Physiology, Genomics, and Biotechnological Applications of Extremophiles

Aparna B. Gunjal
Dr. D.Y. Patil Arts, Commerce, and Science College, Pune, India

Rebecca Thombre
The University of Kent, UK & Blue Marble Space Institute of Science, USA

Javid A. Parray
Department of Higher Education, Government Degree College, Srinagar, India

IGI Global
PUBLISHER of TIMELY KNOWLEDGE

A volume in the Advances in Environmental Engineering and Green Technologies (AEEGT) Book Series

Published in the United States of America by
IGI Global
Medical Information Science Reference (an imprint of IGI Global)
701 E. Chocolate Avenue
Hershey PA, USA 17033
Tel: 717-533-8845
Fax: 717-533-8661
E-mail: cust@igi-global.com
Web site: http://www.igi-global.com

Library of Congress Cataloging-in-Publication Data

Names: Gunjal, Aparna B., 1981- editor. | Thombre, Rebecca S., 1980- editor. | Parray, Javid Ahmad, editor.
Title: Physiology, genomics, and biotechnological applications of extremophiles / Aparna B. Gunjal, Rebecca S. Thombre, Javid A. Parray, editors.
Description: Hershey, PA : Engineering Science Reference, [2022] | Includes bibliographical references and index. | Summary: "The focus of this book is on the advancement in mechanisms of the extremophiles which makes them able to survive under extreme conditions, taking the ecology of the extremophiles into consideration"-- Provided by publisher.
Identifiers: LCCN 2021037631 (print) | LCCN 2021037632 (ebook) | ISBN 9781799891444 (hardcover) | ISBN 9781799891451 (paperback) | ISBN 9781799891468 (ebook)
Subjects: LCSH: Extreme environments--Microbiology. | Ecology.
Classification: LCC QR100.9 .P595 2022 (print) | LCC QR100.9 (ebook) | DDC 578.75/8--dc23/eng/20211004
LC record available at https://lccn.loc.gov/2021037631
LC ebook record available at https://lccn.loc.gov/2021037632

This book is published in the IGI Global book series Advances in Environmental Engineering and Green Technologies (AEEGT) (ISSN: 2326-9162; eISSN: 2326-9170)

British Cataloguing in Publication Data
A Cataloguing in Publication record for this book is available from the British Library.

For electronic access to this publication, please contact: eresources@igi-global.com.

Advances in Environmental Engineering and Green Technologies (AEEGT) Book Series

Sang-Bing Tsai
Zhongshan Institute, University of Electronic Science and Technology of China, China & Wuyi University, China
Ming-Lang Tseng
Lunghwa University of Science and Technology, Taiwan
Yuchi Wang
University of Electronic Science and Technology of China Zhongshan Institute, China

ISSN:2326-9162
EISSN:2326-9170

Mission

Growing awareness and an increased focus on environmental issues such as climate change, energy use, and loss of non-renewable resources have brought about a greater need for research that provides potential solutions to these problems. Research in environmental science and engineering continues to play a vital role in uncovering new opportunities for a "green" future.

The **Advances in Environmental Engineering and Green Technologies (AEEGT)** book series is a mouthpiece for research in all aspects of environmental science, earth science, and green initiatives. This series supports the ongoing research in this field through publishing books that discuss topics within environmental engineering or that deal with the interdisciplinary field of green technologies.

Coverage

- Cleantech
- Green Transportation
- Contaminated Site Remediation
- Sustainable Communities
- Renewable Energy
- Biofilters and Biofiltration
- Radioactive Waste Treatment
- Industrial Waste Management and Minimization
- Green Technology
- Waste Management

IGI Global is currently accepting manuscripts for publication within this series. To submit a proposal for a volume in this series, please contact our Acquisition Editors at Acquisitions@igi-global.com or visit: http://www.igi-global.com/publish/.

The Advances in Environmental Engineering and Green Technologies (AEEGT) Book Series (ISSN 2326-9162) is published by IGI Global, 701 E. Chocolate Avenue, Hershey, PA 17033-1240, USA, www.igi-global.com. This series is composed of titles available for purchase individually; each title is edited to be contextually exclusive from any other title within the series. For pricing and ordering information please visit http://www.igi-global.com/book-series/advances-environmental-engineering-green-technologies/73679. Postmaster: Send all address changes to above address.

Titles in this Series

For a list of additional titles in this series, please visit: www.igi-global.com/book-series

Improving Integrated Pest Management for Crop Protection and Food Security
Muhammad Haseeb (Florida A&M University, USA) Lambert H.B. Kanga (Florida A&M University, USA) and Jawwad A. Qureshi (University of Florida, USA)
Engineering Science Reference • © 2022 • 315pp • H/C (ISBN: 9781799841111) • US $195.00

Modeling and Control of Static Converters for Hybrid Storage Systems
Arezki Fekik (Akli Mohand Oulhadj University, Bouira, Algeria) and Nacereddine Benamrouche (University of Tizi Ouzou, Algeria)
Engineering Science Reference • © 2022 • 355pp • H/C (ISBN: 9781799874478) • US $225.00

Disruptive Technologies and Eco-Innovation for Sustainable Development
Ulas Akkucuk (Boğaziçi University, Turkey)
Engineering Science Reference • © 2022 • 340pp • H/C (ISBN: 9781799889007) • US $195.00

Handbook of Research on Microbial Remediation and Microbial Biotechnology for Sustainable Soil
Junaid Ahmad Malik (Government Degree College, Bijbehara, India)
Engineering Science Reference • © 2021 • 806pp • H/C (ISBN: 9781799870623) • US $295.00

Handbook of Research on Strategic Management for Current Energy Investments
Serhat Yüksel (Istanbul Medipol University, Turkey) and Hasan Dinçer (Istanbul Medipol University, Turkey)
Engineering Science Reference • © 2021 • 422pp • H/C (ISBN: 9781799883357) • US $295.00

Forest Fire Danger Prediction Using Deterministic-Probabilistic Approach
Nikolay Viktorovich Baranovskiy (National Research Tomsk Polytechnic University, Russia)
Engineering Science Reference • © 2021 • 297pp • H/C (ISBN: 9781799872504) • US $195.00

Role of IoT in Green Energy Systems
Vasaki Ponnusamy (Universiti Tunku Abdul Rahman, Malaysia) Noor Zaman (Taylor's University, Malaysia) Low Tang Jung (Universiti Teknologi PETRONAS, Malaysia) and Anang Hudaya Muhamad Amin (Higher Colleges of Technology, UAE)
Engineering Science Reference • © 2021 • 405pp • H/C (ISBN: 9781799867098) • US $195.00

Combating and Controlling Nagana and Tick-Borne Diseases in Livestock
Caleb Oburu Orenge (Department of Veterinary Anatomy and Physiology, Faculty of Veterinary Medicine and Surgery, Egerton University, Kenya)
Engineering Science Reference • © 2021 • 440pp • H/C (ISBN: 9781799864332) • US $185.00

IGI Global
PUBLISHER of TIMELY KNOWLEDGE

701 East Chocolate Avenue, Hershey, PA 17033, USA
Tel: 717-533-8845 x100 • Fax: 717-533-8661
E-Mail: cust@igi-global.com • www.igi-global.com

Table of Contents

Detailed Table of Contents

Extremophiles are extreme nature devotees, mostly bacteria and archaea, which bloom with extreme environmental parameters like temperature, pH, pressure, and salinity. Extremophiles are responsible for the beginning of geographical structures throughout the evolution and establishment of all presently known ecological units. They are classified into several categories like acidophiles, alkaliphiles, psychrophiles, thermophiles, xerophiles, piezophiles/barophiles, halophiles, and many more, as given in this chapter. The subsistence of these microorganisms in extreme environments produces extremolytes and extremozymes that have the potential of valued resources for the enlargement of a bio-based economy. In addition to their solicitations, extremophiles offer treasured information regarding the physiochemical limitations of natural life. This chapter mainly evaluates extremophiles, the classification of extremophiles, and their biotechnological applications in grey, white, and red biotechnologies with the perspective of exploring celestial life.

Halophiles are extremophilic salt-loving microorganisms that can survive in an extremely high level of salinity (10-30% NaCl). They belong to all three groups (i.e., bacteria, archaea, and eukaryotes). Halophiles tolerate high salt concentration due to unique cellular adaptations like salt-in strategy, compatible solute strategy, and enzyme adaptations. The chapter describes the classification, physiology, ecology, and mechanisms of adaptations and biotechnological applications of halophiles.

Alkalophiles are a class of extremophiles capable of survival in alkaline (pH roughly 8.5–11) environments, growing optimally around a pH of 10. At such high pH, the normal cellular functions are detrimentally affected for mesophilic organisms. The alkalophiles successfully manage stability of DNA, plasma membrane, and function of cytosolic enzymes, as well as other unfavorable physiological changes at such an elevated pH. A recent development in NextGen sequencing technology facilitates identifying uncultivable organisms amongst the extreme environments. In recent years, distribution of alkalophiles was reported from Soda Lake, marine environments, saline deserts, and natural thermal vents to natural water bodies. Although alkalophiles were first reported in 1889, their enzymatic and industrial applications still make them an interesting area of research. This chapter provides basic information on environmental distribution, taxonomy, physiology, bioenergetics, and survival mechanism and enzymes produced by alkalophilic organisms.

Microorganisms are the diverse living things present on the Earth. India has numerous unique thermal habitats that comprise several diversity hotspots, such as hot springs, deep oceanic hydrothermal openings, anaerobic biodigesters. The existence of life at high temperatures is quite attractive. At both ends of the temperature range suited with life, only microorganisms can grow and survive. Thermophiles are a typical extremophilic microbes capable of existence in high temperature environments. At such high temperature, the ordinary cellular functions adversely affected for mesophiles. The thermophiles effectively manage instability of the plasma membrane, inactivation of enzymes instability of DNA, as well as other hostile physiological variations at such an elevated temperature. Heat shock proteins (Hsps) have established the most attention in thermophiles under stress condition, which is well described in this chapter. This chapter offers comprehensive information about thermophiles, physiology, metabolism, enzymes of metabolic pathways, and various adaptation mechanisms.

Life on the Earth has evolved in the cold environments. Such cold habitats pose special challenges to the microbes in cold ecosystems, such as minimum metabolic activities, very limited nutrient availability, and often extreme conditions such as pH and salinity apart from temperature. Microbial communities surviving under these extreme conditions must have evolved complex structural and functional adaptations. Prokaryotic adaptations to cold environments are through physiological adaptations by increasing membrane fluidity through large amount of unsaturated fatty acids. These microbes also possess some cold adapted proteins whose steady state levels are maintained. They also produce certain compounds such as polyamines, sugars, polyols, amino acids, and some antifreeze proteins to protect themselves under freezing conditions. They also produce exopolymeric substances that promote adhesion of microbes to moist surfaces to induce biofilm formation which helps getting nutrients and protect the cells from harsh conditions. Antioxidants help destroying toxic reactive oxygen species.

Extremophiles are the mortals that tolerate in the most limiting and aggravating conditions to life. Because of these fantastic ecological criticisms, extremophiles have substituted innumerable intriguing transformations to cell films, proteins, and extracellular metabolites. These stimulatingly regulated usual particles and frameworks as of now play parts in numerous biotechnological fields. Compounds from extremophilic microorganisms as a rule catalyse synthetic responses in non-standard conditions. Such conditions advance accumulation, precipitation, and denaturation, diminishing the movement of most non-extremophilic catalysts, regularly because of the shortfall of adequate hydration. Extremophilic catalysts can go after hydration by means of modifications particularly to their surface through more noteworthy surface charges and expanded sub-atomic movement. These assets have permitted few extremophilic compounds to work within the sight of non-fluid natural solvents, with potential for plan of valuable impetuses.

Extremophiles will be the choice of next generation industrial biotechnology (NGIB) as they are known to be contaminant resistant, but engineering their genomes has always been difficult and time consuming task. CRIPR/Cas (clustered regularly interspaced short palindromic repeat and CRISPR associated proteins) system can be employed for this reason. The genome of an industrially important halophile (i.e., Halomonas) was edited to study a combined effect of four different genes on glucose breakdown and production of poly (3-hydroxybutyrate-co-3-hydroxyvalerate). This editing has resulted in 16-fold increase of 3HV, and the mutants generated by CRIPR/Cas system were significantly effective in synthesizing PHBV. Unfortunately, this system does not always work, specifically in extremophilic microorganisms because Cas9 or Cpf1 are from mesophilic bacteria. Therefore, alternatively, the endogenous CRISPR/Cas system is used for editing the genomes of such organisms. This genome editing of extremophiles will open the doors for developing next generation industrial biotechnology (NGIB).

Extremophiles are the most ancient microbes on the Earth and also a center of attraction for the scientific community for research because of their ability to adapt to extreme habitats. Compatible solutes are among those factors which enable these microorganisms to thrive in such extreme habitats. Under osmotic stress, the majority of extremophiles accumulate specific organic solutes such as amino acids, sugars, polyols, and their derivatives. In addition, proteins in extremophiles are found to be evolved by changing their amino acid composition to alter the hydrophobicity of its core and surface charge to maintain activity. This chapter encompasses a comprehensive study about the role of various compatible solutes in the endurance of microorganisms under extremophilic conditions, synthesis of compatible solutes, nature of extremophilic proteins, and their applications. Furthermore, an attempt has been made to cover various strategies adopted by the scientific community while pursuing research on compatible solutes.

A microorganism dwelling in severe environmental conditions is termed an extremophile. These unfavorable environmental conditions include high salinity, toxin compounds, heavy metals, unfavorable temperature, and extremely acidic and alkaline pH. Microorganisms belonging to prokaryotes include true bacteria and archaea bacteria which prevail in harsh environments. In recent years, extremophilic, basically, archaea bacteria have been reported for their immense potential application in the bioremediation process. Bioremediation is a technique that utilizes microorganisms for the decomposition of organic and inorganic pollutants; anthropogenic activities are the basic cause of soil pollution, water pollution, and air pollution globally. Extremophiles are capable of producing enzymes that are thermolabile and can function normally even in extreme conditions. These enzymes and proteins can be utilized in the bioremediation process under extreme pH, heavy metal stress, and unfavorable temperature conditions. In this chapter, the role of extremophiles in bioremediation is discussed.

and producers of novel enzymes which can be employed in many industries like food, cosmetics, chemical, pharmaceuticals, etc. Currently the researchers have developed keen interest in studying and utilizing the abilities of these extremophiles in food industries. Metabolic pathways and extremozymes are being studied by the researchers and they are trying to utilize its characteristics and also engineer these extremophiles. In food industries, one of the extremophiles, Rhodothermus marinus, which has been an excellent biocatalyst producing lipase as an enzyme, could be utilized to improve to aroma of food and add natural flavour to food. So, the current chapter will deal with the various applications of these extremophiles.

Extremophiles are at center stage of scientific interest owing to their peculiar properties in terms of physiology, ecology, biochemistry, and molecular genetics. The bio-active compounds from extremophiles involve various types of extremolytes. The functional applicability of extremophiles has been far-reaching. Looking to the global scenario medical, pharmaceutical and allied healthcare sectors have a persistent surge for a novel anticancer, antimicrobial, stable drug deliverables, nutraceuticals, fine chemicals, natural antioxidants, and bio-polymers compounds. Genetic engineering tools clubbed with -omics approach enhance and better the chances for applicability of the extremophilic metabolites in varied sectors of red and yellow biotechnology. The chapter provides an insight into the various types of bio-active molecules from extremophiles and their wide biotechnological applicability in the medical and pharmaceutical industry.

Microbial fuel cell (MFC) technology is considered one of the renewable sources of energy for the production of bioelectricity from waste. Due to the depletion of fossil fuels and environmental considerations, MFC haa garnered increasing importance as it is a sustainable and environmentally-friendly method of generation of bioenergy. In MFC, electroactive bacteria (EAB) and biofilms are harnessed to convert organic substances to electrical energy. Extremophiles survive in extreme environments, and they have demonstrated potential applications in microbial electrical systems (MES) and MFC technology. The key limitations of MFC are the low power output and engineering constraints of the fuel cell. Hence, it is imperative to understand the genetics, key metabolic pathways, and molecular mechanisms of the EAB for enhancing the power generation in MFC. This chapter gives a brief overview of the scope and applications of extremophiles in wastewater treatment, bioelectricity, and biohydrogen production using MFC, eventually enhancing the functional efficiency of MFC.

Extremophile and extremozyme capabilities to uphold catalytic actions under extreme situations open up a varied array of biotechnological applications. Extremophiles are a rich supply of biocatalysts used for innumerable purposes. Bioactive molecules and enzymes isolated from organisms inhabiting risky environments being used in biological innovation pipelines and pharmaceutical have positive claims. The species biodiversity has favourable reservoir of the unexploited amalgams with biotechnological significance. Prospective solicitations of extremozymes, chiefly as catalysis of multistep progressions, quorum sensing, bioremediation, biofuel, biodiversity and prospecting, biomining, and genetic technology are explored. To boost the biotechnological uses of extremozymes, research and development efforts are needed to address hurdles such as extremophile culture, gene expression in host cells, and extremozyme bioprocessing. Extremophiles can be a resource for innovative biotechnological comprising industrial biotechnology, agriculture, medical, food, and environmental biotechnology.

Bacterial exopolysaccharides have enormous diversity with valuable characteristics, synthesized by various pathways in extreme conditions like salinity, geothermal springs, or hydrothermal vents. Due to extreme environments, these microorganisms have various adaption principles (e.g., low pH, high temperature, high saltation, and high radiation). Exopolysaccharide is an organic compound produced by most bacteria during fermentation using various carbon sources, resulting in a jelly-like or mass network structure outside the cell wall. This biopolymer has an adherent cohesive layer throughout the cell layer. Hot spring bacterial polysaccharides contain diverse extracellular polymeric substances. With a gain in popularity in applications of thermophilic microbial polysaccharides and its demand in diverse value-added industrial products, this chapter aims to provide valuable information on the physicochemical function and biotechnological applications in the field of food, medical imaging, nano-drugs, bioremediation, cancer, anti-bacterial, tissue engineering, etc.

Preface

Extremophiles are microorganisms that thrive in extremely harsh environments. Discovered just in the early 20th century by Carl Woese and his colleagues, extremophiles have created a paradigm shift in our understanding of microbial life and its physiology under stress. Extremophiles survive under extreme and harsh conditions and it is imperative to understand their physiology and adaptation mechanisms under such conditions. Extremophiles have also garnered immense importance due to its myriad applications, e.g. *Taq polymerase* in PCR reactions and other applications in agriculture, pharmaceutical industry and medicine. With great pleasure we would like to introduce this book entitled *Physiology, Genomics, and Biotechnological Applications of Extremophiles*, focuses on three distinct sections of extremophiles.

The first section of this book includes five chapters that deal with the physiology of extremophiles. Chapter 1 deals with introduction of extremophiles. Chapter 2 focuses on halophiles. Chapter 3 mentions about alkalophiles - their taxonomy, physiology, bioenergetics, survival mechanism and enzymes. Chapter 4 describes about another group of extremophiles i.e., thermophiles, while Chapter 5 described physiology of psychrophiles.

In continuation, the second section of the book includes three chapters that deal with genomics of extremophiles. Chapter 6 describes the biochemistry behind protein adaptation in extremophiles. Chapter 7 focuses on CRISPR/Cas system of extremophiles, while Chapter 8 deals with major compatible solutes and structural adaptation of proteins in extremophiles.

The third section of this book includes eight chapters that deal with the biotechnological applications of extremophiles. Chapter 9 describes the application of extremophiles in bioremediation. Chapter 10 focuses on applications of enzymes from extremophiles. Chapter 11 describes application of extremophiles in sustainable agriculture. Chapter 12 describes application of extremophiles in food industries. Chapter 13 focuses on use of extremophiles in medicine and pharmaceutical industries. Chapter 14 mentions extremophiles in bioenergy production as microbial fuel cells. Chapter 15 describes biotechnological products from extremophiles, and Chapter 16 deals with thermophilic bacterial exopolysaccharides.

We hope this edited book will be a very useful resource for readers, scientists, student's especially undergraduate and postgraduate students, researchers, and Professors from Colleges and Universities for understanding and studying all major aspects of extremophiles.

Aparna B. Gunjal
Dr. D.Y. Patil Arts, Commerce, and Science College, Pune, India

Rebecca Thombre
The University of Kent, UK & Blue Marble Space Institute of Science, USA

Javid Parray
Department of Higher Education, Government Degree College, Srinagar, India

Section 1
Physiology of Extemophiles

Chapter 1
Extremophiles:
Subsistence of an Extreme Nature Enthusiast

Aniruddh Rabari
Department of Microbiology, Gujarat Arts and Science College, Ahmedabad, India

Janki A. Ruparelia
Department of Microbiology, Gujarat Arts and Science College, Ahmedabad, India

Chaitanya Kumar Jha
Department of Microbiology, Gujarat Arts and Science College, Ahmedabad, India

ABSTRACT

Extremophiles are extreme nature devotees, mostly bacteria and archaea, which bloom with extreme environmental parameters like temperature, pH, pressure, and salinity. Extremophiles are responsible for the beginning of geographical structures throughout the evolution and establishment of all presently known ecological units. They are classified into several categories like acidophiles, alkaliphiles, psychrophiles, thermophiles, xerophiles, piezophiles/barophiles, halophiles, and many more, as given in this chapter. The subsistence of these microorganisms in extreme environments produces extremolytes and extremozymes that have the potential of valued resources for the enlargement of a bio-based economy. In addition to their solicitations, extremophiles offer treasured information regarding the physiochemical limitations of natural life. This chapter mainly evaluates extremophiles, the classification of extremophiles, and their biotechnological applications in grey, white, and red biotechnologies with the perspective of exploring celestial life.

DOI: 10.4018/978-1-7998-9144-4.ch001

INTRODUCTION

Extremophiles are extreme nature devotees mostly bacteria and archaea which blooms with extreme environmental parameters. Extremophiles from prokaryotic genera were the first evidence for the life on Earth which are responsible for the beginning of geographical structures throughout the evolution and establishment of all presently known ecological unit. (Pikuta, Hoover et al., 2007)

Extremophiles can endure and bloom in punitive surroundings executed by physical i.e. pressure, radiation, and temperature and natural chemical immoderations i.e. desiccation, oxygen levels, redox potential, salinity and pH. Providentially, there are many extremophiles that flourish in life-threatening surroundings that are found in environment and deals with an outstanding source of auxiliary enzymes in lieu of mesophilic ones (Cullen and MacIntyre, 2016) presently which contests the typical functions of live hood. (Stan-Lotter and Fendrihan, 2012, Shrestha, Chilkoor et al., 2018) to withstand this surroundings they possess temperature stable proteins (thermos, cold stable) enzymes to withstand varying pH and certain secondary metabolites to defend radioactivity based on that they are classified in to various categories as given in Fig. 1 (Irwin, 2020).

Figure 1. Types of extremophiles based on various environmental parameters

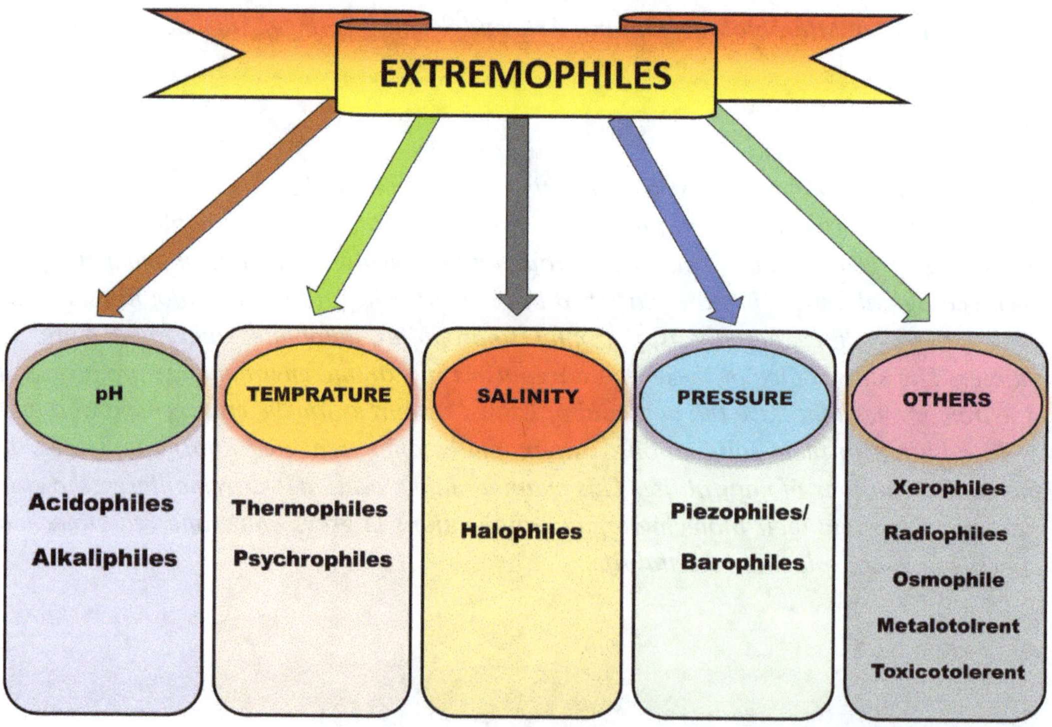

The parameters under which life can survive have been pushed in every direction during the last century, embracing greater swaths of extreme environments. Microorganisms can live in a wide range of environments on Earth, similarly they can tolerate the harsh conditions like space, which include intense radiation, vacuum pressure, drastically fluctuating temperature, and microgravity (Horneck, Klaus et al., 2010). As NASA scientists at Rensselaer Polytechnic Institute found that, during spaceflight on the International Space Station, microbes seems to adapt the space environment in ways "not observed on Earth" and that "may lead to increases in growth and virulence" during travel on the International Space Station.

Now a days many extremophiles and extremozymes have established their potential into large-scale usage in the field of biotechnology (Elleuche, Schroeder et al., 2014) extremophiles have a huge impact on daily basis of life i.e. applications of extremophiles include food, beverage, dairy, textile, detergents and many more...as described in this chapter.

BACKGROUND

Extremophiles sounds amended to the extremes and choose to grow entirely under extreme conditions. This is why these organisms were named "extremophiles" (Latin: extremus, Greek: philiā), which means "extreme-lovers." (Schröder, Burkhardt et al., 2020).

Thomas Dale Brock who is utmost known for discovering hyperthermophiles in Yellowstone National Park's hot springs, while Brock discovered high-temperature bacteria in Yellowstone's Great Fountain region in the late 1960s with his colleague Hudson Freeze and they have extracted a sample named *Thermus aquaticus* (Brock, 1998).

The study of extremophiles began with the publication of "Life at High Temperatures" a 1967's article summarizing Brock's research in the journal Science. Later on *T. aquaticus* was shown to be useful for artificially amplifying DNA segments in 1976 (In Polymerase Chain Reaction). Brock's discoveries aided in the advancement of biology, as well as new breakthroughs in medicine and agriculture, as well as the birth of the disciplined biotechnology (Snyder, 2007).

CLASSIFICATION OF EXTREMOPHILES

Extremophiles are taxonomically broadly distributed and are a functionally assorted group of organisms (Raddadi, Cherif et al., 2015). "Extreme" environments not only denotes temperature, even though heat and cold are the most prominent extremes, but also to extreme pH conditions and high pressure and salinity (Rothschild and Mancinelli, 2001) that includes acidophiles, alkaliphiles, psychrophiles, thermophiles, xerophiles, piezophiles/barophiles, halophiles, metallophiles and radiophiles.

Acidophiles

Acidophiles are rehabilitated to environments with acidic pH values that falls between the ranges of 1 to 5. Generally at pH <2. This group includes bacteria, archaea and some eukaryotic organisms that are found in certain places like sulphuric pools, acidic mine drainage polluted areas, and also in our own stomachs. Some examples of acidophiles are from the genus *Acidobacterium*, *Leptospirillum*, *Picrophilus*, *Ferroplasma* and etc. (Irwin, 2020).

Alkaliphiles

Alkaliphiles are organisms normally bacteria and archaea that optimally grows at pH > 9. They are found in environments like soda lakes, dessert soils and industrial wastes. For example *Chorococcus spp., Synechococcus sp, Ectothiorhodospira, Halorhodospira,* in gram negative mainly *Aeromonas/Vibrio/ Enterobacteria* and *Pseudomonas.* Gram–positive isolates show a prevailing association with *Bacillus spp., Terrabacter* and *Dietzia* (Horikoshi, 2004).

Psychrophiles

Psychrophiles are cold loving organisms which grows at temperature <15 ^{0}C, mostly <5^{0}C .they are commonly found in polar ice, glaciers, fields of snow and in the deep ocean waters. i.e. *Deltaproteobacteria, Firmicutes, Gammaproteobacteria, Bacteroidete, Actinobacteria, Alphaproteobacteria* and etc. (Siddiqui, Williams et al., 2013).

Thermophiles

Thermophiles are high temperature devotees which blooms best at temperature >40 ^{0}C, optimally between 50- 55 ^{0}C. They are found in hot springs, deep-sea hydrothermal vents, volcanic islands, and deep-sea hydrothermal vents. Examples from thermophiles includes Thermus species, *Thermococcus* species, *Aquifex aeolicus, Bacillus stearothermophilus* and etc. (Turner, Mamo et al., 2007).

Xerophiles

It is a kind of extremophilic organisms which grow and reproduce with low water activity generally (aw < 0.7). They are habitat of dry places like desserts, and arid dessert soils. Some commonly found fungus from this group includes *Aspergillus flavus, A.ochraceus, A.carbonarius, Penicillium nordicum* and *P. verrucosum* (Medina, Schmidt-Heydt et al., 2015) other includes *Burkholderiales, Xanthomonadales, actinomycetes* and etc. (Prenafeta-Boldú, Guivernau et al., 2012).

Piezophiles/Barophiles

Piezophiles/Barophiles are kind of organisms that desires to grow at high pressure range up to 110 MPa. They are most commonly found in deep sea environments where pressure is comparatively high. Examples includes *Thermococcus barophilus, Photobacterium profundum, Shewanella violacea, T. barophilus* and *Desulfovibrio hydrothermalis* and etc. (Michoud and Jebbar, 2016).

Halophiles

Halophiles are salt loving organisms which prefers salt concentrations up to 3.4–5 mol L $^{-1}$ NaCl i.e. *Halobacillus, Bacillus, Marinococcus, Salinococcus, Haloarcula, Halococcus, Haloferax* and etc… (DasSarma and Arora, 2001) they are inhabitants of hyper saline environment.

Metalophile

These kind of extremophiles are resistant to high metal concentrations which are found in metal polluted sites that includes bacteria from the genera *Acidithiobacillus, Leptospirillum, Alicyclobacillus, Acidiphilium, Acidimicrobium, Ferrimicrobium, Sulfobacillus* and archaea from the genera *Ferroplasma, Acidiplasma, Sulfolobus, Metallosphaera, and Acidianus* (Raddadi, Cherif et al., 2015).

Radiophile

Radiophiles can resist the high level of radiation (x-rays, gamma rays, UV) as they possess DNA repair mechanisms. They belongs to the *genera Deinococcus, Bacillus, Rubrobacter*, and *Kineococcus*, and *cyanobacteria* including the genera *Nostoc and Chroococcidiopsis* (Gabani and Singh, 2013, Raddadi, Cherif et al., 2015).

Table 1. Classification of extremophiles

Pressure			**Piezotolerant** (0.1–10 MPa)	**Piezophile (10–50 MPa)** *Thermococcus barophilus*	**Hyperpiezophile (>50 MPa)** *Thermococcus piezophilus*
Salinity		Non-halophile (<1.2%) *V. cholerae* *V. mimicus*	Halo tolerant (1.2–2.9%; tolerate Salt concentration up to ≤14.6%) *Halobacterium, Halocalculus,*	Halophile (Grow >8.8% Salt concentration) *Salinibacter ruber*	Extreme halophile (>14.6%, cannot grow < 8.8%) *Halomonas salaria*
Temperature		Psychrophile (Growth < 15 °C) *Arthrobacter sp., Psychrobacter sp*	Mesophile (Growth > 15-45 °C) *Escherichia coli*	Thermophile (Growth > 45-80 °C) *Streptococcus thermophilus*	Hyperthemophile (Growth > 80 °C) *Sulfolobus solfataricus*
pH	Hyperacidophile (Grow at <pH 3) *A.ferrooxidans, A. thiooxidans*	Acidophile (Grow at <pH 5) *Thiobacillus prosperus T. acidophilus,*	Neutrophile (Grow at pH 5–9) *Pseudomonas aeruginosa*	Alkaliphile (Grow at >pH 9) *Natronomonas pharaonis*	Hyperalkaliphile (Grow at >pH 11) *Halorhodospira halochloris*
Water activity			Xerophile (aw < 0.7) *Trichosporonoides nigrescens*		
other	Endoliths (Extreme rockiness) *Ostracoblabe implexis and Lithopythium gangliiforme*	Radiophile (Radiation Loving) *Deinococcus radiodurans*	Metallotolerant (Tolerating high level of Heavy metal Concentration) *Ferroplasma sp. and Cupriavidus metallidurans*	Psychropiezophile (Require 10 MPa pressure and 2-3 °C)	Thermopiezophile (Require 52 MPa pressure and 80 °C) *Pyrococcus yayanosii*
Polyextremophiles	Microorganism that survives or loving more than one extreme conditions combined.				

APPLICATIONS OF EXTREMOPHILES

Biotechnology has a huge prospective particularly when directing for a sustainable bio economy (bio-based economy) (Schröder, Burkhardt et al., 2020). When it comes to applications of extremophiles they are tremendously beneficial in numerous industries like food, dairy, starch liquefaction, baking, brewing, and feed and in medical purpose they are used in pharmaceuticals, diagnosis and PCR when it comes to particular molecular basics. Moreover they are also used in the bio refineries, textile, and detergent making industries. Table. 2. Contain the details of the applications shown in the Fig. 2.

Figure 2. Overview of extremophiles applications in Biotechnology

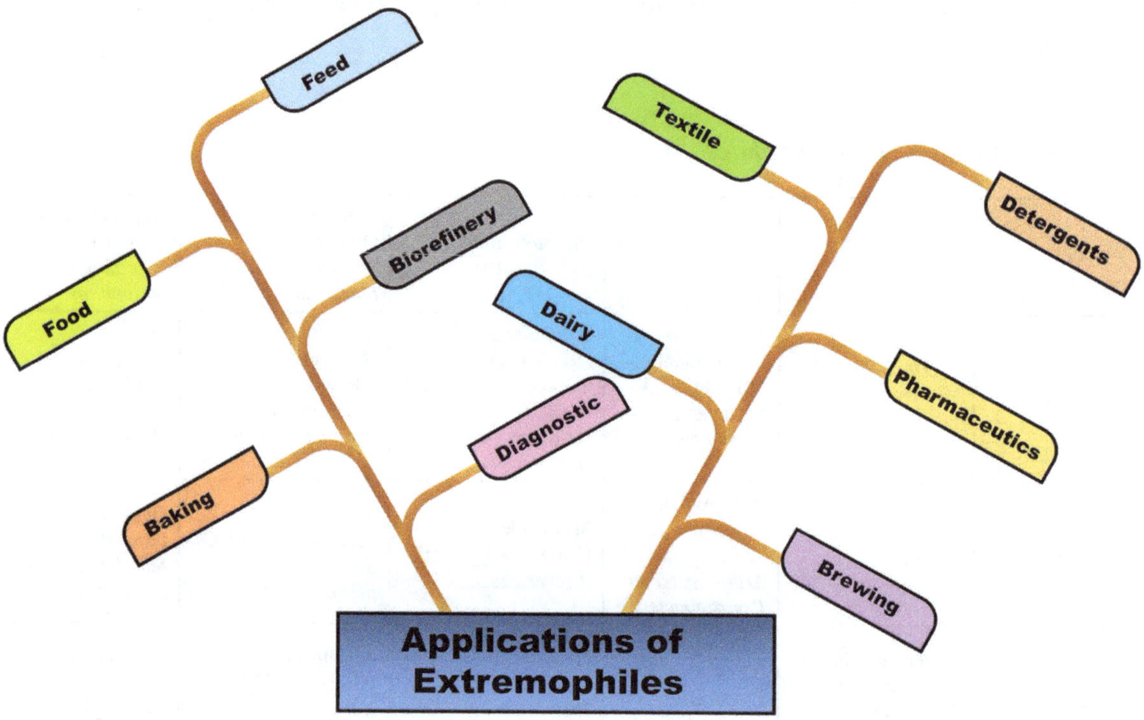

Mainly enzymes of extremophiles carries out complex reactions in an easy way dealing with thrilling conditions which is generally not possible for other organisms. Examples of various extremophiles are given in Table 2. These enzyme's efficacy is usually improved through genetic and/or chemical alteration (Coker, 2016) as well as with the immobilization approaches (Mukhopadhyay, Dasgupta et al., 2015, Ruparelia, 2020), altogether they are intended to make biocatalysts with enhanced stuffs such as improved activity and/or stability to use in definite industrial practises (Gatti-Lafranconi, Natalello et al., 2010).

Table 2. Biotechnological application of extremophiles in daily life

Microorganism	Domain	Habitat	Extreme Condition	Application	Reference
Thermus aquaticus	Bacteria	Hot springs	Thermophile	PCR	(Valones, Guimarães et al., 2009)
Desulfovibrio piezophilus	Bacteria	Wood falls in the Mediterranean Sea	Piezophile	Sulfate-reducing bacterium	(Khelaifia, Fardeau et al., 2011)
Sulfolobus	Archea	sulfur-rich acidic hot springs	Acidophilus	Recovery of metals from the ores	(Sharma, Parashar et al., 2016)
Thermotoga elfii	Bacteria	African oil-producing well	Thermophile	Biofuel Production	(De Vrije, De Haas et al., 2002)
***Thermococcus barophilus* Ch5**	Archaea	Chimmney wall of a deep-sea hydrothermal vent	piezophile	PCR	(Di Giulio, 2005)
Acidithiobacillus	Bacteria	Acid mine Drainage Environment	Acidophilus	Bio mining	(Vera, Schippers et al., 2013)
Halobacterium salinarum	Archaea	Hyper saline lakes, and salterns	Halophile	Carotenoids production	(Schiraldi, Giuliano et al., 2002)
Haloferax alexandrinus	Archaea	Hyper saline lakes	Extreme halophile	Canthaxanthin production	(Asker and Ohta, 2002)
Bacillus acidicola	Bacteria	Acidic Sphagnum peat bog in Wisconsin.	Acidophile	Starch hydrolysis	(Sharma and Satyanarayana, 2012)
Micrococcus antarcticus	Bacteria	Antarctica	Psychrophilic	Detergent, Textile and Bioremediation	(Miao, Hou et al., 2016)
Fomitopsis meliae	Fungi	Gold and gemstone mine site soils	Metallotolerent	Remediation Cu, Pb, Cd and Fe contaminated environments	(Oladipo, Awotoye et al., 2018)
Thermococcus litoralis	Archaea	Submarine thermal springs and Hydrothermal vents	Hyperthermophile	PCR	(Coker, 2016)
Janibacter sp. R02	Bacteria	Antarctica	Thermohalophilic	Pharmaceutical, Leather and Food industry	(Castilla, Panizza et al., 2017)
Erwinia sp. E602	Bacteria	Frozen soil	Psychrophile	Food Industry	(Xia, He et al., 2018)
Paenibacillus tarimensis	Bacteria	Sahara Desert	Thermophile	Textile industry	(Raddadi, Cherif et al., 2013)
Acidianus sulfidivorans	Archaea	solfatara on Lihir Island	Acidophile	Bioremediation	(Plumb, Haddad et al., 2007)
Deinococcus radiodurans	Bacteria	Inside walls of nuclear reactor	Raditolrent, Polyextremophile	Treatment of nuclear energy waste (Bioremediation)	(Brim, McFarlan et al., 2000)
Alkaliphilus metalliredigens	Bacteria	Alkaline borax leachate ponds	Metallotolerent, Alkaliphilic	Useful applications to metal-contaminated alkaline environments	(Roh, Chon et al., 2007)
Bacillus sp	Bacteria	Textile wastewater drain	Alkaliphilic	Treatment and recycling of textile bleaching effluents	(Paar, Costa et al., 2001)
Thioalkalivibrio halophilus	Bacteria	Alkaline hypersaline lake	Alkali-Halophile	Desulfurization of natural gas	(Banciu, Sorokin et al., 2004)
Deinococcus geothermalis	Bacteria	Hot springs, Radioactive places	Raditolrent	Bioremediation of radioactive waste sites	(Ranawat and Rawat, 2017)
Pyrobaculum islandicum	Archaea	Boiling sulfataric and geothermal waters in Iceland	Raditolrent, Polyextremophile	Bioremediation of radioactive waste sites	(Ranawat and Rawat, 2017)
Ferroplasma acidarmanus	Archaea	Heavy metal contaminated sites, Acid mine drainage environment	Metallotolerent, Polyextremophile	Bioleaching of Copper and Arsenic	(Baker-Austin, Dopson et al., 2005)
Cupriavidus metallidurans strain CH34	Bacteria	Decantation tank of zink factory	Metallotolerent	Remediation of heavy metal-contaminated environments	(Nies, 2000)
Pyrococcus abyssi	Archaea	Hydrothermal vent	Piezophile	DNA Polymerase Working in Extreme conditions	(Di Giulio, 2005)
Trichoderma *ghanense*	Fungi	Gold and gemstone mine site soils	Metallotolerent	Remediation Cu, Pb, Cd and Fe contaminated environments	(Oladipo, Awotoye et al., 2018)
Pyrococcus furiosus,	Archaea	Geothermally heated marine sediments	Hyperthermophile	PCR	(Coker, 2016)
Eurotium	Fungi	Dry cured ham and dry beef cecina	Xerophile	Biological control of mites	(Ortiz-Lemus, Campoy et al., 2021)
Halobacterium salinarium	Archaea	Traditionally fermented Thai fish sauce	Halophile	Fermentation of food	(Thongthai, McGenity et al., 1992)
Halalkalicoccus jeotgali	Archaea	Shrimp jeotgal	Halophile	Fermentation of food	(Oren, 2010)
Arthrobacter psychrolactophilus	Bacteria	Soil in the United States	Psychrophile	Biodegradability of organic compounds in the wastewater	(Margesin and Feller, 2010)

FUTURE PROSPECTS

With conventional market success in the DNA polymerase, biofuels, bio mining, and carotenoid sectors of biotechnology, extremophiles and extremozmes have an all-encompassing toe hold in the market place. **(Coker, 2016)** From last few years, consideration for extreme environments has amplified as of interests to isolate and identify formerly unknown extremophilic organisms in pure form and to characterise their different metabolites. Innovative improvements in the cultivation and production of extremophiles derives developments related to the cloning and gene expression in heterologous hosts, will upsurge the numeric enzyme based transformations in chemical, food, pharmaceutical and other industrial applications in near future. Although enzymes and enzymatic proteins of extremophiles are able to defence different kinds of stress responses which includes all of the processes that organisms have developed to survive when they are exposed to environmental challenges, such as heat stress, desiccation, chemical stress, or starvation (Laksanalamai and Robb, 2004) which will deal with new era of biotechnological applications.

CONCLUSION

At present, Biotechnology based on extremophiles has established its way into day-to-day life. where other organisms marks their limit, Extremophiles have made their significant mark with their massive capability of producing extremozymes which are applied in an every aspect of life like in food and drink, either it is production of lactose free milk or in sugar free contents with high nutrition source or if we consider detergent industry (cellulase, lipases) and etc. They have made remarkable impact through surviving at extreme pH, salinity, temperature, pressure, radiation, water activity and many more as described in this chapter. Where at present it is suspected to have existence of radiophiles on space too. Yet innovation in new methodologies for their isolation and to understand their adaptation strategies are still required to open a broad lane for biotechnology.

ACKNOWLEDGMENT

We gratefully acknowledges the scheme of developing high quality research (SHODH), Government of Gujarat, India for the fellowship.

REFERENCES

Asker, D., & Ohta, Y. (2002). Production of canthaxanthin by Haloferax alexandrinus under non-aseptic conditions and a simple, rapid method for its extraction. *Applied Microbiology and Biotechnology*, *58*(6), 743–750. doi:10.100700253-002-0967-y PMID:12021793

Baker-Austin, C., Dopson, M., Wexler, M., Sawers, R. G., & Bond, P. L. (2005). Molecular insight into extreme copper resistance in the extremophilic archaeon 'Ferroplasma acidarmanus' Fer1. *Microbiology*, *151*(8), 2637–2646. doi:10.1099/mic.0.28076-0 PMID:16079342

Banciu, H., Sorokin, D. Y., Galinski, E. A., Muyzer, G., Kleerebezem, R., & Kuenen, J. G. (2004). Thialkalivibrio halophilus sp. nov., a novel obligately chemolithoautotrophic, facultatively alkaliphilic, and extremely salt-tolerant, sulfur-oxidizing bacterium from a hypersaline alkaline lake. *Extremophiles, 8*(4), 325–334. doi:10.100700792-004-0391-6 PMID:15309564

Brim, H., McFarlan, S. C., Fredrickson, J. K., Minton, K. W., Zhai, M., Wackett, L. P., & Daly, M. J. (2000). Engineering Deinococcus radiodurans for metal remediation in radioactive mixed waste environments. *Nature Biotechnology, 18*(1), 85–90. doi:10.1038/71986 PMID:10625398

Brock, T. D. (1998). Early days in Yellowstone microbiology. *ASM News, 64*, 137–140.

Castilla, A., Panizza, P., Rodríguez, D., Bonino, L., Díaz, P., Irazoqui, G., & Giordano, S. R. (2017). A novel thermophilic and halophilic esterase from Janibacter sp. R02, the first member of a new lipase family (Family XVII). *Enzyme and Microbial Technology, 98*, 86–95. doi:10.1016/j.enzmictec.2016.12.010 PMID:28110668

Coker, J. (2016). "Sohail Siddiqui of New South Wales, K." Extremophiles and Biotechnology: Current Uses and Prospects. *F1000 Research, 5*, 396. doi:10.12688/f1000research.7432.1

Coker, J. A. (2016). Extremophiles and biotechnology: Current uses and prospects. *F1000 Research, 5*, 5. doi:10.12688/f1000research.7432.1 PMID:27019700

Cullen, J. J., & MacIntyre, H. L. (2016). On the use of the serial dilution culture method to enumerate viable phytoplankton in natural communities of plankton subjected to ballast water treatment. *Journal of Applied Phycology, 28*(1), 279–298. doi:10.100710811-015-0601-x PMID:26893536

DasSarma & Arora. (2001). *Halophiles*. e LS.

De Vrije, T., De Haas, G., Tan, G., Keijsers, E., & Claassen, P. (2002). Pretreatment of Miscanthus for hydrogen production by Thermotoga elfii. *International Journal of Hydrogen Energy, 27*(11-12), 1381–1390. doi:10.1016/S0360-3199(02)00124-6

Di Giulio, M. (2005). A comparison of proteins from Pyrococcus furiosus and Pyrococcus abyssi: Barophily in the physicochemical properties of amino acids and in the genetic code. *Gene, 346*, 1–6. doi:10.1016/j.gene.2004.10.008 PMID:15716096

Elleuche, S., Schroeder, C., Sahm, K., & Antranikian, G. (2014). Extremozymes—Biocatalysts with unique properties from extremophilic microorganisms. *Current Opinion in Biotechnology, 29*, 116–123. doi:10.1016/j.copbio.2014.04.003 PMID:24780224

Gabani, P., & Singh, O. V. (2013). Radiation-resistant extremophiles and their potential in biotechnology and therapeutics. *Applied Microbiology and Biotechnology, 97*(3), 993–1004. doi:10.100700253-012-4642-7 PMID:23271672

Gatti-Lafranconi, P., Natalello, A., Rehm, S., Doglia, S. M., Pleiss, J., & Lotti, M. (2010). Evolution of stability in a cold-active enzyme elicits specificity relaxation and highlights substrate-related effects on temperature adaptation. *Journal of Molecular Biology, 395*(1), 155–166. doi:10.1016/j.jmb.2009.10.026 PMID:19850050

Horikoshi, K. (2004). Alkaliphiles. *Proceedings of the Japan Academy. Series B, Physical and Biological Sciences*, *80*(4), 166–178. doi:10.2183/pjab.80.166

Horneck, G., Klaus, D. M., & Mancinelli, R. L. (2010). Space microbiology. *Microbiology and Molecular Biology Reviews*, *74*(1), 121–156. doi:10.1128/MMBR.00016-09 PMID:20197502

Irwin, J. A. (2020). Overview of extremophiles and their food and medical applications. Physiological and Biotechnological Aspects of Extremophiles, 65-87.

Khelaifia, S., Fardeau, M.-L., Pradel, N., Aussignargues, C., Garel, M., Tamburini, C., Cayol, J.-L., Gaudron, S., Gaill, F., & Ollivier, B. (2011). Desulfovibrio piezophilus sp. nov., a piezophilic, sulfate-reducing bacterium isolated from wood falls in the Mediterranean Sea. *International Journal of Systematic and Evolutionary Microbiology*, *61*(11), 2706–2711. doi:10.1099/ijs.0.028670-0 PMID:21169465

Laksanalamai, P., & Robb, F. T. (2004). Small heat shock proteins from extremophiles: A review. *Extremophiles*, *8*(1), 1–11. doi:10.100700792-003-0362-3 PMID:15064984

Margesin, R., & Feller, G. (2010). Biotechnological applications of psychrophiles. *Environmental Technology*, *31*(8-9), 835–844. doi:10.1080/09593331003663328 PMID:20662375

Medina, A., Schmidt-Heydt, M., Rodriguez, A., Parra, R., Geisen, R., & Magan, N. (2015). Impacts of environmental stress on growth, secondary metabolite biosynthetic gene clusters and metabolite production of xerotolerant/xerophilic fungi. *Current Genetics*, *61*(3), 325–334. doi:10.100700294-014-0455-9 PMID:25381155

Miao, L.-L., Hou, Y.-J., Fan, H.-X., Qu, J., Qi, C., Liu, Y., Li, D.-F., & Liu, Z.-P. (2016). Molecular structural basis for the cold adaptedness of the psychrophilic β-glucosidase BglU in Micrococcus antarcticus. *Applied and Environmental Microbiology*, *82*(7), 2021–2030. doi:10.1128/AEM.03158-15 PMID:26801571

Michoud, G., & Jebbar, M. (2016). High hydrostatic pressure adaptive strategies in an obligate piezophile Pyrococcus yayanosii. *Scientific Reports*, *6*(1), 1–10. doi:10.1038rep27289 PMID:27250364

Mukhopadhyay, A., Dasgupta, A. K., & Chakrabarti, K. (2015). Enhanced functionality and stabilization of a cold active laccase using nanotechnology based activation-immobilization. *Bioresource Technology*, *179*, 573–584. doi:10.1016/j.biortech.2014.12.070 PMID:25590281

Nies, D. H. (2000). Heavy metal-resistant bacteria as extremophiles: Molecular physiology and biotechnological use of Ralstonia sp. CH34. *Extremophiles*, *4*(2), 77–82. doi:10.1007007920050140 PMID:10805561

Oladipo, Awotoye, Olayinka, Bezuidenhout, & Maboeta. (2018). Heavy metal tolerance traits of filamentous fungi isolated from gold and gemstone mining sites. *Brazilian Journal of Microbiology, 49*, 29-37.

Oren, A. (2010). Industrial and environmental applications of halophilic microorganisms. *Environmental Technology*, *31*(8-9), 825–834. doi:10.1080/09593330903370026 PMID:20662374

Ortiz-Lemus, J. F., Campoy, S., & Martín, J. F. (2021). Biological control of mites by xerophile Eurotium species isolated from the surface of dry cured ham and dry beef cecina. *Journal of Applied Microbiology*, *130*(3), 665–676. doi:10.1111/jam.14839 PMID:32869458

Paar, A., Costa, S., Tzanov, T., Gudelj, M., Robra, K.-H., Cavaco-Paulo, A., & Gübitz, G. (2001). Thermo-alkali-stable catalases from newly isolated Bacillus sp. for the treatment and recycling of textile bleaching effluents. *Journal of Biotechnology*, *89*(2-3), 147–153. doi:10.1016/S0168-1656(01)00305-4 PMID:11500208

Pikuta, E. V., Hoover, R. B., & Tang, J. (2007). Microbial Extremophiles at the Limits of Life. *Critical Reviews in Microbiology*, *33*(3), 183–209. doi:10.1080/10408410701451948 PMID:17653987

Plumb, J. J., Haddad, C. M., Gibson, J. A., & Franzmann, P. D. (2007). Acidianus sulfidivorans sp. nov., an extremely acidophilic, thermophilic archaeon isolated from a solfatara on Lihir Island, Papua New Guinea, and emendation of the genus description. *International Journal of Systematic and Evolutionary Microbiology*, *57*(7), 1418–1423. doi:10.1099/ijs.0.64846-0 PMID:17625168

Prenafeta-Boldú, F. X., Guivernau, M., Gallastegui, G., Viñas, M., de Hoog, G. S., & Elías, A. (2012). Fungal/bacterial interactions during the biodegradation of TEX hydrocarbons (toluene, ethylbenzene and p-xylene) in gas biofilters operated under xerophilic conditions. *FEMS Microbiology Ecology*, *80*(3), 722–734. doi:10.1111/j.1574-6941.2012.01344.x PMID:22380456

Raddadi, N., Cherif, A., Daffonchio, D., & Fava, F. (2013). Halo-alkalitolerant and thermostable cellulases with improved tolerance to ionic liquids and organic solvents from Paenibacillus tarimensis isolated from the Chott El Fejej, Sahara desert, Tunisia. *Bioresource Technology*, *150*, 121–128. doi:10.1016/j.biortech.2013.09.089 PMID:24161550

Raddadi, N., Cherif, A., Daffonchio, D., Neifar, M., & Fava, F. (2015). Biotechnological applications of extremophiles, extremozymes and extremolytes. *Applied Microbiology and Biotechnology*, *99*(19), 7907–7913. doi:10.100700253-015-6874-9 PMID:26272092

Ranawat, P., & Rawat, S. (2017). Radiation resistance in thermophiles: Mechanisms and applications. *World Journal of Microbiology & Biotechnology*, *33*(6), 112. doi:10.100711274-017-2279-5 PMID:28470425

Roh, Y., Chon, C.-M., & Moon, J.-W. (2007). Metal reduction and biomineralization by an alkaliphilic metal-reducing bacterium, Alkaliphilus metalliredigens (QYMF). *Geosciences Journal*, *11*(4), 415–423. doi:10.1007/BF02857056

Rothschild, L. J., & Mancinelli, R. L. (2001). Life in extreme environments. *Nature*, *409*(6823), 1092–1101. doi:10.1038/35059215 PMID:11234023

Ruparelia, J., Saraf, M., & Jha. (2020). Screening and optimization for cellulase production by soil bacterial isolates JRC1 and JRC2. *BBRC, 13*, 159-164.

Schiraldi, C., Giuliano, M., & De Rosa, M. (2002). Perspectives on biotechnological applications of archaea. *Archaea (Vancouver, B.C.)*, *1*(2), 75–86. doi:10.1155/2002/436561 PMID:15803645

Schröder, C., Burkhardt, C., & Antranikian, G. (2020). What we learn from extremophiles. *Chemtexts*, *6*(1), 8. doi:10.100740828-020-0103-6

Sharma, Parashar, & *Satyanarayana. (2016). Acidophilic microbes: biology and applications. Biotechnology* of Extremophiles, 215-241.

Sharma, A., & Satyanarayana, T. (2012). Cloning and expression of acidstable, high maltose-forming, Ca 2+-independent α-amylase from an acidophile Bacillus acidicola and its applicability in starch hydrolysis. *Extremophiles, 16*(3), 515–522. doi:10.100700792-012-0451-2 PMID:22527045

Shrestha, N., Chilkoor, G., Vemuri, B., Rathinam, N., Sani, R. K., & Gadhamshetty, V. (2018). Extremophiles for microbial-electrochemistry applications: A critical review. *Bioresource Technology, 255*, 318–330. doi:10.1016/j.biortech.2018.01.151 PMID:29433771

Siddiqui, K. S., Williams, T. J., Wilkins, D., Yau, S., Allen, M. A., Brown, M. V., Lauro, F. M., & Cavicchioli, R. (2013). Psychrophiles. *Annual Review of Earth and Planetary Sciences, 41*(1), 87–115. doi:10.1146/annurev-earth-040610-133514

Snyder, B. (2007). Why the NSF biology budget should be doubled. *Bioscience, 57*(9), 727–728. doi:10.1641/B570902

Stan-Lotter, H., & Fendrihan, S. (2012). *Adaption of microbial life to environmental extremes*. Springer. doi:10.1007/978-3-211-99691-1

Thongthai, C., McGenity, T., Suntinanalert, P., & Grant, W. (1992). Isolation and characterization of an extremely halophilic archaeobacterium from traditionally fermented Thai fish sauce (nam pla). *Letters in Applied Microbiology, 14*(3), 111–114. doi:10.1111/j.1472-765X.1992.tb00661.x

Turner, P., Mamo, G., & Karlsson, E. N. (2007). Potential and utilization of thermophiles and thermostable enzymes in biorefining. *Microbial Cell Factories, 6*(1), 1–23. doi:10.1186/1475-2859-6-9 PMID:17359551

Valones, M. A. A., Guimarães, R. L., Brandão, L. A. C., Souza, P. R. E., Carvalho, A. A. T., & Crovela, S. (2009). Principles and applications of polymerase chain reaction in medical diagnostic fields: A review. *Brazilian Journal of Microbiology, 40*(1), 1–11. doi:10.1590/S1517-83822009000100001 PMID:24031310

Vera, M., Schippers, A., & Sand, W. (2013). Progress in bioleaching: Fundamentals and mechanisms of bacterial metal sulfide oxidation—part A. *Applied Microbiology and Biotechnology, 97*(17), 7529–7541. doi:10.100700253-013-4954-2 PMID:23720034

Xia, Y., He, L., Mao, J., Fang, P., Ma, X., & Wang, Z. (2018). Purification, characterization, and gene cloning of a new cold-adapted β-galactosidase from Erwinia sp. E602 isolated in northeast China. *Journal of Dairy Science, 101*(8), 6946–6954. doi:10.3168/jds.2018-14605 PMID:29859695

Chapter 2
Halophiles

Aparna Baban Gunjal
Department of Microbiology, Dr. D.Y. Patil Arts, Commerce, and Science College, Pune, India

Nilaja Pradeep Badodekar
https://orcid.org/0000-0001-6356-5066
Logical Life Science Pvt Limited, India

ABSTRACT

Halophiles are extremophilic salt-loving microorganisms that can survive in an extremely high level of salinity (10-30% NaCl). They belong to all three groups (i.e., bacteria, archaea, and eukaryotes). Halophiles tolerate high salt concentration due to unique cellular adaptations like salt-in strategy, compatible solute strategy, and enzyme adaptations. The chapter describes the classification, physiology, ecology, and mechanisms of adaptations and biotechnological applications of halophiles.

INTRODUCTION

Halophiles include diverse group of organisms which include Archaea, Bacteria, and Eukarya and grow in presence of salt (Oren, 2015). The halophiles are isolated mainly from saline soil, water, springs, marshes, brines and lakes. These are classified viz., slight halophiles (0.2-0.5 M salt tolerance), moderate halophiles (0.5-2.5 M salt tolerance), borderline extreme halophiles (2.5-4.0 M salt tolerance) and extreme halophiles (4.0-5.9 M salt tolerance) based on their ability to tolerate salinity (Fig. 1). The ability to tolerate salt depends on parameters such as temperature, pH, nutrients, etc. (Corral et al., 2020). In this way, the halophiles are adapted and limited by specific environmental factors. Halophiles belong to family Halomonadaceae and class Gammaproteobacteria. The halophiles can be aerobes, anaerobes, chemo heterotrophs, photo heterotrophs, or photoautotrophs (Edbeib et al., 2016). The halophiles have been reported to be isolated from Dead Sea of Israel (Wei et al., 2015), lake Urmia near Iran (Mehrshad et al., 2015a), Tuzkoy salt mines of Turkey (Mutlu & Guven, 2015), Great salt lake, Utah in United States of America (Tazi et al., 2014), Rambla Salada, Murcia, Spain (Luque et al., 2014), etc. The sources for isolation of halophiles are represented in Fig. 2.

DOI: 10.4018/978-1-7998-9144-4.ch002

Figure 1. Classification of group of halophiles

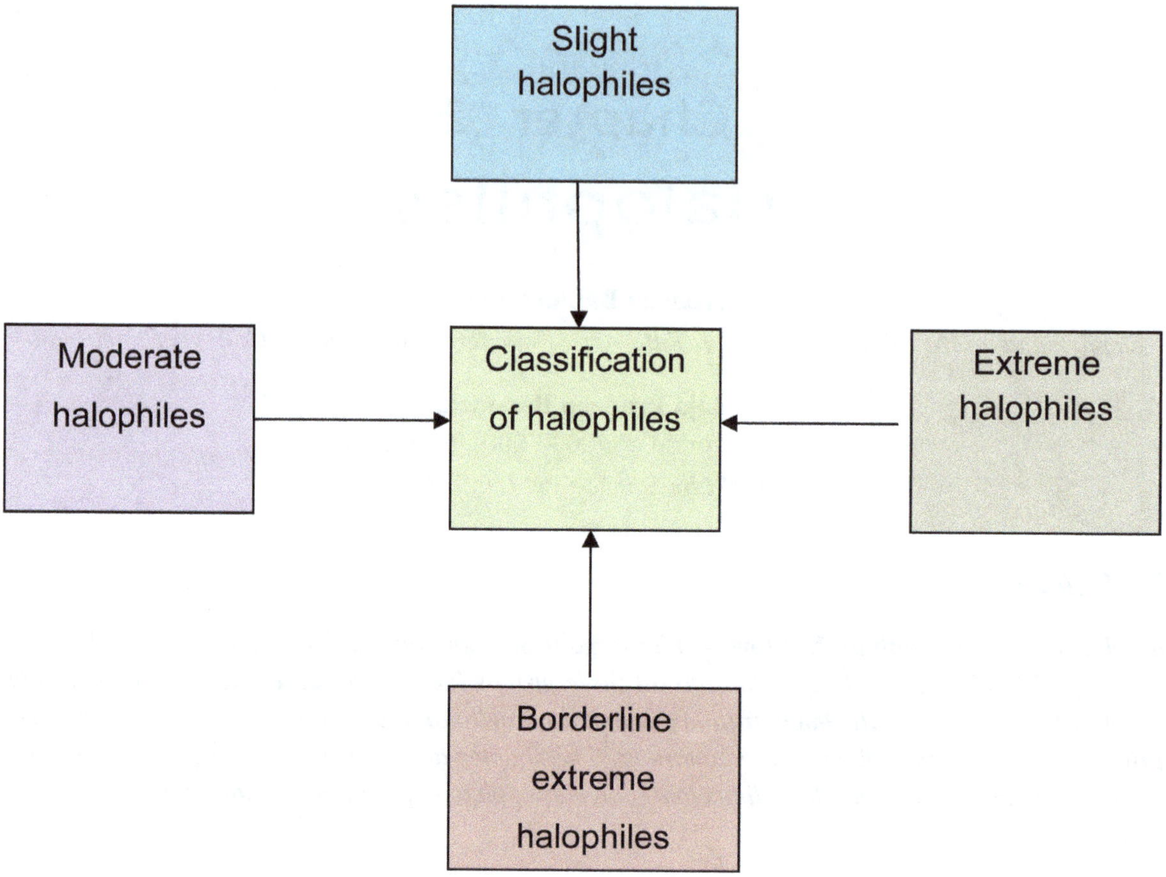

HISTORY OF HALOPHILES

The first halophilic microorganism came into account in 2700 BC (Bass-Becking, 1931) and found in hyper saline area. In 1920s and 1940s halophilic bacteria were isolated from different sources, e.g., fish, animal hides, and anchovies. Elazari isolated various extreme halophiles, *Halobacterium trapanicum* and *Micrococcus morrhuae* and moderate halophiles *Chromohalobacterium marismortui*, *Pseudomonas halestrogus*, and *Flavobacterium halmephium* from the Dead Sea (Edbeib et al., 2016). The ecology, physiology and biochemistry of halophiles have been reported (Oren, 2015). The genome sequencing of halophiles has immense interest. The first genome sequence has been studied of *Halobacterium* NRC-1 (Ng et al., 2000). During the last three years, the genome sequence of eight halophiles has been reported. From these, four halophiles belong to the *Halomonas* genus. The *in-silico* studies help to know exact biology of halophiles (Oren, 2014). The *in-silico* post-genomic and genetic engineering studies have developed new pathways of growth optimization for halophilic microorganisms (Yue et al., 2014).

Figure 2. Sources for isolation of halophiles

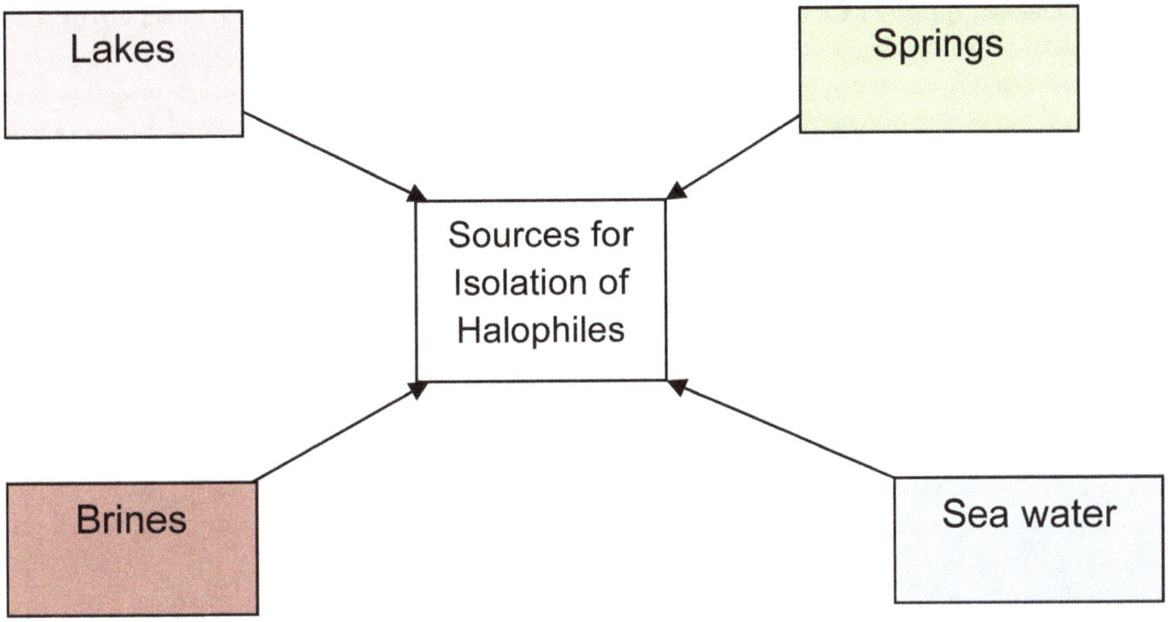

Cellular Adaptation

Microorganisms which cannot tolerate high salt concentration or saline environment tend to lose water because of which cells shrink resulting in complete loss of cell structure and function (Fig. 3a). Halophiles produce osmoprotectants which enhances the osmotic activity of their cytoplasm (Fig. 3b). Some of the halophiles are able to produce excess salt concentration in the cell in order to have equal salt concentrations with that of the environment (Fig. 3c) (Edbeib et al., 2016).

PHYSIOLOGY AND ADAPTATIONS OF HALOPHILES TO THEIR ENVIRONMENTS

The halophiles adapt to high salt concentrations due to various strategies which helps them to maintain osmotic balance. These strategies include cellular adaptations; high salt-in strategy; low-salt, organic solute-in strategy; and enzyme adaptation (Fig. 4).

High Salt-In Strategy

In high-salt-in strategy halophiles produce excess of inorganic ions in order to maintain the salt balance in the environment. The special feature of halophiles is the presence of Cl^- pumps which facilitates transport of Cl^- ions from the environment in the cytoplasm. The amino acids arginine or lysine placed at ends of the channel helps the pumping of Cl^- ions in the cytoplasm uptake (Edbeib et al., 2016). The proton-pump such as bacteriorhodopsin, ATP synthase, and Na^+/H^+ antiporter also helps in this mechanism. Due to this, an electrical potential is created and this helps to transports K^+ ions into the cells due to K^+ uniport mechanism. The electrical potential created must be greater as compared to the diffusion potential of K^+, which enables uptake of Cl^- ions by primary or secondary transporters (Edbeib et al., 2016).

Figure 3. Adaptations of halophiles to extreme saline environments. a) The non-halophile macromolecules are compromised, the water flows outside cell leading to turgor effect. b) Moderate halophiles maintain synthesize compatible organic solutes. c) Extreme halophiles adapt to saline conditions through balance of cellular and outside salt concentrations
Source: Bell, 2012

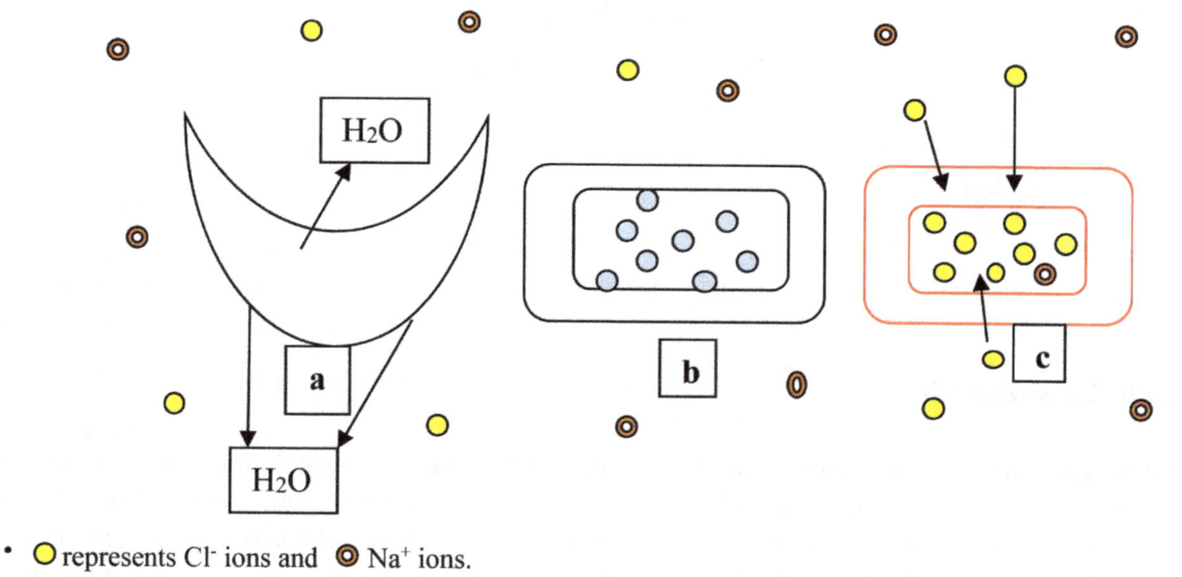

Figure 4. Mechanisms for adaption by halophiles in environment

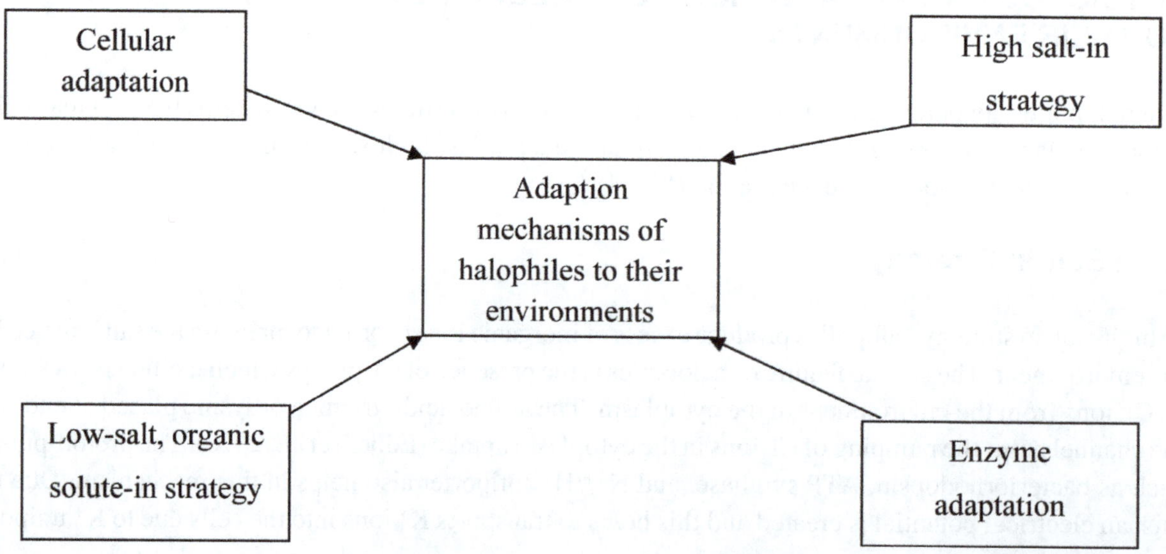

Low-Salt, Organic Solute-In Strategy

The low-salt, organic solute-in strategy makes use of osmolytes. The osmolytes provide protection to the proteins of halophiles from degradation in solvent such as water. Also, these osmolytes enable the halophiles to adapt extreme variations in saline conditions. These osmolytes of halophiles are without positive or negative charges, and water-soluble. The examples of osmolytes in halophiles are betaine, ectoine, hydroxyectoine, glutamine, sucrose and glycine betaine (Edbeib et al., 2016).

Requirement for Salt

Halophilic microorganisms are able to grow under high salt concentration. The ability to tolerate high salt concentration depends on the nutrients in the medium, temperature, etc. Osmoadaptation is the important feature in all halophiles which covers physiological and genetic adaptation under minimum water conditions (Kanekar et al., 2012). Since the cell membranes are permeable to water, so the microorganisms need to maintain their cytoplasm 'isosmotic' so that they are able to survive under the extreme conditions. If a hydrostatic pressure is applied, then the condition under which cytoplasm is maintained becomes 'hyperosmotic'.

Enzyme Adaptation

The enzymes of halophiles have special features viz., ability to tolerate high temperature, remain stable in high salt concentration and pH. These enzymes help the halophiles to grow in presence of high salt concentration, lakes, brines or other saline areas (Edbeib et al., 2016). There are reports on enzymes produced by halophiles viz., amylase, xylanase, protease, cellulase, esterase, amylopullulanase, endo-1, 4-β-xylanase, etc. (Edbeib et al., 2016).

Halophilic Adaptation of Enzymes

The proteins of halophiles adapt "salt-in" strategy to survive under saline environments. These proteins have more of glutamate and aspartate amino acids which are acidic in nature. The proper salt concentration is necessary for these proteins and intracellular organelles of halophiles to function correctly (Kanekar et al., 2012).

Cell Envelopes

The cytoplasmic membrane of halophiles is important. The ability to modify the cytoplasmic membrane with the outside salt concentration is important for the halophiles. The cell membrane of halophiles is composed of phospholipids and glycolipids. Due to the anionic nature of the lipids, they give negative charge to the cell membrane (Kanekar et al., 2012).

Adaptation Strategies of Haloarchaea: Coping with Desiccation, Starvation, and Radiation

Organisms such as Haloarchaea living in extreme environments have to acquire physiological adaptations to cope. Two of the major factors that influence survival are desiccation and nutrition. The salt-in and salt-out strategies utilized help to preserve intracellular water in majority of the cases. However, some species, such as the *Haloquadratum walsbyi* makes use of a protein known as halomucin to help retain water (Bolhuis et al., 2006). Halomucin is a large protein, with an amino acid length of 9159, which helps retain water inside cells through formation of a water-rich capsule. In terms of starvation, numerous microorganisms have been observed to have a higher surface-volume ratio in order to help maximize nutrient usage from the environment. Most recently, Haloarchaea with an increased surface-volume ratio were observed in fluid inclusions found inside halite crystals (Survival of *Halobacteria* within fluid inclusions in salt crystals - Microbiology Society, n.d.). Thus, it can be hypothesized that archaea flatten their cellular structure in order to maximize the surface-volume ratio as a means of dealing with desiccation and starvation. Exposure to radiation leads to double stranded breaks leading to hampered transcription and translation. Haloarchaea are able to survive high doses of radiation that may be attributable to polyploidy. Polyploidy organisms are able to repair double-stranded breaks and give rise to intact chromosomes (Soppa, 2013).

HALOPHILIC MICROORGANISMS

The list of halophilic microorganisms is shown in Table 1.

PHYLOGENETIC DIVERSITY OF HALOPHILES

Halophilic microorganisms are found in three domains of life viz., Archaea, Bacteria, and Eucarya (Fig. 5). The halophiles growing in presence of oxygen - Archaea of the order Halobacteriales and family Halobacteriaceae, are under the group of extremophiles with special features. These are found mainly near Dead Sea, saline lakes such as Lake Magadi, Kenya, saltern ponds, etc. These halophiles show presence of carotenoid pigments (e.g., bacterioruberin, the purple photosynthetic pigment, bacteriorhodopsin, etc.). Due to bacterioruberin pigment, the lakes become red in colour (Oren, 2002). Even the Euryarchaeota includes halophiles which are able to carry methanogenesis process even at NaCl saturation. Within the domain Eucarya, there is less number of halophiles. The green alga *Dunaliella* which is known to produce high amount of carotene is the only one included in Eucaryal microorganism which can grow in presence of high concentration of salt. *Dunaliella* can grow over a wide range of salt concentrations (Oren, 2002). The domain Bacteria includes different types of halophilic and halo tolerant microorganisms, and is widely spread over huge phylogenetic subgroups. Within the domain Bacteria, majority of them are moderate halophiles. The Proteobacteria also include halophilic microorganisms. Halophiles are also seen among the group of Cyanobacteria, e.g. the *Flavobacterium* - Cytophaga branch, spirochetes, and actinobacteria. In the group of *Firmicutes* which are Gram-positive bacteria, halophiles are reported in aerobic as well as anaerobic. The Halanaerobiales includes two families which are Halanaerobiaceae and Halobacteroidaceae. These two families include only anaerobic microorganisms (Oren, 2001).

Halophiles

Table 1. List of halophiles

S. No.	Halophiles
1	*Halobacillus* sp.
2	*Halococcus* sp.
3	*Paenibacillus* sp.
4	*Halomonas* sp.
5	*Halovibrio* sp.
6	*Haloarcula* sp.
7	*Chromohalobacter* sp.
8	*Halobacterium* sp.
9	*Hortaea werneckii*
10	*Halorubrum* sp.
11	*Salinibacter ruber*
12	*Haloferax mediteranei*
13	*Salinivibrio* sp.
14	*Salinibacillus* sp.
15	*Salinicoccus* sp.
16	*Salicola* sp.
17	*Oceanobacillus* sp.
18	*Halalkalicoccus jeotgali*

Figure 5. Halophiles found in three domains of life

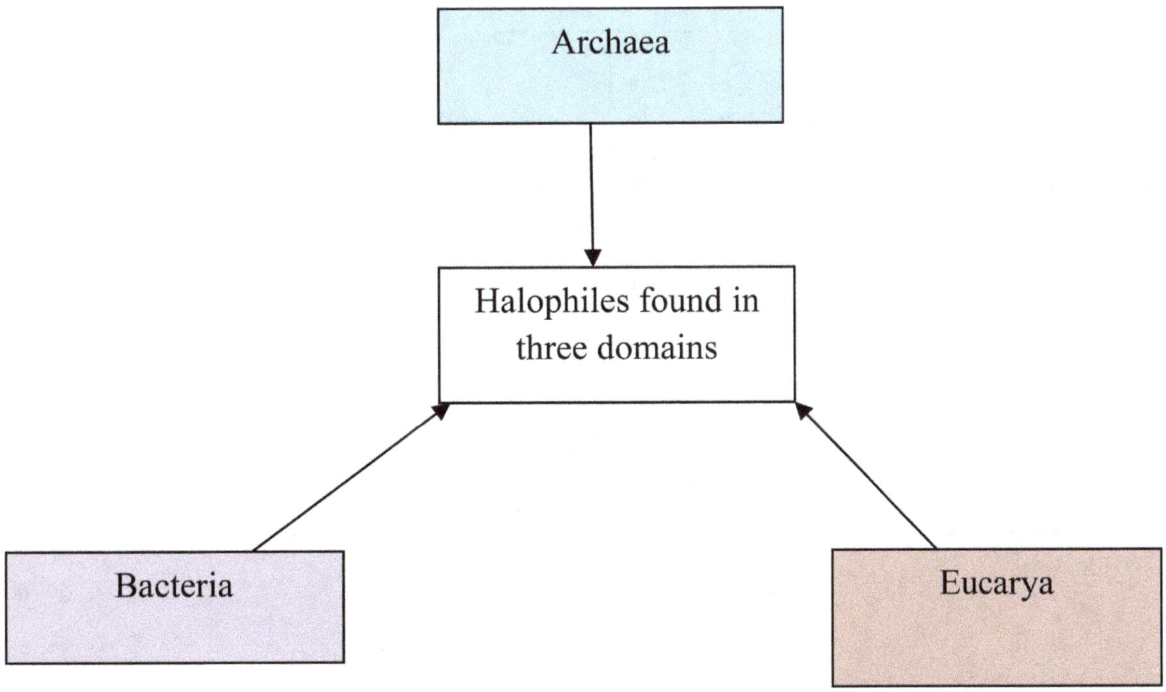

Thalassohalines

Thalassohalines have salt concentration of 35% which is nearly 10 times of sea water. The examples of thalassohaline brines are viz., solar salterns, salt mine, brine springs, etc. (Edbeib et al., 2016). Protozoa, diatoms and algae are found below 10% (w/v) NaCl in solar salterns. The eukaryotes *Dunaliella* spp. and *Artemia salina* are found mostly at very high salt concentration. *Halomonas* sp., *Salinivibrio* sp. and *Flavobacteria* are also reported to be found under the group of thalassohalines (Edbeib et al., 2016).

Athalassohaline

The Great Salt Lake and Dead Sea are the reported hyper saline environments. The human activities have changed the microbial flora and also chemical composition of Great Salt Lake and Dead Sea (Edbeib et al., 2016).

MECHANISMS OF OSMOREGULATION BY HALOPHILES

The different mechanisms of osmoadaptation by halophiles are shown in Fig. 6. The list of different osmolytes is shown in Table 2.

Figure 6. Mechanisms of osmoadaptation by halophiles

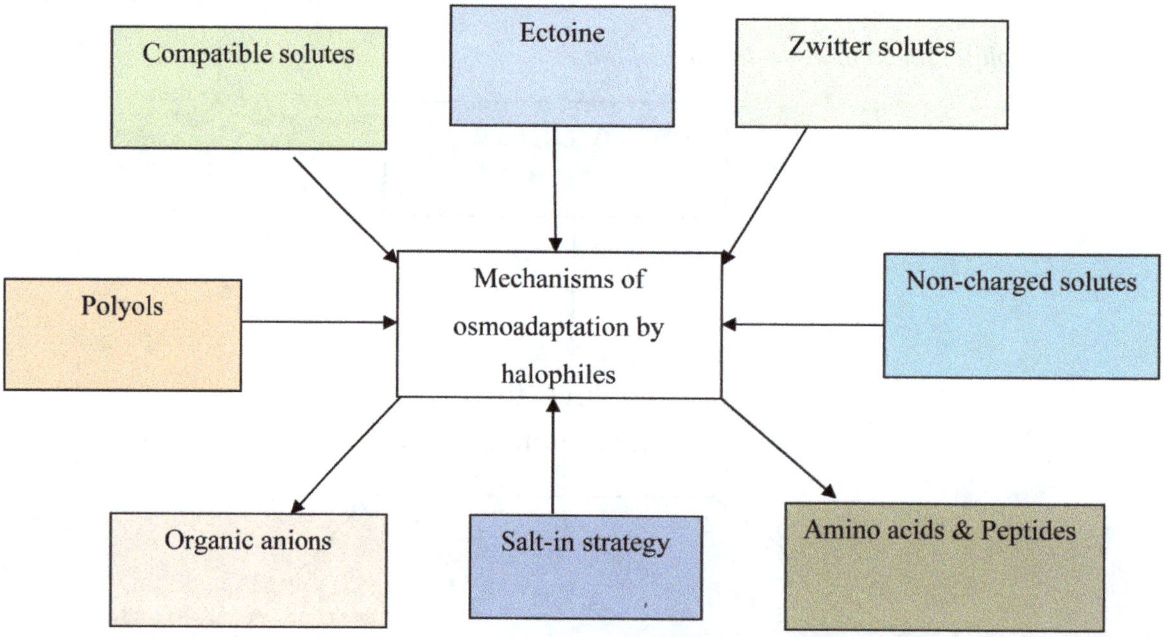

Table 2. List of different osmolytes

Zwitter solutes	Organic anions	Amino acids and peptides	Polyols	Non-charged solutes
betaine	ectoine	Carbamoyl-Lglutamine-1-amide	glucose	trehalose
hydroxyectoine	L-glutamate	N-acetyl glutaminyl glutamine amide	sorbitol	sucrose
β-glutamine			arabitol	α-glucosylglycerol
N-γ-acetyldiaminobutyrate			mannitol	α-Mannosylglyceramide
			glycerol	N-acetylglutaminyl-glutamine amide

Compatible Solute Strategy

Halophiles maintain their cytoplasm with the high salt concentration as they produce osmolytes. These osmolytes or compatible solutes are secreted by the cells itself. Osmolytes are of organic and inorganic types (Nath, 2016). Some of the bacteria do not have intracellular systems and hence are unable to transport water and cannot adjust to osmotic stress. Hence, halophiles maintain internal environment by producing osmolytes (Moghaddam et al., 2016).

Zwitter Solutes

Halophiles have neutral amino acids which serve as osmolytes in saline stress environment. Polar amino acids are useful as intermediates in translation process. The microbial cells produce zwitter solutes derived from amino acids and these serve as osmolytes under high salt concentration environment (Knief et al., 2012). The zwitter ionic solutes include betaine, hydroxyectoine, β-glutamine, N-γ-acetyldiaminobutyrate, etc.

Organic Anions

Due to presence of high intracellular K^+ ions, this results in negative potential in the cell. The anions help to balance this negative potential which enables the halophiles to survive at high salt concentration. Some of the halophiles synthesize ectoine and L-glutamate which are osmolytes (Tanimura et al., 2016). Ectoine and L-glutamate have presence of carboxylate group which gives the negative charge (Pinar et al., 2014). The *Halobacteria* enhance their glutamate levels at high salt concentrations. The negative charge of L-glutamate balances with positive charge of K^+ inside the cell (Borjian et al., 2016).

Amino Acids and Peptides

Proline, glutamine dipeptide and carboxamine are the major osmolytes from the class amino acids and found in halophilic purple sulphur bacteria. Other osmolytes from amino acids include Carbamoyl-Lglutamine-1-amide and N-acetyl glutaminyl glutamine amide (Besse et al., 2015; Raizel et al., 2016).

Polyols

The polyols viz., glucose, sorbitol, arabitol, mannitol and glycerol are used as osmolytes for osmoadaptation by halophiles (Youssef et al., 2014).

Non-Charged Solutes

Non-charged compatible solutes are also used by some halo tolerant and halophilic bacteria for osmoadaptation. Few halo tolerant e.g. yeast *Debaryomyces hatnsenii* and *Hortea werneckii* accumulate glycerol as an osmolyte to protect their cells in extreme conditions (Rhaisa, 2016; Sorokin et al., 2014). The examples of non-charged solutes are trehalose, sucrose, α-glucosylglycerol, α-Mannosylglyceramide and N-acetylglutaminyl-glutamine amide.

Ectoine

Ectoine is synthesized by some halophiles and used as osmolyte to protect under high salt concentration. Some of the halophiles which synthesize ectoine are *Halomonas, Oceanobacillus, Marinococcus* and *Nesterenkonia* (Moghaddam et al., 2016). The pathway for synthesis of ectoine has been studied in *Halomonas* and *Oceanobacillus* (Tanimura et al., 2016).

TAXONOMIC GROUPING OF HALOPHILES

The taxonomic grouping of halophiles is shown in Fig. 7. The halophiles with fully sequenced genomes are *Halomonas salina* CIFRI 1, *Halomonas hydrothermalis* MTCC 5445, *Erythrobacter vulgaris* O1, *Pontibacillus yanchengensis* Y32, *Halomonas zhanjiangensis* DSM 21076, *Gammaproteobacteria* MFB021, *Halomonas titanicae* BH1, *Bacillus* sp. SB49 (Edbeib et al., 2016).

ENZYMES PRODUCTION BY HALOPHILES

The halophiles produce many enzymes which are stable at various salt concentrations or saline environments. The enzymes produced by halophiles include amylase, lipase, xylanase, protease, cellulase, esterase, alcohol dehydrogenase, amylopullulanase, etc. (Fig. 8). These enzymes have numerous biotechnological and industrial applications.

ANTIMICROBIAL RESISTANCE IN HALOPHILES

The problem of antimicrobial resistance (AMR) in non-terrestrial bacteria has seen a surge over the last decades due to increase in the use of antibiotics in industries as well as improper waste disposal. The emergence of antibiotic resistance in halophilic organisms is shown in Fig. 9. The use of antibiotics is widespread in the food industry in order to increase yield and prevent infections. However, the use of excessive antibiotics exerts selection pressure on existing bacterial populations, promoting the growth of

AMR strains. Direct selection pressure as well as conjugal plasmid transfer across distant species due to improper biomedical waste disposal leads to antimicrobial resistance in marine species and halophiles in niches unexposed to antibiotics. Exposure of halophiles, abundant in saline environments, adversely impacts industries relying on biomolecules produced by halophiles as well as poses a threat to human health through their presence in salt-rich foods. The studies concerning halophiles have largely focused on their physiological adaptations and the biomolecules produced as a consequence of the same. Halophiles have emerged as an economically valuable resource being utilized to produce pharmacologically relevant molecules, bioremediation, tanneries, as a source of carotenoids, and so on. In terms of research, halophiles have proven to be unique model organisms to study processes like electron transport system, physiological adaptations of microorganisms to extreme environments, as model organisms for astrobiology, to elucidate protein functions, and several others. However, recent studies reporting the rise of AMR in halophiles is leading to increasing concern due to its widespread impacts.

Figure 7. Taxonomic grouping of halophiles
Source: Kanekar et al., 2012

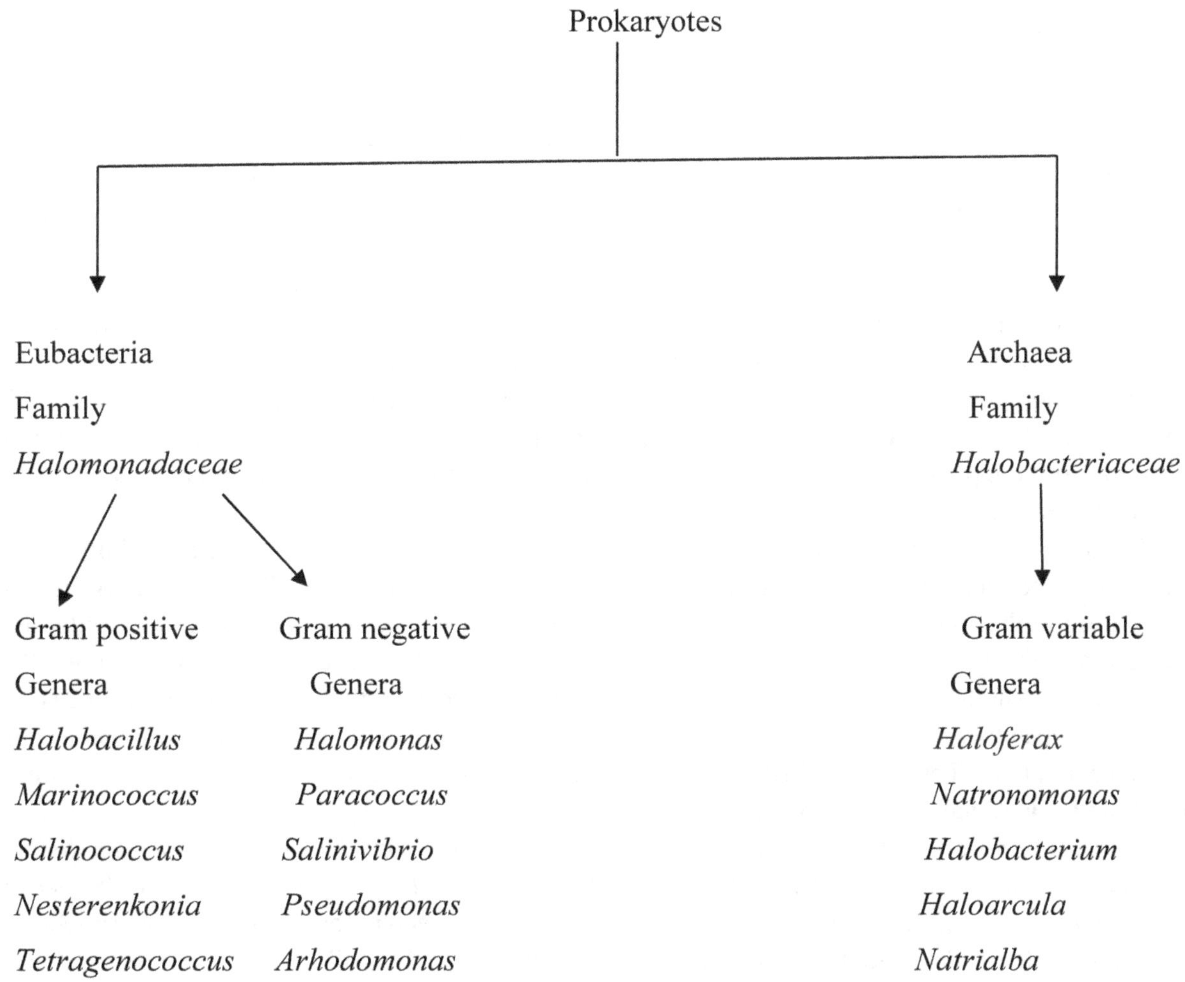

Figure 8. Enzymes produced by halophiles

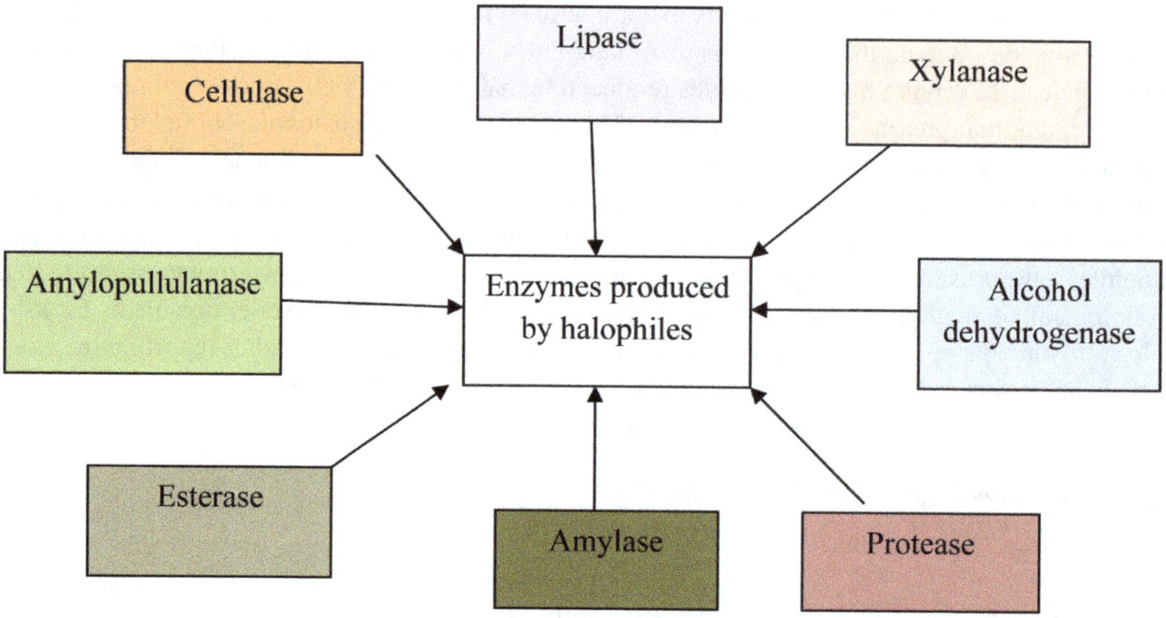

Vibrio parahaemolyticus is a well-known pathogenic organism leading to gastroenteritis in humans upon consumption of contaminated fish or fish products. It is a Gram-negative halophile bacterium that has been recently reported to exhibit AMR due to an increased use of antibiotics in fish farming. The *tdh* and *trh* genes responsible for hemolytic activity confer pathogenicity to *V. parahaemolyticus*; however, among the fish samples tested for the presence of pathogenic strains, <3% tested positive. Antibiotic resistance profiling of *V. parahaemolyticus* revealed ampicillin resistance caused by β-lactamase activity. Thus, ampicillin should no longer be utilized in fish farming. *V. parahaemolyticus* remains susceptible to tetracyclines, chloramphenicol, and nalidixic acid. Even though the presence of the pathogenic strain is not widespread and the organism remains susceptible to several classes of antibiotics, care should be taken to avoid rampant, unregulated use of antibiotics in order to improve yield (Elmahdi et al., 2016; Letchumanan et al., 2015; Tan et al., 2020).

Vibrio vulnificus is a halophile Gram-negative bacterium that causes infections such as cellulitis, gastroenteritis, and septicemia, following consumption of contaminated sea food (*Vibrio vulnificus*: Review of mild to life-threatening skin infections - MDedge Dermatology, n.d.). Examination of *V. vulnificus* revealed a multi-drug resistant profile. The bacterium exhibited resistance to vancomycin, ampicillin, penicillin, cephalothin, and vancomycin; while it was susceptible to kanamycin, tetracycline, and streptomycin (Elmahdi et al., 2016; *Vibrio vulnificus* in aquariums is a novel threat to marine mammals and public health, n.d.). The rise of AMR in bacteria occurring in sea food poses a serious threat to human health with additional economic impacts.

Several other species of marine halophilic bacteria and archea viz., *Halomonas koreensis*, *Salimicrobium flavidum*, *Alkalibacillus almallahensis*, *Salimicrobium salexigens*, *Marinobacteroulmenensis*, *Halomonas smyrnensis*, *Haloarcula sp.*, and *Halovivax sp*. have been reported to exhibit AMR (Shinde & Thombre, 2016). The halophilic bacteria exhibited resistance to the antibiotics ampicillin, bacitracin, chloramphenicol, and ciprofloxacin; while the Haloarchaea exhibited resistance to all antibiotics. Further examination also revealed an increased expression of efflux pumps on nearly all halophiles, which may be indicative of their role in conferring resistance. While antimicrobials such as silver nanoparticles (Shinde & Thombre, 2016) have proven effective against AMR halophiles, further research is required to elucidate the underlying mechanisms and explore additional antimicrobial agents.

In addition to food-borne pathogens, AMR halophiles also have significant detrimental impacts on industries. The presence of AMR halophiles in tanneries leads to degradation of hides cured in brine posing a problem in long-term storage. The halophiles from tanneries exhibited resistance to kanamycin, neomycin, and ampicillin (Ghosh et al., 2010). This may be attributable to a conjugal plasmid transfer from terrestrial bacteria to marine bacteria.

Figure 9. Emergence of antibiotic resistance in halophilic organisms

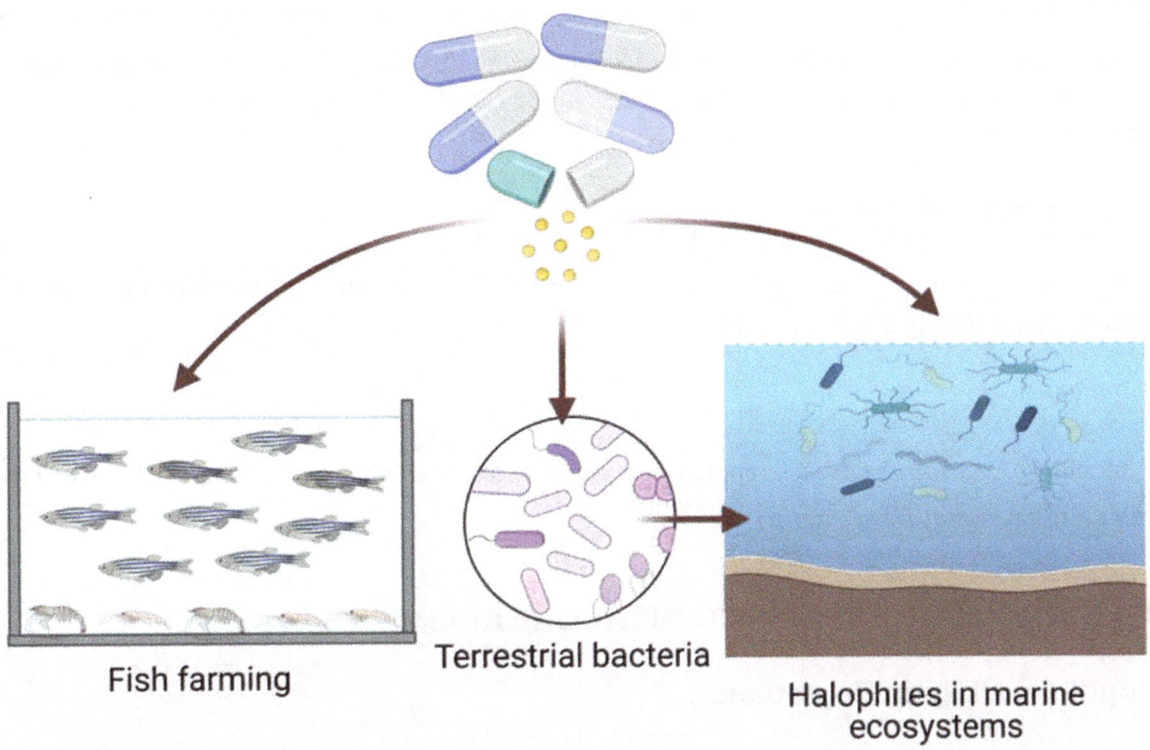

The AMR resistance in halophiles presents an emerging threat to health and economy and thus highlights the need to elucidate resistance mechanisms. The mechanisms of antibiotic resistance in halophilic organisms are shown in Fig. 10. Resistance to antibiotics in halophiles may be attributable to the following factors:

- R-plasmids

Horizontal transfer of plasmids occurs across bacterial species in several ecosystems such as wastewater, marine environments, soil, and lakes. Plasmid transfer has also been shown to occur between genetically distant organisms under a range of environmental factors.

- Expression of efflux pumps

A significant up regulation of efflux pumps was observed in AMR halophiles. Halophiles could potentially utilize any one among the five types of efflux pumps, MATE (Multidrug and toxic efflux) pumps, ABC (ATP binding cassettes) transporters, MF (Major facilitator) pumps, RND (Root Nodulation and Division) pump, and SMR (Small Multidrug Resistance) pumps.

- β-lactamase gene

The β-lactamase gene contributes to resistance against antibiotics by hydrolysing peptide bonds in the beta-lactam ring. This gene is present in the genome of halophilic organisms and could be one of the potential causes of AMR.

- Lack of targetable molecules

Halophile physiology varies significantly from that of other bacteria, thus reducing the number of potentially targetable molecules.

- Polyploidy

Halophilic organisms have been reported to have multiple gene copies, which may help resist DNA damage caused by antibiotics such as nalidixic acid.

APPLICATIONS OF HALOPHILIC MICROORGANISMS

Halophiles in Lignin Degradation

Halophiles play very important role in lignin degradation. *Sagittula stellata*, a marine aerobic bacterium, was reported as first halophile which was able to degrade lignin. Halophiles produce enzyme laccase which helps in lignin degradation (Uthandi et al., 2010). This laccase enzyme is stable at varied salt concentration and temperature. Laccase enzyme is produced by halophiles e.g., *Chromohalobacter* sp., *Bacillus* sp. strain WT, *Aquisalibacillus elongates* (Rezaei et al., 2017), *Bacillus safensis* sp. strain S31 (Siroosi et al., 2018), *Haloferax volcanii* strain SB01, *Pycnoporus sanguineus*, *Digitatispora marina*, *Halocyphina villosa* and *Nia vibrissa* (Pang et al., 2011).

Figure 10. Mechanisms of antibiotic resistance
Source: BioRender, n.d.

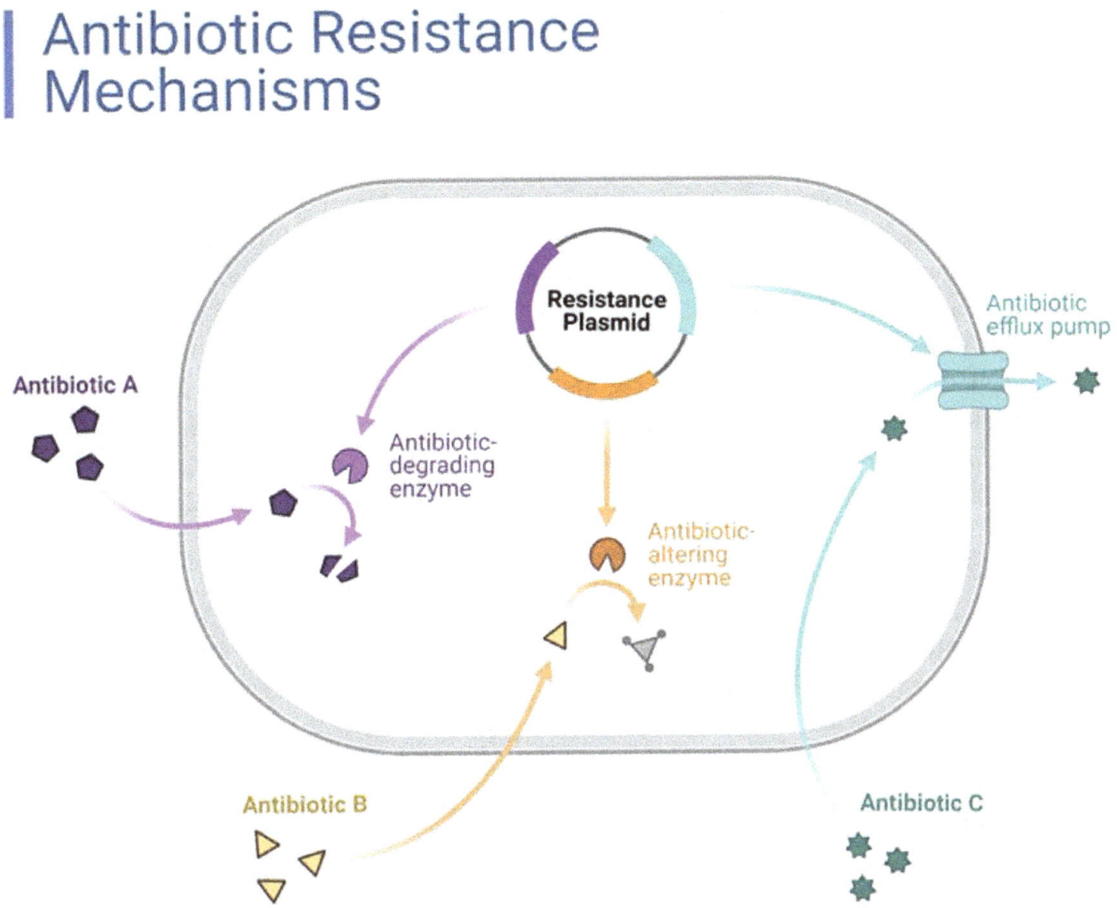

Biodiesel and Methane Production from Halophilic Microalgae

Biodiesel is most important energy source because of many applications (Nejad & Zahedi, 2018). The chemical processes used for the production of biodiesel are costly, time consuming and causes pollution. The production of biodiesel from edible oils is costly. Therefore, use of non-edible oils such as oils from halophilic microalgae is economical and eco-friendly (Abdullah et al., 2015; Yu et al., 2013). Halophilic microalgae are able to grow fast and have high lipid content (about 80% of their weight) which has increased the interest of researchers for their use in the production of biodiesel. *Dunaliella salina* and *Tetraselmis elliptica* microalgae isolated from hyper saline Bardawil (Abomohra et al., 2017) have good content of lipid and fatty acids viz., linoleic, palmitic and oleic acids and hence can be used for the production of biodiesel. The marine microalgae viz., *Nannochloropsis salina* (Bartley et al., 2013) and *A. halophytica* (Miriam et al., 2017) which have the ability to produce lipids also been reported as promising feed stocks for bioenergy production. The production of biodiesel from halophilic microalgae is a very clean and economical technology.

Methane is good source of renewable fuel and can be used to produce heat and electricity. Many of the marine microalgae can be used to produce methane. By hydrolysis, they can convert livestock manure, crop residues, wastes, etc. to methane (CH_4) and carbon dioxide (CO_2) (Amoozegar et al., 2019; Wei et al., 2013).

Applications of Pigments from Halophiles

The halophilic microorganisms produce pigments such as β-carotene, bacteriorhodopsin and bacterioruberin. The halophilic algae *Dunaliella salina* produces β-carotene pigment which has immense applications in food industries as well as medicinal uses such as anti-diabetic, antiviral, anti-inflammatory, etc. (Manjula et al., 2018). Bacteriorhodopsin is another pigment produced by halophilic archaea viz., *Halobacterium salinarium* and *Halobacterium halobium*. It has applications in ocular and computerized devices, photovoltaic cells and artificial retinas (Ashwini et al., 2017). Bacterioruberin pigment is produced by halophilic archaea such as *Halobacterium salinarum, Halorubrum sodomense, Haloarcula vallismortis, Haloarcula japonica, Halococcus morrhuae, Haloferax volcanii*, etc. This pigment has good antioxidant property and also has applications in solar cells (Manjula et al., 2018).

Glycerol and Biohydrogen

The members of order *Halanaerobiales* and green algae *Dunaliella* are known to produce glycerol (Oren, 2017). They breakdown glycerol and form important by-products. Glycerol-based hydrogen production by halophilic microorganisms has been studied from *Halanaerobium saccharolyticum* subsp. *senegalensis*. Biohydrogen can be used for generation of electricity and as fuel.

Ectoine and Hydroxyectoine

Ectoine and hydroxyectoine produced by halophilic actinobacteria have important applications as protective and stabilizing agents for the mammalian cells (Pastor et al., 2010).

Treatment of Saline Wastewaters Using Halophiles

The industrial processes generate huge amount of wastewater with high conc. of salt. Such wastewater contains many toxic compounds, and hence needs to be treated. Anaerobic biodegradation process is effective for the treatment of such wastewater. The halophiles such as *Halobacterium salinarum, Halomonas* spp., *Halanaerobium lacusrosei*, etc. are useful in the anaerobic degradation process for the treatment of saline wastewater (Mohammadipanah et al., 2015).

Therapeutic Compounds from Halophilic Actinobacteria

Halophilic actinobacteria produce many therapeutic compounds viz., anticancer, antitumor, anti-inflammatory, antioxidant, and antimalarial substances. The therapeutic compounds from halophilic actinobacteria are represented in Table 3.

Table 3. Therapeutic compounds from halophilic actinobacteria

Halophilic actinobacteria	Therapeutic compound	Use	Reference
Salinispora tropica	Salinosporamide A	anticancer	(Abdel-Mageed et al., 2010)
Saccharomonospora sp.	Lodopyridone	anticancer activity against the human colon adenocarcinoma cell line HCT-116	(Arasu et al., 2016)
Streptomyces sp.	Cyclomarin A	anti-inflammatory agent	(Arasu et al., 2016)
Streptomyces sp.	Trioxacarcin	antimalarial substance	(Arasu et al., 2016)
Actinokineospora sp.	Actinosporins C and D	antioxidant	(Grkovic et al., 2014)

CONCLUSION

The halophiles are salt loving microorganisms having immense industrial applications. They have various mechanisms to adapt to salt conditions, which includes them under the class of extremophiles and makes them novel. The halophiles are diverse in phylogeny and taxonomy aspects. The halophiles also have antimicrobial resistance and hence can be used for preparation of drugs and therapeutic compounds. More research needs to be carried on the molecular level identification of halophiles and also genes responsible for adaptation by halophiles to extreme salt and other stress conditions.

ACKNOWLEDGMENT

The Figures No. 9 and 10 are done using BioRender (n.d.).

REFERENCES

Abdel-Mageed, W. M., Milne, B. F., Wagner, M., Schumacher, M., Sandor, P., Pathom-aree, W., Goodfellow, M., Bull, A. T., Horikoshi, K., Ebel, R., Diederich, M., Fiedler, H.-P., & Jaspars, M. (2010). Dermacozines, a new phenazine family from deep-sea dermacocci isolated from a Mariana Trench sediment. *Organic & Biomolecular Chemistry*, *8*(10), 2352–2362. doi:10.1039/c001445a PMID:20448892

Abdullah, M. A., Usman, S. M., Shah, A. A., & El-Sayed, H. (2015). Algal biotechnology for bioenergy, environmental remediation and high-value biochemical. In D. Thangadurai & J. Sangeetha (Eds.), *Biotechnology and Bioinformatics* (pp. 301–344). CRC Press.

Abomohra, A. F., El-Sheekh, M., & Hanelt, D. (2017). Screening of marine microalgae isolated from the hyper saline Bardawil lagoon for biodiesel feedstock. *Renewable Energy*, *101*(C), 1266–1272. doi:10.1016/j.renene.2016.10.015

Amoozegar, M. A., Safarpour, A., Noghabi, K. A., Bakhtiary, T., & Ventosa, A. (2019). Halophiles and their vast potential in biofuel production. *Frontiers in Microbiology*, *10*, 1–17. doi:10.3389/fmicb.2019.01895 PMID:31507545

Arasu, M. V., Esmail, G. A., & Abdullah Al-Dhabi, N. (2016). Hyper saline actinomycetes and their biological applications. In D. Dhanasekaran & J. Yi (Eds.), *Actinobacteria - Basics and Biotechnological Applications*. IntechOpen. doi:10.5772/61065

Ashwini, R., Vijayanand, S., & Hemapriya, J. (2017). Photonic potential of Haloarchaeal pigment bacteriorhodopsin for future electronics. *Current Microbiology, 74*(8), 996–1002. doi:10.100700284-017-1271-5 PMID:28573340

Bartley, M. L., Boeing, W. J., Corcoran, A. A., Holguin, F. O., & Schaub, T. (2013). Effects of salinity on growth and lipid accumulation of biofuel microalga *Nannochloropsis salina* and invading organisms. *Biomass and Bioenergy, 54*, 83–88. doi:10.1016/j.biombioe.2013.03.026

Bass-Becking, L. M. (1931). Historical notes on salt and salt-manufacture. *The Scientific Monthly, 32*(5), 434–446.

Bell, E. (2012). *Life at extremes: Environments, organisms, and strategies for survival. CABI Invasive Species 1.* CABI. doi:10.1079/9781845938147.0000

Besse, A., Peduzzi, J., Rebuffat, S., & Carre-Mlouka, A. (2015). Antimicrobial peptides and proteins in the face of extremes: Lessons from archaeocins. *Biochimie, 118*, 344–355. doi:10.1016/j.biochi.2015.06.004 PMID:26092421

BioRender. (n.d.). *Reprinted from Antibiotic resistance mechanisms.* https://app.biorender.com/illustrations/60e5662546850f00acfe0e3a

Borjian, F., Han, J., Hou, J., Xiang, H., & Berg, I. A. (2016). The methyl aspartate cycle in *Haloarchaea* and its possible role in carbon metabolism. *The ISME Journal, 10*(3), 546–557. doi:10.1038/ismej.2015.132 PMID:26241502

Corral, P., Amoozegar, M. A., & Ventosa, A. (2020). Halophiles and their biomolecules: Recent advances and future applications in biomedicine. *Marine Drugs, 18*(1), 33. doi:10.3390/md18010033 PMID:31906001

Edbeib, M. F., Wahab, R. A., & Huyop, F. (2016). Halophiles: Biology, adaptation, and their role in decontamination of hyper saline environments. *World Journal of Microbiology & Biotechnology, 32*(8), 1–23. doi:10.100711274-016-2081-9 PMID:27344438

Elmahdi, S., DaSilva, L. V., & Parveen, S. (2016). Antibiotic resistance of *Vibrio parahaemolyticus* and *Vibrio vulnificus* in various countries: A review. *Food Microbiology, 57*, 128–134. doi:10.1016/j.fm.2016.02.008 PMID:27052711

Ghosh, R., Chattopadhyay, P., Chattopadhyay, B., & Pal, D. (2010). Antibiotic resistance profile of halophilic microorganisms isolated from tannery effluent. *Indian Journal of Biotechnology, 9*(1), 80–86.

Grkovic, T., Abdelmohsen, U. R., Othman, E. M., Stopper, H., Edrada-Ebel, R., Hentschel, U., & Quinn, R. J. (2014). Two new antioxidant actinosporin analogues from the calcium alginate beads culture of sponge-associated *Actinokineospora* sp. strain EG49. *Bioorganic & Medicinal Chemistry Letters, 24*(21), 5089–5092. doi:10.1016/j.bmcl.2014.08.068 PMID:25266784

Kanekar, P., Kanekar, S., Kelkar, A., & Dhakephalkar, P. (2012). Halophiles - Taxonomy, diversity, physiology and applications. In *Microorganisms in Environmental Management: Microbes and Environment* (pp. 1–34). Springer, Netherlands. doi:10.1007/978-94-007-2229-3_1

Knief, C., Delmotte, N., Chaffron, S., Stark, M., Innerebner, G., Wassmann, R., von Mering, C., & Vorholt, J. (2012). Metaproteogenomic analysis of microbial communities in the phyllosphere and rhizosphere of rice. *The ISME Journal*, *6*(7), 1378–1390. doi:10.1038/ismej.2011.192 PMID:22189496

Letchumanan, V., Pusparajah, P., Tan, L. H., Yin, W. F., Lee, L. H., & Chan, K. G. (2015). Occurrence and antibiotic resistance of *Vibrio parahaemolyticus* from shellfish in Selangor, Malaysia. *Frontiers in Microbiology*, *6*, 1417. doi:10.3389/fmicb.2015.01417 PMID:26697003

Luque, R., Bejar, V., Quesada, E., & Llamas, I. (2014). Diversity of halophilic bacteria isolated from Rambla Salada, Murcia (Spain). *Canadian Journal of Microbiology*, *60*(12), 839–846. doi:10.1139/cjm-2014-0342 PMID:25403824

Manjula, D., Jeevitha, P., Ramya, I., Hemapriya, J., & Ravi, A. (2018). Biotechnological applications of halophilic pigments - An overview. *International Journal of Current Microbiology and Applied Sciences*, *7*(7), 4392–4398. doi:10.20546/ijcmas.2018.707.512

Mehrshad, M., Amoozegar, M. A., Makhdoumi, A., Rasooli, M., Asadi, B., Schumann, P., & Ventosa, A. (2015a). *Halovarius luteusgen* nov., sp. nov., a novel extremely halophilic archaeon from Urmia salt lake. *International Journal of Systematic and Evolutionary Microbiology*, *65*(Pt_8), 2420–2425. doi:10.1099/ijs.0.000279 PMID:25899505

Miriam, L. M., Raj, R. E., Kings, A. J., & Visvanathan, M. A. (2017). Identification and characterization of a novel biodiesel producing halophilic *Aphanothece halophytica* and its growth and lipid optimization in various media. *Energy Conversion and Management*, *141*, 91–100. doi:10.1016/j.enconman.2016.05.041

Moghaddam, J. A., Boehringer, N., Burdziak, A., Kunte, H. J., Galinski, E. A., & Schaberle, T. F. (2016). Different strategies of osmoadaptation in the closely related marine Myxobacteria *Enhygromyxa salina* SWB007 and *Plesiocystis pacifica* SIR-1. *Microbiology*, *162*(4), 651–661. doi:10.1099/mic.0.000250 PMID:26842314

Mohammadipanah, F., Hamedi, J., & Dehhaghi, M. (2015). Halophilic bacteria: Potentials and applications in biotechnology. In D. K. Maheshwari & M. Saraf (Eds.), *Halophiles - Biodiversity and Sustainable Exploitation* (pp. 277–321). Springer Publisher. doi:10.1007/978-3-319-14595-2_11

Mutlu, M. B., & Guven, K. (2015). Bacterial diversity in Camalti Saltern, Turkey. *Polish Journal of Microbiology*, *64*(1), 37–45. doi:10.33073/pjm-2015-005 PMID:26094314

Nath, A. (2016). Insights into the sequence parameters for halophilic adaptation. *Amino Acids*, *48*(3), 751–762. doi:10.100700726-015-2123-x PMID:26520112

Nejad, A. S., & Zahedi, A. R. (2018). Optimization of biodiesel production as a clean fuel for thermal power plants using renewable energy source. *Renewable Energy*, *119*, 365–374. doi:10.1016/j.renene.2017.11.087

Ng, W. V., Kennedy, S. P., Mahairas, G. G., Berquist, B., Pan, M., Shukla, H. D., Lasky, S. R., Baliga, N. S., Thorsson, V., Sbrogna, J., Swartzell, S., Weir, D., Hall, J., Dahl, T. A., Welti, R., Goo, Y. A., Leithauser, B., Keller, K., Cruz, R., ... DasSarma, S. (2000). Genome sequence of *Halobacterium* species NRC-1. *Proceedings of the National Academy of Sciences of the United States of America, 97*(22), 12176–12181. doi:10.1073/pnas.190337797 PMID:11016950

Oren, A. (2001). The order Haloanaerobiales. In M. Dworkin, S. Falkow, E. Rosenberg, K. H. Schleifer, & E. Stackebrandt (Eds.), *The Prokaryotes. A Handbook on the biology of bacteria: Eco physiology, isolation, identification, applications* (3rd ed.). Springer-Verlag. doi:10.1007/978-1-4757-2191-1

Oren, A. (2002). Diversity of halophilic microorganisms: Environment, phylogeny, physiology, and applications. *Journal of Industrial Microbiology & Biotechnology, 28*(1), 56–63. doi:10.1038j/jim/7000176 PMID:11938472

Oren, A. (2014). Taxonomy of halophilic Archaea: Current status and future challenges. *Extremophiles, 18*(5), 825–834. doi:10.100700792-014-0654-9 PMID:25102811

Oren, A. (2015). Halophilic microbial communities and their environments. *Current Opinion in Biotechnology, 33*, 119–124. doi:10.1016/j.copbio.2015.02.005 PMID:25727188

Oren, A. (2017). Glycerol metabolism in hyper saline environments. *Environmental Microbiology, 19*(3), 851–863. doi:10.1111/1462-2920.13493 PMID:27511298

Pang, K. L., Chow, R. K., Chan, C. W., & Vrijmoed, L. P. (2011). Diversity and physiology of marine lignicolous fungi in Arctic waters: A preliminary account. *Polar Research, 30*(1), 5859–5865. doi:10.3402/polar.v30i0.5859

Pastor, J., Salvador, M., Argandona, M., Bernal, V., Reina-Bueno, M., Csonka, L., Iborra, J. L., Vargas, C., Nieto, J. J., & Cánovas, M. (2010). Ectoines in cell stress protection: Uses and biotechnological production. *Biotechnology Advances, 28*(6), 782–801. doi:10.1016/j.biotechadv.2010.06.005 PMID:20600783

Pinar, G., Krakova, L., Pangallo, D., Piombino-Mascali, D., Maixner, F., Zink, A., & Sterflinger, K. (2014). Halophilic bacteria are colonizing the exhibition areas of the Capuchin Catacombs in Palermo, Italy. *Extremophiles, 18*(4), 677–691. doi:10.100700792-014-0649-6 PMID:24863363

Raizel, R., Leite, J. S., Hypolito, T. M., Coqueiro, A. Y., Newsholme, P., Cruzat, V. F., & Tirapegui, J. (2016). Determination of the anti-inflammatory and cytoprotective effects of L-glutamine and L-alanine, or dipeptide, supplementation in rats submitted to resistance exercise. *British Journal of Nutrition, 116*(3), 470–479. doi:10.1017/S0007114516001999 PMID:27215379

Rezaei, S., Shahverdi, A. R., & Faramarzi, M. A. (2017). Isolation, one-step affinity purification, and characterization of a polyextremotolerant laccase from the halophilic bacterium *Aquisalibacillus elongatus* and its application in the delignification of sugar beet pulp. *Bioresource Technology, 230*, 67–75. doi:10.1016/j.biortech.2017.01.036 PMID:28161622

Rhaisa, A. (2016). *Studies of Debaryomyces hansenii killer toxin and its effect on pathogenic bloodstream Candida isolates* (Master thesis). University of Nebraska, Lincoln, NE.

Shinde, V., & Thombre, R. (2016). Antibiotic resistance profiling of marine halophilic bacteria and haloarchaea. *Journal of Applied Pharmaceutical Science*, *6*(10), 132–137. doi:10.7324/JAPS.2016.601018

Siroosi, M., Amoozegar, M. A., Khajeh, K., & Dabirmanesh, B. (2018). Decolourization of dyes by a novel sodium azide-resistant spore laccase from a halo tolerant bacterium, *Bacillus safensis* sp. strain S31. *Water Science and Technology*, *77*(12), 2867–2875. doi:10.2166/wst.2018.281 PMID:30065139

Soppa, J. (2013). Evolutionary advantages of polyploidy in halophilic archaea. *Biochemical Society Transactions*, *41*(1), 339–343. doi:10.1042/BST20120315 PMID:23356308

Sorokin, D. Y., Abbas, B., Erik, V. Z., & Muyzer, G. (2014). Isolation and characterization of an obligate chemolithoautotrophic *Halothiobacillus* strain capable of growth on thiocyanate as an energy source. *FEMS Microbiology Letters*, *354*(1), 69–74. doi:10.1111/1574-6968.12432 PMID:24673239

Survival of *Halobacteria* within fluid inclusions in salt crystals. (n.d.). *Microbiology Society*. Retrieved July 7, 2021, from https://www.microbiologyresearch.org/content/journal/micro/10.1099/00221287-134-5-1365?crawler=true

Tan, C. W., Rukayadi, Y., Hasan, H., Thung, T. Y., Lee, E., Rollon, W. D., Hara, H., Kayali, A. Y., Nishibuchi, M., & Radu, S. (2020). Prevalence and antibiotic resistance patterns of *Vibrio parahaemolyticus* isolated from different types of seafood in Selangor, Malaysia. *Saudi Journal of Biological Sciences*, *27*(6), 1602–1608. doi:10.1016/j.sjbs.2020.01.002 PMID:32489301

Tanimura, K., Matsumoto, T., Nakayama, H., Tanaka, T., & Kondo, A. (2016). Improvement of ectoine productivity by using sugar transporter-overexpressing *Halomonas elongate. Enzyme and Microbial Technology*, *89*, 63–68. doi:10.1016/j.enzmictec.2016.03.006 PMID:27233128

Tazi, L., Breakwell, D. P., Harker, A. R., & Crandall, K. A. (2014). Life in extreme environments: Microbial diversity in Great Salt Lake, Utah. *Extremophiles*, *18*(3), 525–535. doi:10.100700792-014-0637-x PMID:24682608

Uthandi, S., Saad, B., Humbard, M. A., & Maupin-Furlow, J. A. (2010). LccA, an archaeal laccase secreted as a highly stable glycoprotein into the extracellular medium by *Haloferax volcanii. Applied and Environmental Microbiology*, *76*(3), 733–743. doi:10.1128/AEM.01757-09 PMID:19966030

Vibrio vulnificus in aquariums is a novel threat to marine mammals and public health. (n.d.). Retrieved July 7, 2021, from https://www.ncbi.nlm.nih.gov/pmc/articles/PMC7169869/

Vibrio vulnificus: Review of Mild to Life-threatening Skin Infections. (n.d.). *MDedge Dermatology*. Retrieved July 7, 2021, from https://www.mdedge.com/dermatology/article/236136/infectious-diseases/*Vibrio-vulnificus*-review-mild-life-threatening-skin

Wei, N., Quarterman, J., & Jin, Y. S. (2013). Marine macro algae: An untapped resource for producing fuels and chemicals. *Trends in Biotechnology*, *31*(2), 70–77. doi:10.1016/j.tibtech.2012.10.009 PMID:23245657

Wei, X., Jiang, Y., Chen, X., Jiang, Y., & Lai, H. (2015). *Amycolatopsis flava* sp. nov., a halophilic actinomycete isolated from Dead Sea. *Antonie van Leeuwenhoek*, *108*(4), 879–885. doi:10.100710482-015-0542-z PMID:26233655

Youssef, N. H., Savage-Ashlock, K. N., McCully, A. L., Luedtke, B., Shaw, E. I., Hoff, W. D., & Elshahed, M. S. (2014). Trehalose / 2-sulfotrehalose biosynthesis and glycine-betaine uptake are widely spread mechanisms for osmoadaptation in the Halobacteriales. *The ISME Journal*, *8*(3), 636–649. doi:10.1038/ismej.2013.165 PMID:24048226

Yu, K. O., Jung, J., Ramzi, A. B., Choe, S. H., Kim, S. W., Park, C., & Han, S. O. (2013). Development of a *Saccharomyces cerevisiae* strain for increasing the accumulation of triacylglycerol as a microbial oil feedstock for biodiesel production using glycerol as a substrate. *Biotechnology and Bioengineering*, *110*(1), 343–347. doi:10.1002/bit.24623 PMID:22886471

Yue, H., Ling, C., Yang, T., Chen, X., Chen, Y., Deng, H., Wu, Q., Chen, J., & Chen, G. Q. (2014). A seawater-based open and continuous process for polyhydroxy alkanoates production by recombinant *Halomonas campaniensis* LS21 grown in mixed substrates. *Biotechnology for Biofuels*, *7*(1), 108–119. doi:10.1186/1754-6834-7-108

Chapter 3
Alkalophiles:
Environmental Distribution, Taxonomy, Physiology, Bioenergetics, Survival Mechanism, and Enzymes

Vishal Arvindlal Mevada
Directorate of Forensic Science, Gandhinagar, India

Urvisha Himmatbhai Beladiya
Veer Narmad South Gujarat University, Surat, India

Himani Rajendrakumar Gandhi
Christ College, Saurashtra University, Rajkot, India

Amitsinh Vijaysinh Mangrola
Shri Alpesh N. Patel Post Graduate Institute of Science and Research, Anand, India

Rajesh Kanjibhai Patel
Veer Narmad South Gujarat University, Surat, India

ABSTRACT

Alkalophiles are a class of extremophiles capable of survival in alkaline (pH roughly 8.5–11) environments, growing optimally around a pH of 10. At such high pH, the normal cellular functions are detrimentally affected for mesophilic organisms. The alkalophiles successfully manage stability of DNA, plasma membrane, and function of cytosolic enzymes, as well as other unfavorable physiological changes at such an elevated pH. A recent development in NextGen sequencing technology facilitates identifying uncultivable organisms amongst the extreme environments. In recent years, distribution of alkalophiles was reported from Soda Lake, marine environments, saline deserts, and natural thermal vents to natural water bodies. Although alkalophiles were first reported in 1889, their enzymatic and industrial applications still make them an interesting area of research. This chapter provides basic information on environmental distribution, taxonomy, physiology, bioenergetics, and survival mechanism and enzymes produced by alkalophilic organisms.

DOI: 10.4018/978-1-7998-9144-4.ch003

INTRODUCTION

Microorganisms are the most common living things on Earth. They are also highly diverse organisms found in almost every corner of the blue planet. Extremophilic microorganisms are a largely unexplored group that can survive in extreme conditions. Among the extremophiles group, Alkaliphiles or alkalophiles can thrive in alkaline (pH 8 to 11) environments. At such a high pH, normal cell functions for mesophilic organisms are adversely affected including inactivation of cytosolic enzymes, instability of DNA and plasma membrane. The adaptations in the genetic and metabolic machinery of these organisms allowed them to thrive in hostile conditions.

Chester (1889) reported bacterium *Sporosarcina pasteurii* (formerly *Bacillus pasteurii*) as the first alkaliphilic organism (Chester, 1897). The organism has capability to generate ammonium carbonate from urea in the presence of ammonia and alkaline environment. Later on in 1934, Vedder reported *Bacillus alcaliphilus* as a second example of isolation of alkalophiles. After the initial discovery, Takahara and Tanabe isolated indigo-reducing alkalophile capable of growing at pH 12.1 in 1960. During 1970-1980s, several studies reported for the applications of alkalophiles industrial applications. However, due to lack of proper methods to classification most of the isolates could not be appropriately classified upto species level. In 1990, Fritzie *et al.,* reclassified such alkalophiles based on their physiological and characteristics up to species level. With the advent of technology, the taxonomic classifications were revised based on DNA G+C mol% values, presence of diaminopimelic acid (DAP) in its cell wall, DNA–DNA hybridization and 16S rRNA gene. Initially, most of the species of *Bacillus* from phylum Fermicutes were proposed in Alkalophiles. However, several research studies have greatly isolated the organisms from phyla Cyanobacteria, Actinobacteria, Proteobacteria, Bacteroidetes, Thermotogae, Spirochaetas, Archaea (Euryarchaeota), and Yeast. The current scenario is still in favor of increased isolation and identification of alkalophiles due to their enormous capability for industrial applications. In recent years, distribution of alkalophiles was reported from soda lake, marine environments, saline deserts, natural thermal vents to natural water bodies. There are innumerous examples of these fascinating organisms have been discovered now and these include primarily prokaryotes (bacteria and archaea) and some eukaryotes (algae, yeast and fungi) (Kumar and Hovik, 2019).

BACKGROUND

Alkalophiles are a class of extremophilic microbes capable of survival in alkaline (pH roughly 8.5–11) environments, growing optimally around a pH of 10. These are generally categorized into two major physiological groups: Alkali-tolerant organisms that show optimal growth in the pH range of 7.0–9.0 but cannot grow above pH 9.5 and alkalophilic organisms that show optimal growth between pH 10.0 and 12.0. Furthermore, the extreme alkalophiles subdivided into facultative alkalophiles, obligate alkalophiles etc. Along with that, alkali tolerant strictly anaerobic strains and obligatory anaerobic species reported in lake Magadi Kenya in 1988 (Norton and Grant, 1988).

As alkalophiles were able to sustain at high temperature under aerobic and anaerobic conditions, the enzymes produced by such organisms were alkalostable and thermostable. Such physiological features allow us to explore new applications in protein engineering and production of thermostable enzymes industries (Reed *et al.,* 2013). Most of the alkalophiles are spore formers, the environmental effects would not directly affect in spore form (Tayyem *et al.,* 2021). However, the organism faces a central problem in pH homeostasis during the growth at extreme pH values. The mechanism behind the survival has been extensively investigated in past decades. In Spite of that, the complete understanding still needs more attention to study undelaying adaptations, bioenergetics and survival mechanisms by extreme alkalophiles.

The alkaline conditions are detrimental to biochemical processes and reproduction compared to normal conditions, as high pH value is extremely harmful to normal cellular processes. Alkalophiles developed several structural and functional adaptations to survive in such alkaline environments. Most of the alkalophiles are difficult to grow in culture media. However, recent developments in genomics make it possible to understand the importance of uncultivable organisms in such extreme environments (Grant and Heaphy, 2010). The genetic structure responsible for these adaptations developed a special interest in fundamental and applied research. The findings from genome and gene sequence revealed several potential applications of the organism grow at higher pH. These organisms deploy several strategies to thrive in extreme habitats; cytosolic acidification, high level of cytochromes, efficient respiratory and cations/proton antiport system, abundant unsaturated branched fatty acids, highly anionic cell surface, efficient protein damage repair system, production of extracellular products to acidify external cell environment etc. These organisms' ability not only supports only growth but also makes them rapid ATP producer compared to non-alkalophilic group of organisms. Various adaptations of the alkalophilic organisms such as acid production, efficient siderophores, highly anionic cell surface, alkaline active enzymes, antibiotics production showing great importance in various fields of Biotechnology (Mamo, 2019).

The main effects due to high pH value are instability of DNA, plasma membrane and cytosolic enzymes present in the cell. There must be specific mechanisms in alkalophilic organisms to sustain in such high pH value. The alkalophilic organisms must either have a specific machinery within them or the acidification mechanism of cytosolic region of the cell. Since few experiments suggest, the cytosolic pH must remain almost neutral. In this case, alkalophiles must have one or more mechanisms of acidifying the cytosol when in the presence of a highly alkaline environment. The present chapter includes some of the major adaptations and bioenergetics of the alkalophiles.

Alkaline enzymes have found major commercial applications in laundry detergents, for effectual food processing, finishing of fabrics, and pulp and paper industries(Bocchini *et al.,* 2003; Priyadarshini *et al.,* 2020; Sehar and Hameed, 2011; Zhao and Eun, 2020). Alkaline protease extensively studied, characterized, and commercialized from *Bacillus* sp. Alkaline proteases are convenient in detergent formulations, silk degumming, food and feed industry, photographic gelatine hydrolysis, leather dehairing, cosmetics, and pharmaceutical preparations (Joshi and Satyanarayana, 2013; Sarkar and K, 2020). The unique therapeutic potential of commercially available alkaline proteases, reported for degradation of prion protein (PrPSc). Amylase operative for biocatalysis of starch-based products (Cha *et al.,* 1998). The highly reported exo-alkaline amylases from alkaliphilic *Bacillus* sp. Alkaline debranching enzymes with amylase like pullulanases, amylopullulanase, neopullulanase, or isoamylase can be effectively applicable in stain removal. They are highly promoted globally at alkaline β-glucanases, as vivacious components of detergents. The activity of cellulase at alkaline pH was stable and not repressed by ingredients of laundry detergents. Alkaline cellulase functions in biostoning to remove pumice stones of denims, and biopolishing through elimination of the rough cellulose lumps formed on fabrics and clothes (Kakkar and Wadhwa, 2021). Cellulase is effective to improve biological de-inking and the activity of recycling paper pulp. Lipase produced by alkalophilic *Pseudomonas sp.* is also extensively used in detergents, pharmaceutical products, and fat processing in meat industries (Yang *et al.,* 2018; Zhao and Eun, 2020). Along with that, alkaline lipase reported for synthesis of biopolymers, biodiesel, pharmaceuticals and other composites. Similarly, alkaliphilic xylanase was studied from soil, kraft pulp, alkaline soda lakes, and pulp and paper industry wastes.

TAXONOMY AND DISTRIBUTIONS

Alkalophiles are categorized into alkali-tolerant, obligate alkalophilic, facultative alkalophilic, and halo-alkalophilic groups depending on the adaptation to a high pH value and / or neutral pH value together with the NaCl concentration (Yumoto *et al.*, 2011). The group of alkali tolerant can grow at pH 9 and their optima pH around 7 but cannot grow at pH greater than 10. *Virgibacillus sediminis* can grow at 6.0 to 10.5 but optimum pH is 7.5 to 8.0 (Chen *et al.*, 2009), *Virgibacillus chiguensis* can survive at pH 5–9 with 7.5 optimum pH but able to grow at pH 10 (Wang *et al.*, 2008) and *Rheinheimera pleomorphica* require pH 6 to 12 for growth but not able to grow above pH 10, and their optimum pH around 7 to 8 (Panda *et al.*, 2020). The Obligate alkaliphiles grow at high basic pH (up to 9) value like, *Alkalibacterium indicireducens, Alkalibacterium iburiense*, and *Bacillus marmarensis* requires a pH range of 9 to 12 for their growth, but cannot grow at pH 7 to 8 (Yumoto *et al.*, 2008); (Nakajima *et al.*, 2005); (Denizci *et al.*, 2010). The facultative alkaliphiles are cluster that rise optimally under harsh alkaline conditions but are also capable of growing near neutral pH. For examples, *Clostridium thermoalcaliphilum* (LI *et al.*, 1994), *Oceanobacillus indicireducens* (Hirota *et al.*, 2013) grow at pH 7 to 12 with optimum pH 10 and *Rhodobaca bogoriensis* (Milford *et al.*, 2000) grow at optimally 9 pH and survive at pH range between 7.5 to 10. *Bacillus* genus is communal and having alkali-tolerant and alkaliphilic attributes for survive under wide range of basic pH e.g., *Bacillus horikoshii* (Nielsen *et al.*, 1995). The haloalkaliphiles are adapted under dual extremities of high salinity and elevated pH (Arayes *et al.*, 2021). *Bacillus arsenic-oselenatis* and *Bacillus selenitireducens* can grow at 8.5 to 10 pH and 6% w/v NaCl concentration (Vyas *et al.*, 2014). After the 2000s, due to the massive potential application and interesting physiology of alkaliphiles and versatility of the alkaline environment for biotechnological applications, novel alkali-philic species are often found. Alkalophiles are widespread in alkaline niches where the pH is above 10. These niches include alkaline soda lakes, desert, rhizosphere, gardens, manure, agricultural soil, Sodium Sulfate Lake, hot baths, sewage sludge from a beverage industry, potato wastewater effluent, and costal area that were found throughout the world harbor environment for diversity of alkalophiles. In Table-1, some important organisms were listed observed worldwide.

Alkalophiles are widely dispersed in eubacteria, eukaryata and archaea domain. The alkaline niches like, Soda lakes, dry and fresh water lakes, soda solonchak, lagoon, steppe epitomize the most stable naturally distributed throughout the world but highly explored and studied in soda saline lakes. About 136 salt bodies are present worldwide. All these are distributed in almost all continents, including Africa (48 sites), Asia (43), America (23), Australia (14) and Europe (8). Maximum African rift valley studied for microbial diversity exploration (Mwatha, 1990; Gouda *et al.*, 2020). Lake Magadi is one of the smallest and most saline lakes in the East Rift Valley (Bernhart Owen *et al.*, 2019). There are several sites available that have not been explored for microbial ecological study. Table 2 contains the list of all the sites available in 5 different continents. Among all the listed sites, some sites explored by researcher, while remaining sites are still not properly explored for any extremophiles diversity study.

Table 1. Worldwide distribution of diverse alkalophilic organism reported in research field

No.	Name of organism	Strain	pH	Temp (°C)	NaCl (W/V)	Source	Reference
1	*Alkalibacterium olivoapovliticus*	WW2- SN4a	8.5	27 - 32	3 - 5	Edible-olive wash-waters	(Ntougias and Russell, 2001)
2	*Bacillus alkalicola*	Zby6	10	37	3	Saline Soda Zhabuye Lake, Tibet, China.	(Zhai *et al.*, 2014)
3	*Bacillus miscanthi*	AK13	8.0 - 9.0	28 - 35	10	Rhizosphere of Miscanthus sacchariflorus	(Shin *et al.*, 2020)
4	*Bacillus okuhidensis*	GTC 854	10.5	45 - 50	10	Okuhida hot spa, Japan	(Li *et al.*, 2002)
5	*Bacillus polygoni*	YN-1	10	29 - 31	5	Indigo balls, Ibaraki, Japan	(Aino *et al.*, 2008)
6	*Bacillus solisalsi*	YC1	7 - 10	35 - 42	15	Sodium sulfate inland Yuncheng Lake, South Shanxi Basin of Shanxi Province, China	(Liu *et al.*, 2009)
7	*Bacillus* sp.	NG-27	9.0 - 10.0	27	-	Decaying organic matter from northern India	(Gupta *et al.*, 2000)
8	*Bacillus vedderi*	JaH	10	40	2.5 – 7.5	bauxite-processing red mud tailing pond.	(Agnew *et al.*, 1995)
9	*Emericellopsis alkalina*	VKPM F1428	10.5		-	The edge of the soda lake Tanatar-2, Altai area, Russia	(Kuvarina *et al.*, 2021)
10	*Exiguobacterium alkaliphilum*	12/1	9.5	35	9.5	Alkaline wastewater drained sludge of a beverage industry, New Delhi, India	(Kulshreshtha *et al.*, 2013)
11	*Exiguobacterium aquaticum*	IMTB-3094	8.0	30	10	Tikkar Tal Lake, Haryana, India	(Raichand *et al.*, 2012)
12	*Exiguobacterium aurantiacum*	DSM 6208	9.5	37	12	Potato wastewater effluent	(Collins *et al.*, 1983; Souza & Deal, 1977)
13	*Exiguobacterium oxidotolerans*	T-2-2	7 - 10	34	12	Drain of a fish processing plant, Hokkaido, Japan	(Yumoto *et al.*, 2004b)
14	*Halomonas alkaliantarctica*	CRSS	9.0	30	10	Saline lake Cape Russell in Antarctica	(Poli *et al.*, 2007)
15	*Nocardiopsis alkaliphila*	YIM-80379	9.5 - 10	28	-	Desert soil sample collected in Egypt	(Hozzein *et al.*, 2004)
16	*Phytoactinopolyspora alkaliphila*	EGI 80629	9.0 - 10.0	30	3 - 5	Xinjiang, north-west China.	(Zhang *et al.*, 2016)
17	*Phytoactinopolyspora halophila*	YIM 96934	7.0 - 8.0	28 - 37	5 - 8	Edge of a saline lake in Xinjiang, north-west China.	(Ding *et al.*, 2019)
18	*Phytoactinopolyspora limicola*	HAJB-30	9.5 -10.0	-	1.0 – 3.0	Soda alkali-saline soil, Heilongjiang, Northeast China	(Wei *et al.*, 2021)
19	*Phytoactinopolyspora mesophila*	XMNu-373	7.0 - 8.0	28 - 37	2 - 5	Mongolia Plateau, Dongwu County, Inner Mongolia Autonomous Region, PR China	(Feng *et al.*, 2020)
20	*Pseudomonas aeruginosa*	-	9.0	50	-	Gut of Penaeus monodon, from Rajakkamangalam marine coastal area, Kanyakumari, Tamilnadu	(Jaradat *et al.*, 2013)
21	*Salinococcus* sp	BAB 3246	9.0-9.5	-	15	Coastal Region of Gujarat, India.	(Mevada *et al., 2017)*
22	*Sodiomyces alkalinus*	CBS 110278	8.7 - 10.5	27	-	Choibalsan area, North-East Mongolia, Shar-Burdiyn lake	(Grum-Grzhimaylo *et al.*, 2013)
23	*Thermococcus acidaminovorans*	DSM 11906	9.0	85	-	Marine hydrothermal Porto di Levante, Vulcano Island, Italy	(Dirmeier *et al.*, 1998)
24	*Thioalkalibacter halophilus*	ALCO 1	8.5	30	0.6	Hypersaline alkaline lakes, South-Western Siberia (Russia)	(Banciu *et al.*, 2008)

Table 2. Environmental site supporting the growth of Alkalophiic organism

No.	Continent	Country	Environment/sites for presence of alkalophiles	Reference
1	Africa	Egypt	Wadi El Natrun	(Gouda *et al.*, 2020)
		Algeria	Shaṭṭ Al-Jarīd	(Brittanica, 2014)
		Ethiopia	Abijatta Lake, Aranguadi (Green Lake), Chitu Lake, Hertale Lake, Metahara Lake, Kilotes Lake, Shala Lake	(Lanzén *et al.*, 2013; Schagerl, 2016)
		Kenya	Bogoria lake, Crater (Sonachi) Lake, Elmenteita Lake, Magadi Lake, Nakuru Lake, Oloidien Lake, Simbi Lake, Taurkana Lake	(Bernhart Owen *et al.*, 2019; Grant and Sorokin, 2011)
		Chad	Bodu Lake, Dijikare Lake, Monboio Lake, Munyanyange Lake, Murumuli Lake, Nunyampaka Lake, Rombou Lake, Yoan Lake	(Kambura, 2016; Schagerl, 2016)
		Libiya	Fezzan Lake	(Schagerl, 2016)
		Uganda	Kikorongo Lake, Mahenga Lake, Nyamunuka Lake, Rukwa North Katwe Lake	(Mwatha, 1990)
		Tanzania	Lake Natron, Lake Embagi, Lake Magad, Lake Manyara, Lake Balangida, Bosotu Crater, Lakes, Lake Kusare, Lake Tulusia, El Kekhooito, Momela Lakes, Lake Lekandiro, Lake Reshitani, Lake Lgarya, Lake Ndutu, Lake Rukwa North	(Bernhart Owen *et al.*, 2019; Kambura, 2016)
		Sudan	Dariba Lake, Malha Crater Lake	(Schagerl, 2016)
2	Europe	Poland	Janikowo (Kuyavia) Soda saline lake	(Kalwasińska *et al.*, 2017)
		UK	Malham Tarn	(Pentecost, 2009)
		Spain	Las Eras	(Sanz-Montero *et al.*, 2019)
		Serbia	Rusanda Lake	(Felföldi *et al.*, 2009)
		Hungary	Böddi-szék, Kelemen-szék, Lake Fehér (Szeged), Lake Neusiedl (Fertő)	(Felföldi *et al.*, 2009; Forró *et al.*, 2017)
3	Asia	India	Chhapra, Chilika Lake, Didwana, Khyagar Lake, Kuchaman, Kushul lake, Little Rann of Kutch, Lonar Lake, Lunkaransar, Mansagar, Namucuo Lake, Pachpadra Lake, Pokaran, Pulicat Lake, Sambhar Lake, Thar Desert, Thob, Tso Kar Salt Lake, Tso Moriri Salt Lake	(Gupta *et al.*, 2015; Kumar *et al.*, 2016; Patel *et al.*, 2015; Singh *et al.*, 2018; Zhu *et al.*, 2020)
		Russia	Barabinskaya Soda solonchak Steppe, Buriatia lake, Kiran soda lake, Kulunda Soda solonchak Steppe, Lake Khatyn, Malyi Kasytui, Torey Lakes, Verkhnee soda lake	(Shapovalova *et al.*, 2009; Sorokin *et al.*, 2008)
		Turkey	Kartsakhi Lake, Lake Salda, Lake Van	(Schagerl, 2016)
		Aksai Chin, India/ China	Lake Surigh Yilganing Kol, Tso Tang Lake, Aksayqin Hu Lake, Lake Hongshan Hu, North Tianshuihai lake, Tianshuihai lake, Pangong Salt Lake, Spanggur Tso (Pongur Tso)	(Wikipedia contributors", 2021)
		China	Badain Jaran Desert, Guozha lake, Lake Zabuye (Drangyer), Mongolia lake, Qinghai Lake	(Itoh *et al.*, 2005; Sorokin *et al.*, 2021)
4	Australia	Australia	Lake Eyre, Lake Werowrap, Lake Corangamite, Red Rock Lake, Lake Chidnup, Lake Torrens, North Blue Lake, Lake Bulla, Mid Blue Lake, Gidgee Lake, Lake Mere, Lake Buchanan, Lake Wyara	(Schagerl, 2016; Timms, 2021)
5	America	US	Alkali Lake, Baldwin Lake, Borax Lake, Mono Lake, Owens Lake, Soap Lake, Soda Lakes (Nevada), Summer Lake, Albert Lake, Lake Lenore, Searles Lake, Deep Springs, Rhodes, Marsh, Harney Lake, Surprise Valley, Pyramid Lake, Walker Lake	(Asao *et al.*, 2011; Grant and Sorokin, 2011; Schagerl, 2016)
		Canada	Goodenough Lake, Manitou Lake	(Grant and Sorokin, 2011; Schultze-Lam *et al.*, 1996)
		Mexico	Lake Alchichica, Lake Texcoco	(Schagerl, 2016)
		Chile	Antofagasta Lake	(Grant and Sorokin, 2011)

CYTOSOLIC ACIDIFICATION

It is very important to maintain the pH of cytoplasm for proper functioning of the intracellular components of the cell. Alkalophiles are normally grow at an elevated alkaline pH, which leads to inadequate cell functioning. However, Alkaliphiles are able to maintain cytosolic acidification by both passive as well as active transport means. In the passive form of transport mechanism, the energy is not required to maintain the pH of cytoplasm. While the Active transport mechanism uses the ATP as an energy molecule to maintain intracellular pH.

The passive mode of acidification is based on cell structure of alkalophilic organisms. Most of the organism contains the acidic polymers composed of residues of aspartic acid, galacturonic acid, gluconic acid, glutamic acid and phosphoric acid. All these residues form a matrix, which is acidic in nature. The acidic matrixes created by acidic residues prevent entry of hydroxide ions and facilitate entry of sodium and hydronium ions from external regions. The importance of these acidic residues already explored with induced mutant alkalophilic organism. The mutant organism lacking the ability to synthesize such residues showing complete loss of ability to grow in alkaline conditions. However, only passive transport mechanism only would not able to maintain pH of intracellular matrix; there must be active transport mechanism work around to maintain the structure

The active mode of acidification relies on Na+/H+ antiporters, which are present in the cell membrane of the alkalophilic organisms. The Na+/H+ antiporter protein of alkalophilic organisms, allows ions H+ and Na+ to move across a membrane in order to change a concentration gradient (Padan *et al.*, 2001). These antiporter proteins in acidophiles recognize a higher pH on the outside of cell, which leads to activation of a homeostatic mechanism to bring the pH level back to optimal range (Padan, 2014). In respiratory cells, H+ ions are first extruded through the electron transport chain and in fermentative cells through the ATPase enzyme. This proton shift creates the proton gradient to facilitate the exchange of Na+/H+. This process accumulates the H+ ions in the cell leads to a lowering of cytosolic pH. Along with that, Na+ dependent solute uptake is crucial in providing sodium for antiport activity as well as nutrients (i.e., sugars, amino acids) or precursors (such as choline) for the purpose of general metabolism. The Na+ can also facilitate nutrient uptake via solute symport, which is necessary for cellular processes. The mutation in the Na+/H+ antiporters via mutation or any other means rendered the neutrophilic nature of the alkalophilic organisms. The alkalophiles combat the high pH concentration in the area by lowering cytosolic pH. The sodium is crucial for the antiporter system to operate in alkalophilic organisms. That is the reason, why most of the alkalophilic organisms are grow in saline environments (Lee *et al.*, 2013).

Bioenergetics and ATP Production

In alkaliphilic organisms, the antiporter system contains huge number of respiratory components such as cytochrome a, b and c. The high concentration of these components may allow the ATP generation via ATPase enzyme present in the cytoplasm of the cell *(Fujisawa et al.,* 2010). There are major two factors are known to associated with proton motive force that drives ATPase in alkalophilic organism; 1. Difference in H+ concentration across membrane and 2. Negatively charged molecules near cell membrane. In general, the cell could not estimate pH across membrane and nearby surface. That requires electron transport chain mechanism and translocation H+ across membrane in respiratory chain for the ATP generation (Figure 1). The ATP generation via respiratory chain in alkalophiles and neutrophils is similar. However, the rates of H+ translocation by alkalophiles were extremely lower compared to neutrophilic organism.

Figure 1. Membrane transport and ATP generation mechanism

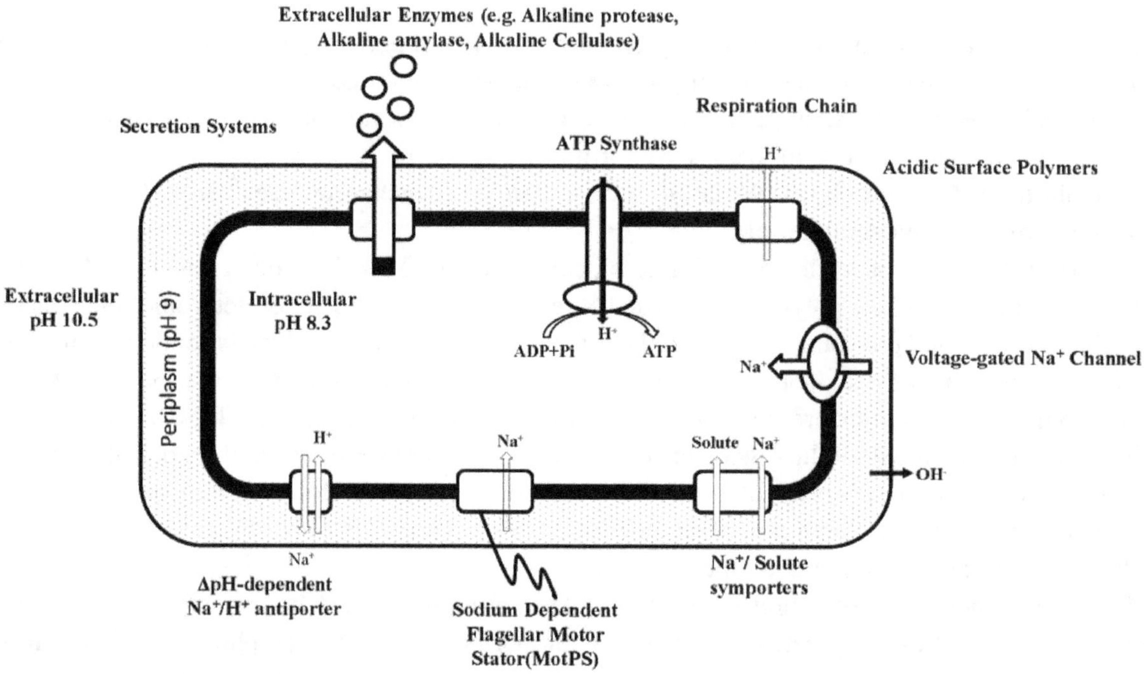

SURVIVAL MECHANISMS

Alkalophiles also known as extremophiles belong to group of special organism, which always need a harsh environment for optimal growth. They use several structural, chemical and defensive mechanisms to grow in extreme environments. This adaptations and mechanism are important in several industrial applications as well as clean garbage near dump sides. The mechanism used by alkalophiles often based on the secretion of enzymes and metabolic products outside of the cell (Singh and Gabani, 2011). This allows the cells adapt stressful conditions from surrounding environment. Some of the important mechanisms are explained here.

1. Ectoine-Mediated Mechanism

Ectoine is the widely occurring and one of the important osmolyte, known as stabilisers protecting cell, proteins and enzymes from the deleterious effect of environmental stress such as alkalinity, high temperature, desiccation, osmotic stress (Lentzen and Schwarz, 2006). Many alkaliphilic halotolerant *Bacillus* sp. were reported for the biosynthesis of ectoine (5-2-methyl-1, 4, 5, 6-tetrahydropyridine- 4-carboxylic acid) (Van-Thuoc *et al.,* 2013). The cluster of *ectABC* genes encode the diaminobutyric acid acetyltransferase (EctA), the diaminobutyric acid aminotransferase (EctB), and the ectoine synthase (EctC), which are responsible for the ectoine biosynthesis (Bursy *et al.,* 2007). Still the mechanism of action for ectoines is not fully understood, but it is believed that the stabilization effect is achieved by the hydration of protein due to the exclusion of ectoines from the protein surface. Ectoine and other compatible solute like trehalose, hydroxyectoine decrease surface area of protein by increasing surface tension resulting in compact fold state of protein (Street *et al.,* 2006). Under the extreme stress condition, the halophiles prone to produce more hydroxyectoine than ectoine, which facilitate more preservation by increasing hydration of protein and rigidity. Microorganism produce hydroxyectoine because ectoines can be degrade into N-acetyldiaminobutyric acid by consuming a hydroxyl group (Van-Thuoc *et al.,* 2013). Zhao *et al.,* have reviewed the production of Ectoine from haloalkaliphiles as an one of the biotechnologically import product (Zhao *et al.,* 2014).

2. Increased Catalytic Activity

Extremophiles produce enzymes which are stable in harsh environment and resist to denaturation. Due to the high pH environment, the catalytic efficiency of in alkalophilic organisms is greatly affected. However, the metaproteomic based discovery of an Enzyme from an Indian Hot Spring suggest the importance of high catalytic efficiency to reduce the damage due to pH (Sander *et al.,* 2021). An optimum activity of enzymes of alkalophilic organisms observed at pH 10 but amylase isolated from Bacillus spp. reported to have hydrolysis capability of starch at pH of 2 as well still active at pH of 12 (Mesbah and Wiegel, 2018). The wide range indicated mechanism of high catalytic activity of alkalophilic enzymes as well. Georlette *et al.,* 2003, suggest adapted DNA ligase has increased activity and high conformational flexibility at active site domain (Georlette *et al.,* 2003). The study on extremophile enzyme suggests that there is an interfacial correlative mechanism between stability, activity, rigidity, and flexibility. In summation enzymes have complex catalytic and regulatory strategies which plays important role in living organisms (Purich, 2010).

3. Amino Acid Accumulation

Accumulation of amino acids near membrane proteins contribute to cation trapping. Secretion of amino acids such as teichuronic acid, teichoic acids, aspartic acid, and phosphoric acid leads to formation of anionic surface in alkalophilic organisms. This negatively charged surface absorbs Na^+ and H^+ from surrounding of cell and repulse hydroxide ions, which facilitates growth in alkaline niches by maintaining pH homeostasis. This property of alkalophiles is also useful in increasing their stability and activity in nonaqueous solvents. In sum, amino acid accumulation participate in passive mechanism of pH homeostasis by regulating osmotic stress (Ling *et al.,* 2018).

4. Activation of the Nuclear Factors

The NF-kB (Nuclear factor-kappaB) is the elementary transcription factor and plays an important role in the cell-cycle progression. Recent developments in sciences facilitate better understanding of nuclear factors. The transciptions factors ChbZIP1 from Alkaliphilic Microalgae *Chlorella* also reported to provide alkaline stability to *Arabidopsis thaliana* (Qu *et al.,* 2021). A wide range of stimuli from alkalophiles including cytokines, environmental particles, oxidative stress, and toxic metals leads to activation of NF-kB (Hayden and Ghosh, 2008). Activation of NF-kB signaling pathway in response to oxidative stress has been frequently observed in certain types of alkalophilic cells (Chen *et al.,* 2001).

5. Overexpression of Heat Shock Protein Genes

Extremophiles overexpress single or collective ubiquitous HSP genes (Heat Shock Protein). Overexpression of these highly conserved HSPs genes, are thought to be induced for survival and play essential role in maintain protein complex machinery, in the growth and viability at high osmotic pressure or in hot condition. Heat Shock Proteins prevent protein mis-folding, aggregation and promote degradation of the irreversibly denatured polypeptides. High temperature exposure and high hydrolytic pressure increase the synthesis rate of heat shock proteins, which mediate folding of molecular chaperones (Koide *et al.,* 2006).

ENZYMES AND THEIR APPLICATIONS

Alkalophiles are known to synthesize a large number of compounds such as extracellular alkaline active enzymes, bioactive substance, siderophores, novel carotenoids and compatible solutes like ectoine, trehalose. Among that the enzyme production shows great potential for application in detergent formulation, environmental protection, textile and fiber processing, paper and pulp processing etc. Unfortunately, Screening of the active compounds from culturable isolated alkalophiles serves as the boundary for the wide application. However, the NGS platform overcomes this problem by exploring the new way to screen biotechnological potential.

Alkaline Protease

Alkaline protease seems to have peaked attention because of its wide application in dairy, detergent industry, leather dehairing, gelatin hydrolysis, silk degumming, silver recovery, cosmetic and pharmaceutical preparations. In natural environment, protease can be produces by plants, animals, and microorganisms. However, it is difficult to take plants and animals as a sustainable proteolytic enzyme sources. Among the different producers, extracellular protease producer microbes are considered with the aim of a large-scale fermentation process. Proteolytic enzymes are basically produced inside cell but their presence found inside (endo-peptidases) as well as outside of cell as exo-peptidases. Both these type of peptidases are having different cell function as per their catalytic property. Horikoshi (1971) reported *Bacillus* sp. no. 221 (*B. clausii* ATCC 21522) as first and highest alkaline protease producer (Horikoshi, 1971). Recently, various novel microorganisms are reported for the alkaline protease production such as, *Bacillus* sp.K-3 (Yin *et al.,* 2019), *Bacillus* sp.HL-8 (Gimza *et al.,* 2019), *Bacillus* sp.HS-4 (Sehar and Hameed, 2011),*Bacillus* sp. N-40 (Sevinc and Demirkan, 2011), *Pseudomonas aerugin* (Ghorbel-Bellaaj *et al.,* 2012), *Streptomyces* sp. GS – 1(Sarkar and K, 2020), *Bacillus licheniformis* MK90 (Hamed *et al.,* 2019). Darwesh *et al.,* 2019, reported alkaline protease production by *Saccharomonospora viridis* strain Hw G550 which shows the nematicidal activity.

In the environment protection, protease found an eco-friendly agent for the degradation of waste because chemical treatment or physical treatment of waste shows hazardous effect for the environment and its surroundings. Researcher proves this method better than physical and chemical method, even though waste management by using protease may be produce useful products (Abdel-Shafy and Mansour, 2018). In the leather industry, protease is very useful for dehairing from skin. Human kind traditionally uses different method for dehairing, such as use of sodium sulfate and lime which have environmental effect including bad odor and air pollution. Besides that, alkaline protease also useful for enhancing quality of leather products. *Streptomyces* sp. GS-1 reported for the alkaline protease production in basal media (BM) supplemented with wheat bran at 45 °C and pH 8.5 which displayed dehairing property (Sarkar and K, 2020). In detergent industry protease cover about 30% enzyme market. Protease enzyme produce from *Bacillus* species have been widely useful in detergent because of their activity and stability at high pH, stability in high temperature, compatibility with other chelating agents and should be active in the presence of inhibitors during the use (Karataş *et al.,* 2013). Nowadays, alkaline protease also apply in medical and health area; one of the hot area of human life. By using proteases from different species scientists are also able to produce different medicine, drugs and vaccines to prevent or treat diseases (Palanivel *et al.,* 2013; Razzaq *et al.,* 2019; Sou *et al.,* 2011). It has been reported that fibrin degradation was achieved by alkaline fibrinolytic proteases. The use of this fibrinolytic enzyme suggests its future use as a cancer drug and in thrombolytic therapy (Jaouadi *et al.,* 2011). Drugs production by using protease have found their application in treating enzyme deficiency and disorder in human kind (Kumar *et al.,* 2012).

Alkaline Amylase

Numerous Amylase- an amylolytic enzyme have been isolated from various alkaline niches and deposited in the literature. Alkaline amylase was the first time produce in Horikoshi-II medium by cultivating alkaliphilic *Bacillus* sp. A-40-2 (Horikoshi, 1971). Subsequently, *Bacillus* sp. NRRL B 3881, *Bacillus* KSM-1876, and several other alkaliphilic microorganisms have been discovered for the starch degrading alkaline amylase production (Ara *et al.,* 1992; Boyer and Ingle, 1972)

Alkaline amylases are hydrolytic enzymes that are stable in an alkaline environment and hydrolyzes starch molecules by cleaving glycosidic bonds to give diverse products including dextrins and smaller polymers of glucose (Windish and Mhatre, 1965). Among the bacteria, *Bacillus* species has been reported widely for the production of amylase. *B. subtilis*, *B. stearothermophilus*, *B. licheniformis*, and *B. amyloliquefaciens* are known to be good producers of alpha amylase (Coradin *et al.,* 2011). Various strategies have applied for the production of recombinant enzyme, such as optimising promoters, ribosome binding site, decreasing the degradation of proteins, fermentation optimization, mutagenesis screening, signal peptides and secretory pathways etc. Transport and transcription levels regulation can enhance alkaline α-amylase production by deleting a putative peptide segment of YwbN'. Insertion of arginine (R) between residues 5 and 6 of YwbN' further increased the protein yield (Yang *et al.,* 2020).

Amylase is classified into Endoamylase, Exoamylase, α-amylase, β – Amylase, and γ- Amylase. These are one of the main enzymes used in industries and play magnificent role in industrial area. Alpha amylase is the important candidate of Starch processing industries, Food industries, Biofuel industries, Detergent industries, Textile industries, Paper industries, and in medical field. The major market of α-amylase is in the starch industry, during the liquefaction process for the hydrolysis of starch (Saini *et al.,* 2017). In food industry amylase can be added to the dough of bread for conversation of starch to dextrin in flour, which followed by yeast fermentation. The addition of amylase in dough increase the rate of fermentation, reduce the viscosity resulting in the improvement in dough rheological property and volume of dough. In addition, amylase also improves the taste, colour and toasting qualities of the bread. *Bacillus subtilis* US586 identified as amylase producer and improved activity in the backing industry (Trabelsi *et al.,* 2019). The textile industries are one of the largest industries in the form of spreading the environment pollution. Application of amylase in textile industry allows the development of eco-friendly technology in fiber processing such as removal of starch sizing agent from woven fabric. Amylase is important component in detergent for removal of starchy foods from cloths (Mojsov *et al.,* 2018). Amylase is also showed important application in medical filed to treat digestive disorder (Pandey *et al.,* 2000). Alkalophilic microorganisms, able to produce amylase are promising tool for the development of sustainable environment by participating in pollution control, and also important in food industry, textile industry and backing industry.

Alkaline Cellulase

In 1984, Horikoshi and his researcher found the Bacillus No. N4 and No. 1139 for the extracellular production of alkaline carboxymethylcellulases (CMCases) (Horikoshi, 2011). Recently certain bacterial strain are identified as a source of Cellulase such as *Bacillus vallismortis* RG-07 (Gaur and Tiwari, 2015), *Pseudomonas* sp. (Kaur *et al.,* 2020), *Bacillus thuringiensis*, *Bacillus licheniformis* (Awasthi *et al.,* 2018), *Bacillus sp.* KSM-S237 (Hakamada *et al.,* 1997), *Bacillus subtilis* BC1 (Dehghanikhah *et al.,* 2020). Cellulase is abundantly found in alkaliphilic microbes and play crucial role in biotechnological applications. Cellulosic material can be hydrolyze by the action of cellulase which cleave β-1,4-glycosidic bonds and resulting the production of glucose (Zainuddin *et al.,* 2015). CMCases aids in various processes for breaking down of cellulosic material in the production of different commercially important products like monomeric sugar production, antibiotic production, single-cell protein (SCP) production which are essential for human kind (Kaur *et al.,* 2020).

In present time cellulase have attracted great attention because of its application in various industrial prospects including food industry, textile industry, and paper and pulp industry, laundry as well as in agriculture. In textile industry cellulase can be applicable for bio-polishing of fabrics, pretreatment of bast fibers, scouring, mercerization, stone finishing and carbonization process (Shah, 2013). Alkaline cellulase is an additive in detergent industry, supplemented into the detergents to enhance the fabric smoothness and removal of soil from the cloths without damaging the fiber of cloth. Cellulase act by passing through the interfibril spaces of cloths to preserve the fabric quality. This process is environment friendly and diminishes the utilization of toxic chemical constituents for washing purpose which prevent humans and environment both from hazardous effect (Niyonzima, 2019). Ito *et al.,* 1989 isolated an alkaliphilic *Bacillus* sp. No. KSM-635 from soil and reported for the production of detergent additive cellulase production at industrial scale. *Bacillus thuringiensis* and *Bacillus licheniformis* are identified as food waste degrader containing 60% cellulose and starch (Awasthi *et al.,* 2018).

Alkaline Lipase

The enzyme can degrade fats into more hydrophilic fatty acids known as lipase. Lipases are ubiquitous in nature and naturally produce by several plants, animals and microorganisms. Lipases from alkalophilic microbes are superior to normal lipase because of their reasonable activity even at wide range of pH, high temperature and high stability in organic solvents. In current scenario, alkaline lipase have versatile industrial potential and attracting more attention (Eddehech *et al.,* 2019). The fat degradation capacity of lipase, put into a member of detergent industry. However, introduction or entry of alkaline lipase in detergent industry is recent but the market value is comparable to proteases. Alkaline lipase is successful to remove hydrophobic stains caused by cosmetics, body fats and oil based foods from cloth due to the high degradation capacity (Horikoshi, 2011). The alkaline lipase also reported in degradation of petroleum hydrocarbons from automobile oil spillage (Sahoo *et al.,* 2020) .

Bacillus sonorensis 4R (Bhosale *et al.,* 2016), *Halobacillus* sp. (Esakkiraj *et al.,* 2016), *Serratia* sp. W3 (Eddehech *et al.,* 2019), *Pseudomonas aeruginosa* HFE733 (Hu *et al.,* 2018) were screened for the production of alkaline lipase. The alkaline lipase from *Pseudomonas aeruginosa* HFE733 reported to degrade oil films on waste water surface. Oil film production causing serious environmental pollution, leading death of any aquatic species by decreasing diffused oxygen level. Based on the finding of Hu *et al.,* 2018, application of alkaline lipase producer *P. aeruginosa* HFE733 may overcome this problem by degrading oil film (Hu *et al.,* 2018). Thus, alkaline lipase can be considered as a potential candidate for environment friendly aspects.

Alkaline Xylanase

Xylan consists of hemicellulose, most abundant natural polymer on the earth planet. The bleaching of lignin from kraft pulp with chlorine often leads to the formation of dioxins. However, xylanases degrade xylans, which bind to lignin and cellulose, and therefore xylanases can bleach kraft pulp without producing dioxins (Damiano *et al.,* 2006) . The Solubility of xylans in water increases under alkaline pH conditions, so alkaline xylanases are suitable for treating xylans (Kumar *et al.,* 2014). For the hydrolysis of xylan, required combine activity of xylanolytic enzyme like β-1,4-endoxylanase, β-xylosidase, αglucuronidase, α-L-arabinofuranosidase, acetyl xylan esterase and phenolic acid esterase. In 1973, Horikoshi and his co-researcher first time isolate xylan producer Alkalophilic *Bacillus* No. C-59-2 from the soil. The purified xylan enzyme from this strain exhibited great stability at pH ranging from 6.0 to 8.0. (Horikoshi, 1973). Alkalphilicity of alkaline xylanase Xyn11A-LC from *Bacillus* sp. SN5 can be improve by saturation mutagenesis at Glu135 residue (Bai *et al.,* 2016). Xylanase have found great potential economic application in diverse biotechnological approach such as in paper and pulp industry for prebleaching of pulp, biofuel production, animal feed, fabric bio-processing, food industry and degumming of plant fabrics (Popa *et al.,* 2016)

Other Applications of Alkaline Enzymes Isolated from Alkalophiles

Keratin, a protein present in human hairs and feathers of the birds. To degrade such compounds keratinases type of proteases are extremely helpful. Previously, and keratinass derived from alkalophilic Bacillus sp. have reported (Kojima *et al.,* 2006; Takami *et al.,* 1992). Alkaline keratinases, which are difficult to break down by proteases due to their high content of disulfide bonds, can effectively break down feathers at an alkaline pH, but keratinases hardly break down human hair. They would be used to recycle waste from feathers. Once keratinase has been found, which can break down human hair, it may be useful to clear clogs in drains. Alkalophiles have traditionally been used in Japan for indigo color reducers as a traditional method. Both facultative and obligate alkaliphiles have been screened from indigo balls by Yumoto *et al.,* (Yumoto *et al.,* 2004a). In addition, alkaliphiles have been used to confirm the quality of Asian yellow alkaline noodles (Saito *et al.,* 2003)

FUTURE RESEARCH DIRECTIONS

The last 50 years of extremophile research have shown limitless capabilities of life on extreme environments. For example, the organisms can grow beyond 100 degree celsius, extreme acidic-alkaline pH, elevated pressure etc. (Merino *et al.,* 2019; Stetter, 1982). Alkalophilic organisms have pushed our understanding of life beyond pH 9. The adaptations, survival mechanism, energy production and molecular stability of cellular components have major applications in the field of Industrial applications. Past several years, the successful industrial products has been given by alkalophilic organism including, alkaline amylase (Ozawa *et al.,* 2007), alkaline protease (Fujiwara *et al.,* 1993), alkaline lipases (Chinnathambi, 2015), alkaline celluloses (Kim *et al.,* 2005), alkaline peroxidase (Ikehata *et al.,* 2005), Gaurdzymes (Olsen and Falholt, 1998), Siderophores (McMillan *et al.,* 2010) etc. Despite the several benefits and industrially important features, the non-cultivable organisms would not grow under *in vitro* conditions. This would limit our understanding to study such uncultivable organisms. With the advent of genome sequencing, transcriptomics and proteomics methods, it would become easier to understand alkalophiles. The multi-omics greatly expands our knowledge to study adaptive mechanisms, metabolic regulations, industrial applications, potential enzymes produced by such organisms (Ece *et al.,* 2018). In addition to that, a recent study also suggests the importance of alkalophiles in antibiotics production (Terra *et al.,* 2018), anti-cancer therapeutics products (VA *et al.,* 2017), IAA production (Goswami *et al.,* 2014). Although the major applications are associated with industrially important products, the alkalophilic organism were shown their importance in environment friendly aspects (such as bioremediation, biodegradation and biocontrol) (Dhakar and Pandey, 2016).

Along with that, The research conducted to use alkalophiles as microbial fuel cells (MFCs) in electronic devices (Logan *et al.,* 2006). As bacteria oxidized organic or inorganic substrate and generate current and generate electrons. These electrons are likely to be transferred to the anode and the movement of electrons creates millivolt currents for powering small microprocessor chips in devices. The psychrophilic alkalophilic *Pseudomonas alcaliphila* MBR produces phenazine-1-carboxylic acid under alkaline conditions (Yumoto *et al.,* 2001). While *Corynebacterium sp.* strain MFC03 uses organic compounds to generate potential redox difference to be use as “Bioelectricity” in Microbial fuel cells.

Actinobacteria are an ecologically important group that play an important role in biogeochemical cycles, bioremediation (Chen *et al.,* 2010), bioweathering (Cockell *et al.,* 2013), plant growth promotion (Palaniyandi *et al.,* 2013), nanoparticle synthesis (Dayma *et al.,* 2019). Along with that, these organism have large array of pharmaceutical applications in antibiotics, anti-inflammatory compounds, antitumor agents and enzyme inhibitor (Shivlata and Satyanarayana, 2015). Prince *et al.,* (2020) reviewed the production of polyhydroxybutyrate for bioplastic production from alkaline cyanobactria from biological wastes (Price *et al.,* 2020). Although, alkalophiles provides several success stories in industrially important products, but are inefficient for antibiotics production. Because most antibiotics are an intracellular product. Obtaining intracellular antibiotics therefore appears difficult. In addition, the antibiotics produced must contains stable molecular components that can remain stable at such a high pH.

As the space explorations are always fascinating to scientists. Astrobiology is the field associated with to search life on planets beyond Earth. Some of the craters on planets and moons in the “Goldilock zones” or habitable zones were suggesting the life of microorganisms. Space explorations has been reported extremely harsh and inhabitable environments such as extreme radiation, extreme temperatures, altered gravity and extreme salinity and nutrients (Thombre *et al.,* 2017). These conditions appear to detrimental to support the growth of life in outer space.

Recent developments in the fields of Genomics revealed several unresolved layers hidden inside the genetic codes. Because the cost of sequencing has been greatly reduced with the development of NextGen sequencing technology. Metagenomics, a science to study all genomic sequencing amongst sample has been revealed uncultivable organism amongst specific alkaline and saline habitats. The current research in culture independent techniques led to in-depth studies into composition, functional capabilities and ecological impact on the environment. Alkalophiles are not limited to just prokaryotic group only. Rather massive numbers of organisms were observed in alkalophilic environment using metagenomics approach. Metagenomics help us gain valuable insights in catabolic, anabolic potentials, genomic features, molecular adaptations, stop codon reassignments and several unique features of organisms at such extreme environment. This also helps understand diversification of microbial lineages and phylogenetic relationship. Other innovative techniques like Fluorescent in situ hybridization with micro autoradiography (MAR-FISH), FISH-NanoSIMS, Raman FISH Spectroscopy, Phylogenetic microarrays with CHIP-SP can serve as potential methods study extreme environments (Hedlund *et al.,* 2014). Finally, as the scientific fraternity gears for the future, the study of alkalophilic organism would help to develop new products with better efficacy and outcome.

CONCLUSION

Alkalophiles have played an important role in biotechnology. Several success stories have been reported for their applications in routine life. Today, alkalophiles are well established in their commercial applications fields such as detergent, paper and pulp, leather and textile industries. The newer technologies in the field of metagenomics, transcriptomics and proteomics have multiple facets to explore alkalophilic organism for unexplored uses in industrial applications. The metagenomics approach imperatively expands the exploration of non-traditional enzymes with new applications as well. However, the main revenue generating products till used is hydrolases, lipase and amylase with a well-established market. Moreover, several other studies have shown the promising potential of alkaliphiles in the production of novel as well as known antibiotics, siderophores and cellular metabolites of great biotechnological importance. On the other hand, the industrial applications of all these substances are still lagging behind for several reasons. One of the important parameter, the cultivation of these alkalophiles is not well-established. Thus, in order to efficiently use the full potentials of alkaliphiles, it is necessary to develop the cultivation system and develop few selected alkalophiles as hosts for heterologous expression, metagenome library construction, metabolic engineering, and other related applications. This chapter is summarizing present and future aspects of alkalophiles and their applications.

ACKNOWLEDGMENT

This research received no specific grant from any funding agency in the public, commercial, or not-for-profit sectors.

REFERENCES

Abdel-Shafy, H. I., & Mansour, M. S. M. (2018). Solid waste issue: Sources, composition, disposal, recycling, and valorization. *Egyptian Journal of Petroleum*, *27*(4), 1275–1290. doi:10.1016/j.ejpe.2018.07.003

Agnew, M. D., Koval, S. F., & Jarrell, K. F. (1995). Isolation and Characterization of Novel Alkaliphiles from Bauxite-Processing Waste and Description of *Bacillus vedderi* sp. nov., a New Obligate Alkaliphile. *Systematic and Applied Microbiology, 18*(2), 221–230. doi:10.1016/S0723-2020(11)80392-9

Aino, K., Hirota, K., Matsuno, T., Morita, N., Nodasaka, Y., Fujiwara, T., Matsuyama, H., Yoshimune, K., & Yumoto, I. (2008). *Bacillus polygoni* sp. nov., a moderately halophilic, non-motile obligate alkaliphile isolated from indigo balls. *International Journal of Systematic and Evolutionary Microbiology, 58*(1), 120–124. doi:10.1099/ijs.0.65193-0 PMID:18175695

Ara, K., Igarashi, K., Saeki, K., Kawai, S., & Ito, S. (1992). Purification and some properties of an alkaline pullulanase from alkalophilic *Bacillus* sp. KSM-1876. *Bioscience, Biotechnology, and Biochemistry, 56*(1), 62–65. doi:10.1271/bbb.56.62

Arayes, M. A., Mabrouk, M. E. M., Sabry, S. A., & Abdella, B. (2021). Diversity and characterization of culturable haloalkaliphilic bacteria from two distinct hypersaline lakes in northern Egypt. *Biologia, 76*(2), 751–761. doi:10.247811756-020-00609-5

Asao, M., Pinkart, H. C., & Madigan, M. T. (2011). Diversity of extremophilic purple phototrophic bacteria in Soap Lake, a Central Washington (USA) Soda Lake. *Environmental Microbiology, 13*(8), 2146–2157. doi:10.1111/j.1462-2920.2011.02449.x PMID:21410624

Awasthi, M. K., Wong, J. W. C., Kumar, S., Awasthi, S. K., Wang, Q., Wang, M., Ren, X., Zhao, J., Chen, H., & Zhang, Z. (2018). Biodegradation of food waste using microbial cultures producing thermostable A-amylase and cellulase under different pH and temperature. *Bioresource Technology, 248*, 160–170. doi:10.1016/j.biortech.2017.06.160 PMID:28709882

Bai, W., Cao, Y., Liu, J., Wang, Q., & Jia, Z. (2016). Improvement of alkalophilicity of an alkaline xylanase Xyn11A-LC from *Bacillus* sp. SN5 by random mutation and Glu135 saturation mutagenesis. *BMC Biotechnology, 16*(1), 77. doi:10.118612896-016-0310-9 PMID:27825339

Banciu, H.L., Sorokin, D.Y., Tourova, T.P., Galinski, E.A., Muntyan, M.S., Kuenen, J.G., & Muyzer, G. (2008). Influence of salts and pH on growth and activity of a novel facultatively alkaliphilic, extremely salt-tolerant, obligately chemolithoautotrophic sufur-oxidizing Gammaproteobacterium *Thioalkalibacter halophilus* gen. nov., sp. nov. from South-Western Sibe. *Extremophiles: Life Under Extreme Conditions, 12*, 391–404. doi:10.1007/s00792-008-0142-1

Bernhart Owen, R., Renaut, R. W., Muiruri, V. M., Rabideaux, N. M., Lowenstein, T. K., McNulty, E. P., Leet, K., Deocampo, D., Luo, S., Deino, A. L., Cohen, A., Sier, M. J., Campisano, C., Shen, C. C., Billingsley, A., Mbuthia, A., & Stockhecke, M. (2019). Quaternary history of the Lake Magadi Basin, southern Kenya Rift: Tectonic and climatic controls. *Palaeogeography, Palaeoclimatology, Palaeoecology, 518*, 97–118. doi:10.1016/j.palaeo.2019.01.017

Bhosale, H., Shaheen, U., & Kadam, T. (2016). Characterization of a Hyperthermostable Alkaline Lipase from *Bacillus sonorensis* 4R. *Enzyme Research, 2016*, 1–11. Advance online publication. doi:10.1155/2016/4170684 PMID:26904276

Bocchini, D. A., Damiano, V. B., Gomes, E., & Da Silva, R. (2003). Effect of *Bacillus circulans* D1 thermostable xylanase on biobleaching of eucalyptus kraft pulp. *Applied Biochemistry and Biotechnology, 105–108*(1-3), 393–402. doi:10.1385/ABAB:106:1-3:393 PMID:12721462

Boyer, E. W., & Ingle, M. B. (1972). Extracellular alkaline amylase from a *Bacillus* species. *Journal of Bacteriology, 110*(3), 992–1000. doi:10.1128/jb.110.3.992-1000.1972 PMID:4624112

Brittanica. (2014). *Chott El-Jarid | lake, Tunisia*. Author.

Bursy, J., Pierik, A. J., Pica, N., & Bremer, E. (2007). Osmotically induced synthesis of the compatible solute hydroxyectoine is mediated by an evolutionarily conserved ectoine hydroxylase. *The Journal of Biological Chemistry, 282*(43), 31147–31155. doi:10.1074/jbc.M704023200 PMID:17636255

Cha, H. J., Yoon, H. G., Kim, Y. W., Lee, H. S., Kim, J. W., Kweon, K. S., Oh, B. H., & Park, K. H. (1998). Molecular and enzymatic characterization of a maltogenic amylase that hydrolyzes and transglycosylates acarbose. *European Journal of Biochemistry, 253*(1), 251–262. doi:10.1046/j.1432-1327.1998.2530251.x PMID:9578484

Chen, F., Castranova, V., & Shi, X. (2001). New insights into the role of nuclear factor-κB in cell growth regulation. *American Journal of Pathology, 159*(2), 387–397. doi:10.1016/S0002-9440(10)61708-7 PMID:11485895

Chen, S., Su, L., Billig, S., Zimmermann, W., Chen, J., & Wu, J. (2010). Biochemical characterization of the cutinases from Thermobifida fusca. *Journal of Molecular Catalysis. B, Enzymatic, 63*(3-4), 121–127. doi:10.1016/j.molcatb.2010.01.001

Chen, Y.-G., Cui, X.-L., Wang, Y.-X., Zhang, Y.-Q., Tang, S.-K., Li, W.-J., Liu, Z.-X., Wen, M.-L., & Peng, Q. (2009). *Virgibacillus sediminis* sp. nov., a moderately halophilic bacterium isolated from a salt lake in China. *International Journal of Systematic and Evolutionary Microbiology, 59*(8), 2058–2063. doi:10.1099/ijs.0.003624-0 PMID:19605714

Chester, F. D. (1897). Report of mycologist: Bacteriological work. *Delaware Agricultural Experiment Station Bulletin, 9*, 38–145.

Chinnathambi, A. (2015). Industrial Important Enzymes from Alkaliphiles – An overview. *Biosciences Biotechnology Research Asia, 12*(3), 2007–2016. doi:10.13005/bbra/1868

Cockell, C. S., Kelly, L. C., & Marteinsson, V. (2013). Actinobacteria—An Ancient Phylum Active in Volcanic Rock Weathering. *Geomicrobiology Journal, 30*(8), 706–720. doi:10.1080/01490451.2012.758196

Collins, M. D., Lund, B. M., Farrow, J. A. E., & Schleifer, K. H. (1983). Chemotaxonomic Study of an Alkalophilic Bacterium, *Exiguobacterium aurantiacum* gen. nov., sp. nov. *Microbiology, 129*(7), 2037–2042. doi:10.1099/00221287-129-7-2037

Coradin, J. H., Braun, A., Viccini, G., de Lima Luz, L. F., Krieger, N., & Mitchell, D. A. (2011). A three-dimensional discrete lattice-based system for modeling the growth of aerial hyphae of filamentous fungi on solid surfaces: A tool for investigating micro-scale phenomena in solid-state fermentation. *Biochemical Engineering Journal, 54*(3), 164–171. doi:10.1016/j.bej.2011.02.012

Damiano, V. B., Ward, R., Gomes, E., Alves-Prado, H. F., & Da Silva, R. (2006). Purification and characterization of two xylanases from alkalophilic and thermophilic *Bacillus licheniformis* 77-2. *Applied Biochemistry and Biotechnology, 129–132*(1-3), 289–302. doi:10.1385/ABAB:129:1:289 PMID:16915648

Darwesh, O. M., El-Hawary, A. S., El Kelany, U. S., & El-Sherbiny, G. M. (2019). Nematicidal activity of thermostable alkaline protease produced by *Saccharomonospora viridis* strain Hw G550. *Biotechnology Reports (Amsterdam, Netherlands), 24*, e00386. doi:10.1016/j.btre.2019.e00386 PMID:31763199

Dayma, P. B., Mangrola, A. V., Suriyaraj, S. P., Dudhagara, P., & Patel, R. K. (2019). Synthesis of bio-silver nanoparticles using desert isolated *Streptomyces intermedius* and its antimicrobial activity. *J. Pharm. Chem. Biol. Sci, 7*, 94–101.

Dehghanikhah, F., Shakarami, J., & Asoodeh, A. (2020). Purification and Biochemical Characterization of Alkalophilic Cellulase from the Symbiotic *Bacillus subtilis* BC1 of the Leopard Moth, Zeuzera pyrina (L.) (Lepidoptera: Cossidae). *Current Microbiology, 77*(7), 1254–1261. doi:10.100700284-020-01938-z PMID:32125446

Denizci, A. A., Kazan, D., & Erarslan, A. (2010). *Bacillus marmarensis* sp. nov., an alkaliphilic, protease-producing bacterium isolated from mushroom compost. *International Journal of Systematic and Evolutionary Microbiology, 60*(7), 1590–1594. Advance online publication. doi:10.1099/ijs.0.012369-0 PMID:19700450

Dhakar, K., & Pandey, A. (2016). Wide pH range tolerance in extremophiles: Towards understanding an important phenomenon for future biotechnology. *Applied Microbiology and Biotechnology, 100*(6), 2499–2510. doi:10.100700253-016-7285-2 PMID:26780356

Ding, Z. G., Ji, Y., Yin, M., Zhao, Y. R., Feng, Y. Z., Chunyu, W. X., & Tang, S. K. (2019). *Phytoactinopolyspora halophila* sp. nov., a halophilic actinomycete isolated from a saline soil. *International Journal of Systematic and Evolutionary Microbiology, 69*(2), 384–389. doi:10.1099/ijsem.0.003153 PMID:30543502

Dirmeier, R., Keller, M., Hafenbradl, D., Braun, F.-J., Rachel, R., Burggraf, S., & Stetter, K.O. (1998). *Thermococcus acidaminovorans sp. nov., a new hyperthermophilic alkalophilic archaeon growing on amino acids*. Academic Press.

Ece, F., Kaya, A., Avci, F. G., Sayar, N. A., Kazan, D., Altinisik Kaya, F. E., Avci, F. G., Sayar, N. A., Kazan, D., Sayar, A. A., & Sariyar Akbulut, B. (2018). What Are the Multi-Omics Mechanisms for Adaptation by Microorganisms to High Alkalinity? A Transcriptomic and Proteomic Study of a Bacillus Strain with Industrial Potential. *OMICS: A Journal of Integrative Biology, 22*(11), 717–732. doi:10.1089/omi.2018.0127 PMID:30457468

Eddehech, A., Zied, Z., Aloui, F., Smichi, N., Noiriel, A., Abousalham, A., & Gargouri, Y. (2019). Production, purification and biochemical characterization of a thermoactive, alkaline lipase from a newly isolated *Serratia* sp. W3 Tunisian strain. *International Journal of Biological Macromolecules, 123*, 792–800. doi:10.1016/j.ijbiomac.2018.11.050 PMID:30439433

Esakkiraj, P., Prabakaran, G., Maruthiah, T., Immanuel, G., & Palavesam, A. (2016). Purification and Characterization of Halophilic Alkaline Lipase from *Halobacillus* sp. *Proceedings of the National Academy of Sciences. India. Section B, Biological Sciences, 86*(2), 309–314. doi:10.100740011-014-0437-1

Felföldi, T., Somogyi, B., Márialigeti, K., & Vörös, L. (2009). Characterization of photoautotrophic picoplankton assemblages in turbid, alkaline lakes of the Carpathian Basin (Central Europe). *Journal of Limnology, 68*(2), 385–395. doi:10.4081/jlimnol.2009.385

Feng, X. M., Zhai, X. H., Liu, S. W., Wang, C. M., Sun, C. H., Zhou, X., Zhang, Y. J., & Li, X. J. (2020). *Phytoactinopolyspora mesophila* sp. Nov., isolated from a saline–alkaline soil. *International Journal of Systematic and Evolutionary Microbiology, 70*(7), 4179–4185. doi:10.1099/ijsem.0.004274 PMID:32558636

Forró, L., Nédli, J., Csata, E., Krízsik, V., Balogh, C., & Tóth, L. G. (2017). Morphometric characteristics and coi haplotype diversity of *Arctodiaptomus spinosus* (copedoda) populations in Soda Pans in Hungary. *Acta Biologica Hungarica, 68*, 279–289. doi:10.1556/018.68.2017.3.5 PMID:28901808

Fujisawa, M., Fackelmayer, O. J., Liu, J., Krulwich, T. A., & Hicks, D. B. (2010). The ATP synthase a-subunit of extreme alkaliphiles is a distinct variant: Mutations in the critical alkaliphile-specific residue Lys-180 and other residues that support alkaliphile oxidative phosphorylation. *The Journal of Biological Chemistry, 285*(42), 32105–32115. doi:10.1074/jbc.M110.165084 PMID:20716528

Fujiwara, N., Masui, A., & Imanaka, T. (1993). Purification and properties of the highly thermostable alkaline protease from an alkaliphilic and thermophilic *Bacillus* sp. *Journal of Biotechnology, 30*(2), 245–256. doi:10.1016/0168-1656(93)90117-6 PMID:7764036

Gaur, R., & Tiwari, S. (2015). Isolation, production, purification and characterization of an organic-solvent-thermostable alkalophilic cellulase from *Bacillus vallismortis* RG-07. *BMC Biotechnology, 15*(1), 19. doi:10.118612896-015-0129-9 PMID:25886936

Georlette, D., Damien, B., Blaise, V., Depiereux, E., Uversky, V. N., Gerday, C., & Feller, G. (2003). Structural and functional adaptations to extreme temperatures in psychrophilic, mesophilic, and thermophilic DNA ligases. *The Journal of Biological Chemistry, 278*(39), 37015–37023. doi:10.1074/jbc.M305142200 PMID:12857762

Ghorbel-Bellaaj, O., Hayet, B. K., Bayoudh, A., Younes, I., Hmidet, N., Jellouli, K., & Nasri, M. (2012). *Pseudomonas aeruginosa* A2 elastase: Purification, characterization and biotechnological applications. *International Journal of Biological Macromolecules, 50*(3), 679–686. doi:10.1016/j.ijbiomac.2012.01.038 PMID:22326423

Gimza, B. D., Larias, M. I., Budny, B. G., & Shaw, L. N. (2019). Mapping the Global Network of Extracellular Protease Regulation in *Staphylococcus aureus*. *MSphere, 4*(5). Advance online publication. doi:10.1128/mSphere.00676-19 PMID:31645429

Goswami, D., Dhandhukia, P., Patel, P., & Thakker, J. N. (2014). Screening of PGPR from saline desert of Kutch: Growth promotion in Arachis hypogea by *Bacillus licheniformis* A2. *Microbiological Research, 169*(1), 66–75. doi:10.1016/j.micres.2013.07.004 PMID:23896166

Gouda, H. A., Moubasher, A. H., Ismail, M. A., & Hussein, N. A. (2020). Alkaliphilic and alkalitolerant fungi in mud and salt crusts collected from Wadi El-Natrun lakes. Journal of Agriculture. *Food and Environment, 01*(02), 29–34. doi:10.47440/JAFE.2020.1205

Grant, W. D., & Heaphy, S. (2010). Metagenomics and recovery of enzyme genes from alkaline saline environments. *Environmental Technology, 31*(10), 1135–1143. doi:10.1080/09593331003646661 PMID:20718296

Grant, W. D., & Sorokin, D. Y. (2011). *Distribution and Diversity of Soda Lake Alkaliphiles*. Extremophiles Handbook. doi:10.1007/978-4-431-53898-1_3

Grum-Grzhimaylo, A. A., Debets, A. J. M., van Diepeningen, A. D., Georgieva, M. L., & Bilanenko, E. N. (2013). *Sodiomyces alkalinus*, a new holomorphic alkaliphilic ascomycete within the Plectosphaerellaceae. Persoonia. *Molecular Phylogeny and Evolution of Fungi, 31*(1), 147–158. doi:10.3767/003158513X673080 PMID:24761040

Gupta, M., Aggarwal, S., Navani, N. K., & Choudhury, B. (2015). Isolation and characterization of a protease-producing novel haloalkaliphilic bacterium *Halobiforma* sp. strain BNMIITR from Sambhar lake in Rajasthan, India. *Annals of Microbiology, 65*(2), 677–686. doi:10.100713213-014-0906-z

Gupta, N., Reddy, V. S., Maiti, S., & Ghosh, A. (2000). Cloning, expression, and sequence analysis of the gene encoding the alkali-stable, thermostable endoxylanase from alkalophilic, mesophilic *Bacillus* sp. strain NG-27. *Applied and Environmental Microbiology, 66*(6), 2631–2635. doi:10.1128/AEM.66.6.2631-2635.2000 PMID:10831448

Hakamada, Y., Koike, K., Yoshimatsu, T., Mori, H., Kobayashi, T., & Ito, S. (1997). Thermostable alkaline cellulase from an alkaliphilic isolate, *Bacillus* sp. KSM-S237. *Extremophiles, 1*(3), 151–156. doi:10.1007007920050028 PMID:9680321

Hamed, A., Khedr, M., & Abdelraof, M. (2019). Molecular characterization of alkaline protease-coding gene from *Bacillus licheniformis* MK90 mutants with biofilm inhibitory activity. *Egyptian Pharmaceutical Journal, 18*(4), 419. doi:10.4103/epj.epj_47_19

Hayden, M. S., & Ghosh, S. (2008). Shared Principles in NF-κB Signaling. *Cell, 132*(3), 344–362. doi:10.1016/j.cell.2008.01.020 PMID:18267068

Hedlund, B. P., Dodsworth, J. A., Murugapiran, S. K., Rinke, C., & Woyke, T. (2014). Impact of single-cell genomics and metagenomics on the emerging view of extremophile "microbial dark matter.". *Extremophiles, 18*(5), 865–875. doi:10.100700792-014-0664-7 PMID:25113821

Hirota, K., Aino, K., Nodasaka, Y., & Yumoto, I. (2013). *Oceanobacillus indicireducens* sp. nov., a facultative alkaliphile that reduces an indigo dye. *International Journal of Systematic and Evolutionary Microbiology, 63*(Pt_4), 1437–1442. doi:10.1099/ijs.0.034579-0 PMID:22843722

Horikoshi, K. (1971). Production of Alkaline Enzymes by Alkalophilic Microorganisms Part II. Alkaline Amylase Produced by *Bacillus* No. A-40-2. *Agricultural and Biological Chemistry, 35*, 1783–1791.

Horikoshi, K. (1971). Production of Alkaline Enzymes by Alkalophilic Microorganisms: Part I. Alkaline Protease Produced by *Bacillus* No. 221. *Agricultural and Biological Chemistry*, *35*(9), 1407–1414. doi :10.1080/00021369.1971.10860094

Horikoshi, K. (1973). *Characters of Bacillus No . C-59-2 The isolated bacteria The charac was that the Table I summarizes the of carbonate an important 37*. Academic Press.

Horikoshi, K. (2011). *Extremophiles Handbook*. Extremophiles Handbook. doi:10.1007/978-4-431-53898-1

Hozzein, W. N., Li, W. J., Ali, M. I. A., Hammouda, O., Mousa, A. S., Xu, L. H., & Jiang, C. L. (2004). *Nocardiopsis alkaliphila* sp. nov., a novel alkaliphilic actinomycete isolated from desert soil in Egypt. *International Journal of Systematic and Evolutionary Microbiology*, *54*(1), 247–252. doi:10.1099/ijs.0.02832-0 PMID:14742488

Hu, J., Cai, W., Wang, C., Du, X., Lin, J., & Cai, J. (2018). Purification and characterization of alkaline lipase production by *Pseudomonas aeruginosa* HFE733 and application for biodegradation in food wastewater treatment. *Biotechnology, Biotechnological Equipment*, *32*(3), 583–590. doi:10.1080/13102818.2018.1446764

Ikehata, K., Buchannan, I., Pickard, M., & Smith, D. (2005). Purification, characterization and evaluation of extracellular peroxidase from two species for aqueous phenol treatment. *Bioresource Technology*, *96*(16), 1758–1770. doi:10.1016/j.biortech.2005.01.019 PMID:16051082

Ito, S., Shdcata, S., Ozaki, K., Kawai, S., Okamoto, K., Inoue, S., Takei, A., Ohta, Y., & Satoh, T. (1989). *Alkaline Cellulase for Laundry Detergents: Production by Bacillus sp. KSM-635 and Enzymatic Properties*. Academic Press.

Itoh, T., Yamaguchi, T., Zhou, P., & Takashina, T. (2005). *Natronolimnobius baerhuensis* gen. nov., sp. nov. and *Natronolimnobius innermongolicus* sp. nov., novel haloalkaliphilic archaea isolated from soda lakes in Inner Mongolia, China. *Extremophiles*, *9*(2), 111–116. doi:10.100700792-004-0426-z PMID:15841343

Jaouadi, B., Abdelmalek, B., Zarai, N., & Bejar, S. 2011. The Bioengineering and Industrial Applications of Bacterial Alkaline Proteases: The Case of SAPB and KERAB. In Progress in Molecular and Environmental Bioengineering - From Analysis and Modeling to Technology Applications. InTech. doi:10.5772/23850

Jaradat, Z., Al Aboudi, A., Shatnawi, M., & Ababneh, Q. (2013). Partial purification and characterization of alkalophilic protease from *Pseudomonas aeruginosa. Journal of Microbiology, Biotechnology and Food Sciences*, *2*, 2455–2461.

Joshi, S., & Satyanarayana, T. (2013). Characteristics and applications of a recombinant alkaline serine protease from a novel bacterium *Bacillus lehensis. Bioresource Technology*, *131*, 76–85. doi:10.1016/j.biortech.2012.12.124 PMID:23340105

Kakkar, P., & Wadhwa, N. (2021). Extremozymes used in textile industry. *Journal of the Textile Institute*, 1–9. doi:10.1080/00405000.2021.1948251

Kalwasińska, A., Felföldi, T., Szabó, A., Deja-Sikora, E., Kosobucki, P., & Walczak, M. (2017). Microbial communities associated with the anthropogenic, highly alkaline environment of a saline soda lime, Poland. *Antonie van Leeuwenhoek*, *110*(7), 945–962. doi:10.100710482-017-0866-y PMID:28382378

Kambura, A. K. (2016). *Metagenomic and metatranscriptomic analysis of bacterial, Archaeal and Fungal communities within the hot springs of Lake Magadi in Kenya* (PhD Thesis). Jomo Kenyatta University of Agrioculture and Technology.

Karataş, H., Uyar, F., Tolan, V., & Baysal, Z. (2013). Optimization and enhanced production of α-amylase and protease by a newly isolated *Bacillus licheniformis* ZB-05 under solid-state fermentation. *Annals of Microbiology*, *63*(1), 45–52. doi:10.100713213-012-0443-6

Kaur, S., Dhanjal, D. S., Singh, S., Chopra, C., & Chopra, R. S. (2020). Isolation and biochemical characterization of an alkalophilic, cold-active cellulase from bacteria isolated from Boniyar region of Kashmir. *Plant Archives*, *20*, 2548–2553.

Kim, J.-Y., Hur, S.-H., & Hong, J.-H. (2005). Purification and characterization of an alkaline cellulase from a newly isolated alkalophilic *Bacillus* sp. HSH-810. *Biotechnology Letters*, *27*(5), 313–316. doi:10.100710529-005-0685-5 PMID:15834791

Koide, T., Vêncio, R. Z. N., & Gomes, S. L. (2006). Global gene expression analysis of the heat shock response in the phytopathogen *Xylella fastidiosa*. *Journal of Bacteriology*, *188*(16), 5821–5830. doi:10.1128/JB.00182-06 PMID:16885450

Kojima, M., Kanai, M., Tominaga, M., Kitazume, S., Inoue, A., & Horikoshi, K. (2006). Isolation and characterization of a feather-degrading enzyme from *Bacillus pseudofirmus* FA30-01. *Extremophiles*, *10*(3), 229–235. doi:10.100700792-005-0491-y PMID:16489414

Kulshreshtha, N. M., Kumar, R., Begum, Z., Shivaji, S., & Kumar, A. (2013). *Exiguobacterium alkaliphilum* sp. nov. isolated from alkaline wastewater drained sludge of a beverage factory. *International Journal of Systematic and Evolutionary Microbiology*, *63*(Pt_12), 4374–4379. doi:10.1099/ijs.0.039123-0 PMID:23838447

Kumar, D. J. M., Rajan, R., Lawrence, L., Priyadarshini, S., Chittybabu, S., & Kalaichelvan, P. T. (2012). Destaining and dehairing capability of partially purified *Bacillus subtilis* protease from optimized fermentation medium. *Asian Journal of Experimental Biological Sciences*, *3*, 613–620.

Kumar, L., Kumar, D., Nagar, S., Gupta, R., Garg, N., Kuhad, R.C., & Gupta, V.K. (2014). Modulation of xylanase production from alkaliphilic *Bacillus pumilus* VLK-1 through process optimization and temperature shift operation. *Biotech*, *3*(4), 345–356. doi:10.1007/s13205-013-0160-2

Kumar, N., & Hovik, A. (2019). Extremophiles : Applications and roles in environmental sustainability. *Environmental Sustainability*, *2*(3), 217–218. doi:10.100742398-019-00082-0

Kumar, V., Saxena, J., & Tiwari, S. K. (2016). Description of a halocin-producing *Haloferax larsenii* HA1 isolated from Pachpadra salt lake in Rajasthan. *Archives of Microbiology, 198*(2), 181–192. doi:10.100700203-015-1175-3 PMID:26659359

Kuvarina, A. E., Gavryushina, I. A., Kulko, A. B., Ivanov, I. A., Rogozhin, E. A., Georgieva, M. L., & Sadykova, V. S. (2021). The emericellipsins a-e from an alkalophilic fungus emericellopsis alkalina show potent activity against multidrug-resistant pathogenic fungi. *Journal of Fungi (Basel, Switzerland), 7*(2), 1–17. doi:10.3390/jof7020153 PMID:33669976

Lanzén, A., Simachew, A., Gessesse, A., Chmolowska, D., Jonassen, I., & Øvreås, L. (2013). Surprising Prokaryotic and Eukaryotic Diversity, Community Structure and Biogeography of Ethiopian Soda Lakes. *PLoS One, 8*(8), e72577. Advance online publication. doi:10.1371/journal.pone.0072577 PMID:24023625

Lee, C., Kang, H. J., von Ballmoos, C., Newstead, S., Uzdavinys, P., Dotson, D. L., Iwata, S., Beckstein, O., Cameron, A. D., & Drew, D. (2013). A two-domain elevator mechanism for sodium/proton antiport. *Nature, 501*(7468), 573–577. doi:10.1038/nature12484 PMID:23995679

Lentzen, G., & Schwarz, T. (2006). Extremolytes: Natural compounds from extremophiles for versatile applications. *Applied Microbiology and Biotechnology, 72*(4), 623–634. doi:10.100700253-006-0553-9 PMID:16957893

Li, Y., Engle, M., Weiss, N., Mandelco, L., & Wiegel, J. (1994). *Clostridium thermoalcaliphilum* sp. nov., an Anaerobic and Thermotolerant Facultative Alkaliphile. *International Journal of Systematic Bacteriology, 44*(1), 111–118. doi:10.1099/00207713-44-1-111 PMID:8123553

Li, Z., Kawamura, Y., Shida, O., Yamagata, S., Deguchi, T., & Ezaki, T. (2002). *Bacillus okuhidensis* sp. nov., isolated from the Okuhida spa area of Japan. *International Journal of Systematic and Evolutionary Microbiology, 52*(4), 1205–1209. doi:10.1099/00207713-52-4-1205 PMID:12148629

Ling, H. L., Rahmat, Z., Bakar, F. D. A., Murad, A. M. A., & Illias, R. M. (2018). Secretome analysis of alkaliphilic bacterium *Bacillus lehensis* G1 in response to pH changes. *Microbiological Research, 215*, 46–54. doi:10.1016/j.micres.2018.06.006 PMID:30172308

Liu, H., Zhou, Y., Liu, R., Zhang, K.-Y., & Lai, R. (2009). Bacillus solisalsi sp. nov., a halotolerant, alkaliphilic bacterium isolated from soil around a salt lake. *International Journal of Systematic and Evolutionary Microbiology, 59*(6), 1460–1464. doi:10.1099/ijs.0.000653-0 PMID:19502335

Logan, B. E., Hamelers, B., Rozendal, R., Schröder, U., Keller, J., Freguia, S., Aelterman, P., Verstraete, W., & Rabaey, K. (2006). Microbial Fuel Cells: Methodology and Technology. *Environmental Science & Technology, 40*(17), 5181–5192. doi:10.1021/es0605016 PMID:16999087

Mamo, G. (2019). Challenges and Adaptations of Life in Alkaline Habitats. *Advances in Biochemical Engineering/Biotechnology, 172*, 85–133. doi:10.1007/10_2019_97 PMID:31273417

McMillan, D. G. G., Velasquez, I., Nunn, B. L., Goodlett, D. R., Hunter, K. A., Lamont, I., Sander, S. G., & Cook, G. M. (2010). Acquisition of iron by alkaliphilic *Bacillus* species. *Applied and Environmental Microbiology, 76*(20), 6955–6961. doi:10.1128/AEM.01393-10 PMID:20802068

Merino, N., Aronson, H. S., Bojanova, D. P., Feyhl-Buska, J., Wong, M. L., Zhang, S., & Giovannelli, D. (2019). Living at the Extremes: Extremophiles and the Limits of Life in a Planetary Context. *Frontiers in Microbiology, 10*, 780. Advance online publication. doi:10.3389/fmicb.2019.00780 PMID:31037068

Mesbah, N., & Wiegel, J. (2018). Improvement of Activity and Thermostability of Agar-Entrapped, Thermophilic, Haloalkaliphilic Amylase AmyD8. *Catalysis Letters, 148*(9), 2665–2674. doi:10.100710562-018-2493-2

Mevada, V., Patel, S., Pandya, J., Joshi, H., & Patel, R. (2017). Whole genome sequencing and annotation of halophilic *Salinicoccus sp.* BAB 3246 isolated from the coastal region of Gujarat. *Genomics Data, 13*, 30–34. doi:10.1016/j.gdata.2017.06.006 PMID:28702355

Milford, A. D., Achenbach, L. A., Jung, D. O., & Madigan, M. T. (2000). *Rhodobaca bogoriensis* gen. nov. and sp. nov., an alkaliphilic purple nonsulfur bacterium from African Rift Valley soda lakes. *Archives of Microbiology, 174*(1-2), 18–27. doi:10.1007002030000166 PMID:10985738

Mojsov, K., Andronikov, D., Janevski, A., Jordeva, S., Kertakova, M., Golomeova, S., Gaber, S., & Ignjatov, I. (2018). Production and application of α-amylase enzyme in textile industry. *Tekstilna industrija, 66*, 23–28.

Mwatha, V.E. (1990). *Microbial Ecology of Kenyan soda lakes*. doi:10.1016/B978-0-12-363522-8.X5001-9

Nakajima, K., Hirota, K., Nodasaka, Y., & Yumoto, I. (2005). *Alkalibacterium iburiense* sp. nov., an obligate alkaliphile that reduces an indigo dye. *International Journal of Systematic and Evolutionary Microbiology, 55*(4), 1525–1530. doi:10.1099/ijs.0.63487-0 PMID:16014476

Nielsen, P., Fritze, D., & Priest, F. G. (1995). Phenetic diversity of alkaliphilic Bacillus strains: Proposal for nine new species. *Microbiology, 141*(7), 1745–1761. doi:10.1099/13500872-141-7-1745

Niyonzima, F. N. (2019). Detergent-compatible bacterial cellulases. *Journal of Basic Microbiology, 59*(2), 134–147. doi:10.1002/jobm.201800436 PMID:30421443

Norton, C. F., & Grant, W. (1988). Survival of halobacteria within fluid inclusions in salt crystals. *Microbiology, 134*(5), 1365–1373. doi:10.1099/00221287-134-5-1365

Ntougias, S., & Russell, N. J. (2001). *Alkalibacterium olivoapovliticus* gen. nov., sp. nov., a new obligately alkaliphilic bacterium isolated from edible-olive wash-waters. *International Journal of Systematic and Evolutionary Microbiology, 51*(3), 1161–1170. doi:10.1099/00207713-51-3-1161 PMID:11414268

Olsen, H. S., & Falholt, P. (1998). The role of enzymes in modern detergency. *Journal of Surfactants and Detergents, 1*(4), 555–567. doi:10.100711743-998-0058-7

Ozawa, T., Endo, K., Igarashi, K., Kitayama, K., Hayashi, Y., Hagihara, H., Kawai, S., Ito, S., & Ozaki, K. (2007). Improvement of the Thermal Stability of a Calcium-free, Alkaline. ALPHA.-Amylase by Site-directed Mutagenesis. *Journal of Applied Glycoscience, 54*(2), 77–83. doi:10.5458/jag.54.77

Padan, E. (2014). Functional and structural dynamics of NhaA, a prototype for Na+ and H+ antiporters, which are responsible for Na+ and H+ homeostasis in cells. Biochimica et Biophysica Acta (BBA) -. *Bioenergetics, 1837*(7), 1047–1062. doi:10.1016/j.bbabio.2013.12.007 PMID:24361841

Padan, E., Venturi, M., Gerchman, Y., & Dover, N. (2001). Na+/H+ antiporters. Biochimica et Biophysica Acta (BBA) -. *Bioenergetics*, *1505*(1), 144–157. doi:10.1016/S0005-2728(00)00284-X

Palanivel, P., Ashokkumar, L., & Balagurunathan, R. (2013). Production, purification and fibrinolytic characterization of alkaline protease from extremophilic soil fungi. *Int J Pharm Bio Sci*, *4*, 101–110.

Palaniyandi, S. A., Yang, S. H., Zhang, L., & Suh, J.-W. (2013). Effects of actinobacteria on plant disease suppression and growth promotion. *Applied Microbiology and Biotechnology*, *97*(22), 9621–9636. doi:10.100700253-013-5206-1 PMID:24092003

Panda, A. N., Ray, L., Mishra, S. R., & Raina, V. (2020). *Rheinheimera pleomorphica* sp. nov., a Novel Alkali-Tolerant Bacteria Isolated from Chilika Lake, India. *Current Microbiology*, *77*(1), 158–165. doi:10.100700284-019-01802-9 PMID:31705390

Pandey, A., Nigam, P., Soccol, C. R., Soccol, V. T., Singh, D., & Mohan, R. (2000). Advances in microbial amylases. *Biotechnology and Applied Biochemistry*, *31*(2), 135–152. doi:10.1042/BA19990073 PMID:10744959

Patel, R., Mevada, V., Prajapati, D., Dudhagara, P., Koringa, P., & Joshi, C. G. (2015). Metagenomic sequence of saline desert microbiota from wild ass sanctuary, Little Rann of Kutch, Gujarat, India. *Genomics Data*, *3*, 137–139. doi:10.1016/j.gdata.2015.01.003 PMID:26484162

Pentecost, A. (2009). The Marl Lakes of the British Isles. *Freshwater Reviews*, *2*(2), 167–197. doi:10.1608/FRJ-2.2.4

Poli, A., Esposito, E., Orlando, P., Lama, L., Giordano, A., de Appolonia, F., Nicolaus, B., & Gambacorta, A. (2007). *Halomonas alkaliantarctica* sp. nov., isolated from saline lake Cape Russell in Antarctica, an alkalophilic moderately halophilic, exopolysaccharide-producing bacterium. *Systematic and Applied Microbiology*, *30*(1), 31–38. doi:10.1016/j.syapm.2006.03.003 PMID:16621401

Popa, A., Israel-Roming, F., & Cornea, C.P. (2016). *Microbial xylanase: A review SEE PROFILE XX*. Academic Press.

Price, S., Kuzhiumparambil, U., Pernice, M., & Ralph, P. J. (2020). Cyanobacterial polyhydroxybutyrate for sustainable bioplastic production: Critical review and perspectives. *Journal of Environmental Chemical Engineering*, *8*(4), 104007. doi:10.1016/j.jece.2020.104007

Priyadarshini, S., Pradhan, S. K., & Ray, P. (2020). Production, characterization and application of thermostable, alkaline α-amylase (AA11) from *Bacillus cereus* strain SP-CH11 isolated from Chilika Lake. *International Journal of Biological Macromolecules*, *145*, 804–812. doi:10.1016/j.ijbiomac.2019.11.149 PMID:31758985

Purich, D.L. (2010). Factors Influencing Enzyme Activity. *Enzyme Kinetics: Catalysis & Control*, 379–484. doi:10.1016/B978-0-12-380924-7.10007-9

Qu, D., Show, P.-L., & Miao, X. (2021). Transcription Factor ChbZIP1 from Alkaliphilic Microalgae *Chlorella* sp. BLD Enhancing Alkaline Tolerance in Transgenic *Arabidopsis thaliana. International Journal of Molecular Sciences*, *22*(5), 2387. doi:10.3390/ijms22052387 PMID:33673599

Raichand, R., Pareek, S., Singh, N. K., & Mayilraj, S. (2012). *Exiguobacterium aquaticum* sp. nov., a member of the genus exiguobacterium. *International Journal of Systematic and Evolutionary Microbiology, 62*(Pt_9), 2150–2155. doi:10.1099/ijs.0.035790-0 PMID:22058319

Razzaq, A., Shamsi, S., Ali, A., Ali, Q., Sajjad, M., Malik, A., & Ashraf, M. (2019). Microbial Proteases Applications. *Frontiers in Bioengineering and Biotechnology, 7*, 110. Advance online publication. doi:10.3389/fbioe.2019.00110 PMID:31263696

Reed, C. J., Lewis, H., Trejo, E., Winston, V., & Evilia, C. (2013). Protein adaptations in archaeal extremophiles. *Archaea (Vancouver, B.C.), 373275*, 1–14. Advance online publication. doi:10.1155/2013/373275 PMID:24151449

Sahoo, R. K., Sahu, A., & Subudhi, E. (2020). Bioremediation of Hydrocarbon Using Bacterial Lipase from Waste Biomass. Iranian Journal of Science and Technology, Transactions A. *Science, 44*(5), 1287–1293. doi:10.100740995-020-00935-y

Saini, R., Singh Saini, H., Dahiya, A., & Harnek Singh Saini, C. (2017). Amylases: Characteristics and industrial applications. *Journal of Pharmacognosy and Phytochemistry, 6*, 1865–1871.

Saito, K., Nishio, Z., Takata, K., Kuwabara, T., Yokota, A., Yamauchi, H., & Oda, Y. (2003). Bacterial Adjunct to Improve the Quality of Alkaline Noodles. Food Science and Technology Research -. *Food Science and Technology Research, 9*(1), 40–44. doi:10.3136/fstr.9.40

Sander, D., Yu, Y., Sukul, P., Schäkermann, S., Bandow, J. E., Mukherjee, T., Mukhopadhyay, S. K., & Leichert, L. I. (2021). Metaproteomic Discovery and Characterization of a Novel Lipolytic Enzyme From an Indian Hot Spring. *Frontiers in Microbiology, 12*, 672727. Advance online publication. doi:10.3389/fmicb.2021.672727 PMID:34149658

Sanz-Montero, M. E., Cabestrero, Ó., & Sánchez-Román, M. (2019). Microbial Mg-rich carbonates in an extreme alkaline lake (Las Eras, Central Spain). *Frontiers in Microbiology, 10*, 148. Advance online publication. doi:10.3389/fmicb.2019.00148 PMID:30800103

Sarkar, G. K. S., & K, S. (2020). Extraction and characterization of alkaline protease from *Streptomyces* sp. GS-1 and its application as dehairing agent. *Biocatalysis and Agricultural Biotechnology, 25*, 101590. doi:10.1016/j.bcab.2020.101590

Schagerl, M. (2016). *Bacteria*. Archaea and Viruses of Soda Lakes, Soda Lakes of East Africa. doi:10.1007/978-3-319-28622-8

Schultze-Lam, S., Ferris, F. G., Sherwood-Lollar, B., & Gerits, J. P. (1996). Ultrastructure and seasonal growth patterns of microbial mats in a temperate climate saline-alkaline lake: Goodenough Lake, British Columbia, Canada. *Canadian Journal of Microbiology, 42*(2), 147–161. doi:10.1139/m96-023

Sehar, S., & Hameed, A. (2011). Extracellular Alkaline Protease by a Newly Isolated *Halophilic Bacillus* sp. *Global Journal of Biotechnology & Biochemistry, 6*, 142–148.

Sevinc, N., & Demirkan, E. (2011). Production of Protease by *Bacillus* sp. N-40 Isolated from Soil and Its Enzymatic Properties. *Journal of Biological and Environmental Sciences, 5*, 95–103.

Shah, S. R. (2013). Chemistry and Applications of Cellulase in Textile Wet Processing. *Research Journal of Engineering Sciences, 2*, 1–5.

Shapovalova, A. A., Khijniak, T. V., Tourova, T. P., & Sorokin, D. Y. (2009). *Halomonas chromatireducens* sp. nov., a new denitrifying facultatively haloalkaliphilic bacterium from solonchak soil capable of aerobic chromate reduction. *Microbiology, 78*(1), 102–111. doi:10.1134/S0026261709010135 PMID:19334604

Shin, B., Park, C., Lee, B. H., Lee, K. E., & Park, W. (2020). *Bacillus miscanthi* sp. Nov., a alkaliphilic bacterium from the rhizosphere of *Miscanthus sacchariflorus. International Journal of Systematic and Evolutionary Microbiology, 70*(3), 1843–1849. doi:10.1099/ijsem.0.003982 PMID:31951191

Shivlata, L., & Satyanarayana, T. (2015). Thermophilic and alkaliphilic Actinobacteria: Biology and potential applications. *Frontiers in Microbiology, 6*, 1014. doi:10.3389/fmicb.2015.01014 PMID:26441937

Singh, H., Kaur, M., Kaur, L., Sharma, S., Mishra, S., Tanuku, N. R. S., & Pinnaka, A. K. (2018). *Bacillus lacus* sp. nov., isolated from a water sample of a salt lake in India. *International Journal of Systematic and Evolutionary Microbiology, 68*(3), 801–809. doi:10.1099/ijsem.0.002588 PMID:29458476

Singh, O. V., & Gabani, P. (2011). Extremophiles: Radiation resistance microbial reserves and therapeutic implications. *Journal of Applied Microbiology, 110*(4), 851–861. doi:10.1111/j.1365-2672.2011.04971.x PMID:21332616

Sorokin, D. Y., Khijniak, T. V., Zakharycheva, A. P., Elcheninov, A. G., Hahnke, R. L., Boueva, O. V., Ariskina, E. V., Bunk, B., Kublanov, I. V., & Evtushenko, L. I. (2021). *Natronoglycomyces albus* gen. Nov., sp. nov, a haloalkaliphilic actinobacterium from a soda solonchak soil. *International Journal of Systematic and Evolutionary Microbiology, 71*(5), 004804. doi:10.1099/ijsem.0.004804 PMID:33999794

Sorokin, I. D., Kravchenko, I. K., Doroshenko, E. V., Boulygina, E. S., Zadorina, E. V., Tourova, T. P., & Sorokin, D. Y. (2008). *Haloalkaliphilic diazotroph*s in soda solonchak soils. *FEMS Microbiology Ecology, 65*(3), 425–433. doi:10.1111/j.1574-6941.2008.00542.x PMID:18616587

Sou, T., Meeusen, E. N., de Veer, M., Morton, D. A. V., Kaminskas, L. M., & McIntosh, M. P. (2011). New developments in dry powder pulmonary vaccine delivery. *Trends in Biotechnology, 29*(4), 191–198. doi:10.1016/j.tibtech.2010.12.009 PMID:21255854

Souza, K. A., & Deal, P. H. (1977). Characteristics of a novel extremely alkalophilic bacterium. *Journal of General Microbiology, 101*(1), 103–109. doi:10.1099/00221287-101-1-103

Stetter, K. O. (1982). Ultrathin mycelia-forming organisms from submarine volcanic areas having an optimum growth temperature of 105 °C. *Nature, 300*(5889), 258–260. doi:10.1038/300258a0

Street, T. O., Bolen, D. W., & Rose, G. D. (2006). A molecular mechanism for osmolyte-induced protein stability. *Proceedings of the National Academy of Sciences of the United States of America, 103*(38), 13997–14002. doi:10.1073/pnas.0606236103 PMID:16968772

Takami, H., Nakamura, S., Aono, R., & Horikoshi, K. (1992). Degradation of Human Hair by a Thermostable Alkaline Protease from Alkaliphilic *Bacillus* sp. No. AH-101. *Bioscience, Biotechnology, and Biochemistry, 56*(10), 1667–1669. doi:10.1271/bbb.56.1667

Tayyem, N. E. S. A., Elhello, K. T. M., & Elmanama, A. A. (2021). Isolation, identification and growth conditions of calcite producing bacteria from urea-rich soil. *African Journal of Microbiological Research*, *15*(1), 37–46. doi:10.5897/AJMR2020.9445

Terra, L., Dyson, P. J., Hitchings, M. D., Thomas, L., Abdelhameed, A., Banat, I. M., Gazze, S. A., Vujaklija, D., Facey, P. D., Francis, L. W., & Quinn, G. A. (2018). A Novel Alkaliphilic *Streptomyces* Inhibits ESKAPE Pathogens. *Frontiers in Microbiology*, *9*, 2458. doi:10.3389/fmicb.2018.02458 PMID:30459722

Thombre, R. S., Bhalerao, A. R., Shinde, V. D., Dhar, S. K., & Shouche, Y. S. (2017). Response of Haloalkaliphilic Archaeon *Natronococcus Jeotgali* RR17 to Hypergravity. *Microgravity Science and Technology*, *29*(3), 191–200. doi:10.100712217-017-9538-9

Timms, B. V. (2021). Drivers restricting biodiversity in Australian saline lakes: A review. *Marine and Freshwater Research*, *72*(4), 462–468. doi:10.1071/MF20205

Trabelsi, S., Ben Mabrouk, S., Kriaa, M., Ameri, R., Sahnoun, M., Mezghani, M., & Bejar, S. (2019). The optimized production, purification, characterization, and application in the bread making industry of three acid-stable alpha-amylases isoforms from a new isolated *Bacillus subtilis* strain US586. *Journal of Food Biochemistry*, *43*(5), 1–13. doi:10.1111/jfbc.12826 PMID:31353531

VA, S., RB, P., PV, P. (2017). Comparative Study of Antimicrobial Potentials of Phospholipid Compound Produced by Halophilic and Alkaliphiles *Bacillus subtilis* isolated from Alkaline Meteorite Crater. *International Journal of Life Sciences (Amravati)*, *5*, 420–424.

Van-Thuoc, D., Hashim, S. O., Hatti-Kaul, R., & Mamo, G. (2013). Ectoine-mediated protection of enzyme from the effect of pH and temperature stress: A study using *Bacillus halodurans* xylanase as a model. *Applied Microbiology and Biotechnology*, *97*(14), 6271–6278. doi:10.100700253-012-4528-8 PMID:23132342

Vyas, K. M., Jadeja, R. N., Patel, D., Devkar, R. V., & Gupta, V. K. (2014). Article. *Polyhedron*, *80*, 20–33. doi:10.1016/j.poly.2013.12.037

Wang, C. Y., Chang, C. C., Ng, C. C., Chen, T. W., & Shyu, Y. T. (2008). *Virgibacillus chiguensis* sp. nov., a novel halophilic bacterium isolated from Chigu, a previously commercial saltern located in southern Taiwan. *International Journal of Systematic and Evolutionary Microbiology*, *58*(2), 341–345. doi:10.1099/ijs.0.64996-0 PMID:18218928

Wei, W., Wang, K., Hu, X., Yang, S., Zeng, C., Hou, Z., Liu, S., Cui, H., & Zhu, L. (2021). *Phytoactinopolyspora limicola* sp. nov., an alkaliphilic actinomycete isolated from a soda alkali-saline soil. *Archives of Microbiology*, *203*(4), 1367–1374. doi:10.100700203-020-02134-0 PMID:33386868

Windish, W. W., & Mhatre, N. S. (1965). Microbial amylases. *Advances in Applied Microbiology*, *7*, 273–304. doi:10.1016/S0065-2164(08)70389-7 PMID:5321878

Yang, H., Ma, Y., Zhao, Y., Shen, W., & Chen, X. (2020). Systematic engineering of transport and transcription to boost alkaline α-amylase production in *Bacillus subtilis*. *Applied Microbiology and Biotechnology*, *104*(7), 2973–2985. doi:10.100700253-020-10435-z PMID:32043188

Yang, S.-J., Ko, J.-A., Kim, H.-S., Jo, M.-H., Lee, H.-N., Park, B.-R., & Kim, Y. (2018). Biochemical Characterization of Alkaliphilic Cyclodextran Glucanotransferase from an Alkaliphilic Bacterium, *Paenibacillus daejeonensis. Journal of Microbiology and Biotechnology, 28*(12), 2029–2035. doi:10.4014/jmb.1810.10007 PMID:30394040

Yin, L., Sun, H., Zhang, H., He, L., Qiu, L., Lin, J., Xia, H., Zhang, Y., Ji, S., Shi, H., & Gao, M. (2019). Quantitatively Visualizing Tumor-Related Protease Activity in Vivo Using a Ratiometric Photoacoustic Probe. *Journal of the American Chemical Society, 141*(7), 3265–3273. doi:10.1021/jacs.8b13628 PMID:30689382

Yumoto, I., Hirota, K., & Kazuaki, Y. (2011). *Extremophiles Handbook.* Extremophiles Handbook. doi:10.1007/978-4-431-53898-1

Yumoto, I., Hirota, K., Nodasaka, Y., Tokiwa, Y., & Nakajima, K. (2008). *Alkalibacterium indicireducens* sp. nov., an obligate alkaliphile that reduces indigo dye. *International Journal of Systematic and Evolutionary Microbiology, 58*(4), 901–905. doi:10.1099/ijs.0.64995-0 PMID:18398191

Yumoto, I., Hirota, K., Nodasaka, Y., Yokota, Y., Hoshino, T., & Nakajima, K. (2004a). *Alkalibacterium psychrotolerans* sp. nov., a psychrotolerant obligate alkaliphile that reduces an indigo dye. *International Journal of Systematic and Evolutionary Microbiology, 54*(6), 2379–2383. doi:10.1099/ijs.0.63130-0 PMID:15545487

Yumoto, I., Hishinuma-Narisawa, M., Hirota, K., Shingyo, T., Takebe, F., Nodasaka, Y., Matsuyama, H., & Hara, I. (2004b). *Exiguobacterium oxidotolerans* sp. nov., a novel alkaliphile exhibiting high catalase activity. *International Journal of Systematic and Evolutionary Microbiology, 54*(6), 2013–2017. doi:10.1099/ijs.0.63129-0 PMID:15545426

Yumoto, I., Yamazaki, K., Hishinuma, M., Nodasaka, Y., Suemori, A., Nakajima, K., Inoue, N., & Kawasaki, K. (2001). *Pseudomonas alcaliphila* sp. nov., a novel facultatively psychrophilic alkaliphile isolated from seawater. *International Journal of Systematic and Evolutionary Microbiology, 51*(2), 349–355. doi:10.1099/00207713-51-2-349 PMID:11321079

Zainuddin, N., Abd Rahman, K. H., Razak Shaari, A., & Mohd Yusof, S. J. H. (2015). Isolation of Potential Alkaliphilic Cellulase-Producer Strains from Limestone Area of Perlis. *Applied Mechanics and Materials, 754–755*, 1054–1058. . doi:10.4028/www.scientific.net/AMM.754-755.1054

Zhai, L., Ma, Y., Xue, Y., Ma, Y. (2014). Bacillus alkalicola sp. nov., An Alkaliphilic, Gram-Positive Bacterium Isolated from Zhabuye Lake in Tibet, China. *Current Microbiology*, 311–316. Https://Doi.Org/10.1007/S00284-014-0576-X

Zhang, Y. G., Lu, X. H., Ding, Y. B., Zhou, X. K., Li, L., Guo, J. W., Wang, H. F., Duan, Y. Q., & Li, W. J. (2016). *Phytoactinopolyspora alkaliphila* sp. Nov., an alkaliphilic actinomycete isolated from a saline-alkaline soil. *International Journal of Systematic and Evolutionary Microbiology, 66*(5), 2058–2063. doi:10.1099/ijsem.0.000992 PMID:26920762

Zhao, B., Yan, Y., & Chen, S. (2014). How could haloalkaliphilic microorganisms contribute to biotechnology? *Canadian Journal of Microbiology, 60*(11), 717–727. doi:10.1139/cjm-2014-0233 PMID:25372346

Zhao, C.-C., & Eun, J.-B. (2020). Shotgun metagenomics approach reveals the bacterial community and metabolic pathways in commercial hongeo product, a traditional Korean fermented skate product. *Food Research International, 131*. doi:10.1016/j.foodres.2020.109030

Zhu, J., Arsovska, B., & Kozovska, K. (2020). Ecology and diversity of Alkaliphilic Cyanobacteria from Soda Lake Lonar, India. *International Journal of Recent Scientific Research, 11*, 37471–37472. doi:10.24327/IJRSR

Chapter 4
Thermophiles:
Physiology, Metabolism, Enzymology, and Adaptation Mechanisms

Amitsinh Mangrola
Shri Alpesh N. Patel Post Graduate Institute of Science and Research, Anand, India

Anjana Ghelani
Shree Ramkrishna Institute of Computer Education and Applied Sciences, Surat, India

Rajesh Kanjibhai Patel
Veer Narmad South Gujarat University, Surat, India

Kunal R. Jain
Sardar Patel University, Vallabh Vidyanagar, India

Pravin Dudhagara
https://orcid.org/0000-0002-4041-6751
Veer Narmad South Gujarat University, Surat, India

Hardik Shah
https://orcid.org/0000-0003-2956-4099
Ganpat University, India

Himani Gandhi
Christ College, Saurashtra University, India

Vishal Mevada
Directorate of Forensic Science, Gandhinagar, India

ABSTRACT

Microorganisms are the diverse living things present on the Earth. India has numerous unique thermal habitats that comprise several diversity hotspots, such as hot springs, deep oceanic hydrothermal openings, anaerobic biodigesters. The existence of life at high temperatures is quite attractive. At both ends of the temperature range suited with life, only microorganisms can grow and survive. Thermophiles are a typical extremophilic microbes capable of existence in high temperature environments. At such high temperature, the ordinary cellular functions adversely affected for mesophiles. The thermophiles effectively manage instability of the plasma membrane, inactivation of enzymes instability of DNA, as well as other hostile physiological variations at such an elevated temperature. Heat shock proteins (Hsps) have established the most attention in thermophiles under stress condition, which is well described in this chapter. This chapter offers comprehensive information about thermophiles, physiology, metabolism, enzymes of metabolic pathways, and various adaptation mechanisms.

DOI: 10.4018/978-1-7998-9144-4.ch004

INTRODUCTION

Extremophiles comprise members from each of the three areas of life such as bacteria, archaea, and eukaryote (Robb *et al.*, 2008). Soon after finding of archaea in 1970s, the phylogenetic tree of life had one more branch of thermophiles and hyperthermophiles in both the domains of Bacteria and Archaea (Amend & Shock, 2001). Prokaryotes are considered as ubiquitous due to their simple dispersal and metabolic adaptability, smaller size, capacity to endure hostile ecosystems, usage of wide range of supplements and capacity to endure hostile ecosystems, hence they form a striking portion of a most of the environments (Kumar *et al.*, 2014). Thermophiles inhabit in tropically heated conditions on the earth. Temperature significantly affects the evolution and distribution of biodiversity and microbial community structure in an ecosystem. Thermophiles are prokaryotic microorganisms show specific attention because of their capacity to endure the denaturing consequence of higher temperatures on biomolecules such as DNA and proteins (Li *et al.*, 2005). Thermophiles are omnipresent and flourish in wide range of environments from marine habitats to hot springs to natural water bodies. In addition to these habitats, thermophiles are now being regularly identified in continental solfataras, heated sediments, mining sites, water heaters to the industrial dumping wastes. The microbes residing extreme environments evolve earlier because of high rate of horizontal gene transfer than those occupying normal habitats (Li *et al.,* 2014).

Temperature is one of the essential factors for the regulation of the activities of microorganisms. Thermophilic microorganisms endure the higher temperatures, yet they require such high temperatures for their optimum growth and survival. Thermophiles are categorized into moderate thermophiles (Topt, 50°C-60°C), extreme thermophiles (Topt, 60°C-80°C), and hyperthermophiles (Topt, 80°C-110°C) (Gupta *et al.,* 2014). The investigation in the Thermophile's arena has become a significant space of exploration and several novel microbial genera and species (Yoneda *et al.,* 2013; Cihan *et al.,*2014). Due to their increased importance, potential applications, and roles in different fields, scientists have concentrated their studies to discover new genus and species across the world (Yoneda *et al.,* 2013; Cihan *et al.,* 2014; Aanniz *et al.,* 2015). To study such a wide range of thermophilic bacteria, both culturable and unculturable methods have been utilized for comprehension study of microbial diversity in hot environments. At present, 16S rRNA sequence-based classification of bacteria appears to be significant for identifying novel taxonomic groups. In addition, more traditional taxonomic features such as G + C contents of DNA, DNA-DNA homology, morphology, and physiological characteristics may be used to separate characters from obtaining larger resolution of classification inside groups of a phylogenetic line of descent. To endure in such a harsh ecosystem, thermophiles produce several unique compounds, for examples, enzymes, stress proteins, chaperones, exopolysaccharides. Moreover, thermophiles have enormous biotechnological potential and industrial applications. From the finding of thermophilic microorganisms, they are the attractive tools for the biologist to understanding the basics physiological adaption and application of their metabolites. In this chapter, we have presented habitats, physiology, metabolism, enzymology and adaptation mechanisms including heat shock proteins and thermostable enzymes of thermophiles isolated from various thermal habitats.

THERMOPHILES FROM VARIOUS THERMAL ENVIRONMENTS IN INDIA

Natural activity has created different types of thermal biological processes where, the heat is formed due to geothermal activity, self-heating. Such habitats reside by thermophiles are different namely thermal springs, deep oceanic hydrothermal outlets, deep marine sediments. Moreover, human-made activity has developed many types of high-temperature biotic processes where the heat is artificially formed. Such environments were occupied by thermophiles (optimal growing temperature between 50°C-65°C) are various markedly anaerobic digestion reactors, garden compost heaps, pressed cooked cheeses (Godon *et al.,* 2020). There are also anthropogenic hot places including compost piles (temperature, generally about 60°C-70 °C but as high as 100 °C), water heaters, industrial processes, and slag heaps (Oshima & Moriya, 2008). Thermophiles are occupants of different habitats such as deep oceanic hydrothermal vents, geothermal springs, and other extreme topographical/geographical locales comprises volcanic habitats, tectonic plates, and compost matters. Various thermal sites and samples includes hot springs, Vegetable waste compost, Anaerobic biogas digestor, Municipal dry vegetable waste was explored. Many thermophilic microbes including bacteria, archaea were detected. In hot springs, *Bacillus, Geobacillus, Paenibacillus, Listeria, Clostridium* were typically found whereas in anaerobic methanogenesis process, Methanogens were found.

Table 1. Representative thermophiles isolated from thermal environments in India

No.	Thermal environments	Location	Predominant genera/common bacteria	Reference (s)
1	Lasundra hot spring	Lasundra, Gujarat	*Bacillus, Geobacillus, Paenibacillus, Listeria, Clostridium*	Mangrola *et al.,* (2015)
2	Vegetable waste compost	Kolkata	*Geobacillus* spp.	Sarkar *et al.,* (2010)
3	Anaerobic biogas digestor	Punjab	Methanogenic bacteria	Singh *et al.,* (2017)
4	Tuwa hot spring	Tuwa, Gujarat	*Bacillus, Paenibacillus, Clostridium, and Geobacillus*	Mangrola *et al.,* (2015)
5	Vegetable waste compost		*Geobacillus* spp.	Varma & Kalamdhad (2015)
6	Unnai hot spring	Unnai, Gujarat	*Bacillus, Paenibacillus, Geobacillus, Clostridium,* and *Anoxybacillus*	Mangrola *et al.,* (2019)
7	Municipal dry vegetable waste	MSW ramp, Koregaon park, Pune.	Methanogens	Wadkar *et al.,* (2013)
8	Tulsi Shyam hot spring	Tulsi Shyam, Gujarat	*Bacillus, Geobacillus, Clostridium,*	Ghelani *et al.,* (2014)

PHYSIOLOGY OF THERMOPHILES

Thermophiles may be Gram positive or negative, spore forming or non-spore forming, and may show an aerobic or anaerobic metabolism pattern. The changes in the physical and chemical conditions such temperature, nutrient availability, light intensity, and pH in the environments develop stress conditions for thermophilic microbes. Due to the presence of Heat shock proteins (HSPs) and other chaperones in the thermophiles, they can withstand stress and survive under extremely heated conditions. In this chapter, we pointed out the growth, cultivation and biochemical basis of heat stability of thermophiles

1. Growth and Cultivation of Thermophiles

Sampling from the different natural and man-made habitats is usually carried out in temperature control containers to avoid the reduction of temperature. Enrichment is a commonly used technique for the isolation and cultivation of thermophiles. The serial dilution of the sample before inoculation is common practice for the isolation of hot spring thermophiles (Pandey *et al.,* 2015). The growth media need to be optimized based on the nature and type of habitats. Trace elements, Elemental sulfur and Casamino Acids are the key ingredients for the isolation of thermophiles (Childers *et al.,*1992). The pH and temperature are set on the habitats or sample's pH and temperature. Seawater is also used for the preparation of media if the sample is of marine origin. The visible growth in liquid media is usually observed after a week and similar time required for the colonial growth on solidified media. Due to the low melting temperature of the agar-agar, gellan gum or polysilicate may be used as solidifying agents (Giavasis *et al.,* 2000). The single-cell culture is prepared using serial dilution, micromanipulator, cell sorters and optical tweezers. Due to different biotopes, thermophiles and hyperthermophiles are adopted to distinct environmental factors, including acidity, oxygen and low salinity within terrestrial solfataric fields or neutrality, high salinity and low redox potentials within the submarine hydrothermal system (Purcell *et al.,* 2007). So, the growth parameters and cultivation requirements of thermophiles are species-specific and vary due to biogeographic variation (Valverde *et al.,* 2012).

Thermophilic microorganisms that survive at temperature beyond mesophilic range of temperature, 25 °C to 40 °C that symbolizes the mainstream of entity. Thermophiles have long remained a gold dust and are known as scarcely available microorganisms requiring elevated temperatures for their optimal growth. An additional term, hyperthermophiles is introduced to recognize microbes that require extreme temperatures. These are inhabitants of the environment with high to extreme temperatures such as hot springs (Panda *et al.,* 2013), volcanic environment (Stathopoulou *et al.,* 2013), fumaroles (Muñoz *et al.,* 2011), mud pots (Castenholz, 1969), geysers (Gaisin *et al.,* 2017), deep-sea hydrothermal vents (Miroshnichenko & Bonch-Osmolovskaya, 2006) and coastal thermal springs (Zelenkina *et al.,* 2009). Thermophiles are phylogenetically related to mesophiles. So, they have the all the properties generally present in mesophiles. These comprise biochemical metabolic pathways, growth yields and metabolic rates, regulatory mechanisms like allosteric regulations, and activation of protein synthesis. The primary difference between and thermophiles and mesophiles are the ability of thermophiles to cultivate at high temperatures.

2. Biochemical Basis of Heat Stability

Adaptations help the organisms to survive in their ecological niche or habitat; adaptations can be morphological, physiological or molecular level. A bacterial cell must be coped up with extremities to survive in harsh conditions. Otherwise, cell damage and destruction may occur, which includes protein unfolding, DNA and RNA denaturation, enzyme inactivation, breakdown of membrane integrity etc. Heat stability and optimal activity at high temperatures are inherent properties of thermophiles' protein and enzymes (Vieille *et al.,* 2001). The biomolecules of the cell, including lipids, nucleic acid, proteins, are very sensitive to heat. But thermophilic biomolecules contain unusual structural components, i.e., *n*-fatty acids, diabolic acid and a glycerol ether lipid in the membrane lipid of *Thermotoga maritime*. The ether lipid may increase stability against heat. Thermophilic archaea contain heat resistant lipid derived from diphytanylglycerol diether or its dimer called di (biphytanyl) diglycerol tetraether. The presence of unusual lipid is reported in thermophilic *Methanopyrus kandleri*, which contains 2,3-di-*O*-Geranylgeranyl-*sn*-glycerol as a dominating ether core lipid in the membrane. Heat stable lipid mono-layer membrane is also common in the few thermophilic bacteria and many archaea. So, thermophiles are evolving habitat-specific adaptation mechanisms to respond effectively in diverse physicochemical conditions of their habitats (Siliakus *et al.*, 2017). Up to 80ºC temperature, the thermophilic bacterial membrane is exclusively composed of ester-lipids and whereas the beyond the 80ºC ether bonds are essential for hyperthermophilic growth (Huber *et al.*, 1992). Hence, the thermophiles maintain the membrane fluidity and permeability at high temperatures. DNA thermal stability can be increased by concentrating the GC content, but it is not universal properties in all thermophiles and hyperthermophiles as *Methanococcus igneus, Pyrococcus furiosus, Acidianus infernus* and many more such thermophiles *exhibit* only 30 to 40% GC content.

Figure 1. Extreme thermophiles and optimum growth temperatures
Source: Zeldes et al., 2015

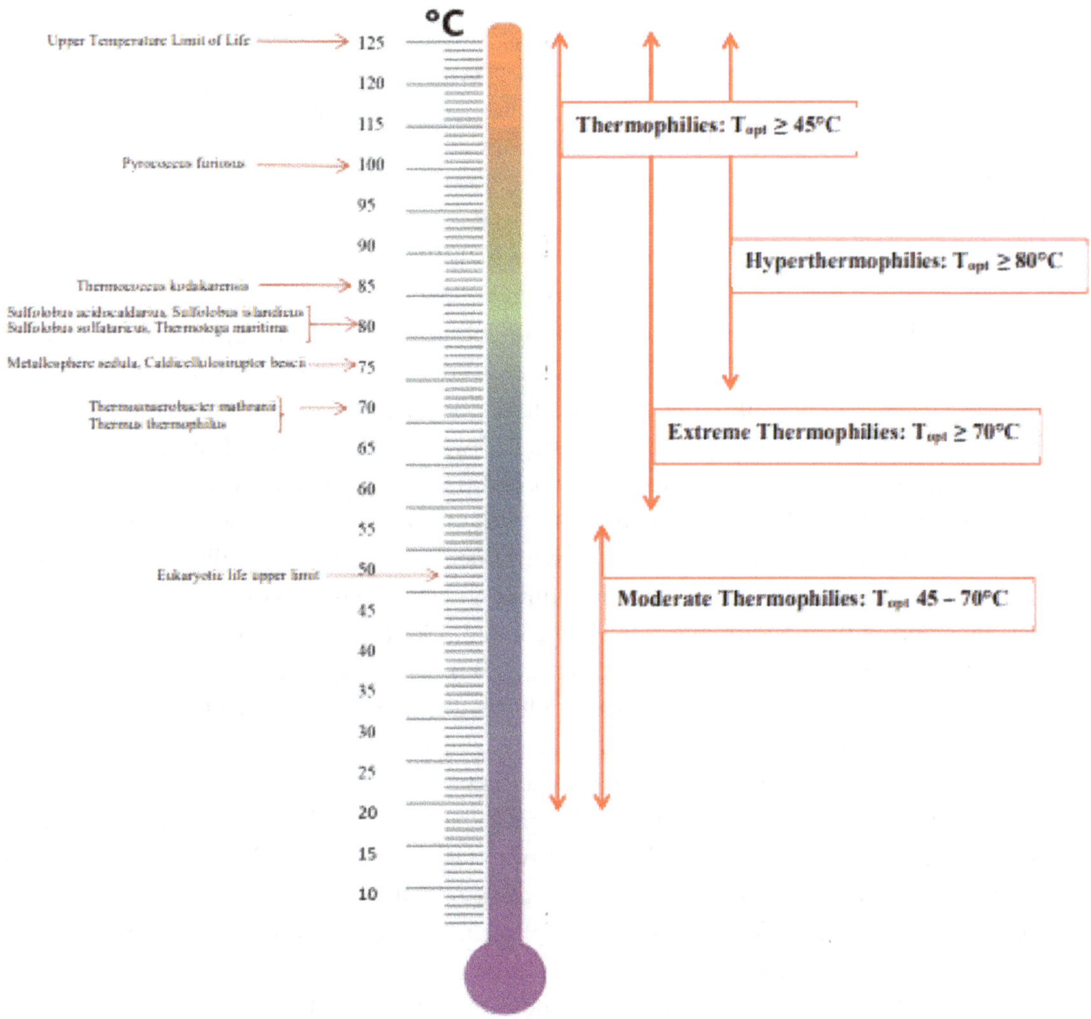

METABOLISM AND ENZYMES OF BIOCHEMICAL PATHWAYS

Thermophiles have been modified their biochemical pathways and well-adopted molecular diversity to survive under hot conditions. In this chapter, we discussed about various metabolic pathways like aerobic and anaerobic and also enzyme involved in the biochemical reactions.

1. Aerobic Metabolism in Thermophiles

The electron donors for basic metabolism of the thermophilic and hyperthermophilic microorganisms, as studied under laboratory conditions are niche depended which usually includes, H2, S, Fe^{2+}, CH_4, H_2S, $S_2O_3^{2-}$, $S_4O_6^{2-}$, sulfide minerals, various amino acids, alcohols, mono-, di- and hydroxyl-carboxylic acids and complex organic substrates (Amend & Shock, 2001). Similarly, the electron acceptors besides the oxygen (for aerobic microbes) it includes S, Fe^{3+}, CO, CO_2, NO, NO_2^-, NO_3^-, N_2O, SO_3^{2-}, SO_4^{2-} and $S_2O_3^{2-}$ (Amend & Shock, 2001). As studied under laboratorial conditions, the major known phylogeny of thermophiles and hyperthermophiles are obligate heterotrophs. Their preferential carbon and energy sources are complex mixtures of carbohydrates and/or polypeptides. Few assimilates CO_2, thus they are strict autotrophs, while others hetero- or autotrophic (i.e., mixotrophic) depending upon the availability of carbon sources (Amend and Shock, 2001). It was observed that all hypertherophiles and many thermophilic microorganisms are chemosynthetic, rather than autotrophic or photosynthetic. They derive cellular energy by the oxidation of inorganic and organic compounds, because native ecotype of these microbes correlates with geochemistry and geophysics of high temperature ecosystem. At such environmental depths sunlight penetration becomes exceedingly low. For thermophiles, the energy yielding mechanisms also depends on sugar (carbohydrate) metabolism like their mesophilic counterparts. In thermophiles entire EM pathway along with their enzymes are decoded (Selig *et al.*, 1997). But for theromophilic bacteria like Geobacillus thermoglucosidasius and Caldicellulosiruptor saccharolyticus, activities of the key enzymes, phosphogluconate dehydratase and 2-keto-3-deoxy-6-phosphogluconate aldolase of ED pathway are not detected (Tang et al. 2009; De Vrije et al. 2007). Under thermophilic conditions oxygen concentration also varies. In *G. thermoglucosidasius* PP pathway operate differently under different oxygen concentration (Tang *et al.*, 2009). Under fermentative conditions with glucose metabolism, oxidative part of the pathway found to have relatively lower activity, with missing of 6-phosphogluconolactonase gene (Tang *et al.*, 2009). However, under oxygen rich environment (i.e., aerobic conditions) two/third of the glucose are shunted to glycolysis pathway and remaining glucose are metabolised through PP pathway. While under microaerophilic conditions, study suggest that, PP pathway and TCA cycle are reduced to half of the original level (Tse & Ma, 2016; Tang *et al.*, 2009). The oxygen concentration also affects the growth rate of the bacteria. *Thermotoga maritima*, a hyperthermophilic bacterium, metabolise glucose through conventional EM pathway (Schröder *et al.*, 1994).

The thermophilic archaea, as observed in bacteria, metabolize carbohydrate through modified EM pathway. The ED and EM pathways are also modified in hyperthermophilic archaea, while the oxidative PP pathway has been replaced by non-oxidative pathway (Van der Oost & Siebers 2007; Schönheit & Schäfer 1995). *Thermococcus stetteri* and *Thermococcus zilligii* have unique glyceraldehyde-3phosphate: ferredoxin oxidoreductase and ADP-dependent phosphofructokinase (Ronimus et al. 1999; Xavier et al. 2000; Kengen et al. 1996). Xavier *et al.*, (2000) during their studies with ^{13}C-glucose-labelling experiment suggests that a novel glycolytic pathway with possible involvement of pentoses might exists in *T. zilligii*. This following pathway is as proposed (Tse & Ma, 2016):

Figure 2.Pentose Phosphoketolase Pathway

Reaction I: Glucose + ADP ⟶ Glucose – 6 – phosphate + AMP (hexokinase)

Reaction II: Glucose – 6 – phosphate + $NADP^+$ ⟶ 6 – phosphogluconate + NADPH + H^+
(glucose – 6 – phosphate dehydrogenase)

Reaction III: 6 – phosphogluconate ⟶ formate + xylulose – 5 – phosphate
(novel lyase)

Reaction IV: Xylulose – 5 – phosphate ⟶ glyceraldehyde – 3 – phosphate + acetyl phosphate
(pentose phosphoketolase)

Reaction V: Glyceraldehyde – 3 – phosphate ⟶ pyruvate ⟶ acetate + CO_2
acetyl phosphate ⟶ acetate
(Pentose Phosphoketolase Pathway)

In the above suggested pathway, according to the Xavier *et al.*, (2000), gluconate-6-phosphate could cleave into formate and xylulose-5-phosphate, reaction catalyse by novel type lyase (as mentioned in Reaction III). While the pentose phosphate can be catabolized through pentose phosphoketolase pathway. The hyperthermophilic archaea such as *Archaeoglobus fulgidus*, *Thermococcales*, *Pyrococcus* and *Desulfurococcales* found to utilize the modified EM pathways with novel enzymes (Figure 3) (Bertoldo & Antranikian 2006; Huber & Stetter 2006; Siebers & Schönheit, 2005; Labes & Schönheit, 2001; Schönheit & Schäfer, 1995; Kengen *et al.* 1994). The enzymes claiming to be novel or unique in thermophilic archaea participating in the modified pathway is ADP-dependent phosphofructokinase, including ADP-dependent kinases and ADP-dependent glucokinase, which in their mesophilic counterparts are found to be depended on ATP (Dörr *et al.*, 2003; Sakuraba *et al.*, 2002; Koga *et al.*, 2000). The metabolic energy derivation from ADP is as equal to ATP-hydrolysis (Hongo *et al.*, 2006). Moreover, at thermophilic temperatures ADP is presence of certain divalent metal ions found to the more stable than ATP (Hongo et al. 2006; Tetas and Lowenstein 1963).

In many thermophilic microbes like *Thermococcales*, *Desulfurococcus amylolyticus* and *Pyrococcus furiosus*, the mesophilic glycolytic enzymes phosphoglycerate kinase and glyceraldehyde-3-phosphate dehydrogenase is replaced by glyceraldehyde-3-phosphate ferredoxin-linked oxidoreductase, a tungsten-containing iron-sulphur protein (Mukund & Adams 1991, 1995; van der Oost *et al.*, 1998; Kletzin *et al.* 1995; Mukund & Adams 1993; Selig *et al.*, 1997). The enzyme catalyses the oxidation reaction converting GA3P into 3-phosphoglycerate and reduced ferredoxin, which in their mesophilic counterpart are two step reactions (de Vos *et al.*, 1998). In above reaction, in thermophilic microbes, phosphoglycerate kinase did not generate ATP; therefore, the substrate-level ATP yield becomes zero (Verhees *et al.*, 2004). The reduction in ATP yield might be because, ferredoxins are more thermo-stable than pyridine nucleotides and intermediate 1,3- biphosphoglycerate is heat liable, whose production can be avoided (Brunner *et al.*, 2001). In *Thermococcus kodakarensis*, the ATP loss during the conversion of phosphoenolpyruvate (PEP) into pyruvate is compensated by replacing it with PEP synthase (Sakuraba *et al.* 2004; Sakuraba

& Ohshima 2002). The PEP synthase uses AMP and a phosphate, to produce an additional 2 ATP per glucose molecule (Imanaka *et al.* 2006; Sakuraba *et al.* 2004; Sakuraba & Ohshima, 2002). Moreover, Imanaka *et al.*, (2006) suggest that, the deletion of PEP synthetase encoding gene from *T. kodakarensis*, decreases the survival rate of the bacterium on the medium containing carbohydrate.

Figure 3. Conventional EM pathway found in mesophilic microorganisms and Modified EM pathway found in thermophilic or hypertheromophilic microorganisms

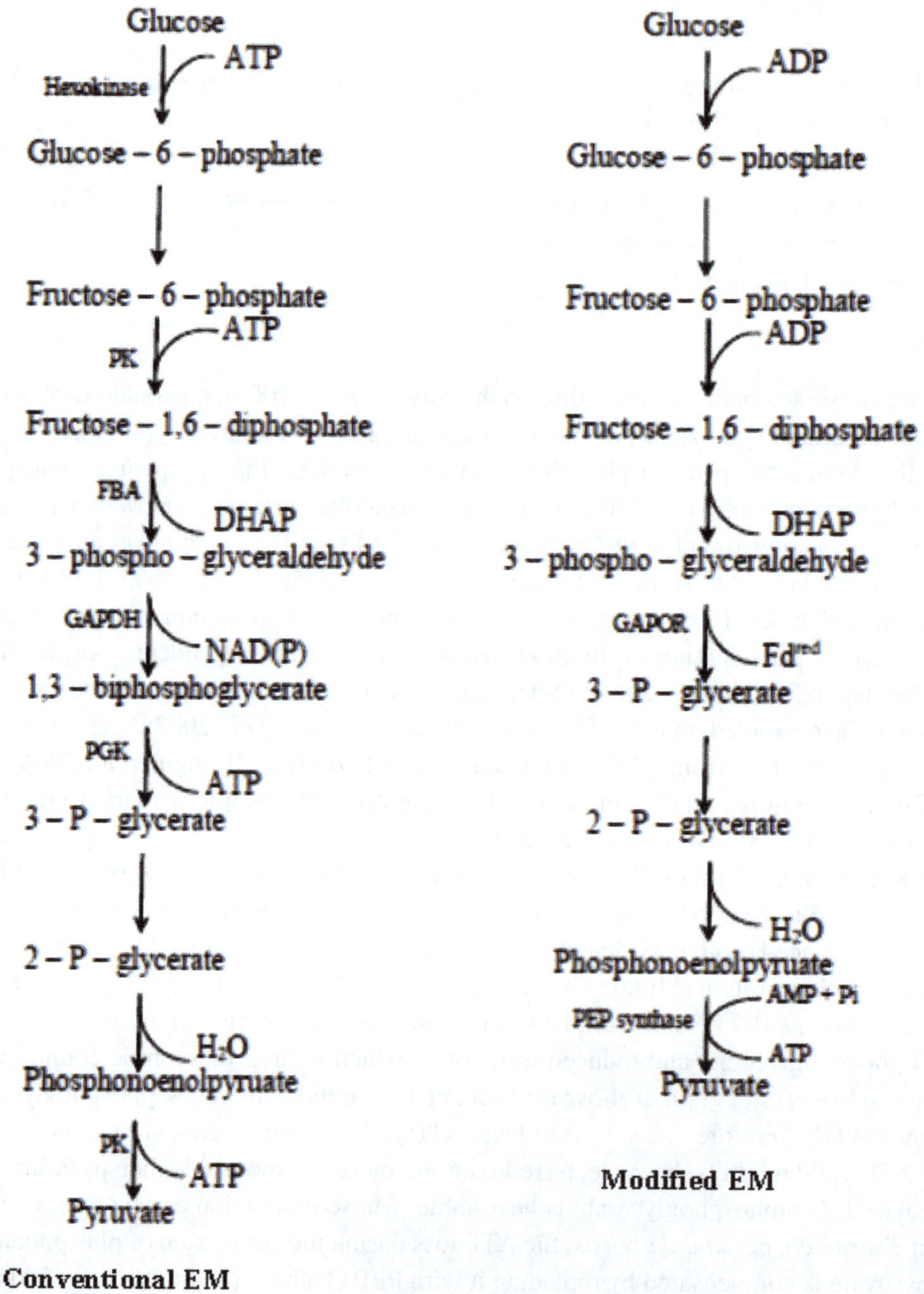

The enzymes of this pathway from mesophilic microbes are either modified or replaces by novel or unique enzymes in modified EM pathways in thermophilic microbes. PFK: phosphofructose kinase, it is modified in thermo/hyperthermopiles; FBA: fructose bisphosphatase aldolase; GAPDH: glyceraldehyde-3-phosphate dehydrogenase, it is modified in hyperthermophilic acidophiles; PGK: phosphoglycerate kinase, it is modified in hyperthermophilies; GAPOR: glyceraldehyde-3-phosphate oxidoreductase is replacing GAPDH and PGK in hyerthermophiles; DAPH: Dihydroxyacetone phosphate (Tse & Ma, 2016). The aerobic acidophilic hyperthermophiles employ the non-phosphorylated ED pathway, where glucose is oxidized into gluconate with the production of 2-keto-3-deoxy-gluconate and novel kinase phosphorylates the glycerate (Siebers & Schönheit, 2005). Looking at the genome sequences of hyperthermophilic archaea, it would be difficult to predict the enzyme homology between mesophilic and thermophilic microorganisms.

2. Anaerobic Metabolism in Thermophiles

Zeikus & Wolfe (1972) isolated an autotrophic methane producer, *Methanobacterium thermoautotrophicum*, from sewage sludge sample. This bacterium is anaerobic in nature and cultivate on a mineral media in presence of H_2 and CO_2 atmospheric environment that is further reduced to CH_4 which acts as a source of carbon for providing energy to the bacterium. In addition to these, bacterium, *Methanobacterium thermoautotrophicum* shows Topt value between 65°C and 70°C and a temperature range from 40 to 75°C. Besides this property, *Methanobacterium thermoautotrophicum* has optimum pH value between 7.2 and 7.6, and the growth was not observed when pH was below 6.0 or above 8.8. The G+C content of DNA is 52% and ribosomes are stable at high temperature (i.e., 80°C). *Methanobacterium thermoautotrophicum* not produced methane when hydrogen is changed with ethanol, formate, methanol, acetate or pyruvate but when exposed with hydrogen and carbon dioxide, it stimulates methane gas productions. Therefore, this bacterium when grown, on all substrates, used only hydrogen as a source of electrons and carbon dioxide as a carbon source. Additionally, Malashenko *et al.*, isolated thermophilic, obligate, methane-oxidizing bacterium *Methylococcus thermophilus*, which cultivates at temperature in the range of 37 to 42°C with an optimum temperature value between 50°C and 56°C. The bacterium utilizes ammonia, nitrite, and nitrate as a nitrogen source. Interestingly, the GC content of the DNA of this bacterium is 63.3% (Malashenko *et al.*, 1975). In the case of aerobic growth, thermophiles require oxygen as a terminal electron acceptor while oxidizing various substrates to harvest electrons. In contrast, anaerobic growth in thermophiles does not require oxygen for their growth requirements. The study of anaerobic thermophiles has numerous advantages such as understanding life under extreme conditions, finding their biotechnological potential, and obtaining hints towards the evolutionary path of life on the earth. Since the primitive earth environment was anaerobic (Kasting, 1993), a study of anaerobic thermophiles can present a wealth of information towards the origins of life and its molecular evolution.

Most of the thermophiles are predominantly obligate anaerobes or facultative aerobes. While seeking answers to the questions as to why the thermophilic microbes grew to be anaerobic led to certain conspicuous factors. These include less water solubility of oxygen under elevated temperatures, presence of reducing gases (H_2 and H_2S), isolated habitat from common environment etc. All of these factors best explain why the thermal environments are anaerobic or low in oxygen. It must be noted that oxygen consumption by aerobic thermophiles in habitat with elevated temperatures is only possible near the water surface. Metabolism in majority of the anaerobic thermophiles is chemoorganoheterotrophic since they utilize organic compounds to fulfil their requirement of energy and carbon. Depending on the substrate available, chemoorganoheterotrophy can be subdivided e.g., cellulolytic, proteolytic, peptidolytic, lipolytic, glycolytic etc. Other metabolic strategies observed in anaerobic thermophiles till date may include chemolithoheterotrophy, chemolithoautotrophy, photoautotrophy, and photoheterotrophy. Observing these strategies adopted, a staggering metabolic diversity can be observed in thermophilic anaerobes.

Phototrophic thermophiles are able to carry out photosynthesis. As stated above, most thermal environments lack oxygen, these thermophilic microbes have to opt for anoxygenic photosynthesis for their requirements. At higher temperature, very few species of anaerobic phototrophs are able to grow or exist. A thermophile, *Chloroflexus aurantiacus* is well known member of microbial mat communities in hot springs that can grow as anoxygenic phototroph (Castenholz & Pierson, 1995). Under anaerobic condition, it grows as photoheterotroph while chemoheterotrophic growth is observed under aerobic conditions in night or darkness. Optimal growth is reported near pH 8 and temperature optima between 52 and 60°C (Pierson & Castenholz, 1974). Molecular assemblies known as chlorosomes are ultimate structures containing chlorophyll c, d or e in the photosynthetic membrane enabling it to efficiently harvest blue-green light (Yakovlev *et al.,*2021). *Chlorobium tepidum*, an inhabitant of hot springs is a green sulfur bacterium requiring sulfide or thiosulfate as a photoreductant and grow as photoautotroph (Madigan *et al.,*1989). Since thiosulfate acts as major electron donor, utilisation of sulfide is much poor in comparison to thiosulfate. At optimal temperature 48°C, its generation time is about 2 hours. Photoheterotrophic growth of *Chlorobium tepidum* is also possible under anaerobic conditions (Wahlund *et al.,*1991). As per the suggestion by Brock and Medigan, thermophilic property of all such phototrophic thermophiles might be of greater significance in defining phylogeny of green sulfur bacteria instead of its phototrophic character (Sirevåg, 2021). Mechanisms behind collection of light energy and its conversion to chemical energy are highly dependent on the variety of pigment contained within photosynthetic reaction centres e.g., bacteriochlorophyll and rhodopsin. Earlier, it was believed that no thermophile exists that contain rhodopsin as a photosynthetic pigment. Recent findings (Shim *et al.,*2021) brings discovery of microbial rhodopsin in notice that is most stable in extreme environments. A proton pumping rhodopsin Tara76 isolated from a thermophilic bacterium Thermus thermophilus shows strong stability against temperature, pH, detergent, salt stress and dehydration stress. It shows higher thermal stability at 80°C, pH 0.02 to 13 and strong resistance to the presence of detergents such as Triton X-100. Such properties of Tara76 ushers in a new direction towards its suitability in biotechnological applications. Mutation studies on thermophilic rhodopsin resulted in further improvement in its thermo-stabilization property.

Thermotogae are anaerobic and organotrophic organism, able to cultivate on a various complex substrate (Conners et al. 2006). These substrates are found in thermal habitats of the globe, including hot springs, petroleum reservoirs, and hydrothermal vents (Ollivier & Cayol 2005; Huber & Hannig, 2006), with some groups cultivate at temperatures up to 90 °C. *Thermotogae* is important microbial ancestry used for biotechnological and industrial relevance. For instant, majority of *Thermotogae* species synthesize hydrogen (Nguyen *et al.,* 2008; Maru *et al.,* 2012). Another important *T. maritima* strain was used for hydrogen gas production through metabolic engineering using in-silico redesign metabolism process (Nogales *et al.,* 2012). Moreover, *T. neapolitana* convert CO_2 and acetate to lactic acid when cultivate in presence of CO_2 (D'Ippolito *et al.*, 2014). *T. maritima* has been investigated for sugar usage by examining the substrate affinity and specificities of sugar transpoters (Nanavati *et al.,* 2005, 2006; Cuneo *et al.*, 2009; Boucher & Noll, 2011; Ghimire-Rijal *et al.,* 2014). Information about substrate specificities, enzymatic activities, and catalytic mechanisms of many of *T. maritima's* glycoside hydrolases are also available (Kleine & Liebl 2006; Comfort *et al.*, 2007; Arti *et al.,* 2012), which has been used, for instance, to engineer an -galactosidase from *T. maritima* into an efficient α-galactosynthase (Cobucci-Ponzano *et al.*, 2011). The transcriptional regulation of glycoside hydrolases and other carbohydrate metabolism-related genes in response to growth on various carbohydrates highlights the differences in carbohydrate utilization, even between closely related *Thermotogae* lineages (Chhabra *et al.*, 2002, 2003; Frock *et al.*, 2012). Moreover, interconnections exist between sugar regulons in *T. maritima's* carbohydrate utilization network, suggesting coordinated regulatory responses to particular types of complex carbohydrates (Rodionov *et al.*, 2013).

Methanogenesis Process

Methanogenesis is a process involving utilisation of carbon dioxide and hydrogen to generate methane gas and falls under the category of chemolithoautotrophic pathways. Among the Archaea, Methanogens are one of the most diverse groupings with genomic DNAs carrying 26 to 68% mol% G+C. To gain energy, they can efficiently utilize acetate, formate, carbon monoxide, carbon dioxide, and methanol (JN, 1992). Thermophilic taxa belonging to families such as Methanococcaceae, Methanocaldococcaceae, Methanothermaceae, and Methanobacteriaceae are well characterized and known for methanogenic potential (Amend & Shock, 2001). The overall reaction is:

$$4H_2 + CO_2 + CH_4 + 2H_2O \tag{1}$$

Methane, a product of methanogenesis is a valuable gas that can be efficiently utilized as a reliable energy source with promising biotechnological potential in treatment of agricultural and domestic wastewater (Hu *et al.*, 2019) as well as generation of hydrogen (Li *et al.*, 2019), a next generation fuel. In pure culture of methanogens, methane can be produced via acetoclastic methanogenesis during conversion of carbon dioxide to methane or by hydrogenotrophic methanogenesis where hydrogen mediates reduction of carbon dioxide to methane. In addition, methylotrophic methanogenesis can also take place where bioconversion of methylated compounds forms methyl-S-CoM which is either broken down further to carbon dioxide and methane or reduced by hydrogen (Lyu *et al.*,2018).

Figure 4. Steps in methanogenic pathway

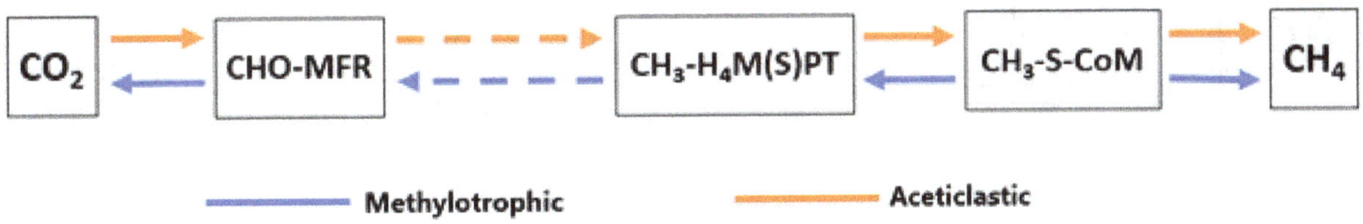

Acetogenic methanogens are obligate anaerobic archaea responsible for the conversion of acetate to methane. *Methanosaeta concilii* and *Methanosaeta* spp. occupy the dominant aceticlastic methanogen in a variety of anaerobic reactors when acetate concentrations are low. RT-PCR based quantitate analysis of community structure in a full-scale digestor supports the existence of microorganisms from the family *Methanosaetaceae*. In such systems, partial pressure generated by hydrogen is considered to be a necessary parameter. Therefore, Hydrogenotrophic methanogens play an essential role in ensuring a stable and efficient process performance (Demirel & Scherer, 2008).

Metabolism in *Clostridium* spp

Metabolic and evolutionary responses of *Clostridium thermocellum* to genetic interventions aimed at improving ethanol production (Holwerda *et al.*,2020). Several species of *Clostridium* are well known for their ability to live in elevated temperatures. *Clostridium thermocellum* can ferment glucose obtained from cellulose hydrolysis to ethanol and acetate (Fig. 4a). The mechanism behind the conservation of ATP in this case is substrate-level phosphorylation. It has been observed that in presence of *Methanobacterium thermoautotrophicum*, transfer of hydrogen at interspecies level promotes excessive production of acetate which in turn conserve additional mole of ATP (Fig. 4b).

Figure 5. (a) Growth of Clostridium thermocellum on cellulose axenically (b) Growth of Clostridium thermocellum in the presence of Methanobacterium thermoautotrophicum
Source: Zinder, 1993

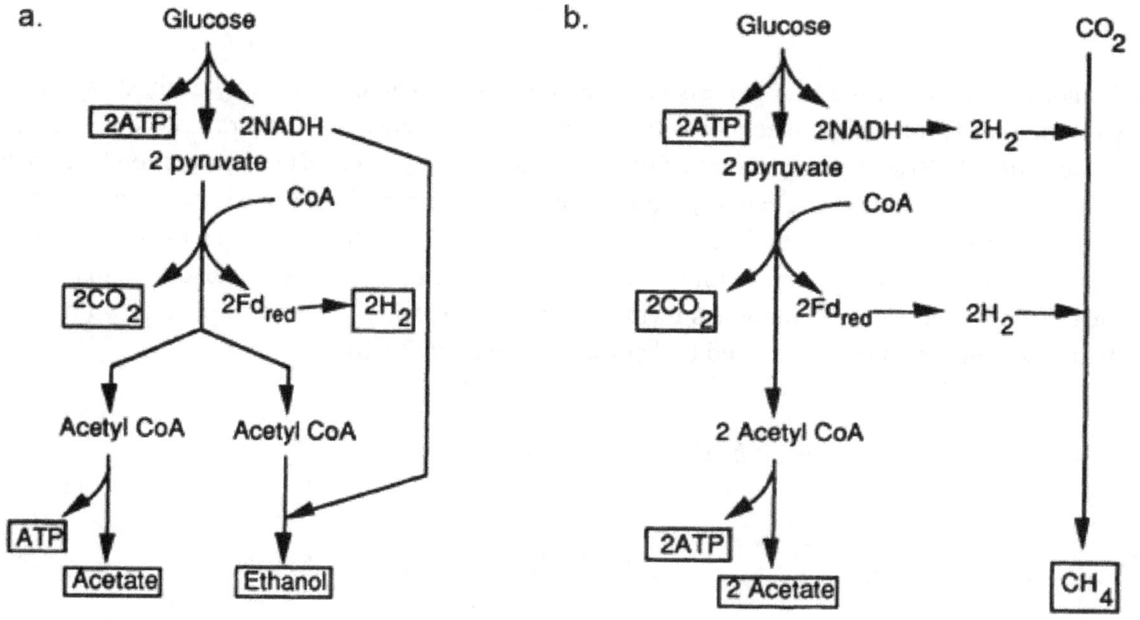

A much faster growth rate can be observed when *Clostridium thermocellum* is grown on hemicellulose instead of cellulosic materials (Zinder, 1993). One more example of a similar organism is *Clostridium Thermosuccinogenes*, a thermophilic anaerobic bacterium that can convert various carbohydrates into acetate and succinate as main fermentation products (Koendjbiharie *et al.*,2018). Bioengineering efforts had been successful in modifying the central fermentative metabolism of *Clostridium thermocellum* targeted to enhance ethanol production. This might be of industrial importance keeping ethanol yield in context (Holwerda *et al.*,2020). Recently, an acetogenic bacteria-based bioethanol production process has attracted considerable attention. Syngas fermentation using microbial catalysts has been hampered by low ethanol yield, which is the biggest obstacle to its commercialization. Metabolic experiments on *Clostridium ljungdahlii* that included growth on carbon monoxide establishes an important link between biomass production and production of ethanol (Liu *et al.,*2020).

ADAPTATION MECHANISMS OF THERMOPHILES

Thermophiles synthesize thermostable proteins or enzymes for their adaption at elevated temperatures. Despite, bacteria of thermophilic origin grow at optimum temperature (Topt), heat-shock reactions are still activated at temperature higher than that of optimum temperatures that required by thermophilic bacteria. Thermophiles developed some modifications compared to mesophilic bacteria for their survival at an elevated temperature (Trivedi *et al.,*2005). The combined hereditary and functional alterations in the cell induced by extreme environmental conflicts referred to as 'Stress responses' (Helmann *et al.,*2001; Tam *et al.,*2006). These microbes survive at high temperature due to the presence of several biomolecules such as nucleic acids, proteins, and lipids. Several mechanisms in the thermophiles, including covalent modifications, extra amino acids, make protein molecules to withstand at high temperature. Many researchers reviewed on structural alteration in the protein biomolecule for adaptability of thermophiles at the extreme temperature (Liu & LiCata, 2013; Chakraborty *et al.,* 2015). More importantly the proteins and enzymes normally denature at higher temperatures. But in thermophilic microorganisms two basic mechanisms have been evolved for stabilizing and preventing the denaturation of proteins at elevated temperatures (Tse & Ma, 2016; Berezovsky and Shakhnovich, 2005). In the first mechanism, three-dimensional protein structures in thermophiles are significantly more compact when compared to homologues proteins in mesophilic microbes, by increasing the number of interactions among the constituting amino acids (Berezovsky & Shakhnovich, 2005). In second mechanism, the structures are not significantly different, but the protein sequence is altered to form few strong interactions to be able to sustain at higher temperatures (Li *et al.,* 2005).

Heat Shock Proteins (HSPs)

The thermophiles have specific heat shock proteins and chaperons for constantly aiding the proteins folding for those who fail to form required conformation at higher temperatures (Feder & Hofmann 1999; Hendrick & Hartl 1993). The heat-shock proteins (HSPs) are varied in structure and mechanism, and are typically classified based on molecular masses of their subunit. Many heat-shock proteins like Hsp10 (GroES), Hsp27 (sHsp), Hsp40 (DnaJ), Hsp60 (Chaperonin, GroEL), Hsp70 (DnaK), Hsp90 (HtpG), Hsp104 (ClpB), and other small HSPs are activated during heat-shock reaction as per their molecular masses (Buchner 1996; Trent, 1996). These chaperones performed the folding of unfolded or nascent polypeptide chain into complete shape of protein and also prevents denatured proteins from mass formation. The well-studied chaperon Hsp60 complex prevents the aggregation of unfolded proteins at elevated temperatures and maintains cell homeostasis by channeling the denatured and misfolded proteins and enzymes into cellular protein degradation systems, a thermosomes (Tse & Ma, 2016; Sterner & Liebl, 2001). Taguchi *et al.,* first isolated and characterized chaperonin from *T. thermophilus* HB8 (Taguchi *et al.,* 1991). In thermophiles, some gene expression products like HSPs perform thermoadaptation functions at higher temperatures and these genes are upregulated at either the transcriptomic or proteomic range because of temperature variations (Wang *et al.,* 2013). The upregulation of HSP genes has been broadly detected at both the transcriptional and proteomics levels in thermophiles. For example, at the transcriptional level, *Pyrococcus furiosus* activated Hsp20 and Hsp60 during a temperature change from 90°C to 105°C (Shockley *et al.,* 2003). At the proteomic level, the upregulation of chaperonin proteins, such as GroES, GroEL, GrpE, and DnaK, when temperature was higher in several thermophiles, such as *Thermotoga maritime* and *Thermoanaerobacter tengcongensis* (Chen *et al.,* 2012; Wang *et al.,* 2007; Wang *et al.,* 2012). The molecular chaperone system of thermophilic archaea is pointedly diverse from that of thermophilic eubacteria (Table 2).

Table 2. Different types of chaperones in thermophiles (Sahlan and Yohda, 2013)

Chaperones	Thermophilic eubacteria	Thermophilic archaea
Hsp100s/AAA proteins	ClpA, ClpB, ClpC, HslU	Absent (Genes for ClpA/B homologues were detected in the genome of *Methanothermobacter thermatotrophicus*)
Hsp90s	HtpG	Absent
Hsp70s	DnaK	Absent
Hsp60s	GroEL	Group II Chaperonin (Thermosome)
Hsp40s	DnaJ	Absent
SHsp	IbpA, IbpB, sHsp	sHsp
Hsp10	GroES	Absent
Prefoldin	Absent	Prefoldin present
Others	Trigger factor PPIase	NAC PPIase

Types of Heat Shock Proteins

1. Small Heat-Shock Proteins (sHsps)

Small heat-shock proteins are the most universal of all chaperone proteins. Structurally, Hsps have α -crystallin domain which is headed by variable N-terminal and C-terminal regions. The chaperone potential of sHsps is dormant when they exist as large oligomeric structures under normal physiological conditions like temperature, pH, and salt concentrations. When physiological condition especially higher temperatures developed, as result these large oligomeric structures of sHsps separates into small oligomers (possibly dimers). In the presence of denatured proteins, separated sHsps produce a bulky stable complex to shield the visitor protein from mass formation.

2. Prefoldin

The prefoldin complex is an approximately 90 kDa molecular chaperone, alternatively known as GimC (genes involved in microtubule biogenesis complex). This chaperon is ubiquitously present in archaea and eukaryotes with their structure similarity, but they are so far unavailable in bacterial species (Geissler *et al.*, 1998; Vainberg *et al.*, 1998). Prefoldin chaperone was first detected in archaea named as *Metha-*

nobacterium thermoautotrophicum (Leroux *et al.*, 1999). The prefoldins are also named as ''holdase'' chaperones and their crystal structure was first determined from the archaeon, *Methanothermobacter thermoautotrophicum* (Siegert *et al.*, 2000; Stirling *et al.*, 2003). Prefoldins are hexameric in nature, composed of two α-subunits and four β- subunits. The body structure of prefoldins made up of two-barrel structure with six lengthen arms like coiled coils extending from it. Each of subunits comprises one or two central β -hairpins, that includes N - and C-terminal lined by coiled-coil spirals. These spirals within each subunit gather into an antiparallel configuration, resulting formation of the packet in the protein structure (Siegert *et al.*, 2000; Ohtaki *et al.*, 2008). It has been recommended that prefoldin chaperone arrest an unfolded protein substrate and transfers it to a group II chaperonin in eukaryotes and archaea (Vainberg *et al.*, 1998). Prior investigations on the prefoldin of *Thermococcus* KS1 strain showed that prefoldin complexes transfer substrate to chaperonins (Class II) through a direct contact process (Iizuka *et al.*, 2008).

Thermal Stability of Proteins

In contrast to salt and pH, temperature influences cells with no distinction between the external and inward cell limits. One major challenge for the thermophilic microorganisms is the cultivation and maintenance its protein integrity under such high temperature. This topic has facilitated attention for long time. In a recent time, proteomic developments deliver more considerable mark for supporting protein stability in thermophiles. Gu & Hilser, (2009) suggested that an adaptation via elevated temperature seems to be focus in the proteins with regulatory proteins and catalytic reactions. Some thermophilic microorganisms own a higher number of disulfide bonds than mesophilic organisms, and it was suggested that some of these organisms utilize disulfide bridges for their protein stabilization (Beeby *et al.*, 2005; Jorda & Yeates, 2011; Mallick, 2002). In thermophiles, utilization of disulfide bridges is astonishing, but, in light of the fact that the intracellular condition is usually in reducing position, preventing the establishment of disulfide linkages. Subsequently, it turns into an inquiry concerning the intracellular condition of thermophilic microbes and the systems of protein disulfide bond arrangement. Afterwards, comparative genomics investigations uncover that a protein of disulfide oxidoreductase was available in thermophilic microbes and that these microbes possess an intracellular disulfide linkage, which is no doubt the important molecular mechanism for intracellular protein disulfide bolding in thermophiles (Beeby *et al.*, 2005, Pedone *et al.*, 2004).

In thermophiles, solutes require to pass three membrane regions like hydrocarbon chain segment near the polar head group (semi-rigid), hydrocarbon tail near the bilayer center (fluid, thus high permeation), and the lipid head group (highly viscous, thus low permeation) for permeation via monopolar diester bilayer membranes.

Thermostable Enzymes

More than 75 proteins or enzymes have been isolated from *Bacillus stearothemophilus* alone. Thermozymes are proteins that reveal greater stability and provide exciting stages for biocatalysis and biotransformation reaction. One of greatest discovery in the field of extremophile is the finding of Taq polymerase, a thermostable DNA polymerase, from *Thermus aquaticus* bacterium which is now a day used in PCR reactions. Thermophiles have developed diverse biochemical, structural and physiological modifications, allowing the catalytic synthesis of enzymes and proteins with interesting physicochemical and structural properties. Extremozyme offers unlimited potentials due to their stability and activity in the presence of heat. These enzymes are especially stable over a wide range of temperatures, and also pH, salt and high concentration organic solvent and surfactants. The great industrial and biotechnological interests of thermophilic enzymes are due to the fact that these enzymes are better suited for harsh industrial processes. There are many advantages of conducting industrial processes at high temperatures, such as the increased solubility of many polymeric substrates, resulting in decreased viscosity, increased bio-availability, faster reaction rate and the decreased risk of microbial contamination. These enzymes have also been used as models for the understanding of thermostability and term activity, which is useful for protein engineering (Sterner and Liebl, 2001; Kumar and Nussinov, 2001).

Thermophilicity of these enzymes is due to the unique enzyme architecture properties (Bezsudnova *et al.,* 2012; Eichler, 2001), including

- Presence of highest arginine, aspartate and glutamate amino acid
- The occurrence of the high frequency of salt bridges
- Compact hydrophobic core and
- A large number of surface ion pairs
- Additional hydrogen bonding and shortening of loops

These strategies, used at differing extents by different thermophilic proteins and enzymes, confer not only higher thermal stability to proteins but also enhanced rigidity and resistance to chemical denaturation. That is why thermostable enzymes are resistant and remains active in solvents and surfactants. Molecular packing is also contributing to the thermostability of enzymes and can be achieved by the occurrence of additional alpha-helices in its structure (Charron *et al.,* 2000).

Thermophilic extremozymes adapting to extremely high temperatures are apparently genetically encoded (Jaenicke and Bohm, 1998). However, different mechanisms of thermostabilisation may be utilized by different bacterial and archaeal enzymes. Thermophiles also produce special proteins known as 'chaperonins', which are thermostable and resistant to denaturation and proteolysis. It is clear that there are a number of features that have been identified as possible contributors to increased enzymes thermostability; however, there are substantial differences between the conclusions reached with different proteins, and there appears to be no universal rule for the structural basis of stability. Acidothermophilic enzymes are structurally similar to thermozymes as they are stable at elevated temperatures. Additionally, such an enzyme contains a high concentration of acid-stable amino acids and hydrophobic amino acids to render the enzyme stable in acidic conditions. Theoretically, the thermostability of enzymes can be calculated by the free energy difference between folded and unfolded enzymes. Extremozymes indicated a lower rate of unfolding and a higher rate of refolding and thereby increase the T_m and decrease the rate of thermal inactivation. Thus, enzyme is thermodynamically and kinetically stabilized at high temperatures.

FUTURE PROSPECTS

Thermophilic enzymes have well known for industrial and biotechnological application in the past two decades. With the advent of next generation high throughput sequencing technologies, gene coding for enzymes of thermophiles can be explored well and also understand various biochemical pathways. Still occult science of heterologous gene expression study is understudy, which is being focus in the future line of this subject area. In order to proficiently usage of complete potential of thermophiles, it is compulsory to develop some thermophilic microbes as a host for gene expression, metagenome library construction metabolic engineering, and other related studies. This chapter is summarizing current and future aspects of thermophiles and their broad range of application in various sectors.

CONCLUSION

It is recognized that microbial life can flourish in extremely harsh environments. This chapter present the overview of physiology, metabolism, enzymes involved in metabolic activities and adaptation mechanisms of thermophiles and hyperthermophiles. The thermo/hyperthermophilic microorganisms are found from various habitats over the planet ranging from marine habitats to hot springs and natural water bodies. These microorganisms have evolved various basic mechanisms like altered cell membrane permeability, high G+C content in their DNA composition, altered protein structures and change in amino acids sequences, etc. as compare to their mesophilic counterparts. Thermophiles also adapts to their environment by changing in the protein folding aided by chaperons and presence of certain thermostable enzymes. The metabolism and metabolic pathways also remain same in the thermo/hyperthermophilic microbes as found in the mesophiles, however, many central pathways like EM and ED pathways are modified along with the few modifications in the participating enzymes.

ACKNOWLEDGMENT

This research received no specific grant from any funding agency in the public, commercial, or not-for-profit sectors.

REFERENCES

Aanniz, T., Ouadghiri, M., & Melloul, M. (2015). Thermophilic bacteria in Moroccan hot springs, salt marshes and desert soils. *Brazilian Journal of Microbiology*, *46*(2), 443–453. doi:10.1590/S1517-838246220140219

Amend, J. P., & Shock, E. L. (2001). Energetics of overall metabolic reactions of thermophilic and hyperthermophilic Archaea and Bacteria. *FEMS Microbiology Reviews*, *25*(2), 175–243. doi:10.1111/j.1574-6976.2001.tb00576.x PMID:11250035

Arti, D., Park, J., Jung, T. Y., Song, H., Jang, M., & Han, N. S. (2012). Structural analysis of α-L-arabinofuranosidase from *Thermotoga maritima* reveals characteristics for thermostability and substrate specificity. *Journal of Microbiology and Biotechnology*, *22*(12), 1724–1730. doi:10.4014/jmb.1208.08043 PMID:23221536

Beeby, M., O'Connor, B. D., Ryttersgaard, C., Boutz, D. R., Perry, L. J., & Yeates, T. O. (2005). The genomics of disulfide bonding and protein stabilization in thermophiles. *PLoS Biology*, *3*(9), 1549–1558. doi:10.1371/journal.pbio.0030309 PMID:16111437

Berezovsky, I. N., & Shakhnovich, E. I. (2005). Physics and evolution of thermophilic adaptation. *Proceedings of the National Academy of Sciences of the United States of America, 102*(36), 12742–12747. doi:10.1073/pnas.0503890102 PMID:16120678

Bertoldo, C., & Antranikian, G. (2006). The order Thermococcales. In M. Dworkin, S. Falkow, E. Rosenberg, K. Schleifer, & E. Stackebrandt (Eds.), *The Prokaryotes* (pp. 69–81). Springer. doi:10.1007/0-387-30743-5_5

Bezsudnova, E. Y., Boyko, K. M., Polyakov, K. M., Dorovatovskiy, P. V., Stekhanova, T. N., Gumerov, V. M., Ravin, N. V., Skryabin, K. G., Kovalchuk, M. V., & Popov, V. O. (2012). Structural insight into the molecular basis of polyextremophilicity of short-chain alcohol dehydrogenase from the hyperthermophilic archaeon *Thermococcus sibiricus. Biochimie, 94*(12), 2628–2638. doi:10.1016/j.biochi.2012.07.024 PMID:22885278

Boucher, N., & Noll, K. M. (2011). Ligands of thermophilic ABC transporters encoded in a newly sequenced genomic region of Thermotoga maritima MSB8 screened by differential scanning fluorimetry. *Applied and Environmental Microbiology, 77*(18), 6395–6399. doi:10.1128/AEM.05418-11 PMID:21764944

Brunner, N. A., Siebers, B., & Hensel, R. (2001). Role of two different glyceraldehyde-3-phosphate dehydrogenases in controlling the reversible Embden-Meyerhof-Parnas pathway in *Thermoproteus tenax*: Regulation on protein and transcript level. *Extremophiles, 5*(2), 101–109. doi:10.1007007920100181 PMID:11354453

Buchner, J. (1996). Supervising the fold: Functional principles of molecular chaperones. *Federation of American Societies for Experimental Biology Journal, 10*(1), 10–19. doi:10.1096/fasebj.10.1.8566529 PMID:8566529

Castenholz, R. W. (1969). The Thermophilic cyanophytes of Iceland and the upper temperature limit. *Journal of Phycology, 5*(4), 360–368. doi:10.1111/j.1529-8817.1969.tb02626.x PMID:27096455

Castenholz, R. W., & Pierson, B. K. (1995). Ecology of Thermophilic Anoxygenic Phototrophs. *Anoxygenic Photosynthetic Bacteria, 2*, 87–103. doi:10.1007/0-306-47954-0_5

Charron, C., Talfournier, F., Isupov, M. N., Littlechild, J. A., Branlant, G., Vitoux, B., & Aubry, A. (2000). The crystal structure of D-glyceraldehyde-3-phosphate dehydrogenase from the hyperthermophilic archaeon *Methanothermus fervidus* in the presence of NADP(+) at 2.1 A resolution. *Journal of Molecular Biology, 297*(2), 481–500. doi:10.1006/jmbi.2000.3565 PMID:10715215

Chen, Z., Wang, Q., Lin, L., Tang, Q., Edwards, J. L., Li, S., & Liu, S. (2012). Comparative evaluation of two isobaric labeling tags, DiART and iTRAQ. *Analytical Chemistry, 84*(6), 2908–2915. doi:10.1021/ac203467q PMID:22404494

Chhabra, S. R., Shockley, K. R., Ward, D. E., & Kelly, R. M. (2002). Regulation of endo-acting glycosyl hydrolases in the hyperthermophilic bacterium *Thermotoga maritima* grown on glucan- and mannan-based polysaccharides. *Applied and Environmental Microbiology, 68*(2), 545–554. doi:10.1128/AEM.68.2.545-554.2002 PMID:11823189

Chhabra, S. R., Shockley, K. R., Conners, S. B., Scott, K. L., Wolfinger, R. D., & Kelly, R. M. (2003). Carbohydrate-induced differential gene expression patterns in the hyperthermophilic bacterium *Thermotoga maritima*. *The Journal of Biological Chemistry*, *278*(9), 7540–7552. doi:10.1074/jbc.M211748200 PMID:12475972

Childers, S. E., Madeline, V., & Noll, K. M. (1992). Improved methods for cultivation of the extremely thermophilic bacterium *Thermotoga neapolitana*. *Applied and Environmental Microbiology*, *58*(12), 3949–3953. doi:10.1128/aem.58.12.3949-3953.1992 PMID:16348824

Cihan, A. C., Cokmus, C., Koc, M., & Ozcan, B. (2014). *Anoxybacillus calidus* sp. nov., a thermophilic bacterium isolated from soil near a thermal power plant. *International Journal of Systematic and Evolutionary Microbiology*, *64*(1), 211–219. doi:10.1099/ijs.0.056549-0 PMID:24052627

Cobucci-Ponzano, B., Zorzetti, C., Strazzulli, A., Carillo, S., Bedini, E., Corsaro, M. M., Comfort, D. A., Kelly, R. M., Rossi, M., & Moracci, M. (2011). A novel α -D-galactosynthase from *Thermotoga maritima* converts β-D-galactopyranosyl azide to α-galacto-oligosaccharides. *Glycobiology*, *21*(4), 448–456. doi:10.1093/glycob/cwq177 PMID:21084405

Comfort, D. A., Bobrov, K. S., Ivanen, D. R., Shabalin, K. A., Harris, J. M., Kulminskaya, A. A., Brumer, H., & Kelly, R. M. (2007). Biochemical analysis of *Thermotoga maritima* GH36 α -galactosidase (TmGalA) confirms the mechanistic commonality of clan GH-D glycoside hydrolases. *Biochemistry*, *46*(11), 3319–3330. doi:10.1021/bi061521n PMID:17323919

Conners, S. B., Mongodin, E. F., Johnson, M. R., Montero, C. I., Nelson, K. E., & Kelly, R. M. (2006). Microbial biochemistry, physiology, and biotechnology of hyperthermophilic *Thermotoga* species. *FEMS Microbiology Reviews*, *30*(6), 872–905. doi:10.1111/j.1574-6976.2006.00039.x PMID:17064285

Cuneo, M. J., Beese, L. S., & Hellinga, H. W. (2009). Structural analysis of semi-specific oligosaccharide recognition by a cellulose-binding protein of *Thermotoga maritima* reveals adaptations for functional diversification of the oligopeptide periplasmic binding protein fold. *The Journal of Biological Chemistry*, *284*(48), 33217–33223. doi:10.1074/jbc.M109.041624 PMID:19801540

D'Ippolito, G., Dipasquale, L., & Fontana, A. (2014). Recycling of carbon dioxide and acetate as lactic acid by the hydrogen-producing bacterium *Thermotoga neapolitana*. *ChemSusChem*, *7*(9), 2678–2683. doi:10.1002/cssc.201402155 PMID:25116280

de Vos, W. M., Kengen, S. W., Voorhorst, W. G., & van der Oost, J. (1998). Sugar utilization and its control in hyperthermophiles. *Extremophiles*, *2*(3), 201–205. doi:10.1007007920050061 PMID:9783166

De Vrije, T., Mars, A., Budde, M., Lai, M., Dijkema, C., De Waard, P., & Claassen, P. (2007). Glycolytic pathway and hydrogen yield studies of the extreme thermophile *Caldicellulosiruptor saccharolyticus*. *Applied Microbiology and Biotechnology*, *74*(6), 1358–1367. doi:10.100700253-006-0783-x PMID:17216445

Demirel, B., & Scherer, P. (2008). The roles of acetotrophic and hydrogenotrophic methanogens during anaerobic conversion of biomass to methane: a review. *Reviews in Environmental Science and Bio/Technology*, *7*(2), 173–190. doi:10.1007/s11157-008-9131-1

Dörr, C., Zaparty, M., Tjaden, B., Brinkmann, H., & Siebers, B. (2003). The Hexokinase of the Hyperthermophile *Thermoproteus tenax* ATP-Dependent Hexokinases and ADP-Dependent glucokinases, two alternatives for glucose phosphorylation in archaea. *The Journal of Biological Chemistry*, *278*(21), 18744–18753. PMID:12626506

Eichler, J. (2001). Biotechnological uses of archaeal extremozymes. *Biotechnology Advances*, *19*(4), 261–278. doi:10.1016/S0734-9750(01)00061-1 PMID:14538076

Feder, M. E., & Hofmann, G. E. (1999). Heat-shock proteins, molecular chaperones, and the stress response: Evolutionary and ecological physiology. *Annual Review of Physiology*, *61*(1), 243–282. doi:10.1146/annurev.physiol.61.1.243 PMID:10099689

Frock, A. D., Gray, S. R., & Kelly, R. M. (2012). Hyperthermophilic Thermotoga species differ with respect to specific carbohydrate transporters and glycoside hydrolases. *Applied and Environmental Microbiology*, *78*(6), 1978–1986. doi:10.1128/AEM.07069-11 PMID:22247137

Gaisin, V. A., Kalashnikov, A. M., Grouzdev, D. S., Sukhacheva, M. V., Kuznetsov, B. B., & Gorlenko, V. M. (2017). *Chloroflexus islandicus* sp. nov., a thermophilic filamentous anoxygenic phototrophic bacterium from a geyser. *International Journal of Systematic and Evolutionary Microbiology*, *67*(5), 1381–1386. doi:10.1099/ijsem.0.001820 PMID:28126046

Geissler, S., Siegers, K., & Schiebel, E. (1998). A novel protein complex promoting formation of functional alpha and gamma-tubulin. *The EMBO Journal*, *17*(4), 952–966. doi:10.1093/emboj/17.4.952 PMID:9463374

Ghelani, A., Patel, R., Mangrola, A., & Dudhagara, P. (2014). Cultivation-independent comprehensive survey of bacterial diversity in Tulsi Shyam Hot Springs, India. *Genomics Data*, *4*, 54–56. doi:10.1016/j.gdata.2015.03.003 PMID:26484176

Ghimire-Rijal, S., Lu, X., Myles, D. A., & Cuneo, M. J. (2014). Duplication of genes in an ATP-binding cassette transport system increases dynamic range while maintaining ligand specificity. *The Journal of Biological Chemistry*, *289*(43), 30090–30100. doi:10.1074/jbc.M114.590992 PMID:25210043

Giavasis, I., Harvey, L. M., & McNeil, B. (2000). Gellan Gum. *Critical Reviews in Biotechnology*, *20*(3), 177–211. doi:10.1080/07388550008984169 PMID:11039329

Godon, J. J., Galès, A., Latrille, E., Ouichanpagdee, P., & Seyer, J.-P. (2020). An "overlooked" habitat for thermophilic bacteria: The phyllosphere. *BioDiscovery*, *23*, 1–14. doi:10.3897/biodiscovery.23.e47033

Gu, J., & Hilser, V. J. (2009). Sequence-based analysis of protein energy landscapes reveals nonuniform thermal adaptation within the proteome. *Molecular Biology and Evolution*, *26*(10), 2217–2227. doi:10.1093/molbev/msp140 PMID:19592668

Gupta, G., Srivastava, S., Khare, S. K., & Prakash, V. (2014). Extremophiles: An overview of microorganism from extreme environment. *International Journal of Agriculture Environment and Biotechnology*, *7*(2), 371–380. doi:10.5958/2230-732X.2014.00258.7

Helmann, J., Wu, M. F. W., Kobel, P. A., Gamo, F.-J., Wilson, M., Morshedi, M. M., Navre, M., & Paddon, C. (2001). Global transcriptional response of *Bacillus subtilis* to heat shock. *Journal of Bacteriology*, *183*(24), 7318–7328. doi:10.1128/JB.183.24.7318-7328.2001 PMID:11717291

Hendrick, J. P., & Hartl, F. (1993). Molecular chaperone functions of heat-shock proteins. *Annual Review of Biochemistry*, *62*(1), 349–384. doi:10.1146/annurev.bi.62.070193.002025 PMID:8102520

Holden, J. F. (2012). Hot Environments. In M. Schaechter (Ed.), *Schmidt TM. Topics in ecological and environmental microbiology* (pp. 313–332). Elsevier Inc.

Holwerda, E. K., Olson, D. G., Ruppertsberger, N. M., Stevenson, D. M., Murphy, S. J. L., Maloney, M. I., Lanahan, A. A., Amador-Noguez, D., & Lynd, L. R. (2020). Metabolic and evolutionary responses of *Clostridium thermocellum* to genetic interventions aimed at improving ethanol production. *Biotechnology for Biofuels, 13*(1), 1-20. doi:10.1186/s13068-020-01680-5

Hongo, K., Hirai, H., Uemura, C., Ono, S., Tsunemi, J., Higurashi, T., Mizobata, T., & Kawata, Y. (2006). A novel ATP/ADP hydrolysis activity of hyperthermostable group II chaperonin in the presence of cobalt or manganese ion. *FEBS Letters*, *580*(1), 34–40. doi:10.1016/j.febslet.2005.11.043 PMID:16343486

Hu, Q., Zhou, N., Gong, K., Liu, H., Liu, Q., Sun, D., Wang, Q., Shao, Q., Liu, H., Qiu, B., & Guo, Z. (2019). Intracellular Polymer Substances Induced Conductive Polyaniline for Improved Methane Production from Anaerobic Wastewater Treatment. *ACS Sustainable Chemistry & Engineering*, *7*(6), 5912–5920. doi:10.1021/acssuschemeng.8b05847

Huber, G., & Stetter, K. O. (1991). *Sulfolobus metallicus*, sp. nov., a novel strictly chemolithoautotrophic thermophilic archaeal species of metal-mobilizers. *Systematic and Applied Microbiology*, *14*(4), 372–378. doi:10.1016/S0723-2020(11)80312-7

Huber, R., & Hannig, M. (2006). Thermotogales. In The Prokaryotes. Springer.

Huber, R., Wilharm, T., Huber, D., Trincone, A., Burggraf, S., König, H., Reinhard, R., Rockinger, I., Fricke, H., & Stetter, K. O. (1992). *Aquifex pyrophilus* gen. nov. sp. nov., represents a novel group of marine hyperthermophilic hydrogen-oxidizing bacteria. *Systematic and Applied Microbiology*, *15*(3), 340–351. doi:10.1016/S0723-2020(11)80206-7

Iizuka, R., Sugano, Y., Ide, N., Ohtaki, A., Yoshida, T., Fujiwara, S., Imanaka, T., & Yohda, M. (2008). Functional characterization of recombinant prefoldin complexes from a hyperthermophilic archaeon, *Thermococcus* sp. strain KS-1. *Journal of Molecular Biology*, *377*(3), 972–983. doi:10.1016/j.jmb.2008.01.070 PMID:18295793

Imanaka, H., Yamatsu, A., Fukui, T., Atomi, H., & Imanaka, T. (2006). Phosphoenolpyruvate synthase plays an essential role for glycolysis in the modified Embden-Meyerhof pathway in *Thermococcus kodakarensis*. *Molecular Microbiology*, *61*(4), 898–909. doi:10.1111/j.1365-2958.2006.05287.x PMID:16879645

Jaenicke, R., & Bohm, G. (1998). The stability of proteins in extreme environments. *Current Opinion in Structural Biology*, *8*(6), 738–748. doi:10.1016/S0959-440X(98)80094-8 PMID:9914256

Reeve, J. N.JN. (1992). Molecular biology of methanogens. *Annual Review of Microbiology*, *46*(1), 165–191. doi:10.1146/annurev.mi.46.100192.001121 PMID:1444253

Jorda, J., & Yeates, T. O. (2011). Widespread disulfide bonding in proteins from thermophilic archaea. *Archaea (Vancouver, B.C.), 2011*, 1–9. doi:10.1155/2011/409156 PMID:21941460

Kasting, J. (1993). Earth's early atmosphere. *Science, 259*(5097), 920–926. doi:10.1126cience.11536547 PMID:11536547

Kawamori, M., Takayama, K., & Takasawa, S. (1987). Production of Cellulases by a Thermophilic Fungus, *Thermoascus aurantiacus* A-131. *Agricultural and Biological Chemistry, 51*(3), 647–654. doi:10.1080/00021369.1987.10868111

Kengen, S., Stams, A. J., & Vos, W. M. (1996). Sugar metabolism of hyperthermophiles. *FEMS Microbiology Reviews, 18*(2-3), 119–137. doi:10.1111/j.1574-6976.1996.tb00231.x

Kengen, S., De Bok, F., Van Loo, N., Dijkema, C., Stams, A., & De Vos, W. (1994). Evidence for the operation of a novel Embden-Meyerhof pathway that involves ADP-dependent kinases during sugar fermentation by *Pyrococcus furiosus. The Journal of Biological Chemistry, 269*(26), 17537–17541. doi:10.1016/S0021-9258(17)32474-2 PMID:8021261

Kleine, J., & Liebl, W. (2006). Comparative characterization of deletion derivatives of the modular xylanase XynA of *Thermotoga maritima. Extremophiles, 10*(5), 373–381. doi:10.100700792-006-0509-0 PMID:16550304

Kletzin, A., Mukund, S., Kelley-Crouse, T. L., Chan, M. K., Rees, D. C., & Adams, M. (1995). Molecular characterization of the genes encoding the tungsten-containing aldehyde ferredoxin oxidoreductase from Pyrococcus furiosus and formaldehyde ferredoxin oxidoreductase from *Thermococcus litoralis. Journal of Bacteriology, 177*(16), 4817–4819. doi:10.1128/jb.177.16.4817-4819.1995 PMID:7642512

Koendjbiharie, J. G., Wiersma, K., & van Kranenburg, R. (2018). Investigating the central metabolism of *Clostridium thermosuccinogenes. Applied and Environmental Microbiology, 84*(13). Advance online publication. doi:10.1128/AEM.00363-18 PMID:29678919

Koga, S., Yoshioka, I., Sakuraba, H., Takahashi, M., Sakasegawa, S., Shimizu, S., & Ohshima, T. (2000). Biochemical characterization, cloning, and sequencing of ADP-dependent (AMP-forming) glucokinase from two hyperthermophilic archaea, *Pyrococcus furiosus* and *Thermococcus litoralis. Journal of Biochemistry, 128*(6), 1079–1085. doi:10.1093/oxfordjournals.jbchem.a022836 PMID:11098152

Kumar, M., Yadav, A. N., Tiwari, R., Prasanna, R., & Saxena, A. K. (2014). Deciphering the diversity of culturable thermotolerant bacteria from Manikaran hot springs. *Annals of Microbiology, 64*(2), 741–751. doi:10.100713213-013-0709-7

Kumar, S., & Nussinov, R. (2001). How do thermophilic proteins deal with heat? *Cellular and Molecular Life Sciences, 58*(9), 1216–1233. doi:10.1007/PL00000935 PMID:11577980

Labes, A., & Schönheit, P. (2001). Sugar utilization in the hyperthermophilic, sulfate-reducing archaeon *Archaeoglobus fulgidus* strain 7324: Starch degradation to acetate and CO_2 via a modified Embden-Meyerhof pathway and acetyl-CoA synthetase (ADP-forming). *Archives of Microbiology, 176*(5), 329–338. doi:10.1007002030100330 PMID:11702074

Ladenstein, R., & Antranikian, G. (1998). (1098). Proteins from hyperthermophiles: Stability and enzymatic catalysis close to the boiling point of water. *Advances in Biochemical Engineering/Biotechnology*, *61*, 37–85. doi:10.1007/BFb0102289 PMID:9670797

Ladenstein, R., & Antranikian, G. (1998). Proteins from hyperthermophiles: Stability and enzymatic catalysis close to the boiling point of water. *Advances in Biochemical Engineering/Biotechnology*, *61*, 37–85. doi:10.1007/BFb0102289 PMID:9670797

Lee, D., Koh, Y., Kim, K., & B., K.-F. M. (1999). *Isolation and characterization of a thermophilic lipase from* Bacillus thermoleovorans *ID-1*. Academic.oup.com

Leroux, M. R., Fandrich, M., Klunker, D., Siegers, K., Lupas, A. N., Brown, J. R., Schiebel, E., Dobson, C. M., & Hartl, F. U. (1999). MtGimC, a novel archaeal chaperone related to the eukaryotic chaperonin cofactor GimC/ prefoldin. *The EMBO Journal*, *18*(23), 6730–6743. doi:10.1093/emboj/18.23.6730 PMID:10581246

Li, H., Liao, Z., Sun, J., Jiang, B., Wang, J., & Yang, Y. (2019). Modelling and simulation of two-bed PSA process for separating H2 from methane steam reforming. *Chinese Journal of Chemical Engineering*, *27*(8), 1870–1878. doi:10.1016/j.cjche.2018.11.022

Li, S. J., Hua, Z.-S., Huang, L., Li, J., Shi, S.-H., Chen, L.-X., Kuang, J.-L., Liu, J., Hu, M., & Shu, W.-S. (2014). Microbial communities evolve faster in extreme environments. *Scientific Reports*, *4*(6205), 1–9. doi:10.1038rep06205 PMID:25158668

Li, W. F., Zhou, X. X., & Lu, P. (2005). Structural features of thermozymes. *Biotechnology Advances*, *23*(4), 271-281. doi:.biotechadv.2005.01.002 doi:10.1016/j

Liao, H., McKenzie, T., & Hageman, R. (1986). Isolation of a thermostable enzyme variant by cloning and selection in a thermophile. *Proceedings of the National Academy of Sciences of the United States of America*, *83*(3), 576–580. doi:10.1073/pnas.83.3.576 PMID:3003740

Liu, Z. Y., Jia, D. C., di Zhang, K., Zhu, H. F., Zhang, Q., Jiang, W. H., Gu, Y., & Li, F. L. (2020). Ethanol metabolism dynamics in *Clostridium ljungdahlii* grown on carbon monoxide. *Applied and Environmental Microbiology*, *86*(14). Advance online publication. doi:10.1128/AEM.00730-20 PMID:32414802

Lyu, Z., Shao, N., Akinyemi, T., & Whitman, W. B. (2018). Methanogenesis. *Current Biology*, *28*(13), R727–R732. doi:10.1016/j.cub.2018.05.021 PMID:29990451

Madigan, M. T., Takigiku, R., Lee, R. G., Gest, H., & Hayes, J. M. (1989). Carbon isotope fractionation by thermophilic phototrophic sulfur bacteria: Evidence for autotrophic growth in natural populations. *Applied and Environmental Microbiology*, *55*(3), 639–644. doi:10.1128/aem.55.3.639-644.1989 PMID:11536609

Malashenko, Y. R., Romanovskaya, V. A., Bogachenko, V. N., & Shved, A. D. (1975). Thermophilic and thermotolerant methane-assimilating bacteria. *Mikrobiologiya*, *44*, 855–862. PMID:1207503

Mallick, P., Boutz, D. R., Eisenberg, D., & Yeates, T. O. (2002). Genomic evidence that the intracellular proteins of archaeal microbes contain disulfide bonds. *Proceedings of the National Academy of Sciences of the United States of America*, *99*(15), 9679–9684. doi:10.1073/pnas.142310499 PMID:12107280

Mangrola, A. V., Dudhagara, P. R., Koringa, P. G., Joshi, C. G., & Patel, R. K. (2018). Metagenomic microbial community profiling of Unnai hot spring by Ion-Torrent based shotgun sequencing. *Microbiology, 87*(1), 143–146. doi:10.1134/S0026261718010113

Mangrola, A. V., Dudhagara, P., Koringa, P., Joshi, C. G., & Patel, R. K. (2015). Shotgun metagenomic sequencing based microbial diversity assessment of Lasundra hot spring, India. *Genomics Data, 4*, 73–75. doi:10.1016/j.gdata.2015.03.005 PMID:26484181

Mangrola, A., Dudhagara, P., Koringa, P., Joshi, C. G., Parmar, M., & Patel, R. (2015). Deciphering the microbiota of Tuwa hot spring, India using shotgun metagenomic sequencing approach. *Genomics Data, 4*, 153–155. doi:10.1016/j.gdata.2015.04.014 PMID:26484204

Maru, B. T., Bielen, A. A. M., Kengen, S. W. M., Constantí, M., & Medina, F. (2012). Biohydrogen Production from Glycerol using *Thermotoga* spp. *Energy Procedia, 29*, 300–307. doi:10.1016/j.egypro.2012.09.036

Miroshnichenko, M. L., & Bonch-Osmolovskaya, E. A. (2006). Recent developments in the thermophilic microbiology of deep-sea hydrothermal vents. *Extremophiles, 10*(2), 85–96. doi:10.100700792-005-0489-5 PMID:16418793

Mukund, S., & Adams, M. W. (1995). Glyceraldehyde-3-phosphate ferredoxin oxidoreductase, a novel tungsten-containing enzyme with a potential glycolytic role in the hyperthermophilic archaeon *Pyrococcus furiosus. The Journal of Biological Chemistry, 270*(15), 8389–8392. doi:10.1074/jbc.270.15.8389 PMID:7721730

Mukund, S., & Adams, M. (1991). The novel tungsten-iron-sulfur protein of the hyperthermophilic archaebacterium, *Pyrococcus furiosus*, is an aldehyde ferredoxin oxidoreductase. Evidence for its participation in a unique glycolytic pathway. *The Journal of Biological Chemistry, 266*(22), 14208–14216. doi:10.1016/S0021-9258(18)98669-2 PMID:1907273

Mukund, S., & Adams, M. (1993). Characterization of a novel tungsten-containing formaldehyde ferredoxin oxidoreductase from the hyperthermophilic archaeon, *Thermococcus litoralis*. A role for tungsten in peptide catabolism. *The Journal of Biological Chemistry, 268*(18), 13592–13600. doi:10.1016/S0021-9258(19)38690-9 PMID:8390467

Muñoz, P. A., Flores, P. A., Boehmwald, F. A., & Blamey, J. M. (2011). Thermophilic bacteria present in a sample from Fumarole Bay, Deception Island. *Antarctic Science, 23*(6), 549–555. doi:10.1017/S0954102011000393

Nanavati, D. M., Nguyen, T. N., & Noll, K. M. (2005). Substrate specificities and expression patterns reflect the evolutionary divergence of maltose ABC transporters in Thermotoga maritima. *Journal of Bacteriology, 187*(6), 2002–2009. doi:10.1128/JB.187.6.2002-2009.2005 PMID:15743948

Nanavati, D. M., Thirangoon, K., & Noll, K. M. (2006). Several archaeal homologs of putative oligopeptide-binding proteins encoded by Thermotoga maritima bind sugars. *Applied and Environmental Microbiology, 72*(2), 1336–1345. doi:10.1128/AEM.72.2.1336-1345.2006 PMID:16461685

Nguyen, T. A. D., Kim, J. P., Kim, M. S., Oh, Y. K., & Sim, S. J. (2008). Optimization of hydrogen production by hyperthermophilic eubacteria, *Thermotoga maritima* and *Thermotoga neapolitana* in batch fermentation. *International Journal of Hydrogen Energy, 33*(5), 1483–1488. doi:10.1016/j.ijhydene.2007.09.033

Nogales, J., Gudmundsson, S., & Thiele, I. (2012). An in silico re-design of the metabolism in Thermotoga maritima for increased biohydrogen production. *International Journal of Hydrogen Energy, 37*(17), 12205–12218. doi:10.1016/j.ijhydene.2012.06.032

Ohtaki, A., Kida, H., Miyata, Y., Ide, N., Yonezawa, A., Arakawa, T., Iizuka, R., Noguchi, K., Kita, A., Odaka, M., Miki, K., & Yohda, M. (2008). Structure and molecular dynamics simulation of archaeal prefoldin: The molecular mechanism for binding and recognition of nonnative substrate proteins. *Journal of Molecular Biology, 376*(4), 1130–1141. doi:10.1016/j.jmb.2007.12.010 PMID:18201719

Ollivier, B., & Cayol, J. (2005). Fermentative, iron-reducing, and nitrate-reducing microorganisms. In B. Ollivier & M. Magot (Eds.), *Petroleum Microbiology* (pp. 71–88). ASM Press. doi:10.1128/9781555817589

Oshima, T., & Moriya, T. (2008). A preliminary analysis of microbial and biochemical properties of high-temperature compost. *Annals of the New York Academy of Sciences, 1125*(1), 338–344. doi:10.1196/annals.1419.012 PMID:18378603

Panda, M. K., Sahu, M. K., & Tayung, K. (2013). Isolation and characterization of a thermophilic *Bacillus* sp. with protease activity isolated from hot spring of Tarabalo, Odisha, India. *Iranian Journal of Microbiology, 5*(2), 159–165. PMID:23825735

Pandey, A., Dhakar, K., Sharma, A., Priti, P., Sati, P., & Kumar, B. (2015). Thermophilic bacteria that tolerate a wide temperature and pH range colonize the Soldhar (95°C) and Ringigad (80°C) hot springs of Uttarakhand, India. *Annals of Microbiology, 65*(2), 809–816. doi:10.100713213-014-0921-0

Pedone, E., Ren, B., Ladenstein, R., Rossi, M., & Bartolucci, S. (2004). Functional properties of the protein disulfide oxidoreductase from the archaeon *Pyrococcus furiosus*: A member of a novel protein family related to protein disulfide-isomerase. *European Journal of Biochemistry, 271*(16), 3437–3448. doi:10.1111/j.0014-2956.2004.04282.x PMID:15291821

Pierson, B. K., & Castenholz, R. W. (1974). A phototrophic gliding filamentous bacterium of hot springs, *Chloroflexus aurantiacus*, gen. and sp. nov. *Archives of Microbiology, 100*(1), 5–24. doi:10.1007/BF00446302 PMID:4374148

Purcell, D., Sompong, U., Lau, C. Y., Barraclough, T. G., Peerapornpisal, Y., & Pointing, S. B. (2007). The effects of temperature, pH and sulphide on community structure of hyperthermophilic streamers in hot springs of northern Thailand. *FEMS Microbiology Ecology, 60*(3), 456–466. doi:10.1111/j.1574-6941.2007.00302.x PMID:17386034

Robb, F., Antranikian, G., Grogan, D., & Driessen, A. (2008). *Thermophiles, biology and technology at high temperatures.* CRC Press.

Rodionov, D. A., Rodionova, I. A., Li, X., Ravcheev, D. A., Tarasova, Y., Portnoy, V. A., et al. (2013). Transcriptional regulation of the carbohydrate utilization network in *Thermotoga maritima. Front. Microbiol., 4*, 244. doi:.00244 doi:10.3389/fmicb.2013

Ronimus, R. S., Koning, J., & Morgan, H. (1999). Purification and characterization of an ADP-dependent phosphofructokinase from *Thermococcus zilligii. Extremophiles, 3*(2), 121–129. doi:10.1007007920050107 PMID:10356998

Sahlan, M., & Yohda, M. (2013). Molecular Chaperones in Thermophilic Eubacteria and Archaea. In Thermophilic Microbes in Environmental and Industrial Biotechnology: Biotechnology of Thermophiles, (pp. 375-394). Springer Science Business Media. doi:10.1007/978-94-007-5899-5

Sakuraba, H., & Ohshima, T. (2002). Novel energy metabolism in anaerobic hyperthermophilic archaea: A modified Embden-Meyerhof pathway. *Journal of Bioscience and Bioengineering, 93*(5), 441–448. doi:10.1016/S1389-1723(02)80090-1 PMID:16233230

Sakuraba, H., Goda, S., & Ohshima, T. (2004). Unique sugar metabolism and novel enzymes of hyperthermophilic archaea. *Chemical Record (New York, N.Y.), 3*(5), 281–287. doi:10.1002/tcr.10066 PMID:14762828

Sakuraba, H., Yoshioka, I., Koga, S., Takahashi, M., Kitahama, Y., Satomura, T., Kawakami, R., & Ohshima, T. (2002). ADP-dependent glucokinase/phosphofructokinase, a novel bifunctional enzyme from the hyperthermophilic archaeon Methanococcus jannaschii. *The Journal of Biological Chemistry, 277*(15), 12495–12498. doi:10.1074/jbc.C200059200 PMID:11856730

Sarkar, S., Banerjee, R., Chanda, S., Das, P., Ganguly, S., & Pal, S. (2010). Effectiveness of inoculation with isolated *Geobacillus* strains in the thermophilic stage of vegetable waste composting. *Bioresource Technology, 101*(8), 2892–2895. doi:10.1016/j.biortech.2009.11.095 PMID:20036532

Schönheit, P., & Schäfer, T. (1995). Metabolism of hyperthermophiles. *World Journal of Microbiology & Biotechnology, 11*(1), 26–57. doi:10.1007/BF00339135 PMID:24414410

Schröder, C., Selig, M., & Schönheit, P. (1994). Glucose fermentation to acetate, CO_2 and H2 in the anaerobic hyperthermophilic eubacterium *Thermotoga maritima*: Involvement of the Embden-Meyerhof pathway. *Archives of Microbiology, 161*(6), 460–470.

Selig, M., Xavier, K. B., Santos, H., & Schonheit, P. (1997). Comparative analysis of Embden-Meyerhof and Entner-Doudoroff glycolytic pathways in hyperthermophilic archaea and the bacterium *Thermotoga. Archives of Microbiology, 167*(4), 217–232. doi:10.1007/BF03356097 PMID:9075622

Shim, J., Soum, V., Kang, K.-W., Chuon, K., Cho, S.-G., Kim, J.-H., Meas, S., Pushkarev, A., Shin, K., & Jung, K.-H. (2021). Discovery of a microbial rhodopsin that is the most stable in extreme environments. *iScience, 24*(6), 102620. Advance online publication. doi:10.1016/j.isci.2021.102620 PMID:34151231

Shockley, K. R., Ward, D. E., Chhabra, S. R., Conners, S. B., Montero, C. I., & Kelly, R. M. (2003). Heat shock response by the hyperthermophilic archaeon *Pyrococcus furiosus. Applied and Environmental Microbiology, 69*(4), 2365–2371. doi:10.1128/AEM.69.4.2365-2371.2003 PMID:12676722

Siebers, B., & Schönheit, P. (2005). Unusual pathways and enzymes of central carbohydrate metabolism in *Archaea. Current Opinion in Microbiology, 8*(6), 695–705. doi:10.1016/j.mib.2005.10.014 PMID:16256419

Siegert, R., Leroux, M. R., Scheufler, C., Hartl, F. U., & Moarefi, I. (2000). Structure of the molecular chaperone prefoldin: Unique interaction of multiple coiled coil tentacles with unfolded proteins. *Cell, 103*(4), 621–632. doi:10.1016/S0092-8674(00)00165-3 PMID:11106732

Siliakus, M. F., van der Oost, J., & Kengen, S. W. M. (2017). Adaptations of archaeal and bacterial membranes to variations in temperature, pH and pressure. *Extremophiles, 21*(4), 651–670. doi:10.100700792-017-0939-x PMID:28508135

Singh, G., Jain, V. K., & Singh, A. (2017). Effect of Temperature and other factors on Anaerobic Digestion Process, responsible for Bio Gas Production. *International Journal of Theoretical and Applied Mechanics., 12*(3), 637–657.

Sirevåg, R. (2021). Thermophilic Phototrophs. *Thermophilic Bacteria*, 195-219. doi:10.1201/9781003068334-8

Stathopoulou, P. M., Savvides, A. L., Karagouni, A. D., & Hatzinikolaou, D. G. (2013). Unraveling the lipolytic activity of thermophilic bacteria isolated from a volcanic environment. *BioMed Research International, 2013*, 1–13. Advance online publication. doi:10.1155/2013/703130 PMID:23738330

Sterner, R., & Liebl, W. (2001). Thermophilic adaptation of proteins. *Critical Reviews in Biochemistry and Molecular Biology, 36*(1), 39–106. doi:10.1080/20014091074174 PMID:11256505

Stirling, P. C., Lundin, V. F., & Leroux, M. R. (2003). Getting a grip on non-native proteins. *EMBO Reports, 4*(6), 565–570. doi:10.1038j.embor.embor869 PMID:12776175

Suzuki, Y., Tsujimoto, Y., Matsui, H., & Watanabe, K. (2006). Decomposition of extremely hard-to-degrade animal proteins by thermophilic bacteria. *Journal of Bioscience and Bioengineering, 102*(2), 73–81. doi:10.1263/jbb.102.73 PMID:17027867

Taguchi, H., Konishi, J., Ishii, N., & Yoshida, M. A. (1991). Chaperonin from a thermophilic bacterium, *Thermus thermophilus*, that controls refolding of several thermophilic enzymes. *The Journal of Biological Chemistry, 266*(33), 22411–22418. doi:10.1016/S0021-9258(18)54588-9 PMID:1682319

Tam, L. T., Antelmann, H., Eymann, C., Albrecht, D., Bernhardt, J., & Hecker, M.Tam le. (2006). Proteome signatures for stress and starvation in *Bacillus subtilis* as revealed by a 2-D gel image color coding approach. *Proteomics, 6*(16), 4565–4585. doi:10.1002/pmic.200600100 PMID:16847875

Tang, Y. J., Sapra, R., Joyner, D., Hazen, T. C., Myers, S., Reichmuth, D., Blanch, H., & Keasling, J. D. (2009). Analysis of metabolic pathways and fluxes in a newly discovered thermophilic and ethanoltolerant *Geobacillus* strain. *Biotechnology and Bioengineering, 102*(5), 1377–1386. doi:10.1002/bit.22181 PMID:19016470

Tetas, M., & Lowenstein, J. M. (1963). the effect of bivalent metal ions on the hydrolysis of adenosine Di-and triphosphate. *Biochemistry, 2*(2), 350–357. doi:10.1021/bi00902a030 PMID:13980762

Trent, J. D. (1996). A review of acquired thermotolerance, heat-shock proteins, and molecular chaperones in archaea. *FEMS Microbiology Reviews, 18*(2-3), 249–258. doi:10.1111/j.1574-6976.1996.tb00241.x

Trivedi, S., RamaRao, S., & Gehlot, H.S. (2005). Nucleic Acid stability in thermophilic prokaryotes: A review. *Journal of Cell & Molecular Biology, 4*, 61–69.

Tse, C., & Ma, K. (2016). Growth and Metabolism of Extremophilic Microorganisms. In Biotechnology of Extremophiles, Grand Challenges in Biology and Biotechnology. Springer. doi:10.1007/978-3-319-13521-2_1

Vainberg, I. E., Lewis, S. A., Rommelaere, H., Ampe, C., Vandekerckhove, J., Klein, H. L., & Cowan, N. J. (1998). Prefoldin, a chaperone that delivers unfolded proteins to cytosolic chaperonin. *Cell, 93*(5), 863–873. doi:10.1016/S0092-8674(00)81446-4 PMID:9630229

Valverde, A., Tuffin, M., & Cowan, D. A. (2012). Biogeography of bacterial communities in hot springs: A focus on the actinobacteria. *Extremophiles, 16*(4), 669–679. doi:10.100700792-012-0465-9 PMID:22711422

van der Oost, J., & Siebers, B. (2007). The glycolytic pathways of Archaea: evolution by tinkering. *Archaea: Evolution, Physiology and Molecular Biology, 22*, 247-260.

van der Oost, J., Schut, G., Kengen, S. M., Hagen, W. R., Thomm, M., & de Vos, W. M. (1998). The ferredoxin-dependent conversion of glyceraldehyde-3-phosphate in the hyperthermophilic archaeon *Pyrococcus furiosus* represents a novel site of glycolytic regulation. *The Journal of Biological Chemistry, 273*(43), 28149–28154. doi:10.1074/jbc.273.43.28149 PMID:9774434

Varma, V. S., & Kalamdhad, A. S. (2015). Evolution of chemical and biological characterization during thermophilic composting of vegetable waste using rotary drum composter. *International Journal of Environmental Science and Technology, 12*(6), 2015–2024. Advance online publication. doi:10.100713762-014-0582-3

Verhees, C., Kengen, S., Tuininga, J., Schut, G., Adams, M., de Vos, W., & Van der Oost, J. (2004). The unique features of glycolytic pathways in archaea. *Biochemistry, 377*(3), 819–822. doi:10.1042/bj3770819

Vieille, C., & Zeikus, G. J. (2001). Hyperthermophilic enzymes: Sources, uses, and molecular mechanisms for thermostability. *Microbiology and Molecular Biology Reviews, 65*(1), 1–43. doi:10.1128/MMBR.65.1.1-43.2001 PMID:11238984

Wadkar, D. V., Modak, P. R., & Chavan, V. S. (2013). Aerobic Thermophilic Composting of Municipal Solid Waste. *International Journal of Engineering Science and Technology, 5*(3), 716–718.

Wahlund, T. M., Woese, C. R., Castenholz, R. W., & Madigan, M. T. (1991). A thermophilic green sulfur bacterium from New Zealand hot springs, *Chlorobium tepidum* sp. nov. *Archives of Microbiology, 156*(2), 81–90. doi:10.1007/BF00290978

Wang, J., Zhao, C., Meng, B., Xie, J., Zhou, C., Chen, X., Zhao, K., Shao, J., Xue, Y., Xu, N., Ma, Y., & Liu, S. (2007). The proteomic alterations of *Thermoanaerobacter tengcongensis* cultured at different temperatures. *Proteomics, 7*(9), 1409–1419. doi:10.1002/pmic.200500226 PMID:17469076

Wang, Q., Cen, Z., & Zhao, J. (2013). The survival mechanisms of thermophiles at high temperatures: An angle of omics. *Physiology (Bethesda, MD), 30*(2), 97–106. doi:10.1152/physiol.00066.2013 PMID:25729055

Wang, Z., Tong, W., Wang, Q., Bai, X., Chen, Z., Zhao, J., Xu, N., & Liu, S. (2012). The temperature dependent proteomic analysis of *Thermotoga maritima. PLoS One*, *7*(10), e46463. doi:10.1371/journal.pone.0046463 PMID:23071576

Xavier, K. B., Martins, L. O., Peist, R., Kossmann, M., Boos, W., & Santos, H. (1996). High-affinity maltose/ trehalose transport system in the hyperthermophilic archaeon *Thermococcus litoralis. Journal of Bacteriology*, *178*(16), 4773–4777. doi:10.1128/jb.178.16.4773-4777.1996 PMID:8759837

Yakovlev, A. G., Taisova, A. S., & Fetisova, Z. G. (2021). Utilization of blue-green light by chlorosomes from the photosynthetic bacterium *Chloroflexus aurantiacus*: Ultrafast excitation energy conversion and transfer. *Biochimica et Biophysica Acta (BBA) -. Bioenergetics*, *1862*(6), 148396. doi:10.1016/j.bbabio.2021.148396 PMID:33581107

Yoneda, Y., Yoshida, T., Yasuda, H., Imada, C., & Sako, Y. (2013). A novel thermophilic, hydrogenogenic, and carboxydotrophic bacterium *Calderihabitans maritimus* gen. nov., sp. nov. from a marine sediment core of an undersea caldera. *International Journal of Systematic and Evolutionary Microbiology*, *63*(Pt_10), 3602–3608. doi:10.1099/ijs.0.050468-0 PMID:23606483

Zeikus, J. G., & Wolfe, R. S. (1972). *Methanobacterium thermoautotrophicus* sp., an anaerobic, autotrophic, extreme thermophile. *Journal of Bacteriology*, *109*(2), 707–715. doi:10.1128/jb.109.2.707-713.1972 PMID:4550816

Zeldes, B. M., Keller, M. W., Loder, A. J., Straub, C. T., Adams, M. W. W., & Kelly, R. M. (2015). Extremely thermophilic microorganisms as metabolic engineering platforms for production of fuels and industrial chemicals. *Frontiers in Microbiology*, *12*(6), 2015–2024. doi:10.3389/fmicb.2015.01209 PMID:26594201

Zelenkina, T., Eshinimayev, B. T., Dagurova, O., Suzina, N., Namsarayev, B. B., & Trotsenko, Y. (2009). Aerobic methanotrophs from the coastal thermal springs of Lake Baikal. *Microbiology*, *78*(4), 492–497. doi:10.1134/S0026261709040134 PMID:19827720

Zinder, S. H. (1993). Physiological Ecology of Methanogens. *Methanogenesis*, 128-206. doi:10.1007/978-1-4615-2391-8_4

Chapter 5
Psychrophiles:
Distribution, Ecology, Physiology, Metabolism, Cold Adapted Enzymes, and Proteins

Shreyas Anantray Bhatt
Hemchandracharya North Gujarat University, India

ABSTRACT

Life on the Earth has evolved in the cold environments. Such cold habitats pose special challenges to the microbes in cold ecosystems, such as minimum metabolic activities, very limited nutrient availability, and often extreme conditions such as pH and salinity apart from temperature. Microbial communities surviving under these extreme conditions must have evolved complex structural and functional adaptations. Prokaryotic adaptations to cold environments are through physiological adaptations by increasing membrane fluidity through large amount of unsaturated fatty acids. These microbes also possess some cold adapted proteins whose steady state levels are maintained. They also produce certain compounds such as polyamines, sugars, polyols, amino acids, and some antifreeze proteins to protect themselves under freezing conditions. They also produce exopolymeric substances that promote adhesion of microbes to moist surfaces to induce biofilm formation which helps getting nutrients and protect the cells from harsh conditions. Antioxidants help destroying toxic reactive oxygen species.

BACKGROUND

Much of the knowledge on the microorganisms inhabiting the cold environments has been accumulated during 1960s and 1970s. Several debatable definitions have been proposed for the range of temperatures for the growth for psychrophiles. It has now been well accepted that the microorganisms defined as psyhrophiles have maximum growth temperature of 20 °C. However, those microbes that grow well at lower temperatures but can also grow at >20 °C are termed as psychrotolerant or also called prychrotrophs. The difficulty faced in defining psychrophiles is because many microorganisms have evolved to withstand the temperature fluctuations and there is no threshold cut off of temperature for the growth

DOI: 10.4018/978-1-7998-9144-4.ch005

of microorganisms. However, proposed the definition of psychrophiles as 'Any organism that grows at 5 °C or below' to distinguish from mesophiles that grow well at 37 °C. However, it is difficult to distinguish psychrotolerant from psychrophiles on the basis of metabolic activities because there are several organisms that can grow above 20 °C temperature and can produce enzymes that can show good activity below 20 °C.

The diversity of psychrophiles comprises vast variety of microorganisms as most of the Earth's surface have temperature favourable for psychrophilic organisms. Moreover, these organisms are also reported from harsh environmental conditions such as high pH, high pressure, nutrient scarcity, high salt concentration, deep sea and ice accretions. The capacity of these organisms to produce different hydrolytic enzymes and antifreeze proteins, antioxidants make them competent to grow and catalyse many organic compounds that help them to survive under inhospitable environments.

INTRODUCTION

The initial colonization of life on the Earth has been reported to be evolved in the cold environments. The largest portion of the global biosphere is represented by the communities growing and surviving at temperatures below 5°C (Siddiqui & Cavicchioli 2006; Margesin & Miteva, 2011). Low to extreme low temperature environment have always been fascinating source for scientists particularly for microbial life existing and for long term survival as well as adaptation to some of the very harsh conditions including frost and permafrost temperatures.

Microbial species inhabiting the cold environment are recognized by different terminology such as "Psychrotrophs", Psychrophiles, Psychrotolerant and Cold active etc., and are exploited for their ability to survive and grow at low temperatures. "Psychrophile" is a generic term used generally to include all organisms inhabiting cold climatic conditions. Psychrophilic microorganisms have been an interesting subject for microbial ecologist, physiologists, geneticists and taxonomists for their biochemical and physiological adaptations through complex processes for protein stability and their catalytic function.

It has been reported that about 80% of the biosphere of the Earth experiences an average of 15°C temperature. However, some of the specialized psychrophilic habitats such as permafrost sediments, Antarctic lakes and deep marine ecosystems have been explored for microbial diversity and found greater diversity than expected.

HABITAT

It is well accepted that wherever there is possibility for existence of microbial life, it is found to exist. Microbes are found in very harsh environments of extremely halophilic conditions especially spore forming bacilli are found in a 250 million year old salt crystal. A bacterial spore has also been revived, cultured and identified from 40 million year old amber . Similarly microbial life has also been reported to be found at several kilometres of depth in the Earth's crust and viable bacterial populations have been discovered at depths of 750 m in Pacific Ocean sites. Bacterial population can also grow and reproduce at ≤ 0 °C in the cloud droplet at high altitude.

Microbes might have existed in permanent ice cap some 14 million years ago and their descendants may live in sub glacial rock crevices, lakes and sediments. Some population of psychrophilic bacteria in deep antarctic ice in the absence of sunlight or oxygen at pressure up to 400 bars, at temperature well below 0°Cand in strongly acidic or saline solutions.

About 10000 years of age in the vein of ice accretion the carbon and energy source can maintain significant numbers of cells that are not multiplying but are actively metabolizing species of *Thiobacillus* can crow at below 0°C and very low pH while fungi belonging to hyper acidophilic group can grow at the least pH. Species of *Acontiumvelatium* has been grown in 2.5 M H_2SO_4. Thus, microorganisms can survive under sub zero temperature, without oxygen, under dark conditions, under pressure up to 400 bars, also tolerate 1 to 2 M of salt concentrations.

The liquid vein habitat has been shown to act as source of water, energy and carbon. However, no species have been reported to tolerate all the inhospitable conditions such as no sunlight, zero oxygen, high pressure of 400 bars, subzero temperature, highly acidic or saline conditions. This implies that those very few surviving in liquid veins in ice may be characteristically different from the known species.

Antarctica

Antarctica is the fifth largest continent. About 98% of it is covered under ice with environment and climatic conditions such as extreme low temperatures, very low humidity, low liquid water availability and high solar radiation. Despite of severity in environmental conditions, microbial communities have adapted and inhabited such environments.

This cold biosphere is also represented by various types of environments viz., deep sea, alpine regions, climatically challenged regions of permafrost as well as polar regions. A diverse group of microbes have been found to proliferate such as bacteria, eukarya and viruses. Some members have limited range of temperatures for growth, maximum being 20 °C, while majority of members have capacity to tolerate warmer climate. Those that grow below 20 °C are called stenopsychrophile while the later members are grouped as eurypsychrophiles. These have been revealed by various genetic methods viz, small subunit rRNA sequencing, FISH and DNA sequencing of whole environmental samples.

Recently, metagenomic and metaproteomic studies have been successfully used for recording diversity, community dynamics and microbial processes driven by psychrophiles. The biochemical, physiological and genetic studies have revealed the cold adaptation of psycharophiles.

Several novel species isolated from terrestrial habitat of antarctica are listed in following table 1.

Maritime and Peninsula: The Antarctic peninsular soil contain complex communities of algae, lichens, mosses etc. The overall macro- and micro- species diversity is law. However, fungi play a significant role as decomposer and thus carryout transformation of nutrients. The dominant–fungal taxa reported include *Basidiomycetes*, *Ascomycetes* and *Oomycetes* (Wicklow and Soderstrom, 1997).

Continental

The oasis regions of continental Antarctica and Queen Mand land soils are dominantly inhabited by gram negative bacteria mainly by the genus *Pseudomonas*. (Shivaji *et al.*, 1989). The other gram negative bacilli strain of *Achromobacter*, *Alkaligenes* and halotolerant – species of *Planococcus* have also been isolated.

Based on phospholipid and lipid profiles, *Psychrobacter* species are reported to belong to the psychrotolerant / psychrophilic and halotolerant group. The ubiquitous distribution of these bacteria in many antarctic marine and terrestrial habitats is attributed to the above characteristics (Bowman *et al.*, 1996).

Table 1. Novel microbial species isolated from soil ecosystems in Antarctica

Species	Taxonomic Affiliation	Source	Growth Range(°C)	Reference
Brevibacillus levickii	*Actinobacteria*	Geothermal soils, Mt Melbourne	15–55	Allen *et al.*, (2005)
Cryobacterium psychrophilum	*Actinobacteria*	Antartica soil	<18	Suzuki *et al.*, (1997)
Friedmanniella Antarctica	*Actinobacteria*	Sandstone, McMurdo Dry Valleys	18-25	Schumann *et al.*, (1997)
Micromonospora endolithica	*Actinobacteria*	Sandstone, McMurdo Dry Valleys	8-39	Hirsch *et al.*, (2004)
Modestobacter multiseptatus	*Actinobacteria*	Soil Asgard Range, Transantarctic Mountains	0-28	Mevs *et al.*, (2000)
Pseudonocardia antarctic	Actinobacteria	Cyanobacterial mat, McMurdo Valley	7-38	Hirsch *et al.*, (2004)
Streptomyces hypolithicus	*Actinobacteria*	Hypolithic community, Miers Valley	22-30	Le Roes-Hill *et al.*, (2009)
Aequorivita sublithincola	Bacteroidetes	Quartz rock	2-25	Bowman and Nichols (2002)
Flavobacterium antarcticum	Bacteroidetes	Penguin habitat, King George Island	5-24	Yi *et al.*, (2005)
Flavobacterium weaverense	Bacteroidetes	Soil, Weaver Peninsula	5-19	Yi and Chun (2006)
Flavobacterium segetis	Bacteroidetes	Penguin habitat, King George Island	5-21	Yi and Chun (2006)
Sejongia antarctica	Bacteroidetes	Penguin habitat, King George Island	4-28	Yi *et al.*, (2005)
Sejongia jeonii	Bacteroidetes	Moss sample, King George Island	4-31	Yi *et al.*, (2005)
Alicyclobacillus pohliae	Firmicutes	Soil, King George Island	0-23	Yu *et al.*, (2008)
Aneurinibacillus terranovensis	Firmicutes	Soil, King George Island	20-55	Allan *et al.*, (2005)
Sporosarcina antarctica	Firmicutes	Soil, King George Island	0-23	Yu *et al.*, (2008)
Pseudoalteromonas antarctica	*γ-Proteobacteria*	Mud, Admiralty Bay, King George Islan	4-30	Bozal *et al.*, (1997)
Pseudomonas guinea	*γ-Proteobacteria*	South Shetland Islands	-4 to 30	Bozal *et al.*, (1997)
Psychrobacter frigidicola	*γ-Proteobacteria*	Penguin colony, Vestfold Hills	-18 to 22	Bowman *et al.*, (1996)
Psychrobacter urativorans	*γ-Proteobacteria*	Penguin colony, Vestfold Hills	-10* to 25	Bowman *et al.*, (1996)
Shewanella livingstonensis	*γ-Proteobacteria*	Sediment, Johnson's Dock, Livingston Island	4-20	Bozal *et al.*, (2002)
Shewanella vesiculosa	*γ-Proteobacteria*	Sediment, Deception Island	-4 to 30	Bozal *et al.*, (2002)
Eukaryotic species				
Cryptococcus albidosimilis	*Basidiomycota*	Soil, Linnaeua Terrace	4 to <20	Vishniac and Kurtzman (1992)
Cryptococcus antarcticus	*Basidiomycota*	Soil, University Valley	4 to <25	Vishniac and Kurtzman (1992)
Cryptococcus consortionis	*Basidiomycota*	Ross Desert	4 – 23	Vishniac (1985)
Cryptococcus lupi	*Basidiomycota*	Dolerite gravel, South Victoria Land	4 to <25	Baharaeen and Vishniac (1982)
Cryptococcus socialis	*Basidiomycota*	Ross Desert	4 -23	Vishniac (1985)
Cryptococcus statzelliae	*Basidiomycota*	Soil, Lichen Valley, Vestvold Hills	<20	Thomas-Hall *et al.*, (2002)
Dioszegia antarctica	*Basidiomycota*	Taylor Valley, South Victoria Land	<20	Connell *et al.*, (2010)
Dioszegia cryoxerica	*Basidiomycota*	Taylor Valley, South Victoria Land	<20	Connell *et al.*, (2010)

Curtsey: Extremophiles Handbook, Ch.6.5, Kirby et al., 2011.

The Dry Valleys

The dry valley of Eastern Antarctica covering about 2% of ice free area is considered to be the coldest and driest desert representing the extreme climatic conditions such as low humidity, very low precipitation and strong winds. The climo – edaphic factors favours the growth of organisms which produces cryoprotactants (Cameron, 1969).

The microbial biomass in these area have been reported to be four magnitude greater than expected (Cowan *et al.,* 2002). Algal species also dominate here that are favoured by climatic conditions and these have been analysed by metagenomic studies (Wood *et al.,* 2008, Pointing *et al.,* 2009). Several other heterotrophs predominantly psychrotrophs are better adapted to dry valley conditions. *Corynebacterium* species also dominate here (Cameron *et al.,* 1972).

Actinobacterial population appears to be dominant phylotype in Antarctic soils. Species of *Arthrobacter*, *Nocordia* and *Streptomycetes* and a noval *Flavobacterum* sp., have been reported.

THE ARCTIC

The arctic circle includes parts of Alaska, Canada, Europe, Greenland, Iceland and Rusia, dominated by arctic ocean. The area include under this region includes tundra, alpine soils, polar deserts and permafrost. Population diversity is sustained by the minerals and other organic compounds from fossil fuels present in the soils of this region (Callaghan *et al.,* 2014).

The tundra region which mostly is a permafrost have quite unique characteristically dwarf plants including shrubs, lichens and mosses. These soils have challenging climatic conditions of low temperature, low pH, and high water content because of drainage limitations. These stress conditions with large organic carbon in tundra region may favour the growth of psychrophiles mainly methanotrophs

Methanogens have been isolated from different temperate soils of various countries such as Western Siberia, Northern Russia, Northern Germany. Different kinds of microorganisms such as bacteria, algae and fungal representatives are reported from polar soils. *Acetobacterium* and *Acenetobacter* species a lipolytic psychrotrophic bacteria have been reported from the specialized habitat of tundra. Members of *Firmicutes*, Flavobacteriaceae and *Moraxellaceae* are found as dominant – species from Little Ank and Spitsbergen capable of growing between 4 °C to 30 °C temperatures.

Cold adapted heterotrophic and lipolytic microbial communities have been reported from the pristine and hydrocarbon contaminated soils of Western Antarctcia having wider pH range from 4.8 to 9.2 with organic matter less than 2%. Here α, β and γ-proteobaeteira were found to be dominating members (Labbe *et al.,* 2007). Apart from these, species of *Pseudomonas*, *Rhodocoans*, *Mycobacterium* and Acinetobacteria have also been found.

Ecology, Distribution and Physiology

The survival of an organism is influenced by climo-edaphic factors among which temperature has very strong influence both directly through its effect on water in liquid form and directly through its influence on the organic molecules within the living cells. Cold environments are colonized by vast variety of microorganisms including bacteria, filamentous fungi, algae, yeasts and archaea.

The challenges faced by the microbes in the cold habitats include reduced growth rates, reduced enzymatic reaction rates, limitations in nutrient availability, extreme conditions of pH, water stress, salinity, hydrostatic pressure and survival under different stress conditions, microorganisms have to evolve a complex range of structural and functional adaptations. Hence, a wide range of metabolic activities are reported through several studies even at sub-zero temperatures in different habitats of cold environments. The distribution of psychrophilic microorganisms in different ecosystems under cold environments are discuss below.

Atmosphere and Clouds

Bacteria have been reported to survive at altitudes of the atmosphere up to the stratosphere (dominated by gram positive bacteria) and mesosphere (41 to 47 km) where temperature up to -100 °C have been reported (Green, 2008; Pearce *et al.,* 2009; Wainwright *et al.,* 2003). Microbial life at this very low temperature is also affected by high UV radiation, very low moisture, oxidative stress, very low nutrients and desiccation. Depending upon the meteorological and seasonal conditions microbes are transported to short and very long distances through aerosols. A model developed for the estimation of bacterial mass transported in the global atmosphere has estimated to be 40 – 1800 Gg (1Gg = 109 g) (Burrows *et al.,* 2009). These microbial cells are further transported in to other different ecosystems after remaining suspended in the atmosphere for days or weeks through precipitation.

The metabolic activity of these microbes and their role in cloud formation through adherence with suspended particulate matters has gained attention of many scientists (Morris *et al.,* 2008). Clouds are indispensable component of atmosphere, and water vapours at higher altitude gets cooled and condensed into water droplets as well as ice crystals. The microbial cells in the cloud water find this as better habitat because the water droplets in the cloud can remain in liquid form even at very low temperature including below zero where these microbial cells can easily metabolize organic pollutants (Sattler *et al.,* 2001; Deguillaume *et al.,*2008). Several bacterial and fungal species have been reported from the atmospheric cloud water (up to10^3 to 10^5 per ml), the species of *Deinococcus aethius* and *Bacillus stratosphericus* have also been reported in the same environment (Ahern *et al.,* 2007; Amato *et al.,* 2005, 2007b). The microbial cells have the ability to play a key role in the formation of ice nuclei through cloud condensation and thus also in precipitation. The cloud condensation is activated at super saturated conditions (Bauer *et al.,* 2002). Cloud condensation efficiency is also enhanced through bio-surfactant activity by the glycoprotein and lipoproteins of the microbial origin (Ekstorm *et al.,* 2010). However, several new techniques have been employed to find out the ecological role of biological ice nucleators in the atmospheric process (Mohler *et al.,* 2007).

Snow

Approximately 35% of the Earth's land surface is covered with snow, predominantly in the Northern hemisphere. Seasonal fluctuation in temperature, oxygen availability, intensive solar radiation including UV irradiation are the key ecological characteristics of snow (Cockell and Cordoba-Jabonero, 2004; Jones 1999). Because of constant fluxes of dust, microbial cells and other biomaterials deposited along with precipitation, also its involvement in formation of glaciers have made snow a habitat just like atmosphere and hence have good impact on soils covered with snow (Hodson *et al.,* 2008; Pérez and Sommaruga, 2007 and Pearce 2011). The dominance of the algal growth as primary producer provides colour to the ice as this environment has intense sunlight with high moisture. (Hohum and Duval, 2001). The dominant bacterial species of *Hymenobacter* is reported in red snow (Fujii *et al.,* 2010). The bacterial abundance in snow cover is influenced by latitude, altitude and seasonal fluctuations. The numbers of bacteria vary between $10^3$and 10^5 per ml and correlate with concentration of Ca^{+2} ion that might serve as carrier against dust particles. The arctic region has higher abundance of microbial communities than antarctic in mountains but increases with altitude. (Carpenter *et al.,* 2000; Liu *et al.,* 2006). Significant diversity among heterotrophic bacteria, cyanobacteria and eukaryotes including psychrotolerant species are reported from this region (Ameto *et al.,* 2007b; Segawa *et al.,* 2005) and also from Tibetian glaciers (Liu *et al.,* 2009b). These microbes being photochemical reactors have been shown to be involved in active exchange of reactive nitrogen species with the atmosphere and thus are contributing significantly in biogeochemical cycling of elements at low temperatures. There is some correlation between the amount of snow cover and bacterial as well as fungal diversity in Alpine Tundra soils (Zinger *et al.,* 2009).

Cryoconite Holes

Cryoconites are specific water filled ice depressions (1m wide and 0.5m deep) with dark material usually found on the surface of snow free glaciers. These are formed by airborne debris which absorbs more solar radiation and melt down in to the ice. Generally these remains open to the atmosphere, however, in antarctica these are covered with ice lid and occasionally they remain connected to the subsurface run off system. Variety of microbial populations have been detected from polar and alpine cryoconite holes such as photosynthetic cyanobacteria and algae, bacteria, viruses, yeasts and metazoa showing greater adaptability to cold environments (Anesio *et al.,* 2007; Christner *et al.,* 2003a; Margesin *et al.,* 2002, Vincent,2007, Zakhia *et al.,*2008). The cold adapted novel species viz., *Pedobacter cryoconitis,* Sphingomonas *glacialis,* and *Mrakiella cryoconiti* have been isolated and reported from glacial habitats (Margesin *et al.,* 2003, Margesin and Fell 2008; Zhang *et al.,* 2011). These glacial mini ecosystems are found to be the 'hot spots' of microbial metabolic activity signifying their active role in the cling of nutrient elements through biogeochemical processes. The photosynthetic microbes are the principle drivers in the primary production which serves as food for the heterotrophic bacteria (Bagshaw *et al.,* 2007; Foreman *et al.,* 2007; Stibal *et al.,* 2007).

Glaciers

Glaciers are the hardest habitat for microorganisms to live as the temperature here remain extremely low up to -56 °C to -12 °C (Priscu *et al.,* 2007). Apart from extreme cold conditions, high pressure, scanty nutrients with low organic matter availability, and absence of solar radiation are other unfavourable conditions making the survival very difficult in these habitats for microorganisms. Despite of these stressful environmental conditions, glaciers are found to be large reservoir of microorganisms for thousands of years. Glaciers with different physical hydrological and geochemical characteristics are said to be the most versatile ecosystems in cryoconite holes and deep sea ice. (Hodson *et al.,* 2008). The diversity and abundance of microorganisms varies with the altitudes and depth of glaciers and ranges between $<10^2$ to 10^6-10^7 cells per ml.

The study of microbial diversity from glaciers is very challenging because it is difficult to drill in deep ice cores; very limited sample volumes with low microbial load are available for cultivation and DNA isolation. Intensive studies on polar and non-polar glaciers have shown vide variety of morphological, physiological and phylogenetic microbial diversity (Miteva, 2008). Major dominating bacterial phylogenetic group comprised *Acinetobacteria, Firmicutes, Proteobacteria* and also included yeasts and fungi, some plant and bacterial viruses and few archea (Simon *et al.,* 2009), with potential of degrading variety of organic compounds. The isolated microbial physiological groups had common characteristics such as ability to grow at very low temperature, with low nutrient availability, pigment production, modifying specific membrane structure and synthesis of cryoprotectants, exopolymers, cold active enzymes and ice binding proteins (Margesin *et al.,* 2005, 2007). The antifreeze proteins also known as ice binding proteins are responsible for inhibition of growth of ice-crystals and thus making the survival of microbes possible under freezing conditions (Christner, 2010 and Raymond *et al.,* 2008). The origin, abundance and composition of microbial communities and their pattern of diversity in the ice layers deposited chronologically under the influence of local and global climatic conditions and composition of glaciers have supported the identification of microbial markers that may help understand the climates of near past.

POLAR AND ALPINE LAKES

Polar and high altitude Alpine lakes represent the majority of cold lakes on Earth which show limnological diversity ranging from fresh water to hyper saline, from highly acidic to highly alkaline, highly aerobic to anaerobic, permafrost to ice free environments (Vincent *et al.,* 2008). Artic region comprises much greater number of lakes compared to Atarctica because of geograpbhy, nature of origin and the environmental conditions (Ryanzhin *et al.,* 2010).

Arctic Lakes

Naturally occurring Arctic lakes cover more than 80,000 sq. km. It is reported that the coastal arctic ecosystem shows microbial activities throughout the year during all seasons. These activities are occurring with different temperature options ranging from 12 °C to 20 °C indicating presence of different phgylogenetic groups (Adams *et al.,* 2010). Among various groups, members of cyanobacteria especially members of *Occilatoriales, Nostocales* and *Chroococales* play a significant and dominant role in high arctic lakes and streams (Bonilla *et al.,* 2005; Jungblut *et al.,* 2010; Tang *et al.,* 1997). As these organisms show tolerance to desiccation, can survive under high solar radiation, resist the freeze-thaw cycles and grow in wider range of temperature, have dominated to other organisms (Tang *et al.,* 1997).

Among the archaeal communities *Euryarchaeota* and *Crenarchaeota* are found to dominate. The *Euryarchaeota* dominate in the ice layers of moderate oxygen and show remarkable nitrification activities because of the presence of ammonia monooxygenase A gene variant. (Pouliot *et al.,* 2009). While *Crenarchaeota* have been recorded exclusively as free living inhabitants of marine waters.

The total area of ice shelves is lower than the antarctica. The main reason of lower ice shelves in Arctic region is global warming. Microbial activities have been reported from Canadian high arctic at lower temperature up to -10°C. *Bacteriodetes, Proteobacteria, Actinobacteria,* represents the dominant bacterial population. While *Euryarchaeota* represent the archaeal population.

Alpine Lakes

The Alpine mountains remain under 1.5 m to 3 m thick ice sheets and for a period that varies between six to nine months in a year. The ice cover that exist in this region is composed of alternating layers of ice and slush and constitute a sandwich like structure (Psenner *et al.,* 1999). This is considered to be the highest productive habitat with higher carbon turnover rates as the nutrients get accumulated through atmospheric deposition and from the catchment area of the lakes. (Waldhuber *et al.,* 2002). The Alpine lakes usually are exposed to high UV irradiation because of very thin layer of atmosphere, they contain very low nutrient elements and the transparency of water is very high (Rose *et al.,* 2009). The composition of microbial population is also affected by climatic and soil factors (Parez and Sommaruga, 2007). The biomass assemblage is also affected by physic-chemical structure of ice cover during its formation and growth (Felip *et al.,* 2002).

The slush layers contained up to 6 x 10^5 cells per ml, while in lake water and snow the same was reported to be 4 x 10^5 cells per ml, where *Alphaproteobacteria* were present in low numbers but *Betaproteobacteria* species dominated in all layers of slush and water (Alfreider *et al.,* 1996). The pelagic microbial communities in the high mountain lakes are recurrent annually, vertically stratified and seasonally variable (Pernthaler *et al.,* 1998). The members of betaproteobacteria have been reported to be dominant throughout the year and contribute to about 24% of total bacterial counts, while alphaproteobacteria constitute about 11% and are dominant during spring. Furthermore, rod shaped archaea have been reported to dominate during Autumn when ice starts getting covered (Pernthaler *et al.,* 1998).

Antarctic Lakes

The covered lakes are usually found in large numbers in Antarctica. There are more than 150 subglacial lakes in this area that are interconnected by subglacial rivers (Hodson *et al.,* 2008; Priscu and Foreman, 2009, Sattler and Storrie Lombardi, 2009). The most interesting subglacial lake is Vostok made of 200 m of accretion ice formed by refreezing on bottom of the 4 km deep ice sheet. The microbiological and biogeochemical study on this lake has revealed significant information of the lake environment which is extremely oligotrophic and highly oxygenated. A large diversity of prokaryotic microbial communities have been reported from very low number of pupulation (only 400 cells per ml) from where a thermophilic proteobacterium, *Hydrogenophilus thermoluteolus* was isolated (Bulat *et al.,* 2004; D'Elia *et al.,* 2008). The lakes of antarctica are permanently covered with ice which include both fresh water and highly saline lake environment and reported to have approximately seven times salinity than the sea water. The ice cover may reach height up to 3-6 metres. These lakes are considered to be oasis of life in a polar desert where microbial diversity and as large as 10^5 to 10^6 cells per ml are reported that include bacteria, archaea, eukaryote and viruses in the ice cover, lake water and sediments (Mosier *et al.,* 2007; Laybourn-Parry and Pearce 2007; Sawstrom *et al.,* 2008; Stingl *et al.,* 2008).

DEEP SEA AND ICE

On the Earth, about 70% of the total surface area covered by ocean and at least half of which is up to 3000 m deep. It is the lowest layer of ocean below thermocline. However, the normal temperature of this extreme environment remains around 2°–4°C, while the hydrothermal vents may have the highest temperature up to 370 °C but this happens very rare. There also occurs in two different zones viz., the abyssal zone between 3000 to 6000 m and the hadal zone with >6000 m, both are experiencing high hydrostatic pressure up to 110MPand are without any light. There is very low availability of nutrients but the oxygen is abundant (~4 ml DO per 1000 ml) at about 6000 m (Lauro and Bartlett, 2008; Nogi, 2008). The deep sea microbial communities may have developed several unique and unusual characteristics to adapt the environmental conditions (Abe, 2004; Deming, 2009b). Most of the isolates belong to psychropiezophiles having growth optima of > 0.1 MPa and cannot be cultivated at more than 20 °C temperature. The ocean microbial life is mainly comprised of bacteria, yeast archaebacteria, and protists. These are responsible for almost 98% of primary production (Whiteman *et al.,* 1998). The microbial abundance in deep sea water is low but the phylogenetic diversity is high which is underestimated through culture based studies (Sogin *et al.,* 2006). The predominating psychrophilic and piezophilic bacteria belong to gammaproteobacteria. Most of the culturable bacteria mainly belong to the genus *Shewanella, Colwellia, Moritella, Marinomonas, Psychromonas, Photobacterium*, and also comprised by some new species (Dang *et al.,* 2009;Nogi, 2008). These organisms' possess very high amount of saturated fatty acids in their cell membranes. These bacteria play a very vital role in biochemical cycling of nutrients in deep sea through characteristic hydrolytic enzymes (Kaneko *et al.,* 2007).

Chemoautotrophic ammonia oxidizing archaea (Crenoarchaeota) in deep sea at the depth of 2000–3000 m play an important role in the transformation of nitrogen. The presence of these archaea in high numbers and ability to carry metabolic reactions at 4°-10°C temperature implies that they might also play vital role in the nitrification process at various depths in the sea water. (Francis *et al.,* 2005; Nakagawa *et al.,* 2007; Kalanetra *et al.,* 2009). Only few members of yeast and fungal species have been reported from deep sea environments. The study of samples collected from the depth of about 15000 - 4000 m have recorded only 32 fungal 18S types (Ban *et al.,* 2007). Mainly the species belonging to *Basidiomycetes* and *Actinomycetes* have been reported from the depth up to 4000 m that belong to different genera such as *Candida, Cryptococcus, Rhodotorula,* and *Torulopsis* (Kutty and Philip, 2008).

The most extreme cold habitat is sea ice which is spread over 30 million sq km in polar oceans (Collins *et al.,* 2010). The temperature here ranges between 0°and -35°C, salinity of 35 -200 psu, high pH and low solar radiation. Abundant microbial communities have been reported in the upper and the lowest layers of sea ice (Bowman and Deming, 2010; Gosink *et al.,* 1993; Junge and Swanson, 2008 and Sullivan and Palmisano, 1984). The primary production in this condition is predominantly carried out by Diatoms.

TERRESTRIAL COLD ENVIRONMENTS

Arctic Soils

In the arctic soils, extreme temperatures, low precipitation, low nutrients low moisture and freeze-thaw cycles are the main challenging factors for microbial metabolism. The dominant microbial communities in these habitats are gram negative bacteria, usually the members of *alpha-*, *beta-* and *gamma- proteobacteria.* The arctic tundra soils contain about 60% of *Pseudomonas.* However, about 30% of the isolates belonged to unclassified bacteria which may play a significant and yet unknown ecological role (Gilichinsky *et al.,* 2008).

Bedrock material has significant effects on the composition of microbial communities while the altitude or vegetation has negligible effects (Mannisto *et al.,* 2007). On the contrary, Wallenstein *et al.,*

(2007), has shown that the bacterial and fungal communities are significantly influenced by the vegetation of the Alaskan soil. Tussock soils having contaminating recalcitrant compounds are reported to be dominated by *Acetobacteria* and *Ascomycota* while shrub soils contain high amounts of available carbon which favours the colonization of Proteobacteria and *Zygomycota.* The arctic bacteria in their cold habitats show higher activities of variety of extracellular enzymes viz., amylases, proteases, lipases and cellulases showing significant ecological role in these habitats (Mannisto and Haggblom, 2006). These also play an important role in the regulation of greenhouse gas methane which is liberated due to climatic warming and this is done by members of *Methanotrophs* (Wartiainen *et al.,* 2003).

Alpine Soils

The 'Alpine' is a region with high altitude and a continuous forests on high mountains. A large variation in temperature is experienced here while frost and ice remain for long time causing freeze-thaw cycles and a very high precipitation is also recorded. Therefore, in these soils a large seasonal variation in microbial communities is also reported (Lipson, 2007). With increase in altitude the environment becomes difficult to be inhabited and so the composition of microbial communities and their activities decrease. But there is significant increase in culturable psychrophilic heterotrophic bacteria and fungi. The culture independent approach has revealed that as the altitude increases the Himalayan soils show decrease in psychrophilic bacterial community with about 73% dominance of *Proteobacteria* and 31% dominance of *Betaproteobactreia.* Similar observations have been reported with culture dependent studies but here the dominant community is of *Gammaproteobacteria.* Studies have shown that microbial diversity have spatial patterns in specific geographical locations with dominance of endemic species (Papke *et al.,* 2003; Souza *et al.,* 2008; Whitaker *et al.,* 2003). However, these depends on how efficient is the analytical methods used for the same (Green *et al.,* 2008; Pearce and Galand 2008; Rammete and Tiedje, 2007).

The cold environments on the Earth, particularly polar regions create ideal ecosystems showing similar ecological characteristics which are geographically separated by climatic conditions (Staley and Gosink, 1999). It is well known that the microbial cells are carried by dust to both the poles, originate from different places among which Asian deserts are the main source. The 16S rRNA studies of soil samples from polar and non-polar glaciers have recorded species form genera viz., *Actinobacteria, Proteobacteria* and *Flavobacrteria* indicating that they all have similar survival mechanism (Chistener *et al.,* 2008b).

Lithic Communities

Microbial colonization are also found in porous translucent rocks which provide trapped moisture for microbial growth (Cockell *et al.,* 2003; Cockell and Stokes, 2004). Three different Lithic communities are classified on the basis of environmental niches where they reside viz., hypoliths, chasmoliths and cryptoendoliths. Hypoliths constitute photosynthetic microbial communities that reside in porous transluscent rocks (Broady 1981b, Thomas, 2005). Most of the green and red algae such as *Leptolyngbya* and *Phormidium* sp. are reported. Heterotrophic bacteria include alpha and gamma – *Proteobacteria, Arthrobacteria, Micrococcuss, Pseudomonas* are also reported (Smith *et al.,* 2000).

Chasmoendoliths mainly reside in cracks of weathering rocks caused by repeated freezing and thawing process in the ice free areas. Chasmoendolithic Communities are normally colonized by endolithic lichen associated with as cyaobacteria such as *Gleocapsa* species. (Nienow & Friedmann, 1993).

Cryptoendolithic organisms colonize in the interstices of crystalline rocks and the microbial colonization is dependent on the geological features of the rock subatrate (Friedmann and Ocampo 1976). Mainly two types of cryptoendolithic communities have been identified viz., lichen and cyanobacterium assemblages and these have clear zonal stratification. Heterotrophic microbes include gram positive cocci and several actinobacterial genera (Siebert& Hirsch 1988).

METABOLISM

Amato Pierre and Brent Christner (2009) have reported that the energy metabolism of psychrophiles under freezing temperature results in increase in the levels of ADP and ATP. The unfavourable conditions created at very low temperature are overcome through increased unsaturation of membrane lipids, increase in enzyme concentration and production of heat shock proteins. The metabolic state of the organism cn easily be evaluated by determining the concentration of ADP and ATP. Although, Proton Motive Force (PMF) plays a significant role in adapting freezing conditions.

The concentration of adenylic acid also have significant effect on metabolism of cold adapted microorganisms. Increased levels of adenylate maintain the optimum metabolic reaction rates by offsetting the decreased rate of enzymatic reactions. Adversely affected metabolic pathways are either repaired or maintained through nigh concentration of ATP and ADP. These are responsible for resumption of normal metabolism under favourable growth conditions.

Studies on physiology of cold adapted microbes have reported the metabolic activities at sub-zero temperatures of -10 °C to -196 °C (Amato *et al.,* 2009, Junge *et al.,* 2004, Mader *et al.,* 2006). Metabolic activities have been reported from solute rich liquid veins existing between grain crystals Price., (2000) and are not restricted to icy microenvironments.

Low temperatures induces the synthesis or uptake of compatible solutes which helps to neutralize the osmotic pressure and also maintain turgor pressure of cell under high salinity condition (Kempf *et al.,* 1998, Ko *et al.,* 1994). At subzero temperatures, the ice formation produces stress on microbial cells through mechanical disruption, osmotic imbalance and oxidative damage which is overcome by the cells by producing antifreeze proteins that prevent recrystallization of water in to ice crystals. To maintain the metabolic reaction within the cell, hydrolysis of ATP into ADP provides the required energy. Cellular ATP concentration is increased at decreasing temperature (Napolitano *et al.,* 2004, Napolitano *et al.,* 2005). Study on *E. coli* mutant have shown that increased ATP levels in the cells have provided tolerance to lower temperature of 0 °C for several days than the wild strain. It has been reported by (Napolitano *et al.,* 2004) that increasing adenylate concentration has positive effect on rate of diffusion between the substrate and the enzyme.

The experiments with *P. cryohalolantis* grown in the temperature range of 15 °C to -20 °C, showed that the concentration of ATP increased with the decrease in temperature. This is due to the cell inhibit ATP utilizing and at the same time stimulate ATP generating pathways at low energy change (Feniouk *et al.,* 2007). However, the ADP concentration was found to be 1.8 to 6.0 times higher than that of ATP and the total adenylate concentration in respiring cells compared to the cells without coupled respiration and was non-significant at freezing temperatures. This implies that for the production of ATP and ADP there may be existing a noncanonical pathway. However, Amato and Christner (2009) have reported that temperature has less effect on the production of ATP than its utilization and this may be the reason of accumulation of ATP in the cell. However, it is also possible that there is a indirect effect of temperature on enhancement in catabolic reactions and suppression of anabolic reactions in the microbial cells. This is also called as physiological regulation due to lower energy change form metabolism (Atkinson., 1968, Feniouk *et al.,* 2007). It is also shown that the higher ATP/ADP ratio in liquid conditions, the hydrolysis of ATP is favoured.

Psychrophilic organisms increase their adenylate concentration with decrease in temperature. This indicate that adenylate may have important role in tolerating cold temperatures. Bakermans *et al.,* (2007) showed that at low temperature (-4 °C), the concentrations of five different proteins that are involved in energy production, transport of products, gene expression as well as protein synthesis were significantly increased. ATP synthase has reported to comprise a submit which is identified as cold induced protein which implies that ATP synthase is highly required for ATP production at subzero temperatures. At freezing temperatures, the microbial cells are partitioned in the interstitial veins existing in the ice crystals and they have been shown to maintain their basal metabolic activity at such low temperatures (Christener., 2002, Junge *et al.,* 2006, Mader *et al.,* 2006, Miteva *et al.,* 2007, Price., 2000). High solute concentration in the liquid fraction of ice make the microbial cells to metabolize and survive under freezing temperatures as well as osmotic pressure (Chen, 1987).

The active biomes in the permanently frozen environments are comprising the significant pool of microorganisms that are capable of remaining metabolically active under frozen conditions (Priscu *et al.,* 2003, Priscu *et al.,* 2008). Bacterial communities have been reported to carry out oxidative phosphorylation at -80 °C temperature (Amato and Christner, 2009).

Cold Adaptation

The Earth's cryosphere with extreme environmental conditions making the life difficult to sustain still harbours much diverse, viable and metabolically active microbial populations belonging to different phylogenetic groups. The common striking feature among all he populations of these microorganisms is their capability to overcome the negative effects of very low temperature by adaptations which may be structural or functional (Gostincar, 2009; Siddiqui and Cavicchioli, 2006).

The ability to sense a very slight change in the temperature makes these microorganisms to adapt to the low temperatures and the very sensitive cold sensor is the cell membrane that behaves as an interface between the internal and external environments. The membrane becomes more rigid at cold temperatures and that activates the membrane associated sensor. The signal is transduced to the regulator to the response and the genes involved in modulating the membrane fluidity are up regulated. This again up-regulates many other genes that are involved in adaptation to the cold conditions (Shivaji and Prakash, 2010). Among the psychrophilic bacteria and archaea the genome sequencing and proteomic studies have shown that the protein synthesis stage is the main up regulating function for growth. While other functions such as RNA and protein folding, maintenance of fluidity in membranes, production and uptake of cryoprotective compounds, regulation of specific metabolic pathways and antioxidant activities are also up-regulated for adapting the organisms to cold conditions (Bowman, 2008; Kurihara and Esaki, 2008; Ng *et al.*, 2010 and Riley *et al.*, 2008).

The lower limit of psychrophiles for growth and reproduction is considered to be -12°C and for maintaining metabolic activity it is -20 °C (Bakermans, 2007). However, significant metabolic activities have also been recorded at subfreezing temperatures but the validation of such reports are awaited. To adopt the cold environment through membrane fluidity, microbes apply various strategies. The most frequently recoded change is the unsaturation of fatty acid composition as observed in most bacteria, fungi and algae. The fatty acid isomerization is other mode of modulation of membrane fluidity where methyl- branched fatty acid composition is increased in polar carotenoids and also the ratio of anteiso to iso-branched fatty acid is increased. The ratio of sterol to phospholipids determines the length of the fatty acid (Russel, 2008).

Cold Active Enzymes

Production of cold active enzymes having better catalytic activity than the mesophilic counterparts under low and moderate temperature is another strategy for adaptation to cold environments by the psychrophiles. These enzymes are heat labile and may get inactivated at temperatures which do not harm the mesophiles. However, the flexibility and dynamics of the active sites are maintained at low temperatures by the psychrophiles, whereas the meso and thermophilic microbes have been severely restricted the molecular motions (D' Amico, *et al.,* 2006; Feller, 2007).

Cold Shock and Heat Shock

A cold shock response is always encountered under sudden temperature changes but the response from psychrophiles is different than meso- and thermo-philes. There are two major mechanisms to overcome the cold shock by the psychrophiles. First, the house keeping proteins synthesis is not repressed and second, the number of cold shock proteins are that are higher and they increase with the severity of cold shock. When the temperature is further lowered, the steady state of these cold shock proteins is maintained by other set of proteins that is also known as cold acclimatization protein, probably produced constitutively.

Antifreeze Proteins and Cryoprotectants

Psychrophilic microorganisms produce several cryoprotective compounds to protect themselves against deleterious effects of ice crystal formation or the intercellular freezing due to extreme low temperatures of environments. AntiFreeze Proteins (AFP) are one of the most produced cryoprotective compounds which either modify the ice crystals or inhibit the growth of ice. This is achieved in two ways, firstly, AFP lower down the freezing point of water prior to freezing and thus altering the melting point and secondly, the AFPs show ice re-crystallization inhibition and cause growth inhibition of large crystals at sub-zero temperatures (Gilbert *et al.,*2004).

Bacteria in cold aquatic environments such as Antarctic or Arctic sea ice have been shown to produce large amounts of exopolymeric compounds (Krembs *et al.,* 2002). The main function of these exopolymeric substances is to facilitate adhesion of bacteria to the wet surfaces resulting in the formation of biofilm which helps to trap the nutrients, helps to protect cells under stress condition and it also mediate biochemical interactions.

Genomics and Proteomics

The complete genome sequenced during 2004-05 of *Desulfotalea psychrophyla* (Rabus *et al.,* 2004), *Colwelliapsychroerythraea*34H (Methe *et al.,* 2005) and *Pseudomonas haloplanktis* TAC125 (Medigue *et al.,* 2005) have revealed that there are several Cold shock proteins (Csps) and Cold acclimatization proteins (Caps) involved in unsaturated fatty acid synthesis. Apart from usual lipid desaturases, two gene clusters that are involved in membrane fluidity / rigidity through steroid degradation have been reported from *P. haloplanktis*. Also from *C. psychroerythraea* a fatty acid cis-trans isomerase, β-keto acyl CoA synthetases and some β-keto acyl carrier proteins have been identified that might enhance membrane fluidity depending upon their cold adaptive activity or their up regulated expression.

The proteomic approach has also been studied for cold adaptation at protein level. Both Csps and Caps have been reported to be expressed by *Arthrobacter globiformis*, a psychrophile (Berger *et al.*, 1996) and also from psychrotrophic species *Aeromonas hydrophila* (Imbert and Gancel, 2004). More than 30 over expressed proteins have been reported from *Bacillus pshychrosaccharolyticus* under psychrophilic environmental conditions. Furthermore, several other proteins involved in transcription, translation, energy metabolism as well as quality control of proteins have been identified from *Methanococcoides burtanii* (Goodchild *et al.*, 2004) at 4 °C temperature. Another enzyme prolyl-cis-trans isomerase was identified by Suzuki *et al.*, (2004) from *Shewanella* species strain SIB1.

Antioxidants

At low temperatures the Reactive Oxygen Species (ROS) rapidly solubilizes and causes significant cell damage. Protection against ROS is thus very important to survive under low temperature. The bacteria detoxify the ROS through production of several antioxidants enzymes such as catalase, superoxide dismutase, oxygen consuming lipid desaturase or they lack the ROS producing pathways (Medigue *et al.*, 2005; Methe *et al.*, 2005)

CONCLUSION

Large are of the Earth is experiencing cold climatic conditions where the average temperature remains between 0 °C and 15 °C. Particularly the Northern and Southern pole have temperature that remains permanently below zero. The microorganisms have adapted to every climatic conditions and have developed the capacity to grow, metabolize and reproduce under inhospitable conditions. The psychrophilic microorganisms are the naturally habitants of the cold environment and they have adapted to the seasonal temperature fluctuations and more particularly the freeze-thaw cycles which have led these microorganisms to evolve different adaptive mechanisms with regard to metabolic reactions, reproduction, survival and protection strategies. These adaptive techniques have enabled microbes to remain active at low and even sub-zero temperatures.

The research in the field of microbial ecology of the cold environment through culture dependent and culture independent molecular techniques has acquired great pace in the present time and thus has led to understand the diversity and ecological role of microbial populations in the cold ecosystems. The modern and emerging field of genomics and proteomics have added new insights into the psychrophilic and also psychrotrophic microbial life. The community structure of microbes has shown that a wide range of gram positive, gram negative bacteria, fungi, yeast, members of archaebacteria, and even cyanobacteria constitute a very large pool of microorganisms and may contain novel and yet to be cultured organisms.

The research in the field of astrobiology includes the diversity and ecology of microbes that permanently reside in the frozen environments such as permafrost and ice is increased in the recent time. This would also give us an insight about the possibility of microbial life in the cryogenic planets of our solar system.

The biogeographical study in microbial ecology would clear the distribution pattern of the psychrophiles and this would also reveal that some new species of the cold environment may be endemic to the particular environment.

REFERENCES

Abe, F. (2004). Piezophysiology of yeast: Occurrence and significance. *Cellular and Molecular Biology, 50*(4), 437–445. PMID:15529753

Adams, H. E., Crump, B. C., & Kling, G. W. (2010). Temperature controls on aquatic bacterial production and community dynamics in arctic lakes and streams. *Environmental Microbiology, 12*(5), 1319–1333. doi:10.1111/j.1462-2920.2010.02176.x PMID:20192972

Ahern, H. E., Walsh, K. A., Hill, T. C. J., & Moffett, B. F. (2007). Fluorescent pseudomonads isolated from Hebridean cloud and rain water produce biosurfactants but do not cause ice nucleation. *Biogeosciences, 4*(1), 115–124. doi:10.5194/bg-4-115-2007

Alfreider, A., Pernthaler, J., Amann, R., Sattler, B., Glockner, F., Wille, A., & Psenner, R. (1996). Community analysis of the bacterial assemblages in the winter cover and pelagic layers of a high mountain lake by in situ hybridization. *Applied and Environmental Microbiology, 62*(6), 2138–2144. doi:10.1128/aem.62.6.2138-2144.1996 PMID:16535341

Allan, R. N., Lebbe, L., Heyrman, J., De Vos, P., Buchanan, C. J., & Logan, N. A. (2005). Brevibacillus levickii sp. nov. and Aneurinibacillus terranovensis sp. nov., two novel thermoacidophiles isolated from geothermal soils of northern Victoria Land, Antarctica. *International Journal of Systematic and Evolutionary Microbiology, 55*(3), 1039–1050. doi:10.1099/ijs.0.63397-0 PMID:15879231

Amato, P., & Christner, B. C. (2009). Energy metabolism response to low-temperature and frozen conditions in Psychrobacter cryohalolentis. *Applied and Environmental Microbiology, 75*(3), 711–718. doi:10.1128/AEM.02193-08 PMID:19060163

Amato, P., Ménager, M., Sancelme, M., Laj, P., Mailhot, G., & Delort, A. M. (2005). Microbial population in cloud water at the Puy de Dôme: Implications for the chemistry of clouds. *Atmospheric Environment, 39*(22), 4143–4153. doi:10.1016/j.atmosenv.2005.04.002

Amato, P., Parazols, M., Sancelme, M., Mailhot, G., Laj, P., & Delort, A. M. (2007b). An important oceanic source of micro-organisms for cloud water at the Puy de Dôme (France). *Atmospheric Environment, 41*(37), 8253–8263. doi:10.1016/j.atmosenv.2007.06.022

Anesio, A. M., Mindl, B., Laybourn-Parry, J., Hodson, A. J., & Sattler, B. (2007). Viral dynamics in cryoconite holes on a high Arctic glacier (Svalbard). *Journal of Geophysical Research, 112*(G4), G04S31. Advance online publication. doi:10.1029/2006JG000350

Atkinson, D. E. (1968). Energy charge of the adenylate pool as a regulatory parameter. Interaction with feedback modifiers. *Biochemistry, 7*(11), 4030–4034. doi:10.1021/bi00851a033 PMID:4972613

Bagshaw, E. A., Tranter, M., Fountain, A. G., Welch, K. A., Basagic, H., & Lyons, W. B. (2007). Biogeochemical evolution of cryoconite holes on Canada glacier, Taylor Valley, Antarctica. *Journal of Geophysical Research. Biogeosciences*, *112*(G4).

Baharaeen, S., & Vishniac, H. S. (1982). Cryptococcus lupi sp. nov., an Antarctic basidioblastomycete. *International Journal of Systematic and Evolutionary Microbiology*, *32*(2), 229-232.

Bakermans, C., Tollaksen, S. L., Giometti, C. S., Wilkerson, C., Tiedje, J. M., & Thomashow, M. F. (2007). Proteomic analysis of Psychrobacter cryohalolentis K5 during growth at subzero temperatures. *Extremophiles*, *11*(2), 343–354. doi:10.100700792-006-0042-1 PMID:17123128

Ban, S., & Trites, A. W. (2007). Quantification of terrestrial haul-out and rookery characteristics of Steller sea lions. *Marine Mammal Science*, *23*(3), 496–507. doi:10.1111/j.1748-7692.2007.00130.x

Bauer, H., Kasper-Giebl, A., Löflund, M., Giebl, H., Hitzenberger, R., Zibuschka, F., & Puxbaum, H. (2002). The contribution of bacteria and fungal spores to the organic carbon content of cloud water, precipitation and aerosols. *Atmospheric Research*, *64*(1-4), 109–119. doi:10.1016/S0169-8095(02)00084-4

Berger, F., Morellet, N., Menu, F., & Potier, P. (1996). Cold shock and cold acclimation proteins in the psychrotrophic bacterium Arthrobacter globiformis SI55. *Journal of Bacteriology*, *178*(11), 2999–3007. doi:10.1128/jb.178.11.2999-3007.1996 PMID:8655472

Bonilla, S., Villeneuve, V., & Vincent, W. F. (2005). Benthic And Planktonic Algal Communities In A High Arctic Lake: Pigment Structure And Contrasting Responses To Nutrient Enrichment 1. *Journal of Phycology*, *41*(6), 1120–1130. doi:10.1111/j.1529-8817.2005.00154.x

Bowman, J. P. (2008). *Genomic analysis of psychrophilic prokaryotes.* In R. Margesin, F. Schinner, J. C. Marx, & C. Gerday (Eds.), *Psychrophiles: from biodiversity to biotechnology* (pp. 265–284). Springer. doi:10.1007/978-3-540-74335-4_16

Bowman, J. P., Cavanagh, J., Austin, J. J., & Sanderson, K. (1996). Novel Psychrobacter species from Antarctic ornithogenic soils. *International Journal of Systematic and Evolutionary Microbiology*, *46*(4), 841–848. PMID:8863407

Bowman, J. P., & Nichols, D. S. (2002). Aequorivita gen. nov., a member of the family Flavobacteriaceae isolated from terrestrial and marine Antarctic habitats. *International Journal of Systematic and Evolutionary Microbiology*, *52*(5), 1533–1541. PMID:12361255

Bowman, J. S., & Deming, J. W. (2010). Elevated bacterial abundance and exopolymers in saline frost flowers and implications for atmospheric chemistry and microbial dispersal. *Geophysical Research Letters*, *37*(13), n/a. doi:10.1029/2010GL043020

Bozal, N., Montes, M. J., Tudela, E., Jiménez, F., & Guinea, J. (2002). Shewanella frigidimarina and Shewanella livingstonensis sp. nov. isolated from Antarctic coastal areas. *International Journal of Systematic and Evolutionary Microbiology*, *52*(1), 195–205. doi:10.1099/00207713-52-1-195 PMID:11837303

Bozal, N., Tudela, E., Rossello-Mora, R., Lalucat, J., & Guinea, J. (1997). Pseudoalteromonas antarctica sp. nov., isolated from an Antarctic coastal environment. *International Journal of Systematic and Evolutionary Microbiology*, *47*(2), 345–351. PMID:9103620

Broady, P. A. (1981b). The ecology of sublithic terrestrial algae at the Vestfold Hills, Antarctica. *British Phycological Journal*, *16*(3), 231–240. doi:10.1080/00071618100650241

Bronwyn, Barnard, Tuffin, & Cowan. (2011). Ecological Distributopn of Microorgnaisms in Terrestrial, Psychrophilic Habitats. In Extremophiles Handbook. Springer. DOI doi:10.1007/978-4-431-53898-1_6.5

Bulat, S. A., Alekhina, I. A., Blot, M., Petit, J. R., Vasileva, L. P., de Angelis, M., Wagenbach, D., Lipenkov, V. Y., Vasilyeva, L. P., Wloch, D., Raynaud, D., & Lukin, V. V. (2004). DNA signature of thermophilic bacteria from the aged accretion ice of Lake Vostok, Antarctica: Implications for searching for life in extreme icy environments. *International Journal of Astrobiology*, *3*(1), 1–12. doi:10.1017/S1473550404001879

Burrows, S. M., Elbert, W., Lawrence, M. G., & Pöschl, U. (2009). Bacteria in the global atmosphere–Part 1: Review and synthesis of literature data for different ecosystems. *Atmospheric Chemistry and Physics*, *9*(23), 9263–9280. doi:10.5194/acp-9-9263-2009

Callaghan, R., Luk, F., & Bebawy, M. (2014). Inhibition of the multidrug resistance P-glycoprotein: Time for a change of strategy? *Drug Metabolism and Disposition: the Biological Fate of Chemicals*, *42*(4), 623–631. doi:10.1124/dmd.113.056176 PMID:24492893

Cameron, R. E. (1969). Cold desert characteristics and problems relevant to other arid lands. *Arid Lands in Perspective.*

Cameron, R. E. (1972). Microbial and ecologic investigations in Victoria valley, southern Victoria Land, Antarctica. *Antarctic Terrestrial Biology*, *20*, 195-260.

Carpenter, E. J., Lin, S., & Capone, D. G. (2000). Bacterial activity in South Pole snow. *Applied and Environmental Microbiology*, *66*(10), 4514–4517. doi:10.1128/AEM.66.10.4514-4517.2000 PMID:11010907

Chen, C. S. (1987). Relationship between water activity and freezing point depression of food systems. *Journal of Food Science*, *52*(2), 433–435. doi:10.1111/j.1365-2621.1987.tb06632.x

Christner, B. C. (2002). Incorporation of DNA and Protein Precursors into Macromolecules by Bacteria at– 15oC. *Applied and Environmental Microbiology*, *68*(12), 6435–6438. doi:10.1128/AEM.68.12.6435-6438.2002 PMID:12450874

Christner, B. C. (2010). Bioprospecting for microbial products that affect ice crystal formation and growth. *Applied Microbiology and Biotechnology*, *85*(3), 481–489. doi:10.100700253-009-2291-2 PMID:19841917

Christner, B. C., Kvitko, B. H. II, & Reeve, J. N. (2003a). Molecular identification of bacteria and eukarya inhabiting an Antarctic cryoconite hole. *Extremophiles*, *7*(3), 177–183. doi:10.100700792-002-0309-0 PMID:12768448

Christner, B. C., Skidmore, M. L., Priscu, J. C., Tranter, M., & Foreman, C. M. (2008b). Bacteria in subglacial environments. In *Psychrophiles: from biodiversity to biotechnology* (pp. 51–71). Springer. doi:10.1007/978-3-540-74335-4_4

Cockell, C., Rettberg, P., Horneck, G., Scherer, K., & Stokes, D. M. (2003). Measurements of microbial protection from ultraviolet radiation in polar terrestrial microhabitats. *Polar Biology*, *26*(1), 62–69. doi:10.100700300-002-0438-z

Cockell, C. S., & Córdoba-Jabonero, C. (2004). Coupling of climate change and biotic UV exposure through changing snow-ice covers in terrestrial habitats. *Photochemistry and Photobiology*, *79*(1), 26–31. PMID:14974712

Cockell, C. S., & Stokes, M. D. (2004). Widespread colonization by polar hypoliths. *Nature*, *431*(7007), 414–414. doi:10.1038/431414a PMID:15386002

Collins, R. E., Rocap, G., & Deming, J. W. (2010). Persistence of bacterial and archaeal communities in sea ice through an Arctic winter. *Environmental Microbiology*, *12*(7), 1828–1841. doi:10.1111/j.1462-2920.2010.02179.x PMID:20192970

Connell, L. B., Redman, R., Rodriguez, R., Barrett, A., Iszard, M., & Fonseca, A. (2010). Dioszegia antarctica sp. nov. and Dioszegia cryoxerica sp. nov., psychrophilic basidiomycetous yeasts from polar desert soils in Antarctica. *International Journal of Systematic and Evolutionary Microbiology*, *60*(6), 1466–1472. doi:10.1099/ijs.0.015412-0 PMID:19671723

Cowan, D. A., Russell, N. J., Mamais, A., & Sheppard, D. M. (2002). Antarctic Dry Valley mineral soils contain unexpectedly high levels of microbial biomass. *Extremophiles*, *6*(5), 431–436. doi:10.100700792-002-0276-5 PMID:12382121

D'Amico, S., Collins, T., Marx, J. C., Feller, G., Gerday, C., & Gerday, C. (2006). Psychrophilic microorganisms: Challenges for life. *EMBO Reports*, *7*(4), 385–389. doi:10.1038j.embor.7400662 PMID:16585939

D'Elia, T., Veerapaneni, R., & Rogers, S. O. (2008). Isolation of microbes from Lake Vostok accretion ice. *Applied and Environmental Microbiology*, *74*(15), 4962–4965. doi:10.1128/AEM.02501-07 PMID:18552196

Dang, H., Zhu, H., Wang, J., & Li, T. (2009). Extracellular hydrolytic enzyme screening of culturable heterotrophic bacteria from deep-sea sediments of the Southern Okinawa Trough. *World Journal of Microbiology & Biotechnology*, *25*(1), 71–79. doi:10.100711274-008-9865-5

Deguillaume, L., Leriche, M., Amato, P., Ariya, P. A., Delort, A.-M., Po¨schl, U., Chaumerliac, N., Bauer, H., Flossmann, A. I., & Morris, C. E. (2008). Microbiology and atmospheric processes: Chemical interactions of primary biological aerosols. *Biogeosciences*, *5*(4), 1073–1084. doi:10.5194/bg-5-1073-2008

Deming, J. W. (2009b). Extremophiles: Cold Environments. In M. Schaechter (Ed.), *Encyclopedia of Microbiology* (pp. 147–158). Elsevier. doi:10.1016/B978-012373944-5.00280-7

Ekström, S., Nozière, B., Hultberg, M., Alsberg, T., Magnér, J., Nilsson, E. D., & Artaxo, P. (2010). A possible role of ground-based microorganisms on cloud formation in the atmosphere. *Biogeosciences*, *7*(1), 387–394. doi:10.5194/bg-7-387-2010

Felip, M., Wille, A., Sattler, B., & Psenner, R. (2002). Microbial communities in the winter cover and the water column of an alpine lake: System connectivity and uncoupling. *Aquatic Microbial Ecology*, *29*(2), 123–134. doi:10.3354/ame029123

Feller, G. (2007). Life at low temperatures: Is disorder the driving force? *Extremophiles, 11*(2), 211–216. doi:10.100700792-006-0050-1 PMID:17160345

Feniouk, B. A., Suzuki, T., & Yoshida, M. (2007). Regulatory interplay between proton motive force, ADP, phosphate, and subunit ϵ in bacterial ATP synthase. *The Journal of Biological Chemistry, 282*(1), 764–772. doi:10.1074/jbc.M606321200 PMID:17092944

Foreman, C. M., Sattler, B., Mikucki, J. A., Porazinska, D. L., & Priscu, J. C. (2007). Metabolic activity and diversity of cryoconites in the Taylor Valley, Antarctica. *Journal of Geophysical Research. Biogeosciences, 112*(G4).

Francis, C. A., Roberts, K. J., Beman, J. M., Santoro, A. E., & Oakley, B. B. (2005). Ubiquity and diversity of ammonia-oxidizing archaea in water columns and sediments of the ocean. *Proceedings of the National Academy of Sciences of the United States of America, 102*(41), 14683–14688. doi:10.1073/pnas.0506625102 PMID:16186488

Friedmann, E. I., & Ocampo, R. (1976). Endolithic blue-green algae in the dry valleys: Primary producers in the Antarctic desert ecosystem. *Science, 193*(4259), 1247–1249. doi:10.1126cience.193.4259.1247 PMID:17837022

Fujii, M., Takano, Y., Kojima, H., Hoshino, T., Tanaka, R., & Fukui, M. (2010). Microbial community structure, pigment composition, and nitrogen source of red snow in Antarctica. *Microbial Ecology, 59*(3), 466–475. doi:10.100700248-009-9594-9 PMID:19847476

Gilbert, J. A., Hill, P. J., Dodd, C. E., & Laybourn-Parry, J. (2004). Demonstration of antifreeze protein activity in Antarctic lake bacteria. *Microbiology, 150*(1), 171–180. doi:10.1099/mic.0.26610-0 PMID:14702410

Gilichinsky, D., Vishnivetskaya, T., Petrova, M., Spirina, E., Mamykin, V., & Rivkina, E. (2008). Bacteria in permafrost. In *Psychrophiles: from biodiversity to biotechnology* (pp. 83–102). Springer. doi:10.1007/978-3-540-74335-4_6

Goodchild, A., Saunders, N. F., Ertan, H., Raftery, M., Guilhaus, M., Curmi, P. M., & Cavicchioli, R. (2004). A proteomic determination of cold adaptation in the Antarctic archaeon, Methanococcoides burtonii. *Molecular Microbiology, 53*(1), 309–321. doi:10.1111/j.1365-2958.2004.04130.x PMID:15225324

Gosink, J. J., Irgens, R. L., & Staley, J. T. (1993). Vertical distribution of bacteria in arctic sea ice. *FEMS Microbiology Ecology, 11*(2), 85–90. doi:10.1111/j.1574-6968.1993.tb05799.x

Gostinčar, C., Grube, M., De Hoog, S., Zalar, P., & Gunde-Cimerman, N. (2009). Extremotolerance in fungi: Evolution on the edge. *FEMS Microbiology Ecology, 71*(1), 2–11. doi:10.1111/j.1574-6941.2009.00794.x PMID:19878320

Green, J. L., Bohannan, B. J., & Whitaker, R. J. (2008). Microbial biogeography: from taxonomy to traits. *Science, 320*(5879), 1039-1043.

Hirsch, P., Mevs, U., Kroppenstedt, R. M., Schumann, P., & Stackebrandt, E. (2004). Cryptoendolithic actinomycetes from Antarctic sandstone rock samples: Micromonospora endolithica sp. nov. and two isolates related to Micromonospora coerulea Jensen 1932. *Systematic and Applied Microbiology, 27*(2), 166–174. doi:10.1078/072320204322881781 PMID:15046305

Hodson, A., Anesio, A. M., Tranter, M., Fountain, A., Osborn, M., Priscu, J., Laybourn-Parry, J., & Sattler, B. (2008). Glacial ecosystems. *Ecological Monographs, 78*(1), 41–67. doi:10.1890/07-0187.1

Hodson, A., Anesio, A. M., Tranter, M., Fountain, A., Osborn, M., Priscu, J., Laybourn-Parry, J., & Sattler, B. (2008). Glacial ecosystems. *Ecological Monographs, 78*(1), 41–67. doi:10.1890/07-0187.1

Hoham, R. W., & Duval, B. (2001). Microbial Ecology of Snow and Freshwater Ice with Emphasis on Snow Algae. In H. G. Jones, J. W. Pomeroy, D. A. Walker, & R. W. Hoham (Eds.), *Snow Ecology* (pp. 168–228). Cambridge University Press.

Jones, H. G. (1999). The ecology of snow-covered systems: A brief overview of nutrient cycling and life in the cold. *Hydrological Processes, 13*(14-15), 2135–2147. doi:10.1002/(SICI)1099-1085(199910)13:14/15<2135::AID-HYP862>3.0.CO;2-Y

Jungblut, A. D., Lovejoy, C., & Vincent, W. F. (2010). Global distribution of cyanobacterial ecotypes in the cold biosphere. *The ISME Journal, 4*(2), 191–202. doi:10.1038/ismej.2009.113 PMID:19890368

Junge, K., Eicken, H., & Deming, J. W. (2004). Bacterial activity at− 2 to− 20 C in Arctic wintertime sea ice. *Applied and Environmental Microbiology, 70*(1), 550–557. doi:10.1128/AEM.70.1.550-557.2004 PMID:14711687

Junge, K., Eicken, H., Swanson, B. D., & Deming, J. W. (2006). Bacterial incorporation of leucine into protein down to− 20 C with evidence for potential activity in sub-eutectic saline ice formations. *Cryobiology, 52*(3), 417–429. doi:10.1016/j.cryobiol.2006.03.002 PMID:16647051

Junge, K., & Swanson, B. D. (2008). High-resolution ice nucleation spectra of sea-ice bacteria: Implications for cloud formation and life in frozen environments. *Biogeosciences, 5*(3), 865–873. doi:10.5194/bg-5-865-2008

Kalanetra, K. M., Bano, N., & Hollibaugh, J. T. (2009). Ammonia-oxidizing Archaea in the Arctic Ocean and Antarctic coastal waters. *Environmental Microbiology, 11*(9), 2434–2445. doi:10.1111/j.1462-2920.2009.01974.x PMID:19601959

Kaneko, R., Hayashi, T., Tanahashi, M., & Naganuma, T. (2007). Phylogenetic diversity and distribution of dissimilatory sulfite reductase genes from deep-sea sediment cores. *Marine Biotechnology (New York, N.Y.), 9*(4), 429–436. doi:10.100710126-007-9003-7 PMID:17497195

Kempf, B., & Bremer, E. (1998). Uptake and synthesis of compatible solutes as microbial stress responses to high-osmolality environments. *Archives of Microbiology, 170*(5), 319–330. doi:10.1007002030050649 PMID:9818351

Ko, R., Smith, L. T., & Smith, G. M. (1994). Glycine betaine confers enhanced osmotolerance and cryotolerance on Listeria monocytogenes. *Journal of Bacteriology, 176*(2), 426–431. doi:10.1128/jb.176.2.426-431.1994 PMID:8288538

Krembs, C., Eicken, H., Junge, K., & Deming, J. W. (2002). High concentrations of exopolymeric substances in Arctic winter sea ice: Implications for the polar ocean carbon cycle and cryoprotection of diatoms. *Deep-sea Research. Part I, Oceanographic Research Papers*, *49*(12), 2163–2181. doi:10.1016/S0967-0637(02)00122-X

Kurihara, T., & Esaki, N. (2008). Proteomic studies of psychrophilic microorganisms. In *Psychrophiles: from biodiversity to biotechnology* (pp. 333–344). Springer. doi:10.1007/978-3-540-74335-4_19

Kutty, S. N., & Philip, R. (2008). Marine yeasts—A review. *Yeast (Chichester, England)*, *25*(7), 465–483. doi:10.1002/yea.1599 PMID:18615863

Labbé, D., Margesin, R., Schinner, F., Whyte, L. G., & Greer, C. W. (2007). Comparative phylogenetic analysis of microbial communities in pristine and hydrocarbon-contaminated Alpine soils. *FEMS Microbiology Ecology*, *59*(2), 466–475. doi:10.1111/j.1574-6941.2006.00250.x PMID:17313586

Lauro, F. M., & Bartlett, D. H. (2008). Prokaryotic lifestyles in deep sea habitats. *Extremophiles*, *12*(1), 15–25. doi:10.100700792-006-0059-5 PMID:17225926

Laybourn-Parry, J., Marshall, W. A., & Madan, N. J. (2007). Viral dynamics and patterns of lysogeny in saline Antarctic lakes. *Polar Biology*, *30*(3), 351–358. doi:10.100700300-006-0191-9

Laybourn-Parry, J., & Pearce, D. A. (2007). The biodiversity and ecology of Antarctic lakes: Models for evolution. *Philosophical Transactions of the Royal Society of London. Series B, Biological Sciences*, *362*(1488), 2273–2289. doi:10.1098/rstb.2006.1945 PMID:17553775

Le Roes-Hill, M., Rohland, J., Meyers, P. R., Cowan, D. A., & Burton, S. G. (2009). Streptomyces hypolithicus sp. nov., isolated from an Antarctic hypolith community. *International Journal of Systematic and Evolutionary Microbiology*, *59*(8), 2032–2035. doi:10.1099/ijs.0.007971-0 PMID:19567563

Lipson, D. A. (2007). Relationships between temperature responses and bacterial community structure along seasonal and altitudinal gradients. *FEMS Microbiology Ecology*, *59*(2), 418–427. doi:10.1111/j.1574-6941.2006.00240.x PMID:17328121

Liu, Y., Yao, T., Kang, S., Jiao, N., Zeng, Y., Shi, Y., ... Huang, S. (2006). Seasonal variation of snow microbial community structure in the East Rongbuk glacier, Mt. Everest. *Chinese Science Bulletin*, *51*(12), 1476–1486.

Liu, Y. Q., Yao, T. D., Jiao, N. Z., Kang, S. C., Xu, B. Q., Zeng, Y. H., Huang, S. J., & Liu, X. B. (2009b). Bacterial diversity in the snow over Tibetan Plateau Glaciers. *Extremophiles*, *13*(3), 411–423. doi:10.100700792-009-0227-5 PMID:19159068

Mader, H. M., Pettitt, M. E., Wadham, J. L., Wolff, E. W., & Parkes, R. J. (2006). Subsurface ice as a microbial habitat. *Geology*, *34*(3), 169–172. doi:10.1130/G22096.1

Männistö, M. K., & Häggblom, M. M. (2006). Characterization of psychrotolerant heterotrophic bacteria from Finnish Lapland. *Systematic and Applied Microbiology*, *29*(3), 229–243. doi:10.1016/j.syapm.2005.09.001 PMID:16564959

Männistö, M. K., Tiirola, M., & Häggblom, M. M. (2007). Bacterial communities in Arctic fjelds of Finnish Lapland are stable but highly pH-dependent. *FEMS Microbiology Ecology, 59*(2), 452–465. doi:10.1111/j.1574-6941.2006.00232.x PMID:17328122

Margesin, R., Dieplinger, H., Hofmann, J., Sarg, B., & Lindner, H. (2005). A cold-active extracellular metalloprotease from Pedobacter cryoconitis—Production and properties. *Research in Microbiology, 156*(4), 499–505. doi:10.1016/j.resmic.2004.12.008 PMID:15862448

Margesin, R., & Fell, J. W. (2008). Mrakiella cryoconiti gen. nov., sp. nov., a psychrophilic, anamorphic, basidiomycetous yeast from alpine and arctic habitats. *International Journal of Systematic and Evolutionary Microbiology, 58*(12), 2977–2982. doi:10.1099/ijs.0.2008/000836-0 PMID:19060092

Margesin, R., & Miteva, V. (2011). Diversity and ecology of psychrophilic microorganisms. *Research in Microbiology, 162*(3), 346–361. doi:10.1016/j.resmic.2010.12.004 PMID:21187146

Margesin, R., Neuner, G., & Storey, K. B. (2007). Cold-loving microbes, plants, and animals—Fundamental and applied aspects. *Naturwissenschaften, 94*(2), 77–99. doi:10.100700114-006-0162-6 PMID:17039344

Margesin, R., Spröer, C., Schumann, P., & Schinner, F. (2003). Pedobacter cryoconitis sp. nov., a facultative psychrophile from alpine glacier cryoconite. *International Journal of Systematic and Evolutionary Microbiology, 53*(5), 1291–1296. doi:10.1099/ijs.0.02436-0 PMID:13130009

Margesin, R., Zacke, G., & Schinner, F. (2002). Characterization of heterotrophic microorganisms in alpine glacier cryoconite. *Arctic, Antarctic, and Alpine Research, 34*(1), 88–93. doi:10.1080/15230430.2002.12003472

Medigue, C., Krin, E., Pascal, G., Barbe, V., Bernsel, A., Bertin, P. N., Cheung, F., Cruveiller, S., D'Amico, S., Duilio, A., Fang, G., Feller, G., Ho, C., Mangenot, S., Marino, G., Nilsson, J., Parrilli, E., Rocha, E. P. C., Rouy, Z., ... Danchin, A. (2005). Coping with cold: The genome of the versatile marine Antarctica bacterium Pseudoalteromonas haloplanktis TAC125. *Genome Research, 15*(10), 1325–1335. doi:10.1101/gr.4126905 PMID:16169927

Methe´, B. A., Nelson, K. E., Deming, J. W., Momen, B., Melamud, E., Zhang, X., Moult, J., Madupa, R., Nelson, W. C., Dodson, R. J., Brinkac, L. M., Daugherty, S. C., Durkin, A. S., DeBoy, R. T., Kolonay, J. F., Sullivan, S. A., Zhou, L., Davidsen, T. M., Wu, M., ... Fraser, C. M. (2005). The psychrophilic lifestyle as revealed by the genome sequence of Colwellia psychrerythraea 34H through genomic and proteomic analyses. *Proceedings of the National Academy of Sciences of the United States of America, 102*(31), 10913–10918. doi:10.1073/pnas.0504766102 PMID:16043709

Mevs, U., Stackebrandt, E., Schumann, P., Gallikowski, C. A., & Hirsch, P. (2000). Modestobacter multiseptatus gen. nov., sp. nov., a budding actinomycete from soils of the Asgard Range (Transantarctic Mountains). *International Journal of Systematic and Evolutionary Microbiology, 50*(1), 337–346. doi:10.1099/00207713-50-1-337 PMID:10826821

Miteva, V. (2008). Bacteria in snow and glacier ice. In *Psychrophiles: from biodiversity to biotechnology* (pp. 31–50). Springer. doi:10.1007/978-3-540-74335-4_3

Miteva, V., Sowers, T., & Brenchley, J. (2007). Production of N 2 O by ammonia oxidizing bacteria at subfreezing temperatures as a model for assessing the N 2 O anomalies in the Vostok ice core. *Geomicrobiology Journal, 24*(5), 451–459. doi:10.1080/01490450701437693

Möhler, O., DeMott, P. J., Vali, G., & Levin, Z. (2007). Microbiology and atmospheric processes: The role of biological particles in cloud physics. *Biogeosciences, 4*(6), 1059–1071. doi:10.5194/bg-4-1059-2007

Morris, C. E., Sands, D. C., Bardin, M., Jaenicke, R., Vogel, B., Leyronas, C., Ariya, P. A., & Psenner, R. (2008). Microbiology and atmospheric processes: An upcoming era of research on bio-meteorology. *Biogeosciences Discussions, 5*(1), 191–212.

Mosier, A. C., Murray, A. E., & Fritsen, C. H. (2007). Microbiota within the perennial ice cover of Lake Vida, Antarctica. *FEMS Microbiology Ecology, 59*(2), 274–288. doi:10.1111/j.1574-6941.2006.00220.x PMID:17092309

Nakagawa, T., Mori, K., Kato, C., Takahashi, R., & Tokuyama, T. (2007). Distribution of cold-adapted ammonia-oxidizing microorganisms in the deep-ocean of the northeastern Japan Sea. *Microbes and Environments, 22*(4), 365–372. doi:10.1264/jsme2.22.365

Napolitano, M. J., & Shain, D. H. (2004). Four kingdoms on glacier ice: Convergent energetic processes boost energy levels as temperatures fall. *Proceedings of the Royal Society of London. Series B, Biological Sciences, 271*(suppl_5), S273–S276. PMID:15503992

Napolitano, M. J., & Shain, D. H. (2005). Distinctions in adenylate metabolism among organisms inhabiting temperature extremes. *Extremophiles, 9*(2), 93–98. doi:10.100700792-004-0424-1 PMID:15490283

Ng, C., De Maere, M. Z., Williams, T. J., Lauro, F. M., Raftery, M., Gibson, J. A. E., Andrews-Pfannkoch, C., Lewis, M., Hoffman, J. M., Thomas, T., & Cavicchioli, R. (2010). Metaproteogenomic analysis of a dominant green sulfur bacterium from Ace Lake, Antarctica. *The ISME Journal, 4*(8), 1002–1019. doi:10.1038/ismej.2010.28 PMID:20237513

Nienow, J. A., & Friedmann, E. I. (1993) Terrestrial lithophytic (rock) communities. In Antarctic Microbiology. Wiley.

Nogi, Y. (2008). Bacteria in the deep sea: psychropiezophiles. In R. Margesin, F. Schinner, J. C. Marx, & C. Gerday (Eds.), *Psychrophiles: from Biodiversity to Biotechnology* (pp. 73–82). Springer. doi:10.1007/978-3-540-74335-4_5

Papke, R. T., Ramsing, N. B., Bateson, M. M., & Ward, D. M. (2003). Geographical isolation in hot spring cyanobacteria. *Environmental Microbiology, 5*(8), 650–659. doi:10.1046/j.1462-2920.2003.00460.x PMID:12871232

Pearce, D. A., Bridge, P. D., Hughes, K. A., Sattler, B., Psenner, R., & Russell, N. J. (2009). Microorganisms in the atmosphere over Antarctica. *FEMS Microbiology Ecology, 69*(2), 143–157. doi:10.1111/j.1574-6941.2009.00706.x PMID:19527292

Pearce, D. A., & Galand, P. E. (2008). *Microbial biodiversity and biogeography. In Polar lakes and rivers, limnology of Arctic and Antarctic aquatic ecosystems*. Oxford University Press.

Pearce, K. E. (2011). Convergence through mobile peer-to-peer file sharing in the Republic of Armenia. *International Journal of Communication*, *5*, 511–528.

Pérez, M. T., & Sommaruga, R. (2007). Interactive effects of solar radiation and dissolved organic matter on bacterial activity and community structure. *Environmental Microbiology*, *9*(9), 2200–2210. doi:10.1111/j.1462-2920.2007.01334.x PMID:17686018

Pernthaler, J., Glöckner, F. O., Unterholzner, S., Alfreider, A., Psenner, R., & Amann, R. (1998). Seasonal community and population dynamics of pelagic bacteria and archaea in a high mountain lake. *Applied and Environmental Microbiology*, *64*(11), 4299–4306. doi:10.1128/AEM.64.11.4299-4306.1998 PMID:9797280

Pointing, S. B., Chan, Y., Lacap, D. C., Lau, M. C., Jurgens, J. A., & Farrell, R. L. (2009). Highly specialized microbial diversity in hyper-arid polar desert. *Proceedings of the National Academy of Sciences of the United States of America*, *106*(47), 19964–19969. doi:10.1073/pnas.0908274106 PMID:19850879

Pouliot, J., Galand, P. E., Lovejoy, C., & Vincent, W. F. (2009). Vertical structure of archaeal communities and the distribution of ammonia monooxygenase A gene variants in two meromictic High Arctic lakes. *Environmental Microbiology*, *11*(3), 687–699. doi:10.1111/j.1462-2920.2008.01846.x PMID:19207564

Price, P. B. (2000). A habitat for psychrophiles in deep Antarctic ice. *Proceedings of the National Academy of Sciences of the United States of America*, *97*(3), 1247–1251. doi:10.1073/pnas.97.3.1247 PMID:10655516

Priscu, J. C., & Christner, B. C. (2003). Earth's icy biosphere. In Microbial diversity and bioprospecting. Washington, DC: ASM Press.

Priscu, J. C., Christner, B. C., Foreman, C. M., & Royston-Bishop, G. (2007). Biological material in ice cores. Encyclopedia of Quaternary Sciences, 2, 1156-1166.

Priscu, J. C., & Foreman, C. M. (2009). Article. In G. E. Likens (Ed.), Encyclopedia of Inland Waters. Lakes of Antarctica (Vol. 2, pp. 555–566). Elsevier Press. doi:10.1016/B978-012370626-3.00038-7

Priscu, J. C., Tulaczyk, S., Studinger, M., Kennicutt, M. C., Christner, B. C., & Foreman, C. M. (2008). *Antarctic subglacial water: origin, evolution and ecology. In Polar lakes and rivers: limnology of Arctic and Antarctic aquatic ecosystems.* Oxford University Press.

Psenner, R., Sattler, B., Wille, A., Fritsen, C. H., Priscu, J. C., Felip, M., & Catalan, J. (1999). Lake ice microbial communities in alpine and antarctic lakes. In R. Margeisn & F. Schinner (Eds.), *Cold-Adapted Organisms* (pp. 17–31). Springer. doi:10.1007/978-3-662-06285-2_2

Rabus, R., Ruepp, A., Frickey, T., Rattei, T., Fartmann, B., Stark, M., Bauer, A., Zibat, A., Lombardot, T., Becker, I., Amann, J., Gellner, K., Teeling, H., Leuschner, W. D., Glockner, F.-O., Lupas, A. N., Amann, R., & Klenk, H.-P. (2004). The genome of Desulfotalea psychrophila, a sulfate-reducing bacterium from permanently cold Arctic sediments. *Environmental Microbiology*, *6*(9), 887–902. doi:10.1111/j.1462-2920.2004.00665.x PMID:15305914

Rammette, A., & Tiedje, J. M. (2007). Biogeography: An emerging cornerstone for understanding prokaryotic diversity, ecology, and evolution. *Microbial Ecology*, *53*(2), 197–207. doi:10.100700248-005-5010-2 PMID:17106806

Raymond, J. A., Christner, B. C., & Schuster, S. C. (2008). A bacterial ice-binding protein from the Vostok ice core. *Extremophiles*, *12*(5), 713–717. doi:10.100700792-008-0178-2 PMID:18622572

Riley, M., Staley, J. T., Danchin, A., Wang, T. Z., Brettin, T. S., Hauser, L. J., Land, M. L., & Thompson, L. S. (2008). Genomics of an extreme psychrophile, Psychromonas ingrahamii. *BMC Genomics*, *9*(1), 1–19. doi:10.1186/1471-2164-9-210 PMID:18460197

Rose, K. C., Williamson, C. E., Saros, J. E., Sommaruga, R., & Fischer, J. M. (2009). Differences in UV transparency and thermal structure between alpine and subalpine lakes: Implications for organisms. *Photochemical & Photobiological Sciences*, *8*(9), 1244–1256. doi:10.1039/b905616e PMID:19707613

Russell, N. J. (2008). Membrane components and cold sensing. In *Psychrophiles: from biodiversity to biotechnology* (pp. 177–190). Springer. doi:10.1007/978-3-540-74335-4_11

Ryanzhin, S. V., Subetto, D. A., Kochkov, N. V., Akhmetova, N. S., & Veinmeister, N. V. (2010). Polar lakes of the World: Current data and status of investigations. *Water Resources*, *37*(4), 427–436. doi:10.1134/S0097807810040019

Sattler, B., Puxbaum, H., & Psenner, R. (2001). Bacterial growth in supercooled cloud droplets. *Geophysical Research Letters*, *28*(2), 239–242. doi:10.1029/2000GL011684

Sattler, B., & Storrie-Lombardi, M. C. (2009). LIFE in Antarctic lakes. In A. K. Bej, J. Aislabie, & R. M. Atlas (Eds.), *Polar Microbiology: The Ecology, Biodiversity, and Bioremediation Potential of Microorganisms in Extremely Cold Environments* (pp. 95–114). Taylor and Francis. doi:10.1201/9781420083880-c4

Säwström, C., Lisle, J., Anesio, A. M., Priscu, J. C., & Laybourn-Parry, J. (2008). Bacteriophage in polar inland waters. *Extremophiles*, *12*(2), 167–175. doi:10.100700792-007-0134-6 PMID:18188502

Schumann, P., Prauser, H., Rainey, F. A., Stackebrandt, E., & Hirsch, P. (1997). Friedmanniella antarctica gen. nov., sp. nov., an LL-diaminopimelic acid-containing actinomycete from Antarctic sandstone. *International Journal of Systematic and Evolutionary Microbiology*, *47*(2), 278–283. PMID:9103610

Segawa, T., Miyamoto, K., Ushida, K., Agata, K., Okada, N., & Kohshima, S. (2005). Seasonal change in bacterial flora and biomass in mountain snow from the Tateyama Mountains, Japan, analyzed by 16S rRNA gene sequencing and real-time PCR. *Applied and Environmental Microbiology*, *71*(1), 123–130. doi:10.1128/AEM.71.1.123-130.2005 PMID:15640179

Shivaji, S., & Prakash, J. S. (2010). How do bacteria sense and respond to low temperature? *Archives of Microbiology*, *192*(2), 85–95. doi:10.100700203-009-0539-y PMID:20049417

Shivaji, S., Rao, N. S., Saisree, L., Sheth, V., Reddy, G. S., & Bhargava, P. M. (1989). Isolation and identification of Pseudomonas spp. from Schirmacher Oasis, Antarctica. *Applied and Environmental Microbiology*, *55*(3), 767–770. doi:10.1128/aem.55.3.767-770.1989 PMID:2930174

Siddiqui, K. S., & Cavicchioli, R. (2006). Cold-adapted enzymes. *Annual Review of Biochemistry*, *75*(1), 403–433. doi:10.1146/annurev.biochem.75.103004.142723 PMID:16756497

Siebert, J., & Hirsch, P. (1988). Characterization of 15 selected coccal bacteria isolated from Antarctic rock and soil samples from the McMurdo-Dry Valleys (South-Victoria Land). *Polar Biology*, *9*(1), 37–44. doi:10.1007/BF00441762 PMID:11538335

Simon, C., Wiezer, A., Strittmatter, A. W., & Daniel, R. (2009). Phylogenetic diversity and metabolic potential revealed in a glacier ice metagenome. *Applied and Environmental Microbiology*, *75*(23), 7519–7526. doi:10.1128/AEM.00946-09 PMID:19801459

Smith, M. C., Bowman, J. P., Scott, F. J., & Line, M. A. (2000). Sublithic bacteria associated with Antarctic quartz stones. *Antarctic Science*, *12*(2), 177–184. doi:10.1017/S0954102000000237

Sogin, M. L., Morrison, H. G., Huber, J. A., Welch, D. M., Huse, S. M., Neal, P. R., Arrieta, J. M., & Herndl, G. J. (2006). Microbial diversity in the deep sea and the underexplored "rare biosphere". *Proceedings of the National Academy of Sciences of the United States of America*, *103*(32), 12115–12120. doi:10.1073/pnas.0605127103 PMID:16880384

Souza, V., Eguiarte, L. E., Siefert, J., & Elser, J. J. (2008). Microbial endemism: Does phosphorus limitation enhance speciation. *Nature Reviews. Microbiology*, *6*(7), 559–564. doi:10.1038/nrmicro1917 PMID:18521074

Staley, J. T., & Gosink, J. J. (1999). Poles apart: Biodiversity and biogeography of sea ice bacteria. *Annual Review of Microbiology*, *53*(1), 189–215. doi:10.1146/annurev.micro.53.1.189 PMID:10547690

Stibal, M., Elster, J., Sabacka, M., & Kastovska, K. (2007). Seasonal and diel changes in photosynthetic activity of the snow alga Chlamydomonas nivalis (Chlorophyceae) from Svalbard determined by pulse amplitude modulation fluorometry. *FEMS Microbiology Ecology*, *59*(2), 265–273. doi:10.1111/j.1574-6941.2006.00264.x PMID:17313577

Stingl, U., Cho, J. C., Foo, W., Vergin, K. L., Lanoil, B., & Giovannoni, S. J. (2008). Dilution-to-extinction culturing of psychrotolerant planktonic bacteria from permanently ice-covered lakes in the McMurdo Dry Valleys, Antarctica. *Microbial Ecology*, *55*(3), 395–405. doi:10.100700248-007-9284-4 PMID:17623231

Sullivan, C. W., & Palmisano, A. C. (1984). Sea ice microbial communities: Distribution, abundance, and diversity of ice bacteria in McMurdo Sound, Antarctica, in 1980. *Applied and Environmental Microbiology*, *47*(4), 788–795. doi:10.1128/aem.47.4.788-795.1984 PMID:16346517

Suzuki, K. I., Sasaki, J., Uramoto, M., Nakase, T., & Komagata, K. (1997). Cryobacterium psychrophilum gen. nov., sp. nov., nom. rev., comb. nov., an obligately psychrophilic actinomycete to accommodate "Curtobacterium psychrophilum" Inoue and Komagata 1976. *International Journal of Systematic and Evolutionary Microbiology*, *47*(2), 474–478. PMID:9103638

Suzuki, Y., Haruki, M., Takano, K., Morikawa, M., & Kanaya, S. (2004). Possible involvement of an FKBP family member protein from a psychrotrophic bacterium *Shewanella* sp. SIB1 in cold adaptation. *European Journal of Biochemistry*, *271*(7), 1372–1381. doi:10.1111/j.1432-1033.2004.04049.x PMID:15030488

Tang, E. P., Tremblay, R., & Vincent, W. F. (1997). Cyanobacterial Dominance Of Polar Freshwater Ecosystems: Are High-Latitude Mat-Formers Adapted To Low Temperature? 1. *Journal of Phycology*, *33*(2), 171–181. doi:10.1111/j.0022-3646.1997.00171.x

Thomas, D. N. (2005). Photosynthetic microbes in freezing deserts. *Trends in Microbiology, 13*(3), 87–88. doi:10.1016/j.tim.2004.11.002 PMID:15737723

Thomas-Hall, S., Watson, K., & Scorzetti, G. (2002). Cryptococcus statzelliae sp. nov. and three novel strains of Cryptococcus victoriae, yeasts isolated from Antarctic soils. *International Journal of Systematic and Evolutionary Microbiology, 52*(6), 2303–2308. PMID:12508901

Vincent, W. F. (2007). Cold tolerance in cyanobacteria and life in the cryosphere. In *Algae and cyanobacteria in extreme environments* (pp. 287–301). Springer. doi:10.1007/978-1-4020-6112-7_15

Vincent, W. F., Hobbie, J. E., & Laybourn-Parry, J. (2008). Introduction to the limnology of high-latitude lake and river ecosystems. Polar lakes and rivers: limnology of Arctic and Antarctic aquatic ecosystems, 1-24. doi:10.1093/acprof:oso/9780199213887.003.0001

Vishniac, H. S. (1985). Cryptococcus socialis sp. nov. and Cryptococcus consortionis sp. nov., Antarctic basidioblastomycetes. *International Journal of Systematic and Evolutionary Microbiology, 35*(1), 119–122. PMID:11540820

Vishniac, H. S., & Kurtzman, C. P. (1992). Cryptococcus antarcticus sp. nov. and Cryptococcus albidosimilis sp. nov., basidioblastomycetes from Antarctic soils. *International Journal of Systematic and Evolutionary Microbiology, 42*(4), 547–553.

Wainwright, M., Wickramasinghe, N. C., Narlikar, J. V., & Rajaratnam, P. (2003). Microorganisms cultured from stratospheric air samples obtained at 41 km. *FEMS Microbiology Letters, 218*(1), 161–165. doi:10.1111/j.1574-6968.2003.tb11513.x PMID:12583913

Waldhuber, S., Sattler, B., Semmler, J., Wille, A., & Psenner, R. (2002, November). Proceedings of the 2nd Eurpean Workshop on Exoas- trobiology. Comparative 16S rDNA analysis of extremophiles in the winter cover of a high mountain lake. In *Exo-Astrobiology* (Vol. 518, pp. 489-490). Academic Press.

Wallenstein, M. D., McMahon, S., & Schimel, J. (2007). Bacterial and fungal community structure in Arctic tundra tussock and shrub soils. *FEMS Microbiology Ecology, 59*(2), 428–435. doi:10.1111/j.1574-6941.2006.00260.x PMID:17313585

Wartiainen, I., Hestnes, A. G., & Svenning, M. M. (2003). Methanotrophic diversity in high arctic wetlands on the islands of Svalbard (Norway)-denaturing gradient gel electrophoresis analysis of soil DNA and enrichment cultures. *Canadian Journal of Microbiology, 49*(10), 602–612. doi:10.1139/w03-080 PMID:14663494

Whitaker, R. J., Grogan, D. W., & Taylor, J. W. (2003). Geographic barriers isolate endemic populations of hyperthermophilic archaea. *Science, 301*(5635), 976–978. doi:10.1126cience.1086909 PMID:12881573

Whitman, W. B., Coleman, D. C., & Wiebe, W. J. (1998). Prokaryotes: The unseen majority. *Proceedings of the National Academy of Sciences of the United States of America, 95*(12), 6578–6583. doi:10.1073/pnas.95.12.6578 PMID:9618454

Wicklow, D. T., & Soderstrom, B. E. (1997). *Environmental and microbial relationships.* Springer.

Wood, S. A., Rueckert, A., Cowan, D. A., & Cary, S. C. (2008). Sources of edaphic cyanobacterial diversity in the Dry Valleys of Eastern Antarctica. *The ISME Journal*, *2*(3), 308–320. doi:10.1038/ismej.2007.104 PMID:18239611

Yi, H., & Chun, J. (2006). Flavobacterium weaverense sp. nov. and Flavobacterium segetis sp. nov., novel psychrophiles isolated from the Antarctic. *International Journal of Systematic and Evolutionary Microbiology*, *56*(6), 1239–1244. doi:10.1099/ijs.0.64164-0 PMID:16738098

Yi, H., Oh, H. M., Lee, J. H., Kim, S. J., & Chun, J. (2005). Flavobacterium antarcticum sp. nov., a novel psychrotolerant bacterium isolated from the Antarctic. *International Journal of Systematic and Evolutionary Microbiology*, *55*(2), 637–641. doi:10.1099/ijs.0.63423-0 PMID:15774636

Yi, H., Yoon, H. I., & Chun, J. (2005). Sejongia antarctica gen. nov., sp. nov. and Sejongia jeonii sp. nov., isolated from the Antarctic. *International Journal of Systematic and Evolutionary Microbiology*, *55*(1), 409–416. doi:10.1099/ijs.0.63273-0 PMID:15653910

Yu, Y., Xin, Y. H., Liu, H. C., Chen, B., Sheng, J., Chi, Z. M., Zhou, P.-J., & Zhang, D. C. (2008). Sporosarcina antarctica sp. nov., a psychrophilic bacterium isolated from the Antarctic. *International Journal of Systematic and Evolutionary Microbiology*, *58*(9), 2114–2117. doi:10.1099/ijs.0.65838-0 PMID:18768614

Zakhia, F., Jungblut, A. D., Taton, A., Vincent, W. F., & Wilmotte, A. (2008). Cyanobacteria in cold ecosystems. In R. Margesin, F. Schinner, J. C. Marx, & C. Gerday (Eds.), *Psychrophiles: from biodiversity to biotechnology* (pp. 121–135). Springer. doi:10.1007/978-3-540-74335-4_8

Zhang, D. C., Busse, H. J., Liu, H. C., Zhou, Y. G., Schinner, F., & Margesin, R. (2011). Sphingomonas glacialis sp. nov., a psychrophilic bacterium isolated from alpine glacier cryoconite. *International Journal of Systematic and Evolutionary Microbiology*, *61*(3), 587–591. doi:10.1099/ijs.0.023135-0 PMID:20382784

Zinger, L., Shahnavaz, B., Baptist, F., Geremia, R. A., & Choler, P. (2009). Microbial diversity in alpine tundra soils correlates with snow cover dynamics. *The ISME Journal*, *3*(7), 850–859. doi:10.1038/ismej.2009.20 PMID:19322246

Section 2
Genomics of Extremophiles

Chapter 6
Biochemistry Behind Protein Adaptations in Extremophiles

Kailas V. Fuldeore
Serum Institute of India, India

Nafisa Patel
Naran Lala College of Professional and Applied Sciences, Navsari, India

Aradhana Hitesh bhai Bavarva
Naran Lala College of Professional and Applied Sciences, Navsari, India

Vrushali Ashok Wagh
Naran Lala College of Professional and Applied Sciences, Navsari, India

ABSTRACT

Extremophiles are the mortals that tolerate in the most limiting and aggravating conditions to life. Because of these fantastic ecological criticisms, extremophiles have substituted innumerable intriguing transformations to cell films, proteins, and extracellular metabolites. These stimulatingly regulated usual particles and frameworks as of now play parts in numerous biotechnological fields. Compounds from extremophilic microorganisms as a rule catalyse synthetic responses in non-standard conditions. Such conditions advance accumulation, precipitation, and denaturation, diminishing the movement of most non-extremophilic catalysts, regularly because of the shortfall of adequate hydration. Extremophilic catalysts can go after hydration by means of modifications particularly to their surface through more noteworthy surface charges and expanded sub-atomic movement. These assets have permitted few extremophilic compounds to work within the sight of non-fluid natural solvents, with potential for plan of valuable impetuses.

DOI: 10.4018/978-1-7998-9144-4.ch006

INTRODUCTION

Extremophiles are the microorganisms that are equipped for enduring and flourishing in conditions recently thought to be unwelcoming or unequipped for supporting life. A huge portion of the organic entities living conditions have adjusted such conditions, for example, pressing factor and temperature like remote ocean or the soluble pH and saline. Such conditions are extremely astonishing for organic entities are exceptionally specific with explicit protein variations, for example, chaperone frameworks or chemicals fit for working in the climate without denaturing. Because of such conditions these creatures are able to work below conditions in which mesophilic proteins may not. Proteins and catalysts disengaged from extremophiles are considered helpful for an assortment of uses, because of their incomparable properties to work in hostile conditions. Current utilization of diverse proteins sourced from extremophiles have as of now for assorted as atomic science reagents (Terpe, 2013) or as clothing cleansers (Ito et al., 1998). The biotechnological along with mechanical interest for constant compounds working in unforgiving functional conditions has flooded.

From the massive majority of the spots, we can disconnect the extremophiles and furthermore can be used as a part of bioremediation of contaminated climate (Zhuang et al., 2010). Biosphere contains numerous extremophilic microorganisms with catalysts fit for working in unfavourable conditions (Gomes & Steiner, 2004; Hough & Danson, 1999). Microorganisms which fill in outrageous conditions have been a significant wellspring of steady and important catalysts (Adams et al., 1995; DasSarma et al., 2010; Kaul & Asano, 2012). The compounds likewise called "extremozymes", play out analogous capacities as their non-outrageous partners, yet they can catalyse such responses in conditions which repress or denature the less outrageous structures. The immense mainstream of the proteins got from extremophiles shows poly extremophilicity, for instance steadiness and action in more than one outrageous condition, embracing high salt, basic pH, low temperature, and non-fluid medium (Bowers et al., 2009; Pire et al., 2004). According to a phylogenetic perspective, extremophiles have a place with the realm of Archaea, one of the three species of life, not withstanding the areas of Bacteria and eukaryotes. Archaea creates in very limits conditions like warmth, chilly, corrosive, base, saltiness, pressing factor, and radiation.

PROTEIN ADAPTATION MECHANISMS IN EXTREMOPHILES

Extremophiles have custom to alternate their enzymes in order that those continue to be functionally energetic in great situations at which an enzyme from non-extremophile may want to have different-sensible aggregated, prompted or denatured (Figure 1).

Radiation Adaptation by Extremophiles

Microorganisms that are profoundly impervious to undeniable degrees of ionizing and bright radiation are called radiophiles. Two kinds of radiations where organic entities' reactions fundamentally in (gamma) radiation and UV radiation - have been considered.

Ionizing radiation is dependable fundamentally for twofold abandoned breaks in the genome of creatures. Nonetheless, it has additionally been displayed to harm the proteins and lipids and instigate industrious oxidative pressure (Slade & Radman, 2011). Because of such a state of ionizing, radio safe living beings have fostered all, or a blend of, the accompanying methodologies: novel and versatile DNA fix components, cancer prevention agent and enzymatic safeguard frameworks, and a consolidated nucleoid.

Figure 1. Protein adaptations mechanisms in extremophiles
Source: Kumar et al., 2018

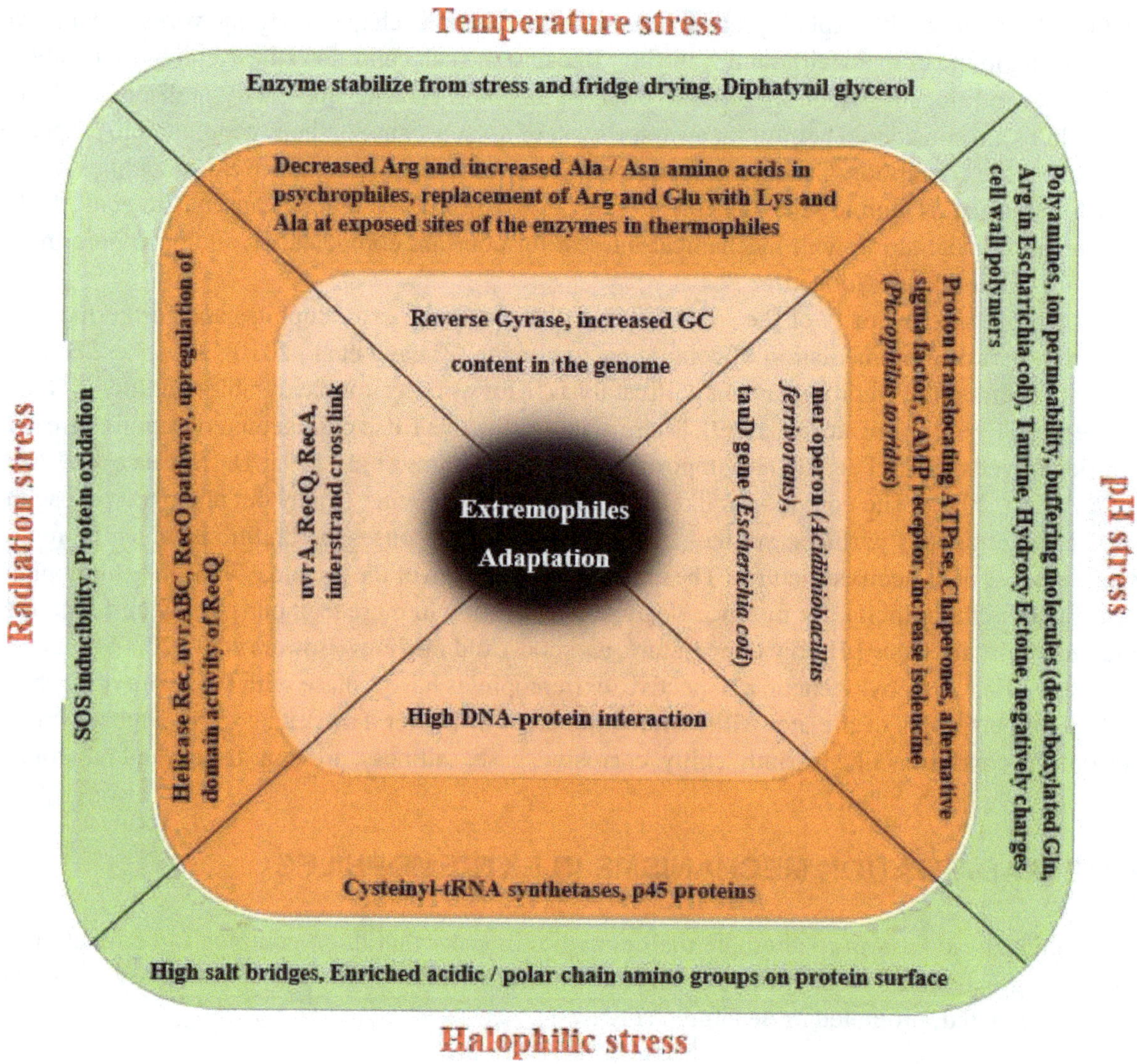

Quick and exact fix of genomes is fundamental in enduring dosages of ionizing radiation. This has been demonstrated to be refined using the nucleotide extraction fix pathway (uvrA1B), base extraction fix pathway (ung and mutY), and homologous recombination pathway (recA, ruvA, ddrA, and pprA) in *Deinococcus radiodurans* (Pavlopoulou et al., 2016) and single-abandoned restricting proteins in *Halobacterium* sp. NRC-1 (Rfa-like qualities) (DeVeaux et al., 2007) and *Deinococcus radiodurans* (DdrB and SSB) (Cox et al., 2010; Lockhart & DeVeaux, 2013).

In some particularly delicate species, protein harm causes demise before twofold abandoned breaks begin to shape. It has been proposed that, particularly for bacterial species, the job of protein harm in cell demise because of ionizing radiation is disparaged. For instance, *D. radiodurans* cells contain a few oxidative pressure counteraction and resilience components: cell cleaning via disposal of oxidized macromolecules, particular insurance of proteins against oxidative harm, and camouflage of reactive oxygen species creation. A condensed nucleoid has likewise been displayed to advance the effectiveness/precision of DNA fix (Minsky et al., 2006), besides to restrict the dissemination (Daly et al., 2007) of radiation-created DNA parts. Unrelated to ionizing radiation where DNA harm is essentially twofold abandoned breaks, UV radiation harms DNA in more unobtrusive manners through the development of cyclobutene pyrimidine dimers (that is, thymine dimers) and pyrimidine (6-4) pyrimidine photoproducts (that is, 6-4 photograph items). These two kinds of harm represent generally 80% of photolesions prompted by UV radiation (Jones & Baxter, 2017). Nonetheless, cyclobutene pyrimidine dimers are unquestionably more various and ordinarily dwarf pyrimidine (6-4) pyrimidine photoproducts. To fix these DNA sores, life forms utilize a mix of photoreactivation (phr) qualities, nucleotide extraction fix (uvr ABCD, xpf, and rad), base extraction fix (mutY and nth), and homologous recombination (recA and radA/51) (Jones & Baxter, 2017). Furthermore, organic entities have advanced a set-up of photoprotection gadgets to shield themselves from nonstop openness to UV radiation. These integrates carotenoids, superoxide dismutase and hydro peroxidases, quality duplication by means of polyploidy, and genome organization (that is, decrease in the number of bi pyrimidine sequences) (Jones & Baxter, 2017). Nonetheless, as in ionizing radiation, responsive oxygen species obstruction with ordinary metabolic cycles is a more common reason for cell demise (Quintana-Cabrera et al., 2012).

Temperature Adaptations by Extremophiles

Thermophiles are living beings fit for enduring temperatures outside the mesophilic range-in the scope of 40°C to 85°C and these conditions incorporate, aqueous vents or geothermally warmed mines, cave frameworks and warm underground aquifers. Various organically fundamental atoms like proteins, lipids or nucleic acids, are unprotected to embarrassment because of the outrageous warmth, therefore thermophilic organic entities have fostered various methodologies to redress. Catalysts sourced from thermophiles are generally esteemed for their thermotolerance, the most eminent model maybe is the DNA polymerase sourced from *Thermophilus aquaticus* usually alluded to as Taq polymerase, which empowered the course of PCR which is a basic supporting of atomic science (Saiki et al., 1988). There is a scope of additional thermostable chemicals that can detached a possibility of substrates from chitin, lignin, cellulose just as an assortment of protein, lipid and polysaccharides. These catalysts have effectively tracked down various modern uses in paper-production, feed handling and cleanser businesses (Elleuche et al., 2015). Various thermophilic organisms have shown the capacity to create novel mixtures, for example, the thermofiles from *Talaromyces thermophilus* with novel enemy of nematode movement (Elleuche et al., 2015). Actinobacteria from warm springs in Teng Chong, China was examined and found to have a different cluster of non-ribosomal peptide union (NRPS) and polyketide blend (PKS) potentials, proposing the chance of innovative mixtures being found from a thermophilic climate (Elleuche et al., 2015).

While warm vents and underground aquifers are viewed as the absolute most outrageous conditions on Earth, a few creatures can flourish in these threatening areas where most life would die. Among these are thermophiles and hyperthermophiles. These two offer comparative transformations to get by in these limits, they contrast in their temperature development ideal. Hyperthermophiles can become ideally up to 105°C, while thermophiles are delegated developing between 50°C and 70°C. At such outrageous temperatures, proteins coming up short on the vital transformations go through irreversible unfurling, uncovering the hydrophobic centres, which causes conglomeration (Tomazic & Klibanov, 1988). Thermophilic proteins have a few transformations that enable the protein to hold construction and capacity in limits of temperature. Probably the most noticeable are expanded number of huge hydrophobic deposits, disulphide bonds, and ionic associations (Champdore et al., 2007).

From the get go, psychrophile has frequently been characterized as a life form with an ideal temperature of under 15°C. In any case, this definition has various issues: it is discretionary, doesn't represent eukaryotes, and treats cells as simple warm units. Others have taken on the term's euro psychrophile and steno psychrophile to allude to a "wide" or "slender" scope of expansion at low temperatures. All things considered, some accept that the utilization of these terms doesn't "push" scientists enough to look for "valid" psychrophiles - those creatures ready to develop well under 0°C. Psychrophiles have been disengaged from an assortment of normal and man-made conditions (Champdore et al., 2007). Microorganisms are for the most part subject to the temperature of their current circumstance and as an outcome should discover approaches to adjust to the constraint set on them by temperature.

The catalysts from thermophiles show high potential in cleanser, food, feed, starch, material, cowhide, mash and paper and drug businesses. They have additionally been utilized as models for understanding the reasons conferring thermostability and thermoactivity, to give treasured prompts protein designing. A few three-dimensional constructions have been tackled and contrasted and those of mesophilic partners, with a definitive objective of clarifying the systems basic thermostability. In an extensive report, it was tracked down that expanded ionic cooperation and hydrogen bonds (surface charge), expanded protein centre hydrophobicity, and less uncovered thermolabile amino acids give soundness on the thermophilic proteins (Gupta et al., 2014).

Regardless, this definition has various issues: it is optional, doesn't address eukaryotes, and treats cells as basic warm units. Others have accepted the terms eurypsychrophile and steno psychrophile to suggest a "wide" or "confined" extent of improvement at low temperatures. Creatures are generally dependent upon the temperature of their present situation and as a result ought to find ways to deal with acclimate as far as possible put on them by temperature.

The mixtures from thermophiles display significant potential in cleaning agent, food, feed, starch, material, calfskin, squash paper and medication associations. They have similarly been used as models for understanding the reasons offering thermostability and thermoactivity, to give important prompts protein planning. A couple three-dimensional plans have been tended to and differentiated and those of mesophilic accomplices, with an authoritative target of clarifying the instruments essential thermostability. In a careful report, it was found that extended ionic cooperation and hydrogen bonds (surface charge), extended protein community hydrophobicity, and less revealed thermolabile amino acids give adequacy on the thermophilic proteins (Gupta et al., 2014).

Halophilic Adaptations by Extremophiles

Salt influences the dissolvability, steadfastness, and variation of a protein, which finally impacts its ability to work. Commonly halophilic microorganisms and eukaryotes, prevent the section of the inorganic salts like NaCl into the cell and integrate slight typical particles, known as osmolytes, to change the osmotic squeezing factor (Mevarech et al., 2000). Halophilic Archaea, not withstanding, make due by taking in high assemblies of inorganic salts, requiring their proteins to pass on varieties that license them to remain consistent and valuable. The lower openness of water can cause hydrophobic amino acids in a protein to lose hydration and aggregate. Thusly, high salt obsessions build up hydrophobic correspondences in a protein. Salt moreover interferes with the electrostatic associations between charged amino acids (Mevarech et al., 2000).

Bioinformatics examination of halophilic proteins has shown that their progressions also dependably contain less serine. Serine is at partner with water anyway not at matching charged particles, so serine is less significant for proteins at high salt obsessions (Zhang & Ge, 2013).

Little is contemplated the changes in their archaeal accomplices, but changes in the proportion of certain archaeal have all the earmarks of being huge. Various instruments of change are overall not remarkable, yet several reports recommend that changes are not a direct result of a specific get-together of characteristics/impetuses however rather are the outcome of an overall change in processing like how the salt-in halophiles have progressed an acidic proteome to keep proteins dissolvable and dynamic at high salinities instead of altering the surge a few qualities (Amrani et al., 2014; Vannier et al., 2015).

Halophiles exploit idiosyncratic transformation systems to make due in hyper saline territories. Halophilic archaea aggregate salts (NaCl or KCl) up to the fixations, isotonic with the climate, to keep an osmotic equilibrium. To adapt to such high salt focus, their proteins procure a generally huge number of contrarily charged amino corrosive deposits on their surface to forestall precipitation (Gupta et al., 2014).

This is because of the presence of negative charges on their surface that permits a higher surface hydration, so that forestalling high saline fixation causes protein total wonders. Acidophiles can inhabit very corrosive pH esteems, typically illegal for cell life. Furthermore, this class of organic entities can brace the film against the threatening external climate, by a biofilm to forestall the dissemination into the cell, for instance, or through the consolidation of unsaturated fats to ensure the cell. Other methodology is addressed by a functioning component dependent on a siphon that launches hydrogen particles out of the cell, assisting with keeping consistent the inward pH (D'Auria et al., 2000; D'Auria et al., 1999; D'Auria et al., 1998). Alkaliphiles occupy amazingly fundamental pH esteems (10-12). Like acidophiles, they have components of pH guideline to stay near impartiality.

Pressure Adaptations by Extremophiles

Microorganisms that prefer high-pressure conditions are termed piezophiles (barophiles). Piezophiles are living beings that live under amazingly high hydrostatic pressing factor frequently in different limits, similar to high or low temperature. Their regular environment is somewhere down in the sea, under outrageous tension, and in the outrageous warmth of aqueous vents or in the cold of the sea.

A few piezophiles have had the option to be refined regularly requiring complex frameworks to keep up with the pressing factors required utilizing gadgets, for example, water powered siphons and complex gas frameworks to keep up with pressures as high as 38 MPa (Pettit, 2011; Zhang et al., 2015). The utilization of particular gear to culture is hard for some labs to secure and work securely due to the pressures included; along these lines many investigations have gone to non-refined procedures, for example, genomics to study these specific organic entities.

Extremophile organic entities are regularly found in profound lakes, for example, Baikal (1.6 km) and Tanganyika (1.5 km), the sea, and subsurface networks (Rosenbaum et al., 2012). To have a thought of how much pressing factor is included, one should recollect that hydrostatic pressing factor increments generally 10.5 kPa per meter profundity while lithostatic (overburden). One of the way proteins can adapt to pressure is to shape multimeric proteins. The piezophile protein, TET3 peptidase (TET3) from *Pyrococcus horikoshii,* structures a prudent dodecamer, as opposed to a barrel-formed multimer, and shows expanded strength at high pressing factor (Rosenbaum et al., 2012).

Piezophiles do have numerous variations to fundamentally accomplish temperature limits- both thermophilic and psychrophilic (Hay et al., 2009), there are various variations which are remarkably adjusted for the high pressing factor conditions. These recollect thick hydrophobic centres for proteins as well as a penchant to frame multimeric proteins to manage the pressing factor, and these transformations have been as of late explored (Reed et al., 2013). This piezophiles have uniquely adjusted cell films to adjust to the pressing factor, for example, fusing polyunsaturated unsaturated fats (Usui et al., 2012), but the instruments behind this still can't seem to be clarified (Jebbar et al., 2015).

Piezophillic proteins could assume a part in high pressing factor cleansing of food sources, for certain catalysts being urgent to taste for example certain proteolytic proteins being significant in the aging of cheeses, and it is conceivable with additional examination piezophiles could fill this job (Rosenbaum et al., 2012).

pH Adaptations by Extremophiles

Acidophiles are organisms capable of flourishing in environmental conditions with less pH, such as hot springs, human made niches such as mines waste drainage systems (Simonato et al., 2006). It includes *A. ferrooxidans, A. thiooxidans.*

Acidophiles are often gathered with thermophiles, as most acidophilic conditions are likewise considered thermophilic (Baker-Austin & Dopson, 2007). One of the critical variations of acidophilic living beings is various transporter protein system situated on the cell membrane to help maintain cytosolic pH levels (Baker-Austin & Dopson, 2007). As far as applications, bioleaching is the cycle whereby microorganisms debase metal minerals to eliminate the metal particles into arrangement, where they can be reaped permitting beneficial metal extraction from low-grade minerals(Elleuche et al., 2014).

Acidophiles are specially esteemed for their capacity to bioleach a scope of metal particles including substantial metals typically poisonous for most life forms (Dopson et al., 2003). In terms of applications, bioleaching is the process whereby microorganisms degrade metal ores to remove the metal ions into solution, where they can be reaped permitting productive metal extraction from low grade minerals. Acidophiles are especially esteemed for their capacity to bioleach a scope of metal particles including substantial metals regularly poisonous for most life forms (Dopson et al., 2003). Because of the capability of acidophiles to lessen substantial metal particles into the climate, corrosive mine seepage frameworks can represent an issue for encompassing water frameworks. These corrosive mine waste frameworks regularly contain other perilous natural mixtures (like aliphatic mixtures) which can likewise be debased by acidophiles (Gemmell & Knowles, 2000) and it is conceivable using acidophiles that these conditions might be bioremediated (Rani et al., 2009).

Alkaliphiles are on the far edge of the pH range, enduring outrageous degrees of alkalinity, for example > pH 9. Alkaliphiles includes *Halorhodospira halochloris, Natronomonas pharaonis,* and *Thiohalospira alkaliphila.* Quite possibly the most widely recognized normally happening antacid environment are the soft drink lakes which can be emphatically basic (pH 10-12) just as profoundly saline, contingent upon the singular lake (Tindall et al., 1984). Alkaliphilic life forms customarily keep up with lower inside pH levels contrasted with the general climate by means of adjusted cell films and Na+/H+ carrier proteins, which assist with keeping up with the proton rationale power inside the cell (Krulwich,

1995). Alkaliphilic chemicals ordinarily have a wider scope of pH movement having the option to work under impartial and antacid pH conditions which has driven industry to use these catalysts for different purposes, for example, dehairing in conceal restoring (Horikoshi, 1999), just as more normal uses, for example, cleansers (Fujinami & Fujisawa, 2010). Soluble cellulases sourced from algophilic *Bacillus* sp. have effectively been utilized as cleanser added substances (Ito et al., 1998).

CONCLUSION

Extremophiles are discerning microorganisms, which have capability to flourish in scaring atmosphere together with several pressure conditions involving dangerous assortments of elements like temperature, pH, pressure, saltiness, most extreme radiation, slight groupings of water and supplements. Extremophiles have capacity to adjust their sub-atomic assortment and biochemical pathways, because of this they can persevere through such sorts of reformatory conditions. Extremophiles showed significant role to modify their enzymes therefore these microorganisms endure functionally vigorous in adversative conditions and opposite to this enzyme of non-extremophilic microorganisms showed aggregation, precipitation and denaturation. Relative genomics showed that extremophiles are capable to have the matchless arrangement of proteins and qualities that grant them through biochemical apparatus vital for prosper in interesting conditions. These proteins are going about as shields for extremophiles against a broad scope of energizing conditions for instance: radiation, temperature, pH, pressing factor, medications and synthetics. This multipurpose system of extremophiles assists with forming tricks to adjust the proteins and qualities which have huge helpful potential alongside business esteems.

REFERENCES

Adams, M. W., Perler, F. B., & Kelly, R. M. (1995). Extremozymes: Expanding the limits of biocatalysis. *Bio/technology, 13*(7), 662–668. PMID:9634802

Amrani, A., Bergon, A., Holota, H., Tamburini, C., Garel, M., Ollivier, B., Imbert, J., Dolla, A., & Pradel, N. (2014). Transcriptomics reveal several gene expression patterns in the piezophile *Desulfovibrio hydrothermalis* in response to hydrostatic pressure. *PLoS One, 9*(9), e106831. doi:10.1371/journal.pone.0106831 PMID:25215865

Baker-Austin, C., & Dopson, M. (2007). Life in acid: pH homeostasis in acidophiles. *Trends in Microbiology, 15*(4), 165–171. doi:10.1016/j.tim.2007.02.005 PMID:17331729

Bowers, K. J., Mesbah, N. M., & Wiegel, J. (2009). Biodiversity of poly-extremophilic Bacteria: Does combining the extremes of high salt, alkaline pH and elevated temperature approach a physico-chemical boundary for life? *Saline Systems, 5*(1), 1–8. doi:10.1186/1746-1448-5-9 PMID:19930649

Champdore, M. D., Staiano, M., Rossi, M., & D'Auria, S. (2007). Proteins from extremophiles as stable tools for advanced biotechnological applications of high social interest. *Journal of the Royal Society, Interface, 4*(13), 183–191. doi:10.1098/rsif.2006.0174 PMID:17251151

Cox, M. M., Keck, J. L., & Battista, J. R. (2010). Rising from the Ashes: DNA Repair in *Deinococcus radiodurans*. *PLOS Genetics*, *6*(1), e1000815. doi:10.1371/journal.pgen.1000815 PMID:20084105

D'Auria, S., Di Cesare, N., Gryczynski, Z., Gryczynski, I., Rossi, M., & Lakowicz, J. R. (2000). A thermophilic apoglucose dehydrogenase as nonconsuming glucose sensor. *Biochemical and Biophysical Research Communications*, *274*(3), 727–731. doi:10.1006/bbrc.2000.3172 PMID:10924344

D'Auria, S., Herman, P., Rossi, M., & Lakowicz, J. R. (1999). The fluorescence emission of the apoglucose oxidase from Aspergillus niger as probe to estimate glucose concentrations. *Biochemical and Biophysical Research Communications*, *263*(2), 550–553. doi:10.1006/bbrc.1999.1330 PMID:10491329

D'Auria, S., Moracci, M., Febbraio, F., Tanfani, F., Nucci, R., & Rossi, M. (1998). Structure-function studies on β-glycosidase from *Sulfolobus solfataricus*. Molecular bases of thermostability. *Biochimie*, *80*(11), 949–957. doi:10.1016/S0300-9084(00)88892-6 PMID:9893955

Daly, M. J., Gaidamakova, E. K., Matrosova, V. Y., Vasilenko, A., Zhai, M., Leapman, R. D., Lai, B., Ravel, B., Li, S.-M. W., Kemner, K. M., & Fredrickson, J. K. (2007). Protein oxidation implicated as the primary determinant of bacterial radioresistance. *PLoS Biology*, *5*(4), e92. doi:10.1371/journal.pbio.0050092 PMID:17373858

DasSarma, P., Coker, J. A., Huse, V., & DasSarma, S. (2010). Halophiles, biotechnology. Encyclopedia of industrial biotechnology, bioprocess, bioseparation, and cell technology, 2769-2777.

DeVeaux, L. C., Müller, J. A., Smith, J., Petrisko, J., Wells, D. P., & DasSarma, S. (2007). Extremely radiation-resistant mutants of a halophilic archaeon with increased single-stranded DNA-binding protein (RPA) gene expression. *Radiation Research*, *168*(4), 507–514. doi:10.1667/RR0935.1 PMID:17903038

Dopson, M., Baker-Austin, C., Koppineedi, P. R., & Bond, P. L. (2003). Growth in sulfidic mineral environments: Metal resistance mechanisms in acidophilic micro-organisms. *Microbiology*, *149*(8), 1959–1970. doi:10.1099/mic.0.26296-0 PMID:12904536

Elleuche, S., Schäfers, C., Blank, S., Schröder, C., & Antranikian, G. (2015). Exploration of extremophiles for high temperature biotechnological processes. *Current Opinion in Microbiology*, *25*, 113–119. doi:10.1016/j.mib.2015.05.011 PMID:26066287

Elleuche, S., Schroeder, C., Sahm, K., & Antranikian, G. (2014). Extremozymes—Biocatalysts with unique properties from extremophilic microorganisms. *Current Opinion in Biotechnology*, *29*, 116–123. doi:10.1016/j.copbio.2014.04.003 PMID:24780224

Fujinami, S., & Fujisawa, M. (2010). Industrial applications of alkaliphiles and their enzymes–past, present and future. *Environmental Technology*, *31*(8-9), 845–856. doi:10.1080/09593331003762807 PMID:20662376

Gemmell, R. T., & Knowles, C. J. (2000). Utilisation of aliphatic compounds by acidophilic heterotrophic bacteria. The potential for bioremediation of acidic wastewaters contaminated with toxic organic compounds and heavy metals. *FEMS Microbiology Letters*, *192*(2), 185–190. doi:10.1111/j.1574-6968.2000.tb09380.x PMID:11064193

Gomes, J., & Steiner, W. (2004). The biocatalytic potential of extremophiles and extremozymes. *Food Technology and Biotechnology*, *42*(4), 223–225.

Gupta, G. N., Srivastava, S., Khare, S. K., & Prakash, V. (2014). Extremophiles: An overview of micro-organism from extreme environment. *International Journal of Agriculture Environment and Biotechnology*, *7*(2), 371–380. doi:10.5958/2230-732X.2014.00258.7

Hay, S., Evans, R. M., Levy, C., Loveridge, E. J., Wang, X., Leys, D., Allemann, R. K., & Scrutton, N. S. (2009). Are the catalytic properties of enzymes from piezophilic organisms pressure adapted? *ChemBioChem*, *10*(14), 2348–2353. doi:10.1002/cbic.200900367 PMID:19681091

Horikoshi, K. (1999). Alkaliphiles: Some applications of their products for biotechnology. *Microbiology and Molecular Biology Reviews*, *63*(4), 735–750. doi:10.1128/MMBR.63.4.735-750.1999 PMID:10585964

Hough, D. W., & Danson, M. J. (1999). Extremozymes. *Current Opinion in Chemical Biology*, *3*(1), 39–46. doi:10.1016/S1367-5931(99)80008-8 PMID:10021406

Ito, S., Kobayashi, T., Ara, K., Ozaki, K., Kawai, S., & Hatada, Y. (1998). Alkaline detergent enzymes from alkaliphiles: Enzymatic properties, genetics, and structures. *Extremophiles*, *2*(3), 185–190. doi:10.1007007920050059 PMID:9783164

Jebbar, M., Franzetti, B., Girard, E., & Oger, P. (2015). Microbial diversity and adaptation to high hydrostatic pressure in deep-sea hydrothermal vents prokaryotes. *Extremophiles*, *19*(4), 721–740. doi:10.100700792-015-0760-3 PMID:26101015

Jones, D. L., & Baxter, B. K. (2017). DNA repair and photoprotection: Mechanisms of overcoming environmental ultraviolet radiation exposure in halophilic archaea. *Frontiers in Microbiology*, *8*, 1882. doi:10.3389/fmicb.2017.01882 PMID:29033920

Karan, R., Capes, M. D., & DasSarma, S. (2012). Function and biotechnology of extremophilic enzymes in low water activity. *Aquatic Biosystems*, *8*(1), 1–15. doi:10.1186/2046-9063-8-4 PMID:22480329

Kaul, P., & Asano, Y. (2012). Strategies for discovery and improvement of enzyme function: State of the art and opportunities. *Microbial Biotechnology*, *5*(1), 18–33. doi:10.1111/j.1751-7915.2011.00280.x PMID:21883976

Kletzin, A. (2007). General characteristics and important model organisms. *Archaea: molecular and cellular biology*, 14-92.

Krulwich, T. A. (1995). Alkaliphiles: Basicmolecular problems of pH tolerance and bioenergetics. *Molecular Microbiology*, *15*(3), 403–410. doi:10.1111/j.1365-2958.1995.tb02253.x PMID:7783613

Lockhart, J. S., & DeVeaux, L. C. (2013). The essential role of the Deinococcus radiodurans ssb gene in cell survival and radiation tolerance. *PLoS One*, *8*(8), e71651. doi:10.1371/journal.pone.0071651 PMID:23951213

Mevarech, M., Frolow, F., & Gloss, L. M. (2000). Halophilic enzymes: Proteins with a grain of salt. *Biophysical Chemistry*, *86*(2-3), 155–164. doi:10.1016/S0301-4622(00)00126-5 PMID:11026680

Minsky, A., Shimoni, E., & Englander, J. (2006). Ring-like nucleoids and DNA repair through error-free nonhomologous end joining in Deinococcus radiodurans. *Journal of Bacteriology*, *188*(17), 6047–6051. doi:10.1128/JB.01951-05 PMID:16923869

Pavlopoulou, A., Savva, G. D., Louka, M., Bagos, P. G., Vorgias, C. E., Michalopoulos, I., & Georgakilas, A. G. (2016). Unraveling the mechanisms of extreme radioresistance in prokaryotes: Lessons from nature. *Mutation Research/Reviews in Mutation Research*, *767*, 92–107. doi:10.1016/j.mrrev.2015.10.001 PMID:27036069

Pettit, R. K. (2011). Culturability and secondary metabolite diversity of extreme microbes: Expanding contribution of deep sea and deep-sea vent microbes to natural product discovery. *Marine Biotechnology (New York, N.Y.)*, *13*(1), 1–11. doi:10.100710126-010-9294-y PMID:20437069

Pire, C., Marhuenda-egea, F. C., Esclapez, J., Alcaraz, L., Ferrer, J., & José Bonete, M. (2004). Stability and enzymatic studies of glucose dehydrogenase from the archaeon Haloferax mediterranei in reverse micelles. *Biocatalysis and Biotransformation*, *22*(1), 17–23. doi:10.1080/10242420310001643179

Quintana-Cabrera, R., Fernandez-Fernandez, S., Bobo-Jimenez, V., Escobar, J., Sastre, J., Almeida, A., & Bolaños, J. P. (2012). γ-Glutamylcysteine detoxifies reactive oxygen species by acting as glutathione peroxidase-1 cofactor. *Nature Communications*, *3*(1), 1–8. doi:10.1038/ncomms1722 PMID:22395609

Rani, A., Souche, Y. S., & Goel, R. (2009). Comparative assessment of in situ bioremediation potential of cadmium resistant acidophilic Pseudomonas putida 62BN and alkalophilic Pseudomonas monteilli 97AN strains on soybean. *International Biodeterioration & Biodegradation*, *63*(1), 62–66. doi:10.1016/j.ibiod.2008.07.002

Reed, C. J., Lewis, H., Trejo, E., Winston, V., & Evilia, C. (2013). Protein adaptations in archaeal extremophiles. *Archaea*. PMID:24151449

Rosenbaum, E., Gabel, F., Durá, M. A., Finet, S., Cléry-Barraud, C., Masson, P., & Franzetti, B. (2012). Effects of hydrostatic pressure on the quaternary structure and enzymatic activity of a large peptidase complex from *Pyrococcus horikoshii. Archives of Biochemistry and Biophysics*, *517*(2), 104–110. doi:10.1016/j.abb.2011.07.017 PMID:21896270

Saiki, R. K., Gelfand, D. H., Stoffel, S., Scharf, S. J., Higuchi, R., Horn, G. T., Mullis, K. B., & Erlich, H. A. (1988). Primer-directed enzymatic amplification of DNA with a thermostable DNA polymerase. *Science*, *239*(4839), 487–491. doi:10.1126cience.239.4839.487 PMID:2448875

Schleper, C., Jurgens, G., & Jonuscheit, M. (2005). Genomic studies of uncultivated archaea. *Nature Reviews. Microbiology*, *3*(6), 479–488. doi:10.1038/nrmicro1159 PMID:15931166

Simonato, F., Campanaro, S., Lauro, F. M., Vezzi, A., D'Angelo, M., Vitulo, N., Valle, G., & Bartlett, D. H. (2006). Piezophilic adaptation: A genomic point of view. *Journal of Biotechnology*, *126*(1), 11–25. doi:10.1016/j.jbiotec.2006.03.038 PMID:16780980

Slade, D., & Radman, M. (2011). Oxidative stress resistance in Deinococcus radiodurans. *Microbiology and Molecular Biology Reviews*, *75*(1), 133–191. doi:10.1128/MMBR.00015-10 PMID:21372322

Terpe, K. (2013). Overview of thermostable DNA polymerases for classical PCR applications: From molecular and biochemical fundamentals to commercial systems. *Applied Microbiology and Biotechnology, 97*(24), 10243–10254. doi:10.100700253-013-5290-2 PMID:24177730

Tindall, B. J., Ross, H. N. M., & Grant, W. D. (1984). Natronobacterium gen. nov. and Natronococcus gen. nov., two new genera of haloalkaliphilic archaebacteria. *Systematic and Applied Microbiology, 5*(1), 41–57. doi:10.1016/S0723-2020(84)80050-8

Tomazic, S. J., & Klibanov, A. M. (1988). Mechanisms of irreversible thermal inactivation of Bacillus alpha-amylases. *The Journal of Biological Chemistry, 263*(7), 3086–3091. doi:10.1016/S0021-9258(18)69038-6 PMID:3257756

Usui, K., Hiraki, T., Kawamoto, J., Kurihara, T., Nogi, Y., Kato, C., & Abe, F. (2012). Eicosapentaenoic acid plays a role in stabilizing dynamic membrane structure in the deep-sea piezophile Shewanella violacea: A study employing high-pressure time-resolved fluorescence anisotropy measurement. *Biochimica et Biophysica Acta (BBA)- Biomembranes, 1818*(3), 574–583. doi:10.1016/j.bbamem.2011.10.010 PMID:22037146

Vannier, P., Michoud, G., Oger, P., Marteinsson, V., & Jebbar, M. (2015). Genome expression of *Thermococcus barophilus* and *Thermococcus kodakarensis* in response to different hydrostatic pressure conditions. *Research in Microbiology, 166*(9), 717–725. doi:10.1016/j.resmic.2015.07.006 PMID:26239966

Wright, P. C., Westacott, R. E., & Burja, A. M. (2003). Piezotolerance as a metabolic engineering tool for the biosynthesis of natural products. *Biomolecular Engineering, 20*(4-6), 325–331. doi:10.1016/S1389-0344(03)00042-X PMID:12919816

Zhang, G., & Ge, H. (2013). Protein hypersaline adaptation: Insight from amino acids with machine learning algorithms. *The Protein Journal, 32*(4), 239–245. doi:10.100710930-013-9484-3 PMID:23592219

Zhang, Y., Li, X., Bartlett, D. H., & Xiao, X. (2015). Current developments in marine microbiology: High-pressure biotechnology and the genetic engineering of piezophiles. *Current Opinion in Biotechnology, 33*, 157–164. doi:10.1016/j.copbio.2015.02.013 PMID:25776196

Zhuang, X., Han, Z., Bai, Z., Zhuang, G., & Shim, H. (2010). Progress in decontamination by halophilic microorganisms in saline wastewater and soil. *Environmental Pollution, 158*(5), 1119–1126. doi:10.1016/j.envpol.2010.01.007 PMID:20163899

Chapter 7
Genome Editing and CRISPR/Cas System of Extremophiles and Its Applications

Suneeta Gireesh Panicker
Dr. D.Y. Patil Arts, Commerce, and Science College, Pune, India

ABSTRACT

Extremophiles will be the choice of next generation industrial biotechnology (NGIB) as they are known to be contaminant resistant, but engineering their genomes has always been difficult and time consuming task. CRIPR/Cas (clustered regularly interspaced short palindromic repeat and CRISPR associated proteins) system can be employed for this reason. The genome of an industrially important halophile (i.e., Halomonas) was edited to study a combined effect of four different genes on glucose breakdown and production of poly (3-hydroxybutyrate-co-3-hydroxyvalerate). This editing has resulted in 16-fold increase of 3HV, and the mutants generated by CRIPR/Cas system were significantly effective in synthesizing PHBV. Unfortunately, this system does not always work, specifically in extremophilic microorganisms because Cas9 or Cpf1 are from mesophilic bacteria. Therefore, alternatively, the endogenous CRISPR/Cas system is used for editing the genomes of such organisms. This genome editing of extremophiles will open the doors for developing next generation industrial biotechnology (NGIB).

INTRODUCTION

The Nobel Prize of year 2020 in Chemistry 2020 was awarded to Emmanuelle Charpentier and Jennifer Doudna for discovering the genetic scissors called CRISPR/Cas9: **C**lustered **R**egularly **I**nterspaced **S**hort **P**alindromic **R**epeat - CRISPR-associated sequences (CRISPR-Cas).

DOI: 10.4018/978-1-7998-9144-4.ch007

The Discovery of the CRISPR-Cas System in Prokaryotes

In 1993, for the first time, CRISPRs were observed in archaea, named *Haloferax mediterranei* (Mojica et al., 1993) and then subsequently as the life science moved into the genomic era, these were detected in more and more number of bacteria and archaea genomes. The powerful CRISPR-Cas9 system discovery initiated with the identification of repeated genome structures present in bacteria and Archaea. An unusual repeated structure in the *Escherichia coli* genome was reported in 1987. This sequence consisted of five extremely homologous sequences of 29 base pairs (bp) along with a reverse complementary (dyad symmetry) sequence of 14 bp that were interspersed by variable spacer sequences of 32 bp (Ishino et al., 1987). Later, the halophilic Archaea *Haloferax mediterranei*, showed the presence of similar, repeated structures in its genome as shown in figure 1. It had 14 nearly conserved sequences of 30 bp, repeated at regular distances (Mojica et al., 1993). Subsequently, these types of repeats were revealed through bioinformatics analyses of various prokaryotes. It was thus shown that such repeats possessed a characteristic feature of short, partially palindromic sequences in clusters which were separated by exclusive interfering elements of fixed length. It was also observed that these intervening sequences differed from organism to organism, but had an ancestral origin and high biological relevance (Mojica et al., 2000). These arrays of repeated sequences are called **C**lustered **R**egularly **I**nterspaced **S**hort **P**alindromic **R**epeats, abbreviated as **CRISPR** (Jansen et al., 2002). The analysis of the unique, non-repetitive sequences in CRISPR showed they matched with the genetic code of different viruses, pointing towards the protective immunity against viruses. This led to a hypothesis that if a bacterium overcomes a virus infection and succeeds in surviving then it will add a piece of that virus genetic code to its genome, like a memory of that infection. In the efforts of understanding the function of CRISPR, researchers discovered special genes called CRISPR-associated, abbreviated as *cas*. This group of genes were located only in the CRISPR-containing prokaryotes and were always found adjacent to CRISPRs. The *cas* genes encoded proteins that had helicase and nuclease activity, advocating their role in DNA metabolism or gene expression (Jansen et al., 2002). In the coming years, a number of Cas protein subfamilies were discovered (Haft et al., 2005; Makarova et al., 2006). Cas proteins control the three functional stages of CRISPR adaptive defence. These are adaptation, processing, and interference as shown in figure 2.

Figure 1. CRISPR structure in prokaryotes

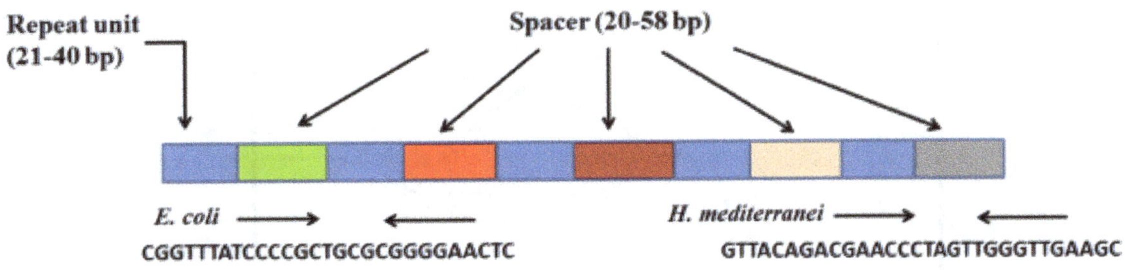

Figure 2. Three functional stages of CRISPR adaptive defence

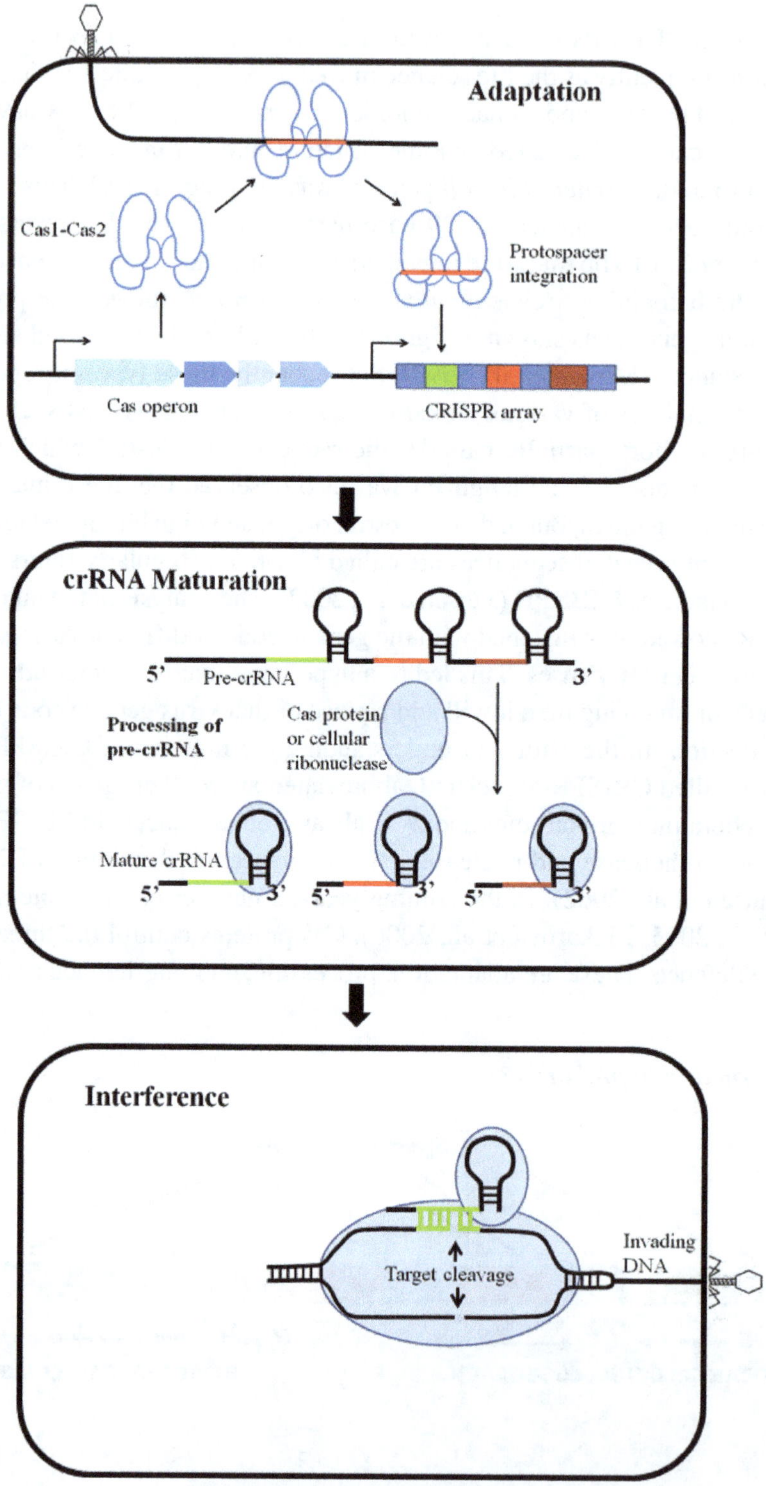

The adaptation stage was best studied in a type I-E CRISPR system of *E. coli*. In this system, a stable complex was formed between two strongly conserved nucleases (Cas1 and Cas2) (Nuñez et al., 2014). Two Cas1 dimers were linked to one Cas2 dimer and this is considered as the minimum requirement for the *de novo* spacer acquisition (Yosef et al., 2012). crRNA processing is the second stage (Bhaya et al., 2011), which involves shortening of the long crRNA. When the crRNA is transcribed from its gene, the tracrRNA (trans-activating CRISPR RNA which has complementary sequences for palindromic sequences) binds to it. The scissor protein Cas9 also joins this complex. This complex signals the RNase III to cleave the long crRNA into a smaller piece which consists of the spacer or the non-repetitive sequence along with the palindrome repeat RNA sequence bound to tracrRNA and all of these bound in the Cas9 protein, forming a ribonucleoprotein complex (RNP). During the final stage of CRISPR-Cas-meditated immunity that is interference, the RNP complex can recognize the incoming foreign DNA (or RNA) via base pairing of crRNA and target nucleic acids (Brouns et al., 2008). The complementary DNA sequence recognised by a crRNA is called as a protospacer. Almost all CRISPR-Cas systems mediate both adaptation and interference by specifically recognizing a short sequence located at the 5 -flanking region of the protospacers, called the protospacer adjacent motif (PAM) (Wang et al., 2015; Yamano et al., 2017). The CRISPR-Cas system, with its memory component, resembles the adaptive immune system of vertebrates, with a very critical difference that the immune system in animal is not inheritable. Makarova et al., 2011, on the basis of the diversity and their erratic & ubiquitous distribution suggested that CRISPR-Cas system emerged in an ancient ancestral archaea and then it was spread horizontally to the bacteria.

CRISPR-CAS SYSTEM AS PROKARYOTIC ACQUIRED IMMUNE SYSTEM

A CRISPR-Cas system function in immune system of prokaryotes was experimentally demonstrated in 2007. These experiments were carried out in lactic acid bacterium *Streptococcus thermophiles* (Barrangou et al., 2007). A sequence from phage known to infect *S. thermophiles* was inserted into the spacer part of CRISPR that was detected in S. thermophiles. This used CRISPR was shown to make the bacterium resistant to the phage used in this experiment and this was proven by the loss of resistance from the bacterium on deletion of the corresponding protospacer sequence from the phage genome. Furthermore, it was also experimentally proven that the plasmids which carried the sequences matching the CRISPR spacers were not allowed to get transformed into the bacterium by the CRISPR-Cas system (Marraffini and Sontheimer, 2008).

Next, van der Oost's group demonstrated that the CRISPR genes transcribed the CRISPR RNA which were processed and made functional by the cooperative association of Cas proteins that are produced from the genes located next to the CRISPR (Brouns et al., 2008). These findings were subsequently confirmed by reconstituting the CRISPR-Cas system of *S. thermophilus* in E. coli. This transformed E. coli demonstrated heterologous protection against phage infection and the plasmid transformation (Sapranauskas et al., 2011). The *in vitro* capacity of the purified Cas9-CRISPR RNA (crRNA) complex to cleave the target DNA was there after demonstrated by many researchers (Gasiunas et al., 2012; Jinek et al., 2012). In this way CRISPR-Cas system was then widely recognised as prokaryotic acquired immunity system (Horvath and Barrangou, 2010; Wiedenheft et al., 2012). By then numerous and hugely varied Cas proteins were discovered. These proteins were shown to have functions at different stages of CRISPR immunity. They possessed nucleic acid manipulation capacities such as helicases, nucleases, and polymerases (Charpentier et al., 2015; Tamulaitis et al., 2017). In brief, the most widely known types of CRISPR-Cas systems contain highly conserved Cas1 and Cas2. These two Cas proteins bind to each other to form a complex to play a role in the adaptation module and thus are necessary for the insertion of new spacers in the CRISPR system. During the stage of the expression of the CRISPR genes, the transcribed pre-crRNA needs to be processed into the mature crRNAs and this is performed by the type-specific Cas endonucleases. In the interference stage, the effector Cas endonucleases is bound to the crRNAs to form a complex that is required to recognize and cut the target DNA or RNA in a sequence-dependent manner. The Cas proteins taking part in the expression and interference stages differ in different CRISPR-Cas system and the same enzyme may function in both the stages but this is not the case with the adaptation module Cas proteins.

Classification of Crispr-Cas Systems

The CRISPR-Cas systems, depending on the encoded effector proteins, are recently classified into two classes, 1 and 2 (Shmakov et al., 2017). Class 1 CRISPR-Cas system consists of 4 to 7 Cas proteins in an uneven stoichiometry that form functional multisubunit effector complexes. This class 1 system is widely present in archaea and bacteria, which include all hyperthermophiles, consisting of nearly 90 per cent of all known CRISPR-cas systems. Class 2 contributes to the remaining 10 per cent and is known to use a multidomain single effector protein. Class 2 is mostly observed only in bacteria (Burstein et al., 2017). Class 1 mainly includes, types I, III, and IV while class 2 consists of types II, V, and VI. These types are distinguished by the presence of characteristic signature proteins: type I has Cas3, type II has Cas9, and type III has Cas10. The type I effector complexes are called as CRISPR-associated complex for antiviral defense (Cascade) and that of type III are known as Csm/Cmr complexes. These two complexes are structurally same and have an evolutionary relationship (Venclovas, 2016; Koonin et al., 2017). The type IV systems remain functionally uncharacterized and they do not have the Cas1 and Cas2 proteins of adaptation module (Makarova et al., 2015). The studies of Class-2, Type-II CRISPR-Cas system in S. thermophiles and Streptococcus pyogenes showed that this system consists of four cas genes. Three of these (cas1, cas2, csn2), were shown to be involved in in spacer acquisition. The fourth one (cas9), which was early known as cas5 and csn1, was shown to be necessary for interference (Barrangou et al., 2007) and this was proved when the target DNA was prevented from cleavage by the inactivated cas9 gene (Garneau et al., 2010). Furthermore, in the experiments with heterologous CRISPR-Cas system of S. thermophiles in E. coli, parts of the system were inactivated in order to study the role of each component in the system. These experiments clearly stated that the interference step in CRISPR immunity is solely carried out by Cas9 protein alone. It was also shown that the Cas9 protein possesses two nuclease domains, HNH and RuvC that are mandatory for this interference activity (Sapranauskas et al., 2011).

PROTECTION OF BACTERIAL CHROMOSOMAL DNA FROM THE INDIGENOUS CRISPR

The spacers in the CRISPR were shown to cleave the DNA of matching sequences, so the next question that arose from this fact was that how the bacteria's own DNA did get protection from its own the CRISPR system? Sequences around protospacers were studied extensively to answer this question. Protospacers are the sequences in the phage genomes that contribute to the spacers in CRISPR system.

Studies revealed that there were short sequence motifs just at a distance of two nucleotides from protospacer sequences (Bolotin et al., 2005; Horvath et al., 2008). These important sequence motifs were further named as protospacer adjacent motifs or PAMs (Mojica et al., 2009). This was demonstrated by studying the phage resistance to CRISPR in S. thermophilus. The phages that were able to survive bacterial immunity were isolated and analysed in these studies. It was shown that the phages which were able to escape the CRISPR immunity had acquired mutations in the PAMs, concluding that these short sequences play a crucial role in targeting (Deveau et al., 2008). More such studies further demonstrated that the PAM sequences are not only necessary for target interference but are required for the integrating the new spacer sequences into CRISPRs (Wang et al., 2015; Anders et al., 2014).

tracrRNA IN crRNA MATURATION

To study the maturation of crRNA after its transcription, differential RNA sequencing was used to characterize small, non-coding RNA molecules in *S. pyogenes* (Deltcheva et al., 2011). Based on expression of pre-crRNA and mature crRNA molecules, an active CRISPR locus was identified in this process. Along with these results, surprisingly, lots of RNA molecules transcribed from the opposite strand of the CRISPR array and which were 210 bp upstream of the CRISPR locus were identified. These RNA transcripts of 25 nucleotides (nt) length and nearly perfectly complementarity (1-nt mismatch) to the repeat regions of the CRISPR locus, were thus named as trans-encoded small RNA (tracrRNA). They were thought to base pair with pre-crRNA (Deltcheva et al., 2011) and form a RNA duplex region that included the processing sites of the two RNAs; pre-crRNA and tracrRNA. This pairing also suggested that it was necessary for co-processing of these two RNAs upon pairing (Figure 3). Additionally, it was also noted that the co-processed duplex of tracrRNA and pre-crRNA possessed short 3′ overhangs like those made by the endoribonuclease RNase III. Finally it was found that Cas9 protein was also involved in processing of this duplex. It was shown that Cas9 acts as an anchor to assist the base pairing between tracrRNA and pre-crRNA and in turn supports the host RNase III protein for recognition and cleavage (Deltcheva et al., 2011). Experimentally it was proven that the cleavage of the target DNA by Cas9 was triggered upon addition of tracrRNA in an in vitro reaction. The scientists thus concluded that the tracrRNA had two crucial roles: activating the processing of pre-crRNA by RNase III enzyme and successively triggering the cleavage of crRNA-guided DNA by Cas9 (Deltcheva et al., 2011; Jinek et al., 2012).

AN EPOCH-MAKING EXPERIMENT THAT GAVE CRISPR-CAS9 GENOME EDITING TOOL

Emmanuelle Charpentier and Jennifer Doudna collaborated for the studies on usage of crRNA to direct the sequence specificity of the nuclease. After their discovery of the crucial role of tracr RNA, they worked on the utterly necessary regions of tracrRNA and crRNA for the cleavage of target DNA by the enzyme Cas9. Activating motifs were found in tracrRNA. A seed region of about 10 nt in the proximal region of PAM sequence of the target strand was also determined by the researchers and was proved to be clearly important for the recognition of target. Using all their discoveries and the related knowledge the two scientists then tried to figure out how to use CRISPR-Cas system as the genetic scissors in a simplified manner. In this attempt they fused the tracrRNA and CRISPR-RNA into a single molecule and called it as a guide RNA. They then used this simplified genetic scissors to undertake an epoch-making experiment: whether these scissors can be used to cut the DNA at a location specified by the researchers. They took a gene and selected five different places where the gene should be cleaved. Then they changed the CRISPR part of the scissors to match its code to the code where the cuts were to be made. As a result they were successful in cutting the DNA molecules at exactly the right places. They published their discovery of the CRISPR/ Cas9 genetic scissors in 2012 and soon after that various research groups demonstrated that this tool can be used to manipulate the DNA of human and mice origin (Jinek et al., 2012).

Genome Editing Mechanism

CRISPR- Cas9, a class 2, type II genomic editing system is most widely used as it requires only Cas9 nuclease in contrast to complex multiprotein complexes, which are required for other CRISPR systems. Cas9 from *Streptococcus pyogenes* (Cas9Spy) is widely used Cas9 nuclease, although *Streptococcus thermophilus* nuclease is also used in some cases. Cas9 protein is a crRNA-dependent endonuclease and contains two unrelated nuclease domains; HNH and RuvC. HNH is used to cleave the target DNA strand and the RuvC is used to cut the complementary strand (Ishino et al., 2018). The system consists of the Cas9 nuclease, trans-activating CRISPR RNA (tracrRNA), and crRNA. crRNAs consists of spacers and their adjacent Direct Repeats (DRs). crRNA and tracrRNA are fused into a chimeric single guide RNA (sgRNA) (Jinek et al., 2012; Hsu et al., 2014). The target specificity is determined by the complementarity of sgRNA and the target DNA and also by the PAM motifs. sgRNA complexes with the Cas9 and guides it to the specific sequences on the exogenous target DNA.

The Cas9/sgRNA complex scans the target DNA sequence to find PAM sequence and uncoils the DNA in the priming region (10 bp immediately after the PAM) (Szczelkun et al., 2014). After successful complementary base pairing between sgRNA sequence and target DNA, the HNH-domain of Cas9 creates a cut in the target strand, and the RuvC-domain cuts the non-target strand (Anders et al., 2014; Nishimasu et al., 2015) and is seen in figure 3. Thus double-stranded break (DSB) is introduced by the CRISPR-Cas9 system in the target DNA, leading in a gap in that DNA (Chneiweiss et al., 2017). The gap is repaired by the endogenous repair system. In eukaryotic genome editing techniques, the gap may be repaired by non-homologous end-joining (NHEJ), in which the cut DNA is ligated without a homologous DNA template, directly or by introducing an insertion or deletion. As a result of insertions and deletions, frameshifts can occur and cause loss of gene function (Kosicki et al., 2018). Hence this NHEJ system can be used to create gene knockouts or to completely remove a target gene. Homologous Directed Repair (HDR), is an alternative repair mechanism used to fill the gap. This method involves filling the gap with a new sequence complementary to a specifically introduced DNA template. HDR method is highly accurate but has lower frequency than NHEJ (Maruyama et al., 2015. With increase in its frequency, HDR can be used both for adding and reversing target mutation in the genome. It can also be utilized to insert complete genes in the genome using a template (Lino et al., 2018). Occasional NHEJ repair are known to occur in HDR, in spite of presence of the template but NHEJ inhibitors are used to overcome this problem (Yu et al., 2015).

With the help of this powerful genome editing tool, scientists are able to manipulate the DNA sequences in a wide variety of cells and organisms. These days it is used to modify the DNA of cells and model organisms with a basic intention of understanding the function of the various genes and their regulation in the genetic disorders.

Figure 3. Mechanism of action of crRNA, tracrRNA and Cas9 complex to cleave the target DNA

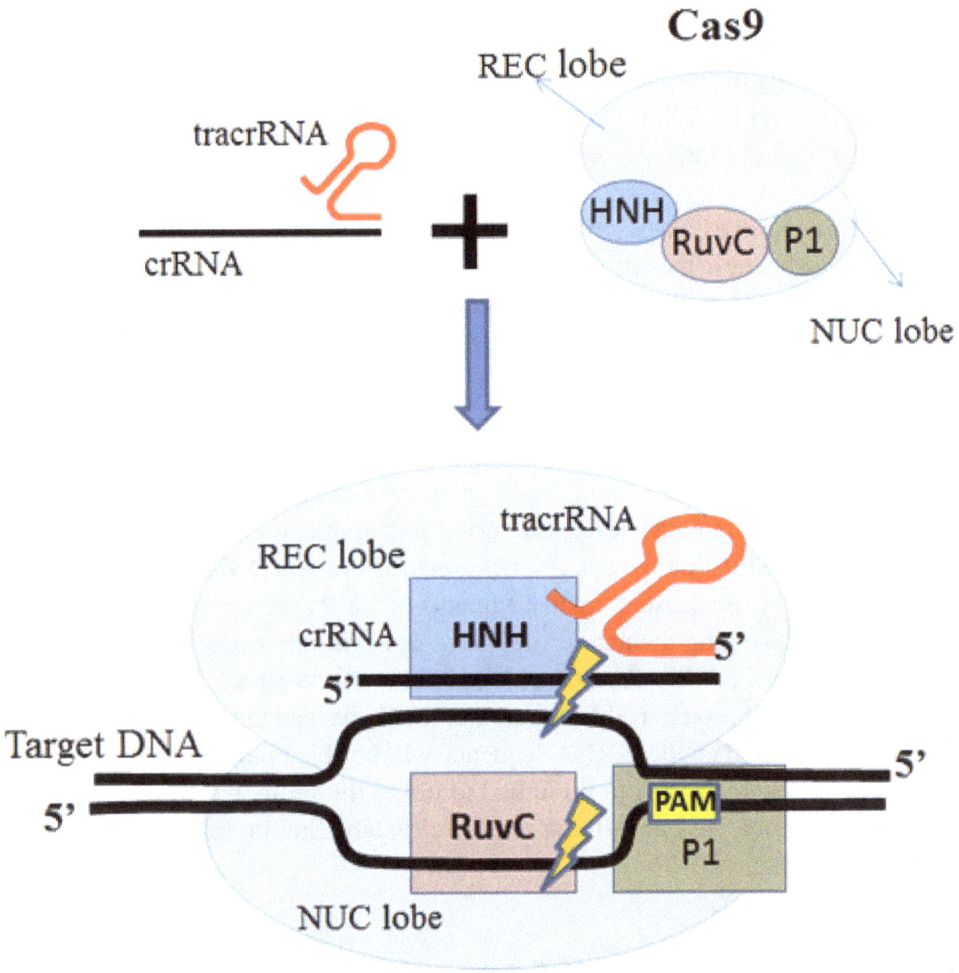

Application of Crispr-Cas Tools in Editing Genomes of Bacteria and Archaea

The growth of a bioindustry is hugely influenced by microbial engineering. The CRISPR-Cas genome editing tool proves to be heavily beneficial for generating economically important strains. After its discovery this editing method has been rapidly and remarkably adapted for practical editing of many genomes including eukaryotic cells. The technology is gradually emerging and diversifying for gene silencing, genome editing, and genome-wide screening of essential genes in bacterial and archaeal genomes (Jiang et al., 2015; Gophna et al., 2017). For example, heterologous recombineering was coupled with CRISPR-Cas system using linear single-stranded (single-stranded DNA recombineering [SSDR]) or double-stranded (double-stranded DNA recombineering [DSDR]) DNA templates and was successfully employed in E. coli (Mougiakos et al., 2016). In archaea, with the help of endogenous CRISPR-Cas systems, gene silencing was established in *Sulfolobus solfataricus*, *Sulfolobus islandicus*, and *Haloferax volcanii* ((Peng et al., 2017). Recently, bacterial Cas9 protein was used to edit the genome of a mesophilic archaeon *Methanosarcina acetivorans* (Nayak and Metcalf, 2017).

OTHER APPLICATIONS OF CRISPR-CAS9

Other than genome editing, due to their remarkable diversity, these loci were used as genetic markers for identification and typing of the species, much before their actual function was determined. For example, *Yersinia pestis* (Cui et al., 2008), *Salmonella* spp. (Liu et al., 2011), *Mycobacterium tuberculosis* (Kamerbeek et al., 1997) and *Corynebacterium diphtheria* (Mokrousov et al., 2005) typing is done by using CRISPR.

CRISPR-Cas9 can be used as a new novel mechanism of antimicrobial action to control the antibiotic-resistant pathogenic bacteria by cutting their genomes. This has been tried with antibiotic-resistant *Staphylococcus* spp. which infected the mice skin, got killed using CRISPR-Cas9 (Citorik et al., 2014). Intestinal infection by pathogenic *E. coli* was also reported to be prevented by CRISPR-Cas9 (Chen et al., 2013). In spite of the technical issues like delivery methods, for which active research is being carried out and expected to produce ways to overcome such challenges in very near future, CRISPR-Cas can be a powerful therapeutic agent. The CRISPR-Cas system can be also used for preventing different industrially valuable bacterial strains by imparting phage resistance in them. The beneficial bacteria of fermented food industry acquire phage infection during the production process. Such important strains can be protected using CRISPR-Cas system. As mentioned above, the DNA cleavage activity of Cas9 is mediated by HNH and RuvC nuclease domains. Mutants of Cas9 that were devoid of cleavage activity (dCas9) were created by changing the active site amino acids in order to study the regulation of gene expression. The complex of CRISPR-dCas9 is able to bind to the target DNA but cannot cut it. Thus by tagging the green fluorescent protein (GFP) tag to the dCas9, we can label a particular position, as the CRISPR-dCas9 complex based on the sgRNA sequence will be able bind to the target sequence (Bikard et al., 2014). Apart from this, dCas9 can be linked to either the promoter region or the open reading frame of a gene and carry out live intracellular site-specific labelling in order to artificially control the gene expression (Gilbert et al., 2013).

EXTREMOPHILES IN BIOTECHNOLOGY

Theromophiles have more tendencies to possess CRISPR loci than mesophiles (Nelson et al. 1999). Thermophilic organisms possess several advantages over mesophilic organisms, which make them of particular interest when being used as production hosts in different industries (Olson et al., 2015). For instance, thermophiles are capable to grow and ferment at thermophilic temperatures, this ability helps in reducing the costs of cooling in fermentation (Kambam and Henson, 2010) and also leads to increase in the solubility of the substrate and product (Ma et al., 2014). This thermophilic property of these organisms also helps in reducing the mesophile contamination risk and nonsterilized fermentation can be carried when they are used as production hosts, thus reducing the sterilization costs (Ouyang et al., 2013). Furthermore, when thermophiles or their products are used for enzymatic lignocellulose degradation, an optimum temperature can be used to allow efficient simultaneous saccharification and fermentation (Bhalla et al., 2013). However, the use nonmodel thermophiles in fermentation can be restricted due to the lack of a powerful genome editing tool as those used for mesophilic model organisms are not functional in these thermophiles (Taylor et al., 2011). Similarly the high salt concentrations and alkaline pH strongly inhibits the growth of non-halophilic microorganisms (Don et al., 2006; Quillaguamán et al., 2010), thus facilitating open non-sterile fermentation (Johnson et al., 2009), reducing the cost and troubles of sterilization and decontamination. Even more, organisms those are able to grow in marine environment, like *Halomonas,* can be used to save the fresh water for different purposes (Tan et al., 2011). Next generation industrial biotechnology (NGIB) will promote the use of extremophiles but molecular engineering in these organisms is usually difficult. Thus, to move the NGIB to a high level, CRISPR like technologies can help in convenient engineering of extremophiles. *Halomonas bluephagenesis* and *H. campaniensis* can be industrially used continuously in open and unsterile conditions for the production of biological materials (Chen et al., 2017). Thus the two extremophiles have demonstrated their capability as a skeleton for NGIB.

CRISPR/CAS SYSTEM OF EXTREMOPHILES AND ITS APPLICATIONS

CRISPR Distribution in the Genomes of Extremophiles

A comparative study has shown that actinobacteria has more CRISPRs than crenarchaeota and hyperthermophilic bacteria. Psychrophilic archaea have more CRISPRs than other psycrophiles and these are equally distributed throughout their genomes. Even the crenarchaeota and proteobacteria show equal spread of CRISPRs in their genomes (Chellapandi and Ranjani, 2015).

The genome of Caldivirga maquilingensis contains many CRISPRs and the genome of *Kosmotoga olearia* has an extensive CRISPR system consisting of 84 spacers. Similarly, Thermotoga lettingae shows the presence of a classic CRISPR system with 53 spacers. The genomic survey carried out in this study demonstrated that out of the 12 extremophiles, distinctive CRISPR-Cas systems were present in three thermophilic archaea (*Clostridium maquilingensis, Thermococcus gammotolerans, Thermococcus sibiricus*), and two thermophilic bacteria (*Petrotoga mobilis, Thermotoga* lettingae).

Design and Assembly of CRISPR-CAS Systems in Thermophiles

The cluster of the Cas gene are mostly located downstream of CRISPR locus, after about few hundred of basepairs. These genes are highly organised in the genomes, generally in the order of cas5–cas3–cas1–cas2. The subtypes genes that include cmr4, cmr5, cmr1 and cmr6 are also clustered. The cas gene cluster in *T. gammotolerans* has been shown to be in the order of cas6–cas5– cas3–cas4–cas1. Similar order was observed in *P. mobilis* with an additional cas2 at the end (cas6–cas5–cas3–cas4–cas1–cas2). This type of order of the cas genes appears like a superoperon. In contrast, *T. lettingae* showed a varied organization and order of cas genes. Its genome has csh1 and csh2 genes in between the cas2 and cas5 (Chellapandi and Ranjani, 2015).

CAS Family Genes

The CAS system of thermophilic archaea commonly shows the presence of endoribonucleases, CRISPR-associated protein, CRISPR-associated autoregulator, CRISPR-associated helicase, HD-like nuclease, and RAMP associated protein. Only *T. gammotolerans* shows the existence of RecB family exonuclease. *C. maquilingensis* marks the presence of nine cas-related and cmr genes. Both *P. mobilis* and *T. gammotolerans* consists of a CAS system with cas1–cas6 cluster without cas2 in their genome.

This cluster is known to encode endoribonucleases, nuclease, exonuclease, and helicase. *T. lettingae* has a cas gene familiy (cas1– cas6), gene clusters of cmr and csh, and a csx11 gene in its genome. The function of csh1, csh2 and csx11 genes is not yet clear. DEATH box helicase, DUF364, and RAMP superfamily functions were significantly annotated (Chellapandi and Ranjani, 2015) among the eight identified genes.

CRISPR/Cas Editing in Halophile

CRISPR-Cas9 editing in Halophilic *Halomonas bluephagenesis* was done using two plasmids: one for encoding Cas9 protein and other for sgRNA and donor DNA. A low copy number pSEVA321 plasmid was generated by adding the native *S. pyogenes* cas9 gene promoter (CAS), expressing the Cas9 gene. Another high copy number pSEVA241 plasmid was constructed by adding the J23119 promoter expressing the sgRNA and donor DNA containing 500–1000 bp homologous arms. These two plasmids were then transferred into *H. bluephagenesis* TD01 via conjugation for genome editing. In this experiment, phaC gene was targeted and hence the sgRNA pQ03-G1 was designed accordingly. As a result 75% editing efficiency was achieved. This was confirmed by colony PCR and DNA sequencing. In next experiment when the phaP1 gene was targeted using sgRNA pQ85-G26 editing efficiency reached 100%. Furthermore, to prove the robustness of the CRISPR/Cas9 tool in editing *H. bluephagenesis* TD01 genome, 13 different sgRNAs targeting 13 genes and 9 sgRNAs targeting intergenic regions (IR) were constructed. The results showed that the editing efficiency varied from 12.5% to 100.0% (Table 1 and S4). The limitations of the homologous recombination system in *H. bluephagenesis*, did not make simultaneous multiple site editing possible, which can be easily done in *Escherichia coli* (Jiang et al., 2015) and *Corynebacterium glutamicum* (Cho et al., 2017). Thus some manipulations of the homologous recombination system in *H. bluephagenesis* might help in overcoming this issue and will be able to save time and work required in engineering the halophiles. To date, conjugation is the only method for DNA transfer in *H. bluephagenesis*. Hence the sgRNA plasmid has to contain the donor DNA. This results in overexpression of large exogenous gene(s) in the bacteria and may cause metabolic stress. Thus, if the sgRNA plasmid could be transferred by electroporation or chemical method, the donor DNA could be inserted in the chromosome as a part of DNA and not as a plasmid. This will help in reducing the overexpression and the metabolic stress caused due it. A successful example of genome editing using CRISPR-Cas9 system is deletion of multiple genes to study glucose catabolism in *H. bluephagenesis*.

The CRISPR/Cas system is a different type of defence system which is adaptive, hereditary and it can identify the invading foreign sequences by a sequence specific technique. This technique needs CRISPR RNA which complements with the invading foreign nucleic acid to identify it along with a PAM sequence. One recent study identified such PAM sequences for halophilic archaeon *Haloferax volcanii*. They also found that many motifs were functional in activating the defence reaction. On the other hand, the comparative sequence data showed that very few PAM sequences are required to select the protospacers from the invading DNA, suggesting that protospacer selection needs more precise requirements than the defence action. The CRISPR-repeat sequences when compared by sequenced haloarchaea showed that the repeat sequence of half or more species were conserved and carried same type of CRISPR/Cas system (Maier et al., 2012).

CRSIPR-Cas Genome Editing in Thermophiles

Production of biofuels and different natural products, through microbial fermentation of renewable resources is increasing with a great speed. Nevertheless, the production cost of such protocols at extreme conditions urge for improvised and efficient strains. Thermophilic organisms have the ability to grow and ferment at increased temperatures, thus can decrease the production along with an additional advantage of using nonsterile conditions. Even more, the manipulation of sporulation gene (such as sigF) regulation, can offer high capability for commercialization.

For editing the genome of very useful and powerful thermophilic organisms, some genetic tools are available but these are not highly efficient to give 100% results (Olson et al., 2015). The Cas9 protein from *S. pyogenes* (SpCas9) is not active at higher temperatures (at or above 42°C). In an attempt to overcome this problem a thermostable, ThermoCas9 from CRISPR-Cas type-IIC system of a thermophilic bacterium *Geobacillus thermodenitrificans* T1230 was isolated and used instead of the *S. pyogenes* Cas9 (Mougiakos et al., 2017). This thermostable Cas9 protein is active between 20 and 70°C *in vitro*. The thermostability of this protein is shown to be affected by sgRNA-structure. ThermoCas9 was employed for editing the genome of a thermophilic *Bacillus smithii* ET 138, which is an industrially beneficial thermophile. Genome editing and silencing in this bacteria was carried out at 55°C (Mougiakos et al., 2017). Additionally, the researchers applied ThermoCas9 for editing the genome of mesophile *Pseudomonas putida* KT2440 for no CRISPR-Cas9 editing tool was available at that time (Aparicio et al., 2016) and thus proved the broad temperature range and applicability of this thermo stable Cas9.

It was observed that the CRISPR-Cas9 or Cpf1 system was not always equally effective in editing the genomes of bacterial and archaeal cells, more specifically the harder ones like extremophilic microorganisms. An alternative for this hindrance was to use the endogenous CRISPR-Cas systems of such prokaryotic cells. With this alternative strategy, there remains only the need for a shuttle vectors expressing the sgRNA matching the target site and carrying the repairing donor DNA. The endogenous ribonuclease will be able to process the CRISPR RNA. A group of scientists performed genome editing in the thermophilic archaeon *Sulfolobus islandicus* by using the endogenous type I and III systems and achieved high-efficiency editing (Li et al., 2016). The endogenous CRISPR-Cas system of S. islandicus was used for deletion, insertion and site-specific mutagenesis, to manipulate it genome (Li et al., 2016). This was more of straightforward approach than the classic one (using the Cas9 from *S. pyogenes*). They also successfully attempted to manipulate the genome of another *Sulfolobus* strain (*S. islandicus* HVE10/4), which is widely used as host to study the viruses by using the similar approach of using the endogenous Type I-A CRISPR-Cas system (Guo et al., 2011).

A study by Mougiakos et al., 2017, showed that spCas9 was not active at 42°C and above in *Bacillus smithii* ET 138. They attempted to tightly control its activity not by using an inducible promoter, but by changing the temperature of organism's cultivation. They performed gene deletion, gene disruption, and a gene insertion by recruiting plasmid mediated homologous recombination template for introducing the wanted changes to the genetic material at 45°C. The elimination of nonedited cells was carried out by counter selection of spCas9 at 37°C. The temperature range for the growth of ET 138 is between 37 and 65 °C and this organism is known to utilize both C5 and C6 sugars (Bosma et al., 2015). The researchers used this tool to consecutively deleted the sporulation gene *sigF* and the pyruvate dehydrogenase complex subunit gene *pdhA* in the ldhL mutant. As result they obtained a triple mutant strain that was sporulation-deficient did not produce L-lactate and acetate. Such sporulation-deficient strains are industrially important and desired for in fermentations for safety reasons. Thus, for the first time, a temperature-controlled recombination/counter selection tool was used for genome editing of a moderate thermophile.

After the discovery of a thermostable Cas9, a ThermoCas9-based genome editing system for *B. smithii* ET 138, was developed by the same group. They combined homologous recombination and ThermoCas9-based counter selection at 55⁰C (Mougiakos et al., 2017). The requirement of G-rich PAM sequences by SpCas9 and C-rich PAM sequences by ThermoCas9 however, may not be always available at the target regions of editing, more specifically in AT-rich areas of the genomes. Contrastingly, the Cas12a requires a T-rich PAM sequence to cleave the target DNA sequence (Zetsche et al., 2015). Hence, the inclusion of thermostable Cas12a-based tools can be used to extend the genome editing systems for thermophilic organisms.

The *in vitro* temperature tolerance and stability of three different Cas12a orthologs was studied by a group of researchers. They used AsCas12a from *Acidaminococcus* sp. BV3L6, FnCas12a from *Francisella tularensis* subsp. *novicida* U112, and LbCas12a from *Lachnospiraceae* bacterium ND2006 for their experiments. For their subsequent *in vivo* studies that involved the production of a genome editing system, specifically for the moderate thermophiles, they selected FnCas12a nuclease. The homologous recombination based editing at higher temperatures was merged with FnCas12a-based counter selection at lower temperatures with an intention to achieve faster, precise and efficient deletions of genes in *B. smithii* and thus, decreased the editing process time to 2–3 days. This was the first attempt to use Cas12a-based genome editing method for facultative thermophiles. They demonstrated that the thermostability of Cas12a nucleases was strongly influenced by their accompanying guide RNA. FnCas12a nuclease was found to be the most thermostable among all Cas12a nucleases tested, but it was difficult to specify the reason for its higher thermostability as both sequence and structural homology of LbCas12a and AsCas12a with FnCas12a were comparatively high (35% sequence identity with AsCas12a and 42% with LbCas12a) (Mohanraju et al., 2021).

Clostridium thermocellum possess a powerful lignocellulose-solubilizing capacity. This characteristic of *C. thermocellum* can be explored for biofuel production but due to the limitation of genetic techniques in this organism, the improvement in the biofuel production is lagging behind. Thus to overcome this issue a group of scientists used a native Type I–B and heterologous Type II CRISPR-Cas systems. The genome of *C. thermocellum* was edited by repurposing the native Type I–B system. Three various thermophilic Cas9 variants (Type II) were used for this editing. Another Cas9 isolated from *Geobacillus stearothermophilus* which was named as GeoCas9 was also found active in this organism. They were successful in introducing a nonsense mutation into pyrF by employing CRISPR-mediated homology directed repair. In both editing cases they came across a limiting step which was homologous recombination between the repair template and the genome. This limiting step was overcome by exploring three novel thermophilic recombinases. As a result, they demonstrated that exo/beta homologs that were isolated from *Acidithiobacillus caldus*, were actively functional in *C. thermocellum*. Initially, only 40% genome editing efficiency was achieved at the pyrF locus when the Type I–B system was used to engineer the strain. Later, this efficiency was increased to 71% by using the recombineering machinery. Similarly, only 12.5% genome editing efficiency was obtained with Type II GeoCas9 system which was then increased to 94% when recombineering machinery was expressed. Thus by adding the recombinases to the thermophilic CRISPR system (either Type I–B or Type II) the scientists were able to develop a novel, efficient CRISPR editing tool. This made it possible to engineer *C. thermocellum* in a better way to increase the lignocellulose degradation and biofuel production (Walker et al., 2020).

Mini-metagenomic sequencing was used by a group of researchers to locate a different CRISPR-Cas9 system in *Ignavibacterium*, a hyperthermophilic bacterium from Yellowstone National Park's Lower Geyser Basin. The temperature of this hot spring is known to surpass 90 °C (Yu et al., 2017). They found that the IgnaviCas9 was active *in vitro*, at temperatures up to 100°C. This property helps the enzyme to cleave the DNA beyond the 44°C limitation of *Streptococcus pyogenes* Cas9 (SpyCas9) and the 70°C limitation of both *Geobacillus stearothermophilus* Cas9 (GeoCas9) and *Geobacillus thermodenitrificans* T12 Cas9 (ThermoCas9). The potential of this enzyme was applied to prepare a bacterial RNA-seq library that can be used to remove the undesired cDNA from 16s ribosomal rRNA without any additional step. Addition of such hyperthermo stable IgnaviCas9 to the CRISPR-Cas9 tool system proves to be extremely beneficial, exciting and also broadens the temperature range of the system (Schmidt et al., 2019).

Cas9-based tools for engineering thermophilic bacteria were highly researched and SpCas9- and thermostable ThermoCas9/GeoCas9-based systems were found suitable for genome editing in these bacteria (Mougiakos et al., 2017; Ganguly et al., 2020). Additionally, a Cas12a-based genome editing tool was shown to offer different benefits over Cas9-based system. Cas12a recognizes TTTV PAM sequences, which raises the target site number in a genome and thus leads to precise selection of the cutting region. This type of tool can be useful for creating point mutations and when the target sequences are short intergenic regions. The Cas12a can be a very good option where the inherent toxicity to SpCas9 in some organisms, has been reported. The pre-crRNA processing activity of Cas12a can be exploited for faster and facile multiplexing (Adiego-Perez et al., 2019).

Thermoanaerobacter ethanolicus can utilize both pentose and hexose and hence can be a very good option for thermophilic ethanol fermentations. This organism can produce lactate, acetate, hydrogen, and ethanol from sugars that are obtained from the degradation of plant carbohydrate polymer at temperatures above 65°C. To engineer the function of few important enzymes in ethanol formation, a thermostable Cas9-based system for genome editing was developed. Thymidine kinase gene (tdk), acetaldehydealcohol dehydrogenase gene (adhE), and redox sensing protein gene (rsp) were the three target genes. These genes were successfully edited using the thermostable Cas9-based system for genome editing. Editing included gene knockout and knock-in of a small DNA fragment. 77% genome-editing efficiency was achieved after optimization of the transformation strategies. Their *in vivo* results showed that RSP played a crucial role in energy metabolism regulation. Thus they concluded that an efficient way to identify the genes of biosynthetic and energy metabolism pathway is the genetic system (Le et al., 2021).

CaldoCas9 was another thermostable Cas9 protein identified and developed for genome editing in thermophiles using CRISPR-Cas9 system. The researchers designed a guide RNA which was associated with this CaldoCas9. Using these two in their in vitro experiments, they were able to demonstrate the target specific nuclease activity up to 65°C of this pair. After complete characterization of the PAM sequence specificity of CaldoCas9, it was shown that this system preferred 5-NNNNGNMA. This powerful system was then inserted into a plasmid vector to facilitate its delivery and use in the extreme thermophile *Thermus thermophiles* for editing its genome at 65°C. This vector was then used to knock out genes on the bacterial chromosome and megaplasmid of *T. thermophiles* and succeeded in achieving mutants at about 90% frequency. They also showed that they could cure this plasmid construct from the mutants, in order to carry out a next cycle of genome editing. This was the 1st CRISPR-Cas9 based genome editing report in the extreme thermophile *T. thermophiles* and might help to develop such genome editing tools for other extremophiles. They have even deposited their vector construct in the Addgene plasmid repository to facilitate genome editing in other extreme thermophiles (Adalsteinsson et al., 2021).

Multiple CRISPR-Cas systems with varied compositions and mechanisms are demonstrated by many researchers. One such group was able to show that a hyperthermophilic euryarchaeon, *Thermococcus kodakarensis* (Tko) has both type A & B Csa and Cst CRISPR-Cas system. They carried out RNA deep sequencing and northern analysis to analyse the expression and composition of crRNAs from the three CRISPRs in Tko. The results of this analysis revealed the presence of an eight nucleotide conserved sequence tag at the 5' end of the crRNAs that were associated with these CRISPR-Cas systems. The organism was then challenged with a plasmid. This plasmid consisted of the sequences that were targets for the endogenous crRNAs. As a result they could observe CRISPR-Cas-mediated plasmid silencing. This silencing was influenced by the crRNA complementary sequence on the plasmid and also by another sequence element that was apparently adjoining the PAM sequence. Furthermore it was clear that the silencing was independent of the target sequence orientation in the plasmid, and must have occurred at the DNA level, probably due to DNA degradation. Additionally, they successfully engineered the chromosomal CRISPR locus to express customized crRNAs directed against the plasmid and were able to perform directed silencing of an invader plasmid. Their results widely demonstrate the feasibility of CRISPR engineering approach to construct infection resistant prokaryotic strains for industrial use (Elmore et al., 2013).

The earlier study of the hyperthermophilic bacterium *Thermotoga maritima* MSB8 genome revealed that organisms of this group undergo huge transfer of their genes with the archaeal domain (Nelson et al. 1999; Nesbo and Doolittle, 2003). This hypothesis was further validated by comparative genome hybridization (CGH). The researchers studied genome flexibility and lateral gene transfer in the *Thermotogales* (Mongodin et al., 2005). The investigation showed that the metabolic diversity of the organisms of this species was contributed by the loss and gain of various genes. Surprisingly, this gene transfer was not associated with any mobile elements or special repeated sequences but was due to the eight distinct CRISPRs present on their chromosome. These CRISPRs were characterized by the presence of 30-bp repeat element interspersed with a specific sequence of almost same size. Furthermore, a deeper analysis of the short spacers at five loci from five different strains of *T. neapolitana*, showed that they can be grouped into three distinct groups: strain RQ7, strain VMA1/L2B, and a third group with strains NS-E, LA10, and LA4. Similar conclusion was drawn previously by a group of researchers and they stated that hierarchical clustering of CGH data for *T. neapolitana* strains and *T. maritima* strain MSB8, after comparing found to be different from that of the phylogenetic comparison, which compares the 16S rRNA genes of the same strains (Mongodin et al., 2005). Thus, by using the CRISPR spacer sequence analysis we can add that data to 16S rRNA analysis and reconstruct the relationship between these strains. The spacer sequences of strain RQ7 was compared with strain VMA1/L2B and it was shown that the two, share portions of two DNA joints. The strain RQ7 was also shown to have the same two DNA joints as that of the NS-E, LA10, and LA4 strains. As a conclusion of this study, RQ7- strain appears to be the ancestral link between these three *T. neapolitana* strain groups (DeBoy et al., 2006).

The type III-A ststems contains Csm named interference complex which is multisubunit consisting of Cas proteins and a CRISPR RNA (crRNA). This Csm complex is involved in targeting of both RNA and transcriptionally active DNA with the help of guide RNAs but the mechanism is not yet very clear. To get insights in the mechanism a group of scientists overexpressed the five components of the Type III-A Csm complex (TthCsm) along with a defined crRNA sequence in *E. coli* cells. These components were from *Thermus thermophiles*. They then purified the TthCsm complexes from *E. coli* cells. The complexes were found to be highly efficient at 65°C than at 37°C in targeting complementary ssRNA. They also noticed that the single-stranded DNA was cleaved in a sequence-independent by TthCsm and this cleavage was activated by recognizing the complementary ssRNA and did not require complementarity between the first 8 nucleotides (5′ tag) of the crRNA and the 3′ flanking region of the ssRNA. When the histidine-aspartate (HD) nuclease domain of the TthCsm subunit was mutated, the DNA cleavage was destroyed. DNA cleavage was triggered by RNA binding but not by cleavage. Thus, this study presents a new model that stimulates the cleavage of exposed ssDNA in the cell by binding of an ssRNA target to the Csm complex, similar to the RNA polymerase unwinding double-stranded DNA (dsDNA) during transcription. Their findings demonstrated a controllable, thermostable system to deeply study the targeting mechanism with the help of cryo-electron microscopy and x-ray crystallography (Liu et al., 2017).

Advanced sensitive techniques for comparing the DNA sequences and for predicting the protein structures were used by a group of researchers to perform a comparative study of clusters of genes which are grouped as "dark matter islands", in archaeal genomes. These clusters include nearly 20 percent of their genomes and are shown to highly diverse and heterogenous. Particularly, two different, alleged CRISPR-Cas systems that were not demonstrated earlier were identified. It was observed that these variants of CRISPR-Cas were unique. They were seen in only one genome each and very few proteins expressed from these regions were found to be similar in their sequences with the identified Cas protein families studied in Archaeal Clusters of Orthologous Genes (arCOGs). In spite of these different properties, they both consisted of the CRISPR-Cas characters. The Cas1 gene, which is important for the insertion of the spacer, was found missing in this system and so it was thought that this system was incapable of integrating the new spacers into the CRISPR cassettes. Additionally, they observed that though the Cas1 gene was absent in *Thermococcus onnurineus*, it instead, consisted of intact type III-A CRISPR-Cas system (TON_0892-TON_0898) plus presence of some more CRISPR repeat arrays at different regions in the genome. To conclude, these scientists stated that it would be very interesting to study such CRISPR-Cas systems, where either the system functions without spacer insertion or applies another method of integration is not clear. This group also studied *Ignisphaera aggregans* for a different CRISPR-Cas system. The identified CRISPR-Cas system of this organism showed the presence of Cas6 and Cas4 family genes and a gene similar to Csa3 that expressed an anticipated transcriptional regulator consisting of a HTH domain and ligand-binding domain (Makarova et al. 2011a, b). All the three proteins expressed from this locus exhibited restricted resemblance with Csm3 protein. One more protein identified in the HHpred search was found to be extremely deviated form of Cas10 protein in which all catalytic sites stay undamaged, suggesting that the CRISPR-Cas system of *Ignisphaera* is very different form of Type III (Makarova et al., 2014).

CONCLUSION AND FUTURE DIRECTIONS

The CRISPR-Cas systems in bacteria function like genomic sensors to acquire the foreign DNA fragments and confer immunological memory and adaptive immunity in them. The gRNA-directed CRISPR-Cas tool is a recent method of choice for genome editing of various organisms. This system truly holds great potential for genome/gene(s) editing (Puchta, 2017; Tang et al., 2017) but also has some limitations. One of them is temperature sensitivity of the Cas9 proteins. Such limitations can be taken care by understanding the genome of extremophile for the CRISPR-Cas editing tool. Another is off-target activity of CRISPR that can lead to undesired and unrealized pathological outcomes. Alternate specific and more stringent CRISPR-Cas from various other strains of archaea or bacteria can mitigate these off target effects. Improvising of the specificity can also lead to cost efficiency. Upregulation of gene transcription can be achieved by utilizing multiple transcriptional activators and thus the CRISPR efficiency can be enhanced. Multiple sgRNAs were used to recruit many dCas9 activators on the promoter to activate the transcription of genes. Thus, by using this natural genome editing tool ethically and responsibly, we can employ this for the human welfare and benefits of scientific research (Bosley et al., 2015; Mathews et al., 2015). Just eight years after their discovery, this genome editing tool has drastically remodelled life sciences.

REFERENCES

Adalsteinsson, B. T., Kristjansdottir, T., Merre, W., Helleux, A., Dusaucy, J., Tourigny, M., Fridjonsson, O., & Hreggvidsson, G. O. (2021). Efficient genome editing of an extreme thermophile, *Thermus thermophilus,* using a thermostable Cas9 variant. *Scientific Reports*, *11*(1), 9586. doi:10.103841598-021-89029-2 PMID:33953310

Adiego-Perez, B., Randazzo, P., Daran, J. M., Verwaal, R., Roubos, J., Daran-Lapujade, P., & van der Oost, J. (2019). Multiplex genome editing of microorganisms using CRISPR-Cas. *FEMS Microbiology Letters*, *366*(8), fnz086. doi:10.1093/femsle/fnz086 PMID:31087001

Anders, C., Niewoehner, O., Duerst, A., & Jinek, M. (2014). Structural basis of PAM-dependent target DNA recognition by the Cas9 endonuclease. *Nature*, *513*(7519), 569–573. doi:10.1038/nature13579 PMID:25079318

Aparicio, T., Jensen, S. I., Nielsen, A. T., de Lorenzo, V., & Martínez-García, E. (2016). The Ssr protein (T1E_1405) from *Pseudomonas putida* DOT-T1E enables oligonucleotide-based recombineering in platform strain *P. putida* EM42. *Biotechnology Journal*, *11*(10), 1309–1319. doi:10.1002/biot.201600317 PMID:27367544

Barrangou, R., Fremaux, C., Deveau, H., Richards, M., Boyaval, P., Moineau, S., Romero, D., & Philippe Horvath, P. (2007). CRISPR provides acquired resistance against viruses in prokaryotes. *Science*, *315*(5819), 1709–1712. doi:10.1126cience.1138140 PMID:17379808

Barrangou, R., Fremaux, C., Deveau, H., Richards, M., Boyaval, P., Moineau, S., Romero, D. A., & Horvath, P. (2007). CRISPR provides acquired resistance against viruses in prokaryotes. *Science*, *315*(5819), 1709–1712. doi:10.1126cience.1138140 PMID:17379808

Bhalla, A., Bansal, N., Kumar, S., Bischoff, K. M., & Sani, R. K. (2013). Improved lignocellulose conversion to biofuels with thermophilic bacteria and thermostable enzymes. *Bioresource Technology*, *128*, 751–759. doi:10.1016/j.biortech.2012.10.145 PMID:23246299

Bhaya, D., Davison, M., & Barrangou, R. (2011). CRISPR-Cas systems in bacteria and archaea: Versatile small RNAs for adaptive defense and regulation. *Annual Review of Genetics*, *45*(1), 273–297. doi:10.1146/annurev-genet-110410-132430 PMID:22060043

Bikard, D., Euler, C. W., Jiang, W., Nussenzweig, P. M., Goldberg, G. W., Duportet, X., Fischetti, V. A., & Marraffini, L. A. (2014). Exploiting CRISPR-Cas nucleases to produce sequence-specific antimicrobials. *Nature Biotechnology*, *32*(11), 1146–1150. doi:10.1038/nbt.3043 PMID:25282355

Bolotin, A., Quinquis, B., Sorokin, A., & Ehrlich, S. D. (2005). Clustered regularly interspaced short palindrome repeats (CRISPRs) have spacers of extrachromosomal origin. *Microbiology (Reading)*, *151*(Pt 8), 2551–2561. doi:10.1099/mic.0.28048-0 PMID:16079334

Bosley, K. S., Botchan, M., Bredenoord, A. L., Carroll, D., Charo, R. A., Charpentier, E., Cohen, R., Corn, J., Doudna, J., Feng, G., Greely, H. T., Isasi, R., Ji, W., Kim, J. S., Knoppers, B., Lanphier, E., Li, J., Lovell-Badge, R., Martin, G. S., ... Zhou, Q. (2015). CRISPR germline engineering—The community speaks. *Nature Biotechnology*, *33*(5), 478–486. doi:10.1038/nbt.3227 PMID:25965754

Bosma, E. F., van de Weijer, A. H. P., Daas, M. J. A., van der Oost, J., de Vos, W. M., & van Kranenburg, R. (2015). Isolation and screening of thermophilic *Bacilli* from compost for electrotransformation and fermentation: Characterization of *Bacillus smithii* ET 138 as a new biocatalyst. *Applied and Environmental Microbiology*, *81*(5), 1874–1883. doi:10.1128/AEM.03640-14 PMID:25556192

Brouns, S. J., Jore, M. M., Lundgren, M., Westra, E. R., Slijkhuis, R. J., Snijders, A. P., Dickman, M. J., Makarova, K. S., Koonin, E. V., & van der Oost, J. (2008). Small CRISPR RNAs guide antiviral defense in prokaryotes. *Science*, *321*(5891), 960–964. doi:10.1126cience.1159689 PMID:18703739

Brouns, S. J. J., Jore, M. M., Lundgren, M., Westra, E. R., Slijkhuis, R. J. H., Snijders, A. P. L., Dickman, M. J., Makarova, K. S., Koonin, E. V., & Van der Oost, J. (2008). Small CRISPR RNAs guide antiviral defense in prokaryotes. *Science*, *321*(5891), 960–964. doi:10.1126cience.1159689 PMID:18703739

Burstein, D., Harrington, L. B., Strutt, S. C., Probst, A. J., Anantharaman, K., Thomas, B. C., Doudna, J. A., & Banfield, J. F. (2017). New CRISPR-Cas systems from uncultivated microbes. *Nature*, *542*(7640), 237–241. doi:10.1038/nature21059 PMID:28005056

Charpentier, E., Richter, H., van der Oost, J., & White, M. F. (2015). Biogenesis pathways of RNA guides in archaeal and bacterial CRISPR-Cas adaptive immunity. *FEMS Microbiology Reviews*, *39*(3), 428–441. doi:10.1093/femsre/fuv023 PMID:25994611

Chellapandi, P., & Ranjani, J. (2015). Knowledge-based discovery for designing CRISPRCAS systems against invading mobilomes in thermophiles. *Systems and Synthetic Biology*, *9*(3), 97–106. doi:10.100711693-015-9176-8 PMID:26279704

Chen, B., Gilbert, L. A., Cimini, B. A., Schnitzbauer, J., Zhang, W., Li, G. W., Park, J., Blackburn, E. H., Weissman, J. S., Qi, L. S., & Huang, B. (2013). Dynamic imaging of genomic loci in living human cells by an optimized CRISPR/Cas system. *Cell*, *155*(7), 1479–1491. doi:10.1016/j.cell.2013.12.001 PMID:24360272

Chen, X., Yu, L., Qiao, G., & Chen, G. (2018). Reprogramming *Halomonas* for industrial production of chemicals. *Journal of Industrial Microbiology & Biotechnology*, *45*(7), 545–554. doi:10.100710295-018-2055-z PMID:29948194

Chneiweiss, H., Hirsch, F., Montoliu, L., Müller, A. M., Fenet, S., Abecassis, M., Merchant, J., Baertschi, B., Botbol-Baum, M., Houghton, J. A., Kritikos, M., Mifsud, J., Bartnik, E., Rath, J., Druml, C., Friedrich, B., Carvalho, A. S., Lanzerath, D., & Raymond, A. S. (2017). Fostering responsible research with genome editing technologies: A European perspective. *Transgenic Research*, *26*(5), 709–713. doi:10.100711248-017-0028-z PMID:28730514

Cho, J. S., Choi, K. R., Prabowo, C. P. S., Shin, J. H., Yang, D., Jang, J., & Lee, S. Y. (2017). CRISPR/Cas9-coupled recombineering for metabolic engineering of *Corynebacterium glutamicum. Metabolic Engineering*, *42*, 157–167. doi:10.1016/j.ymben.2017.06.010 PMID:28649005

Citorik, R. J., Mimee, M., & Lu, T. K. (2014). Sequence-specific antimicrobials using efficiently delivered RNA-guided nucleases. *Nature Biotechnology*, *32*(11), 1141–1145. doi:10.1038/nbt.3011 PMID:25240928

Cui, Y., Li, Y., Gorge, O., Platonov, M. E., Yan, Y., Guo, Z., Pourcel, C., Dentovskaya, S. V., Balakhonov, S. V., Wang, X., Song, Y., Anisimov, A. P., Vergnaud, G., & Yang, R. (2008). Insight into microevolution of *Yersinia pestis* by clustered regularly interspaced short palindromic repeats. *PLoS One*, *3*(7), e2652. doi:10.1371/journal.pone.0002652 PMID:18612419

DeBoy, R. T., Mongodin, E. F., Emerson, J. B., & Nelson, K. E. (2006). Chromosome evolution in the Thermotogales: Large-scale inversions and strain diversification of CRISPR sequences. *Journal of Bacteriology*, *188*(7), 2364–2374. doi:10.1128/JB.188.7.2364-2374.2006 PMID:16547022

Deltcheva, E., Chylinski, K., Sharma, C., Gonzales, K., Chao, Y., Pirzada, Z., Eckert, M., Vogel, J., & Charpentier, E. (2011). CRISPR RNA maturation by *trans*-encoded small RNA and host factor RNase III. *Nature*, *471*(7340), 602–607. doi:10.1038/nature09886 PMID:21455174

Deveau, H., Barrangou, R., Garneau, J. E., Labonté, J., Fremaux, C., Boyaval, P., Romero, D. A., Horvath, P., & Moineau, S. (2008). Phage response to CRISPR-encoded resistance in Streptococcus thermophilus. *Journal of Bacteriology*, *190*(4), 1390–1400. doi:10.1128/JB.01412-07 PMID:18065545

Don, T. M., Chen, C. W., & Chan, T. H. (2006). Preparation and characterization of poly (hydroxyalkanoate) from the fermentation of *Haloferax mediterranei. Journal of Biomaterials Science. Polymer Edition*, *17*(12), 1425–1438. doi:10.1163/156856206778937208 PMID:17260512

Elmore, J. R., Yokooji, Y., Sato, T., Olson, S., Claiborne, V. C., Glover, I. III, Graveley, B. R., Atomi, H., Terns, R. M., & Terns, M. P. (2013). Programmable plasmid interference by the CRISPR-Cas system in *Thermococcus kodakarensis. RNA Biology*, *10*(5), 828–840. doi:10.4161/rna.24084 PMID:23535213

Ganguly, J., Martin-Pascual, M., & van Kranenburg, R. (2020). CRISPR interference (CRISPRi) as transcriptional repression tool for *Hungateiclostridium thermocellum* DSM 1313. *Microbial Biotechnology*, *13*(2), 339–349. doi:10.1111/1751-7915.13516 PMID:31802632

Garneau, J. E., Dupuis, M. È., Villion, M., Romero, D. A., Barrangou, R., Boyaval, P., Fremaux, C., Horvath, P., Magadán, A. H., & Moineau, S. (2010). The CRISPR/Cas bacterial immune system cleaves bacteriophage and plasmid DNA. *Nature*, *468*(7320), 67–71. doi:10.1038/nature09523 PMID:21048762

Gasiunas, G., Barrangou, R., Horvath, P., & Siksnys, V. (2012). Cas9-crRNA ribonucleoprotein complex mediates specific DNA cleavage for adaptive immunity in bacteria. *Proceedings of the National Academy of Sciences of the United States of America*, *109*(39), E2579–E2586. doi:10.1073/pnas.1208507109 PMID:22949671

Gilbert, L. A., Larson, M. H., Morsut, L., Liu, Z., Brar, G. A., Torres, S. E., Stern-Ginossar, N., Brandman, O., Whitehead, E. H., Doudna, J. A., Lim, W. A., Weissman, J. S., & Qi, L. S. (2013). CRISPR-mediated modular RNA-guided regulation of transcription in eukaryotes. *Cell*, *154*(2), 442–451. doi:10.1016/j.cell.2013.06.044 PMID:23849981

Gophna, U., Allers, T., & Marchfelder, A. (2017). Finally, archaea get their CRISPR-Cas toolbox. *Trends in Microbiology*, *25*(6), 430–432. doi:10.1016/j.tim.2017.03.009 PMID:28391963

Guo, L., Brugger, K., Liu, C., Shah, S. A., Zheng, H., Zhu, Y., Wang, S., Lillestol, R. K., Chen, L., Frank, J., Prangishvili, D., Paulin, L., She, Q., Huang, L., & Garrett, R. A. (2011). Genome analyses of Icelandic strains of Sulfolobus islandicus, model organisms for genetic and virus-host interaction studies. *Journal of Bacteriology*, *193*(7), 1672–1680. doi:10.1128/JB.01487-10 PMID:21278296

Haft, D. H., Selengut, J., Mongodin, E. F., & Nelson, K. E. (2005). A guild of 45 CRISPR-associated (Cas) protein families and multiple CRISPR/Cas subtypes exist in prokaryotic genomes. *PLoS Computational Biology*, *1*(6), e60. doi:10.1371/journal.pcbi.0010060 PMID:16292354

Horvath, P., & Barrangou, R. (2010). CRISPR/Cas, the immune system of bacteria and archaea. *Science*, *327*(5962), 167–170. doi:10.1126cience.1179555 PMID:20056882

Horvath, P., Romero, D. A., Coûté-Monvoisin, A. C., Richards, M., Deveau, H., Moineau, S., Boyaval, P., Fremaux, C., & Barrangou, R. (2008). Diversity, activity, and evolution of CRISPR loci in Streptococcus thermophilus. *Journal of Bacteriology*, *190*(4), 1401–1412. doi:10.1128/JB.01415-07 PMID:18065539

Hsu, P. D., Lander, E. S., & Zhang, F. (2014). Development and applications of CRISPR-Cas9 for genome engineering. *Cell*, *157*(6), 1262–1278. doi:10.1016/j.cell.2014.05.010 PMID:24906146

Ishino, Y., Krupovic, M., & Forterre, P. (2018). History of CRISPR-Cas from Encounter with a Mysterious Repeated Sequence to Genome Editing Technology. *Journal of Bacteriology*, *200*(7), 5429. doi:10.1128/JB.00580-17 PMID:29358495

Ishino, Y., Shinagawa, H., Makino, K., Amemura, M., & Nakata, A. (1987). Nucleotide sequence of the iap gene, responsible for alkaline phosphatase isozyme conversion in *Escherichia coli,* and identification of the gene product. *Journal of Bacteriology*, *169*(12), 5429–5433. doi:10.1128/jb.169.12.5429-5433.1987 PMID:3316184

Jansen, R., Embden, J. D., Gaastra, W., & Schouls, L. M. (2002). Identification of genes that are associated with DNA repeats in prokaryotes. *Molecular Microbiology*, *43*(6), 1565–1575. doi:10.1046/j.1365-2958.2002.02839.x PMID:11952905

Jiang, Y., Chen, B., Duan, C., Sun, B., Yang, J., & Yang, S. (2015). Multigene editing in the *Escherichia coli* genome via the CRISPR-Cas9 system. *Applied and Environmental Microbiology*, *81*(7), 2506–2514. doi:10.1128/AEM.04023-14 PMID:25636838

Jinek, M., Chylinski, K., Fonfara, I., Hauer, M., Doudna, J. A., & Charpentier, E. (2012). A programmable dual-RNA-guided DNA endonuclease in adaptive bacterial immunity. *Science*, *337*(6096), 816–821. doi:10.1126cience.1225829 PMID:22745249

Johnson, K., Jiang, Y., Kleerebezem, R., Muyzer, G., & Van Loosdrecht, M. C. M. (2009). Enrichment of a mixed bacterial culture with a high polyhydroxyalkanoate storage capacity. *Biomacromolecules*, *10*(4), 670–676. doi:10.1021/bm8013796 PMID:19193058

Kambam, P. K. R., & Henson, M. A. (2010). Engineering bacterial processes for cellulosic ethanol production. *Biofuels*, *1*(5), 729–743. doi:10.4155/bfs.10.46

Kamerbeek, J., Schouls, L., Kolk, A., van Agterveld, M., van Soolingen, D., Kuijper, S., Bunschoten, A., Molhuizen, H., Shaw, R., Goyal, M., & van Embden, J. (1997). Simultaneous detection and strain differentiation of *Mycobacterium tuberculosis* for diagnosis and epidemiology. *Journal of Clinical Microbiology*, *35*(4), 907–914. doi:10.1128/jcm.35.4.907-914.1997 PMID:9157152

Koonin, E. V., Makarova, K. S., & Zhang, F. (2017). Diversity, classification and evolution of CRISPR-Cas systems. *Current Opinion in Microbiology*, *37*, 67–78. doi:10.1016/j.mib.2017.05.008 PMID:28605718

Kosicki, M., Tomberg, K., & Bradley, A. (2018). Repair of double-strand breaks induced by CRISPR-Cas9 leads to large deletions and complex rearrangements. *Nature Biotechnology*, *36*(8), 765–771. doi:10.1038/nbt.4192 PMID:30010673

Le, Y., Fu, Y., & Sun, J. (2021). Genome editing of the anaerobic thermophile Thermoanaerobacter ethanolicus using thermostable Cas9. *Applied and Environmental Microbiology*, *87*, e01773–e20. PMID:33067194

Li, Y., Pan, S., Zhang, Y., Ren, M., Feng, M., Peng, N., Chen, L., Liang, Y. X., & She, Q. (2016). Harnessing Type I and Type III CRISPR-Cas systems for genome editing. *Nucleic Acids Research*, *44*(4), e34. doi:10.1093/nar/gkv1044 PMID:26467477

Lino, C. A., Harper, J. C., Carney, J. P., & Timlin, J. A. (2018). Delivering CRISPR: A review of the challenges and approaches. *Drug Delivery*, *25*(1), 1234–1257. doi:10.1080/10717544.2018.1474964 PMID:29801422

Liu, F., Kariyawasam, S., Jayarao, B. M., Barrangou, R., Gerner-Smidt, P., Ribot, E. M., Knabel, S. J., & Dudley, E. G. (2011). Subtyping *Salmonella enterica* serovar enteritidis isolates from different sources by using sequence typing based on virulence genes and clustered regularly interspaced short palindromic repeats (CRISPRs). *Applied and Environmental Microbiology*, *77*(13), 4520–4526. doi:10.1128/AEM.00468-11 PMID:21571881

Liu, T. Y., Iavarone, A. T., & Doudna, J. A. (2017). RNA and DNA targeting by a reconstituted *Thermus thermophiles* Type III-A CRISPR-Cas system. *PLoS One*, *12*(1), 1–20. doi:10.1371/journal.pone.0170552

Ma, K., Maeda, T., You, H., & Shirai, Y. (2014). Open fermentative production of l-lactic acid with high optical purity by thermophilic Bacillus coagulans using excess sludge as nutrient. *Bioresource Technology*, *151*, 28–35. doi:10.1016/j.biortech.2013.10.022 PMID:24201025

Maier, L. K., Fischer, S., Stoll, B., Brendel, J., Pfeiffer, F., Dyall-Smith, M., & Marchfelder, A. (2012). The immune system of halophilic archaea. *Mobile Genetic Elements*, *2*(5), 228–232. doi:10.4161/mge.22530 PMID:23446883

Makarova, K., Aravind, L., Wolf, Y., & Koonin, E. (2011). Unification of Cas protein families and a simple scenario for the origin and evolution of CRISPR-Cas systems. *Biology Direct*, *6*(1), 38. doi:10.1186/1745-6150-6-38 PMID:21756346

Makarova, K. S., Aravind, L., Wolf, Y. I., & Koonin, E. V. (2011a). Unification of Cas protein families and a simple scenario for the origin and evolution of CRISPR-Cas systems. *Biology Direct*, *6*(1), 38. doi:10.1186/1745-6150-6-38 PMID:21756346

Makarova, K. S., Grishin, N. V., Shabalina, S. A., Wolf, Y. I., & Koonin, E. V. (2006). A putative RNA-interference-based immune system in prokaryotes: Computational analysis of the predicted enzymatic machinery, functional analogies with eukaryotic RNAi, and hypothetical mechanisms of action. *Biology Direct*, *1*(1), 7. doi:10.1186/1745-6150-1-7 PMID:16545108

Makarova, K. S., Haft, D. H., Barrangou, R., Brouns, S. J., Charpentier, E., Horvath, P., Moineau, S., Mojica, F. J., Wolf, Y. I., Yakunin, A. F., van der Oost, J., & Koonin, E. V. (2011b). Evolution and classification of the CRISPR-Cas systems. *Nature Reviews. Microbiology*, *9*(6), 467–477. doi:10.1038/nrmicro2577 PMID:21552286

Makarova, K. S., Wolf, Y. I., Alkhnbashi, O. S., Costa, F., Shah, S. A., Saunders, S. J., Barrangou, R., Brouns, S. J., Charpentier, E., Haft, D. H., Horvath, P., Moineau, S., Mojica, F. J., Terns, R. M., Terns, M. P., White, M. F., Yakunin, A. F., Garrett, R. A., van der Oost, J., ... Koonin, E. V. (2015). An updated evolutionary classification of CRISPR-Cas systems. *Nature Reviews. Microbiology*, *13*(11), 722–736. doi:10.1038/nrmicro3569 PMID:26411297

Makarova, K. S., Wolf, Y. I., Forterre, P., Prangishvili, D., Krupovic, M., & Koonin, E. V. (2014). Dark matter in archaeal genomes: A rich source of novel mobile elements, defense systems and secretory complexes. *Extremophiles*, *18*(5), 877–893. doi:10.100700792-014-0672-7 PMID:25113822

Marraffini, L. A., & Sontheimer, E. J. (2008). CRISPR interference limits horizontal gene transfer in staphylococci by targeting DNA. *Science*, *322*(5909), 1843–1845. doi:10.1126cience.1165771 PMID:19095942

Maruyama, T., Dougan, S. K., Truttmann, M. C., Bilate, A. M., Ingram, J. R., & Ploegh, H. L. (2015). Increasing the efficiency of precise genome editing with CRISPR-Cas9 by inhibition of nonhomologous end joining. *Nature Biotechnology*, *33*(5), 538–542. doi:10.1038/nbt.3190 PMID:25798939

Mathews, D. J., Chan, S., Donovan, P. J., Douglas, T., Gyngell, C., Harris, J., Regenberg, A., & Lovell-Badge, R. (2015). CRISPR: A path through the thicket. *Nature*, *527*(7577), 159–161. doi:10.1038/527159a PMID:26560284

Mohanraju, P., Mougiakos, I., Albers, J., Mabuchi, M., Fuchs, R., Curcuru, J., van Kranenburg, R., Brett Robb, G., & van der Oost, J. (2021). Development of a Cas12a-Based Genome Editing Tool for Moderate Thermophiles. *The CRISPR Journal*, *4*(1), 82–91. doi:10.1089/crispr.2020.0086 PMID:33538626

Mojica, F. J., Díez-Villaseñor, C., Soria, E., & Juez, G. (2000). Biological significance of a family of regularly spaced repeats in the genomes of Archaea, Bacteria and mitochondria. *Molecular Microbiology*, *36*(1), 244–246. doi:10.1046/j.1365-2958.2000.01838.x PMID:10760181

Mojica, F. J. M., Díez-Villaseñor, C., García-Martínez, J., & Almendros, C. (2009). Short motif sequences determine the targets of the prokaryotic CRISPR defence system. *Microbiology (Reading)*, *155*(Pt 3), 733–740. doi:10.1099/mic.0.023960-0 PMID:19246744

Mojica, M. J., Juez, G., & Rodríguez-Valera, F. (1993). Transcription at different salinities of *Haloferax mediterranei* sequences adjacent to partially modified PstI sites. *Molecular Microbiology*, *9*(3), 613–621. doi:10.1111/j.1365-2958.1993.tb01721.x PMID:8412707

Mokrousov, I., Narvskaya, O., Limeschenko, E., & Vyazovaya, A. (2005). Efficient discrimination within a *Corynebacterium diphtheriae* epidemic clonal group by a novel macroarray-based method. *Journal of Clinical Microbiology*, *43*(4), 1662–1668. doi:10.1128/JCM.43.4.1662-1668.2005 PMID:15814981

Mongodin, E. F., Hance, I. R., DeBoy, R. T., Gill, S. R., Daugherty, S., Huber, R., Fraser, C. M., Stetter, K., & Nelson, K. E. (2005). Gene transfer and genome plasticity in *Thermotoga maritima*, a model hyperthermophilic species. *Journal of Bacteriology*, *187*(14), 4935–4944. doi:10.1128/JB.187.14.4935-4944.2005 PMID:15995209

Mougiakos, I., Bosma, E. F., de Vos, W. M., van Kranenburg, R., & van der Oost, J. (2016). Next generation prokaryotic engineering: The CRISPR-Cas toolkit. *Trends in Biotechnology*, *34*(7), 575–587. doi:10.1016/j.tibtech.2016.02.004 PMID:26944793

Mougiakos, I., Mohanraju, P., Bosma, E., Vrouwe, V., Finger, M., Naduthodi, M., Gussak, A., Brinkman, R., van Kranenburg, R., & Oost, J. (2017). Characterizing a thermostable Cas9 for bacterial genome editing and silencing. *Nature Communications*, *8*(1), 1647. doi:10.103841467-017-01591-4 PMID:29162801

Nayak, D. D., & Metcalf, W. W. (2017). Cas9-mediated genome editing in the methanogenic archaeon *Methanosarcina acetivorans. Proceedings of the National Academy of Sciences of the United States of America*, *114*(11), 2976–2981. doi:10.1073/pnas.1618596114 PMID:28265068

Nelson, K. E., Clayton, R. A., Gill, S. R., Gwinn, M. L., Dodson, R. J., Haft, D. H., Hickey, E. K., Peterson, J. D., Nelson, W. C., Ketchum, K. A., McDonald, L., Utterback, T. R., Malek, J. A., Linher, K. D., Garrett, M. M., Stewart, A. M., Cotton, M. D., Pratt, M. S., Phillips, C. A., ... Fraser, C. M. (1999). Evidence for lateral gene transfer between archaea and bacteria from genome sequence of Thermotoga maritima. *Nature*, *399*(6734), 323–329. doi:10.1038/20601 PMID:10360571

Nesbo, C. L., & Doolittle, W. F. (2003). Targeting clusters of transferred genes in *Thermotoga maritima. Environmental Microbiology*, *5*(11), 1144–1154. doi:10.1046/j.1462-2920.2003.00515.x PMID:14641594

Nishimasu, H., Cong, L., Yan, W. X., Ran, F. A., Zetsche, B., Li, Y., Kurabayashi, A., Ishitani, R., Zhang, F., & Nureki, O. (2015). Crystal Structure of *Staphylococcus aureus* Cas9. *Cell*, *162*(5), 1113–1126. doi:10.1016/j.cell.2015.08.007 PMID:26317473

Nuñez, J. K., Kranzusch, P. J., Noeske, J., Wright, A. V., Davies, C. W., & Doudna, J. A. (2014). Cas1-Cas2 complex formation mediates spacer acquisition during CRISPR-Cas adaptive immunity. *Nature Structural & Molecular Biology*, *21*(6), 528–534. doi:10.1038/nsmb.2820 PMID:24793649

Olson, D. G., Sparling, R., & Lynd, L. R. (2015). Ethanol production by engineered thermophiles. *Current Opinion in Biotechnology*, *33*, 130–141. doi:10.1016/j.copbio.2015.02.006 PMID:25745810

Ouyang, J., Ma, R., Zheng, Z., Cai, C., Zhang, M., & Jiang, T. (2013). Open fermentative production of L-lactic acid by Bacillus sp. strain NL01 using lignocellulosic hydrolyzates as low-cost raw material. *Bioresource Technology*, *135*, 475–480. doi:10.1016/j.biortech.2012.09.096 PMID:23127843

Peng, N., Han, W., Li, Y., Liang, Y., & She, Q. (2017). Genetic technologies for extremely thermophilic microorganisms of Sulfolobus, the only genetically tractable genus of crenarchaea. *Science China. Life Sciences*, *60*(4), 370–385. doi:10.100711427-016-0355-8 PMID:28251462

Puchta, H. (2017). Applying CRISPR/Cas for genome engineering in plants: The best is yet to come. *Current Opinion in Plant Biology*, *36*, 1–8. doi:10.1016/j.pbi.2016.11.011 PMID:27914284

Quillaguamán, J., Guzmán, H., Van-Thuoc, D., & Hatti-Kaul, R. (2010). Synthesis and production of polyhydroxyalkanoates by halophiles: Current potential and future prospects. *Applied Microbiology and Biotechnology*, *85*(6), 1687–1696. doi:10.100700253-009-2397-6 PMID:20024541

Sapranauskas, R., Gasiunas, G., Fremaux, C., Barrangou, R., Horvath, P., & Siksnys, V. (2011). The *Streptococcus thermophilus* CRISPR/Cas system provides immunity in *Escherichia coli*. *Nucleic Acids Research*, *39*(21), 9275–9282. doi:10.1093/nar/gkr606 PMID:21813460

Schmidt, S. T., Yu, F. B., Blainey, P. C., May, A. P., & Quake, S. R. (2019). Nucleic acid cleavage with a hyperthermophilic Cas9 from an uncultured Ignavibacterium. *Proceedings of the National Academy of Sciences of the United States of America*, *116*(46), 23100–23105. doi:10.1073/pnas.1904273116 PMID:31659048

Shmakov, S., Smargon, A., Scott, D., Cox, D., Pyzocha, N., Yan, W., Abudayyeh, O. O., Gootenberg, J. S., Makarova, K. S., Wolf, Y. I., Severinov, K., Zhang, F., & Koonin, E. V. (2017). Diversity and evolution of class 2 CRISPR-Cas systems. *Nature Reviews. Microbiology*, *15*(3), 169–182. doi:10.1038/nrmicro.2016.184 PMID:28111461

Szczelkun, M. D., Tikhomirova, M. S., Sinkunas, T., Gasiunas, G., Karvelis, T., Pschera, P., Siksnys, V., & Seidel, R. (2014). Direct observation of R-loop formation by single RNA-guided Cas9 and Cascade e_ector complexes. *Proceedings of the National Academy of Sciences of the United States of America*, *111*(27), 9798–9803. doi:10.1073/pnas.1402597111 PMID:24912165

Tamulaitis, G., Venclovas, C., & Siksnys, V. (2017). Type III CRISPR-Cas immunity: Major differences brushed aside. *Trends in Microbiology*, *25*(1), 49–61. doi:10.1016/j.tim.2016.09.012 PMID:27773522

Tan, D., Xue, Y. S., Aibaidula, G., & Chen, G. Q. (2011). Unsterile and continuous production of polyhydroxybutyrate by *Halomonas* TD01. *Bioresource Technology*, *102*(17), 8130–8136. doi:10.1016/j.biortech.2011.05.068 PMID:21680179

Tang, X., Lowder, L. G., Zhang, T., Malzahn, A. A., Zheng, X., Voytas, D. F., Zhong, Z., Chen, Y., Ren, Q., Li, Q., Kirkland, E. R., Zhang, Y., & Qi, Y. (2017). A CRISPR-Cpf1 system for efficient genome editing and transcriptional repression in plants. *Nature Plants*, *3*(3), 17018. doi:10.1038/nplants.2017.18 PMID:28211909

Taylor, M. P., van Zyl, L., Tuffin, I. M., Leak, D. J., & Cowan, D. A. (2011). Genetic tool development underpins recent advances in thermophilic whole-cell biocatalysts. *Microbial Biotechnology*, *4*(4), 438–448. doi:10.1111/j.1751-7915.2010.00246.x PMID:21310009

Venclovas, C. (2016). Structure of Csm2 elucidates the relationship between small subunits of CRISPR-Cas effector complexes. *FEBS Letters*, *590*(10), 1521–1529. doi:10.1002/1873-3468.12179 PMID:27091242

Walker, J., Lanahan, A., Zheng, T., Toruno, C., Lynd, L., Cameron, J., Olson, D., & Eckert, C. (2020). Development of both type I–B and type II CRISPR/Cas genome editing systems in the cellulolytic bacterium *Clostridium thermocellum*. *Metabolic Engineering Communications*, *10*, e00116. doi:10.1016/j.mec.2019.e00116 PMID:31890588

Wang, J., Li, J., Zhao, H., Sheng, G., Wang, M., Yin, M., & Wang, Y. (2015). Structural and Mechanistic Basis of PAM-Dependent Spacer Acquisition in CRISPR-Cas Systems. *Cell, 163*(4), 840–853. doi:10.1016/j.cell.2015.10.008 PMID:26478180

Wiedenheft, B., Sternberg, S. H., & Doudna, J. A. (2012). RNA-guided genetic silencing systems in bacteria and archaea. *Nature, 482*(7385), 331–338. doi:10.1038/nature10886 PMID:22337052

Yamano, T., Zetsche, B., Ishitani, R., Zhang, F., Nishimasu, H., & Nureki, O. (2017). Structural Basis for the Canonical and Non-canonical PAM Recognition by CRISPR-Cpf1. *Molecular Cell, 67*(4), 633–645. doi:10.1016/j.molcel.2017.06.035 PMID:28781234

Yosef, I., Goren, M. G., & Qimron, U. (2012). Proteins and DNA elements essential for the CRISPR adaptation process in *Escherichia coli. Nucleic Acids Research, 40*(12), 5569–5576. doi:10.1093/nar/gks216 PMID:22402487

Yu, C., Liu, Y., Ma, T., Liu, K., Xu, S., Zhang, Y., Liu, H., La Russa, M., Xie, M., Ding, S., & Qi, L. S. (2015). Small molecules enhance CRISPR genome editing in pluripotent stem cells. *Cell Stem Cell, 16*(2), 142–147. doi:10.1016/j.stem.2015.01.003 PMID:25658371

Yu, F. B., Blainey, P. C., Schulz, F., Woyke, T., Horowitz, M. A., & Quake, S. R. (2017). Microfluidic-based mini-metagenomics enables discovery of novel microbial lineages from complex environmental samples. *eLife, 6*, e26580. doi:10.7554/eLife.26580 PMID:28678007

Zetsche, B., Gootenberg, J. S., Abudayyeh, O. O., Slaymaker, I. M., Makarova, K. S., Essletzbichler, P., Volz, S. E., Joung, J., van der Oost, J., Regev, A., Koonin, E. V., & Zhang, F. (2015). Cpf1 is a single RNA-guided endonuclease of a class 2 CRISPR-Cas system. *Cell, 163*(3), 759–771. doi:10.1016/j.cell.2015.09.038 PMID:26422227

Chapter 8
Major Compatible Solutes and Structural Adaptation of Proteins in Extremophiles

Hardik Shah
https://orcid.org/0000-0003-2956-4099
Ganpat University, India

Khushbu Panchal
Ganpat University, India

Amisha Panchal
Ganpat University, India

ABSTRACT

Extremophiles are the most ancient microbes on the Earth and also a center of attraction for the scientific community for research because of their ability to adapt to extreme habitats. Compatible solutes are among those factors which enable these microorganisms to thrive in such extreme habitats. Under osmotic stress, the majority of extremophiles accumulate specific organic solutes such as amino acids, sugars, polyols, and their derivatives. In addition, proteins in extremophiles are found to be evolved by changing their amino acid composition to alter the hydrophobicity of its core and surface charge to maintain activity. This chapter encompasses a comprehensive study about the role of various compatible solutes in the endurance of microorganisms under extremophilic conditions, synthesis of compatible solutes, nature of extremophilic proteins, and their applications. Furthermore, an attempt has been made to cover various strategies adopted by the scientific community while pursuing research on compatible solutes.

DOI: 10.4018/978-1-7998-9144-4.ch008

INTRODUCTION

Over the previous decades, the scientific community had explored various extreme environments for the hunt of microorganisms that can thrive in them. Because of their ability to sustain life in extreme habitats, these microorganisms are entitled extremophiles. Extremophiles can be considered as an important key to the evolutionary history on earth since such habitats are existing on earth from the time of evolution. These organisms can survive in extreme hot niches, acidic, alkaline, salt solutions, and ice whereas some have the unique ability to grow in toxic waste (Alavi et al., 2020), organic solvents (Isken & de Bont, 1998), and heavy metals (Alavi et al., 2020). Extreme conditions mentioned above are intolerable for other earthly life-forms. The word 'extreme' in itself implies that they are capable of growing and surviving in uncommon conditions. These microorganisms utilize various strategies for adaptation to extreme habitats. Proteins and other molecules of extremophilic microorganisms can be great resources for mankind with unique biotechnological importance (Raddadi et al., 2015). The organisms are categorized as extremotolerant and extremophiles which indicates that former organisms can tolerate extreme environments while later love such harsh conditions respectively (Verma et al., 2020).

Most of these organisms fall under the domain of Bacteria and Archaea. This includes those that love the extremely high temperature and extremely low temperature known as Thermo/Hyperthermophiles and psychrophiles respectively, those that survive in high pH and low pH termed as Acidophiles and Alkalophiles respectively, organisms survive in high salt concentration known as Halophiles, those that thrive at high atmospheric pressure called barophiles and organisms which can withstand a high level of toxic agents are termed, toxitolerants (Rampelotto, 2013). As physicochemical parameters can be extremely high where these microorganisms are capable to live, they are also called polyextremophiles (Chela-Flores, 2013). A common example of polyextremophiles is thermoacidophiles which can survive in extremely high temperatures as well as they require acidic pH for survival. In addition to being extremely alkaline and acidic simultaneously, many hot springs are rich in metal content. Several hypersaline lakes are extremely alkaline, while the deep oceans are usually cold and oligotrophic (very low nutrient content) and exposed to high pressure (Strazzulli et al., 2020).

ADAPTATION TO THE EXTREME ENVIRONMENT

For any cell with smaller dimensions of micrometer-scale and a higher surface-to-volume ratio, surviving in an extreme environment could be a critical challenge. In the case of hyperthermophiles, to survive high temperatures for a lifetime, all the cell components must be either heat resistant or had adopted an alternative mechanism that aids in thermal stability. Therefore, they are evolved with the strategies to survive in extremely high temperatures (Berezovsky & Shakhnovich, 2005). Biomolecules required in the cell such as lipids, nucleic acid, and proteins must be heat resistant for survival in high temperatures. Structural and functional proteins available in the cells must be thermostable. Additionally, the process of protein folding must be synchronized in a manner to fold it appropriately immediately followed by protein synthesis to exert thermostability. Reverse gyrase is a type I DNA topoisomerase found in thermophiles, that stabilize DNA throughout the processes involving DNA such as replication, transcription, etc. (Garnier et al., 2021). Heat stability of the protein in hyper/thermophiles is due to the increased number of salt bridges and the highly hydrophobic interior of the protein. Membranes of these microbes are found to be rich in saturated fatty acids. In some thermophiles, special lipids are present such as hopanoids

(Hippchen et al., 1981). Archeal hyperthermophiles have hydrocarbons of various lengths composed of repeating units of 5-6 compound phytans bonded by ether linkage to glycerophosphate. They also possess histones phylogenetically related to the eukaryotic core histones, such as H2A and H2B (Sabath et al., 2013). As indicated by *In vitro* studies, binding with histones increased the melting temperature of purified DNA significantly. The strategy behind the stabilization of RNA secondary structure, the stem areas have an increased content of GC bases and it is further strengthened by post-transcriptional modifications (Podar et al., 2020) The enzymes isolated from the hyper/thermophiles are highly stable at high temperatures, for example, an amylase obtained from *Pyrococcus woesei* exhibits optimal activity at 130°C (Koch et al., 1991). The function of heat shock proteins seems to become essential for the growth of hyperthermophiles at higher temperatures.

At the upper-temperature border of the growth of hyperthermophiles, the function of heat-shock proteins appears to become essential. Molecular studies on crude extract of *Pyrodictium occultum* at extreme temperatures about 108°C, 80% fraction of the soluble protein consisted of a heat-inducible molecular chaperone designated as thermosome (Sabath et al., 2013).

Halophiles remain constantly exposed to the environment with higher concentrations of salt for a lifelong period. In hypersaline conditions, high osmolarity results in loss of water in the external medium which can be deleterious to most of the cells since it may result in dehydration. Halophiles prevent such loss of water by adapting special strategies (Cycil et al., 2020) Adaption to elevated salt concentration begins with cell wall structure where cell wall of halophiles lacks the peptidoglycan and possesses ether-linked lipid. Various proteins and enzymes such as Archean type RNA polymerase can maintain rigidity even when exposed to higher salt concentrations, which allows increased endurance during contact with broad range of salt concentrations (Farraj et al., 2020).

The first strategy to adjust in the environments with high salt concentrations is 'high salt-in strategy' which involves the accumulation of potassium and chloride molar concentration. As the protein should maintain its proper conformation and activity at near-saturating salt concentration, this strategy is helpful for adaptation of the intracellular enzymatic machinery in the presence of salt (Ruginescu et al., 2020). Alternatively, halophiles also adapt to the salt stress either by producing a large amount of internal solute or by containing a solute extracted from the outside environment. The best-known example is *Halobacterium salinarum* that maintains a high intracellular concentration of KCl. Thus, the enzymes present in its cytoplasm will be able to function only if there is a high intracellular salt concentration (Dennis, 2020) (Ruginescu et al., 2020)

Another strategy for molecular adaptation in halophiles is biosynthesis or accumulation of organic compatible solutes. Also, in such bacteria, the concentration of K^+ ions inside the cell is higher than the concentration of Na^+ ions outside the cell which acts as a solute and it will exclude salt from their cytoplasm as much as possible. Hence, it maintains cell integrity and does not interfere with normal enzymatic activity (Imhoff, 2020).

COMPATIBLE SOLUTES

The term "Compatible solutes" was coined by Brown. The substance which is compatible with cellular metabolism and its accumulation in extremophiles will make them adapted to osmotic stress and protect macromolecules as well as a cell against stress conditions such as high temperature, high salt concentration, and desiccation. Thus, the mechanism by which solutes can stabilize enzymes and proteins as well as the utilization of these solutes to protect macromolecules and cells from heating, and freezing is of great research interest nowadays.

A most interesting property of the extremophilic cell is to adapt and survive in external media such as high temperature and high salt concentration. In the high salt concentration, water availability to the cell is inversely proportional to dissolved solute concentration. So, to survive in extremely high salt concentration or high temperature the organisms will accumulate the small organic solutes which will counteract the external osmotic pressure (Harishchandra et al., 2010). These small organic molecules are also called osmolytes or compatible solutes. Most topics of research interest about extremophiles are based on the occurrence of these osmolytes, study about their biosynthetic pathways, and mechanism of their regulation in external osmotic pressure. There is a consensus among studies that the accumulation of solutes has a major role in osmotic balance and it increases protein stability as well. They act as a chemical chaperone in cells and have a thermostabilizing capacity which can be exploited for various biotechnological applications (Schulz et al., 2017)

Mostly, compatible solutes are produced by halophiles and hyper/thermophiles such as sugars and their derivatives (Zahid et al., 2015), amino acids and their derivatives (Imhoff & Rodriguez-Valera, 1984), polyols and derivatives (Rice et al., 2019), betaines and ectoines (Crowley, 2017). Majorly bacteria and eukaryotes usually will accumulate neutral compatible solutes whereas archaea will produce negatively charged solutes. In some examples of archaea, they can modify many of the same neutral or zwitterionic solutes to negatively charged solutes. Compatible solutes are either synthesized by the cell or transported into the cell from an external source. The key role of these molecules is to modulate individual enzyme activities, assist in the maintenance of turgor pressure, cell volume, and electrolyte concentration. All these functions provide a valuable reason to entitle it as 'compatible solutes' (Cheng et al., 2020) Compatible solutes fall into three categories: i) zwitterionic solutes, ii) noncharged solutes, and iii) anionic solutes.

1. Zwitterionic Solutes

The accumulation of Intracellular K^+ leads to the osmotic balance across the membrane and also stabilizes the turgor pressure. As the salinity of the growth medium is increased, the incoming charges of K^+ are not compensated by the accumulation of Cl^-. Instead, it is neutralized by the accumulation of organic solutes such as amino acids and derivatives like α-glutamate. The most of zwitterionic solutes produced are betaine, ectoine, hydroxyectoine, Nε-acetyl-β-lysine, and, β-glutamine. As intracellular reserves of nutrients and energy, some organisms can accumulate the mixtures of both carbohydrate and nitrogen-containing solutes e.g., *Halomonas socia* NY-011, a moderate halophile can accumulate glutamate as a major solute along with the synthesis of secondary compatible solutes such as ectoine, hydroxyectoine, and glycine betaine (Peng et al., 2020). In most halophiles and thermophiles when the amount of fixed nitrogen is high the ectoine is accumulated and intracellular trehalose is lower down or completely absent (Vargas et al., 2006). But during Nitrogen starvation, ectoine is replaced by trehalose and partially replaces by glycine betaine. Thus, based on Nitrogen availability the activity of trehalose and glycine betaine synthesis is followed (Galinski & Herzog, 1990).

Table 1. Occurrence and importance of zwitterionic compatible solutes

Zwitterionic Compatible Solutes	Occurrence and Importance
Betaine 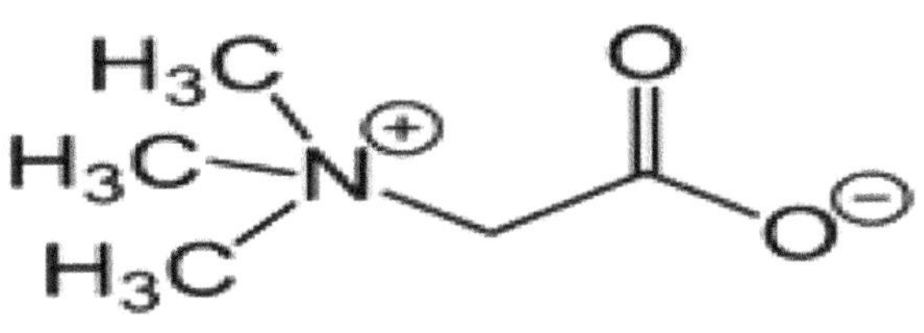	The universal solute, found in most halophilic bacteria in phylogenetic diversity. It is transported from a complex medium and concentration varies with external NaCl. The synthesis of betaine is carried out either by oxidation of choline or by methylation of glycine. As a compatible solute, betaine is responsible for the ability of extremophiles to stabilize cellular proteins. The presence of this compatible solute is observed in mostly halotolerant and halophilic microbes such as *Thioalkalivibrio versutus, Actinopolyspora halophila, Halorhodospira halochloris*, etc. (Berben et al., 2017; Deole & Hoff, 2020)
Ectoine 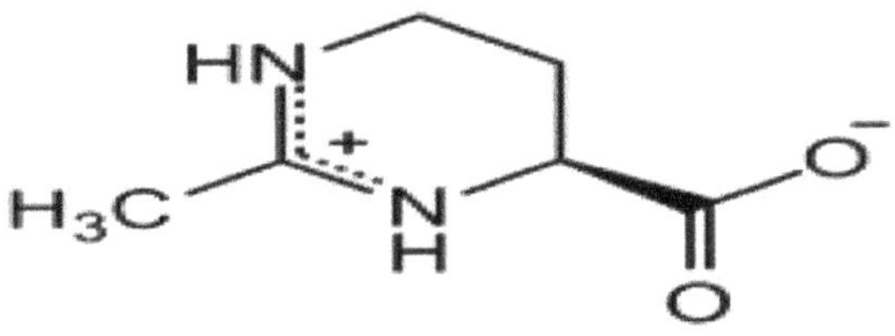	Ectoine is known as Cyclic tetrahydro pyrimidine (1,4,5,6-tetrahydro-2-methyl-4-pyrimidine carboxylic acid). The concentration of intracellular ectoine is increased by an increase in extracellular NaCl concentration. Hence, ectoine is a major compatible solute of aerobic chemoheterotrophic bacteria, alkaline, hypersaline Monolake, halophilic methylotrophic bacteria, etc. Examples of microorganisms includes *Spororsarcina pasteurii, Thioalkalimicrobium aerophilum, Chromohalobacter israelensis, Halomonas elongate, Methylophaga alcalica*, etc. (Harishchandra et al., 2011; Vargas et al., 2006)
Hydroxyectoine 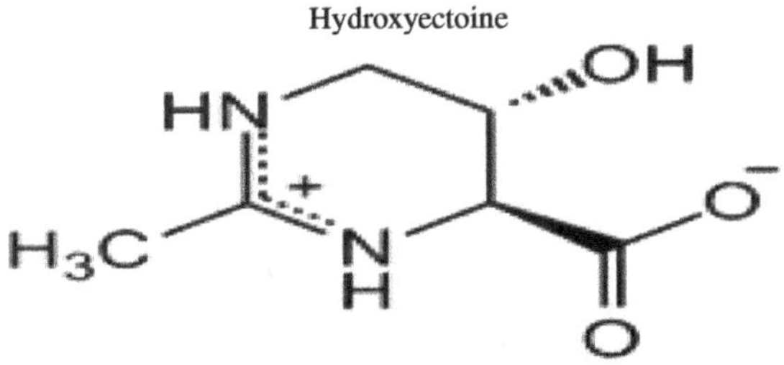	The accumulation of Hydroxyectoine is similar to the same as ectoine because it is a derivative of ectoine. It is a rare naturally occurring compatible solute known as (4S, 5S)-2-methyl-5-hydroxy- 1,4,5,6-tetrahydrpyrimidine-4-carboxylic acid. Hydroxyectoine accumulation by the organism protects own self from the detrimental effect of water efflux, cell dehydration, and considerable drop in turgor under hyperosmotic conditions. Typically, hydroxyectoine is synthesized by *Halomonas elongate* and *Nocardiopsis halophila* (Galinski, 1993; Waditee-Sirisattha et al., 2016)
Nε-acetyl-β-lysine 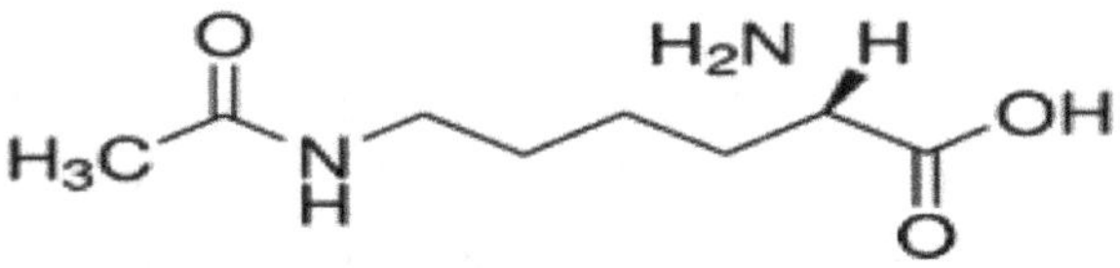	Most methanogenic organisms can accumulate several β-amino acids for osmotic balance. When the external NaCl concentration increases the two β-amino acids have been shown to accumulate. Mostly wide range of mesophilic and a few thermophilic methanogens accumulate a high amount of Nε-acetyl-β-lysine (Jadhav et al., 2018).
β-glutamine 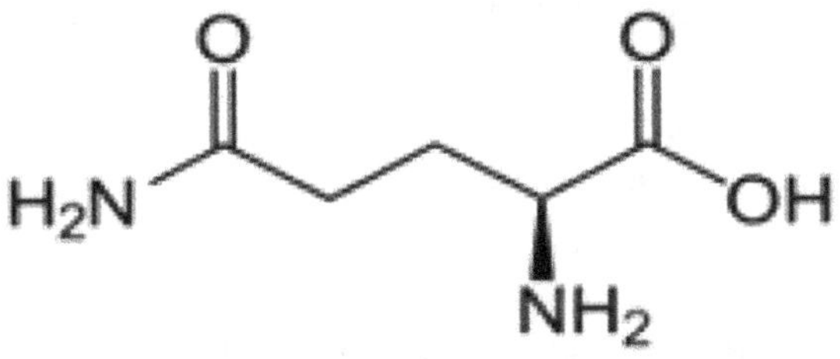	β -glutamine is neutral at physiological pH and majorly accumulated at high osmolality. It is also described as endogenous osmolyte accumulated in *Corynebacterium* spp. and its β-form is accumulated in the pool of halophilic methanogenic archaea. Most methanohalophilic organisms can accumulate β-glutamine along with Nε-acetyl-β-lysine and Betaine (Goude et al., 2004; Jadhav et al., 2018).

2. Non-Charged Solutes

Mostly non-charged solutes are α-glycosylglycerol, α-mannosylglyceramide, trehalose, sucrose, N-α-carbamoyl-L-glutamine 1-amide, and N-acetylglutaminylglutamine amide. All kinds of Halophilic and Halotolerant organisms can produce this osmolyte. The osmotic balance is carried out by few carbohydrates because of having a reducing end and are chemically reactive, and in a protein-rich environment, these non-charged solutes will react with surface amino groups. So, to avoid this, the glycosidic bond is formed with the reactive end of sugar and either glycerol and glyceramide. In halophilic eukaryotic algae and fungi, glycerols and other polyols are produced for osmotic adaptation. Mostly in Halophilic methanogens for example *Methanohalophilus* sp. widely contain glycine betaine which is widespread in nature and also accumulate β-glutamine, β-glutamate, and proline group of β-amino acids and derivatives (Cheng et al., 2020)

Uncharged Solutes have two classes that is Amino acids and peptides called as osmolytes: (i) a carboxamine, and (ii) acetylate glutamine dipeptide. These amino acid derivatives, modifications mask the α-amino and α-carboxyl groups. Halophilic Phototrophic bacterium such as *Ectothiorhodospira marismortui* accumulate amino acid derivative N-α-Carbamoyl-L-glutamine 1-amide. Most halophilic purple sulfur bacteria can synthesize dipeptides like N-acetylglutaminylglutamine amide.

3. Organic Anions

Mostly Halophilic cells have negative potential inside and high intracellular K^+. So, the negatively charged solutes can counteract osmotic pressure by maintaining intracellular K^+. Almost at low external NaCl, many bacteria and archaea accumulate L-α-Glutamate as a compatible solute, whereas methanogenic organisms at high NaCl will convert anionic glutamate isomers to the zwitterionic solute Nε acetyl-β-lysine for osmotic balance. Mostly Anionic solutes have carboxylate which supplies a negative charge or contains phosphate or sulfate group (Empadinhas & Costa, 2011).

β-Glutamate: most methanogens accumulate β-glutamate and α-glutamate for osmotic balance. In NMR experiments turnover for α- and β- glutamate shows that β- glutamate is ideal compatible solute because it will remain relatively static with compared to α-glutamate. In thermophilic and halophilic organisms such as *Methanothermococcus thermolithotrophicus*, with an increase in external salt concentration both the α- and β- glutamate concentration will increase. Also, when the NaCl concentration will lower down it will accumulate negatively charged glutamates and thus the total intracellular glutamate occurs at a comparable concentration to the intracellular K^+ (Goude et al., 2004).

β-Hydroxybutyrate and derivatives: Organisms such as *Methylarcula marina* and *Methylarcula terricola* and *Photobacterium profundum* SS9 used Soluble poly-β-hydroxybutyrates as carbon reservoirs in cells and it will be accumulated in moderate concentration. The concentration of β-hydroxybutyrates at fixed hydrostatic pressure increase with increasing external NaCl and its polymer functional at osmolyte. As the intracellular level responds to hydrostatic pressure, β-hydroxybutyrate is termed as piezolytes (Youssef et al., 2013).

Anionic Polyols and Carbohydrates: In bacteria at high intracellular concentrations the negatively charged carbohydrate such as α-glycosylglycerate and α- mannosylglycerate are produced (Goude et al., 2004). Mostly the accumulation of α-Mannosylglycerate is higher in the exponential phase and it will decrease when it enters the stationary phase. In the case of temperature stress conditions, there will be an accumulation of mannosylglycerate; the increased NaCl leads to the accumulation of neutral α-mannoglyceramide. In the NMR experiment, its turnover is measured which shows roughly twice as slow as α-glutamate and 2-4 times faster than zwitterions i.e., betaine and Nε-acetyl-β-lysine (Empadinhas & Costa, 2011).

Table 2. Occurrence and importance of non-charged compatible solutes

Uncharged Compatible Solute	Occurrence and Importance
Trehalose 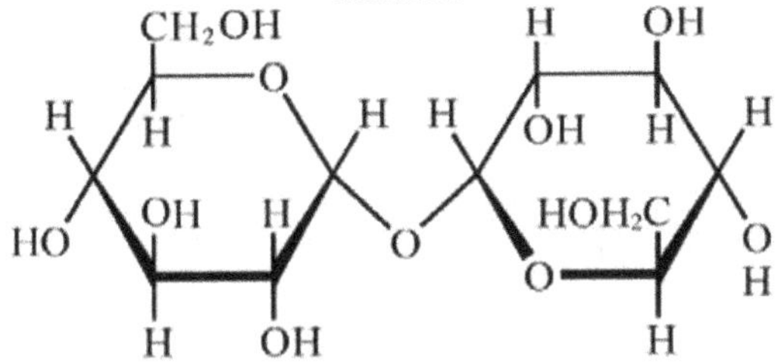	The trehalose is a non-reducing disaccharide used by microorganisms to counteract and act as an osmolyte. Trehalose is mainly accumulated when the NaCl concentration is less. Trehalose is mostly produced by mainly halotolerant and halophilic such as *Pyrobaculumb aerophilum*; *Actinopolyspora halophila*; etc. (Cheng et al., 2020)
Sucrose 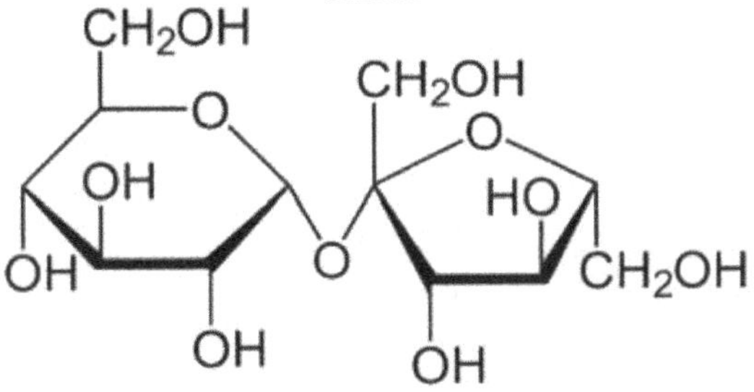	Sucrose the major osmolyte can be transported by some halotolerant and halophilic organisms and this can enhance the growth in higher NaCl. It is critical for stationary phase survival and could regulate metabolic pathways that are active under nutritional stress conditions. Sucrose is a major osmoprotectant in plants and synthesis in both cyanobacteria and plants.
N-α-carbamoyl-L-glutamine 1-amide 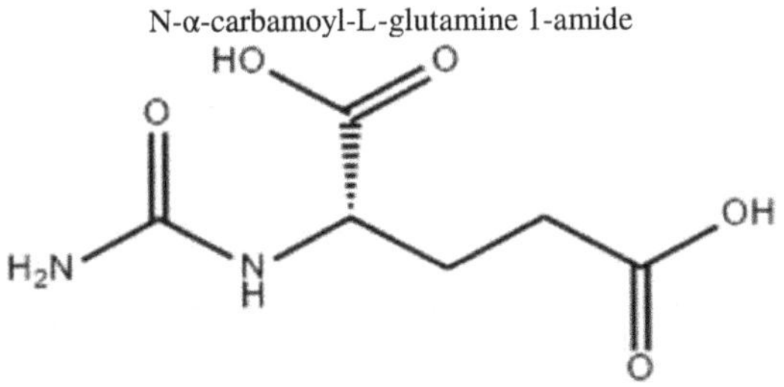	N-α-carbamoyl-L-glutamine 1-amide (CGA) is reported to be accumulated in *Ectothiorhodospira marismortui*. Accumulation of CGA leads to increased osmotic balance in these microbes against surrounding medium (Fischel & Oren, 1993).
N-acetylglutaminylglutamine amide 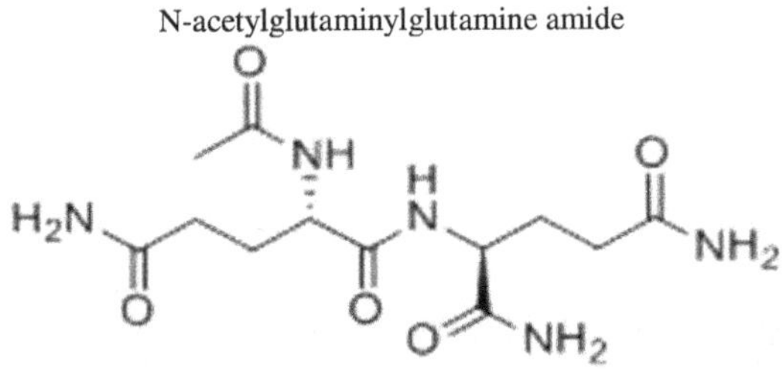	N-acetylglutaminylglutamine amide (NAGGN) is observed to play key role in osmoprotection of cells in various microorganisms such as *Sinorhizobium meliloti*. Similar role of NAGGN is observed in certain other microbes facing osmotic stress (Sagot et al., 2010).

Table 3. Occurrence and importance of organic anions as compatible solutes

Anionic Compatible Solute	Occurrence
L-α-Glutamate 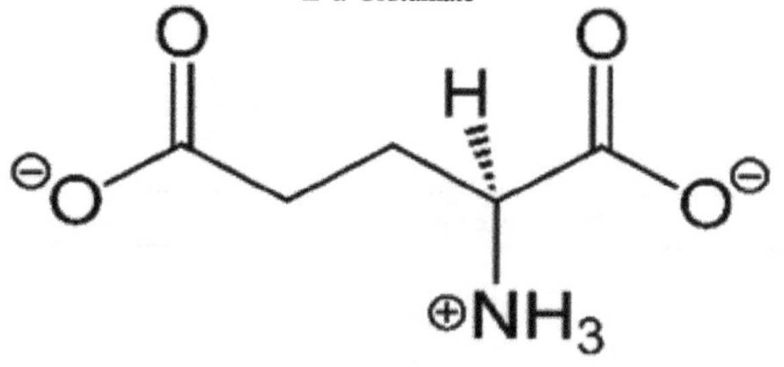	Most halotolerant and methanogenic halophiles such as *Halomonas elongate* and *Halobacterium salinarum* can synthesize this solute (Goude et al., 2004).
β-Glutamate 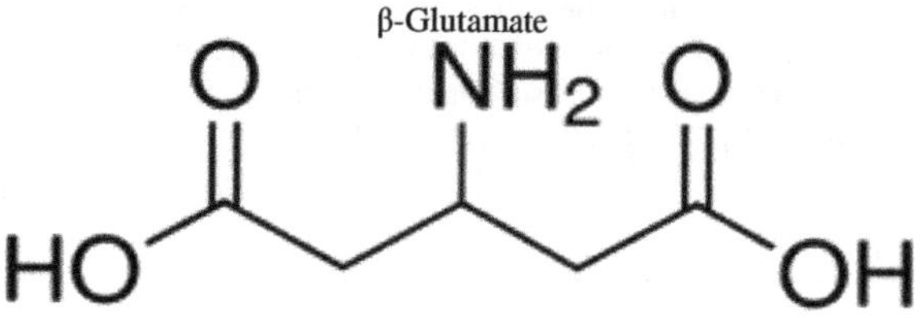	Halotolerant and halophilic microorganisms such as *Methanothermococcus thermolithotrophicus* and *Nacardiopsis halophila* are able to accumulate β- glutamate (Goude et al., 2004).
Hydroxybutyrate 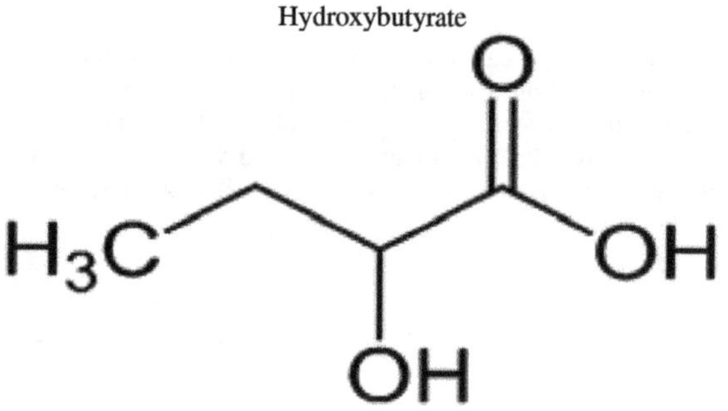	The higher accumulation of Hydroxybutyrate is reported in *Photobacterium profundum* (Youssef et al., 2013).
Poly β-Hydroxybutyrate 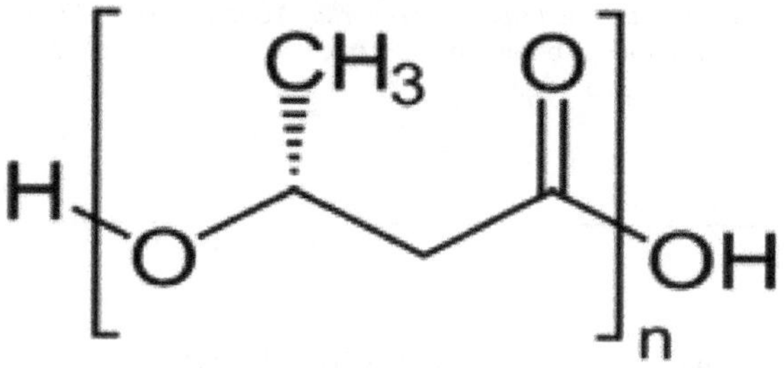	Apart from *Photobacterium profundum*, Poly β-Hydroxybutyrate is produced by *Methylarcula marina* and *M. terricola* (Youssef et al., 2013).
α-glycosylglycerate 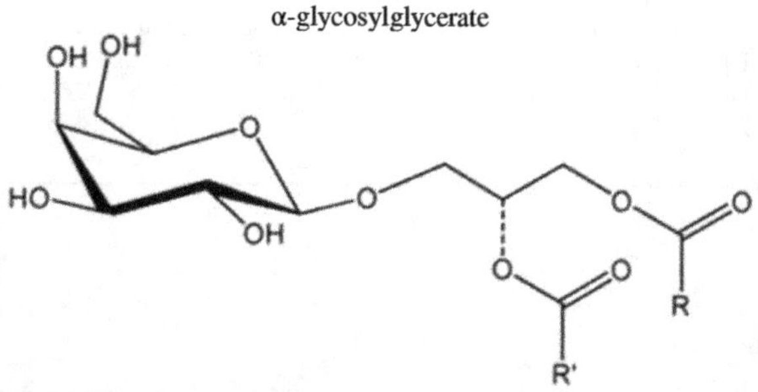	Thermophilic bacteria such as *Thermus thermophilus* and *Rhodothermus marinus*, as well as halophilic microorganisms such *Agmenellum quadruplicatum and Stenotrophomonas maltophilia* can accumulate α-glycosylglycerate as a solute (Goude et al., 2004).

METHODS FOR DETECTION OF COMPATIBLE SOLUTES

The methods used to detect solutes include H-NMR spectroscopy, two-dimensional experiments, HPLC, Refractive index detection, anion-exchange chromatography, and pulse amperometric detection, etc.

H-NMR spectroscopy is an example of a method being used to detect and quantify solutes in cell cultures without extraction while HPLC is used to quantify specific solutes. Refractive index detection is used for a specific class of molecules while the anion-exchange chromatography plus pulse amperometric detection method is extremely sensitive and used to detect solutes after hydrolytic cleavage of the pyrimidine ring (Welsh, 2000)

Accumulation of osmolytes in bacterial cells grown using a saline nutrient medium with varying concentrations of salt and a range of organic sources has been evaluated in *H. aquamarina*, *H. dabanensis*, *Halobacillus* sp. and *Sediminibacillus* sp. using H-NMR spectroscopy. NMR ^{1}H spectra revealed the production of both ectoine and betaine under influence of elevated salt concentrations in *H. aquamarina*. Betaine was observed to be a sole osmolyte produced by *Sediminibacillus* sp. *Halobacillus* sp. and *H. dabanensis* (Amasha, 2018).

Study of ectoine production by *Halomonas salina* BCRC17875 in the presence of various concentrations of salt, carbon source, and nitrogen source applying optimized cultural conditions was carried which was followed by FAB-MS analysis (Fast Atom Bombardment Mass Spectroscopy) and ^{1}H NMR. The utilization of both of these techniques aided the identification and characterization of ectoine produced by *Halomonas salina*. Upon application of FAB-MS and ^{1}H NMR, it was concluded that the yield of ectoine was having the potential to commercialize (W. C. Chen et al., 2018).

Similar studies on *Halomonas socia* NY-011, a moderately halophilic microbe, implemented ^{13}C NMR analysis to identify compatible solutes synthesized during growth in a saline environment. Comparison of the chemical shift δ (ppm) values with published standard spectrums provide in-depth details about the compatible solutes synthesized by *Halomonas socia* NY-011 such as glutamate, ectoine, hydroxyectoine, and glycine betaine even in minute quantities (Peng et al., 2020).

A comprehensive study on the synthesis of compatible solutes by *Virgibacillus halodenitrificans* PDB-F2 shown a preference of hydroxyectoine over ectoine as compatible solutes under osmotic shock. HPLC analysis of cellular extracts was reported as an essential tool in the detection of ectoine and hydroxyectoine (Zhou et al., 2017).

INTERACTION OF COMPATIBLE SOLUTES WITH PROTEINS

Compatible solutes play a vital role in the cell exposed to high osmotic concentration by directly competing for interaction with the surface of proteins with water molecules. In response to this energetically unfavourable interaction, the osmolyte trapped on the surface of the protein is expelled, making the protein hydrated. The presence of osmolytes may result in increased osmotic pressure which in turn triggers compactly folded protein since it has less area exposed to osmotic stress as compared to denatured protein (X. Chen et al., 2019). Having a look at the thermodynamic aspect here, in the presence of osmolytes, the free energy level of the native protein in the unfolded state is increased. The protein folds more compactly due to an increase in energy. The polypeptide chains within proteins interact with water and osmolytes differently. The interaction between osmolytes and polypeptide backbone is less favourable than the same between water and polypeptide (Arsiccio et al., 2020). In native proteins, the polypeptide backbone is more exposed, which lets other solutes interact, shifting the energy equilibrium required for protein folding (Jadhav et al., 2018).

Figure 1. Overview of adaptations in extremophilic proteins

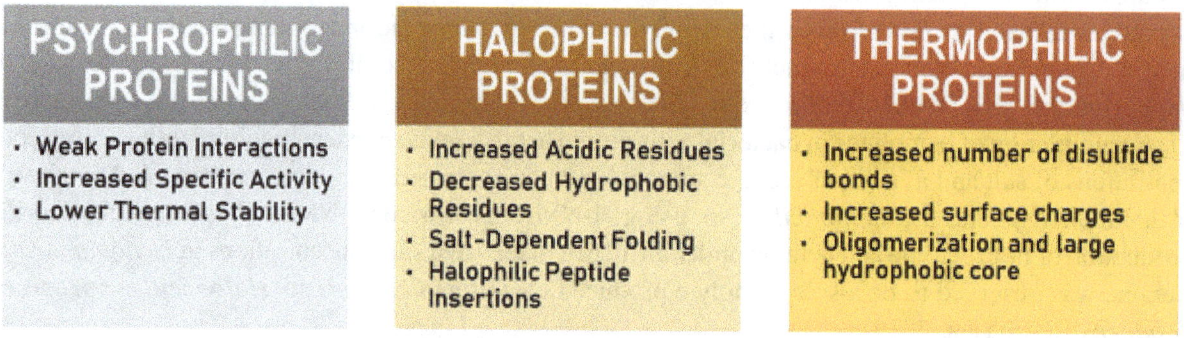

STRUCTURAL ADAPTATION OF PROTEINS IN EXTREMOPHILES

Organisms that acclimatized in extremely salty environments are known as halophilic organisms that are found in salt lakes, pools of evaporating seawater, solar salterns, and other hypersaline environments. They are classified into three categories according to their requirement of salt concentration, respectively known as halotolerant, moderately halophile, and extreme halophile. They exhibit both Eubacterial and Archaeal domains of life. In Eubacteria, halophiles are a very heterogeneous group. In Archaea, however, halophilic is strictly limited to the members of the Haloarchaea class and the 'Nanohaloarchaeota' subphylum (Loukas et al., 2018). Halophilic Archaea are of special interest among halophiles since they dominate the most hypersaline environments on Earth (Oren, 2015).

1. Structural Adaptation in Thermophilic Proteins

Natural habitats for thermophilic microbes may include natural as well as man-made environments. Examples of natural habitats include deep-sea hydrothermal vents, mud pots, geysers, volcanic environments, fumaroles, coastal thermal springs, and hot springs. Manmade heated environments such as compost piles slag heaps, certain industrial processes, and water heaters can also provide a suitable environment for the growth of thermophiles (Castenholz, 1969; Gaisin et al., 2017; Miroshnichenko & Bonch-Osmolovskaya, 2006; Muñoz et al., 2011; Panda et al., 2013; Stathopoulou et al., 2013) Protein structure is stabilized due to a complex pattern of bonding between all the amino acids in it. Thermophilic microbes live at higher temperature values at which proteins of mesophilic microorganisms may get denatured. Proteins of thermophiles adapt to various strategies so that the conformation of proteins remains functional e.g., Increased Number of Disulfide Bonds, Increased Surface Charges, Increased Salt-Bridging and Oligomerization of Large Hydrophobic Core.

Increased number of disulfide bonds: In determining the overall structure of a protein, disulfide bridging between cysteine residues plays a crucial role. Organisms in all realms of life have adapted their own specific mechanisms to maintain proper bridging, some of which are favourable while others may completely inactivate enzymes. It has been shown that these structural elements play a significant role in thermostable enzymes since they appear to possess a function in maintaining quaternary structure in thermophilic proteins. Effects of disulfide bonds on the hydrolytic activity of xylanase produced by *Talaromyces thermophiles* F1208 are well documented. Mutation studies involving introducing and breaking down different disulfide bonds in the structure of xylanase confirmed that hydrolysis characteristics of xylanases are greatly influenced due to the existence of disulfide bonds. Introducing or breaking down the disulfide bond in enzyme may end up with a variety of results such as a change in optimal pH for hydrolytic activity, decrease in the optimum temperature, reduction in thermal stability, changes in substrate specificity, etc. (Fan et al., 2020)

Increased surface charges: It has been observed that when polar uncharged surface residues are replaced by polar charged residues on the protein surface, the overall stability is increased due to several factors. Reduction in protein stability can be observed when polar residues like glutamine and asparagine are deaminated at higher temperatures (Fukuchi & Nishikawa, 2001). A major reason behind the increased thermal stability of a thermophilic protein is reported to be the reduced number of thermolabile amino acids in its structure. Taking an example, thermophilic ribosomal protein from *Thermococcus celer* can have increased thermal capacity while mutations are carried out in its structure at favourable positions where charged residues are located. In addition, the replacement of surface charges with alanine resulted in decreased thermal capacity (Chi-Fung Lee et al., 2005).

Oligomerization and large hydrophobic core: Both of these strategies have been found to increase the thermal stability of enzymes at elevated temperatures. The results of these strategies include preventing hydrophobic residues from getting exposed to solvents, ensure the hydrophobic core is packed tightly, enhance the rigidity of each subunit upon oligomerization. Novel structural adaptations are revealed upon a study on *Ignicoccus hospitalis* and *Pyrobaculum aerophilum* where acetyl-CoA synthetases are observed in the octameric state favouring increased rigidity of its subunits as compared to its mesophilic counterpart which forms mono or dimer (Bräsen et al., 2005; Mayer et al., 2012). A study of thermostable amylase from *Pyrococcus furiosus* indicated the absence of oligomerization property while comparing with earlier examples. In this case, a novel domain at N-terminal is reported with the ability to act as a structural lid to the active site (Park et al., 2013).

2. Structural Adaptation in Halophilic Proteins

Proteins present in these types of organisms are significantly affected by salt. The high Salt concentration is responsible for less solubility, destabilization of protein, and protein disfunction. Evolutionary studies show that they developed stratagems for acclimatization in high salt concentrations. In Haloarchaea, the salt-in strategy is observed which involves cytoplasmic accumulation of K^+ ions. A study on extremely halophilic *Proteobacterium halorhodospira halophila* had reported the role of K^+ concentration in the medium altering the use of osmoprotectant. *H. halophila* was observed to accumulate KCl in molar concentration when hypersaline media with substantial K^+ concentrations is utilized for growth (Deole & Hoff, 2020). In most cases, Haloarchaea having the ability to accumulate high concentrations of cytoplasmic K^+ while exporting Na^+ ions to the extracellular space for maintenance of osmotic pressure of cell (Oren, 2013). There are numerous research studies carried out to explore the mechanism of salt tolerance which depicted the production of salt-tolerant enzymes by halophilic and halotolerant microorganisms. The organisms devise protection against a high concentration of Na^+. Dombrowski and Reiser and Tasch have studied the isolation of viable organisms from ancient rock salts (Fendrihan et al., 2006). In another way, most halophilic organisms having a strategy for the synthesis of compatible solutes like betaine and ectoine in the cytoplasm which are responsible for osmoregulation (Gunde-Cimerman et al., 2018). The composition of the cell envelope and outer membrane modified as ionic strength of the outer membrane is changed. The electrostatic interactions between charged amino acids are also affected by salt (Karan et al., 2012). High salt concentrations are not appropriate for protein function in the case of non-halophilic proteins because of alteration in hydrophobic and electrostatic interactions which are important to maintain its functional stability and appropriate folding. Modifications in these properties are responsible for protein destabilization, unfolding, and aggregation of protein and ultimately leads to precipitation.

Increased Acidic Residues: There are remarkable variances among halophilic and non-halophilic proteins. One of the most notable differences observed is on protein's surface which shows a large increase in acidic residues, like glutamic and aspartic acid (Zhang et al., 2013). These acidic residues may have many possible roles. It is described that if the protein's surface is having an increased negative charge, proteins can directly compete for water molecules with ions present due to which the protein stays stable in solution (Britton et al., 2006; Karan et al., 2012). This characteristic is due to the crystal structures of halophilic proteins which having the water-binding ability with these acidic surface residues (Mevarech et al., 2000). One special case must be noticed here for β-galactosidase from *Heloferax lacusprofundi* showing exceptional stability in elevated NaCl concentration up to 4 M, It is reported that the composition of amino acid and its acidic residues are responsible for this feature. The additional stability is achieved through compensatory mechanisms such as fewer hydrogen bonds with salt bridges in higher amounts (Karan et al., 2012). Furthermore, Bioinformatics studies have also been carried out for halophilic proteins. This study has shown that the amount of serine is very less in their sequences. Notably, Serine plays an important role with water interactions but not at competing with charged ions, so it is believed that serine is not much necessary for halophilic proteins (Zhang & Ge, 2013). It is not well understood how protein and cations interact with each other because crystal structures of halophilic proteins are unable to distinguish between salt and water. Hydration dynamics study in folded and unfolded states of halophilic protein suggested that halophilic proteins do not have increased waters of hydration due to their greater negative charge (Qvist et al., 2012). Furthermore, Molecular structure homology simulations of halophilic dihydrofolate reductases reveal similar hydrogen-bonding networks which are similar to the non-halophilic proteins (Kastritis et al., 2007). Such findings can raise questions on the nature of acidic residues in keeping halophilic proteins soluble. Following a study on halophilic adaptation of proteins, it has been observed that crystalizing conditions for proteins involve salting-out conditions that result in improved water binding and exclusion of salt (Madern et al., 2000). The γ-carbonic anhydrase (γ-CA) data was obtained from Discovery Deep of Red Sea brine pool using culture-independent methods. Further expression of γ-CA in the bioengineered strain of *Halobacterium* sp. and analysis of its X-ray crystallographic data indicated that the protein possesses increased charged residues, hydrogen bonds and salt bridges which can be the most important features illustrating extremophilicity (Vogler et al., 2020). The acidic residues also play a key role in protein flexibility as halophilic proteins have a large number of negative charges on their surface (Mevarech et al., 2000).

Decreased Hydrophobic Residues: Along with *in vitro* studies, certain bioinformatical data is also available on halophiles. Bioinformatics based studies on the sequence of halophilic proteins reported that they also contain different hydrophobic residues which play important role in adaption in harsh environments. Siglioccolo et al. studied halophilic protein structure by using the known crystal structures of 15 pairs of halophilic and non-halophilic proteins. The studies determined that the portion of halophilic proteins having hydrophobic contact in the core, exposed to molar concentrations of inorganic salt, is consistently smaller than other proteins found in mesophiles (Siglioccolo et al., 2011). An explanation behind this is that lower hydrophobic contact in the core may counterbalance the increased strength of hydrophobic interactions in high salt concentrations (Siglioccolo et al., 2011). Based on structural models constructed from homology modelling of halophilic dihydrofolate reductase along with approaches toward visual inspection of core and surface of models shown, it has been proposed that dihydrofolate reductase can maintain its fold under the influence of high salt concentrations due to the presence of weak hydrophobic cores and sharing a higher portion of negatively charged surfaces (Kastritis et al., 2007).

Salt-Dependent Folding: Diverse research data are available stating reliance of halophilic proteins on salt to fold and are useful to understand halophilic protein adaptation (Müller-Santos et al., 2009). The fluorescence spectroscopy and circular dichroism-based study on the cysteinyl tRNA synthetase in *H. salinarum* NRC-1 show that at increasing salt concentrations, the enzyme not only folds but also protein's stability increases and becomes resistant towards thermal denaturation (Reed et al., 2014). A consensus set of ten different α-amino acids has been identified which are entitled as "prebiotic" and are: Aspartic acid, Isoleucine, Glutamic acid, Proline, Alanine, Serine, Threonine, Glycine, Leucine, and Valine. Proteins with such consensus were found to have a requirement of high salt concentrations to go for cooperative folding and exhibit substantial acidification of pI. In a halophilic environment, these prebiotic comprise a set of amino acids available in the foldable configuration. In addition, it can also be suggested that the existence of these prebiotic amino acids is vital for the biogenesis of halophilic proteins (Longo et al., 2013).

Halophilic Peptide Insertions: Adaptations to high salt levels are not always found throughout the protein sequence. Insertion of a peptide in a protein has been known to increase the halophilicity of the protein. Such insertions can be best defined as small protein sequences containing a number of amino acids as little as 2 amino acids and as much as 30 amino acids. There may be a major proportion of acidic amino acids in the insert but may also include the noteworthy presence of proline and glycine. There is no consensus on the exact purpose of the extra sequences, but most studies indicate that they may increase the flexibility of the protein when exposed to halophile conditions (Brininger et al., 2018). As seen in cysteinyl-tRNA synthetase from halophilic bacteria, these insertions usually contain a large number of acidic amino acids. Adding a catalytic insertion increased enzyme efficiency greatly (Evilia & Hou, 2006). *Haloarcula marismortui* serinyl-tRNA synthetase also has an insertion sequence, presumably contributing to better flexibility (Zaccai et al., 1989). There was evidence that ferredoxin from the same organism contained 15 negatively charged amino acids on its N-terminus. Adding this insertion to the enzyme is thought to increase its solvent-accessible surface area. They may function in a number of ways and might be used to inject evolutionary adaptations that make a protein more halophilic (Bianca-Lucia Marg et al., 2004)

Deletion of a motif representing insertion peptide has been found to drastically reduce the amino-acylation ability of halophilic tRNA synthetase while the presence of the same is observed to express greater enzyme stability at low salt and strong salt dependency (Evilia & Hou, 2006). Dependence of halophilic proteins on salt for proper folding

3. Structural Adaptation in Psychrophilic Proteins

Psychrophiles (also known as cryophiles) are a class of extremophiles, they can grow, survive and proliferate at low temperatures (temperatures below 20°C). Especially, these organisms can produce cold-active enzymes which having tremendous biotechnological potential as well as informative models for fundamental research into the structure and function of the protein (Yusof et al., 2021). The majority of studies have been done on genome sequences, proteomic, and transcriptomic which suggest various adaptive features for protein synthesis and protein folding at very low temperatures (Feller, 2013). In order with bacterial and eukaryotic proteins, much research has been done on psychrophilic protein adaptations (Collins et al., 2014). Nevertheless, archaeal organisms are also habituated in cold environments. Some research studies have been also done on archaeal proteins (Dong & Chen, 2012). For a growing cell, very low temperature is not suitable especially in protein functions as at a low temperature protein activity is affected (Feller, 2010). At low temperatures, enzyme activity decreases because mean kinetic energy also lowers; the protein conformational movements also become slower and therefore enzymatically less efficient (Cavicchioli et al., 2000). Also, at low temperatures, the enzyme's activity reduces as the energy barrier of activation for catalysis becomes too great for a protein. The adaptation of proteins in psychrophiles allows them to thrive in cold temperatures, even though these proteins function best at temperatures above their physiological temperatures (Feller, 2010). At low temperatures, psychrophilic proteins are more flexible, because of their high activity level, and they can move and change shape more easily (Collins et al., 2014). Psychrophilic proteins are subjected to a thermodynamic challenge, as their molecular motion is diminished as entropy and enthalpy decrease in the influence of subordinate temperatures (Brininger et al., 2018).

Weak Protein Interactions: Psychrophilic proteins are flexible by nature, so a lower energy barrier is found between the various protein conformations (Feller, 2010). These protein conformations are just because of differences in amino acid composition between proteins of psychrophiles and mesophiles. Furthermore, in cold-active proteins, the stabilizing interactions are typically weakened or removed which is found in normal protein. According to Feller's review on cold- and heat active enzymes, the psychrophilic proteins adaptations were summarized as follows: (i) at cold temperature notable increase found in the number of glycine residues, which is helpful in psychrophilic proteins conformational change, (ii) proline residues are reduced from loop regions, which provide conformational rigidity, (iii) the amount of arginine residues are reduced, which responsible for salt bridge and hydrogen bond formation, (iv) there is a reduction in the size of nonpolar residues in the protein core, that leads for weaker hydrophobic interactions (Feller, 2010). A study has been done based on these features, which includes proteins from the archaeal cold-adapted halophile *Halorubum lacusprofundi* which shows a decrease in amino acids such as hydrophobic tryptophan, and hydrogen bond-forming residues, like glutamic acid. *H. lacusprofundi* also possesses *β*-galactosidase enzyme, in which an increasing hydrophobicity is observed on the protein surface, which replaced anionic electrostatic interactions which are normally found on halophilic proteins (DasSarma et al., 2013; Karan et al., 2013). Furthermore, some data have been reported that psychrophilic methanogens proteins also showing these types of amino acid trends (Thomas & Cavicchioli, 1998). In 2003, a study has been done by Saunders and co-workers and they examined genomes from several archaeal methanogens across a wide range of optimal growth temperatures. A three-dimensional protein model was developed from two draft genome sequences of psychrophilic methanogens, M*ethanogenium frigidum,* and M*ethanococcoides burtonii,* followed by comparative studies between modelled proteins from mesophilic and thermophilic counterparts. As expected, the number of charged residues on the amino acid surface is decreased and the amount of glutamine and threonine is increased in cold-adapted proteins. In this way, the surface of the protein is repelled from charges without causing aggregation by making the surface too hydrophobic (Saunders et al., 2003). The surface charge is the major contributing factor behind the feature of cold adaptation in DNA ligases obtained from psychrophilic microorganisms P*sychromonas* sp. and A*liivibrio salmonicida.* Surface-exposed patches with higher hydrophobicity were identified via sequence comparison and homology modelling of DNA ligases from both these microorganisms (Berg et al., 2019).

Increased Specific Activity: A psychrophilic enzyme has a highly flexible structure. Due to this, at low temperatures, a psychrophilic enzyme shows higher catalytic activity as compared to the same enzyme from a mesophile. In fact, at low temperatures, a psychrophilic enzyme shows the specific activity ($kc_{at)}$ about characteristically 10 times superior to a mesophilic enzyme (Georlette et al., 2003). An observable increase in the size of the binding site in psychrophilic proteins is a typical observation that can be made to explain the greater $kca_{t.}$ The psychrophilic enzymes are having several mechanisms for the enlargement of the substrate-binding area while there is no impact on catalytic residues (Feller, 2010). The increased substrate-binding area can be achieved by a variety of mechanisms such as deletion of loops near the binding site (Russell et al., 1998), the presence of strategic residues of glycine near the active sites (Feller, 2010), and pulling the protein backbone out to increase substrate accessibility (Aghajari et al., 2003). As a result, substrates are not able to bind as well to a psychrophilic enzyme, and, therefore, the Michaelis-Menten constant (*Km*) $_{of}$ psychrophilic enzymes is high (D'Amico et al., 2006; Feller, 2010). At low temperature, enzyme activity improves as enzymes having poor substrate affinity because (Feller, 2010)

Low**er Thermal Stability: As** a result of weaker interactions between amino acid residues, the psychrophilic protein cannot be "frozen" in a specific conformation, allowing molecular motions needed for catalysis. These types of weaker interactions are responsible for the low stability of protein; thus, cold-adapted proteins unfold at lower temperatures than mesophilic proteins (D'Amico et al., 2001; Georlette et al., 2003). Notably, data shows that a single transition is responsible for the thermal unfolding of psychrophilic proteins. As cold-adapted proteins have weaker interactions, their stability is reduced, and due to fewer stabilizing interactions, local unfolding greatly damages the protein. These types of attributes have been reported in an archaeal cold-shock protein of Met*hanogenium frigidum, w*hich was shown to be less stable at its optimal temperature than its mesophilic homolog from E. *coli (G*iaquinto et al., 2007). In the recent study, data was reported for structural and mutational analysis of adenylate kinases using psychrophilic Bac*illus globisporus an*d mesophilic Bac*illus subtilis. T*hese data provide structural ideas about the thermal stability of adenylate kinases of psychrophiles and mesophiles as well as emphasize the need for hydrophobic interactions in the thermal stability of a protein (Moon et al., 2019). Thermal denaturation experiments on cytochrome c' of a psychrophiles She*wanella violacea an*d She*wanella benthica by* circular dichroism spectral measurements depict the contribution of proline and lysine at the molecular surface in the higher stability while the zilch influence of alanine in the stability. The difference in amino acid composition for both of these microorganisms is also responsible for the dissimilar psychrophilic behaviour of cytochrome c'. Piezophile is an additional characteristic of She*wanella violacea an*d She*wanella benthica. H*ence, the contribution of proline, lysine and alanine residues is either in psychrophilism or piezophilicm or both is not very much clear and requires further investigations (Suka et al., 2019). Substitution of alanine with serine in cytochrome c6 o_f psychrophilic diatoms decreases thermal stability proving the significance of respective amino acid residues in protein stabilization (Wilson et al., 2020).

BIOTECHNOLOGICAL APPLICATIONS OF OSMOLYTES

Osmolytes obtained from extremophilic microorganisms can be of importance in varieties of areas. Possible applications of betaines include as a cryoprotectant during storage for an extended period (Cleland et al., 2004; Sui et al., 2018), osmoprotectant in cosmetics to prevent denaturation of skin proteins (Desmarais et al., 1997), anticoagulant to decrease the probability of heart attacks, inducing stress tolerance in non-halophilic organisms (*US6855734B2 - Glycine Betaine and Its Use - Google Patents,* n.d.). Similarly, ectoine can also be used as an osmoprotectant to save skin from the effects of ultraviolet radiations along with application in PCR to enhance amplification of GC rich regions (Schnoor et al., 2004). Sucrose and glycerol can be applied to protect normal cells during cancer treatment from toxic chemicals and plasma. Trehalose has potential application to prevent denaturation of proteins and enzymes by providing them with thermostability (Borges et al., 2002). Glycine betaine and urea were combinedly studied for their effect on human telomerase RNA where they were found effective in maintaining the stability of primary and tertiary structures in RNA (Jeffrey J. Schwinefus et al., 2007; Lambert & Draper, 2007). In addition, proline has been found to enhance the stabilization of secondary and tertiary structures of RNA (Lambert & Draper, 2007). Molecular methods are evolving continuously over the past decade which in turn gives brighter scope of conducting research in further depth. Transformation of betaine genes from different microorganisms to one which cannot synthesize any compatible solute can be beneficial in inducing stress resistance. Although, certain limitations were reported such as limited accumulation of osmoprotectants in the host cell. Hence, further investigations can be conducted to overcome such limitations (Jadhav et al., 2018).

APPLICATIONS OF EXTREMOPHILIC PROTEINS

Extremophilic proteins have a wide scope of catalysing various reactions that cannot be operated in a normal environment. Taq polymerase, a thermophilic enzyme is well known for extensive application in numerous molecular biology techniques. Similarly, extremophilic proteins may have an ample number of applications that can benefit mankind in either way.

Hydrolytic enzymes from extremophilic microorganisms such as amylase are of boundless importance in the food, medical, analytical, textile, and pulp industries. Thermophilic acidophilic, as well as psychrophilic amylase, can be of high demand in the food processing and starch industry (Kour et al., 2019). Protease produced by psychrophilic bacteria can have potential application in biodegradation of human waste rich in proteins while lipase from psychrophilic origin can be of great significance in the food industry such as preparing lipase modified butterfat, ripening of cheese, hydrolysis of cheese fat (Bruins et al., 2001). Other enzymes finding importance are xylanase and laccase in paper and pulp processes, Glutamate dehydrogenase in processes related to food, transaminase and nitrilase for drug synthesis in the pharmaceutical industry, lipase present in detergents, etc. (Atalah et al., 2019).

Ice structure proteins or thermal hysteresis proteins, also known as antifreeze proteins (AFPs), are non-collaborative biological antifreeze that can penetrate ice and inhibit ice growth and crystallization. A new generation of antifreeze is being developed for use in improving freeze tolerance of crops and aquaculture fish, gas hydrate inhibition in the petroleum industry, and surface coatings for preventing/decreasing ice formation on materials (Dolev et al., 2016; Voets, 2017). Oppositely, ice nucleation proteins (INPs) promote the formation of ice and an artificial snow production technology is currently being developed based on INP. Other potential applications of INPs includes minimizing freezing expenses in food, beverages, microfluidic devices and cloud seeding for climate control (Collins & Margesin, 2019).

Further applications of extremophilic proteins include the development of biosensors. An experiment to utilize thermophilic proteins to develop an enzyme biosensor based on glucokinase from *Bacillus stearothermophilus* for detection of glucose which utilizes the sulfhydryl-reactive fluorophore IA-INS [2-(4-(iodoacetamido) anilino) naphthalene-6-sulfonic acid] is successfully demonstrated (Vasudevan & Jayshree, 2020).

CONCLUSION

Compatible solutes impart additional benefits to extremophilic microorganisms by providing stability to their cellular components. These solutes have various applications starting from osmoprotectant in common cosmetic preparation to stabilizing PCR protocols. While focusing on extremophilic enzymes, due to their multiple potential applications in diverse processes, extremozymes have become highly sought after in agriculture, medicine, and the food industry, over the past decade. Other extremophilic proteins such as INPs, AFPs and protein chaperons are explored for their potential biotechnological applications and will continue for a long period till the technological advancements in molecular techniques permit in-depth study. All the products produced by extremophiles are like wonders created by nature and are being continuously explored and exploited by humans for all possible applications.

REFERENCES

Aghajari, N., van Petegem, F., Villeret, V., Chessa, J.-P., Gerday, C., Haser, R., & van Beeumen, J. (2003). Crystal structures of a psychrophilic metalloprotease reveal new insights into catalysis by cold-adapted proteases. *Proteins*, *50*(4), 636–647. doi:10.1002/prot.10264 PMID:12577270

Alavi, S., Rafieyan, S., Yavari-Bafghi, M., & Amoozegar, M. A. (2020). Extremophiles: A Powerful Choice for Bioremediation of Toxic Oxyanions. *Microbial Bioremediation & Biodegradation*, 203–249. doi:10.1007/978-981-15-1812-6_9

Amasha, R. H. (2018). Use of NMR to determine compatible solutes in halophilic bacteria isolated from highly saline areas. *Applied Microbiology: Open Access*, *4*. Advance online publication. doi:10.4172/2471-9315-C3-017

Arsiccio, A., Ganguly, P., la Cortiglia, L., Shea, J.-E., & Pisano, R. (2020). ADD Force Field for Sugars and Polyols: Predicting the Additivity of Protein–Osmolyte Interaction. *The Journal of Physical Chemistry B*, *124*(36), 7779–7790. doi:10.1021/acs.jpcb.0c05345 PMID:32790371

Atalah, J., Cáceres-Moreno, P., Espina, G., & Blamey, J. M. (2019). Thermophiles and the applications of their enzymes as new biocatalysts. *Bioresource Technology*, *280*, 478–488. doi:10.1016/j.biortech.2019.02.008 PMID:30826176

Berben, T., Overmars, L., Sorokin, D. Y., & Muyzer, G. (2017). Comparative Genome Analysis of Three Thiocyanate Oxidizing Thioalkalivibrio Species Isolated from Soda Lakes. *Frontiers in Microbiology, 0*(FEB), 254. doi:10.3389/fmicb.2017.00254 PMID:28293216

Berezovsky, I. N., & Shakhnovich, E. I. (2005). Physics and evolution of thermophilic adaptation. *Proceedings of the National Academy of Sciences of the United States of America, 102*(36), 12742–12747. doi:10.1073/pnas.0503890102 PMID:16120678

Berg, K., Leiros, I., & Williamson, A. (2019). Temperature adaptation of DNA ligases from psychrophilic organisms. *Extremophiles, 23*(3), 305–317. doi:10.1007/s00792-019-01082-y

Borges, N., Ramos, A., Raven, N. D., Sharp, R. J., & Santos, H. (2002). Comparative study of the thermostabilizing properties of mannosylglycerate and other compatible solutes on model enzymes. *Extremophiles, 6*(3), 209–216. doi:10.1007/s007920100236

Bräsen, C., Urbanke, C., & Schönheit, P. (2005). A novel octameric AMP-forming acetyl-CoA synthetase from the hyperthermophilic crenarchaeon Pyrobaculum aerophilum. *FEBS Letters, 579*(2), 477–482. doi:10.1016/j.febslet.2004.12.016 PMID:15642362

Brininger, C., Spradlin, S., Cobani, L., & Evilia, C. (2018). The more adaptive to change, the more likely you are to survive: Protein adaptation in extremophiles. *Seminars in Cell & Developmental Biology, 84*, 158–169. doi:10.1016/j.semcdb.2017.12.016 PMID:29288800

Britton, K. L., Baker, P. J., Fisher, M., Ruzheinikov, S., Gilmour, D. J., Bonete, M.-J., Ferrer, J., Pire, C., Esclapez, J., & Rice, D. W. (2006). Analysis of protein solvent interactions in glucose dehydrogenase from the extreme halophile <em>Haloferax mediterranei</em>. *Proceedings of the National Academy of Sciences of the United States of America, 103*(13), 4846–4851. doi:10.1073/pnas.0508854103 PMID:16551747

Bruins, M. E., Janssen, A. E. M., & Boom, R. M. (2001). Thermozymes and their applications. *Applied Biochemistry and Biotechnology, 90*(2), 155–186. doi:10.1385/ABAB:90:2:155

Castenholz, R. W. (1969). The thermophilic cyanophytes of Iceland and the upper temperature limit. *Journal of Phycology, 5*(4), 360–368. doi:10.1111/j.1529-8817.1969.tb02626.x PMID:27096455

Cavicchioli, R., Thomas, T., & Curmi, P. M. G. (2000). Cold stress response in Archaea. *Extremophiles, 4*(6), 321–331. doi:10.1007/s007920070001

Chela-Flores, J. (2013). Polyextremophiles. *Summary and Conclusions*, 611–615. Advance online publication. doi:10.1007/978-94-007-6488-0_29

Chen, W. C., Hsu, C. C., Lan, J. C. W., Chang, Y. K., Wang, L. F., & Wei, Y. H. (2018). Production and characterization of ectoine using a moderately halophilic strain Halomonas salina BCRC17875. *Journal of Bioscience and Bioengineering, 125*(5), 578–584. doi:10.1016/j.jbiosc.2017.12.011 PMID:29331525

Chen, X., Bhandari, B., & Zhou, P. (2019). Insight into the effect of glycerol on stability of globular proteins in high protein model system. *Food Chemistry, 278*, 780–785. doi:10.1016/j.foodchem.2018.11.117 PMID:30583443

Cheng, H., Sun, Y., Chang, H., Cui, F., Xue, H., Shen, Y., Wang, M., & Luo, J. (2020). Compatible solutes adaptive alterations in Arthrobacter simplex during exposure to ethanol, and the effect of trehalose on the stress resistance and biotransformation performance. *Bioprocess and Biosystems Engineering, 43*(5), 895–908. doi:10.1007/s00449-020-02286-9

Cleland, D., Krader, P., McCree, C., Tang, J., & Emerson, D. (2004). Glycine betaine as a cryoprotectant for prokaryotes. *Journal of Microbiological Methods*, *58*(1), 31–38. doi:10.1016/j.mimet.2004.02.015 PMID:15177901

Collins, T., D'Amico, S., Marx, J.-C., Feller, G., & Gerday, C. (2014). Cold-Adapted Enzymes. *Physiology and Biochemistry of Extremophiles*, 165–179. doi:10.1128/9781555815813.ch13

Collins, T., & Margesin, R. (2019). Psychrophilic lifestyles: mechanisms of adaptation and biotechnological tools. *Applied Microbiology and Biotechnology, 103*(7), 2857–2871. doi:10.1007/s00253-019-09659-5

Crowley, E. (2017). Compatible Solute Ectoine Review: Protection Mechanisms and Production Methods. [*Journal of Undergraduate Studies at Trent*, *5*(1), 32–39. https://ojs.trentu.ca/ojs/index.php/just/article/view/79

Cycil, L. M., DasSarma, S., Pecher, W., & McDonald, R. (1567). AbdulSalam, M., & Hasan, F. (2020). Metagenomic Insights Into the Diversity of Halophilic Microorganisms Indigenous to the Karak Salt Mine, Pakistan. *Frontiers in Microbiology*, *0*. Advance online publication. doi:10.3389/fmicb.2020.01567 PMID:32793134

D'Amico, S., Gerday, C., & Feller, G. (2001). Structural Determinants of Cold Adaptation and Stability in a Large Protein. *The Journal of Biological Chemistry*, *276*(28), 25791–25796. doi:10.1074/jbc.M102741200 PMID:11325973

D'Amico, S., Sohier, J. S., & Feller, G. (2006). Kinetics and Energetics of Ligand Binding Determined by Microcalorimetry: Insights into Active Site Mobility in a Psychrophilic α-Amylase. *Journal of Molecular Biology*, *358*(5), 1296–1304. doi:10.1016/j.jmb.2006.03.004 PMID:16580683

DasSarma, S., Capes, M. D., Karan, R., & DasSarma, P. (2013). Amino Acid Substitutions in Cold-Adapted Proteins from Halorubrum lacusprofundi, an Extremely Halophilic Microbe from Antarctica. *PLoS One*, *8*(3), e58587. doi:10.1371/journal.pone.0058587 PMID:23536799

Dennis, P. P. (2020). The Molecular Biology of Halophilic Archaebacteria. The Biology of Halophilic Bacteria, 255–287. doi:10.1201/9781003069140-9/MOLECULAR-BIOLOGY-HALOPHILIC-ARCHAEBACTERIA-PATRICK-DENNIS

Deole, R., & Hoff, W. D. (2020). A potassium chloride to glycine betaine osmoprotectant switch in the extreme halophile Halorhodospira halophila. *Scientific Reports, 10*(1), 1–10. doi:10.1038/s41598-020-59231-9

Desmarais, D., Jablonski, P. E., Fedarko, N. S., & Roberts, M. F. (1997). 2-sulfotrehalose, a novel osmolyte in haloalkaliphilic archaea. *Journal of Bacteriology*, *179*(10), 3146–3153. doi:10.1128/jb.179.10.3146-3153.1997 PMID:9150208

Dolev, M. B., Braslavsky, I., & Davies, P. L. (2016). *Ice-Binding Proteins and Their Function*. doi:10.1146/annurev-biochem-060815-014546

Dong, X., & Chen, Z. (2012). Psychrotolerant methanogenic archaea: Diversity and cold adaptation mechanisms. *Science China Life Sciences, 55*(5), 415–421. doi:10.1007/s11427-012-4320-0

Empadinhas, N., & da Costa, M. S. (2011). Diversity, biological roles and biosynthetic pathways for sugar-glycerate containing compatible solutes in bacteria and archaea. *Environmental Microbiology, 13*(8), 2056–2077. doi:10.1111/j.1462-2920.2010.02390.x PMID:21176052

Evilia, C., & Hou, Y.-M. (2006). Acquisition of an Insertion Peptide for Efficient Aminoacylation by a Halophile tRNA Synthetase. *Biochemistry, 45*(22), 6835–6845. doi:10.1021/bi0521386 PMID:16734420

Fan, G., Wu, Q., Li, Q., Sun, B., Ma, Y., Wu, K., Wang, C., Teng, C., Yang, R., & Li, X. (2020). Impact of the disulfide bond on hydrolytic characteristics of a xylanase from Talaromyces thermophiles F1208. *International Journal of Biological Macromolecules, 164*, 1748–1757. doi:10.1016/j.ijbiomac.2020.07.270 PMID:32777426

Farraj, D. A. al, Hadibarata, T., Yuniarto, A., Alkufeidy, R. M., Alshammari, M. K., & Syafiuddin, A. (2020). Exploring the potential of halotolerant bacteria for biodegradation of polycyclic aromatic hydrocarbon. *Bioprocess and Biosystems Engineering, 43*(12), 2305–2314. doi:10.1007/s00449-020-02415-4

Feller, G. (2010). Protein stability and enzyme activity at extreme biological temperatures. *Journal of Physics Condensed Matter, 22*(32), 323101. doi:10.1088/0953-8984/22/32/323101 PMID:21386475

Feller, G. (2013). Psychrophilic Enzymes: From Folding to Function and Biotechnology. *Scientifica, 2013*, 1–28. doi:10.1155/2013/512840 PMID:24278781

Fendrihan, S., Legat, A., Pfaffenhuemer, M., Gruber, C., Weidler, G., Gerbl, F., & Stan-Lotter, H. (2006). Extremely halophilic archaea and the issue of long-term microbial survival. *Life in Extreme Environments*, (2-3), 125–140. doi:10.100711157-006-0007-y PMID:21984879

Fischel, U., & Oren, A. (1993). Fate of compatible solutes during dilution stress in Ectothiorhodospira marismortui. *FEMS Microbiology Letters, 113*(1), 113–118. doi:10.1111/j.1574-6968.1993.tb06497.x

Fukuchi, S., & Nishikawa, K. (2001). Protein surface amino acid compositions distinctively differ between thermophilic and mesophilic bacteria. *Journal of Molecular Biology, 309*(4), 835–843. doi:10.1006/jmbi.2001.4718 PMID:11399062

Gaisin, V. A., Kalashnikov, A. M., Grouzdev, D. S., Sukhacheva, M., Kuznetsov, B. B., & Gorlenko, V. M. (2017). Chloroflexus islandicus sp. nov., a thermophilic filamentous anoxygenic phototrophic bacterium from a geyser. *International Journal of Systematic and Evolutionary Microbiology, 67*(5), 1381–1386. doi:10.1099/ijsem.0.001820 PMID:28126046

Galinski, E. A. (1993). Compatible solutes of halophilic eubacteria: molecular principles, water-solute interaction, stress protection. *Experientia, 49*(6), 487–496. doi:10.1007/BF01955150

Galinski, E. A., & Herzog, R. M. (1990). The role of trehalose as a substitute for nitrogen-containing compatible solutes (Ectothiorhodospira halochloris). *Archives of Microbiology, 153*(6), 607–613. doi:10.1007/BF00245273

Garnier, F., Couturier, M., Débat, H., & Nadal, M. (2021). Archaea: A Gold Mine for Topoisomerase Diversity. *Frontiers in Microbiology, 0*, 1083. doi:10.3389/fmicb.2021.661411 PMID:34113328

Georlette, D., Damien, B., Blaise, V., Depiereux, E., Uversky, V. N., Gerday, C., & Feller, G. (2003). Structural and Functional Adaptations to Extreme Temperatures in Psychrophilic, Mesophilic, and Thermophilic DNA Ligases *. *The Journal of Biological Chemistry, 278*(39), 37015–37023. doi:10.1074/jbc.M305142200 PMID:12857762

Giaquinto, L., Curmi, P. M. G., Siddiqui, K. S., Poljak, A., DeLong, E., DasSarma, S., & Cavicchioli, R. (2007). Structure and function of cold shock proteins in archaea. *Journal of Bacteriology, 189*(15), 5738–5748. doi:10.1128/JB.00395-07 PMID:17545280

Goude, R., Renaud, S., Bonnassie, S., Bernard, T., & Blanco, C. (2004). Glutamine, glutamate, and α-glucosylglycerate are the major osmotic solutes accumulated by Erwinia chrysanthemi strain 3937. *Applied and Environmental Microbiology, 70*(11), 6535–6541. doi:10.1128/AEM.70.11.6535-6541.2004 PMID:15528516

Gunde-Cimerman, N., Plemenitaš, A., & Oren, A. (2018). Strategies of adaptation of microorganisms of the three domains of life to high salt concentrations. *FEMS Microbiology Reviews, 42*(3), 353–375. doi:10.1093/femsre/fuy009 PMID:29529204

Harishchandra, R. K., Sachan, A. K., Kerth, A., Lentzen, G., Neuhaus, T., & Galla, H.-J. (2011). Compatible solutes: Ectoine and hydroxyectoine improve functional nanostructures in artificial lung surfactants. *Biochimica et Biophysica Acta (BBA) - Biomembranes, 1808*(12), 2830–2840. doi:10.1016/j.bbamem.2011.08.022 PMID:21889490

Harishchandra, R. K., Wulff, S., Lentzen, G., Neuhaus, T., & Galla, H. J. (2010). The effect of compatible solute ectoines on the structural organization of lipid monolayer and bilayer membranes. *Biophysical Chemistry, 150*(1–3), 37–46. doi:10.1016/j.bpc.2010.02.007 PMID:20206435

Hippchen, B., Röll, A., & Poralla, K. (1981). Occurrence in soil of thermo-acidophilic bacilli possessing ω-cyclohexane fatty acids and hopanoids. *Archives of Microbiology, 129*(1), 53–55. doi:10.1007/BF00417180

Imhoff, J. F. (2020). Osmotic Adaptation in Halophilic and Halotolerant Microorganisms. *The Biology of Halophilic Bacteria*, 211–253. doi:10.1201/9781003069140-8

Imhoff, J. F., & Rodriguez-Valera, F. (1984). Betaine is the main compatible solute of halophilic eubacteria. *Journal of Bacteriology, 160*(1), 478–479. doi:10.1128/jb.160.1.478-479.1984 PMID:6148337

Isken, S., & de Bont, J. A. M. (1998). Bacteria tolerant to organic solvents. *Extremophiles, 2*(3), 229–238. doi:10.1007/s007920050065

Jadhav, K., Kushwah, B., & Jadhav, I. (2018). Insight into Compatible Solutes from Halophiles: Exploring Significant Applications in Biotechnology. *Microbial Bioprospecting for Sustainable Development*, 291–307. doi:10.1007/978-981-13-0053-0_16

Jeffrey, J. (2007). Human Telomerase RNA Pseudoknot and Hairpin Thermal Stability with Glycine Betaine and Urea: Preferential Interactions with RNA Secondary and Tertiary Structures†. *Biochemistry, 46*(31), 9068–9079. doi:10.1021/bi602637v PMID:17630773

Karan, R., Capes, M. D., DasSarma, P., & DasSarma, S. (2013). Cloning, overexpression, purification, and characterization of a polyextremophilic β-galactosidase from the Antarctic haloarchaeon Halorubrum lacusprofundi. *BMC Biotechnology, 13*(1), 1–11. doi:10.1186/1472-6750-13-3

Karan, R., Capes, M. D., & DasSarma, S. (2012). Function and biotechnology of extremophilic enzymes in low water activity. *Aquatic Biosystems, 8*(1), 4. doi:10.1186/2046-9063-8-4 PMID:22480329

Kastritis, P. L., Papandreou, N. C., & Hamodrakas, S. J. (2007). Haloadaptation: Insights from comparative modeling studies of halophilic archaeal DHFRs. *International Journal of Biological Macromolecules, 41*(4), 447–453. doi:10.1016/j.ijbiomac.2007.06.005 PMID:17675150

Koch, R., Spreinat, A., Lemke, K., & Antranikian, G. (1991). Purification and properties of a hyperthermoactive α-amylase from the archaeobacterium Pyrococcus woesei. *Archives of Microbiology, 155*(6), 572–578. doi:10.1007/BF00245352

Kour, D., Rana, K. L., Kaur, T., Singh, B., Chauhan, V. S., Kumar, A., Rastegari, A. A., Yadav, N., Yadav, A. N., & Gupta, V. K. (2019). Extremophiles for Hydrolytic Enzymes Productions: Biodiversity and Potential Biotechnological Applications. *Bioprocessing for Biomolecules Production*, 321–372. doi:10.1002/9781119434436.ch16

Lambert, D., & Draper, D. E. (2007). Effects of Osmolytes on RNA Secondary and Tertiary Structure Stabilities and RNA-Mg2+ Interactions. *Journal of Molecular Biology, 370*(5), 993–1005. doi:10.1016/j.jmb.2007.03.080 PMID:17555763

Lee, C.-F., Makhatadze, G. I., & Wong, K.-B. (2005). Effects of Charge-to-Alanine Substitutions on the Stability of Ribosomal Protein L30e from Thermococcus celer†. *Biochemistry, 44*(51), 16817–16825. doi:10.1021/bi0519654 PMID:16363795

Longo, L. M., Lee, J., & Blaber, M. (2013). Simplified protein design biased for prebiotic amino acids yields a foldable, halophilic protein. *Proceedings of the National Academy of Sciences of the United States of America, 110*(6), 2135–2139. doi:10.1073/pnas.1219530110 PMID:23341608

Loukas, A., Kappas, I., & Abatzopoulos, T. J. (2018). HaloDom: A new database of halophiles across all life domains. *Journal of Biological Research (Greece), 25*(1), 2. Advance online publication. doi:10.118640709-017-0072-0 PMID:29796383

Madern, D., Ebel, C., & Zaccai, G. (2000). Halophilic adaptation of enzymes. *Extremophiles, 4*(2), 91–98. doi:10.1007/s007920050142

Marg, B.-L., Schweimer, K., Sticht, H., & Oesterhelt, D. (2004). A Two-α-Helix Extra Domain Mediates the Halophilic Character of a Plant-Type Ferredoxin from Halophilic Archaea†,‡. *Biochemistry, 44*(1), 29–39. doi:10.1021/bi0485169 PMID:15628843

Mayer, F., Küper, U., Meyer, C., Daxer, S., Müller, V., Rachel, R., & Huber, H. (2012). AMP-forming acetyl coenzyme a synthetase in the outermost membrane of the hyperthermophilic crenarchaeon ignicoccus hospitalis. *Journal of Bacteriology*, *194*(6), 1572–1581. doi:10.1128/JB.06130-11 PMID:22247508

Mevarech, M., Frolow, F., & Gloss, L. M. (2000). Halophilic enzymes: Proteins with a grain of salt. *Biophysical Chemistry*, *86*(2), 155–164. doi:10.1016/S0301-4622(00)00126-5 PMID:11026680

Miroshnichenko, M. L., & Bonch-Osmolovskaya, E. A. (2006). Recent developments in the thermophilic microbiology of deep-sea hydrothermal vents. *Extremophiles, 10*(2), 85–96. doi:10.1007/s00792-005-0489-5

Moon, S., Kim, J., Koo, J., & Bae, E. (2019). Structural and mutational analyses of psychrophilic and mesophilic adenylate kinases highlight the role of hydrophobic interactions in protein thermal stability. *Structure and Dynamics*, *6*(2), 024702. doi:10.1063/1.5089707 PMID:31111079

Müller-Santos, M., de Souza, E. M., Pedrosa, F. de O., Mitchell, D. A., Longhi, S., Carrière, F., Canaan, S., & Krieger, N. (2009). First evidence for the salt-dependent folding and activity of an esterase from the halophilic archaea Haloarcula marismortui. *Biochimica et Biophysica Acta (BBA) -. Molecular and Cell Biology of Lipids*, *1791*(8), 719–729. doi:10.1016/j.bbalip.2009.03.006 PMID:19303051

Muñoz, P. A., Flores, P. A., Boehmwald, F. A., & Blamey, J. M. (2011). Thermophilic bacteria present in a sample from Fumarole Bay, Deception Island. *Antarctic Science*, *23*(6), 549–555. doi:10.1017/S0954102011000393

Oren, A. (2013). Life at high salt concentrations, intracellular KCl concentrations, and acidic proteomes. *Frontiers in Microbiology*, *4*(NOV). Advance online publication. doi:10.3389/fmicb.2013.00315 PMID:24204364

Oren, A. (2015). Halophilic microbial communities and their environments. *Current Opinion in Biotechnology*, *33*, 119–124. doi:10.1016/j.copbio.2015.02.005 PMID:25727188

Panda, M. K., Sahu, M. K., & Tayung, K. (2013). Isolation and characterization of a thermophilic Bacillus sp. with protease activity isolated from hot spring of Tarabalo, Odisha, India. *Iranian Journal of Microbiology*, *5*(2), 159–165. https://pubmed.ncbi.nlm.nih.gov/23825735 PMID:23825735

Park, J. T., Song, H. N., Jung, T. Y., Lee, M. H., Park, S. G., Woo, E. J., & Park, K. H. (2013). A novel domain arrangement in a monomeric cyclodextrin-hydrolyzing enzyme from the hyperthermophile Pyrococcus furiosus. *Biochimica et Biophysica Acta (BBA) - Proteins and Proteomics*, *1834*(1), 380–386. doi:10.1016/j.bbapap.2012.08.001 PMID:22902546

Peng, S., Kai, M., Yang, X., Luo, Y., & Bai, L. (2020). Study on the osmoregulation of "Halomonas socia" NY-011 and the degradation of organic pollutants in the saline environment. *Extremophiles, 24*(6), 843–861. doi:10.1007/s00792-020-01199-5

Podar, P. T., Yang, Z., Björnsdóttir, S. H., & Podar, M. (2020). Comparative Analysis of Microbial Diversity Across Temperature Gradients in Hot Springs From Yellowstone and Iceland. *Frontiers in Microbiology*, *0*, 1625. doi:10.3389/fmicb.2020.01625 PMID:32760379

Qvist, J., Ortega, G., Tadeo, X., Millet, O., & Halle, B. (2012). Hydration Dynamics of a Halophilic Protein in Folded and Unfolded States. *The Journal of Physical Chemistry B, 116*(10), 3436–3444. doi:10.1021/jp3000569 PMID:22329545

Raddadi, N., Cherif, A., Daffonchio, D., Neifar, M., & Fava, F. (2015). Biotechnological applications of extremophiles, extremozymes and extremolytes. *Applied Microbiology and Biotechnology, 99*(19), 7907–7913. doi:10.1007/s00253-015-6874-9

Rampelotto, P. H. (2013). Extremophiles and Extreme Environments. *Life 2013, 3*(3), 482–485. doi:10.3390/life3030482

Reed, C. J., Bushnell, S., & Evilia, C. (2014). Circular Dichroism and Fluorescence Spectroscopy of Cysteinyl-tRNA Synthetase from Halobacterium salinarum ssp. NRC-1 Demonstrates that Group I Cations Are Particularly Effective in Providing Structure and Stability to This Halophilic Protein. *PLoS One, 9*(3), e89452. doi:10.1371/journal.pone.0089452 PMID:24594651

Rice, T., Zannini, E., Arendt, E. K., & Coffey, A. (2019). *A review of polyols – biotechnological production, food applications, regulation, labeling and health effects.* doi:10.1080/10408398.2019.1625859

Ruginescu, R., Gomoiu, I., Popescu, O., Cojoc, R., Neagu, S., Lucaci, I., Batrinescu-Moteau, C., & Enache, M. (2020). Bioprospecting for Novel Halophilic and Halotolerant Sources of Hydrolytic Enzymes in Brackish, Saline and Hypersaline Lakes of Romania. *Microorganisms, 8*(12), 1903. doi:10.3390/microorganisms8121903

Russell, R. J. M., Gerike, U., Danson, M. J., Hough, D. W., & Taylor, G. L. (1998). Structural adaptations of the cold-active citrate synthase from an Antarctic bacterium. *Structure (London, England), 6*(3), 351–361. doi:10.1016/S0969-2126(98)00037-9 PMID:9551556

Sabath, N., Ferrada, E., Barve, A., & Wagner, A. (2013). Growth Temperature and Genome Size in Bacteria Are Negatively Correlated, Suggesting Genomic Streamlining During Thermal Adaptation. *Genome Biology and Evolution, 5*(5), 966–977. doi:10.1093/gbe/evt050 PMID:23563968

Sagot, B., Gaysinski, M., Mehiri, M., Guigonis, J.-M., le Rudulier, D., & Alloing, G. (2010). Osmotically induced synthesis of the dipeptide N-acetylglutaminylglutamine amide is mediated by a new pathway conserved among bacteria. *Proceedings of the National Academy of Sciences of the United States of America, 107*(28), 12652–12657. doi:10.1073/pnas.1003063107 PMID:20571117

Saunders, N. F. W., Thomas, T., Curmi, P. M. G., Mattick, J. S., Kuczek, E., Slade, R., Davis, J., Franzmann, P. D., Boone, D., Rusterholtz, K., Feldman, R., Gates, C., Bench, S., Sowers, K., Kadner, K., Aerts, A., Dehal, P., Detter, C., Glavina, T., ... Cavicchioli, R. (2003). Mechanisms of Thermal Adaptation Revealed From the Genomes of the Antarctic Archaea Methanogenium frigidum and Methanococcoides burtonii. *Genome Research, 13*(7), 1580–1588. doi:10.1101/gr.1180903 PMID:12805271

Schnoor, M., Voß, P., Cullen, P., Böking, T., Galla, H. J., Galinski, E. A., & Lorkowski, S. (2004). Characterization of the synthetic compatible solute homoectoine as a potent PCR enhancer. *Biochemical and Biophysical Research Communications, 322*(3), 867–872. doi:10.1016/j.bbrc.2004.07.200 PMID:15336543

Schulz, A., Stöveken, N., Binzen, I. M., Hoffmann, T., Heider, J., & Bremer, E. (2017). Feeding on compatible solutes: A substrate-induced pathway for uptake and catabolism of ectoines and its genetic control by EnuR. *Environmental Microbiology, 19*(3), 926–946. doi:10.1111/1462-2920.13414 PMID:27318028

Siglioccolo, A., Paiardini, A., Piscitelli, M., & Pascarella, S. (2011). Structural adaptation of extreme halophilic proteins through decrease of conserved hydrophobic contact surface. *BMC Structural Biology, 11*(1), 1–12. doi:10.1186/1472-6807-11-50

Stathopoulou, P. M., Savvides, A. L., Karagouni, A. D., & Hatzinikolaou, D. G. (2013). Unraveling the lipolytic activity of thermophilic bacteria isolated from a volcanic environment. *BioMed Research International, 2013*, 1–13. Advance online publication. doi:10.1155/2013/703130 PMID:23738330

Strazzulli, A., Cobucci-Ponzano, B., Iacono, R., Giglio, R., Maurelli, L., Curci, N., Schiano-di-Cola, C., Santangelo, A., Contursi, P., Lombard, V., Henrissat, B., Lauro, F. M., Fontes, C. M. G. A., & Moracci, M. (2020). Discovery of hyperstable carbohydrate-active enzymes through metagenomics of extreme environments. *The FEBS Journal, 287*(6), 1116–1137. doi:10.1111/febs.15080 PMID:31595646

Sui, X., Wen, C., Yang, J., Guo, H., Zhao, W., Li, Q., Zhang, J., Zhu, Y., & Zhang, L. (2018). Betaine Combined with Membrane Stabilizers Enables Solvent-Free Whole Blood Cryopreservation and One-Step Cryoprotectant Removal. *ACS Biomaterials Science & Engineering, 5*(2), 1083–1091. doi:10.1021/acsbiomaterials.8b01286 PMID:33405798

Suka, A., Oki, H., Kato, Y., Kawahara, K., Ohkubo, T., Maruno, T., Kobayashi, Y., Fujii, S., Wakai, S., Lisdiana, L., & Sambongi, Y. (2019). Stability of cytochromes c′ from psychrophilic and piezophilic Shewanella species: implications for complex multiple adaptation to low temperature and high hydrostatic pressure. *Extremophiles, 23*(2), 239–248. doi:10.1007/s00792-019-01077-9

Thomas, T., & Cavicchioli, R. (1998). Archaeal cold-adapted proteins: Structural and evolutionary analysis of the elongation factor 2 proteins from psychrophilic, mesophilic and thermophilic methanogens. *FEBS Letters, 439*(3), 281–286. doi:10.1016/S0014-5793(98)01375-1 PMID:9845338

US6855734B2 - Glycine betaine and its use - Google Patents. (n.d.). Retrieved August 15, 2021, from https://patents.google.com/patent/US6855734B2/en

Vargas, C., Jebbar, M., Carrasco, R., Blanco, C., Calderón, M. I., Iglesias-Guerra, F., & Nieto, J. J. (2006). Ectoines as compatible solutes and carbon and energy sources for the halophilic bacterium Chromohalobacter salexigens. *Journal of Applied Microbiology, 100*(1), 98–107. doi:10.1111/j.1365-2672.2005.02757.x PMID:16405689

Vasudevan, N., & Jayshree, A. (2020). Extremozymes and Extremoproteins in Biosensor Applications. Encyclopedia of Marine Biotechnology, 1711–1736. doi:10.1002/9781119143802.ch72

Verma, S., Kumar, R., Kumar, P., Sharma, D., Gahlot, H., Sharma, P. K., & Meghwanshi, G. K. (2020). Cloning, Characterization, and Structural Modeling of an Extremophilic Bacterial Lipase Isolated from Saline Habitats of the Thar Desert. *Applied Biochemistry and Biotechnology, 192*(2), 557–572. doi:10.1007/s12010-020-03329-3

Voets, I. K. (2017). From ice-binding proteins to bio-inspired antifreeze materials. *Soft Matter, 13*(28), 4808–4823. doi:10.1039/C6SM02867E PMID:28657626

Vogler, M., Karan, R., Renn, D., Vancea, A., Vielberg, M.-T., Grötzinger, S. W., DasSarma, P., DasSarma, S., Eppinger, J., Groll, M., & Rueping, M. (2020). Crystal Structure and Active Site Engineering of a Halophilic γ-Carbonic Anhydrase. *Frontiers in Microbiology*, *0*, 742. doi:10.3389/fmicb.2020.00742 PMID:32411108

Waditee-Sirisattha, R., Kageyama, H., Takabe, T., Waditee-Sirisattha, R., Kageyama, H., & Takabe, T. (2016). Halophilic microorganism resources and their applications in industrial and environmental biotechnology. *AIMS Microbiology, 2*(1), 42–54. doi:10.3934/microbiol.2016.1.42

Welsh, D. T. (2000). Ecological significance of compatible solute accumulation by micro-organisms: From single cells to global climate. *FEMS Microbiology Reviews*, *24*(3), 263–290. doi:10.1111/j.1574-6976.2000.tb00542.x PMID:10841973

Wilson, M., Tillery, L., Tabaie, E., Beery, G., Preusker, J., Bhola, I., & Frato, K. (2020). Alanine to serine substitutions drive thermal adaptation in a psychrophilic diatom cytochrome c6. *JBIC Journal of Biological Inorganic Chemistry, 25*(3), 489–500. doi:10.1007/s00775-020-01777-0

Youssef, N. H., Savage-Ashlock, K. N., McCully, A. L., Luedtke, B., Shaw, E. I., Hoff, W. D., & Elshahed, M. S. (2013). Trehalose/2-sulfotrehalose biosynthesis and glycine-betaine uptake are widely spread mechanisms for osmoadaptation in the Halobacteriales. *The ISME Journal, 8*(3), 636–649. doi:10.1038/ismej.2013.165

Yusof, N. A., Hashim, N. H. F., & Bharudin, I. (2021). Cold Adaptation Strategies and the Potential of Psychrophilic Enzymes from the Antarctic Yeast, Glaciozyma antarctica PI12. *Journal of Fungi, 7*(7), 528. doi:10.3390/jof7070528

Zaccai, G., Cendrin, F., Haik, Y., Borochov, N., & Eisenberg, H. (1989). Stabilization of halophilic malate dehydrogenase. *Journal of Molecular Biology, 208*(3), 491–500. doi:10.1016/0022-2836(89)90512-3 PMID:2795658

Zahid, N., Schweiger, P., Galinski, E., & Deppenmeier, U. (2015). Identification of mannitol as compatible solute in Gluconobacter oxydans. *Applied Microbiology and Biotechnology, 99*(13), 5511–5521. doi:10.1007/s00253-015-6626-x

Zhang, G., & Ge, H. (2013). Protein Hypersaline Adaptation: Insight from Amino Acids with Machine Learning Algorithms. *The Protein Journal, 32*(4), 239–245. doi:10.1007/s10930-013-9484-3

Zhang, G., huihua, G., & Yi, L. (2013). Stability of halophilic proteins: From dipeptide attributes to discrimination classifier. *International Journal of Biological Macromolecules*, *53*, 1–6. doi:10.1016/j.ijbiomac.2012.10.031 PMID:23142140

Zhou, Y., Sun, Z. Y., Li, H., Qian, C. J., Wu, X., Tang, H. Z., Ali, F., & di Liu, Y. (2017). Investigation of compatible solutes synthesis and transport of Virgibacillus halodenitrificans PDB-F2 with complete genome analysis. *International Biodeterioration & Biodegradation*, *122*, 165–172. doi:10.1016/j.ibiod.2017.05.005

Section 3
Biotechnological Applications of Extremophiles

Chapter 9
Extremophiles and Their Application in Bioremediation

Himanshu Pandey
Dr. Yashwant Singh Parmar University of Horticulture and Forestry, India

Devendra Singh
Motilal Nehru National Institute of Technology, Allahabad, India

Vinay Kumar Dhiman
Dr. Yashwant Singh Parmar University of Horticulture and Forestry, India

Vivek Kumar Dhiman
Himachal Pradesh University, Shimla, India

Devendra Pandey
Central Institute for Subtropical Horticulture, Lucknow, India

ABSTRACT

A microorganism dwelling in severe environmental conditions is termed an extremophile. These unfavorable environmental conditions include high salinity, toxin compounds, heavy metals, unfavorable temperature, and extremely acidic and alkaline pH. Microorganisms belonging to prokaryotes include true bacteria and archaea bacteria which prevail in harsh environments. In recent years, extremophilic, basically, archaea bacteria have been reported for their immense potential application in the bioremediation process. Bioremediation is a technique that utilizes microorganisms for the decomposition of organic and inorganic pollutants; anthropogenic activities are the basic cause of soil pollution, water pollution, and air pollution globally. Extremophiles are capable of producing enzymes that are thermolabile and can function normally even in extreme conditions. These enzymes and proteins can be utilized in the bioremediation process under extreme pH, heavy metal stress, and unfavorable temperature conditions. In this chapter, the role of extremophiles in bioremediation is discussed.

DOI: 10.4018/978-1-7998-9144-4.ch009

INTRODUCTION

Many living organisms are able to interact with pollutants through a biological method so as to degrade the harmful pollutants in a very slow process; depending upon their capability, the rate of this process can be helpful in controlling the pollution. Due to the involvement of organisms in remediation, their biological processes rate, lower costing, no management issues making bioremediation an efficient and ideal biological process compared to physical and chemical methods of remediation. Numerous bacteria, fungi, and some plants are able to neutralize the pollutants through degradation to elementary constituents or to non-harmful form (Azubuike et al., 2016). Bioremediation is the process of detoxification of pollutants present in the environment and the hazardous waste material components into mild or attenuated or less toxic components by means of biological metabolic processes, i.e., degradation, mineralization, immobilization, removal, transformation by the action of living organisms primarily microorganisms, the enzymes they produce or by some green plants as a treatment sites contamination to the most original form (Azubuike et al., 2016; Dua et al., 2002). It can be applied to clear contamination in soil, groundwater, ponds, waste sludge, or streams using microbial action of natural or genetically modified microbes. A proper favorable environment for optimized growth can be provided via engineering designs; thus, the technique can be implanted in contaminated sites. Figure 1 describes the basic principles of bioremediation.

SIGNIFICANCE OF BIOREMEDIATION

Bioremediation is becoming a major decontamination method due to its approach in most human toxic waste treatment. Pollutants are becoming direct or indirect issues of public health, environmental contamination, and the cost of treating the industrial, municipal, industrial, agriculture waste varies on their treatment methods. All these wastes are responsible for polluting water resources and soil, which are directly related to all living life. As water and soil are directly related to each other, there is the exchange of everything between the two, whether it's the organic decomposed plants, animals, or the leaching of minerals to direct contact of aquatic life to land life. Bioremediation is an inexpensive methodology critically needed for the removal of pollutants (Azubuike et al., 2016). Bioremediation is effective in removing petroleum industry pollutants that are organic hydrocarbons such as aromatic, alkanes, and alkenes compounds. Subsurface and soil contamination due to leakage of pipelines, oil spills, automobile industries, manufacturing, etc.

Agriculture is also untouched from contaminating terrestrial, aquatic life forms. Due to the application of different pesticides in different forms, i.e., powder, aqueous solutions, emulsions, sprays, etc., they enter the environment in different forms. Structurally they are inorganic and organic compounds mostly comprising Cl, P, N, and other complex forms (Dua et al., 2002). They are stable in the environment for many years or even decades, causing ecological instability and health issues. Fertilizers, on the other hand, pollute the surface or groundwater by leaching or runoff. Eutrophication of this runoff causes intoxication in aquatic life; phosphate, nitrates, and nitrites are the leached runoff of agriculture (Dua et al., 2002). Bioremediation helps in the transformation, assimilation of these chemicals and also providing the breakdown products of these complex chemicals to plants in simpler form in some cases (Azubuike et al., 2016).

Figure 1.

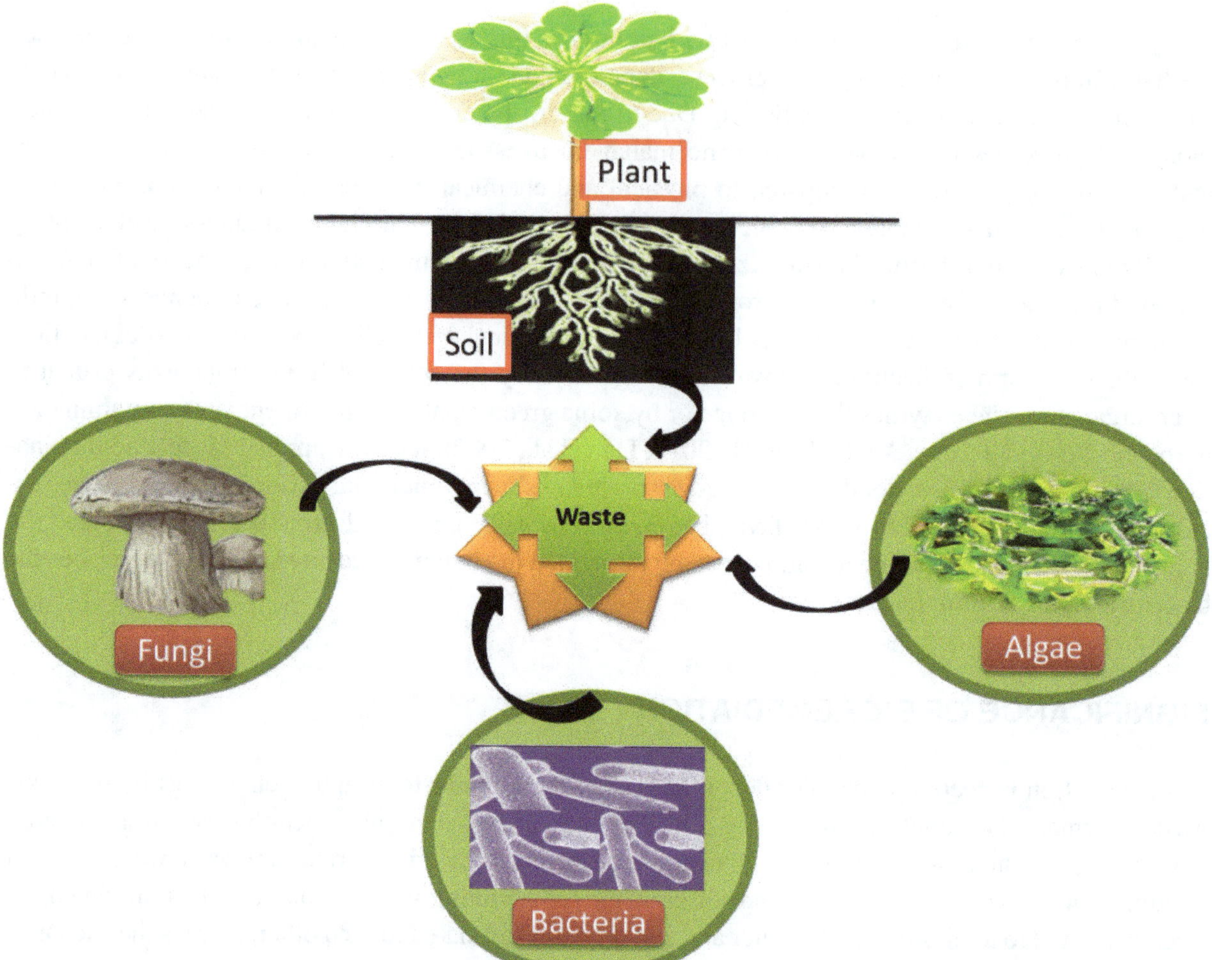

Soil consists of majorly persistent pollutants that need to be degraded through bioremediation. Various human practices that are responsible for the environmental accumulation of persistent pollutants include the process of mining and smeltering of raw material of sites, i.e., metals and ores mining, landfills, overuse of pesticides in farmlands, industrial and municipal wastes, including sludge-slurry with rich effluents, manufacturing plants of gases, nuclear waste dumping sites, ammunition waste dump sites, electronic wastes components, petroleum oil waste materials, hospital waste, etc (Dua et al., 2002). Different types of chemical pollutants are present in the environment, which includes PAHs (poly-aromatic hydrocarbons), NACs (nitroaromatic compounds of metals and industrial solvents), polynuclear aromatic hydrocarbons (PAHs), aromatic (e.g., benzene, ethylbenzene, toluene, xylene) chloroaromatics, s-triazines, biphenyls, polycyclic aromatics, polychlorinated biphenyls, hydrocarbon of halogenated petroleum, pesticides, etc. are liable to microbiological degradation (Dua et al., 2002).

Mechanism of Bioremediation

It may be reconsidered that bioremediation leads to degrading and detoxifying the harmful pollutants within the environment via a natural metabolic process of living organisms such as bacteria, algae, fungi, or plants into simpler constituents of metabolic bi-products that are less toxic or can be further converted to non-toxic or mineralized constituents, or as biomass or to conversion to CO_2, H_2O (Azubuike et al., 2016). Microbes dominate and play a crucial role in the bioremediation process due to the ubiquitous and diverse natural environment they harbor, hence mostly unidentified, which can potentially degrade pollutants. Some microorganisms indigenous to contaminated sites are isolated and brought to the same site with further optimization or sometimes introducing other microbes to the site to enhance the bioremediation in the area (Dua et al., 2002). The transformation of these toxic pollutants requires the degradation and utilization of complex compounds for growth purposes through metabolic reactions. The more complex compound is, the slower the process becomes due to physical and chemical properties of compounds, type and solubility, the bioavailability of pollutants to microbes, moreover the population and capability of microbes also influence the process. Total degradation of compounds requires a favorable environment, effective degradation, potential microorganisms that are efficient in the process. Hence it is important to provide an operative and controlled environment for efficient and fast degradation of pollutants (Azubuike et al., 2016).

In general, bioremediation can be trusted upon three methods, i.e., natural attenuation, biostimulation, bioaugmentation. Natural attenuation (passive or intrinsic remediation) is solely based on the ability of the native microbial community for pollutant degradation, and biostimulation is based on providing the nutritional balance as inorganic or organic nutrition for faster functioning of the native microbial community. Bioaugmentation is the specialized aided microbial community to remove pollutants (Sarkar et al., 2020). In situ and Ex-situ are the two contaminants treatment processes. In situ, bioremediation involves a treatment process, namely referred to as on-site or original place or within the contaminated site. It requires a large area, is less expensive, and requires less labor intensity, but conditions are difficult to maintain due to open contaminated sites. It involves numerous procedures, including bioaugmentation, bioventing, biosparging, biostimulation, phytoremediation, etc (Das and Kazy, 2014). On the other side, ex-situ bioremediation involves a treatment process referred to off-site or away from the original contaminated site as in a laboratory or controlled condition. It involves moving away the water or soil contaminated with pollutants and providing them with conditions favorable for degradation as shown in the Table 1.

Bioremediation Exploitation in Extremophiles

The diversity of microorganisms varies with different environmental conditions; they have evolved to utilize the robustness of enzymes and biocatalysts, providing changes necessary for their survival. Sensing the changes in the environment around them and utilizing the extremozyme accordingly with specificity is essential, as the metabolic changes provide endurability, adaptation, and sustain according to nutrition. Extremozymes are stable at high temperature, ionic states, and variable range of pH, solvents as proteins present within the molecular level of these enzymes vary according to conditions required for survival (in acidic or alkaline pH, under high pressure, low to high temperature, halophilic, etc. enzymes work optimally requiring such conditions) (Dumorné et al., 2017). Such enzymes can be exploited or engineered and modified with a biotechnological perspective by changing the biology for industrial remediation purposes. Due to this unique capability of enduring multiple stress variations, these microorganisms are named extremophiles. Almost every habitat (hydrothermal vents, sea-ice, volcanic mud, volcanic hot-spring vents, deserts, deep anoxic seas, ocean and lakes, nuclear contaminations, oligotrophic environments, and many others) that is incapable of dwelling life makes an exception for these microorganisms (Merino et al., 2019). Microorganisms surviving in harsh environmental conditions are most likely to metabolize and sustain in remediation sites. They have the potential to utilize different nutrient sources, biomass beneath the contamination sites, as many microorganisms are found beneath the deep layers of earth. The investigations of these sites, engineering the microbes' potential to endure and adapt contamination area for bioremediation, can be achieved by identifying, selecting, and exploiting the extremophiles (Table 2).

Table 1. Different Bioremediation methods

Bioremediation methods [a,b]	Categories	Methods details	Advantages	Disadvantages
In-situ Bioremediation	Biosparging	Induced air flow under pressure is injected to increase biological degradation by naturally occurring microbes	Minimal disturbance to saturated zone, easy to operate	Suitable under uniform-permeable soils, aquifers
	Biostimulation	Nutrients are supplied to optimize the growth and activity of the natural microbial population	Effective for volatile compounds, both soil and water are treatable, natural attenuation	Environmental constraints, i.e., site-specific
	Bioaugmentation	Genetically modified and induced microbes are aided to the targeted site	Non-invasive in the site operation	Monitoring is difficult
	Bioventing	Injected air and nourishment is aided by wells sufficient to maintain microbial activity	Cost-effective operation	Environmental constrain
Ex-situ Bioremediation	Bioreactors	Microbes aid the biological degradation in a closed container or a tank under favorable condition	Degradation rate and progression is optimum, same as the growth of microbes	A problem in mass transfer, i.e., uniform distribution of substrate (pollutant)
	Biopiles	Both composting, as well as landfilling is involved simultaneously	Effective against soil pollutants, less water pollution in soil	Some of the adsorbed pollutants are inaccessible to microbes limiting their bioavailability (pollutants in water or soil that are unavailable to biota)
	Landfarming	Waste is tilled in different layers of topsoil amended with water and nutrients	Cost-effective operation	The area is required for tilling
	Biocomposting	Aerobic microbes decomposition in a higher reaction rate and temperature	Less toxic gas and pollutants after composting	Requires human handling

Where [a] and [b] denote the reference Nandal, 2015 and Tyagi and Kumar, 2021

Table 2. Range of conditions optimum for extremophiles habitat

Lower to Higher (based on parameter range)[a]					
pH (0 - 14)	Extreme-acidophiles (< pH 3)	Acidophilic (< pH 5)	Neutrophiles (pH 5–9)	Alkaliphiles (> pH 9)	Hyper-alkaliphiles (> pH 11)
Temperature(°C)		Cryophiles (< 20°C)	Mesophilic (20 to 45°C)	Thermophiles (45 to 80°C)	Superthermophiles (> 80°C)
Salinity (as% NaCl (w/v))		Non-halophiles (< 1.2%)	Salt-tolerants (1.2 to 2.9%; tolerate ≤ 14.6%)	Halophilic (> 8.8%)	Extreme halophiles (> 14.6%, cannot grow < 8.8%)
Pressure (MPa)			Piezo-tolerant / baro-tolerant (0. 1 to 10 MPa)	Piezophiles (10 to 50 MPa)	Hyperpiezophiles / hyperbarophiles (> 50 MPa)
Water activity (aw)			Xerotolerant (a_w < 0.7)		
Poly-extremophile	Tolerance under multiple conditions				

Where [a] denotes the reference Merino et al., 2019

IDENTIFICATION OF EXTREMOPHILES IN BIOREMEDIATION

Pollution directed environment includes mostly industrialized areas, mining sites, dumping areas, petroleum wells, and other practices which produce xenobiotics as in the agriculture sector. Many of the microorganisms present in these ecologies potentially have adapted to unnatural and altered ecosystems. Metabolism of these microbes is able to convert pollutants to a more stable or harmless compound that can be further used by other organisms or should remain in a low toxic state. With the advent of research and development in the field of molecular biology, many tools and techniques have been developed for the identification and functioning of the biological system in microorganisms (Tailliez et al., 2002). Our understanding has expanded on how to categorize microorganisms based on their functionality; at the gene level, detection of genes responsible for a peculiar function essential for the expression of proteins and enzymes that work in multi-stress conditions can be exploited based on interest. Different techniques used are based on ribosomal RNA present in prokaryotes as 5S, 16S, and 23S, i.e., in bacteria and archaea. 16 rDNA, 16 rRNA sequencing, hybridization (FISH) and probe (Isotopes probing) techniques, different PCR based techniques (Real-Time PCR, RAPD, etc.), ARDRA, T-RFLP, DNA fingerprinting (DGGE, TGGE) analyzing ITS, oligonucleotide region for sequencing (Tailliez et al., 2002; Zhao, et al., 2010). As studies are ongoing to identify more new diversity of microorganisms, their essence in producing life on earth cannot be unnoticed. The functionality of microorganisms is the basis of the functionality of the earth's ecosystem with other higher organisms as well as with the earth's geography. Their survival in extreme conditions, including the industrial, petroleum polluted water and lands, radioactive, heavy metal enriched sites, and other polluted landmarks, have been of great interest to scientists and engineers. Identification of these microbes in polluted sites is based on physiological, molecular, growth aspects as well as their degradation potential on different complex pollutants. The only limitation of applying the extremophiles in remediation is the environment that often needs to be favorable so that these microorganisms can grow and metabolize pollutants to simpler products (Merino et al., 2019). Table 3 shows the categorization of extremophiles on the basis of their habitat.

Table 3. Extremophiles categorized based on their habitat

Extremophiles based on different categories [a]		
Acidophiles	Organisms that grow under optimum pH < 3, i.e., under acidic conditions.	*Thiobacillus* sp., *Ferroplasma* sp., *Acidithiobacillus ferrooxidans, Acidobacteria, Picrophilus oshimae, P. torridus, Acidovorax facilis*
Alkaliphiles	Organisms that grow above optimum pH > 9, i.e., above alkaline condition.	*Natronomonas, Virgibacillus chiguensis, Bacillus horikoshii, B. pseudofirmus*
Halophiles	Organisms that require a minimum of 0.2 M salt (>5% NaCl (w/v)) for growth.	*Halococcus, Halobacterium salinarum, Halobacterium* sp., *Haloferax sp., Haloterrigena*
Thermophiles	Organisms that inhabit at temperatures of 60 - 80°C.	*Thermoplasma volcanium, T. acidophilum, Methanocaldococcus jannaschii*
Hyperthermophiles	Organism growing at an optimum temperature of 80°C or above.	*Methanopyrus kandleri, Archaeoglobus sp., Nanoarchaeum, Ignicoccus sp., Pyrodictium* strain 121, *Pyrolobus*
Psychrophiles	Organisms are requiring an optimum temperature of < 20°C - 15 °C or lower.	*Arthrobacter flavus, A. glacialis, Micrococeus cryophilus, Pseudomonas aeruginosa*
Piezophiles	Organisms that grow optimum in a high (> 60 MPa) hydro-static pressure.	*Methanothermococcus thermolithotrophicus, Pyrococcus yayanosil, Methanocaldococcus jannaschii, Methanothermococcus sp.*
Anoxiphiles	Organisms inhabiting O_2 deprived conditions	*Ferroglobus*
Oligotrophic	Organisms that are able to inhabit nutrient-deprived conditions	*Pelagibacter ubique*
Endolithes	Organisms that inhabit inside rocks, shells, corals, or inside pores within rocks grains.	*Schizothrix lacustris, Solentia sp., Cytophaga, Nitrospirae, Chloroflexi, Hyella gigas, H. racemus*
Hypolithes	Organisms that inhabit underneath the rock soil interface in extreme desert areas.	*Chroococci diopsis, Gloeocapsa sp.*
Metallotolerant	Organisms that are able to tolerate high levels of heavy metals in their environment, including metals like Cu, Ca, As, Zn, etc.	*Ferroplasma sp., Cupriavidus metallidurans, Halbacterium sp., Ralstonia sp., Halococcussa lifodinae, Haloferax sp.*
Radioresistant	Organisms are resisting a higher level of ionizing radiation.	*Deinococcus perariditoris, Deinococcus radiodurans, Thermococcus gammatolerance, Rubrobacter sp.*
Toxitolerant	Organisms are withstanding damaging mediators such as benzene, toluene, or water within the nuclear reactor core.	*Pseudomonas putida*
Xerophiles	Organisms are inhabiting lower water activity, such as in the extreme halophilic or endolithic environment.	*Artemia salina, Streptomyces bulli*
Polyextremophiles	Organisms inhabiting multi-stress conditions (pressure, nutrient-deficient environment or temperature, etc. simultaneously)	*Deinococcus radiodurans, Picrophilus torridus, P. oshima, Natronobacterium sp.,*

Where [a] denotes the reference Seckbach et al., 2011

Extremophiles in Bioremediation of Hydrocarbons

Hydrocarbons are present in land and water due to natural as well as anthropogenic processes leading to pollutants gathering. Hydrocarbons composed of aliphatic, aromatic, or branched chains constitute petroleum basis of modern energy fuel, other organic compounds (Prince 1993), elements like O (1%), S(0.5%), N(0.5%), and metallogenic hydrocarbons of Fe, Ni, Cu or mixed can be found in smaller amount

(<0.1%). Annual consumption of petroleum has increased day by day for transportation, automobile industry, machinery based on diesel-petroleum, other crude products such as gasoline or its use in burning as kerosene fuel to fueling gas for cooking. As they are hydrophobic in nature, anthropogenic processes lead to their gathering around the solid surface and soil, groundwater, and sea-ocean water. Human activities like transport and road traffic and land use to accidental spillage during petroleum industry activities, leakage from containers during transport, oil refineries, mishandling, or mismanagement error during its consumption or use are the major factor for hydrocarbon release (Souza et al., 2014; Prince et al., 2014). Its side product hydrocarbons (coal, crude oil, petroleum, natural gas) derivatives are formed from planktons, algae, plant, animal dead remains trapped beneath earth's layers over million years in geological cycles comprising of intense heat, temperature, the pressure under anaerobic condition. These hydrocarbons are found in the sediment of source rocks and move upward through migration from permeable strata and get accumulated in impermeable dense rock layers, "the trap." If migrated fluid (crude oil or natural gas) doesn't encounter the impermeable rock layers, they tend to flow to the surface or deposited on the ocean floor, therefore, leaching out as natural gas or liquid form.

Prokaryotes, including archaea and bacteria, are soul utilizer of hydrocarbons, as reported by several researchers. Hydrocarbon is the carbon and energy source of these microorganisms, thereby are part of bioremediation processes occurring in remediation (Dean-Ross, Moody, & Cerniglia, 2002). An environment with many hydrocarbons is bound to provide rich carbon in marine or surface petroleum sites as well as high polyaromatic hydrocarbon inhabiting the microbes adapted to these surroundings (Zhang et al., 2012). PAHs are toxic, mutagenic, and carcinogenic complex pollutants prior to causing health issues, marking them as primary target pollutants that need to be degraded (Prasad 2016). These tolerant communities will thrive in any similar hydrocarbon-rich environment as the native niche; once exploited, these communities can be used as an augmented template by engineers for remediation purposes and bio-transformation of contamination areas. Hydrocarbon-rich environments are an inhabitant of archaea as reported in dam and dikes containing tailing materials, the soil of refueling station, oil spill deep within the horizon, petroleum waste, etc. the archaea phyla Euryarchaeota and Crenarchaeota predominates (Sutton et al., 2013). Euryarchaeota is diverse in its extreme nature as halophilic, thermophilic, and methanogenic, whereas Crenarchaeota are hyperthermophilic utilizing sulfur as an energy source (Woese et al., 1990). Methanogenic archaea are belonging to Euryarchaeota, i.e., *Methanocella* sp., *Methanobacterium* sp., *Methanomicrobiales*, *Methanosaeta* are found in oil refinery (Das and Kazy, 2014; Gao et al., 2015; Sarkar et al., 2016).

Oil alkanes are converted into methane and CO_2 by methanogens (*Methanomicrobia, Methanoculleus*) (hydrogenotrophic), syntropic (*Smithella I*), and fermentative bacteria, and sometimes to acetic acid and hydrogen by *Methanosaeta* (acetotroph) (Meijer et al., 1999). H_2 is reduced by sulfur-reducing bacteria. *Acidaminobacter, Propionibacteriaceae, Kosmotoga, Sediminibacterium, Thermanaerovibrio, Acetobacterium, Desulfuromonas, Synergistetes, Fusibacter,* etc. many degrading communities of bacteria are found in the oil, natural gas, and coal sites (Dwason et al., 2012; Kryachko et al., 2012; Baea et al., 2006; Dipippo et al., 2009). Dolfing (2014) also estimated the higher population of methanogens such as hydrogenotrophs over acetoclastic in crude oil biodegradation. Roy et al., (2018) studied the diversity of microbial communities in the sludge of different refineries (native location in Assam, India) by isolating and characterizing sludge samples in different carbon media(maltose, lactose, sucrose, mannitol, galactose, inositol), alkanes (dodecane, hexadecane, cyclohexane, pentadecane) thereby analyzing their potential for remediation of total petroleum hydrocarbons (TPH) in sludge, thereby categorizing them based on 16 rRNA gene sequencing and amplification. The isolates belong to obligate/facultative anaerobes, fermentative, syntrophic, methanogenic, CO_2 assimilating, hydrocarbon-degrading, some xenobiotics degrading communities indigenous to sludge found are *Methanobacterium, Methanosaeta, Desulfitobacter, Desulfosporosinus, Pseudomonas, Coprothermobacter, Bacillus, Rhodobacter, Achromobacter,* etc. Eight native isolates were biostimulated and bioaugmentated in killed sludge as a consortium augmented with P, N, and N+P and compared with microcosms without nutrient amendment.N+P amended consortium was found to be most efficient in degrading total petroleum hydrocarbons (TPH), i.e., 75%, followed by N amended consortium (63.3%), and P amended consortium (57.3%) due to cooperative effect of native communities. Moreover, bacteria involved in the consortium were analyzed for PICRUSt (Phylogenetic investigation of communities by reconstruction of unobserved states), thereby showing a higher allocation of genes involved in nutrient transport-utilization and metabolism pathways related to hydrocarbon utilization. Some substrates that are degraded by Extremophiles are mention in the table 4.

Table 4. Hydrocarbons degradation by Extremophiles

Substrate	Bacteria involved in transforming organic compounds [a]
Coal-oil	*Pseudomonas sp & Ochrobactrum sp Stenotrophomonas maltophila*
Fossil fuel	*Acinetobacter sp*
Alkanes & organic-aromatic hydrocarbons	*Georgenia sp, Bacillus sp, Stappia sp* and *Isolptericola sp*
Heptadecane, coal-oil, naphthalene	*Pseudomonas, Marinobacter* and *Cycloclasticus*
Phenanthrene, Anthracene, Fluorene	*Ochrobactrum sp, Enterobactercloceae, and Stenotrophomonas maltophila sp*
Aniline	*Dietzianatronolimnaea*
Pyrene, Benzoanthracene, Chrysene	*Novosphingobium pentaromativorans*
Alkanes	*Alcanivorax dieselolei*
Benzo(e)Pyrene	*Achromobacter sp* and *Rhodanobacter sp*
Coal- oil	*Halomonassp.*
Rock- oil	*Cellulomonas* sp., *Dietziamaris, Halomonas eurihalina*
Benzene, Xylene	*Marinobacter sp*
Benzene, toluene	*Alcanivorax sp*
Organic compounds	*Arthrobacter and Micrococcus*
Phenathrene, Anthracene	*Matelella*
Benzene, Xylene, toluene	*Planococcus*
Naphthalene, Phenanthrene, Pyrene	*Halorubrum ezzemoulense, Halobacterium piscisalsi* and *Halobacterium salinarium*
Petroleum compounds	*Marinobacter sp, Erwiniaananas sp* and *Bacillus sp*
Benzene, toluene	*Bacillus sp, Acidovoraxdelafieldi, Halobacillussalinus,*
Benzol, phenylmethane	*Arhodomonas sp*
Diesel oil	*Halobacterium salinarum*

Where [a] denotes the reference Arulazhagan et al., 2017

Figure 2.

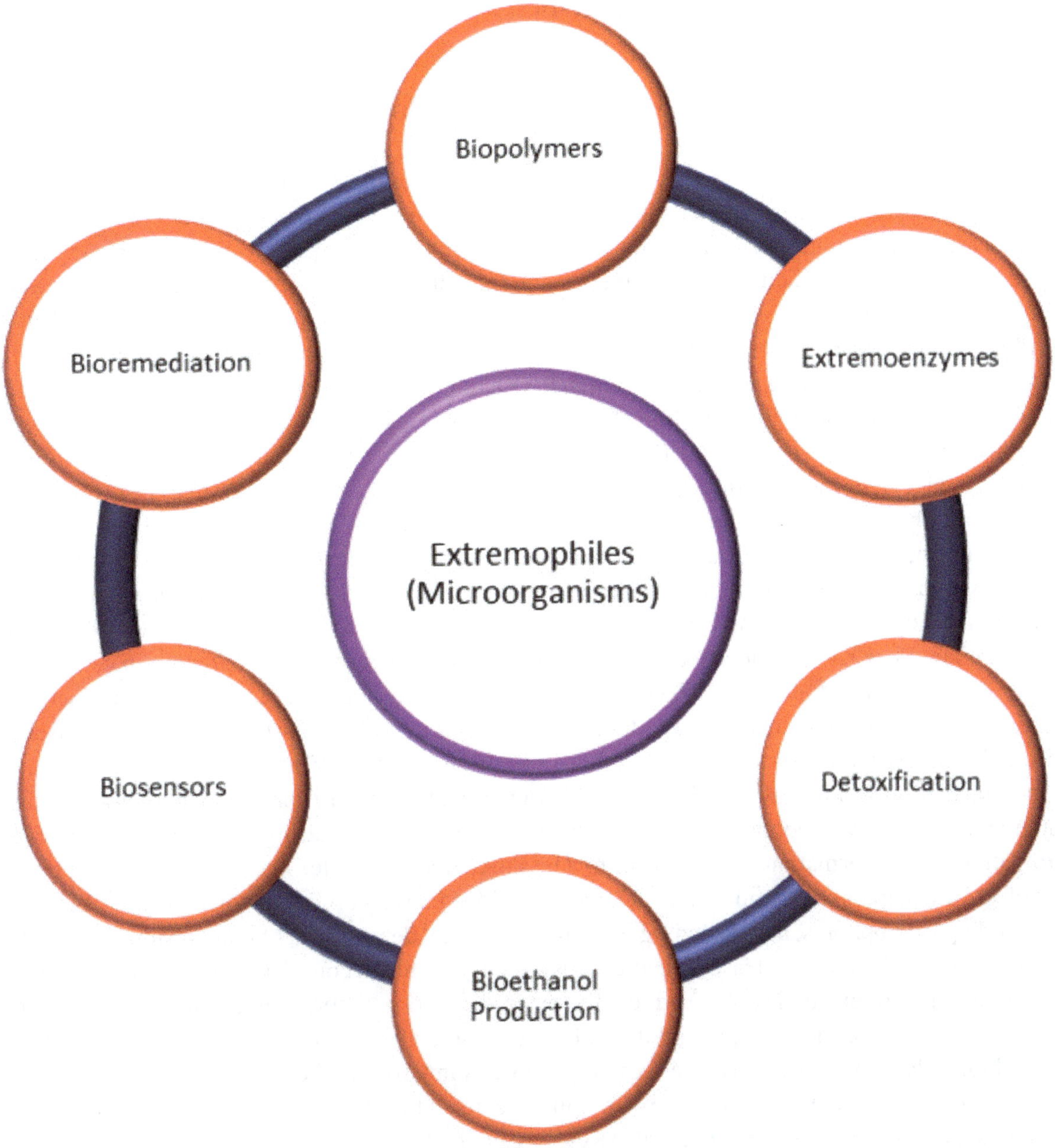

Park and Park (2018) studied the alkanes degradation in crude oils, suggesting the degradation takes in extreme environments of aerobic, anaerobic, low-high temperature, salt, acidic pH by archaea and bacteria. These microbes involve various metabolic pathways involving different survival mechanisms (exopolysaccharides, chaperons, proteins, solute gradient inside the cell, etc., for protection) as well as in secreting enzymes for alkanes degradation. *Archaeoglobus fulgidus* (S reducing thermophile) secrete alkylsuccinate synthase for long-chain alkanes, *Candidatus Syntrophoarchaeum butanivorans* (thermophilic) oxidize under special condition (anaerobically) by enzyme alkyl-coenzyme (Park and Park,

2018). Thus, extremophiles' application for hydrocarbon remediation is limitless; with technological advancement, the approach for bioremediation varies considerably. Figure 2 summarizes the application of extremophiles in various sectors.

Extremophiles in Bioremediation of Heavy Metals

Heavy metals have a density of 5 g cm^{-3} or above, are present in soil naturally or by anthropogenic practices. They are naturally found in volcanic ash or dust, volcanic eruption causes them to move upward into the earth's crust, and atmosphere along with gases and heat, rocks enriched with heavy metals, weathering of these rocks adds the heavy metals enrichment to the soils. Anthropogenic practices due to humans causing the addition of heavy metals may be due to mining-smelting activities (e.g., Cd, As, Hg, Pb), use of metal-based pesticides (e.g., Zn, Cu, Pb, As, Se, Cd) in agriculture, bioaccumulation and leaching, their processing and purification in the metallurgic industry (e.g., Co, Hg, Cr, As, Cd, Cu, Zn, Ni) for further technological use like electronics, dumping and non-recycling of electronic waste, emission during fossil fuel combustion (Fe, Ni, Cu), use in weapon and military training (Pb, Cd, Cr, Be, etc.), its presence in sewage. All these practices are the major source of soil pollution as heavy metals are persistent, non-biodegradable in the environment. Metals essential for living beings and plants include Fe, Zn, Mg, Cu, Co, Mo for the functioning of life processes in trace amount. They are used for stabilizing biomolecules, are the present complex structure of pigment, enzyme co-factors. Metals, when absorbed as macronutrients or traces, consumed in excess, become harmful or toxic (Goyeret al., 1997). Other metals like Hg, Pb, Cr, Ar produce reactive oxygen species (ROS) that cause oxidation in cells leading to their damage (Pinto et al., 2003). Excess of heavy metals is the cause of instability, malfunctioning of various systems in the body. Mutation, teratogens, cancer, immune disorders, neurological disorders is due to ROS, circulatory system (Kim et al., 2015; Korashy et al., 2017).

Bioremediation is an ideal technique that is efficient and necessary for removing heavy metals in contaminated areas. Microorganisms reduce or recover the heavy metal in the contamination site by assimilation or transformation to biomass or to a less toxic/reactive state. These metal ions can be used as an electron acceptor or assimilated inside microbes. Mechanisms used to involve the use of chelators (iron-binding siderophores, metal-binding proteins), microbial degradation of plant residues, animal tissues, and microbial biomass releases organic acids which form metal complex, volatilization in mercury and selenium metal ions by MerA enzyme (Barkay et al., 2003). Absorption, accumulation of metals, transformation (sequestration or precipitation of metal), mineralization (recovery of metal) resulting in a lower bioavailability of metal. Hence these are the steps involved in the remediation of metal pollutants from the environment. *Citrobacter* sp. immobilizes metal ions for generating biofilms to bind more metal ions, precipitating insoluble minerals (Keasling et al., 2000).

Many bacteria like thermo-acidophilic archaea that are lithoautotroph are usually metal resistant and show resistance to metals due to the presence of genes involved in (metal efflux transport, metallochaperons (trafficking metal ions) efflux system, i.e., *Archaeoglobus fulgidis, Sulfolobus solfataricus, S. acidocaldarius, Metallosphaera cuprina* (Schelert et al., 2013) are resistant to Cu due to *cop genes* specially *copA* involved in efflux system, *S. solfataricus, Metallosphaera sedula, Leptospirillum ferriphilum* consists of merH, A, I genes that provide resistance to Hg (Aurenik et al., 2008; Mi et al., 2011; Schelert et al., 2013). Various halo-archaea bacteria are utilized for clearing up toxic metals contaminated areas and wastewater. Popescu and Dumitru (2009) found 2 strains of *Haloferax* capable of biosorption of metals Pb, Cr, Ni, Zn in the saline medium. *Halobacterium noricense* was found to absorb Cd (Showalter et al., 2016), *H. salinarum* was also able to absorb Ca^{2+} ion, assisting them in forming aggregates of cells (Kawakami et al., 2007), *Halobacterium* sp. are also resistant to high Manganese (Mn) (Naikand Furtado, 2014).

Role of Extremophiles in Bioremediation Through Extremozymes

The enzyme synthesized by extremophilic microorganisms, well known as extremozymes, are kind of proteins that can withstand high temperature, extreme pH, and degrade harmful organic substances, thus could possibly be utilized for industrial bio-transformations (Schiraldi and De Rosa, 2002). Enzymes such as amylases, xylanases, and proteases, etc., are isolated from an extremophilic microorganism that includes psychrophiles, hyperthermophiles, and acidophiles which help in the degradation of different types of the polymer may play a significant role in industry related to detergent, food, pulp, and paper. Extremozymes are also involved in the synthesis of different other classes of enzymes like cellulases, pectinases, proteases, lipases, phytases, esterases, catalases, keratinases, and peroxidases. Owing to the different characteristics of these extremozymes, they are supposed to narrow down the gap between industries of biological origin and chemical-based (Gunjal et al., 2021, Taylor et al., 2011).

The protein biocatalysts, unlike normal enzymatic reactions, are named "extremozymes" due to their resist, resilient nature of catalyzing a biochemical reaction at the condition that normally degrade or inactivate the enzymes. These include extremely high-low temperature, pH, the difference in an organic solvent that is abnormal for biocatalyst; hence they can be used in the biotransformation of various pollutants (Gunjalet al., 2021, Schiraldi and De Rosa, 2002). These robust enzymes are present in hyperthermophiles, acidophiles, psychrophiles and have can depolymerize xenobiotics, toxic pesticides, reduce or oxidize metal ions to stabilize and transform them. Enzymes that stabilize in extreme conditions are also operatable in remediation sites depending upon their specificity and efficiency to degrade pollutants. Hence, more of these enzymes need to be exploited as novel biocatalysts, properties like easily extractable, purified through downstream processing, requiring no cofactor for its activity, which is operatable under varied conditions in bioremediation sites (Peeples, 2014). Further enzymes that aid in the conversion process is functional without cofactor regeneration and can be produced through large-scale fermentation in a cost-effective way with minimal purification requirement. Extremozyme that are used in bioremediation can be listed as lipases esterases, peroxidases, nitrilases, laccases, monooxygenases, dioxygenases, cellulases, arylsulfotransferase, l-Haloacid Dehalogenase. Pectinase. Few enzymes that are known to perform independently without requiring a high amount of co-factors are laccase, phenoloxidase, phosphotriesterase (Oves et al., 2016).

Pectinases are used to treat activated sludge wastewater (Gundala et al., 2017) by alkalophilic bacteria like *Bacillus* sp. (Tanabe et al., 1987) can be augmented with citrus fruit for increasing activity of pectinolytic microbes. Cellulases consist of endoglucanase (responsible for free cellulose fiber chain), exoglucanase (remove cellobiose units), β-glucosidase (hydrolyze cellobiose to glucose subunit) (Karigar et al., 2011). *Acidothermus cellulolyticus, Bacillus subtilis, B. licheniformis, Cellulomonas biazotea, Pseudomonas Cellulosa, Acetivibrio cellulolyticus, Clostridium thermocellum, C. cellulolyticum, Cellulomonas fimi, Thermomonospora fusca, Trichoderma ressei, Aspergillus* sp. (Mojsov, 2012) are some cellulase producing microorganisms (mostly bacteria and fungi). Most of the alkaliphilic extremophiles are able to decolorize and reduce the dyes for lesser textile industry effluents (Table 5) (Gunjalet al., 2021).

Table 5. Showing the target substrate of different microorganism

Types and Conditions	Extremozyme	Microorganism	Target substrate
Thermophiles and Superthermophiles (High temperature)	Phosphotriesterase-like lactonase Laccase phenolase Phosphotriesterase-like lactonase	*Sulfolobus sulfataricus, Thermoascu saurantiacus* *Thermusthermophilus HB27, Geobacillus stearothermophilus*	Organo-phosphates aromatic organic hydrocarbons Phenolic compounds of organic origin Organo-phosphates
Acidophilous (Low pH)	Phosphotriesterase-like lactonase	*Sulfolobus sulfataricus*	Organo-phosphates

Laccases are is classified as an oxidoreductase type enzyme that oxidizes substrate and forms water by reducing oxygen. Laccases are high stability at mesophilic temperature range, pH 5.5-9.0, highly active when Cl is available (Fanget al., 2012). The enzyme is used by microorganisms to degrade hydrocarbons by oxidizing the poly forms of various phenols, amines, aryl diamines, and other polyaromatic hydrocarbons. It can also depolymerize and oxidize (-OCH_3 group) lignin which are a form of polyquinones, polyphenols present in plants. PAHs are transformed into simpler CO_2 and H_2O by these enzymes. Laccases have application in degrading and decolorizing azo dyes in textile—*Pseudomonas resinovorans*, alkali, and halotolerant bacteria are able to degrade azo dyes like Bezactiv Blue, Tubantin Brown. Xenobiotics like pentachlorophenol, 2,4,6-trichlorophenol, 2,4- dichlorophenol are degraded by *Trametes pubescens* via laccases (Gunjalet al., 2021, Oves et al., 2016).

Pesticides are a mixture of substances (natural, organic, or synthetic) that are involved in the prevention, suppression, destruction of insects and pests or any other pest/pathogens (harmful) from causing disease in plants. The commonly used pesticides are organophosphorus (chlorpyrifos, parathion, glyphosate, diazinon, methyl parathion, monocrotophos, etc.), high toxicity, non-specific target toxicity. These pesticides inhibit the regulation of the nervous system due to enzyme inhibition essential for the functioning of the nervous system. Organochlorine pesticides, which are persistent, highly toxic, bioaccumulation in mammal tissues, constitute dichloro-diphenyl-trichloroethane (DDT), lindane. Enzymes such as organophosphate hydrolase, phosphotriesterase catalyze encoded by opd gene in *Agrobacterium radiobacter, Flavobacterium, Brevundimonas diminuta, Alteromonas* sp. to detoxify pesticides (Horne et al., 2002). Bacterial enzymes due to substrate specificity can be considered as a detoxifying agent for these pesticides. Phosphotriesterase produced by archaea *Sulfolobus solfataricus* due to there is able to degrade ester bond in pesticides (Merone et al., 2005). Pesticides like carbamates, pyrethroids can be converted into -COOH and -OH via hydrolysis in water.

ROLE OF RADIATION RESISTANCE MICROORGANISM

Extremolytes that are resistant to radiation can be utilized for nuclear waste management and bioremediation purpose. In *Shewanella sp* and *Geobacter sp,* cytochrome enzymes (Lloyd et al., 2003) were found effective in reducing soluble uranium (U) radio-isotopes into non-radioactive species. Further, it was reported Nickel-Iron hydrogenase enzyme reduced Tc (VIII) in *Desulfovibrio sp* (Luca et al., 2001). Fujimoto & Morita (2006) was able to obtain a pure culture of Halomonas sp. (a new strain) that cleared up Technetium from solution by insolubilization of that element. Kim et al., (2012) reported that

Geobacter sp. and *Rhodoferax sp* can be utilized to transform radioisotopes via the enzymatic method into non-radioactive isotope in a similar way as discussed by the above procedure.

Radioactive waste can also be reduced by indirect enzyme action and could be applied to remediate soil pollution by radionuclides. For this purpose, metal-reducing and sulfur-reducing extremophiles can be utilized for reducing soluble radioactive waste. This method involves the oxidation of hydrocarbons, while in another process, hydrogen is used to reduce Iron or Sulfur in another form, i.e., sulfate. These different forms were further reduced by different insoluble compounds (van Hullebusch et al., 2005). Examples of such microorganisms include sulfur-reducing bacteria known as *Mycobacterium flavescens* that are able to synthesize natural substances like organic acids, siderophores & secondary metabolites when grown under the presence of certain heavy metals.

Radiation-tolerant microorganisms play a vital role in human therapeutics, medicines, biotechnology, and bioremediation of harmful and radioactive substances. Because of the capability of radio tolerant extremophiles to thrive in a high radiation environment, the different types of secondary metabolites and extremozymes synthesized by them, which can be useful in human therapeutics, as well as biodegradation of radionucleotide the, are of universal importance. Despite of all the information regarding radiation-tolerant bacteria and their potential application in various fields still, there is a need to discover much more so to further exploit their efficiency in the area of medicine and environmental biotechnology.

CONCLUSION

Extremophiles are the microorganism that is able to thrive in harsh environmental conditions. Due to their ability to survive in the extreme environment, they can be broadly classified into three major groups, which include halophiles (salt-tolerant), thermophiles (high temperature tolerant) acidophiles (low pH tolerant). They play a crucial role in understanding the ecological relationship and can be useful in metagenomic studies. Their properties like surviving in water scarcity, growing under metal stress (i.e., high concentration of heavy metal), establishing itself under the higher concentration of pesticides, poly hydroxyl alkanoate, radioactive substances, and highly alkaline conditions make them suitable for the purpose of bioremediation process. Bioremediation which involves the utilization of microorganisms is termed microbial remediation. As bacteria can be cultured easily and due to their rapid multiplication rate, they can be utilized for different purposes like microbial remediation and biotechnological purpose, including gene cloning and genetic engineering. Extremophiles provide an alternative and cheap technology for clearing up the environment, further making it pollution-free. As they are a crucial part of the food chain in which their major role is decomposition and degradation of compounds to provide energy thir role in cleaning up the environment cannot be overlooked.

REFERENCES

Arulazhagan, P., Mnif, S., Rajesh Banu, J. R., Huda, Q., & Jalal, M. A. B. (2017). HC-0B-01: Biodegradation of hydrocarbons by extremophiles. In *Environmental Footprints and Eco-Design of Products and Processes* (pp. 137–162). Springer.

Auernik, K. S., Maezato, Y., Blum, P. H., & Kelly, R. M. (2008). The genome sequence of the metal-mobilizing, extremely thermophilic archaeon Metallosphaera sedula provides insight into bioleaching-associated metabolism. *Applied and Environmental Microbiology, 74*(3), 682–692. doi:10.1128/AEM.02019-07 PMID:18083856

Azubuike, C. C., Chikere, C. B., & Okpokwasili, G. C. (2016). Bioremediation techniques classification based on site of application: Principles, advantages, limitations and prospects. *World Journal of Microbiology & Biotechnology, 32*(11), 180. doi:10.100711274-016-2137-x PMID:27638318

Bae, H. S., Moe, W. M., Yan, J., Tiago, I., da Costa, M. S., & Rainey, F. A. (2006). Propionicicella superfundia gen. nov., sp. nov, a color solvent-tolerant propionate forming, facultative anaerobic bacterium isolated from contaminated groundwater. *Systematic and Applied Microbiology, 29*(5), 404–413. doi:10.1016/j.syapm.2005.11.004 PMID:16338112

Barkay, T., Miller, S. M., & Summers, A. O. (2003). Bacterial mercury resistance from atoms to ecosystems. *FEMS Microbiology Reviews, 27*(2–3), 355–384. doi:10.1016/S0168-6445(03)00046-9 PMID:12829275

Das, R., & Kazy, S. K. (2014). Microbial diversity, community composition and metabolic potential in hydrocarbon contaminated oily sludge: Prospects for in situ bioremediation. *Environmental Science and Pollution Research International, 21*(12), 7369–7389. doi:10.100711356-014-2640-2 PMID:24682711

Dawson, K. S., Strąpoć, D., Huizinga, B., Lidstrom, U., Ashby, M., & Macalady, J. L. (2012). Quantitative fluorescence in situ hybridization analysis of microbial consortia from a biogenic gas field in Alaska's Cook Inlet basin. *Applied and Environmental Microbiology, 78*(10), 3599–3605. doi:10.1128/AEM.07122-11 PMID:22427501

Dean-Ross, D., Moody, J., & Cerniglia, C. E. (2002). Utilization of mixtures of polycyclic aromatic hydrocarbons by bacteria isolated from contaminated sediment. *FEMS Microbiology Ecology, 41*(1), 1–7. doi:10.1111/j.1574-6941.2002.tb00960.x PMID:19709233

Dipippo, J. L., Nesbø, C. L., Dahle, H., Doolittle, W. F., Birkland, N. K., & Noll, K. M. (2009). Kosmotogaolearia gen. nov., sp. nov., a thermophilic, anaerobic heterotroph isolated from an oil production fluid. *International Journal of Systematic and Evolutionary Microbiology, 59*(12), 2991–3000. doi:10.1099/ijs.0.008045-0 PMID:19643902

Dolfing, J. (2014). Thermodynamic constraints on syntrophic acetate oxidation. *Applied and Environmental Microbiology, 80*(4), 1539–1541. doi:10.1128/AEM.03312-13 PMID:24480834

Dua, M., Singh, A., Sethunathan, N., & Johri, A. K. (2002). Biotechnology and bioremediation: Successes and limitations. *Applied Microbiology and Biotechnology, 59*(2–3), 143–152. PMID:12111139

Dumorné, K., Córdova, D. C., Astorga-Eló, M., & Renganathan, P. (2017). Extremozymes: A potential source for industrial applications. *Journal of Microbiology and Biotechnology, 27*(4), 649–659. doi:10.4014/jmb.1611.11006 PMID:28104900

Fang, Z. M., Li, T. L., Chang, F., Zhou, P., Fang, W., Hong, Y. Z., Zhang, X. C., Peng, H., & Xiao, Y. Z. (2012). A new marine bacterial laccase with chloride-enhancing, alkaline-dependent activity and dye decolorization ability. *Bioresource Technology, 111*, 36–41. doi:10.1016/j.biortech.2012.01.172 PMID:22377476

Fujimoto, K., & Morita, T. (2006). Aerobic removal of technetium by a marine Halomonas strain. *Applied and Environmental Microbiology, 72*(12), 7922–7924. doi:10.1128/AEM.00819-06 PMID:17056690

Gao, Y. C., Wang, J. N., Guo, S. H., Hu, Y. L., Li, T. T., Mao, R., & Zeng, D. H. (2015). Effects of salinization and crude oil contamination on soil bacterial community structure in the Yellow River Delta region, China. *Applied Soil Ecology, 86*, 165–173. doi:10.1016/j.apsoil.2014.10.011

Goyer, R. A. (1997). Toxic and essential metal interactions. *Annual Review of Nutrition, 17*(1), 37–50. doi:10.1146/annurev.nutr.17.1.37 PMID:9240918

Gundala, P., & Chinthala, P. (2017). Extremophilic pectinases. In R. K. Sani & R. N. Krishnaraj (Eds.), *Extremophilic enzymatic processing of lignocellulosic feedstocks to bioenergy* (p. 155). Springer. doi:10.1007/978-3-319-54684-1_9

GunjalAparna, B., Waghmode Meghmala, S., & Patil Neha, N. (2021). Role of extremozymes in bioremediation. *Research Journal of Biotechnology, 16*, 3.

Horne, I., Sutherland, T. D., Harcourt, R. L., Russell, R. J., & Oakeshott, J. G. (2002). Identification of an opd (organophosphate degradation) gene in an Agrobacterium isolate. *Applied and Environmental Microbiology, 68*(7), 3371–3376. doi:10.1128/AEM.68.7.3371-3376.2002 PMID:12089017

Karigar, C. S., & Rao, S. S. (2011). Role of microbial enzymes in the bioremediation of pollutants: A review. *Enzyme Research, 2011*, 805187. doi:10.4061/2011/805187 PMID:21912739

Kawakami, Y., Hayashi, N., Ema, M., & Nakayama, M. (2007). Effects of divalent cations on Halobacterium salinarum cell aggregation. *Journal of Bioscience and Bioengineering, 104*(1), 42–46. doi:10.1263/jbb.104.42 PMID:17697982

Keasling, J. D., Van Dien, S. J., Trelstad, P., Renninger, N., & McMahon, K. (2000). Application of polyphosphate metabolism to environmental and biotechnological problems. Biochemistry. *Biokhimiia (Moscow, Russia), 65*(3), 324–331. PMID:10739475

Kim, H. S., Kim, Y. J., & Seo, Y. R. (2015). An overview of carcinogenic heavy metal: Molecular toxicity mechanism and prevention. *Journal of Cancer Prevention, 20*(4), 232–240. doi:10.15430/JCP.2015.20.4.232 PMID:26734585

Kim, S. J., Koh, D. C., Park, S. J., Cha, I. T., Park, J. W., Na, J. H., Roh, Y., Ko, K. S., Kim, K., & Rhee, S. K. (2012). Molecular analysis of spatial variation of iron-reducing bacteria in riverine alluvial aquifers of the Mankyeong River. *Journal of Microbiology (Seoul, Korea), 50*(2), 207–217. doi:10.100712275-012-1342-z PMID:22538648

Korashy, H. M., Attafi, I. M., Famulski, K. S., Bakheet, S. A., Hafez, M. M., Alsaad, A. M. S., & Al-Ghadeer, A. R. M. (2017). Gene expression profiling to identify the toxicities and potentially relevant human disease outcomes associated with environmental heavy metals exposure. *Environmental Pollution, 221*, 64–74. doi:10.1016/j.envpol.2016.10.058 PMID:27916491

Kryachko, Y., Dong, X., Sensen, C. W., & Voordouw, G. (2012). Compositions of microbial communities associated with oil and water in a mesothermic oil field. *Antonie van Leeuwenhoek, 101*(3), 493–506. doi:10.100710482-011-9658-y PMID:22038128

Lloyd, J. R., Leang, C., Hodges Myerson, A. L., Coppi, M. V., Cuifo, S., Methe, B., Sandler, S. J., & Lovley, D. R. (2003). Biochemical and genetic characterization of PpcA, a periplasmic c-type cytochrome in Geobacter sulfurreducens. *The Biochemical Journal, 369*(1), 153–161. doi:10.1042/bj20020597 PMID:12356333

Luca, D., Denier van der Gon, A. W., Anita, V., Ponjee, M. W. G., Brongersma, H. H., & Popa, G. (2001). Surface nitridation processes and non-linear behaviour of the reactive magnetron discharge with titanium target. *Vacuum, 61*(2–4), 163–167. doi:10.1016/S0042-207X(00)00474-7

Matte-Tailliez, O. M., Brochier, C., Forterre, P., & Philippe, H. (2002). Archaeal phylogeny based on ribosomal proteins. *Molecular Biology and Evolution, 19*(5), 631–639. doi:10.1093/oxfordjournals.molbev.a004122 PMID:11961097

Meijer, W. G., Nienhuis-Kuiper, M. E., & Hansen, T. A. (1999). Fermentative bacteria from estuarine mud: Phylogenetic position of Acidaminobacter hydrogenoformans and description of a new type of gram-negative, propionigenic bacterium as Propionibacter pelophilus gen. nov., sp. nov. *International Journal of Systematic Bacteriology, 49*(3), 1039–1044. PMID:10425761

Merino, N., Aronson, H. S., Bojanova, D. P., Feyhl-Buska, J., Wong, M. L., Zhang, S., & Giovannelli, D. (2019). Living at the extremes: Extremophiles and the limits of life in a planetary context. *Frontiers in Microbiology, 10*, 780. doi:10.3389/fmicb.2019.00780 PMID:31037068

Merone, L., Mandrich, L., Rossi, M., & Manco, G. (2005). A thermostable phosphotriesterase from the archaeon Sulfolobus solfataricus: Cloning, overexpression and properties. *Extremophiles: Life under Extreme Conditions, 9*(4), 297–305.

Mi, S., Song, J., Lin, J., Che, Y., Zheng, H., & Lin, J. (2011). Complete genome of Leptospirillum ferriphilum ML-04 provides insight into its physiology and environmental adaptation. *Journal of Microbiology (Seoul, Korea), 49*(6), 890–901. doi:10.100712275-011-1099-9 PMID:22203551

Mojsov, K. (2012). Microbial cellulases and their applications in textile processing. *International Journal of Marketing and Technology, 2*(11), 12–29.

Naik, S., & Furtado, I. (2014). Equilibrium and kinetics of adsorption of Mn+2 by haloarchaeon Halobacterium sp. GUSF (MTCC3265). *Geomicrobiology Journal, 31*(8), 708–715. doi:10.1080/01490451.2013.859769

Nandal, M., Solanki, P., Rastogi, M., & Hooda, R. (2015). Bioremediation: A sustainable tool for environmental management of oily sludge. *Nature Environment and Pollution Technology, 14*(1), 181.

Oves, M., Saghir Khan, M., & Huda Qari, A., NadeenFelemban, M., & Almeelbi, T. (2016). Heavy metals: Biological importance and detoxification strategies. *Journal of Bioremediation & Biodegradation, 7*(2), 1–15.

Park, C., & Park, W. (2018). Survival and energy producing strategies of alkane degraders under extreme conditions and their biotechnological potential. *Frontiers in Microbiology, 9*, 1081. doi:10.3389/fmicb.2018.01081 PMID:29910779

Peeples, T. L. (2014). Bioremediation using extremophiles. In *Microbial biodegradation and bioremediation* (pp. 251–268). Elsevier. doi:10.1016/B978-0-12-800021-2.00010-8

Pinto, E., Sigaud-Kutner, T. C. S., Leitao, M. A. S., Okamoto, O. K., Morse, D., & Colepicolo, P. (2003). Heavy metal–induced oxidative stress in algae 1. *Journal of Phycology, 39*(6), 1008–1018. doi:10.1111/j.0022-3646.2003.02-193.x

Popescu, G., & Dumitru, L. (2009). Biosorption of some heavy metals from media with high salt concentrations by halohilicarchaea. *Biotechnology and Biotechnological Equipment, 23*(sup1), 791–795.

Prince, R. C. (1993). Petroleum spill bioremediation in marine environments. *Critical Reviews in Microbiology, 19*(4), 217–242. doi:10.3109/10408419309113530 PMID:8305136

Prince, R. C., & Butler, J. D. (2014). A protocol for assessing the effectiveness of oil spill dispersants in stimulating the biodegradation of oil. *Environmental Science and Pollution Research International, 21*(16), 9506–9510. doi:10.100711356-013-2053-7 PMID:23943003

Roy, A., Dutta, A., Pal, S., Gupta, A., Sarkar, J., Chatterjee, A., Saha, A., Sarkar, P., Sar, P., & Kazy, S. K. (2018). Biostimulation and bioaugmentation of native microbial community accelerated bioremediation of oil refinery sludge. *Bioresource Technology, 253*, 22–32. doi:10.1016/j.biortech.2018.01.004 PMID:29328931

Sarkar, J., Kazy, S. K., Gupta, A., Dutta, A., Mohapatra, B., Roy, A., Bera, P., Mitra, A., & Sar, P. (2016). Biostimulation of indigenous microbial community for bioremediation of petroleum refinery sludge. *Frontiers in Microbiology, 7*, 1407. doi:10.3389/fmicb.2016.01407 PMID:27708623

Sarkar, J., Roy, A., Sar, P., & Kazy, S. K. (2020). Accelerated bioremediation of petroleum refinery sludge through biostimulation and bioaugmentation of native microbiome. In *Emerging technologies in environmental bioremediation* (pp. 23–65). Elsevier. doi:10.1016/B978-0-12-819860-5.00002-X

Schelert, J., Rudrappa, D., Johnson, T., & Blum, P. (2013). Role of MerH in mercury resistance in the archaeon Sulfolobus solfataricus. *Microbiology, 159*(6), 1198–1208. doi:10.1099/mic.0.065854-0 PMID:23619003

Schiraldi, C., & De Rosa, M. (2002). The production of biocatalysts and biomolecules from extremophiles. *Trends in Biotechnology, 20*(12), 515–521. doi:10.1016/S0167-7799(02)02073-5 PMID:12443873

Seckbach, J., Oren, A., & Stan-Lotter, H. (Eds.). (2013). *Polyextremophiles. Cellular Origin, Life in Extreme Habitats and Astrobiology*. Springer.

Sharma, A., Singh, S. B., Sharma, R., Chaudhary, P., Pandey, A. K., Ansari, R., Vasudevan, V., Arora, A., Singh, S., Saha, S., & Nain, L. (2016). Enhanced biodegradation of PAHs by microbial consortium with different amendment and their fate in-situ condition. *Journal of Environmental Management, 181*, 728–736. doi:10.1016/j.jenvman.2016.08.024 PMID:27558829

Showalter, A. R., Szymanowski, J. E. S., Fein, J. B., & Bunker, B. A. (2016). An x-ray absorption spectroscopy study of Cd binding onto a halophilic archaeon. *Journal of Physics: Conference Series, 712*(1), 012079. doi:10.1088/1742-6596/712/1/012079

Souza, E. C., Vessoni-Penna, T. C., & de Souza Oliveira, R. P. (2014). Biosurfactant-enhanced hydrocarbon bioremediation: An overview. *International Biodeterioration & Biodegradation, 89*, 88–94. doi:10.1016/j.ibiod.2014.01.007

Sutton, N. B., Maphosa, F., Morillo, J. A., Abu Al-Soud, W., Langenhoff, A. A., Grotenhuis, T., Rijnaarts, H. H., & Smidt, H. (2013). Impact of long-term diesel contamination on soil microbial community structure. *Applied and Environmental Microbiology*, *79*(2), 619–630. doi:10.1128/AEM.02747-12 PMID:23144139

Tanabe, H., Yoshihara, K., Tamura, K., Kobayashi, Y., Akamatsu, I., Niyomwan, N., & Footrakul, P. (1987). Pretreatment of pectic wastewater from orange canning process by an alkalophilic Bacillus sp. *Journal of Fermentation Technology*, *65*(2), 243–246. doi:10.1016/0385-6380(87)90173-7

Taylor, M. P., van Zyl, L., Tuffin, I. M., Leak, D. J., & Cowan, D. A. (2011). Genetic tool development underpins recent advances in thermophilic whole-cell biocatalysts. *Microbial Biotechnology*, *4*(4), 438–448. doi:10.1111/j.1751-7915.2010.00246.x PMID:21310009

Tyagi, B., & Kumar, N. (2021). Bioremediation: Principles and applications in environmental management. In Bioremediation for environmental sustainability (pp. 3–28). Elsevier.

van Hullebusch, E. D., Peerbolte, A., Zandvoort, M. H., & Lens, P. N. (2005). Sorption of cobalt and nickel on anaerobic granular sludges: Isotherms and sequential extraction. *Chemosphere*, *58*(4), 493–505. doi:10.1016/j.chemosphere.2004.09.017 PMID:15620741

Woese, C. R., Kandler, O., & Wheelis, M. L. (1990). Towards a natural system of organisms: Proposal for the domains Archaea, Bacteria and Eucarya. *Proceedings of the National Academy of Sciences of the United States of America*, *87*(12), 4576–4579. doi:10.1073/pnas.87.12.4576 PMID:2112744

Zhang, D. C., Mörtelmaier, C., & Margesin, R. (2012). Characterization of the bacterial archaeal diversity in hydrocarbon-contaminated soil. *The Science of the Total Environment*, *421–422*, 184–196. doi:10.1016/j.scitotenv.2012.01.043 PMID:22386232

Zhao, L., Ma, Z., Luan, Y., Lu, A., Wang, J., & Pan, L. (2011). Molecular methods of studying microbial diversity in soil environments. In. *IFIP Advances in Information and Communication Technology International Conference on Computer and Computing Technologies in Agriculture*. Springer. 10.1007/978-3-642-18336-2_10

KEY TERMS AND DEFINITIONS

Biodegradation: A natural process of breakdown of materials by using microbes.

Bioremediation: A process which uses micro-organism to treat the pollutants.

Enzymes: Biological catalysts accelerate the reaction.

Extremophiles: An organism that grows optimally in extreme conditions.

Microorganisms: They are the microscopic organisms that may exist as single-celled or in cell clusters form.

Pollutant: Contaminants that introduced into the environment and brought adverse effects.

Transformation: A process of changing from one form to another.

Chapter 10
Industrial Applications of Enzymes From Extremophiles

Prashant Satishbhai Arya
Gujarat University, India

Shivani Maheshkumar Yagnik
Christ College, India

Rakeshkumar Ramanlal Panchal
https://orcid.org/0000-0001-8715-8553
Gujarat University, India

Kiransinh Narendrasinh Rajput
Gujarat University, India

Vikram Hiren Raval
https://orcid.org/0000-0001-8912-1850
Gujarat University, India

ABSTRACT

Extremophilic microorganisms have developed a variety of molecular tactics to exist in extreme environments. Researchers are fascinated by extremophiles and unearth various enzymes from these fascinating microbes. Extremozymes are astonishing biocatalysts with distinctive properties of catalysis and stability under a multitude of daunting conditions of salt, pH, organic solvents, and temperature, which open up new possibilities for biocatalysis and biotransformation and outcompetes mesophilic counterparts. Biotechnological implications include simple, immobilized, as well as whole-cell applications. Stability in organic solvents adds to the asymmetric catalysis and thereby exemplifies the applicability of extremozymes and in fostering biobased economies. Marine, cold-adapted enzymes, and those that help in the removal of a toxic hazardous substance from the environment are obvious choices for food industries and bioremediation. The major area of application and research emphasis includes textile, detergents, food, dairy, agriculture, and environmental remediation.

DOI: 10.4018/978-1-7998-9144-4.ch010

INTRODUCTION: EXTREMOPHILES AND BIOCATALYSTS ASSOCIATED INDUSTRIES

The current and future applications of extreme environment tolerating microorganisms and their enzymes in the field of biotechnology are discussed in this chapter. The microbial biotechnology has enormous advantages be it in the field of environment, product manufacturing or industrial processes for value added commercial products. The microbial approach to these processes has given it a new and improved eco-friendly dimension. This is now applied for management of different type of wastes generated by various industries. Microbial bioengineering contributes to the development of long-term technologies with a variety of process and market advantages. Microbial biotechnology has a number of advantages, including the ability to produce existing and novel products in a sustainable manner, as well as a reduce dependence on non - renewable fuels and many other resources, which improves industrial economics. Since 1970s, biotechnology and enzymology in particular has had a considerable influence on many sectors including healthcare, medicine, food, agriculture, environment and synthesis of fine chemicals. Until enzymes that could tolerate organic and biphasic conditions were discovered, "white biotechnology" was primarily based on aqueous enzymology (Vashist and Sharma, 2018).

The usage of enzymes extracted from microbes is beneficial to hundreds of industrial processes and products. However, the bulk of enzymes now on the market are made with mesophilic enzymes, which are frequently inhibited under extreme condition in industrial processes (Raddadi et al., 2015). Additionally, bio catalytic stability is important for cost reduction since enzymes that are stable enough to endure industrial conditions can be used for several cycles of the bio catalytic process, resulting in cost savings. The discovery of extremophilic bacteria opens the door to the synthesis of extremozymes that are stable in a variety of environment, which might be beneficial in industry applications. Enzymes which catalyze reactions in non-physiological environments and/or using non-natural molecules can also be found in extreme environments (Littlechild, 2015). As a result of their resilience to extreme physico-chemical conditions, extremozymes have grown at a rapid pace for their use in various industrial processes.

Extremophilic bacteria, on the other hand, have extremely flexible metabolisms and unique structural characteristics in their bio macromolecules that allow them to survive and grow in these hostile environments (Dalmaso et al., 2015). These extremophilic organisms have a wide range of functions and are taxonomically diverse too. The main categories of extremophiles and some examples are listed below in Table 1. These include the microbes that tolerate temperature extremities, specifically the psychrophiles, hyperthermophiles and thermophiles; alkaliphiles and acidophiles, which tolerate pH extremes; barophiles (piezophiles) which survive optimally under high pressure; halophiles, which thrive in high-salinity environments; xerophiles, which tolerate low water activity, and radioresistant organisms, which tolerate high levels of radioactivity that pose lethal effects on almost all organisms.

According to Dewan (2014), the market for industrial enzymes was expected to grow at an annual average growth rate of 8% in the next 5 years, reaching US$ 7,100 million by 2018. Microbes that produce novel hydrolytic enzymes, with biotechnological potential and remarkable activity at low temperatures are now being sought. Due to their biodegradability and exceptional stability, extremophilic microbes are a source of extremozymes with a wide range of commercial applications (Dumorne, et al., 2017). As biocatalysts, extremozymes are stable and active even within extreme environmental conditions traditionally thought to be incompatible with biology. Cold-tolerant, alkali-tolerant, acid-tolerant and salt-tolerant extremozymes are just a few of the resistant biomolecules that have been made possible by the use of extremozymes. Thus the quest of novel enzymes with unique catalytic features and improved stability remain a topmost priority in enzyme research (Raddadi et al., 2015).

Table 1. Major classification and subtypes of extremophiles based on their environmental habitat

Environmental conditions	Survival and defensive strategies	Types of extremophile	Growth condition	Environment/source/ geographical location	Reference/s
Temperature	Stabilization of enzymes from stress and freeze drying; protection of oxidative protein damage; reduction of VLS in immunotoxin therapy,	Hyperthermophile	Temperatures of 80°C or above are ideal for growth.	Volcano, East Pacific, Porto di Levante, Submarine Hydrothermal vents,	Dalmaso et al., 2015, Hamdan, 2018, Zhu et al., 2020
		Thermophile	Temperatures of 60-80°C or above are ideal for growth.	Yellowstone National Park, US, Grand Prismatic Spring, Hot Spring,	
	reduction in the packing of acyl chains in the cell membranes; increased flexibility in protein structure; Translation of cold-evolved enzymes; cold shock and nucleic acid binding proteins	Psychrophile	Temperatures of 10°C or less are ideal for growth, with a maximum of 20°C.	Ice, snow, Antarctic ice and Arctic Ocean	
pH	Maintaining a circumneutral intracellular pH; constant pumping of protons in and out of cytoplasm; acidic polymers of the cell membrane; passive regulation of the cytoplasmic pools of polyamines and low membrane permeability	Acidophile	pH optimal for growth is 4-5 or less.	Acid mine drainage, volcanic springs	Sharma et al., 2012
		Alkaliphile	Growth at pH values above 8	Soda Lakes	Horikoshi, 1999
Salinity	Protection of skin immune cells from UV radiation; enzyme stabilization against heating, freezing, and drying; protection of the skin barrier against water loss and drying out; block of UVA induced ceramide release in human keratinocytes	Halophile	Salt-loving organisms require at least 1 M of salt for growth	High salt concentration such as salter pond brines and natural salt lakes, Soda Lakes, ocean,	Coker, 2016
Pressure	Maintaining a circumneutral intracellular pH; constant pumping of protons in and out of cytoplasm; acidic polymers of the cell membrane; passive regulation of the cytoplasmic pools of polyamines and low membrane permeability	Barophile	Microbe that thrives at hydrostatic pressures of at least 40 MPa	Antarctic ice, Mariana Trench, Deep ocean,	Horikoshi, 1998
Radiation	DNA repair mechanisms, produce the UV-absorbing compounds under UV stress	Radiophile	Organisms that can tolerate large doses of ionizing radiation.	UV radiation, Sunlight,	Singh and Gabani, 2010
Desiccation	Accumulation and expression of osmo-protectants, alternative carbon sources, alternative energy consuming and production path, DNA repair mechanism, production of extracellular polymeric substances (EPSs)	Xerophile	Microbe with of strong desiccation resistance and low water activity growth	Rock, surfaces, Desert, Atacama, Desert	Lebre et al., 2017

GEO-ECOLOGICAL CLASSIFICATION & ABUNDANCE

Microbial communities in extreme habitats have undergone the physiological adaptations to the daunting stress of low/high temperature, salinity, and chemicals. Recently, possible applications of microbial communities from such habitats have been focused and diverse sectors viz. agriculture, food, textile, cosmetics and medicine are profited. Various ecosystems such as volcanoes, hot springs, Sulphur springs, thermal vents, salt lakes, soda lakes, Dead Sea, solar salterns, acid mine drainage, deep sea sediments, and many other such environments harbor potential extremophilic microbes with unexplored bio catalytic potential. Extremophiles are living organisms that can survive and reproduce in diverse geo ecological niches with multitude of extremities of pressure, pH, temperature, redox potential, radiation and salinity (Figure 1) (Purohit et al., 2014; Kour et al., 2019).

The Saltern Ecosystem

Thalassohaline hypersaline habitats are defined as those that result from the evaporation of sea water and have halite (NaCl) concentrations of more than 10% (m/w). Similar ecological niches are found throughout the globe and exist in form of marine ponds, salt marshes, soda lakes and manmade salterns (Raval et al., 2013; Singh et al., 2013). They are made up of a number of shallow ponds found in tropical, subtropical, and temperate climates around the world. As the seawater evaporates in these conditions, the NaCl concentration increases. With their consecutive evaporation ponds, salt-pan mud, and wooden fences, the solar salterns Secovlje, present a reasonably simple habitat that is ideal for halotolerant and halophilic microorganism research (Zajc et al., 2012). At first, it was thought that archaea, bacteria, and eukaryotes dominated the hypersaline waters of salterns. Diverse halotolerant and halophilic microbes were first isolated from the sun salterns of Secovlje, and then from salterns all over the world.A majority of halotolerant and strictly halophilic fungi have been isolated from hypersaline environments including seawater. Previously, xerophilic fungi that can grow at low water activity were solely identified from domestic environments, as they can spoil food that has been preserved by reducing the biologically accessible water through drying, freezing, or solute addition. Food-borne organisms that can grow at low aw are thought to have a general xerophilic character and so would not be found in natural hypersaline habitats.

Figure 1. Geo-ecological niche and their of extremophilic diversity

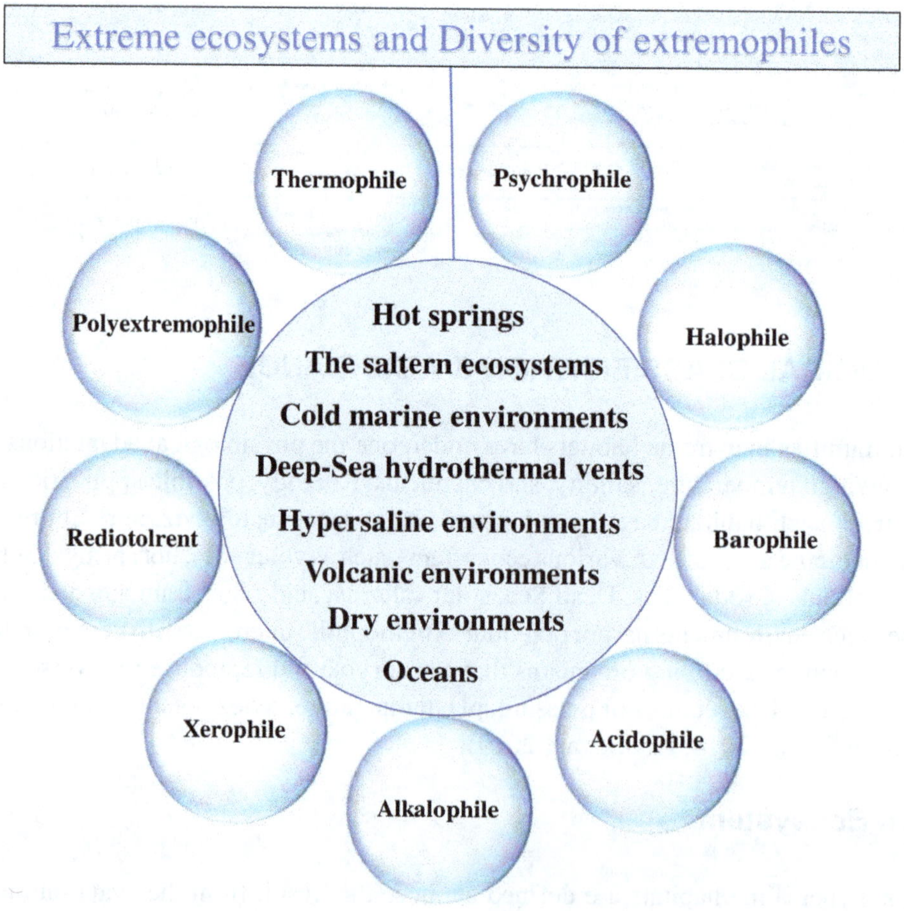

Hot Springs

The sites where warm or hot groundwater emerges from the Earth are known as hot springs. Thomas Brock's groundbreaking work in identifying *Thermus aquaticus* from a thermal environment (Brock and Freeze, 1969) has completely changed our perspective of the microbial richness of hot springs (Arya et al., 2020a, Narsing Rao et al., 2021). At the phylum level, Firmicutes were dominant followed by *Actinobacteria*, *Proteobacteria*, and *Bacteroidetes*. At the genus level, *Bacillus* is dominant followed by *Brevibacillus*, *Paenibacillus*, *Oceanibaculum*, *Micromonospora*, *Microbacterium*, *Terrimonas*, *Aneurinibacillus*, *Actinocorallia*, *Sphingomonas*, *Micrococcus*, and *Rhodoligotrophos* (Narsing Rao et al., 2021). Moreover, Solfataric fields are the most important biotopes thermoacidophilic microbes. Most of thermoacidophilic microbes that usually lived in solfataric fields belong to the archaea including the genera *Acidianus*, *Desulfurolobus*, *Metallosphaera*, *Stygiolobus*, *Sulfolobus*, *Sulfurisphaera*, *Sulfurococcus*, *Thermoplasma*, and *Picrophilus*. Microbes belong to bacteria were also found including *Acidithiobacillus*, *Acidimicrobium*, *Sulfobacillus* and *Hydrogenobaculum* genera (Bertoldo et al., 2004).

Deep-Sea Hydrothermal Vents (DSHV)

The ocean covers over three-quarters of the Earth's surface area, with an average depth of 3800 meters, meaning that deep-sea ecosystems make up the great majority of our planet. Deep sea has been one of the world's most enigmatic and unexplored ecosystems, and it contains a unique microbial community which is significant to biogeochemical cycles (Sogin et al., 2006). The absence of light, as well as the existence of atypical low temperature and high hydrostatic pressure, signify the habitat. In some environments, such as DSHV (temperatures over 400°C) and deep hypersaline anoxic basins (DHABs) (high salinity, abysses up to 11 km deep and high pressures), these climatic conditions become even more difficult for normal microbes. Deep-sea microbes are the living entities able to survive and reproduce under extreme geochemical and physiological environments detrimental to the conventional species. Prokaryotes, which comprise microbes from archaea and bacteria domains, make up the great bulk of deep-sea extremophiles (Harrison et al., 2013). The limited application of extremozymes is due to special nutritional requirements and extreme growth requirements of parent microbe that restricts their isolation and maintenance. However, the ever-rising number of extremophilic genes and metagenomes that are now accessible via next-generation sequencing provide us an ever-increasing resource novel extremozymes for the non-cultivable microbes (Jin et al., 2019). Many of the extremophiles from deep sea environments survive under more than one extreme condition and hence often termed as polyextremophiles. Prokaryotes such as *Pyrococcus horikoshii* and *Thermococcus profundis*, endure such conditions in deep sea environments, also tend to dominate the hydrothermal vents (Fang et al., 2010).

Cold Marine Environments

A meagre thought of "cold marine habitats," takes us to the North Arctic and South Antarctic, ice covered glaciers with scarce or practically no life forms. The temperature of marine water also lowers with depth as latitude and seasons change. Temperatures as low as 4–5°C can be found near the thermocline (a body of water where temperature fluctuates rapidly with depth) (Sullivan and Baross, 2007). Psychrophiles and psychrotrophs are distinguished by their growth temperature, which determines whether they are cold-loving or cold-tolerant microorganisms. Lower temperatures are preferred by organisms known as psychrophiles, which grow between 0-20°C (optimum at 15°C), while psychrotrophs, which can endure cold temperatures and have an ideal growth temperature of 20°C. *Moritella, Photobacterium, Colwellia, Shewanella*, and *Psychromonas* are the most often cultivated psychrophilic organisms (Fang et al., 2010). Psychrophilic microorganisms may be discovered from both the archaea and bacteria domains, belonging to diverse genera, such as *Colwellia, Arthrobacter, Listeria, Marinobacter, Methanogenium, Halomonas, Glaciecola, Hyphomonas, Moritella, Planococcus, Psychrobacter, Psychroserpens, Shewanella* and *Sphingomonas* (Poli et al., 2017).

Dry Environments

Climate change has been linked to an increase in dry conditions in many locations owing to poor rainfall, extreme heat, and drought. All living species require water to survive but Arid habitats, such as deserts, are thought to be life's dry limit (Bull and Asenjo, 2013; Bhatt et al., 2017). Extremophile xerotolerant creatures can thrive in habitats with a moisture content of less than 0.75 (Lebre et al., 2017). Other elements, such as low water activity, high and low temperatures, low organic carbon contents, high salinity, and strong radiation, exacerbate the xeric conditions, limiting the life of organisms. Nevertheless, xerotolerant microorganisms are capable to inhabit this extreme environment. Using culture-independent methods, the frequency of microbial communities such as *Cyanobacteria, Chloroflexi, Proteobacteria, Gemmatimonadetes, Firmicutes, Planctomycetes, Actinobacteria*, and *Thermomicrobia* phyla in the surface sediment of the hyper-arid core was found to be limited (Piubeli et al., 2015). In the Atacama Desert, Actinobacteria are most prevalent cultivable phylum including genus *Microlunatus, Prauserella, Microcella, Arthrobacter, Streptomyces, Nocardia, Propionibacterium*, and *Kocuria* (Bull and Asenjo, 2013).

Hypersaline Environments

Hypersaline ecosystems are habitats with salinities that are significantly greater than seawater. They are classified as either thalassohaline or athalassohaline based on whether they originated from seawater or not. Extremophilic organisms known as halophiles have adapted in living in highly hypersaline conditions (Raval et al., 2018). The most accepted definition of halophiles is bacteria that grow best at Na^+ concentration higher than 0.2 M (Oren, 2006). These organisms are divided into three groups based on their optimum salt concentration for growth: extreme halophiles that grow with 3.4–5.1 M (20-30%, w/v) NaCl; moderate halophiles with 0.85–3.4 M (3-25% w/v) NaCl; and slight halophiles with 0.2–0.85 M (1-5% w/v) NaCl concentration (DasSarma and DasSarma, 2015). Marine salterns are niche for a wide range of halophilic and halotolerant microbes that grow over the salt concentration gradient. The bottoms of many hypersaline ponds are covered by purple and green sulphur and non-sulfur bacteria (Caumette, 1993). A variety of sulfur oxidizing, sulfate reducing, methanogenic and homoacetogenic varieties of bacteria and archaea are found in anoxic mats and sediments. On the contrary, many aerobic representatives from archaeal generas viz. *Halobacterium, Natronobacterium, Haloferax, Haloarcula* and bacteria from *Salinivibrio, Pseudomonas, Alteromonas, Alcaligens Acinetobacter* and *Flavobacterium* have been reported from similar ecological niches. From approximately 4 M NaCl to saturation (>5.1 M NaCl), halophilic archaea dominate the brine pools and most other microbial activity ceases (Oren, 2006). With ease maintenance and control being better than methanogenic and hyperthermophilic archaea, representatives of the genera *Halobacterium, Haloferax*, and *Haloarcula*, have now become useful models for investigating the archaeal domain. Hydrolases, such as DNAases, lipases, amylases, gelatinases, and proteases, are one of the most significant groups of enzymes produced by halophilic microorganisms because they can function under environments that cause most proteins either denature or precipitate out. Halophilic hydrolases are usually thermostable and pH-adjustable. If catalytic reactions are required to function in physical and chemical variables, such as in the presence of surfactant, metal ions, organic solvents, at high temperature and salt content, hydrolases produced from halophilic bacteria can be used (Raval et al., 2015; Zheng et al., 2016).

Volcanic Environments

Volcanism may be defined as "the manifestation at the surface of a planet or satellite of internal thermal processes through the emission of solid, liquid, or gaseous products". Environments resulting from volcanic activities are diverse, ranging from acidic hot springs to deep ocean basaltic habitats. Since volcanic environments, especially active fumaroles, are considered analogous to some of the earliest environments on Earth, the study of their microbial diversity may provide insights into the origin of life and extraterrestrial life as suggested by Medrano-Santillana and his coworkers (2017). However, only a few studies revealing the prokaryotes (bacteria and archaea) diversity of soil fumaroles, steam deposits and steam aerosols have been conducted. Fumaroles have been identified as a diversity hotspot for extremophile microorganisms, with the potential to uncover new species and metabolic capabilities, according to Benson and co-workers (2011). The volcanic niches harbor diverse microbial flora such as *Massilia*, *Methylobacterium*, *Pseudomonas costantinii* and *C. niabensis*, *Acidobacterium capsulatum*, *Ktedonobacter racemifer*, *T. phaeum*, *Pseudonocardia yunnanensis*, *M. aerilata*, *Nocardia xishanensis*, *Thermoanaerobacter mathranii*, *Desulfobacterium anilini*, *Planktothricoides raciborskii*, *D. thiodismutans*, *D. anilini* and *T. sulfurigignens*), iron reducers (*T. ferriorganovorum*), syntrophic acetate-oxidizers (*T. phaeum*), chemo-organotrophs (*N. alkalitolerans*), phototrophs (*P. raciborskii*), acetogens (*Moorella thermoacetica*), and methanogens (*M. thermautotrophicus*) (Medrano-Santillana et al., 2017).

DIVERSITY OF EXTREMOPHILES

The diversity of microorganisms in harsh environments has been comprehensively researched in recent years, with a focus on culture-dependent and culture-independent approaches. Extremophiles from a wide range of natural extreme habitat have been characterized, and potential biological uses have been developed, arising from effective microbial strain and its hydrolytic enzymes. Extremophilic bacteria, such as acidophiles, alkalophiles, halophiles, psychrophiles, thermophiles, and xerophiles, have a diverse microbial variety. Microbes and their enzymes from severe habitats could be useful in agriculture, food, detergent, and other industries. The Table 2 shows narrative description on diversity of extremophilic microorganisms and their industrial applications.

Psychrophiles

A great proportion of the earth's biosphere (>85%) permanently experiences temperatures below 5°C. Psychrophiles are bacteria that are cold-adapted or cold-loving and have key growth temperatures minimum, optimum, and maximum) at or below 0°C, 15°C, and 20°C, respectively, while microorganisms that withstand cold temperatures with a higher growth optimum, and maximum (above 25°C) are called 'psychro-tolerant'. The largest coverage of these cold environments is successfully colonized by a wide diversity of extremophilic microorganisms, including bacteria, archaea, yeasts, filamentous fungi, and algae. Forster was first to report the discovery of bacteria capable of growing at 0°C, particularly in seawater and saltwater fish, in 1887. The ability to thrive at such low temperatures requires a vast array of adaptations to maintain the metabolic rates and sustain growth compatible with life in these severe environmental conditions. Owing to the functionally prominent attributes an extensive research on psychrophilic microbes is underway. Majority of this focuses on their survival strategies, novel adaptations and biochemical as well as genetic basis for such adaptability (Hamdan, 2018).

Table 2. Extremophiles, extremozymes and their industrial applications

Extreme condition	Microbial diversity	Enzymes	Industrial applications	Reference/s
Thermophile	*Geobacillus thermodenitrificans; Thermus aquaticus* (YT-1), *Anoxybacillus* sp., GXS-BL, *Caldicellulosiruptor bescii, Anoxybacillus thermarum* FRM-RBK02, *Cellulomonas fimi* ATCC484, *Geobacillus proteiniphilus, Thermoanaerobacterium, Pyrococcus, Aeribacillus pallidus, Consortium* TC-5, *Clostridium* sp. strain WST, *Geobacillus kaustophilus, Janibacter* sp. strain R02, *Bacillus mojavensis* SO-10	Lipase, protease, amylases, pullulanase, glucoamylases, cellulases, xylanases, DNA polymerase, Esterase, chitinase	Detergent additive, Hydrolysis of macromolecules, food and feed, brewing and baking, Production of biofuels, Flavor modification, Processing and textile industry, Molecular biology	Dalmaso et al., 2015; Zhu et al., 2020
Psychrophile	*Achromobacteria, Alcaligenes, Altermonas, Aquaspirillum, Arthrobacter, Bacillus, Bacteroides, Brevibacterium, Clostridium, Colwellia, Cytophaga, Flavobacterium, Gelidibacter, Methanococcoides, Methanogenium, Methanosarcina, Microbacterium, Micrococcus, Moritella, Octadecabacter, Phormidium, Photobacterium, Polaribacter, Polaromonas, Pseudomonas, Psychroserpens, Shewanella, Psychrobacter*	Proteases Amylase Cellulases Dehydrogenases Catalase, Esterase Lipase, Pectinase β-Galactosidase	Detergents industry Food industry Bakery industry Preparation of animal feed Textiles industry Biosensor	Hamdan, 2018
Acidophile	*Ferroplasma acidarmanus, Sulfolobus solfataricus, Teratosphaeria acidotherma, Alicyclobacillus acidocaldarius, Acidothermus cellulolyticus, Bacillus aerophilus, Bacillus atrophaeus, Bacillus nanhaiensis, Pseudomonas chlororaphis, Staphylococcus devriesei*	Amylase glucoamylase Proteases Cellulases Oxidases	Processing industry Food and Feed industry Desulfurization of coal Mining industry	Sharma et al., 2012; Kour et al., 2019
Alkaliphile	*Alkaliphilus transvaalensis, Bacillus krulwichiae, Nesterenkonia lacusekhoensis, Bacillus sp., Bacillus alcalophilus, Micrococcus, Pseudomonas, Streptomyces, Alkalibacterium psychrotolerans, Arthrobacter ramosus, Bacillus halodurans* TSEV1, *B. halodurans* PPKS-2, *Halorubrum lacusprofundi, Burkholderia cepacia* ST-200	Proteases, Cellulases Pullulanase Amylase, Xylanases β-Glucanase Endocellulase	Detergents industry, Food industry, Feed preparation, Fermentation of beer and wine, Breadmaking, Fruit juice processing	Horikoshi, 1999; Bhattacharya et al., 2017; Kour et al., 2019
Halophile	*Haloarcula, Halomonas, Halobacterium, Halorhabdus, Halobacillus, Halothermothrix, Natronococcus, Marinococcus, Micrococcus,*	Proteases, Lipase, Dehydrogenases β-Galactosidase, xylanase, amylase	Peptides synthesis, Biocatalysis in organic media, Asymmetric chemical synthesis. Detergents industry, Food industry	Coker, 2016
Barophile	*Moritella* spp; *Alphaproteobacterium, Photobacterium, Shewanella, Colwellia, Psychromonas* and *Moritella*	To be defined	Food processing and antibiotic production	Horikoshi, 1998; Oger and Cario, 2014
Radiophile	*Deinococcus depolymerans; D. guangriensis; D. radiodurans; D. wulumuqiensis; D. gradis; Stenotrophomonas* sp. (YLP1); *D. misasensis; Aeromonas eucrenophila* (UVR4); *Cellulosimicrobium cellulans* (UVP1); *Roultella planticola* (UVR1); *B. stratosphericus* (UVR3); *Bacillus pumilus* (UVP4); *Enterobacter* sp. (UVP3); *Micrococcus yunnanensis* (UV20HR)	DNA repair Enzymes Dehydrogenase Oxido-reductase	Therapeutic industry Bioremediation Bioleaching	Singh and Gabani, 2010

Cold habitats dominate the vast majority of our planet, covering three-quarters of the earth's surface, and span from the Arctic to the Antarctic and from high-mountain regions represented by the deep sea (90% of the ocean volume), followed by snow (35% of land surface), permafrost (24% of land surface), sea ice (13% of the earth's surface) and finally glaciers (10% of land surface) (Sullivan and Baross, 2007). Psychrophilic bacteria denote very important members of the sea ice habitat, including many unique taxa. Representatives of the family Vibrionaceae are among the most commonly reported bacteria to populate almost all extreme environments (D'Amico et al., 2006). Nevertheless, a wide range of phylogenetic diversity within the genera *Achromobacteria, Alcaligenes, Octadecabacter, Phormidium, Altermonas, Aquaspirillum, Moritella, Photobacterium, Colwellia, Cytophaga, Polaromonas, Polaribacter, Gelidibacter, Methanococcoides, Methanogenium, Methanosarcina, Psychrobacter,* and *Vibrio* have been reported to be psychrophilic in nature. In general, fungi are relatively rare in deep sea habitats compared to bacteria. Fungal isolates reported in frozen environments belong mainly to the genera *Rhodotorula,*

Penicillium, *Ustilago*, *Alternaria*, *Aureobasidium*, *Cladosporium*, *Geomyces*, *Ulocladium*, *Valsa*, and *Verticillium* (Hamdan, 2018).

Thermophiles

Thermophiles are divided into two main categories: thermophiles, which need temperatures of 50°C or higher for optimum growth, and hyperthermophiles, which have an optimum temperature for growth over 80°C. However, thermophilic extremophiles exhibit a remarkable range of operational temperatures, between 0°C and 120°C, and at pH values between of-0-12. This means that many thermophiles are in fact polyextremophiles, i.e. thermo-alkaliphiles or thermoacidophiles, and they exhibit molecular adaptations that allow them to exist in these harsh environments. Thermophiles span all three domains of life, but the majority belong to either bacteria or archaea. Which group dominates depends on the temperature: bacteria dominate between 50-90°C, whereas, at temperatures over 90°C, archaea dominate, so the majority of hyperthermophiles are archaea (Sysoev et al., 2021). Thermophilic microorganisms have been isolated from a variety of environments over the years, including domestic laundry and hot water heaters, compost piles, silage, slag heaps, and industrial processes, as well as natural habitats such as geothermally heated terrestrial hot springs, heated soils, submarine hydrothermal systems and geysers (Irwin, 2020). Due to the robust catalytic feature and thermal resistivity a variety of industrial bioprocesses employ thermophilic enzymes for a range of commercial applications. Many thermophile microorganisms produce industrially significant extremozymes, which belong into numerous categories such as bacteria domains (cyanobacteria, *Clostridium*, *Thermus*, *Bacillus*, *Thiobacillus*, *Actinobacteria*, green and purple bacteria and numerous other genres) and the archaea domains (*Pyrococcus*, *Thermococcus*, *Sulfolobus*, *Thermoplasma*, and *Methanogens*) (Dalmaso et al., 2015). Thermophilic microorganisms were first identified in the 1960s, when Thomas Brock isolated *Thermus aquaticus*, now famous as the origin of Taq DNA polymerase essential for the polymerase chain reaction, from Yellowstone National Park in the US (Brock and Freeze, 1969). Moreover, Gurumurthy and Neelagund (2012) described thermophilic *Geobacillus* sp. which synthesized hyper-thermostable α-amylase isolated from thermal springs. Several thermophilic bacteria and their enzymes have been reported to have significant industrial uses and these include thermophilic *Anoxybacillus* sp., GXS-BL, *Caldicellulosiruptor bescii*, *Anoxybacillus thermarum* FRM-RBK02, *Cellulomonas fimi* ATCC484, *Geobacillus proteiniphilus*, *Thermoanaerobacterium*, *Pyrococcus*, *Aeribacillus pallidus*, *Consortium* TC-5, *Clostridium* sp. strain WST, *Geobacillus kaustophilus*, *Janibacter* sp. strain R02, *Bacillus mojavensis* SO-10, *Caldicellulosiruptor*, *Anoxybacillus flavithermus* and *Bacillus licheniformis* (Zhu et al., 2020).

Halophiles

Microorganisms known as halophiles require high salt concentrations to grow. The four main categories of halophiles are mild halophiles (0.2 M), moderately halophiles (0.5–2.5 M), probable extreme halophiles (2.5–4.0 M), and extreme halophiles (4.0–5.9 M) based on their ideal NaCl concentration for survival. To survive in these environments, halophilic species employ a variety of adaption strategies. Only a few examples include altered protein electrostatic potential, regulating osmotic pressure with glycine-betaine and ectoine, absorption of Cl^- and K^+ into cells by transporters, and coordinated action of bacteriorhodopsin and ATP synthase (Karan et al., 2012a). Halophilic microbes and its enzymes are employed in a variety of areas, like food industry, production of solar salt, textiles and leather industries, environmental bioremediation, and pharmaceutical industry. The biological characteristics of halophile-produced compounds, enzymes, and suitable solutes might have implications in fine chemicals, pharmaceuticals, and bio implants (Jin et al., 2019). The stability of these enzymes in solvents like benzene, toluene, and chloroform, which are commonly employed in various industrial process, is one of their key selling features. Halophiles are a significant source of novel enzymes and well known genera are *Halobacterium*, *Acinetobacter*, *Bacillus*, *Haloferax*, *Halorhabdus*, *Micrococcus*, *Halobacillus*, and *Halothermothrix*, have been reported (Coker, 2016). The nuclease from *Micrococcus varians* is reported to degrade RNA at 60°C in presence of 12% w/v salt. It is also utilized for synthesis of 50 guanylic acid, a flavouring agent. Polysaccharide hydrolytic enzymes from halotolerant *Haloarcula* sp. strain LLSG7 have higher stability in presence of organic solvent (Karan et al., 2012b; Li and Yu, 2013). Interestingly, some halophilic biocatalysts have polyextremophilic ability, they withstand high/low pH, temperature, high salt, and non-aqueous catalysis, suggesting that can be applicable in diversified biotech industries (Raval et al., 2014; Akal et al., 2019).

Alkaliphiles

Alkaliphiles are creatures that thrive in alkaline environments (pH > 9), with optimal growth occurring at pH 10. Obligate and facultative alkaliphiles are the two primary physiological types of extremophiles. Alkaliphiles coexist with neutrophilic bacteria at slightly alkaline pH and can flourish under high alkaline environment. Alkaliphilic microorganisms of many types, including bacteria from the genus *Bacillus*, *Micrococcus* and *Streptomyces* and eukaryotes such as yeast and filamentous fungus have been isolated form alkaline habitats, including hyper-saline lakes (e.g., Lake Natron, Tanzania) and soda lakes (e.g., Lake Mono, CA, United States) (Horikoshi, 1999), highly alkaline wastes generating industries (e.g., indigo dye plants) or soils with high alkalinity (e.g., estuaries with extended evaporation periods, alkaline cracks in clay particles). Alkaliphilic microbes have been investigated as potential new sources for a variety of biotechnological uses, including wastewater treatment (e.g., dye-containing effluents). Textile sludge is prominent for high salt and alkaline pH along with the presence of hazardous dyes. (Prasad and Rao, 2013). Bhattacharya and his co-workers described that in hostile conditions, several alkaliphiles have been discovered and investigated as biocatalysts for remediation of dye-containing effluents. *Nesterenkonia lacusekhoensis* EMLA3 degrades the toxic azo dye methyl red in the presence of high salt concentrations and heavy metals (Ni (II), Cr (VI), and Hg(II)). Ammonia has been extracted by activity of alkaliphilic bacteria from N-rich saline wastewater, which is mostly generated by coke factories, fuel refineries, and the fertilizer industry. Ammonia pollution in effluents has been cleaned up using chemolithoautotrophic alkalophilic bacteria. Bacteria that convert all ammonia to nitrate could be valuable for waste water treatment and ecosystem engineering (Lawson and Lucker, 2018).

Acidophiles

Acidophiles are microbes that survive in environments with a pH < 3-4. Acidophilic proteins have not been studied for adaptability. The endo-β-glucanase from *Sulfolobus solfataricus* is found to be stable at a pH of 1.8. Nonetheless, endo-glucanase are generally unstable under acidic environments. *Ferroplasma acidiphilum* alpha-glucosidase has been found to be stable at low pH (Sharma et al., 2012). In recent research, Pikuta et al. (2007) found that carboxylesterase in *Ferroplasma acidiphilum* had a pH optimum of around 2 while other cytoplasmic enzymes activity reduced at pH>5. Acidophilic enzymes produce multi-enzyme complexes at cytoplasmic pH and have a lot of promise for biotechnological and commercial uses in biofuel and ethanol generation. Cellulolytic and xylanolytic enzymes are employed in a high-temperature, high-acidity environment to aid hydrolyze cellulolytic resources and make them more manageable (Dumorne et al., 2015).

Piezophiles/Barophiles

Piezophiles (formerly known as barophiles) are microorganisms that thrive in high-pressure niche, such as deep-sea bacteria or archaea, and are of interest to different biotechnology industries. Greater hydrostatic pressure is one of the physical characteristics that plays a significant role in the distribution of life on Earth in deep-ocean environment. The oceans cover 95% of biosphere and provide an average depth of 3800 metres and a pressure of 380 atmosphere (atm) or 38 MPa. Pressures of 700 to 1100 atm (70 to 110 MPa) prohibit most microbes from growing in the deepest sections of the sea. In addition, the temperature in deep water is generally between 1-3 °C. Yet, significant pressure and temperature may be observed in hydrothermal vent, and marine microorganisms may be exposed to temperature and pressure ranging from 1–300°C and 1–1100 atm (0.1–110 MPa), respectively (Abe and Horikoshi, 2001). The Mariana Trench is the world's deepest ocean hole, with a maximum depth of 11 kilometres and atypical pressure of 1100 atmospheres (110 MPa). Organisms that would survive at normal pressure and temperature, as well as stringent piezophiles, such as *Moritella yayanosii* and *Shewanella benthica*, survive between 700 and 800 atm (70-80 MPa), but not less than 500 atm (50 MPa) (Kato, 2011). Psychrophilic bacteria make up the majority of bacterial piezophiles, which are divided into five genera viz. *Photobacterium*, *Shewanella*, *Colwellia*, *Psychromonas,* and *Moritella*, while the majority of piezophilic Archaea are (hyper)thermophiles like, *Thermococcales* (Oger and Cario, 2014). Owing to the fact the piezophilic enzymes have a wide range of commercial uses but still limited study has been done on them. Abe and Horikoshi (2001) revealed that alpha-amylase from piezophilic produces trisaccharide instead of maltobiose and tetra-saccharide utilizing malto-oligosaccharide as a substrate at high pressure and low energy. This reaction has a lot of industrial and biotech potential, especially in the food and detergent industries.

Radiophiles

Environmental extremes include high levels of ionizing radiation. This has an effect on cells by either directly destroying DNA or by releasing reactive oxygen radicals, which can induce DNA mutations or strand breaks. However, some organisms can survive these conditions referred as radioresistant. These microorganisms are unusual in that they can survive in both ionizing and nonionizing radiation environments, which would be harmful to others. *Deinococcus radiodurans* is known for its ability to withstand supra-lethal ionizing radiation and UVR (1000 Jm^{-2}) (Yuan et al., 2009). Endolithic cyanobacteria have also found a way to protect themselves from the harmful effects of UVR. Most micro-organisms produce photoprotective pigments that are able to absorb UVR much like melanin does in humans. The strain of cyanobacterium *Chroococcidiopsis* exhibits resistance against ionizing radiation because of efficient DNA repair mechanisms (Billi et al., 2000). Major ultraviolet radiation (UVR)–resistant extremophilic microorganisms *Deinococcus guangriensis*, *D. wulumuqiensis*, *D. xibeiensis*; *D. gobiensis*; *D. gradis*; *D. Prochlorococcus* MED4, *Stapylococcus* sp., *Exiguobaceterium* sp., *Acinetobacter* sp., *Bacillus* sp., *Micrococcus* sp., *Sphyngomonas* sp., *Bacillus horneckiae* and *Microbacterium maritypicum* (Singh and Gabani, 2010).

Xerophiles

Xerophiles are microorganisms that can survive and grow in arid conditions with water activity a_w<0.75 by forming spores that help them mitigate environmental stress. Adaptive mechanisms are connected to water loss prevention and increased water retention through the accumulation of compatible solutes, production of extracellular polymeric substances (EPSs), adaptations on the cell membrane to retain intracellular water, and synthesis of DNA repair proteins (Lebre et al., 2017). These unique adaptations allow xerophiles to be used in microbial electrochemical systems or in next-generation industrial biotechnology, where they can be used for treating long-chain fatty acids, cellulose, chitin, rubbers, or other compounds (Chen and Jiang, 2018).

FUNCTIONAL ATTRIBUTES AND BIOTECHNOLOGICAL IMPLICATIONS OF EXTREMOZYMES

More than 500 items are being made with enzymes, and around 150 manufacturing industries benefit from the usage of microorganism-derived enzymes. Furthermore, there are about 3000 enzymes recognized, with around 65% of them being hydrolases utilized in the industry viz. paper & pulp, starch processing, detergent, textile, and almost 25% being used in food processing. According to research, the variety of extremophile bacteria may be more than previously thought (Figure 2) (Dalmaso et al., 2015; Kour et al., 2019). Nature provides a variety of temperatures, salinities, pH, and pressures in severe deep-sea habitats, which can be used to find novel and possibly efficient enzymes. Which are much more appropriate for industrial uses, and that has been discovered that extremozymes generated by extremophilic microbes have a wide range of industrial applications due to its elevated activity and extreme stability. Excellent stability and enzymatic activity, extremeozymes are viable alternatives to traditional biotechnology approaches in the industries such as detergent, feed & food, agricultural, textile and leather, paper & pulp, pharmaceutical, beverage, and biomining (Figure 3) (Raddadi et al., 2015).

Figure 2. Distribution of extremozymes for diverse biotechnological applications

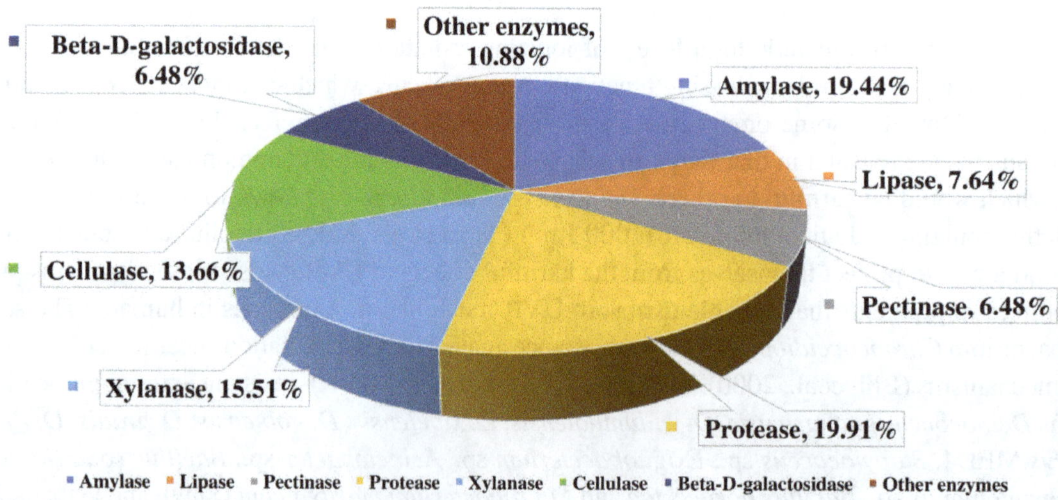

Figure 3. Biotechnological implications of extremozymes

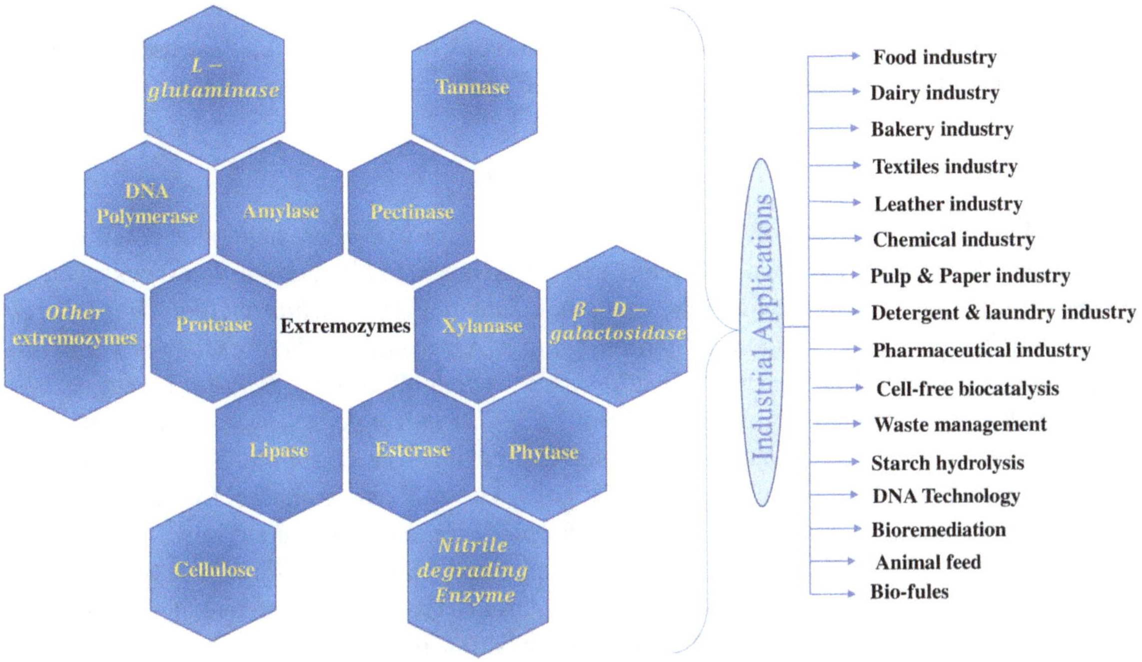

Amylase

Diastase, an amylase enzyme, was first reported by a French chemist, Anselme Payen (Hill and Needham 1970). The term amylase was used initially to entitle enzymes accomplished of hydrolyzing α-1, 4-glucosidic bonds of amylose, amylopectin, glycogen, and their degradation products. The three major α-, β-, and γ-amylases are subtypes of amylases used very specifically for catalysis. The hydrolysis of α-1, 4 glycosidic bonds in starch by α-amylase results in glucose, dextrin, wherein dextrin is limited. The process of scattering indissoluble starch particles in aqueous solution monitored by partial hydrolysis by means of thermostable amylases is referred to as liquefaction. *Bacillus amyloliquefaciens* amylase was the first liquefying alpha-amylase on a large scale (Shukla et al., 2015). Asoodeh et al., (2010), stated a rare acidophilic α-amylase produced from newly isolated *Bacillus* sp. In the food processing industries, thermostable, acidophilic alpha-amylases were commonly used. In the starch industry, the natural pH of starch slurry is generally around 4.5. Since these activities are often done on an alkaline pH in both cold and hot, thermo and alkaline stable α -amylases are being used as a progressive in detergents for washing processes. Alkaline and thermotolerant α-amylases have been cleansed from *Bacillus* species, *B. licheniformis* and *Bacillus halodurans*. The *B. subtilis* JS-2004 strain has been shown to generate large quantities of thermostable -amylase with properties that make it suitable for use in the starch and food industries (Asgher et al., 2007). Zhang and Zeng (2008), reported an actinomycetes isolated from the deep sea sediment which is identified as *Nocardiopsis* sp. and produces α-amylase. For centuries, enzymes in particular malt and microbial α-amylases have been broadly used in the baking industry (Hamer, 1995). In bread and rolls to provide higher volume to these products, these enzymes were used to provide improved color. Cereal enzymes from barley malt and microbial enzymes from fungi and bacteria have been used in baking industry. The use of α-amylase for the production of low viscosity, high molecular weight starch for covering of paper is reported. Since 1975, α-amylases have been used in powder laundry detergents. Currently, 90% of all liquid detergents contain α-amylase (Kottwitz et al., 1994). With the introduction of new frontiers in biotechnology, the scope of amylase uses has extended into several other fields, such as clinical, medicinal, and analytical chemistry.

Proteases

An enzyme that performs the proteolysis: catabolism of protein by hydrolysis of peptide bonds is referred to as proteases. Proteases can be found in all types of organisms, including prokaryotes, eukaryotes, and viruses (Raval et al., 2015). Several microbial strains including fungi (*Penicillium griseofulvin, Aspergillus niger, Aspergillus flavus, Chrysosporium keratinophilum, Aspergillus melleu, Fusarium graminarum, Scedosporium apiosermum*) and bacteria (*Bacillus proteolyticus, B. subtilis, Bacillus firmus, B. amyloliquefaciens, Bacillus alcalophilus, B. licheniformis, Bacillus thuringiensis*) were reported to produce proteases (Yadav et al., 2018). Organic solvents stable protease from *Oceanobacillus* sp. (GQ162111) also were reported by Pandey and his co-workers (2012). The cold-active anaerobic bacteria producing extracellular proteases can potentially be employed for biodegradation of organic wastes rich in protein, such as waste from humans. Alam et al., (2005), reported the psychrotrophic proteolytic bacterium *Clostridium* sp. SPA3 isolated from the lake sediment of Antarctica forming maximum cell mass of 5–10 ∘C and produced extracellular protease. An aerobic Gram-positive, thermophilic in nature *Bacillus* species (P-OOlA) has been isolated from an alkaline hot spring by Atalo and Gashe (1993).

Pectinase

The enzymes hydrolyzing these pectic substances are broadly known as pectinases, and include polygalacturonases, pectin esterases, pectin lyases, and pectate-lyases, depending on their mode of action. The enzymes are widely circulated in higher plants and microorganisms. They are of primary significance for plants as they support cell wall extension and softening of some plant tissues. They also support in keeping ecological balance by causing decomposition and recycling of waste plant materials. Other important markers of pectinolytic enzymes include plant pathogenicity and rotting of fruits and vegetables (Fraissinet-Tachet and Fevre, 1996). Pectinases are produced predominantly from the genus *Bacillus, Aspergillus* and *Pseudomonas*. Pectinases have been utilized in a variety of traditional industrial processes over the years, including textile processing, plant fiber processing, tea, coffee, oil extraction, industrial wastewater treatment, and the treatment of pectinaceous materials, among others (Ricard and Reid, 2004). Fruit juice extraction and clarifying is the most common industrial application of pectinases that increase in fruit juice volume. Pectinases, in combination with amylases, lipases, cellulases, and hemicellulases, were used to safely and environmentally remove sizing agents from cotton, substituting hazardous caustic soda. Biotechnological degumming using pectinases in combination with xylanases presents an ecofriendly and economic alternative (Hoondal et al., 2002).

Cellulase

Cellulases break down the -1, 4-D-glucan bonds in cellulose to produce glucose, cellobiose, and cello-oligosaccharides as main products. Because of its numerous applications in cotton processing, paper recycling, juice extraction, detergent enzymes, and animal feed additives, cellulases are now the third most widely used industrial enzyme on the planet. There are several industries involved in production of cellulase for textile, detergent, paper industries, and other industries (Sukumaran et al., 2009). De Siqueira and his team (2010) reported in the production of cellulase microbes has been used including aerobic and anaerobic bacteria; white rot and soft rot fungi and anaerobic fungi. The majority of cellulases used in industry come from filamentous fungi like *Fusarium*, *Humicola*, *Trichoderma*, *Penicillium*, and *Phanerochaete*. Among the fungi, *Myceliophthora thermophila* while among the thermophilic, aerobic bacteria, only a few actinomycetes are actively cellulolytic, mostly *Thermomonospora curvata* and *Thermoactinomyces cellulosae*. Cellulose-producing psychrotrophic bacteria belonging to the genera *Paenibacillus* and *Pseudomonas* have been isolated from the cold environments in the Western Himalaya. Cellulases have been used in detergents since the early 1990s. Cellulases remove cellulose microfibrils generated during the production and washing of cotton-based fabrics during textile cleaning (Kasana and Gulati, 2011).

Xylanases

Xylanases are glycosidases that comprise endo-1, 4-b-xylanase and β-xylosidase and catalyzing the endo-hydrolysis of 1, 4-b- D-xylosidic linkages in xylan present in hemicelluloses of plants, converting them into the monomeric sugars. The function is performed with the assistance of certain other hydrolytic enzymes such as acetyl xylan esterase, α-L-arabino-furanosidase, α-glucuronidase, and phenolic acid including ferulic and p-coumaric acid esterase (Thomas et al., 2017). Xylanases have also been obtained from microbes of extreme environments such as thermophiles including *Thermoascus aurantiacus*, *Bacillus stearothermophilus*, *Dictyoglomus thermophilum*, *Caldicellulosiruptor* sp., *Thermobifida alba*, *Chaetomium thermophilum*, *Clostridium thermocellum*, *Geobacillus Nonomuraea flexuosa*, *Rhodothermus marinus*, *Thermotoga maritime*; psychrophiles such as *Cryptococcus adeliae*, *Alternaria alternate*, *Clostridium* sp., *Penicillium* sp., *Phoma* sp., *Coprinus psychromorbidus*, *Flavobacterium frigidarium*, and *Pseudoalteromonas haloplanktis* (Butt et al., 2008; Kour et al., 2019). Xylanases are used in a variety of industries. These are used in baking industry to improve the strength of the gluten and to increase dough tolerance, to make dough softer, reducing the sheeting work requirements and considerably increases the volume of the baked bread (Harbak and Thygesen 2002). Xylanases, in combination with other enzymes including cellulases, amylases, and pectinases, improves the yield of juices by means of liquefaction of fruit and vegetables; stabilization of the fruit pulp; enhances the recovery of aromas, edible dyes, essential oils, hydrolysis of substances, mineral salts, pigments vitamins, reduces the viscosity, that hamper the physical or chemical clearing of the juice, or which otherwise may lead to the cloudiness in the concentrate. Xylanases are also used to cause the hydrolysis of arabino xylans to lower oligosaccharides, which diminishes the viscosity of beer thereby eliminating its muddy aspect. Xylanase is utilized in biscuit-making, to make cream crackers lighter and improve the texture, palatability, and uniformity of the wafers (Juturu and Wu, 2012).

Lipase/ Esterases

Esterases and lipases are the most widely used biocatalysts in fine chemical applications, largely because of the advantages of these catalysts for the production of optically pure compounds. Large libraries of thermophilic esterases and lipases have been developed by a number of screening and enzyme discovery methods (Amoozegar et al., 2006). Lipolytic activity has been demonstrated in *Aspergillus* sp., *Acremo-*

nium strictum, *Cunninghamella verticillata*, *Humicola lanuginose*, *Lipomyces starkeyi*, yeasts including *Candida antarctica*, *Pichia burtonii*, *Rhodotorula glutinis*, *Saccharomyces lipolytica*, *Yarrowia lipolytica* and bacterial sources of lipases include *Acinetobacter* sp., *B. coagulans*, *B. stearothermophilus*, *Bacillus thermocatenulatus*, *Burkholderia glumae*, *Chromobacterium viscosum*, *Lactobacillus* sp., *Pseudomonas cepacia* further they have also been obtained from actinomycetes including *Streptomyces fradiae* and other *Streptomyces* sp. (Kour et al., 2019). An esterase from *Bacillus licheniformis*, *Bacillus acidocaldarius*, *Pyrococcus* was expressed from an *E. coli* recombinant strain that increased their organic solvent, pH and thermal stability. Three recombinant cold-adapted lipases are derived from the psychrophillic strain *Moxarella furiosus* TA144, which have a high affinity for butyrate instead of longer-chain esters (especially in one of the lipases) (Ikeda and Clark, 1998). Lipases are used in the food industry. One of the most significant components of meals is fats and oils. Lipase-modified butter fat is used in a wide range of food processing applications. Lipase-mediated food items include chocolates with cocoa butter replacements, bread, structured lipids such as human milkfat replacers, low-calorie health oils, and nutraceuticals. Commercially manufactured lipases are being used to develop flavor in dairy products, as well as in the food processing such as baked goods, beer, fruits, meat, dairy, and vegetables. In the past, phospholipases were employed to cure egg yolks that is useful for processing of custard, mayonnaise, baby foods, dressings, and in dough preparation, sauces, like Hollandaise, Béarnaise, and cafe de Paris. Lipases are often used to hydrolyze milkfat, speed up the ripening of cheese, and lipolyze butter, fat, and cream. In addition, it is utilized to enhance the flavor of bread items and can increase texture and softness (Ray, 2012).

Phytases

Phytases are a class of phosphatases, which possess the ability to release at least one phosphate from phytate in vitro. Microbe-derived phytases are among the most promising for biotechnological applications. Extracellular phytate-degrading enzymes have been reported in the molds and yeast, but in bacteria, these enzymes are mainly cell-associated, with the exception of *Bacillus* and *Enterobacter*, periplasmic in *E. coli*, whereas phytate-degrading activity in *Selenomonas ruminantium* and *Mitsuokella multiacidus* has been revealed to be associated with the outer membrane (Kour et al., 2019). Phytases have also been reported from thermophilic fungi, including *A. niger*, *Humicola lanuginose*, *M. thermophila*, *Rhizomucor pusillus*, *Sporotrichum thermophile*, *T. aurantiacus*, *Fusarium verticillioides*, *Penicillium*, *Rhizopus* and *Rhizoctonia* sp. (Ushasree et al., 2017). The phytase activity has been also reported from *Y. lipolytica*, *Candida tropicalis*, *Zygosaccharomyces bisporus*, *Williopsis saturnus*, *Zygosaccharomyces priorionus* and *Schizosaccharomyces octosporus*. Further, phytases possess various industrial applications. It has been revealed to be an exceptional bread-making improver. It has been known that addition of commercial fungal phytase from *A. niger* in the dough ingredients containing fiber formulation speeds up proofing, improves bread shape, enhances specific volume, adds to softness to the crumb. Phytases have been chiefly used as animal feed additive in diets mainly for swine and poultry, and to some extent for fish. The first commercial phytase products was launched into market in 1991 (Afinah et al., 2010).

β-D-galactosidase/Lactase

β-D-galactosidase (also referred to as lactase) The hydrolysis of 1,4-D-galactosidic bonds is catalysed by this enzyme. Lactose is hydrolyzed by this enzyme in glucose and galactose, that are sugars and is sweeter, more soluble, and easier to digest than lactose. β-D-galactosidase has a wide range of industrial uses including dairy, fermentation, and food industries. It has been produced from bacterial sources, including *Lactobacillus fermentum*, *Saccharomyces cerevisiae*, *Propionibacterium acidipropionici*, *Kluyveromyces marxianus*, *Streptococcus thermophilus*, *Kluyveromyces lactis*, *E. cloacae*, *Bifidobacterium longum*, *Beijerinckia indica*, *Bacillus circulans* and fungal including *Paecilomyces aerugineus*, *Penicillium chrysogenum*, and *Aspergillus aculeatus* (Kour et al., 2019). *B. stearothermophilus*, *Pyrococcus woesei*, *Thermus* sp. have been reported to produce thermostable β-galactosidase. Psychrophiles have been shown to have cold-active and cold-adapted β -galactosidase from *A. psychrolactophilus* and *P. haloplanktis*, thermostable β-galactosidase from *A. niger*, a novel acidophilic intracellular β-galactosidase from *Teratosphaeria acidotherma*. The enzyme is used chiefly in industrial-scale production of milk and dairy products. Milk and dairy products are treated with lactase to reduce their lactose content and also to deal with the problems of lactose insolubility and lack of sweetness. Furthermore, this treatment could make milk, an appropriate diet, available to a significant number of lactose intolerant people and children. Whey hydrolysis also transforms lactose into useful products such as sweet syrup, which can be used in a variety of operations in the baking, confectionary, dairy, and soft-drink industries (Van De Voorde et al., 2014).

Glutaminase

Glutaminase is also an amidase enzyme that releases ammonia when it transforms glutamine to glutamate. Microbes, plants, and animal tissues can all provide these. The source of this enzyme has been identified as a variety of bacterial and fungal strains including *A. oryzae*, *Aerobacter aerogenes*, *Actinomucorelegans* sp., *Candida scottii*, *Cryptococcus laurentii*, *Enterobacter cloacae*, *Erwinia aroideae*, *Escherichia coli*, *Hansenula sp.*, *Klebsiella aerogenes*, *Proteus morganii*, *Providencia* sp., *P. aeruginosa*, *Rhodotorula* sp., *Serratia marcescens*, *Streptomyces enissocaesilis*, *Torulopsis candida, Vibrio* sp. and *Xanthomonas juglandis* (Kour et al., 2019). Also, glutaminase is one of the chief enzymes controlling the taste of fermented foods, including soy sauce, especially in Asian countries. By raising the glutamic acid concentration in fermented foods, it improves their flavor and aroma. Glutamic acid and aspartic acid, well known as important amino acids, contributes to fine taste, umami and sharp taste, sour simultaneously making contribution to nutritional effects to food. Salt-tolerant glutaminase obtained from *Stenotrophomonas maltophilia* is extensively used in Japanese soy sauce fermentation (Patel et al., 2021).

Nitrile-Degrading Enzymes

Nitrile-degrading enzymes are of considerable importance in industrial bio-transformations, with several examples of commercial implementation. A thermophilic *Bacillus* sp. capable of transforming aliphatic nitriles, cyclic nitriles and dinitriles (Graham et al., 2000). Interestingly, in the presence of urea or chloroacetone, amidase activity was inhibited and the amide intermediate accumulated, as demonstrated in continuous bioreactor experiments. Another thermophilic nitrile hydratase was identified in *Bacillus pallidus* Dac521 (Cramp and Cowan, 1999). Only a few aliphatic substrates were hydrolyzed by the enzyme while none of the cyclic, hydroxy-, di-, or aromatic nitriles substrates were hydrolyzed. Benzonitrile reduced the activity in an irreversible manner. During the last few years, the halophilic archaea *Haloferax mediterranei* has been used as haloarchaea model to look for potential uses of the enzymes supporting denitrification in this species. This microorganism is able to assimilate nitrate and nitrite because of the presence of nitrate and nitrite reductases. It tolerates high nitrate or nitrite concentrations, which makes it a very interesting microorganism for wastewater or brines bioremediation applications (Martinez-Espinosa et al., 2015)

Other Extremozymes

Many other enzymes have been identified from extremophiles. Libraries of thermostable dehydrogenases have been identified that are useful in stereoselective transformation of ketones to alcohols. An NADP alcohol dehydrogenase from *P. furiosus* is reported to be thermostable under anaerobic conditions, but very labile in the presence of oxygen (Ma and Adams, 1999). Ethanol, 2-phenylethanol, phenylacetaldehyde, tryptophol, 1, 3-propanediol, acetaldehyde, and methyl glyoxal are among the alcohols and aldehydes used by *P. furiosus* alcohol dehydrogenase. Kinetic investigations revealed that catalyzing aldehyde reduction with NADPH as the electron source was preferred. Two thermophilic glutamate dehydrogenases (native and recombinant) were found to be stabilized by pressure up to 750 atm (Sun et al., 1999). A cold-adapted phosphatase from *Shewanella* sp. was expressed in *E. coli*, exhibiting the same high catalytic activity at low temperature as the native phosphatase. Alanine racemase from *Bacillus psychrosaccharolyticus* showed the highest catalytic activity at 0°C, while extremely labile over 35°C (Okubo et al., 1999).

CONCLUDING REMARKS AND PERSPECTIVES

In recent years, the extremozymes from extremophilic microbes have been in high demand in the agriculture, medicine, and food industry, due to their potential applications in diverse processes. Here in the present chapter a bird's eye view is presented on diversity of extremophilic microbes, their occurrence in nature and lastly a summation of hydrolytic enzymes secreted by them are noted. These have quite diverse potential applications in areas such as agriculture, bioconversion of hemicellulose, biodegradation, bioethanol production, biorefinery, chemical industry, composting, dairy industry, detergent industry, feed supplement, food industry, feed industry, glucose feedstock from cellulose, improving rumen digestion, leather industry, molecular biology, paper and pulp industry, peptide synthesis, pharmaceutical industry, and therapeutic agent. The scope to use extremozymes for biotechnological applications is widening with time. This potential usage in health-related and food related areas is sufficient to search for novel and more efficient extremozymes.

DECLARATIONS OF INTEREST

None of the authors have any potential conflict of interest. All authors have read the manuscript and give their consent for publication.

ACKNOWLEDGMENT

Authors PSA, RRP, KNR and VHR acknowledge the infrastructural support provided by Department of Microbiology & Biotechnology, Gujarat University, Ahmedabad. Authors SMY and VHR kindly acknowledge the infrastructural support from Christ College and Department of Biosciences, Saurashtra University, Rajkot respectively. VHR heartily acknowledge the junior & senior research fellowships by Department of Biotechnology (DBT), Science and Engineering Research Board (SERB) Young Scientist Fellowship, and Universities Grant Commission (UGC)-Basic Science Research (BSR) Start-Up Research Grant No.F.30-521/2020. The Research fellowship under the "SHODH - ScHeme of Developing High-Quality Research" by Department of Higher Education, Government of Gujarat Ref No: 201901380034 to PSA is also heartily acknowledged.

REFERENCES

Abe, F., & Horikoshi, K. (2001). The biotechnological potential of piezophiles. *Trends in Biotechnology*, *19*(3), 102–108. doi:10.1016/S0167-7799(00)01539-0 PMID:11179803

Afinah, S., Yazid, A. M., Anis Shobirin, M. H., & Shuhaimi, M. (2010). Phytase: Application in food industry. *International Food Research Journal*, *17*(1), 13–21.

Akal, A. L., Karan, R., Hohl, A., Alam, I., Vogler, M., Grötzinger, S. W., Eppinger, J., & Rueping, M. (2019). A polyextremophilic alcohol dehydrogenase from the Atlantis II Deep Red Sea brine pool. *FEBS Open Bio*, *9*(2), 194–205. doi:10.1002/2211-5463.12557 PMID:30761247

Alam, S. I., Dube, S., Reddy, G. S. N., Bhattacharya, B. K., Shivaji, S., & Singh, L. (2005). Purification and characterisation of extracellular protease produced by *Clostridium* sp. from Schirmacher oasis, Antarctica. *Enzyme and Microbial Technology*, *36*(5-6), 824–831. doi:10.1016/j.enzmictec.2005.01.011

Amoozegar, M. A., Salehghamari, E., Khajeh, K., Kabiri, M., & Naddaf, S. (2008). Production of an extracellular thermohalophilic lipase from a moderately halophilic bacterium, Salinivibrio sp. strain SA-2. *Journal of Basic Microbiology*, *48*(3), 160–167. doi:10.1002/jobm.200700361 PMID:18506896

Arya, P., Trivedi, M., Ramani, N., Patel, K., & Rajput, K. N. (2020a). Isolation, Production and Applications of Alkaline Protease from Hot Springs Bacterial Isolates. *Bioscience Biotechnology Research Communications*.

Asoodeh, A., Chamani, J., & Lagzian, M. (2010). A novel thermostable, acidophilic α-amylase from a new thermophilic "*Bacillus* sp. *Ferdowsicous*" isolated from Ferdows hot mineral spring in Iran: Purification and biochemical characterization. *International Journal of Biological Macromolecules*, *46*(3), 289–297. doi:10.1016/j.ijbiomac.2010.01.013 PMID:20109486

Atalo, K., & Gashe, B. A. (1993). Protease production by a thermophilic *Bacillus* species (P-001A) which degrades various kinds of fibrous proteins. *Biotechnology Letters*, *15*(11), 1151–1156. doi:10.1007/BF00131207

Benson, C. A., Bizzoco, R. W., Lipson, D. A., & Kelley, S. T. (2011). Microbial diversity in nonsulfur, sulfur and iron geothermal steam vents. *FEMS Microbiology Ecology*, *76*(1), 74–88. doi:10.1111/j.1574-6941.2011.01047.x PMID:21223339

Bertoldo, C., Dock, C., & Antranikian, G. (2004). Thermoacidophilic microorganisms and their novel biocatalysts. *Engineering in Life Sciences*, *4*(6), 521–532. doi:10.1002/elsc.200402155

Bhatt, H. B., Begum, M. A., Chintalapati, S., Chintalapati, V. R., & Singh, S. P. (2017). Desertibacillus haloalkaliphilus gen. nov., sp. nov., isolated from a saline desert. *International Journal of Systematic and Evolutionary Microbiology*, *67*(11), 4435–4442. doi:10.1099/ijsem.0.002310 PMID:28920841

Bhattacharya, A., Goyal, N., & Gupta, A. (2017). Degradation of azo dye methyl red by alkaliphilic, halotolerant *Nesterenkonia lacusekhoensis* EMLA3: Application in alkaline and salt-rich dyeing effluent treatment. *Extremophiles*, *21*(3), 479–490. doi:10.100700792-017-0918-2 PMID:28255636

Billi, D., Friedmann, E. I., Hofer, K. G., Caiola, M. G., & Ocampo-Friedmann, R. (2000). Ionizing-radiation resistance in the desiccation-tolerant cyanobacterium *Chroococcidiopsis. Applied and Environmental Microbiology*, *66*(4), 1489–1492. doi:10.1128/AEM.66.4.1489-1492.2000 PMID:10742231

Brock, T. D., & Freeze, H. (1969). *Thermus aquaticus* gen. n. and sp. n., a nonsporulating extreme thermophile. *Journal of Bacteriology*, *98*(1), 289–297. doi:10.1128/jb.98.1.289-297.1969 PMID:5781580

Bull, A. T., & Asenjo, J. A. (2013). Microbiology of hyper-arid environments: Recent insights from the Atacama Desert, Chile. *Antonie van Leeuwenhoek*, *103*(6), 1173–1179. doi:10.100710482-013-9911-7 PMID:23564162

Butt, M. S., Tahir-Nadeem, M., Ahmad, Z., & Sultan, M. T. (2008). Xylanases and their applications in baking industry. *Food Technology and Biotechnology*, *46*(1), 22–31.

Caumette, P. (1993). Ecology and physiology of phototrophic bacteria and sulfate-reducing bacteria in marine salterns. *Experientia*, *49*(6), 473–481. doi:10.1007/BF01955148

Chen, G. Q., & Jiang, X. R. (2018). Next generation industrial biotechnology based on extremophilic bacteria. *Current Opinion in Biotechnology*, *50*, 94–100. doi:10.1016/j.copbio.2017.11.016 PMID:29223022

Coker, J. A. (2016). Sohail Siddiqui of New South Wales, K. *Extremophiles and Biotechnology: Current Uses and Prospects. F1000 Research*, *5*, 396. doi:10.12688/f1000research.7432.1

Cramp, R. A., & Cowan, D. A. (1999). Molecular characterisation of a novel thermophilic nitrile hydratase. *Biochimica et Biophysica Acta (BBA)-. Protein Structure and Molecular Enzymology*, *1431*(1), 249–260. doi:10.1016/S0167-4838(99)00010-2 PMID:10209297

D'Amico, S., Marx, J. C., Gerday, C., & Feller, G. (2003). Activity-stability relationships in extremophilic enzymes. *The Journal of Biological Chemistry*, *278*(10), 7891–7896. doi:10.1074/jbc.M212508200 PMID:12511577

Dalmaso, G. Z. L., Ferreira, D., & Vermelho, A. B. (2015). Marine extremophiles: A source of hydrolases for biotechnological applications. *Marine Drugs*, *13*(4), 1925–1965. doi:10.3390/md13041925 PMID:25854643

DasSarma, S., & DasSarma, P. (2015). Halophiles and their enzymes: Negativity put to good use. *Current Opinion in Microbiology*, *25*, 120–126. doi:10.1016/j.mib.2015.05.009 PMID:26066288

de Kato, C. (2011). High pressure and prokaryotes. In Extremophiles Handbook. Academic Press.

de Siqueira, F. G., de Siqueira, E. G., Jaramillo, P. M. D., Silveira, M. H. L., Andreaus, J., Couto, F. A., Batista, L. R., & Ferreira Filho, E. X. (2010). The potential of agro-industrial residues for production of holocellulase from filamentous fungi. *International Biodeterioration & Biodegradation*, *64*(1), 20–26. doi:10.1016/j.ibiod.2009.10.002

Dewan, S. S. (2014). *Global markets for enzymes in industrial applications*. BCC Research.

Dumorne, K., Cordova, D. C., Astorga-Elo, M., & Renganathan, P. (2017). Extremozymes: A potential source for industrial applications. *Journal of Microbiology and Biotechnology*, *27*(4), 649–659. doi:10.4014/jmb.1611.11006 PMID:28104900

Fang, J., Zhang, L., & Bazylinski, D. A. (2010). Deep-sea piezosphere and piezophiles: Geomicrobiology and biogeochemistry. *Trends in Microbiology*, *18*(9), 413–422. doi:10.1016/j.tim.2010.06.006 PMID:20663673

Forster, J. (1887). Ueber einige Eigenschaften leuchtender Bakterien. *Centr. Bakteriol. Parasitenk.*, *2*, 337–340.

Fraissinet-Tachet, L., & Fevre, M. (1996). Regulation by galacturonic acid of pectinolytic enzyme production by *Sclerotinia sclerotiorum*. *Current Microbiology*, *33*(1), 49–53. doi:10.1007002849900073 PMID:8661689

Graham, D., Pereira, R., Barfield, D., & Cowan, D. (2000). Nitrile biotransformations using free and immobilized cells of a thermophilic *Bacillus* spp. *Enzyme and Microbial Technology*, *26*(5-6), 368–373. doi:10.1016/S0141-0229(99)00169-6 PMID:10713209

Gunde-Cimerman, N., Zalar, P., de Hoog, S., & Plemenitaš, A. (2000). Hypersaline waters in salterns–natural ecological niches for halophilic black yeasts. *FEMS Microbiology Ecology*, *32*(3), 235–240. PMID:10858582

Gurumurthy, D. M., & Neelagund, S. E. (2012). Molecular characterization of industrially viable extreme thermostable novel alpha-amylase of *Geobacillus* sp Iso5 Isolated from geothermal spring. *Journal of Pure & Applied Microbiology*, *6*, 1759–1773.

Hamdan, A. (2018). Psychrophiles: Ecological significance and potential industrial application. *South African Journal of Science*, *114*(5-6), 1–6. doi:10.17159ajs.2018/20170254

Hamer, R. J. (1995). Enzymes in the baking industry. In *Enzymes in food processing* (pp. 190–222). Springer. doi:10.1007/978-1-4615-2147-1_6

Harbak, L., & Thygesen, H. V. (2002). Safety evaluation of a xylanase expressed in *Bacillus subtilis*. *Food and Chemical Toxicology*, *40*(1), 1–8. doi:10.1016/S0278-6915(01)00092-8 PMID:11731030

Harrison, J. P., Gheeraert, N., Tsigelnitskiy, D., & Cockell, C. S. (2013). The limits for life under multiple extremes. *Trends in Microbiology*, *21*(4), 204–212. doi:10.1016/j.tim.2013.01.006 PMID:23453124

Hill, R. (1970). *The chemistry of life: eight lectures on the history of biochemistry*. CUP Archive.

Hmidet, N., Bayoudh, A., Berrin, J. G., Kanoun, S., Juge, N., & Nasri, M. (2008). Purification and biochemical characterization of a novel α-amylase from *Bacillus licheniformis* NH1: Cloning, nucleotide sequence and expression of amyN gene in Escherichia coli. *Process Biochemistry*, *43*(5), 499–510. doi:10.1016/j.procbio.2008.01.017

Hoondal, G., Tiwari, R., Tewari, R., Dahiya, N. B. Q. K., & Beg, Q. (2002). Microbial alkaline pectinases and their industrial applications: A review. *Applied Microbiology and Biotechnology*, *59*(4), 409–418. PMID:12172603

Horikoshi, K. (1998). Barophiles: Deep-sea microorganisms adapted to an extreme environment. *Current Opinion in Microbiology*, *1*(3), 291–295. doi:10.1016/S1369-5274(98)80032-5 PMID:10066496

Horikoshi, K. (1999). Alkaliphiles: Some applications of their products for biotechnology. *Microbiology and Molecular Biology Reviews*, *63*(4), 735–750. doi:10.1128/MMBR.63.4.735-750.1999 PMID:10585964

Ikeda, M., & Clark, D. S. (1998). Molecular cloning of extremely thermostable esterase gene from hyperthermophilic archaeon *Pyrococcus furiosus* in Escherichia coli. *Biotechnology and Bioengineering*, *57*(5), 624–629. doi:10.1002/(SICI)1097-0290(19980305)57:5<624::AID-BIT15>3.0.CO;2-B PMID:10099242

Irwin, J. A. (2020). Overview of extremophiles and their food and medical applications. In *Physiological and Biotechnological Aspects of Extremophiles* (pp. 65–87). Academic Press. doi:10.1016/B978-0-12-818322-9.00006-X

Jin, M., Gai, Y., Guo, X., Hou, Y., & Zeng, R. (2019). Properties and applications of extremozymes from deep-sea extremophilic microorganisms: A mini review. *Marine Drugs*, *17*(12), 656. doi:10.3390/md17120656 PMID:31766541

Juturu, V., & Wu, J. C. (2012). Microbial xylanases: Engineering, production and industrial applications. *Biotechnology Advances*, *30*(6), 1219–1227. doi:10.1016/j.biotechadv.2011.11.006 PMID:22138412

Karan, R., Capes, M. D., & DasSarma, S. (2012a). Function and biotechnology of extremophilic enzymes in low water activity. *Aquatic Biosystems*, *8*(1), 1–15. doi:10.1186/2046-9063-8-4 PMID:22480329

Karan, R., Kumar, S., Sinha, R., & Khare, S. K. (2012b). Halophilic microorganisms as sources of novel enzymes. In *Microorganisms in sustainable agriculture and biotechnology* (pp. 555–579). Springer. doi:10.1007/978-94-007-2214-9_25

Kottwitz, B., Upadek, H., & Carrer, G. (1994). Application and benefits of enzymes in detergents. *Chimica Oggi*, *12*(11-12), 21–24.

Kour, D., Rana, K. L., Kaur, T., Singh, B., Chauhan, V. S., Kumar, A., Rastegari, A.A., Yadav, N., Yadav, A.N. & Gupta, V. K. (2019). Extremophiles for hydrolytic enzymes productions: biodiversity and potential biotechnological applications. *Bioprocessing for biomolecules production*, 321-372.

Lawson, C. E., & Lücker, S. (2018). Complete ammonia oxidation: An important control on nitrification in engineered ecosystems? *Current Opinion in Biotechnology*, *50*, 158–165. doi:10.1016/j.copbio.2018.01.015 PMID:29414055

Lebre, P. H., De Maayer, P., & Cowan, D. A. (2017). Xerotolerant bacteria: Surviving through a dry spell. *Nature Reviews. Microbiology, 15*(5), 285–296. doi:10.1038/nrmicro.2017.16 PMID:28316329

Li, X., & Yu, H. Y. (2013). Halostable cellulase with organic solvent tolerance from *Haloarcula* sp. LLSG7 and its application in bioethanol fermentation using agricultural wastes. *Journal of Industrial Microbiology & Biotechnology, 40*(12), 1357–1365. doi:10.100710295-013-1340-0 PMID:24037323

Littlechild, J. A. (2015). Enzymes from extreme environments and their industrial applications. *Frontiers in Bioengineering and Biotechnology, 3*, 161. doi:10.3389/fbioe.2015.00161 PMID:26528475

Ma, K., & Adams, M. W. (1999). An unusual oxygen-sensitive, iron-and zinc-containing alcohol dehydrogenase from the hyperthermophilic archaeon *Pyrococcus furiosus. Journal of Bacteriology, 181*(4), 1163–1170. doi:10.1128/JB.181.4.1163-1170.1999 PMID:9973342

Martinez-Espinosa, R. M., Richardson, D. J., & Bonete, M. J. (2015). Characterisation of chlorate reduction in the haloarchaeon *Haloferax mediterranei. Biochimica et Biophysica Acta (BBA)- General Subjects, 1850*(4), 587–594. doi:10.1016/j.bbagen.2014.12.011 PMID:25512066

Medrano-Santillana, M., Souza-Brito, E. M., Duran, R., Gutierrez-Corona, F., & Reyna-Lopez, G. E. (2017). Bacterial diversity in fumarole environments of the Paricutín volcano, Michoacán (Mexico). *Extremophiles, 21*(3), 499–511. doi:10.100700792-017-0920-8 PMID:28289847

Narsing Rao, M. P., Dong, Z. Y., Luo, Z. H., Li, M. M., Liu, B. B., Guo, S. X., Hozzein, W. N., Xiao, M., & Li, W. J. (2021). Physicochemical and Microbial Diversity Analyses of Indian Hot Springs. *Frontiers in Microbiology, 12*, 368. doi:10.3389/fmicb.2021.627200 PMID:33763045

Nowlan, B., Dodia, M. S., Singh, S. P., & Patel, B. K. C. (2006). Bacillus okhensis sp. nov., a halotolerant and alkalitolerant bacterium from an Indian saltpan. *International Journal of Systematic and Evolutionary Microbiology, 56*(5), 1073–1077. doi:10.1099/ijs.0.63861-0 PMID:16627657

Oger, P., & Cario, A. (2014). The high pressure life of piezophiles. *Biologie Aujourd'hui, 208*(3), 193–206. doi:10.1051/jbio/2014023 PMID:25474000

Okubo, Y., Yokoigawa, K., Esaki, N., Soda, K., & Kawai, H. (1999). Characterization of Psychrophilic Alanine Racemase from *Bacillus psychrosaccharolyticus. Biochemical and Biophysical Research Communications, 256*(2), 333–340. doi:10.1006/bbrc.1999.0324 PMID:10080917

Oren, A. (2006). Life at high salt concentrations. *The Prokaryotes, 3*, 263-282.

Pandey, S., Rakholiya, K. D., Raval, V. H., & Singh, S. P. (2012). Catalysis and stability of an alkaline protease from a haloalkaliphilic bacterium under non-aqueous conditions as a function of pH, salt and temperature. *Journal of Bioscience and Bioengineering, 114*(3), 251–256. doi:10.1016/j.jbiosc.2012.03.003 PMID:22727739

Patel, N. Y., Baria, D. M., Yagnik, S. M., Rajput, K. N., Panchal, R. R., & Raval, V. H. (2021). Bioprospecting the future in perspective of amidohydrolase L-glutaminase from marine habitats. *Applied Microbiology and Biotechnology, 105*(13), 5325–5340. Advance online publication. doi:10.100700253-021-11416-6 PMID:34236482

Pikuta, E. V., Hoover, R. B., & Tang, J. (2007). Microbial extremophiles at the limits of life. *Critical Reviews in Microbiology*, *33*(3), 183–209. doi:10.1080/10408410701451948 PMID:17653987

Piubeli, F., de Lourdes Moreno, M., Kishi, L. T., Henrique-Silva, F., García, M. T., & Mellado, E. (2015). Phylogenetic profiling and diversity of bacterial communities in the Death Valley, an extreme habitat in the Atacama Desert. *Indian Journal of Microbiology*, *55*(4), 392–399. doi:10.100712088-015-0539-3 PMID:26543264

Poli, A., Finore, I., Romano, I., Gioiello, A., Lama, L., & Nicolaus, B. (2017). Microbial diversity in extreme marine habitats and their biomolecules. *Microorganisms*, *5*(2), 25. doi:10.3390/microorganisms5020025 PMID:28509857

Prasad, A. A., & Rao, K. B. (2013). Aerobic biodegradation of Azo dye by *Bacillus cohnii* MTCC 3616; an obligately alkaliphilic bacterium and toxicity evaluation of metabolites by different bioassay systems. *Applied Microbiology and Biotechnology*, *97*(16), 7469–7481. doi:10.100700253-012-4492-3 PMID:23070653

Purohit, M. K., Raval, V. H., & Singh, S. P. (2014). Haloalkaliphilic bacteria: molecular diversity and biotechnological applications. In Geomicrobiology and Biogeochemistry (pp. 61-79). Springer.

Raddadi, N., Cherif, A., Daffonchio, D., Neifar, M., & Fava, F. (2015). Biotechnological applications of extremophiles, extremozymes and extremolytes. *Applied Microbiology and Biotechnology*, *99*(19), 7907–7913. doi:10.100700253-015-6874-9 PMID:26272092

Raval, V. H., Bhatt, H. B., & Singh, S. P. (2018). Adaptation strategies in halophilic bacteria. In *Extremophiles* (pp. 137–164). CRC Press. doi:10.1201/9781315154695-7

Raval, V. H., Pillai, S., Rawal, C. M., & Singh, S. P. (2014). Biochemical and structural characterization of a detergent-stable serine alkaline protease from seawater haloalkaliphilic bacteria. *Process Biochemistry*, *49*(6), 955–962. doi:10.1016/j.procbio.2014.03.014

Raval, V. H., Purohit, M. K., & Singh, S. P. (2013). Diversity, population dynamics and biocatalytic potential of cultivable and non-cultivable bacterial communities of the saline ecosystems. In *Marine Enzymes for Biocatalysis* (pp. 165–189). Woodhead Publishing. doi:10.1533/9781908818355.2.165

Raval, V. H., Purohit, M. K., & Singh, S. P. (2015). Extracellular proteases from halophilic and haloalkaliphilic bacteria: Occurrence and biochemical properties. In *Halophiles* (pp. 421–449). Springer. doi:10.1007/978-3-319-14595-2_16

Ray, A. (2012). Application of lipase in industry. *Asian. The Journal of Pharmacy Technology*, *2*(2), 33–37.

Ricard, M., & Reid, I. D. (2004). Purified pectinase lowers cationic demand in peroxide-bleached mechanical pulp. *Enzyme and Microbial Technology*, *34*(5), 499–504. doi:10.1016/j.enzmictec.2003.12.005

Sharma, A., Kawarabayasi, Y., & Satyanarayana, T. (2012). Acidophilic bacteria and archaea: Acid stable biocatalysts and their potential applications. *Extremophiles*, *16*(1), 1–19. doi:10.100700792-011-0402-3 PMID:22080280

Shukla, R. J., & Singh, S. P. (2015). Production optimization, purification and characterization of α-amylase from thermophilic Bacillus licheniformis TSI-14. *Starch*, *67*(7-8), 629–639. doi:10.1002tar.201500046

Singh, O. V., & Gabani, P. (2011). Extremophiles: Radiation resistance microbial reserves and therapeutic implications. *Journal of Applied Microbiology*, *110*(4), 851–861. doi:10.1111/j.1365-2672.2011.04971.x PMID:21332616

Singh, S. P., Thumar, J. T., Gohel, S. D., Kikani, B., Shukla, R., Sharma, A., & Dangar, K. (2013). Actinomycetes from marine habitats and their enzymatic potential. In *Marine Enzymes for Biocatalysis* (pp. 191–214). Woodhead Publishing. doi:10.1533/9781908818355.2.191

Sogin, M. L., Morrison, H. G., Huber, J. A., Welch, D. M., Huse, S. M., Neal, P. R., Arrieta, J. M., & Herndl, G. J. (2006). Microbial diversity in the deep sea and the underexplored "rare biosphere". *Proceedings of the National Academy of Sciences of the United States of America*, *103*(32), 12115–12120. doi:10.1073/pnas.0605127103 PMID:16880384

Sukumaran, R. K., Singhania, R. R., Mathew, G. M., & Pandey, A. (2009). Cellulase production using biomass feed stock and its application in lignocellulose saccharification for bio-ethanol production. *Renewable Energy*, *34*(2), 421–424. doi:10.1016/j.renene.2008.05.008

Sullivan, W. T. III, & Baross, J. (Eds.). (2007). *Planets and life: the emerging science of astrobiology*. Cambridge University Press. doi:10.1017/CBO9780511812958

Sun, M. M., Tolliday, N., Vetriani, C., Robb, F. T., & Clark, D. S. (1999). Pressure-induced thermostabilization of glutamate dehydrogenase from the hyperthermophile *Pyrococcus furiosus*. *Protein Science*, *8*(5), 1056–1063. doi:10.1110/ps.8.5.1056 PMID:10338016

Sysoev, M., Grötzinger, S. W., Renn, D., Eppinger, J., Rueping, M., & Karan, R. (2021). Bioprospecting of Novel Extremozymes From Prokaryotes—The Advent of Culture-Independent Methods. *Frontiers in Microbiology*, 12. PMID:33643258

Thomas, L., Joseph, A., Singhania, R. R., Patel, A. K., & Pandey, A. (2017). Industrial enzymes: xylanases. In *Current developments in biotechnology and bioengineering* (pp. 127–148). Elsevier. doi:10.1016/B978-0-444-63662-1.00006-3

Ushasree, M. V., Vidya, J., & Pandey, A. I. (2017). Other enzymes: phytases. In *Current developments in biotechnology and bioengineering* (pp. 309–333). Elsevier. doi:10.1016/B978-0-444-63662-1.00014-2

Van De Voorde, I., Goiris, K., Syryn, E., Van den Bussche, C., & Aerts, G. (2014). Evaluation of the cold-active *Pseudoalteromonas haloplanktis* β-galactosidase enzyme for lactose hydrolysis in whey permeate as primary step of d-tagatose production. *Process Biochemistry*, *49*(12), 2134–2140. doi:10.1016/j.procbio.2014.09.010

Vashist, S., & Sharma, R. (2018). Why Settle for Mediocre, When Extremophiles Exist? In *Extremophiles in Eurasian Ecosystems: Ecology, Diversity, and Applications* (pp. 435–451). Springer. doi:10.1007/978-981-13-0329-6_16

Yadav, A. N., Verma, P., Kumar, S., Kumar, V., Kumar, M., Sugitha, T. C. K., Singh, B. P., Saxena, A. K., & Dhaliwal, H. S. (2018). Actinobacteria from rhizosphere: molecular diversity, distributions, and potential biotechnological applications. In *New and future developments in microbial biotechnology and bioengineering* (pp. 13–41). Elsevier. doi:10.1016/B978-0-444-63994-3.00002-3

Yuan, M., Zhang, W., Dai, S., Wu, J., Wang, Y., Tao, T., Chen, M., & Lin, M. (2009). *Deinococcus gobiensis* sp. nov., an extremely radiation-resistant bacterium. *International Journal of Systematic and Evolutionary Microbiology, 59*(6), 1513–1517. doi:10.1099/ijs.0.004523-0 PMID:19502345

Zajc, J., Zalar, P., Plemenitaš, A., & Gunde-Cimerman, N. (2012). The mycobiota of the salterns. In *Biology of Marine Fungi* (pp. 133–158). Springer. doi:10.1007/978-3-642-23342-5_7

Zhang, J. W., & Zeng, R. Y. (2008). Purification and characterization of a cold-adapted α-amylase produced by *Nocardiopsis* sp. 7326 isolated from Prydz Bay, Antarctic. *Marine Biotechnology (New York, N.Y.), 10*(1), 75–82. doi:10.100710126-007-9035-z PMID:17934774

Zheng, B., Jiang, X., Xu, Z., Fang, Y., & Li, L. (2016). Characterization of a novel metallo-β-lactamases fold hydrolase from *Pelagibacterium halotolerans*, a marine halotolerant bacterium isolated from East China Sea. *Extremophiles, 20*(1), 37–44. doi:10.100700792-015-0795-5 PMID:26560315

Zhu, D., Adebisi, W. A., Ahmad, F., Sethupathy, S., Danso, B., & Sun, J. (2020). Recent development of extremophilic bacteria and their application in biorefinery. *Frontiers in Bioengineering and Biotechnology, 8*, 483. doi:10.3389/fbioe.2020.00483 PMID:32596215

Chapter 11
Application of Extremophiles in Sustainable Agriculture

Julius Eyiuche Nweze
Biology Centre, Czech Academy of Sciences, Ceske Budejovice, Czech Republic

Justus Amuche Nweze
https://orcid.org/0000-0003-3481-040X
Biology Centre, Czech Academy of Sciences, Ceske Budejovice, Czech Republic

Shruti Gupta
https://orcid.org/0000-0001-9974-9449
Biology Centre, Czech Academy of Sciences, Ceske Budejovice, Czech Republic

ABSTRACT

With the increasing demands for foods and other agriculture-based products, sustainable agricultural practices are the cornerstone for improving low-input agricultural production. In contrast to crop production, plant-microorganism interaction (PMI) plays a crucial role. PMI significantly raises productivity as well as maintaining the overall health of the crop. During harsh and extreme physiological conditions, plant-associated extremophilic microbes (PAEM) are known to contribute to crop production, survivability, and fitness. Thus, the application of extremophiles either in the form of biofertilizer or biopesticides is highly beneficial. Extremophiles have been adapted to withstand diverse harsh environmental conditions. They possess unique mechanisms at the molecular level to produce enormous potential extremozymes and bioactive compounds. Consequently, extremophiles represent the foundation of efficient and sustainable agriculture. This chapter introduces the significance and application of plant-associated extremophilic microbes in sustainable agriculture.

DOI: 10.4018/978-1-7998-9144-4.ch011

INTRODUCTION

Unfavorable soil conditions counted as a serious threat to agriculture and its productivity. This condition includes variation in soil pH from alkalinity to acidity, elevated temperature, drought, salinity, and involvement of several chemical and heavy metal contaminants in the soil ecosystem. However, several practices have been implemented in order to overcome these stressed conditions and increase agricultural productivity, like the development of stress-tolerant plants, several chemical components, etc. Recently, extremophiles including fungi (AMF), rhizospheric bacteria, and archaea (PGPR and PGPA) have been reported in plants to coping the environmental stress potentially (Akinola & Babalola, 2020).

An extremophile is a term used to describe organisms that live in environments that could be considered "extreme" due to their exceptional living conditions. These organisms thrive and love 'extreme' conditions such as pH, temperature, salinity, etc (Rampelotto, 2013). Also, some of the organisms can survive in extreme metal or chemical conditions (e.g. reduced content of oxygen). However, the vast majority of extremophiles are unicellular organisms from the bacterial and archaeal domains of life. There are many plant-associated extremophilic organisms under various abiotic and biotic stresses which have been documented including bacteria, archaea, and eukarya domains (Rampelotto, 2013) (Fig. 1). It is obvious that these extreme ecosystems harbor unique biodiversity of organisms with the ability to adapt to a variety of environmental stress. Some adaptive traits have allowed these extremophiles to thrive optimally in one or more extreme environments. Poly-extremophiles, on the other hand, thrive in a variety of environmental stresses (Merino et al., 2019).

These organisms could be implemented asbio-inoculum by agricultural practitioners in order to conquer soil stresses (Akinola & Babalola, 2020) (Fig. 1). Under abiotic/biotic stresses, plant-associated extremophiles were reportedly involved in plant growth and adaptations. Thus, these plant-associated microbes are termed "plant-associated extremophilic microbes" (Yadav, 2017) which simply can be considered as "PAEM". Although, all domains from archaea to eukarya are known to involve in PAEM. Additionally, these practices are effectively working as biofertilizers, and biocontrol agents not ended by soil pollution and known to maintain soil health, fertility (Igiehon et al., 2019) and cost-efficient in sustainable agricultural practices (Yadav, 2017).

EXTREMOPHILES: AN OVERVIEW

Extremophiles are divided into groups based on the type of habitat in which they may thrive and reproduce, as shown in Figure 2. These include temperature extremes tolerant - psychrophiles, thermophiles, and hyperthermophiles; xerophiles (tolerate low water activity), barophiles/piezophiles (thrive under high pressures), halophiles (high salinity), extreme pH lovers - acidophiles and alkaliphiles, and radioresistant organisms (Kaushik et al., 2021; Merino et al., 2019). Some extremophiles, such as thermoalkaliphiles and halophilic alkali thermophiles, are adapted to numerous types of extreme environments (poly-extremophiles), further complicating the classifications. *Sulfolobus acidocaldaricus* is a polyextremophilic archaeon that thrives at pH 3 and 80 degrees Celsius (Rastädter et al., 2021). *Paenibacillus* and *Bacillus* spp., which exist in hot springs (India) with temperatures ranging from twenty to eighty degrees Celsius and pH values of 5-14, are examples of microbes that can withstand a wide range of pH as well as high temperatures. The majority of these pH-tolerant microorganisms are neutrophiles (Hussain et al., 2020).

Figure 1. Overview management of sustainable agriculture by plant-associated extremophiles under environmental stresses

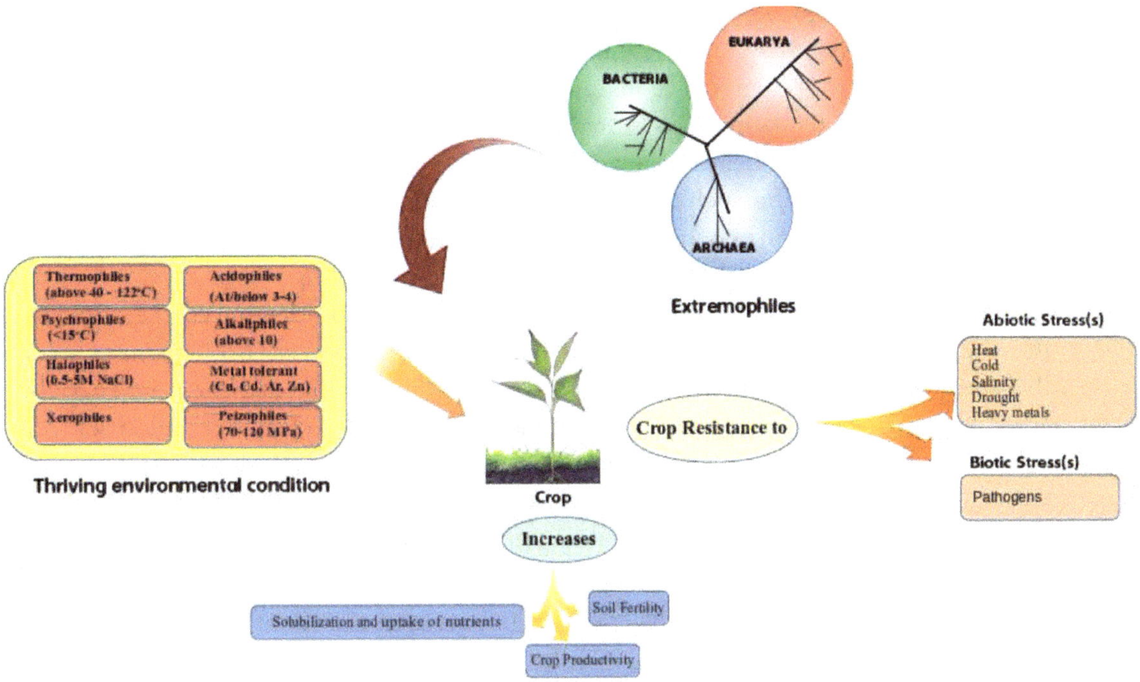

Figure 2. Major classes of extremophiles

Condition	Category	Growth conditions
Temperature	Psychrophile	<15 oC
	Thermophile	50-80 oC
	Hyperthermophile	>80 oC
Water availability	Xerophile	Low water activity
Hydrostatic pressure	Barophile (Piezophile)	>10 MPa
Salinity	Halophile	[NaCl] >0.2 M
pH	Acidophile	pH <3
	Alkaliphile	pH >9
Radiation	Radioresistant	High ionizing radiation
Oxygen tension	Anaerobe	Low/no oxygen
Chemicals	Metal tolerant	High metal concentration

THERMOPHILES/HYPERTHERMOPHILES

Initially, thermophilic anaerobes with chemo-lithoautotrophic/heterotrophic metabolism and capable of being supported by hydrothermal energy sources dominated early life on Earth. Thermophiles and hyperthermophiles are organisms that thrive in temperatures >44°C and >80°C, respectively. They can be found in several natural ecosystems such as hydrothermal vents, volcanic deposits, and hot springs, among other places (compost facilities and anaerobic reactor) (Arras et al., 2019). Their extremozymes are responsible for their survival in such environments. Extreme heat does not cause the amino acids in these enzymes to lose their form and misfold, allowing them to continue to function properly (Kohli et al., 2020).

The adaptation of their thermostable proteins through amino acid modifications in their fundamental structure, thereby thermally enhancing their stabilities, is a common process used by these organisms to protect their cellular structural components at high temperatures (Zeldes et al., 2015). Thermophilic proteins have a higher proportion of alpha-helical amino acid residues and shorter amino acid lengths. Upregulation of the glycolysis pathway (which provides immediate energy), heat shock proteins (HSPs) (Chaperone-assisted protein folding), DNA damage repair mechanisms, cell component stabilization with compatible solutes, andmembrane stabilization with polyamines and branched-chain fatty acids, are all important mechanisms use by thermophiles to withstand high temperature (Zeldes et al., 2015; Pedone et al., 2020).

Psychrophiles

Psychrophiles (also known as Cryophiles) are organisms that survive in temperatures as low as 5 degrees Fahrenheit. They can be found in permafrost, polar ice, cold soils, alpine snowpacks, and cold ocean water, and they belong to bacteria, eukarya, and archaea kingdoms (Salwan & Sharma, 2020). Some organisms possess special enzymes (Extremozymes) which are very active at this extreme temperature and that enables them to survive in such extreme environments. The synthesis of low-temperature active proteins has been reported as well as the presence of fatty acids that contain cyclopropane, unsaturated fatty acids, and short-chain-fatty-acids (maintain membrane fluidity) in their plasma membranes, which serve to protect the cells from the cold stress (Coker, 2019). Other survival mechanisms include increased production of chaperones and cold-shock proteins (CSPs) (which protect RNA and protein synthesis), synthesis of antifreeze proteins (AFPs) (which creates thermal hysteresis), and accumulation of mannitol and other cryoprotectants (defend against UV radiation) (Raymond-Bouchard & Whyte, 2017).

Xerophile

Climate change has been linked to an increase in dry conditions in many locations due to poor rainfall, high temperatures, and drought. Xerophiles thrive in excessively dry environments that might be exceedingly cold or hot, such as the Great Basin, the Antarctic, and the Atacama Desert, among other places. Some xerophiles, like their psychrophilic counterparts, can use trehalose in place of water, which protects membranes and other structures during dry periods (Coker, 2019).

In arid habitats, xerophiles use evasion of environmental stress and adaptive processes to survive. The development of spores during the evasion of a dry environment means that cells are converted to a non-replicative viability state. Adaptive mechanisms are possible through the production of extracellular polymeric substances (EPS), the buildup of osmoprotectants (trehalose, glycine betaine, L-glutamate), the synthesis of DNA-repair proteins to prevent water loss and increase water retention, and the cell membrane modifications to prevent intracellular loss of water (Koul et al., 2021; Coker, 2019).

Barophile (Piezophile)

Barophiles are organisms that thrive at pressures of 400 atmospheres or more. One of the examples of a high-pressure environment is the deep-sea habitat, such as dee lakes and ocean floors (pressure can exceed 380 atm). Subsurface rocks with significant lithostatic pressures are another example. Obligate barophiles are barophiles that cannot thrive outside of their high-pressure environments (Salwan & Sharma, 2020). Barotolerants are those who can live under high pressures and in less harsh environments. An obligate barophile is a Gram-negative proteobacterium called *Halomonas salaria*, which needs a pressure of 1000 atm. They are able to survive by maintaining their membrane's phospholipids fluidity. That is why the pressure gradient between the interior and exterior of the cell, as well as the external environment, is compensated by this fluidity (Shukla et al., 2020; Salwan & Sharma, 2020).

Halophile

Microorganisms from all three domains of life can be found in environments with varying salt concentrations, such as oceans, saline lakes, coastal, salars, and polar ice areas.Halophiles are organisms that must have salt to survive. At salinities greater than 1.5M, prokaryotic bacteria predominate. Halophiles are classified as extreme halophiles, borderline extreme halophiles, moderate halophiles or minor halophiles based on the salinity of >4.0–5.9 M, >2.5–4.0 M, 0.5–2.5 M or 0.2 M, respectively (Zhang et al., 2018).

Minimizing cellular water loss using their unique metabolic properties is the initial step in overcoming the problems of hypersaline environmental conditions. To accomplish this, halophiles accumulate cytoplasmic solutes through several processes. So, to expel Na+ and consume K^+, halophilic archaea use an ATP-powered Na+/K+ pump or Na+/K+-ATPase. Halotolerant bacteria use glycerol, betaine and ectoine as a suitable solute to maintain osmotic pressure equilibrium (Zhang et al., 2018; Shrestha et al., 2018).

Acidophiles

Acidophiles, also known as acidophilic organisms, are organisms that survive in highly acidic environments. These species can be found in Archaea, Bacteria, and Eukarya, among other branches of the tree of life.Extreme acidophilic microorganisms thrive at pH 3 or below, and moderate acidophiles thrive at pH 3 to 5. While microorganisms that are acid-tolerant have an optimal pH of > 5 but are nevertheless active in low pH conditions (Sharma et al., 2016).

A near-neutral cytoplasmic pH is frequently maintained in acidophiles to protect acid-labile cellular components, which necessitates the creation of a significant pH gradient. There are three reported ways/strategies by which they adapt to an acidic environment, which includes the use of proton flux pumping systems to maintain the pH, improved DNA and protein repair systems, and the cell membranepermeability reduction to prevent the protons entry into the cytoplasm (Sharma et al., 2016; Ando et al., 2021).

Alkaliphile

Alkaliphiles are organisms that can survive and thrive in an alkaline environment with a pH of 8.5 to 11. These can be found in alkaline environments such as carbonate-rich soils, the Yellowstone National Park's Octopus Spring, Soda Lake in California's Carrizo Plain National Monument, as well as in California's Eastern Sierra's Mono Lake. Obligate alkaliphiles (optimal growth pH range of 10.0 - 12.0), facultative alkaliphiles (pH range of 7.0–9.5), and haloalkaliphiles are the three main physiological types of alkaliphilic organisms based on their alkalinity need for growth and survival (dual extremities of alkaline pH and high salt) (Mamo & Mattiasson, 2016).

Alkaliphiles adapt by employing electrogenic secondary cation/proton antiporters to maintain cytoplasmic pH homeostasis and H+ absorption. To avoid environmental alkalinity, another option is to have a protective layer of acidic chemicals outside the cell made up of teichuronic acid, teichuronopeptide, and acidic amino acids. Others have evolved pH-stable enzymes such as cytochrome C on the outer membrane, which regulates proton transfer across the membrane, and phosphoserine aminotransferase, which improves hydrophobic contacts and negatively charged amino acids at the interface to improve alkaline stability (Mamo & Mattiasson, 2016; Kulkarni et al., 2019).

Metallophyte

A metallophyte is a plant that can resist high levels of heavy metals, such as lead. Plants are classed as "obligate metallophytes" (those that can only survive in the presence of specific metals) and "facultative metallophytes" (those that can only survive in the presence of a specific metal). Due to phytotoxicity, metalliferous soils provide relatively restricted habitats for plants, resulting in significant selection pressures (Salwan & Sharma, 2020). Heavy-metal plant communities are made up of genetically changed ecotypes that have developed unique tolerances to heavy metals like copper, lead, nickel, cadmium, zinc, and arsenic through microevolutionary processes. The bioavailable percentage of metal (loids) in the soil, as well as the kind of mineralization, define metal tolerance. Expectedly, at extremely high soil metal concentrations, even metal-tolerant genotypes are incapable of evolving excessive tolerances to multiple heavy metals at the same time, especially in polymetallic soils (lan R. Alford et al., 2010; Midhat et al., 2019).

POTENTIAL APPLICATION OF PAEM

Biocontrol Potential

Extremophilic organisms are characterized by the conditions under which they can live and develop while maintaining a complex and dynamic metabolic pathway (Horikoshi, 2007). Through direct and indirect mechanisms, some of these microorganisms may live within the plant rhizosphere, causing serious damage to their hosts or protecting them from pathogens (Torracchi et al., 2020). Their survival depends on the ability to generates a rich source of biomolecules with diverse biological activities that enable them to thrive in extreme environments (Anwar et al., 2020). The protective microorganisms themselves and some of their biomoleculescan provide the large biotechnological potential for biocontrol agents in agriculture (HÃggblom & Margesin, 2005) (Fig. 2). Because of the excretion of a variety of secondary metabolites (antibiotics, quorum-sensing interfering agents, and siderophores), biocontrol agents either inhibit or restrict the growth of plant pathogens.Plant pathogens may also be controlled indirectly, such as by competition for space and nutrients, or by impeding their chemical communication (Köhl et al., 2019). Biopesticides made from mesophilic microorganisms are currently a well-established technology. Regrettably, the efficacy of commercially available biopesticides is limited in many parts of the world due to environmental factors. These biocontrol agents are not active at low orhigh temperatures. In high mountains such as the Himalayas or the Andes, for example, the adaptation and survival of biological agents may be a major limiting factor (Pandey & Yarzábal, 2018). That is why researchers are increasingly interested in studying the microbial diversity of permanently cold environments, with the intention of isolating psychrophiles with biocontrol abilities. Though still in their infancy, the cold-adapted microorganisms (psychrophiles) have been demonstrated to have high biocontrol capability (Fig. 2). For instance,*Pseudomonas* is one of the most researched biocontrol psychrophiles (Moreno & Rojo, 2014). It makes antimicrobial substances (organic acids, hydrogen cyanide, biosurfactants, enzymes, and iron-chelating compounds) (Tapia-Vázquez et al., 2020; Kaur et al., 2006), boosts plant systemic resistance to pathogens (Ogata-Gutiérrez et al., 2016), and reduces cold temperatures and other environmental stresses from impacting on plants (Ortiz-Ojeda et al., 2017). Others are involved in hyperparasitism, such as a psychrotolerant strain of *Trichoderma atroviride*. These stains are capable of inhibiting the growth of pathogens that can cause serious systemic damages to the terrestrial organs and aerial of part plants in cold regions by producing highly active chitinases (McBeath, 2002; Shanmugam et al., 2014). For low-temperature habitats, psychrotrophic microorganisms could be useful in agriculture as bio-inoculants and biocontrol agents (Yadav, 2017).

The biocontrol ability of thermophilic bacteria and fungi also have been reported, especially those producing chitinases (Figure 4). These microbial chitinases are known to hydrolyze the cell walls of many fungi and the chitinous gut epithelium of insects. Chitinases from *Thermomyces lanuginosus* (thermophilic fungus) have been demonstrated to cause high mortality in second instar larvae of the African sugarcane borer (*Eldana saccharina*), other fungi, and stalk-boring lepidopterans (Okongo et al., 2019). The thermophilic mould*Myceliopthora thermophila* produces thermostable, acid-tolerant, and organic solvent stable chitinases that inhibit the growth of phytopathogenic fungi like *Curvularia lunata* and *Fusarium oxysporum*. Thus, these microorganisms could be applied as biofungicide in the tropics during summer (Dua et al., 2016). *Bacillus licheniformis* and *Bacillus stearothermophilus*, which display high antagonism against various soil-borne fungal plant diseases, are two other potential thermophilic biocontrol candidates, especially in stressed rhizosphere conditions. They have a strong tolerance to high temperature, pH and salt concentrations, and antibiotic resistance properties (Rajashree & S., 2018). Another new bacterial strain XT1 (*Bacillus methylotrophicus*) isolated from an extreme halophilic environment has shown very good properties as a plant biopesticide. It also proved effective against some plant pathogens such as necrotrophic plant pathogen *Botrytis cinerea* and verticillium wilt on olive trees (Toral et al., 2018; Castro et al., 2020).

In greenhouse conditions for instance, Lactobacillus plantarum TC92 and PM411 have been reported to be effective in controlling multiple pathogens, such as Xanthomonas arboricola pv. pruni, Pseudomonas syringae pv. actinidiae, and Xanthomonas fragariae, that affect Prunus, kiwifruit, and strawberry respectively. The pH lowering action and the generation of lactic acid were the inhibitory mechanisms of these extremophiles (Daranas et al., 2018). In the field conditions, the two major rice diseases, bacterial panicle blight and sheath blight, caused by Burkholderia glumae and Rhizoctonia solani respectively, were significantly suppressed by the antagonistic activities of some plant associated strains of Bacillius sp (Shrestha et al., 2016). Another two-year field experiments showed the efficacy of Trichoderma viride, Pseudomonas fluorescens, and Glomus fasciculatum in controlling the spiral nematode, Helicotylenchus multicinctus, that infest banana (Jayakumar & Seenivasan, 2019; Saravanan et al., 2020; Jonathan et al., 2004).

Figure 3. Potential applications of extremophiles in sustainable agriculture

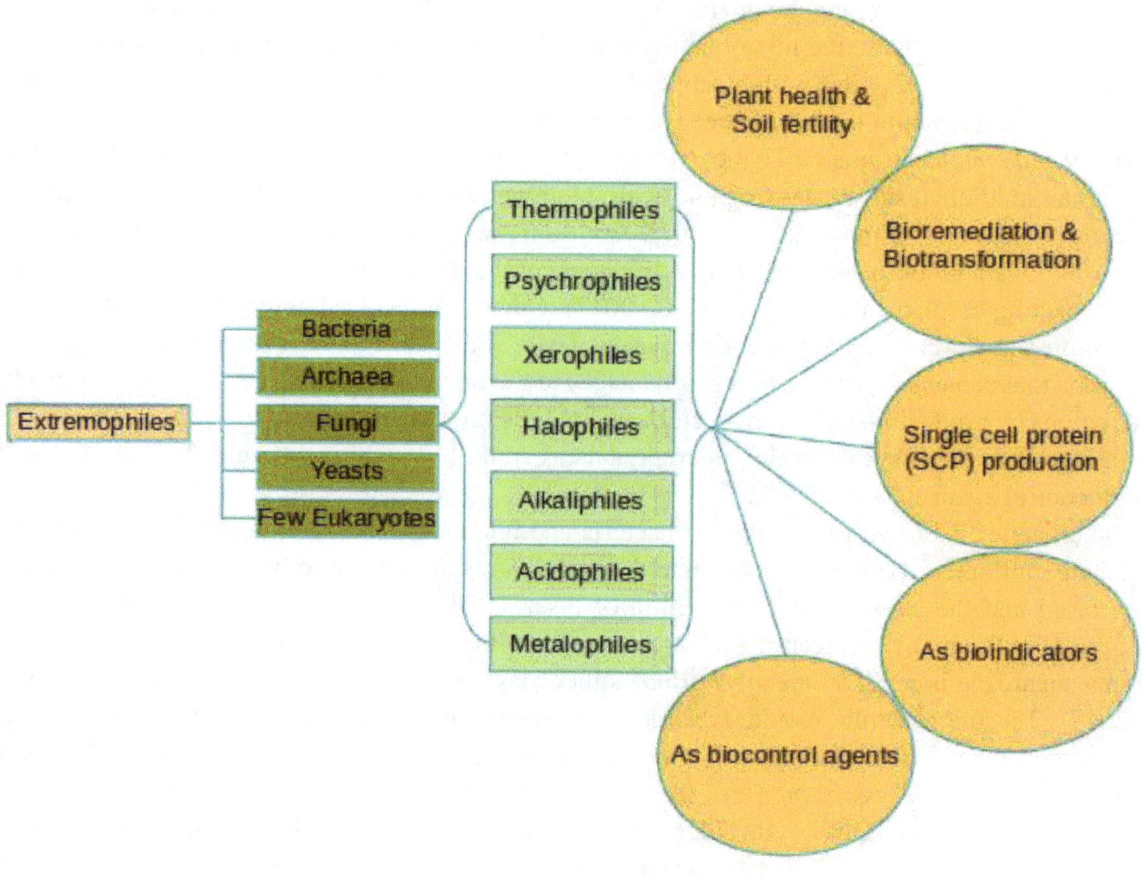

As Bioindicators

Microbial communities are vital for the soil productivity and health and have a lot of promise as environmental bioindicators (Lixia, 2007). Strong correlations exist between their relative abundances and soil pH, salinity, or temperature. These connections between some soil conditions and different soil microbial groups not only show their ecological aspects, but also how specific bacterial taxa might reflect the effect of some anthropogenic stresses. For instance, the differential abundances of the family Pirellulaceae provide strong evidence as biological indicators of the soil pH (Hermans et al., 2017). It's been hypothesized that critical microbial communities connected with pH changes could be exploited as bioindicators for either future plant growth or acid generation conditions that prevent it (Hottenstein et al., 2019).

Salinization is a known danger to soil fertility all over the world, and community salt tolerant microorganisms have been closely correlated with soil salinity. These specific bacterial taxa have higher relative abundances in communities with high salt tolerance, suggesting that they may be employed as bioindicators of soil salinity (Rath et al., 2018) (Figure 4). They could be used as bioindicators to detect salinity problems in the soil ecosystem, the level of salinity before using low-quality aquatic water for irrigating agricultural crops, especially in the arid and semi-arid areas or where salinity is variable due to flooding (Zaghloul et al., 2020). For example, the differential abundance of *Haliangium, Altererythrobacter* or *Pirellula* could be explored as potential bioindicators soil salinity (Ezeokoli et al., 2020).

Enzymes from extremophiles can serve as good bioindicators in soils affected by heavy metals, stress conditions, differentxenobiotic substances, and management practices (Rao et al., 2014). Because enzymes respond quickly to environmental changes, assessing soil enzymes is one of the most sensitive and reliable ways to measure soil health (Utobo & Tewari, 2015). Epiphytic microorganisms like *Pseudomonas, Agrobacterium, Pantoea*, and *Methylobacterium* are the most adaptable in nature because they can withstand strong UV radiation and extreme temperatures (40 °C–55 °C). They can be found in the phyllosphere of a variety of crops that are grown in both normal and hard environments (Verma et al., 2019; Meena et al., 2011). Just like epiphytic bacteria, lichen are also sensitive monitoring organisms that can detect temporal climate variation (Stapper & John, 2015).

Based on field investigation, the effects of long-term mining activities on the agricultural soil quality in the Chinese Yunnan Province showed that microbial oxidoreductases activities indicate the degree of heavy metal pollution, and its accuracy could be supplemented with hydrolases (for C, N, P and S recycling) (Yang et al., 2016).

PLANT HEALTH AND SOIL FERTILITY

Plant-associated microorganisms have been found in plants growing under a number of abiotic stresses. Some of these microorganisms are extremophilic bacteria with a wide range of plant-growth-promoting properties. Some major plant symbionts are nitrogen-fixing bacteria, plant growth-promoting bacteria, and mycorrhizal fungi. They help plants adapt to extreme pH, salinity, temperature, and drought stress (Alsharif et al., 2020). For instance, rhizobacteria can promote plant growth through mobilization of soil nutrients (such as potassium and phosphate solubilization, iron sequestration, and nitrogen fixation), regulation of phytohormones and growth regulators (1-aminocyclopropane-1-carboxylic acid deaminase and indole-3-acetic acid) (Bokhari et al., 2019). These microorganisms, especially thermophiles, may be applied as biofertilizers for crops improvements and soil health for sustainable agriculture in arid ecosystems. (Yadav, 2017). The halophilic bacterial strain,*Bacillus methylotrophicus,* produces an organic solution to heal plants and help them grow (biostimulant).This strain has the capacity to fix nitrogen, solubilize organic and inorganic phosphate, produce siderophore, enzymes and volatile metabolites (Torres et al., 2020).Therefore, evaluating the physiological,biochemical and molecular characteristics of plants/microorganisms inhabiting naturally under adverse circumstances could improve crop plants for the adverse conditions and might be a promising solution for the future of agriculture (Dikilitas et al., 2021).

Extremophiles have the potential to improve soil fertility, water retention, and plant protection, as well as boost pollinator health (such as bees) and overall agricultural conditions. Psychrophilic and psychrotolerant microorganisms, for example, play a significant part in the polar ecosystem's biomass production, nutrient turnover, and litter breakdown processes (HÃggblom & Margesin, 2005). These bacteria are important in agriculture, especially in temperate climates where short growing seasons and freezing temperatures mark the agroecosystems, putting both plant and microbial life under stress from the cold (Mishra et al., 2010).At both the individual and cellular level, a group of Antarctic halophiles can effectively reduce the physiological effect of saline stress on salt-susceptible crops like lettuce (Acuña-Rodríguez et al., 2019). Other Antarctic rhizospheric bacteria such as *Colobanthus quitensis* and *Arthrobacter sp.* are important in the survival and physiology of plants under salt stress (Gallardo-Cerda et al., 2018; Torres-Díaz et al., 2016).

Figure 4. Functional and potential applications of extremophiles in sustainable agriculture

Microorganisms	Class of extremophile	Functions	Potential application	References
Pseudomonas	Psychrophiles	Produces antimicrobial substances for boosting plant systemic resistance to pathogens	Biocontrols	Moreno et al. (2014)
Trichoderma atroviride	Psychrotolerant	Capable of inhibiting the growth of pathogens causing severe damages to the aerial and terrestrial organs of plants	Biocontrols	Shanmugam et al. (2014)
Thermomyces lanuginosus	Thermophiles	Produces chitinases for hydrolysing the chitinous gut epithelium of insects and cell walls of many fungi that attack plants	Biocontrols	Okongo et al. (2019)
Myceliopthora thermophila	Thermophiles	Produces an acid tolerant, thermostable and organic solvent stable chitinases that inhibits the growth of phytopathogenic fungi	Biocontrols	Dua et al. (2016)
Bacillus licheniformis and *Bacillus stearothermophilus*	Thermophiles	Have high antagonism against various soil-borne fungal plant diseases, especially in stressed rhizosphere conditions	Biocontrols	Rajashree et al. (2018)
Bacillus methylotrophicus	Halophiles	Effective against some necrotrophic plant pathogen *Botrytis cinerea* and verticillium wilt on olive trees Produces organic solution to heal plants and help them grow	Biocontrols Biostimulants	Toral et al. (2018; Castro et al. (2020) Torres et al. (2020)
Pirellulaceae	Acidophiles	Relative abundance provides a strong evidence of the soil pH	Bioindicators	Hermans et al. (2017)
Haliangium, Altererythrobacter and *Pirellula*	Halophiles	Relative abundance provides a strong evidence of the soil salinity	Bioindicators	Ezeokoli et al. (2020)
Pseudomonas, Agrobacterium, Pantoea, Methylobacterium, and lichens	Thermophiles	Sensitive for detecting temporal climatic variation	Bioindicators	Verma et al. (2019); Stapper et al. (2015)
Colobanthus quitensis and *Arthrobacter*	Halophilse	Important for plant physiology and survival under controlled conditions of salt stress	Plant health	Torres-Díaz et al. (2016)
Mycobacterium, Gordona, Rhodococcus, and Nocardia	Acidopiles; thermophiles	For bioremediation of soil contaminated with high molecular weight polycyclic aromatic hydrocarbon	Bioremediation	Santos et al. (2007)
Geobacilli	Thermophiles	Produce biosurfactants that act as solubilizing agents	Bioremediation	Perfumo et al. (2006)

Under greenhouse conditions, *Trichoderma* strains have shown the capability to solubilize and increase phosphate uptake (up to 141%) and promote soybean plant growth from 2.1% to 41.1% (*Bononi* et al., 2020). A field study of plant growth-promoting activities of actinobacteria in Morocco has shown that they have high levels of phosphorus and potassium solubilization (8.56 - 12.39 mg/mL) and could grow under nitrogen-free conditions. They have a high level of indole-3-acetic acid (6.70 to 75.54 µg/mL) and siderophores (138.92 µg/mL) producing abilities (*N*afi*s* et al., 2019).

BIOREMEDIATION OF CONTAMINATED AGRICULTURAL SOIL

Agricultural soil contaminated with pollutants is a global concern, and their removal requires biological means that are eco-friendly, potentially cost-effective, and sustainable. Such pollutants include various organic, inorganic, recalcitrant chemicals, toxic heavy metals, xenobiotics, fertilizers, polyaromatic hydrocarbons, agrochemicals, dyes, and pesticides (Khatoon et al., 2021). Due to their adaptation to extreme environments, unique defense mechanisms, and possession of robust enzymatic and biocatalytic systems, extremophiles can be used for the bioremediation of these toxic compounds (Shukla & Singh, 2020). For instance, extremophiles such as *Mycobacterium*, *Gordona*, *Rhodococcus*, and *Nocardia*, have a great potential in bioremediation of soil environment contaminated with high molecular weight polycyclic aromatic hydrocarbon (Kästner et al., 1994; Santos et al., 2007). The degradation of hydrocarbon-contaminated soil is most effective at high temperatures (around 60°C), as demonstrated by thermophilic bacteria, Geobacilli, which can produce biosurfactants that act as solubilizing agents (Perfumo et al., 2006). For bioremediation of heavy metal contaminated soil, thermophilic microorganisms have distinctive cell wall structures, metabolic and enzymatic properties that could facilitate the interactions and degradation of heavy metals (Mir et al., 2021). Soil contaminated with engine oil/diesel, kerosene, or fuel and amended with organic (cow dung and poultry litter) or inorganic (NPK and urea) fertilizer can be remediated and restore its natural characteristics by a microbial consortium containing *E.coli, Proteus, Klebsiella,*and *Pseudomonas sp.* (Nduka et al., 2012). Bioaugmentation with various extremophilic microorganisms such as *Geobacillus thermoparaffinivorans* IR2, *Bacillus licheniformis*, and *Geobacillus stearothermophilus* IR4 improves the decontamination of long alkyl substances (Elumalai et al., 2017). Psychrotrophiles and halophiles have also demonstrated superior performance in the remediation of organic hydrocarbon contaminants (Lin et al., 2009).

AN ADDITIONAL MODE OF AGRICULTURE

Protein from traditional sources is insufficient to meet the current and future human consumption needs. Single-cell proteins from petroleum and cellulosic waste fermentation are possible sources of extra protein (Ritala et al., 2017). Microalgae can provide an alternate form of food production for humans and animals on land with little or no agricultural values, avoiding competing with traditional agricultural resources (Navarro et al., 2016). As a result, extremophilic microalgae have the potential to play a key role in the commercialization of microalgae-based bioproducts and agriculture in the future (Varshney et al., 2015). For the production of SCP fromcellulosic wastes, thermophilic actinomyces appear to be the most effective organisms.Other extremophilic microorganisms that can be used in this regard are *Lactobacillus acidophilus*, *Rhodobacter capsulatus*, *Saccharomyces cerevisiae*, *Spirulina*, *Chlorella*, *Saccharomyces boulardii*, *Polysporous*, and *Trichoderma* (Kwatra et al., 2021).

CONCLUSION AND FUTURE DIRECTION

Environmental stresses are known to create unfavorable soil ecosystems for agriculture. Under stressed conditions, microorganisms from the extreme environment are significantly important to improve the health and productivity of the agriculture system. Several technologies have been implemented in this direction including extremophilic microbial bio-inoculum. The bio-inoculum of extremophiles, colonizing the plant's rhizosphere, phyllosphere, and internal tissues. Moreover,plant-associated extremophilic microbes (PAEM) potentially solubilizing the nutrients (bioavailability of minerals), releasing the phytohormones, protecting against the invasion of pathogens (antagonistic behavior), and toxic materials, and are known to provide tolerance against drought, cold, and salinity stress. Thus, PAEM could be remarkably helpful against both abiotic and biotic stresses in plants. Also, enhances crop productivity without compromising environmental health with minimal cost input. However, it is highly recommended to focus more on research and field practices in this area in order to promote eco-friendly with the least cost input sustainable agriculture practices. However, these practices can be achieved by targeting microbial engineering and integrated with biomolecular science. Adopting the extremophiles in sustainable agriculture could help in alleviating environmental stress and enhances crop productivity.

REFERENCES

Acuña-Rodríguez, I. S., Hansen, H., Gallardo-Cerda, J., Atala, C., & Molina-Montenegro, M. A. (2019). Antarctic Extremophiles: Biotechnological Alternative to Crop Productivity in Saline Soils. *Frontiers in Bioengineering and Biotechnology, 7*. doi:10.3389/fbioe.2019.00022

Akinola, S. A., & Babalola, O. O. (2020). The importance of adverse soil microbiomes in the light of omics: Implications for food safety. *Plant Soil and Environment, 66*(9), 421–430. doi:10.17221/118/2020-PSE

Alford, I., Pilon-Smits, E. A. H., & Paschke, M. W. (2010). Metallophytesa view from the rhizosphere. *Plant and Soil, 337*(1-2), 33–50. doi:10.1007/s11104-010-0482-3

Alsharif, W., Saad, M. M., & Hirt, H. (2020). Desert Microbes for Boosting Sustainable Agriculture in Extreme Environments. *Frontiers in Microbiology, 11*. doi:10.3389/fmicb.2020.01666

Ando, N., Barquera, B., Bartlett, D. H., Boyd, E., Burnim, A. A., Byer, A. S., Colman, D., Gillilan, R. E., Gruebele, M., Makhatadze, G., Royer, C. A., Shock, E., Wand, A. J., & Watkins, M. B. (2021). The Molecular Basis for Life in Extreme Environments. *Annual Review of Biophysics, 50*(1), 343–372. doi:10.1146/annurev-biophys-100120-072804 PMID:33637008

Anwar, U. B., Zwar, I. P., & de Souza, A. O. (2020). Biomolecules produced by extremophiles microorganisms and recent discoveries. In *New and Future Developments in Microbial Biotechnology and Bioengineering* (pp. 247–270). Elsevier. doi:10.1016/B978-0-444-64301-8.00012-3

Arras, W., Hussain, A., Hausler, R., & Guiot, S. R. (2019). Mesophilic, thermophilic and hyperthermophilic acidogenic fermentation of food waste in batch: Effect of inoculum source. *Waste Management (New York, N.Y.), 87*, 279–287. doi:10.1016/j.wasman.2019.02.011 PMID:31109527

Bokhari, A., Essack, M., Lafi, F. F., Andres-Barrao, C., Jalal, R., Alamoudi, S., & Razali, R. (2019). Bioprospecting desert plant Bacillus endophytic strains for their potential to enhance plant stress tolerance. *Scientific Reports, 9*(1). doi:10.1038/s41598-019-54685-y

Castro, D., Torres, M., Sampedro, I., Martínez-Checa, F., Torres, B., & Béjar, V. (2020). Biological Control of Verticillium Wilt on Olive Trees by the Salt-Tolerant Strain Bacillus velezensis XT1. *Microorganisms, 8*(7), 1080. doi:10.3390/microorganisms8071080

Coker, J. A. (2019). Recent advances in understanding extremophiles. *F1000Research, 8*, 1917. doi:10.12688/f1000research.20765.1

Daranas, N., Roselló, G., Cabrefiga, J., Donati, I., Francés, J., Badosa, E., & Spinelli, F. (2018). Biological control of bacterial plant diseases withLactobacillus plantarumstrains selected for their broad-spectrum activity. *Annals of Applied Biology, 174*(1), 92–105. doi:10.1111/aab.12476

Dikilitas, M., Karakas, S., Simsek, E., & Yadav, A. N. (2021). Microbes from Cold Deserts and Their Applications in Mitigation of Cold Stress in Plants. In *Microbiomes of Extreme Environments* (pp. 126–152). CRC Press. doi:10.1201/9780429328633-7

Dua, A., Joshi, S., & Satyanarayana, T. (2016). Recombinant exochitinase of the thermophilic mould Myceliopthora thermophila BJA: Characteristics and utility in generating N-acetyl glucosamine and in biocontrol of phytopathogenic fungi. *Biotechnology Progress, 33*(1), 70–80. doi:10.1002/btpr.2370

Elumalai, P., Parthipan, P., Karthikeyan, O. P., & Rajasekar, A. (2017). *Enzyme-mediated biodegradation of long-chain n-alkanes (C32 and C40) by thermophilic bacteria. Biotech, 7(2).*

Ezeokoli, O. T., Bezuidenhout, C. C., Maboeta, M. S., Khasa, D. P., & Adeleke, R. A. (2020). Structural and functional differentiation of bacterial communities in post-coal mining reclamation soils of South Africa: bioindicators of soil ecosystem restoration. *Scientific Reports, 10*(1). doi:10.1038/s41598-020-58576-5

Gallardo-Cerda, J., Levihuan, J., Lavín, P., Oses, R., Atala, C., Torres-Díaz, C., & Cuba-Díaz, M. (2018). Antarctic rhizobacteria improve salt tolerance and physiological performance of the Antarctic vascular plants. *Polar Biology, 41*(10), 1973–1982. doi:10.1007/s00300-018-2336-z

Hãggblom, M., & Margesin, R. (2005). Microbial life in cold ecosystems. FEMS Microbiology Ecology, 53(1), 1–2. doi:10.1016/j.femsec.2005.03.006

Hermans, S. M., Buckley, H. L., Case, B. S., Curran-Cournane, F., Taylor, M., & Lear, G. (2017). Bacteria as Emerging Indicators of Soil Condition. Applied and Environmental Microbiology, 83(1). doi:10.1128/AEM.02826-16

Horikoshi, K. (2007). Past present and future of extremophiles. *Extremophiles, 12*(1), 1–2. doi:10.1007/s00792-007-0127-5

Hottenstein, J. D., Neilson, J. W., Gil-Loaiza, J., Root, R. A., White, S. A., Chorover, J., & Maier, R. M. (2019). Soil Microbiome Dynamics During Pyritic Mine Tailing Phytostabilization: Understanding Microbial Bioindicators of Soil Acidification. *Frontiers in Microbiology, 10*. doi:10.3389/fmicb.2019.01211

Hussain, A., Ahmad, M., Nafees, M., Iqbal, Z., Luqman, M., Jamil, M., & Maqsood, A. (2020). Plant-growth-promoting Bacillus and Paenibacillus species improve the nutritional status of Triticum aestivum L. PLOS ONE, 15(12), e0241130. doi:10.1371/journal.pone.0241130

Igiehon, N. O., Babalola, O. O., & Aremu, B. R. (2019). Genomic insights into plant growth promoting rhizobia capable of enhancing soybean germination under drought stress. *BMC Microbiology, 19*(1). doi:10.1186/s12866-019-1536-1

Jayakumar, J., & Seenivasan, N. (2019). Evaluation of Biocontrol Agents and Organic Amendments for the Management of Root Knot Nematode and Spiral Nematode in Banana. *International Journal of Current Microbiology and Applied Sciences, 8*(5), 613–621. doi:10.20546/ijcmas.2019.805.071

Kästner, M., Breuer-Jammali, M., & Mahro, B. (1994). Enumeration and characterization of the soil microflora from hydrocarbon-contaminated soil sites able to mineralize polycyclic aromatic hydrocarbons (PAH). *Applied Microbiology and Biotechnology, 41*(2), 267–273. doi:10.1007/BF00186971

Kaur, R., Macleod, J., Foley, W., & Nayudu, M. (2006). Gluconic acid: An antifungal agent produced by Pseudomonas species in biological control of take-all. *Phytochemistry, 67*(6), 595–604. doi:10.1016/j.phytochem.2005.12.011

Kaushik, S., Alatawi, A., Djiwanti, S. R., Pande, A., Skotti, E., & Soni, V. (2021). Potential of Extremophiles for Bioremediation. Microbial Rejuvenation of Polluted Environment. Springer Singapore. 10.1007%2F978-981-15-7447-4_12

Khatoon, H., Rai, J. P. N., & Jillani, A. (2021). Role of fungi in bioremediation of contaminated soil. In *Fungi Bio-Prospects in Sustainable Agriculture Environment and Nano-technology* (pp. 121–156). Elsevier. doi:10.1016/B978-0-12-821925-6.00007-1

Köhl, J., Kolnaar, R., & Ravensberg, W. J. (2019). Mode of Action of Microbial Biological Control Agents Against Plant Diseases: Relevance Beyond Efficacy. *Frontiers in Plant Science, 10*. doi:10.3389/fpls.2019.00845

Kohli, I., Joshi, N. C., Mohapatra, S., & Varma, A. (2020). Extremophile An Adaptive Strategy for Extreme Conditions and Applications. *Current Genomics, 21*(2), 96–110. doi:10.2174/1389202921666200401105908

Koul, B., Chaudhary, R., & Taak, P. (2021). Extremophilic microbes and their application in bioremediation of environmental contaminants. In *Microbe Mediated Remediation of Environmental Contaminants* (pp. 115–128). Elsevier. doi:10.1016/B978-0-12-821199-1.00012-2

Kulkarni, S., Dhakar, K., & Joshi, A. (2019). Alkaliphiles. In *Microbial Diversity in the Genomic Era* (pp. 239–263). Elsevier. doi:10.1016/B978-0-12-814849-5.00015-0

Kwatra, B., Yadav, P., Balasubramaniam, S., Garg, S., & Koyande, T. (2021). Reviewing major microbes as a single cell protein. *International Journal of Medical and Biomedical Studies, 5*(3). doi:10.32553/ijmbs.v5i3.1842

Lin, X., Yang, B., Shen, J., & Du, N. (2009). Biodegradation of Crude Oil by an Arctic Psychrotrophic Bacterium Pseudoalteromomas sp. P29. *Current Microbiology, 59*(3), 341–345. doi:10.1007/s00284-009-9440-9

Lixia, Z. (2007). Soil microbial characteristics as bioindicators of soil health. Biodiversity Science, 15(2), 162. doi:10.1360/biodiv.060290

Mamo, G., & Mattiasson, B. (2016). Alkaliphilic Microorganisms in Biotechnology. In *Biotechnology of Extremophiles* (pp. 243–272). Springer International Publishing. doi:10.1007/978-3-319-13521-2_8

McBeath, J. B. (2002). Snow mold-plant-antagonist interactions: Survival of the fittest under the snow. *The Plant Health Instructor. Science and Society*. Advance online publication. 10.1094%2Fphi-i-2002-1010-01

Meena, K. K., Kumar, M., Kalyuzhnaya, M. G., Yandigeri, M. S., Singh, D. P., Saxena, A. K., & Arora, D. K. (2011). Epiphytic pink-pigmented methylotrophic bacteria enhance germination and seedling growth of wheat (Triticum aestivum) by producing phytohormone. *Antonie van Leeuwenhoek, 101*(4), 777–786. doi:10.1007/s10482-011-9692-9

Merino, N., Aronson, H. S., Bojanova, D. P., Feyhl-Buska, J., Wong, M. L., Zhang, S., & Giovannelli, D. (2019). Living at the Extremes: Extremophiles and the Limits of Life in a Planetary Context. *Frontiers in Microbiology, 10*. doi:10.3389/fmicb.2019.00780

Midhat, L., Ouazzani, N., Hejjaj, A., Ouhammou, A., & Mandi, L. (2019). Accumulation of heavy metals in metallophytes from three mining sites (Southern Centre Morocco) and evaluation of their phytoremediation potential. *Ecotoxicology and Environmental Safety, 169*, 150–160. doi:10.1016/j.ecoenv.2018.11.009

Mir, M. Y., Hamid, S., Rohela, G. K., Parray, J. A., & Kamili, A. N. (2021, March). *Composting and Bioremediation Potential of Thermophiles*. Wiley. doi:10.1002/9781119547976.ch7

Mishra, P. K., Joshi, P., Bisht, S. C., Bisht, J. K., & Selvakumar, G. (2010). Cold-Tolerant Agriculturally Important Microorganisms. In *Plant Growth and Health Promoting Bacteria* (pp. 273–296). Springer Berlin Heidelberg. doi:10.1007/978-3-642-13612-2_12

Moreno, R., & Rojo, F. (2014). Features of pseudomonads growing at low temperatures: another facet of their versatility. *Environmental Microbiology Reports, 6*(5), 417–426. doi:10.1111/1758-2229.12150

Nafis, A., Raklami, A., Bechtaoui, N., Khalloufi, F. E., Alaoui, A. E., Glick, B. R., & Hafidi, M. (2019). Actinobacteria from Extreme Niches in Morocco and Their Plant Growth-Promoting Potentials. *Diversity, 11*(8), 139. doi:10.3390/d11080139

Navarro, F., Forján, E., Vázquez, M., Montero, Z., Bermejo, E., Castaño, M. Á., & Toimil, A. (2016). Microalgae as a safe food source for animals: nutritional characteristics of the acidophilic microalgaCoccomyxa onubensis. *Food & Nutrition Research, 60*(1), 30472. doi:10.3402/fnr.v60.30472

Nduka, J. K., Umeh, L. N., & Okerulu, I. O. (2012). Utilization of Different Microbes in Bioremediation of Hydrocarbon Contaminated Soils Stimulated With Inorganic and Organic Fertilizers. *Journal of Petroleum & Environmental Biotechnology, 3*(2). doi:10.4172/2157-7463.1000116

Ogata-Gutiérrez, K., Alvarado, D., Chumpitaz-Segovia, C., & Zúñiga-Dávila, D. (2016). Characterization of Plant Growth Promoting Rhizobacteria Isolated from the Rhizosphere of Peruvian Highlands Native Crops. *International Journal of Plant & Soil Science, 11*(1), 1–8. doi:10.9734/IJPSS/2016/24573

Okongo, R. N., Puri, A. K., Wang, Z., Singh, S., & Permaul, K. (2019). Comparative biocontrol ability of chitinases from bacteria and recombinant chitinases from the thermophilic fungus Thermomyces lanuginosus. *Journal of Bioscience and Bioengineering, 127*(6), 663–671. doi:10.1016/j.jbiosc.2018.11.007

Ortiz-Ojeda, Ogata-Gutiérrez, & Zúñiga-Dávila. (2017). Evaluation of plant growth promoting activity and heavy metal tolerance of psychrotrophic bacteria associated with maca (emLepidium meyenii/em Walp.) rhizosphere. *AIMS Microbiology, 3*(2), 279–292. doi:10.3934/microbiol.2017.2.279

Pandey, A., & Yarzábal, L. A. (2018). Bioprospecting cold-adapted plant growth promoting microorganisms from mountain environments. *Applied Microbiology and Biotechnology, 103*(2), 643–657. doi:10.1007/s00253-018-9515-2

Pedone, E., Fiorentino, G., Bartolucci, S., & Limauro, D. (2020). Enzymatic Antioxidant Signatures in Hyperthermophilic Archaea. *Antioxidants, 9*(8), 703. doi:10.3390/antiox9080703

Perfumo, A., Banat, I. M., & Marchant, R. (2006). The use of thermophilic bacteria in accelerated hydrocarbon bioremediation. In *Environmental Problems in Coastal Regions VI.* WIT Press. doi:10.2495/CENV060071

Rajashree, R. P., & S., G. B. (2018). Isolation and characterization of thermophilic bacteria from different habitats and their assessment for antagonism against soil-borne fungal plant pathogens. *African Journal of Microbiology Research, 12*(24), 556–566. doi:10.5897/AJMR2015.7511

Rampelotto, P. (2013). Extremophiles and Extreme Environments. *Life, 3*(3), 482–485. doi:10.3390/life3030482

Rao, M. A., Scelza, R., Acevedo, F., Diez, M. C., & Gianfreda, L. (2014). Enzymes as useful tools for environmental purposes. *Chemosphere, 107*, 145–162. doi:10.1016/j.chemosphere.2013.12.059

Rastädter, K., Wurm, D. J., Spadiut, O., & Quehenberger, J. (2021). Physiological Characterization of Sulfolobus acidocaldarius in a Controlled Bioreactor Environment. *International Journal of Environmental Research and Public Health,* 18. PMID:34064179

Rath, K. M., Fierer, N., Murphy, D. V., & Rousk, J. (2018). Linking bacterial community composition to soil salinity along environmental gradients. *The ISME Journal, 13*(3), 836–846. doi:10.1038/s41396-018-0313-8

Raymond-Bouchard, I., & Whyte, L. G. (2017). From Transcriptomes to Metatranscriptomes: Cold Adaptation and Active Metabolisms of Psychrophiles from Cold Environments. In Psychrophiles: From Biodiversity to Biotechnology. Springer International Publishing. 10.1007%2F978-3-319-57057-0_18

Ritala, A., Häkkinen, S. T., Toivari, M., & Wiebe, M. G. (2017). Single Cell ProteinState-of-the-Art Industrial Landscape and Patents 20012016. *Frontiers in Microbiology, 8*. doi:10.3389/fmicb.2017.02009

Salwan, R., & Sharma, V. (2020). Physiology of extremophiles. Physiological and Biotechnological Aspects of Extremophiles. Elsevier. 10.1016%2Fb978-0-12-818322-9.00002-2

Santos, R., Fernandes, J., Fernandes, N., Oliveira, F., & Cadete, M. (2007). Mycobacterium parascrofulaceum in Acidic Hot Springs in Yellowstone National Park. *Applied and Environmental Microbiology, 73*(15), 5071–5073. doi:10.1128/AEM.00353-07

Saravanan, P., Ilavarasan, N., Karthikeyan, A., & Padmanaban, B. (2020). Management of Root Knot Nematode Meloidogyne incognita in Banana (Cv. Robusta) through Biocontrol Agents and Neem Cake. *Indian Journal of Agricultural Research*. doi:10.18805/IJARe.A-5507

Shanmugam, V., Chugh, P., & Sharma, P. (2014). Cold-tolerant Trichoderma species for the managementof Fusarium wilt of tomato plants. *Annals of Microbiology, 65*(1), 543–551. doi:10.1007/s13213-014-0890-3

Sharma, A., Parashar, D., & Satyanarayana, T. (2016). Acidophilic Microbes: Biology and Applications. In Biotechnology of Extremophiles (pp. 215–241). Springer International Publishing. doi:10.1007/978-3-319-13521-2_7

Shrestha, B. K., Karki, H. S., Groth, D. E., Jungkhun, N., & Ham, J. H. (2016). Biological Control Activities of Rice-Associated Bacillus sp. Strains against Sheath Blight and Bacterial Panicle Blight of Rice. PLOS ONE, 11(1), e0146764. doi:10.1371/journal.pone.0146764

Shrestha, N., Chilkoor, G., Vemuri, B., Rathinam, N., Sani, R. K., & Gadhamshetty, V. (2018). Extremophiles for microbial-electrochemistry applications: A critical review. *Bioresource Technology, 255*, 318–330. doi:10.1016/j.biortech.2018.01.151

Shukla, A. K., & Singh, A. K. (2020). Exploitation of Potential Extremophiles for Bioremediation of Xenobiotics Compounds: A Biotechnological Approach. *Current Genomics, 21*(3), 161–167. doi:10.2174/1389202921999200422122253

Shukla, P. J., Bhatt, V. D., Suriya, J., & Mootapally, C. (2020, August). *Marine Extremophiles*. Wiley. doi:10.1002/9781119143802.ch74

Stapper, N. J., & John, V. (2015). Monitoring climate change with lichens as bioindicators. Pollution atmosphérique (N226). INIST-CNRS. 10.4267%2Fpollution-atmospherique.4936

Tapia-Vázquez, I., Sánchez-Cruz, R., Arroyo-Domínguez, M., Lira-Ruan, V., Sánchez-Reyes, A., Rayo Sánchez-Carbente, M. del, & Padilla-Chacón, D. (2020). Isolation and characterization of psychrophilic and psychrotolerant plant-growth promoting microorganisms from a high-altitude volcano crater in Mexico. *Microbiological Research, 232*, 126394. doi:10.1016/j.micres.2019.126394

Toral, L., Rodríguez, M., Béjar, V., & Sampedro, I. (2018). Antifungal Activity of Lipopeptides From Bacillus XT1 CECT 8661 Against Botrytis cinerea. *Frontiers in Microbiology, 9*. doi:10.3389/fmicb.2018.01315

Torracchi, J. E., Morel, M. A., Tapia-Vázquez, I., Castro-Sowinski, S., Batista-García, R. A., & R., L. A. Y. (2020). Fighting plant pathogens with cold-active microorganisms: biopesticide development and agriculture intensification in cold climates. *Applied Microbiology and Biotechnology, 104*(19), 8243–8256. doi:10.1007/s00253-020-10812-8

Torres, M., Llamas, I., Torres, B., Toral, L., Sampedro, I., & Béjar, V. (2020). Growth promotion on horticultural crops and antifungal activity of Bacillus velezensis XT1. *Applied Soil Ecology, 150*, 103453. doi:10.1016/j.apsoil.2019.103453

Torres-Díaz, C., Gallardo-Cerda, J., Lavin, P., Oses, R., Carrasco-Urra, F., Atala, C., & Acuña-Rodríguez, I. S. (2016). Biological Interactions and Simulated Climate Change Modulates the Ecophysiological Performance of Colobanthus quitensis in the Antarctic Ecosystem. PLOS ONE, 11(10), e0164844. doi:10.1371/journal.pone.0164844

Utobo, E. B., & Tewari, L. T. L. (2015). Soil enzymes as bioindicators of soil ecosystem status. *Applied Ecology and Environmental Research, 13*(1). Advance online publication. 10.15666%2Faeer%2F1301_147169

Varshney, P., Mikulic, P., Vonshak, A., Beardall, J., & Wangikar, P. P. (2015). Extremophilic micro-algae and their potential contribution in biotechnology. *Bioresource Technology, 184*, 363–372. doi:10.1016/j.biortech.2014.11.040

Verma, P., Yadav, A. N., Khannam, K. S., Mishra, S., Kumar, S., Saxena, A. K., & Suman, A. (2019). Appraisal of diversity and functional attributes of thermotolerant wheat associated bacteria from the peninsular zone of India. *Saudi Journal of Biological Sciences, 26*(7), 1882–1895. doi:10.1016/j.sjbs.2016.01.042

Yadav, A. N. (2017). Beneficial role of extremophilic microbes for plant health and soil fertility. *Journal of Agricultural Science and Botany, 1*(1). doi:10.35841/2591-7897.1.1.30-33

Yadav, A. N. (2017). Extreme Cold Environments: A Suitable Niche for Selection of Novel Psychrotrophic Microbes for Biotechnological Applications. *Advances in Biotechnology & Microbiology, 2*(2). doi:10.19080/AIBM.2017.02.555584

Yang, J. S., Yang, F. L., Yang, Y., Xing, G. L., Deng, C. P., Shen, Y. T., & Luo, L. Q. (2016). A proposal of core enzyme bioindicator in long-term Pb-Zn ore pollution areas based on topsoil property analysis. *Environmental Pollution, 213*, 760–769. doi:10.1016/j.envpol.2016.03.030

Zaghloul, A., Saber, M., Gadow, S., & Awad, F. (2020). Biological indicators for pollution detection in terrestrial and aquatic ecosystems. *Bulletin of the National Research Centre, 44*(1). doi:10.1186/s42269-020-00385-x

Zeldes, B. M., Keller, M. W., Loder, A. J., Straub, C. T., Adams, M. W. W., & Kelly, R. M. (2015). Extremely thermophilic microorganisms as metabolic engineering platforms for production of fuels and industrial chemicals. *Frontiers in Microbiology, 6*. doi:10.3389/fmicb.2015.01209

Zhang, X., Lin, Y., & Chen, G.-Q. (2018). Halophiles as Chassis for Bioproduction. *Advanced Biosystems, 2*(11), 1800088. doi:10.1002/adbi.201800088

Chapter 12
Application of Extremophiles in Food Industries

Ankita Pradeepkumar Leanwala
Naran Lala College of Professional and Applied Sciences, Navsari, India

ABSTRACT

Extremophiles have adapted themselves at extreme environmental conditions like high or low temperature, pH, salinity, and pressure. Extremophiles may be either acidophilic, alkaliphilic, halophilic, thermophilic, psychrophilic, oligotrophic, endolithic, and xerophilic. There extremozymes are found to be biocatalysts and producers of novel enzymes which can be employed in many industries like food, cosmetics, chemical, pharmaceuticals, etc. Currently the researchers have developed keen interest in studying and utilizing the abilities of these extremophiles in food industries. Metabolic pathways and extremozymes are being studied by the researchers and they are trying to utilize its characteristics and also engineer these extremophiles. In food industries, one of the extremophiles, Rhodothermus marinus, which has been an excellent biocatalyst producing lipase as an enzyme, could be utilized to improve to aroma of food and add natural flavour to food. So, the current chapter will deal with the various applications of these extremophiles.

INTRODUCTION

Organisms are found to be omnipresent in universe. Many organisms have adapted themselves to various environmental conditions like low temperature (psychrophiles), high temperature (thermophiles and hyperthermophiles), high salinity (halophile), low pH (acidophiles), high pH (alkaliphiles), low nutrient concentration (oligotrophs), low water activity (xerophile), heavy metal concentrations (metallotolerant), high pressure (barophile or piezophile), high radiations (radioresistant), low oxygen concentrations, (Gomes and Steiner, 2004; Cowan *et. al.*, 2015), high antibiotic concentrations, high concentration of carbon (Capnophile), etc. Such organisms are termed as extremophiles. The term extremophile was given by MacElroy in 1974. Taxonomic classification includes prokaryotes, eukaryotes, bacteria and archaea. (Woese *et. al.*, 1990; Zhang, 2018). They have adapted themselves to survive in ecological niches that are unsuitable for others, for example, deep-sea hydrothermal vents, hot and cold deserts, soda lakes,

DOI: 10.4018/978-1-7998-9144-4.ch012

inland saline systems, solar salterns, environments highly contaminated with nuclear waste or heavy metals, as well as lithic or rock environments and hot springs. (Raddadi *at. el.,* 2015).

They are found to be potential source of biomolecules which can be utilized for many industrial purposes. Nowadays they have been in the limelight as the researchers have found them to be a treasurer of industrially utilizable metabolites and its products. But only 1-2% of the extremophiles have been utilized uptil now and in future the expanding research may help to exploit them. (Gomes and Steiner, 2004).

Extremophiles have been found to produce extremozymes which are protein that can function under extreme conditions. These enzymes possess some unique characteristics like extreme thermal stability, resistance against chemical denaturants such as detergents, chaotropic agents, organic solvents and extremes of pH. (Gupta *et. al.,* 2014). Proteases, pectinases, keratinases, cellulases, amylases, xylanases, lipases, esterases, catalases, peroxidases, phytases, etc. are few examples of extremozymes produced them. They can be a good biocatalyst and can be employed in various biotechnological processes. (Gomes and Steiner, 2004). More than 300 different enzymes have been identified and have been employed in industrial and biotechnological processes. (Singh *et. al.,* 2021)

According to *Raveendran et. al. 2018* microorganisms such as bacteria, yeast, fungi and their enzymes are widely used in the food industries. These microbial enzymes are used because they are found to be more stable than plant and animal enzymes. They are utilized to improve the texture and taste of in food industry.

BACKGROUND

Extremophiles are a group of organisms that survive under extreme environmental conditions such as high or low temperature, pH, salinity, and pressure. These organisms have evolved and developed strategies and mechanisms to survive under extreme conditions. These extremophiles mainly belong to the genus *Acidithiobacillus, Arthrobacter, Bacillus, Caldicellulosiruptor, Clostridium, Coprothermobacter, Enterobacter, Geobacillus, Micrococcus, Paenibacillus, Penicillium, Picrophilus, Pseudoalteromonas,* and *Thermobifida.* (Zhu *et. al.*, 2020)

Recently researchers are utilizing the enzymes produced by the extremophiles and have still developed keen interest in searching novel enzymes produced by them. Many research companies have been working to develop the strategies to genetically modify or design the extremophiles to take maximum advantage for utilizing extremozymes produced by them in biotechnological processes and food industries. For improvement in food and the food products constant efforts are made by the food industries to employ these enzymes. Thermophiles, psychrophiles, acidophiles and alkaliphiles, halophiles, piezophiles, toxitolerant and radiophile are the extremophiles which can be employed in the food industries.

Thermophiles

Thermophiles can be classified into moderate, extreme and hyperthermophiles. Moderate thermophiles grows at optimum 50–60 °C, extreme thermophiles grows at optimum 60–80 °C whereas hyperthermophiles grows at optimum 80–110 °C. Thermophylic microorganisms are good source of thermostable enzymes. (Singh *et. al.*, 2011). Thermophilic enzymes can increase reaction rates during high temperature processing of food. More than 40 extremophilic enzymes, characterized from the hot springs of Yellowstone National Park are found to be active at high temperatures and have also been found to have various applications in food processing specifically, starch processing, and ethanol production. (Khan & Sathya, 2018).

Psychrophiles

Such type of organisms are found at the low temperature environments of the Earth, including upper atmosphere, ocean deeps, polar regions, glaciers, shallow subterranean regions, refrigerated appliances and on and in plants and animals inhabiting cold regions. (Gupta *et. al.*, 2014). The psychrophilic enzymes produced by cold-adapted microorganisms possess a high catalytic efficiency at low temperature that makes them potentially good to use in food industry. (Kumar *et al.,* 2011). Moreover psychrophilic enzymes are found to have high catalytic activity and stability at low temperature. This property makes them as potential candidates for food processing. By employing these enzymes in processing of food it not only minimizes the microbial contamination but also helps in preventing the deterioration of raw materials. (Joseph *et. al.*, 2019). The major application of these enzymes includes processes such as lactose hydrolysis, lactose free milk and cheese, clarification of fruit juices, protein and milk processing, etc. Cold-active β-galactosidase is found to hydrolyse lactose to glucose and galactose at refrigerated temperature. Thus, this enzyme is employed to produce lactose-free milk-derived foods for lactose-intolerant people. (Hamdan, 2018). *Hoyoux et al.* patented a cold-active β-galactosidase from an Antarctic psychrophile, *Pseudoalteromonas haloplanktis,* for its capacity to hydrolyse lactose during milk storage at low temperatures. Similarly cold active pectinase is employed in retaining the quality and nutritional vales of fruit juices. (*Adapa et. al.,* 2014). In wine industries cold active pectinases which is derived from yeast and fungi increases the production and helps in retention of volatile compounds. This helps to maintain the aroma of wines. (*Singh et. al.,* 2012). In making of bread cold-active xylanases which can be obtained from the Antarctic bacterium *Pseudoalteromonas haloplanktis* hydrolyse ß-1,4-xylan present in all flours and ultimately improves the quality of bread. (Struvay & Feller, 2012).

Acidophiles and Alkaliphiles

Acidophiles thrives under acidic conditions with a pH optimum for growth at, or below, pH 3 whereas alkaliphiles thrives in alkaline environments at a pH of 8 or more. (Rothschild and Mancinelli, 2001; Gomes and Steiner, 2004). Certain alkaliphiles produces various extremozymes like xylanases, pectinases, cyclodextrins, xylanases and proteases. (Table 1). Among these enzymes protease is found to be an industrially important enzyme. Proteases are generally active from pH 7.0–11.0. Bacillus produces all the major subtilisins which can be applied in industries. (Sarethy *et. al.,* 2011). In the food industry, during processing of soy sauce alkaline and neutral proteases from fungi are used. Proteolytic modification of soy proteins by these enzymes improves the functional properties. This consequently results in soluble hydrolysates which show better solubility and lowers the bitterness. (Rao *et. al.*, 1998). They are employed in pulp bleaching in food industry. α – Amylase is mainly produced from *Bacillus.* α – amylases act on the a-1,4 bonds between adjoining glucose units leading to the formation of glucose, maltose, and maltotriose. This extracellular enzyme is used in the food industry for brewing, baking and processing of fruit juice. (Sarethy *et. al.*, 2011).

Halophiles

They are the extremophiles which grow in optimum saline environments with most species requiring more than 2.0 M NaCl for growth and survival. A special protein layer is coated on Halophiles which blocks excessive salt from entering its cells. (Singh et. al., 2021). Natural pigment producing organisms have been employed to color the food stuff in textile industries.

Piezophiles

Microorganisms that can survive under high-pressure are termed as piezophiles or barophiles. These piezophyle inhabit in the deep sea or oceans. Mostly they belong to the genera *Shewanella, Pyrococcus, Methanococcus,* and *Moritella* (*Kato et al.,* 1998; *Kato and Nogi,* 2001). The enzymes produced by piezophiles are found to be stable at high pressure. In food industries high pressure is required for for processing, sterilization and packing. High pressures are utilized in industries to induce gel formation, starch granules, the denaturation or coagulation of proteins or the transition of lipid phases. The use of high pressure leads to better flavor and color preservation as compared to the use of high temperature. Thus, these enzymes have been majorly utilized in food industries. (Gupta *et. al.,* 2014).

Toxitolerant

Proteases and lipases produced by toxitolerants are used in esterification. (Gupta *et. al.*, 2014).

Radiophile

These organisms are able to survive under high or extreme pressure. Yet not much information is available regarding their utilization in food industries.

The enzymes secreted by extremophiles are summarized in **Table: 1** along with their applications in food industries:

Apart from enzymes others compounds can even be utilized in food industries. Halophillic archaea and algae produces a natural pigment carotenoid. (Schiraldi *et. al.*, 2002). Canthaxanthin is derived from carotenoids which might be utilized in food industry as food dye and feed additive. The strain of Halophilic archaea *Haloferax alexandrinus* are the producers of Canthaxanthin. (Coker, 2016). It is also called as Food Orange 8, Carophyll Red. It is a lipid-soluble antioxidant used as a food dye. It is used as a colourant in BBQ sauce, drinks, dressings and other products. As a feed additive, it is used in poultry farming, fishing and crustaceans.

Dunaliella salina, the halophilic alga is the major source for β-carotene. β-carotene is a red/orange pigment which is the primary colorant in pumpkins, carrots and halophilic microorganisms. It is commercially employed to increase the dry weight which results in 30–40 g dry weight/m2 per day. (DasSarma *et. al.,* 2009). Primarily it is found to be an excellent food supplement. It is lipid as well as water soluble which makes it excellent as an additive in the baking process (e.g. food coloring) and emulsions (e.g. confectionery and prepared foods) (Coker, 2016).

Table 1. Extremophiles used in food industries

Extremophiles	Enzymes produced	Use of the enzymes	Ref.
Thermophiles (Moderate thermophiles 50-60°C, thermophiles 60-80°C Hyper thermophiles >80°C)	Amylases	Glucose and fructose for sweeteners, have application in the liquefaction of gelatinized starch and saccharification.	Coker, 2016 Khan & Sathya. 2018.
	Proteases	Baking, cheese making, brewing, and baking. Tenderizes and enhances flavour of meat during refrigerated storage.	He *et. al.*, 2004; Coker, 2016; Khan & Sathya, 2018.
	Lipases, esterases	Hydrolysis in food and feed	Gupta *et. al.*, 2014
	Dehydrogenases	Oxidation reactions	Gupta *et. al.*, 2014
	Glycosyl hydrolases	Dairy products	Gupta *et. al.*, 2014
	Xylanase	Bread making	Jiang *et. al.*, 2005
	Lipases, pullulanase, and amylopullulanase	Brewing and baking.	Satyanarayana *et. al., 2005*
	a-Amylases, glucoamylase, a-glucosidase, pullulanase, amylopullulanase and xylose/glucose isomerases	Starch processing, and glucose and fructose for sweeteners	Satyanarayana *et. al., 2005*
Psychrophile (Low temperature <15°C)	Protease dehydrogenases	Cheese maturation, dairy production and food applications	Dumorne *et. al.*, 2017
	Lipases and protease	Cheese manufacture	Satyanarayana *et. al., 2005*
	β-galactosidase	Lactose hydrolysis, Lactose free milk and cheese.	Khan & Sathya, 2018; Satyanarayana *et. al., 2005*
	Glycocosidases which includes (amylases, proteases, and xylanases)	In bakery industry improve texture and flavour	Khan & Sathya, 2018
	Pectinases	Retains flavour and improves clarification during food processing.	Khan & Sathya, 2018
	Proteases	Meat tenderizing	Satyanarayana *et. al., 2005*
Acidophile (Low pH<2-3)	Amylases, glucoamylases	Starch processing	Gupta *et. al.*, 2014
	Proteases, cellulase	Feed components	Gupta *et. al.*, 2014
	Beta Mannanase	Juice clarification, coffee viscosity reduction	Harnpicharnchai *et. al.*, 2016
Alkaliphiles (High pH<9)	Xylanases	Used in bread making and other bakery products and feed additive industries have found interest in the functional food industry for the synthesis of xylo oligosaccharides and ferulic acid.	Khan & Sathya, 2018.
	α - Amylase	brewing, baking, processing of fruit juice	Sarethy *et. al.*, 2011
	Pectinases	Pulp bleaching	Gupta *et. al.*, 2014
	Cyclodextrins	Foodstuffs	Satyanarayana *et. al., 2005*
	Xylanases and proteases	Pulp bleaching	Satyanarayana *et. al., 2005*
Halophile (High Salt concentration e.g. 2–5 M NaCl)	Proteases	Peptide synthesis	Gupta *et. al.*, 2014
	Amylases	high-fructose corn syrup	Coker, 2016
	Whole microorganism	producing poly (β -glutamic acid; PGA) & poly (β-hydroxy butyric acid; PHB)	Gupta *et. al.*, 2014
	Nucleases, amylases, proteases	Flavouring agents	*Satyanarayana et. al. 2005*
	g-Linoleic acid, b-carotene and cell extracts	Health foods, dietary supplements, food colouring, and feedstock	*Satyanarayana et. al. 2005*
Piezophile (High pressure up to 130 Mpa)	Whole microorganism	Food processing and antibiotic production	Gupta *et. al.*, 2014
		utilized in food	Gupta *et. al.*, 2014
	Whole microorganism	Microbially enhanced oil recovery process	Satyanarayana *et. al., 2005*
Toxitolerant	Proteases, lipases	Esterification	Gupta *et. al.*, 2014
(High levels of toxic reagents/ organic solvents)	Cyclodextrin glucanotransferase	Utilized in food industry	Gupta *et. al.*, 2014
Metalophile	Whole microorganism	Biomineralization	Gupta *et. al.*, 2014
(High metal concentration)			
Radiophile	NI	NI	Dumorne *et. al.*, 2017
(High radiation levels)			

NI = No Information

EXPLORING EXTREMOPHYLES FROM ENVIRONMENT

Recent studies made by the researcher's have proved that the cultivable techniques have only being able to cultivate only 0.1% of the total diversity of microorganisms. Thus, newer techniques based on PCR amplifications can be widely employed to explore the hidden potential of the extremophiles. These includes ribosomal intergenic spacer analysis (RISA), denaturing gradient gel electrophoresis (DGGE), temperature gradient gel electrophoresis (TGGE), single-strand conformation polymorphism (SSCP), random amplified polymorphic DNA (RAPD), and amplified ribosomal DNA restriction analysis (ARDRA). These methods generate the fingerprints which enables us to review the composition of the community. (Kent, 2002; Fierer 2006; Pikuta *et. al.*, 2007). This will ultimately help to check the hidden extremophiles which are non-cultivable and their potential of producing biomolecules or enzymes can be employed in industrial sectors in large scale.

FUTURE RESEARCH DIRECTIONS

The diversity of extremophilic organisms in soils is much higher, exceeding our knowledge. Due to the extremozymes having commercial applications as well as economically advantageous, researchers have been found to focus on exploring the new species for novel extremophiles. The advancement in molecular techniques and also the increasing availability of environmental rDNA sequences, metagenomic approaches and bioinformatics have proved to be boon to the society in exploring the diversity of extremophiles. The novel enzymes still to be revealed can bring vast improvement in the field of food industry. Future research will be made through metabolomics to explore and engage the extremophiles for their incomparable application in food industries.

CONCLUSION

Research in this field will be everlasting and expanding and interesting. Extremozymes are going to be extremely valuable in the food industries. Though the current problem is to produce the metabolites from extremophiles in large scale. Collaboration between industries, academia and government can jointly eliminate this problem and play an important role in employing these extremozymes widely.

ACKNOWLEDGMENT

I would like to acknowledge Dr. Aparna Gunjal, Assistant Professor, Department of Microbiology, Dr. D.Y. Patil, Arts and Commerce and Science College, Pimpri, Pune, Maharashtra to have faith in me and give me the opportunity to enlighten the subject.

REFERENCES

Adapa, V., Ramya, L. N., Pulicherla, K. K., & Rao, K. S. (2014). Cold active pectinases: Advancing the food industry to the next generation. *Applied Biochemistry and Biotechnology, 172*(5), 2324–2337. doi:10.100712010-013-0685-1 PMID:24390855

Coker, J. A. (2016). Extremophiles and biotechnology: Current uses and prospects. *F1000 Research, 5*, 5. doi:10.12688/f1000research.7432.1 PMID:27019700

Cowan, D. A., Ramond, J. B., Makhalanyane, T. P., & De Maayer, P. (2015). Metagenomics of extreme environments. *Current Opinion in Microbiology, 25*, 97–102. doi:10.1016/j.mib.2015.05.005 PMID:26048196

DasSarma, P., Coker, J. A., Huse, V., & DasSarma, S. (2009). *Halophiles, industrial applications. Encyclopedia of industrial biotechnology: bioprocess, bioseparation, and cell technology*, 1-43.

Dumorné, K., Córdova, D. C., Astorga-Eló, M., & Renganathan, P. (2017). *Extremozymes: A potential source for industrial applications*. Academic Press.

Fierer, N., & Jackson, R. B. (2006). The diversity and biogeography of soil bacterial communities. *Proceedings of the National Academy of Sciences of the United States of America, 103*(3), 626–631. doi:10.1073/pnas.0507535103 PMID:16407148

Gomes, J., & Steiner, W. (2004). The biocatalytic potential of extremophiles and extremozymes. *Food Technology and Biotechnology, 42*(4), 223–225.

Gupta, G. N., Srivastava, S., Khare, S. K., & Prakash, V. (2014). Extremophiles: An overview of microorganism from extreme environment. *International Journal of Agriculture Environment and Biotechnology, 7*(2), 371–380. doi:10.5958/2230-732X.2014.00258.7

Hamdan, A. (2018). Psychrophiles: Ecological significance and potential industrial application. *South African Journal of Science, 114*(5-6), 1–6. doi:10.17159ajs.2018/20170254

Harnpicharnchai, P., Pinngoen, W., Teanngam, W., Sornlake, W., Sae-Tang, K., Manitchotpisit, P., & Tanapongpipat, S. (2016). Production of high activity Aspergillus niger BCC4525 β-mannanase in Pichia pastoris and its application for mannooligosaccharides production from biomass hydrolysis. *Bioscience, Biotechnology, and Biochemistry, 80*(12), 2298–2305. doi:10.1080/09168451.2016.1230003 PMID:27648762

He, H., Chen, X., Li, J., Zhang, Y., & Gao, P. (2004). Taste improvement of refrigerated meat treated with cold-adapted protease. *Food Chemistry, 84*(2), 307–311. doi:10.1016/S0308-8146(03)00242-5

Jiang, Z., Li, X., Yang, S., Li, L., & Tan, S. (2005). Improvement of the breadmaking quality of wheat flour by the hyperthermophilic xylanase B from Thermotoga maritima. *Food Research International, 38*(1), 37–43. doi:10.1016/j.foodres.2004.07.007

Joseph, B., Kumar, V., & Ramteke, P. W. (2019). Psychrophilic Enzymes: Potential Biocatalysts for Food Processing. In Enzymes in Food Biotechnology (pp. 817-825). Academic Press.

Kato, C., Li, L., Nogi, Y., Nakamura, Y., Tamaoka, J., & Horikoshi, K. (1998). Extremely barophilic bacteria isolated from the Mariana Trench, Challenger Deep, at a depth of 11,000 meters. *Applied and Environmental Microbiology, 64*(4), 1510–1513. doi:10.1128/AEM.64.4.1510-1513.1998 PMID:9546187

Kato, C., & Nogi, Y. (2001). Correlation between phylogenetic structure and function: Examples from deep-sea Shewanella. *FEMS Microbiology Ecology, 35*(3), 223–230. doi:10.1111/j.1574-6941.2001.tb00807.x PMID:11311432

Kent, A. D., & Triplett, E. W. (2002). Microbial communities and their interactions in soil and rhizosphere ecosystems. *Annual Review of Microbiology, 56*(1), 211–236. doi:10.1146/annurev.micro.56.012302.161120 PMID:12142496

Khan, M., & Sathya, T. A. (2018). Extremozymes from metagenome: Potential applications in food processing. *Critical Reviews in Food Science and Nutrition, 58*(12), 2017–2025. doi:10.1080/10408398.2017.1296408 PMID:28605203

Kumar, L., Awasthi, G., & Singh, B. (2011). Extremophiles: A novel source of industrially important enzymes. *Biotechnology, 10*(2), 121–135. doi:10.3923/biotech.2011.121.135

MacElroy, R. D. (1974). Some comments on the evolution of extremophiles. *Bio Systems, 6*(1), 74–75. doi:10.1016/0303-2647(74)90026-4

Pikuta, E. V., Hoover, R. B., & Tang, J. (2007). Microbial extremophiles at the limits of life. *Critical Reviews in Microbiology, 33*(3), 183–209. doi:10.1080/10408410701451948 PMID:17653987

Raddadi, N., Cherif, A., Daffonchio, D., Neifar, M., & Fava, F. (2015). Biotechnological applications of extremophiles, extremozymes and extremolytes. *Applied Microbiology and Biotechnology, 99*(19), 7907–7913. doi:10.100700253-015-6874-9 PMID:26272092

Rao, M. B., Tanksale, A. M., Ghatge, M. S., & Deshpande, V. V. (1998). Molecular and biotechnological aspects of microbial proteases. *Microbiology and Molecular Biology Reviews, 62*(3), 597–635. doi:10.1128/MMBR.62.3.597-635.1998 PMID:9729602

Rothschild, L. J., & Mancinelli, R. L. (2001). Life in extreme environments. *Nature, 409*(6823), 1092–1101. doi:10.1038/35059215 PMID:11234023

Sarethy, I. P., Saxena, Y., Kapoor, A., Sharma, M., Sharma, S. K., Gupta, V., & Gupta, S. (2011). Alkaliphilic bacteria: Applications in industrial biotechnology. *Journal of Industrial Microbiology & Biotechnology, 38*(7), 769–790. doi:10.100710295-011-0968-x PMID:21479938

Satyanarayana, T., Raghukumar, C., & Shivaji, S. (2005). Extremophilic microbes: Diversity and perspectives. *Current Science*, 78–90.

Schiraldi, C., Giuliano, M., & De Rosa, M. (2002). Perspectives on biotechnological applications of archaea. *Archaea (Vancouver, B.C.), 1*(2), 75–86. doi:10.1155/2002/436561 PMID:15803645

Singh, L., Sharma, G., Awasthi, G., & Kumar, L. (2021). *The Impact of extremophilic Enzyme: Widely Application in Industrial uses*. Academic Press.

Singh, P., Hamid, B., Lone, M. A., Ranjan, K., Khan, A., Chaurse, V. K., & Sahay, S. (2012). Evaluation of pectinase activity from the psychrophilic fungal strain Truncatella angustata-BPF5 for use in wine industry. *Journal of Endocytobiosis and Cell Research, 22*, 57–61.

Singh, P., Hamid, B., Lone, M. A., Ranjan, K., Khan, A., Chaurse, V. K., & Sahay, S. (2012). Evaluation of pectinase activity from the psychrophilic fungal strain Truncatella angustata-BPF5 for use in wine industry. *Journal of Endocytobiosis and Cell Research, 22*, 57–61.

Singh, S. K., Singh, S. K., Tripathi, V. R., Khare, S. K., & Garg, S. K. (2011). A novel psychrotrophic, solvent tolerant Pseudomonas putida SKG-1 and solvent stability of its psychro-thermoalkalistable protease. *Process Biochemistry, 46*(7), 1430–1435. doi:10.1016/j.procbio.2011.03.012

Struvay, C., & Feller, G. (2012). Optimization to low temperature activity in psychrophilic enzymes. *International Journal of Molecular Sciences, 13*(9), 11643–11665. doi:10.3390/ijms130911643 PMID:23109875

Woese, C. R., Kandler, O., & Wheelis, M. L. (1990). Towards a natural system of organisms: Proposal for the domains Archaea, Bacteria, and Eucarya. *Proceedings of the National Academy of Sciences of the United States of America, 87*(12), 4576–4579. doi:10.1073/pnas.87.12.4576 PMID:2112744

Zhang, X., Li, S. J., Li, J. J., Liang, Z. Z., & Zhao, C. Q. (2018). Novel natural products from extremophilic fungi. *Marine Drugs, 16*(6), 194. doi:10.3390/md16060194 PMID:29867059

Zhu, D., Adebisi, W. A., Ahmad, F., Sethupathy, S., Danso, B., & Sun, J. (2020). Recent development of extremophilic bacteria and their application in biorefinery. *Frontiers in Bioengineering and Biotechnology, 8*, 483. doi:10.3389/fbioe.2020.00483 PMID:32596215

Chapter 13
Application of Extremophiles in Medicine and Pharmaceutical Industries

Nidhi Yogeshbhai Patel
https://orcid.org/0000-0001-5451-3491
Gujarat University, India

Shivani Maheshkumar Yagnik
Christ College, India

Dhritiksha Mansukhbhai Baria
Gujarat University, India

Vikram Hiren Raval
https://orcid.org/0000-0001-8912-1850
Gujarat University, India

ABSTRACT

Extremophiles are at center stage of scientific interest owing to their peculiar properties in terms of physiology, ecology, biochemistry, and molecular genetics. The bio-active compounds from extremophiles involve various types of extremolytes. The functional applicability of extremophiles has been far-reaching. Looking to the global scenario medical, pharmaceutical and allied healthcare sectors have a persistent surge for a novel anticancer, antimicrobial, stable drug deliverables, nutraceuticals, fine chemicals, natural antioxidants, and bio-polymers compounds. Genetic engineering tools clubbed with -omics approach enhance and better the chances for applicability of the extremophilic metabolites in varied sectors of red and yellow biotechnology. The chapter provides an insight into the various types of bio-active molecules from extremophiles and their wide biotechnological applicability in the medical and pharmaceutical industry.

INTRODUCTION

DOI: 10.4018/978-1-7998-9144-4.ch013

Ocean embraces about 80% of the minuscule life forms that are highly diverse and less estimated, yet have been a major area of interest (Kateb G., 2021). Marine waters have an enormous quantity as well as a range of assorted compounds/metabolites derived from the biological origin and found to contribute to human welfare. The marine bioactive compounds are explored for copious biotechnological uses (Rathinam & Sani, 2018). Additionally, they have astonishing peculiarities in terms of the synthesis of various bioactive compounds and/or antimicrobial substances (Olaniyan & Adetunji, 2021) that may serve as a model for futuristic biotechnological applications.

Microorganisms that are capable to alter their cellular mechanism in response to surrounding extremities are regarded as extremophiles. They are the remarkable ones that endure in extreme surroundings such as high salinity, temperature, pH, osmotic pressure, ion concentration, and radiation (Hagagy et al., 2021). The extremophiles possess more than one adaptability property due to which they are utmost explored and are also referred to as polyextremophiles (Raval et al., 2013; Raval et al., 2015; Di Donato et al., 2018; Patel, 2020). The "survival strategies" adopted by extremophiles also show a profound effect on their potential applications (Daoud & Ali, 2020). Extremophiles have become a focus of scientific interest owing to their peculiar properties in terms of physiology, ecology, biochemistry, and molecular genetics. Furthermore, the potentiality of bioactive compounds from unexplored and immense surroundings that gain key interest amongst the scientific community nowadays (Kushveer et al., 2021; Patel et al., 2021). They possess distinct adaptation mechanisms that make them fascinating to explore the properties of bioactive compounds from extremophiles (Baria et al., 2020). The chapter is shaped in a way to give insights into types of bioactive molecules from extremophiles and their comprehensive biotechnological applicability in the medicinal and pharmaceutical industry.

DIVERSITY AND ECO-PHYSIOLOGICAL ADAPTATIONS OF EXTREMOPHILES

Extremophiles are diverse microorganisms and endure extremities, show peculiarities for persistence and progression. The acquaintance of bioactive molecules, enzymes and novel compounds from extremophiles exemplify and render enhanced attention (Patel et al., 2021). However, the higher life forms are unable to withstand extremities when compared with the bacterial and archaeal counterparts. Concerning eco-physiological adaptations, extremophilic microbes are well classified via extreme ambiences and the geophysical constraints viz. temperature, pH, salinity, ionizing radiation, and osmotic pressure. Based upon peculiarities, robustness, and diversity many adaptation strategies has been employed by extremophiles viz. high salt-in, salting-out, compatible solute synthesis, sodium pump, chloride pump, osmoregulation, pH, etc. (Amoozegar et al., 2019).

Eco-Physiological Adaptation of Extremophiles

Temperature

Temperature theatres an imperative role in the growth of extremophiles. Any trivial change can affect the diversity and type of microbial system residing therein drastically. The extremophiles that acclimatize their cellular machinery in response to wavering temperature are regarded as thermophiles (Joyce, 2021). Generally, they are adapted to endure in surrounding 15°C & above 85°C are referred to as thermophiles.

Based upon the range of temperature they are categorized into hyperthermophiles, thermophiles, mesophiles, psychrophiles (Chung et al., 2020).

Hyperthermophiles are the ones that obligately need high temperatures around 80ºC for growth. The first hyperthermophilic life form *Sulfolobus acidocaldarius* was reported from Yellowstone National Park in 1972. Major reports thereafter have shown the vast archaeal and bacterial diversity. The report from the Gulf of California on thermophilic organism *Methanopyrus kandleri* showed the optimum growth temperature is 110ºC. However, the same species of *Methanopyrus kandleri was* obtained from the Kairei vent field, the Central Indian Ridge, and was able to tolerate temperature up to 122ºC. Additionally, *Thermus aquaticus, Pyrococcus furiosus, Bacillus stearothermophilus, Clostridium cellulovorans, Coprothermobacter proteolyticus, Ferdowsicous, Geobacillus thermoleovorans,* etc. have been reported as thermophilic microorganisms. By employing step-wise thermal adaptation strategies the organisms adapted to grow and survive with a wide range of temperatures. As revealed to thermophiles, mesophiles, psychrophiles, and the hyperthermophiles seem to possess the t-RNA gene viz. genetic makeup with rich GC content that renders stable intramolecular linkages and confers thermal stability (Atalah et al., 2019).

Thermophiles and mesophiles are those that need the best temperature for persistence and growth, for mesophiles i.e. 15-40°C; whereas in the case of thermophiles i.e. 60-80°C. As compared to hyperthermophiles, the most common are thermophiles, and few according to species vary from enteric bacteria viz. *Bacillus*, true bacteria, etc.; chemical process cluster includes eubacteria, purple bacteria; Archaeal cluster includes *Pyrococcus, Sulfolobus, Thermococcus*, etc. Most usual surroundings worldwide wherever thermophiles endure best vary from 80°C to 122°C that involves hydrothermal vents and plight springs includes species i.e. *Thermocrinis, Hyperthermus, Pyrococcus*, etc. (Clarke, 2014). Due to supermolecule binding properties, presence of reverse gyrase sequence, polyamines, the thermophiles distinguish from mesophiles in terms of structural stability by forming peculiar organic compound chains, additionally, heat shock proteins impart thermostability (Taylor, 2010).

Psychrophiles fancy lower temperatures for growth and survival. From all the acute conditions on Earth, psychrophiles are lavish in terms of diversity, biomass, and distribution. Across the globe, nearly 80% region is glaciers with persistent temperatures ~5°C. Below a thousand meters depth, the ocean temperature is 2–4°C; alpine, glacier-like Polar Regions and soil are among the other cold regions. These regions are well inhabited by robust microbes, hence psychrophiles are plentiful. The cryo-enzymes from *Shewanella* species, from the Antarctic regions, are well documented (Shukla et al., 2020). *Psychrobacter* is the prominent genera along with species from *Halorubrum, Methanococcoides, Photobacterium, Sphingopyxis, Polaribacter*, and *Octadecabacter* (Casanueva, 2010). Psychrophiles demonstrate membrane flexibility by saturated or unsaturated carboxylic acid, a decrease in size of a charged cluster of lipids, leads to alterations within the lipid composition (Guan et al., 2013). Additionally, by maintaining membrane fluidness in the cell the cold-adapted organisms confer protection against crystallization and vasoconstrictive properties (Suyal et al., 2021).

pH

The major biological processes ensue a neutral pH range, but few distinctive organisms differ from this and grow at extreme values on the pH scale. Conditional shifts in pH values give rise to microbes that are stringent and can be classified as acidophiles, neutrophiles, and alkalophilic (Chen, 2021).

Acidophiles are microorganisms that can tolerate and grow at pH < 3. Generally, acidophiles can encompass many surroundings but the foremost are acid mines, acidic industrial effluents, and other leachates, etc. (Hedrich & Schippers, 2021). Amongst these the most abundantly reported are sulfur oxidizers, species that belongs to a prokaryotic and eukaryotic group. Major adaptations strategies employed by the cell is pH homeostasis, values around pH 4.6 equilibrate by acidophiles. By confining the entry of protons through cytoplasmic membrane they can withstand lower pH surrounding. In the case of *Acidithiobacillus ferrooxidans*, the shift in pH from 4.5 to 1.0 leads to reduce the pore size of the membrane. To restraint, the proton entrance is done by employing proton influx inhibition through a chemosmotic gradient. A similar mechanism is also reported in many acidophilic bacteria viz. *Leptospirillum ferriphilum, Acidithiobacillus thiooxidans,* and *Acidithiobacillus caldus*, etc. (Feng et al., 2021). This is due to sudden alterations in pH that the organisms release protons and organic acids by proton motive force. Additionally, acidophiles harbour unique cell stability strategies i.e. chaperones molecules and novel enzymes to compensate for the DNA damage that occurs at lower pH. (Mirete et al., 2017)

Microbes growing facultatively at highly alkaline pH > 8 are regarded as alkalophiles. Naturally, two alkali containing environments occur i.e. one with high Ca^{2+} that embraces groundwater and low Ca^{2+} that embraces soda lakes, deserts. (Postec et al., 2021). The high Ca^{2+} containing surroundings found naturally are California, Yugoslavia, Jordan, Turkey, etc. Few other alkaline conditions are due to serpentinization during industrial processes, which release Fe^{2+}; soda lakes embrace the foremost alkaline environment with higher Na_2CO_3 concentrations. The most abundantly occurring bacteria are *Bacillus* sp. and *Cyanobacteria.* Few reported alkalophiles are *Anaerobranca horikoshii, Anaerobranca gottschalkii, Thermococcus alcaliphilus, Thermococcus acidoaminivorans, Thermococcus alcaliphilus,* and *Thermococcus acidoaminivorans* are isolated from Yellow stone national park and Bogoriae lake Kenya respectively (Shukla et al., 2020). The adaptation strategies employed include the presence of acidic polymers in the cell wall i.e. glutamic acid, galacturonic acid, gluconic acid, aspartic acid, hexosamines, and other amino acids (Horikoshi, 2016).

Salinity

The organism, which requires salinity for growth and metabolism, are considered as halophiles. Based on the salt needs and adaptation strategies employed the halophiles are categorized as slight (growth range, 0-0.3 M salt concentration), moderate (0.1-1.0 M salt concentration) to extreme (optimum 0.2-2.0 M salt concentration) halophiles. The halotolerant can withstand a high amount of salinity. Halophiles are found to be present in all 3 domains: Bacteria, Archaea, and Eukarya (Singh et al., 2013). *Proteobacteria, Firmicutes, Spirochaetes, Cyanobacteria,* and *Bacteriodetes*, etc. are the various phyla to which halophiles belong. *Thalassohaline* (Seawater originated and NaCl as a prevailing salt) and *Athalassohaline* (Non-sea-water originated mainly prevails sodium, magnesium, potash, borax salts) are the two main types of hypersaline habitat (DasSarma & DasSarma, 2015). The most usual examples of *Athalassohaline* surrounding are the Dead Sea, alkaline soda lakes, carbonate springs, alkaline soil, Great Salt Lake, solar lake, salt pans, Organic Lake, deep lake, etc. Amongst extremophiles, the utmost studied group is halophiles, they are peculiar in adaption strategies towards salinity has been a key focused area of research. *Haloquadratum, Halobacterium, Natronococcus, Halomonas* spp, *Flavobacterium, Natronobacterium, Pseudomonas, Volcaniella, Halovibrio* sp., *Paracoccus* sp. etc. are few reported halophilic microorganisms (Raval et al., 2018).

Ionizing Radiation

The extremophiles that are adapted and resistant to lethal mutagenic UV and ionizing radiation are generally referred to as radiophiles (Santra & Banerjee, 2021). There are very scarce reports of radiophiles until now only one bacteria viz. *Deinococcus* sp. is reported. However, in 1956 Anderson first reported a strain of *Deinococcus radiodurans* from sterilized X-ray cans. Additionally, many species of *Deinococcus* reported are *Deinococcus radiophilus, Deinococcus proteolyticus, Deinococcus radiopugnans, and Deinobacter grandis* (Kaushik et al., 2021). Adaptation strategies employed by bacteria are mainly of two types i.e. excision repair and recombination repair of DNA molecules. The ill effect of radiation leads to oxidative stress of vital biomolecules in the living system. The microbial system able to survive radiation is generally referred to as radioresistant. The reported cyanobacteria show the adaptive strategy towards the toxic effect of UVR due to efficient DNA repair mechanisms. By employing modern tools and techniques the biosynthetic pathway of radiation responsive and DNA repair enzymes could be modified in such a way that it permits another organism to probably thrive in extreme radiation (Rajpurohit et al., 2021). The metabolites synthesize from gene responsible for protection against radiation that seems to play a crucial role as anticancer drugs and potent antibiotics, etc. UV-absorbing substances are synthesized under UV stress by *Exiguobacterium* sp., and *Stenotrophomonas maltophilia.*

Osmotic Pressure

The microorganisms obligate requirements of pressure are generally referred to as barophiles/piezophiles. The optimum range for growth of barophiles reported is 70-80 MPa and 100-200 MPa. The organisms disparagingly survive if pressure drops to around 50 MPa. The core habitat for barophiles is generally the bottom of the ocean where extremities in terms of hydrostatic pressure and low temperature exist i.e. (1-2°C). Barosensitive microorganisms reported from deep oceans involve *Vibrio marinus, Shewanella* sp. PT99, *Shewanella benthica, Colawellia* spp. etc. Apart from mentioned habitat barophiles are also reported from Marina Trench below the 10,898-meter depth where extremities about 1100 atm pressure and 2°C temperature found. However, it seems to be very challenging to control the set of growth parameters for barophiles, main peculiar feature for isolation of barophiles is the darkness due to the occurrence of high pressure, lower temperature, and lack of light in the ocean (Gil et al., 2021). Barophilic systems are perceptive to UV light and gamma radiation, but still, the mechanism is not very clear. If in the case of barophiles pressure drops to lower then it leads to membrane fragmentation, nucleotide structural modification, and plasmolysis. Additionally, protonation and deprotonation reduce cell motility a key factor for the survival of barophiles. Another survival mechanism is hemoviscous i.e. enhanced number of polyunsaturated fatty acids over mono-unsaturated ones. Scarce reports on applications of barophiles but most commonly the biocatalysts obtained can be used as a cryoprotectant with a possible application in food preservation, molecular biology, medicine, and pharmaceuticals (Yadav et al., 2021).

VARIOUS STRATEGIES IN A NUTSHELL

Strategies Adopted by Halophiles

Salt Adaptation

The "high-salt-in strategy" found in *Halobacteriaceae* and *Halanaerobiales* (Firmicutes); accumulate salt instead of organic solutes. Extremely halophilic and aerobic red *Salinibacter ruber* (Bacteroidetes) which was isolated from the saltern crystallizer seawater (Choi et al., 2020). To cope with excessive salinity and water stress, halophiles use two basic methods. One technique is to use excessive salt. Some halophiles raise internal osmolarity by accumulating K^+ ions in the cytoplasm, which necessitates the expenditure of two ATP molecules for each K^+. Another is the 'low-salt-in' approach, which is prevalent in Bacteria and Eukarya involves the cells accumulating suitable solutes. Sugars, alcohols, amino acids, N-acetylated diamino acids, glycine betaine, ectoine, and hydroxyectoine are examples of tiny compounds that may be absorbed from the environment or produced intracellularly and can function as stress protectants. They keep proteins from 'salting out' in the cell. This method is used by halotolerant and moderately halophilic species, while severe halophiles also employ the 'high salt in' approach (Raval et al., 2018).

Sodium Pump

To adapt for high external Na^+ concentrations, archaea and a few bacteria have developed an active Na^+ and K^+ antiport, resulting in the exclusion of Na^+ from the cell. The H^+ gradient can also be utilized for Na^+ excretion or ATP production. The findings on salt adaptation suggest that it may have been one of the first adaptations during evolution (Pal et al., 2019).

Chloride Pump

The anaerobic *Halanaerobiales* and the aerobic highly halophilic *Salinibacter rubber* bacteria both have a high demand for chloride. Instead of organic osmotic solutes, these organisms collect inorganic salts. As a result, chloride plays a particular role in halo-adaptation in several halophilic bacteria (Purohit et al., 2014; Martin & McMinn, 2018).

Compatible Solutes in Osmoregulation

The fundamental mechanism developed by extremophiles to overcome osmotic stress is osmoregulation. The osmolytes such as compatible solutes are involved in osmoregulation, which protects cells and promotes growth, is the most commonly used response to hyperosmotic stress (Peng et al., 2020). The moderate halophiles and halotolerant organisms either balance the high external osmotic pressure by producing or take up certain organic molecules as their compatible solutes. Many halophilic methanogenic archaea/ bacteria accumulate organic compatible solutes as a strategy to sustain osmotic imbalance. Several compatible solutes comprising betaine, glycine, ectoines, sugar, sugar alcohols, and amino acid derivatives are synthesized. The majority of these are either zwitterionic or uncharged. Organic solutes in the cell help the microbial cell to adapt in an extensive salt concentration range (Raval et al., 2013, Jawahar et al., 2019). Due to the high ionic composition of hypersaline environments, the osmotic pressure relative to the cytoplasm is extremely high. Obligatory halophilic archaeon *Methanohalophilus portucalensis* and *Halomonas elongata* maintain a high concentration of glycine betaine as compatible solutes from the surrounding. The transportation of glycine betaine in *Halobacillus halophillus* is influenced by intracellular Cl^- concentrations. Under high NaCl concentrations, the haloalkaliphilic, sulfur-oxidizing bacteria *Thioalkalivibrio versutus* produced glycine betaine as the major organic compatible solute. To balance the external osmotic pressure, certain organisms produce sugars like trehalose or sucrose as compatible solutes (Sharshar et al., 2019). Sucrose and 5-oxo-1-proline are accumulated by haloalkaliphilic methanotrophs. At high NaCl concentrations, haloalkaliphilic methanotrophs accumulate sucrose and 5-oxo-1-proline, in addition to ectoine production.

Unusual solutes such as N-acetyl-lysine, β-amino acids, di-myoinositol phosphate, and mannosylglycerate are produced by archaea. In these circumstances, solute absorption takes precedence over *de-novo* synthesis. The microbes can thrive in high-saline environments. The build-up of the suitable solute, glycine betaine, which is an N-trimethyl derivative of glycine, allows the organisms to live at high salt concentrations and cold temperatures (Raval et al., 2018). The secondary uptake system opuD and two binding-protein-dependent transport systems, opuA and opuC (proU) are three transport mechanisms for glycine betaine in *Bacillus subtilis*. The build-up of the suitable solute glycine betaine is largely responsible for the organism's capacity to withstand both high salt and cold temperatures. Glycine betaine absorption in gram-positive bacteria and even plants has been widely investigated genetically (Singh et al., 2013; Kaczmarek et al., 2021).

Survival Strategies in Acidophiles/Alkaliphiles

Several investigations have proved that cell walls can shield the cell from alkaline conditions and help to maintain a neutral cytoplasmic pH. Alkaliphilic *Bacillus* spp. include acidic polymers such as gluconic acid, galacturonic acid, glutamic acid, phosphoric acid, and aspartic acid and in addition to peptidoglycan in cell wall composition. The capacity of the cell surface to absorb sodium and hydronium ions while repelling hydroxide ions may be due to the negative charges on the acidic non-peptidoglycan components. The presence of sodium ions in the environment appears to be necessary for efficient solute transport via the membranes of alkaliphilic *Bacillus* spp. Thus, haloalkaliphilic bacteria require a high salt concentration and alkaline pH, for stronger adaptation mechanisms (Purohit et al., 2014). Acidophiles employ several pH-control strategies, including limiting/passive proton entry into the cytoplasmic membrane and purging protons (Mirete et al., 2017). They also have a highly impermeable cell membrane that prevents active proton pumping from allowing protons to enter the cytoplasm. *Picrophilus oshinaeis* is a well-studied example of bacteria that can thrive at pH 0.7 while having an internal pH of 5. (Zhu et al., 2020).

Survival Strategies in Thermophiles

The elevated temperatures result in the denaturation of proteins, and the destruction of intracellular linkages that are detrimental to organisms. Extremophiles have several survival adaptations Figure 1 viz. secretion of chaperones or heat shock proteins often called thermosomes to prevent protein denaturation at high temperatures (Annamalai et al., 2016). Thermophiles have unique hydrogen bonds that combine with hydrophobicity to resist protein unfolding. Additionally, the formation of salt bridges and the increased number of disulfide bonds contribute to thermo-tolerance (Chakravorty et al., 2017). Several other mechanisms for thermo-resistance include oligomerization, structural compactness, hydrophobic contacts between subunits, and glycosylation. Synthetic biology approaches such as directed evolution and site-directed mutagenesis are specifically applied to study the thermostability of target proteins/enzymes, which is critical for their usage in the bioenergy sector (Adesioye et al., 2018).

Survival Strategies in Psychrophiles

Psychrophiles can survive in extremely cold temperatures majorly due to their cellular cold-adaptability mechanisms, which include the regulation of small RNA-binding proteins, cold-shock proteins, and extracellular polymeric substances (EPS) to defend cells from mechanical disruption caused by low temperatures. Furthermore, the existence of plasmids, transposable/mobile genetic elements associated with the production of unsaturated fatty acids, and the presence of G+C-rich areas encoding tRNAs, elongation factors, and RNA polymerases in the genome of psychrophiles increase their cold adaptation. Furthermore, strong translational and posttranslational processing capability in psychrophiles may be required for their development at low temperatures (De Maayer et al., 2014). Psychrophilic enzymes have more structural flexibility, less thermal stability, and higher specific activity towards a cold environment.

Figure 1. Various strategies of extremophiles

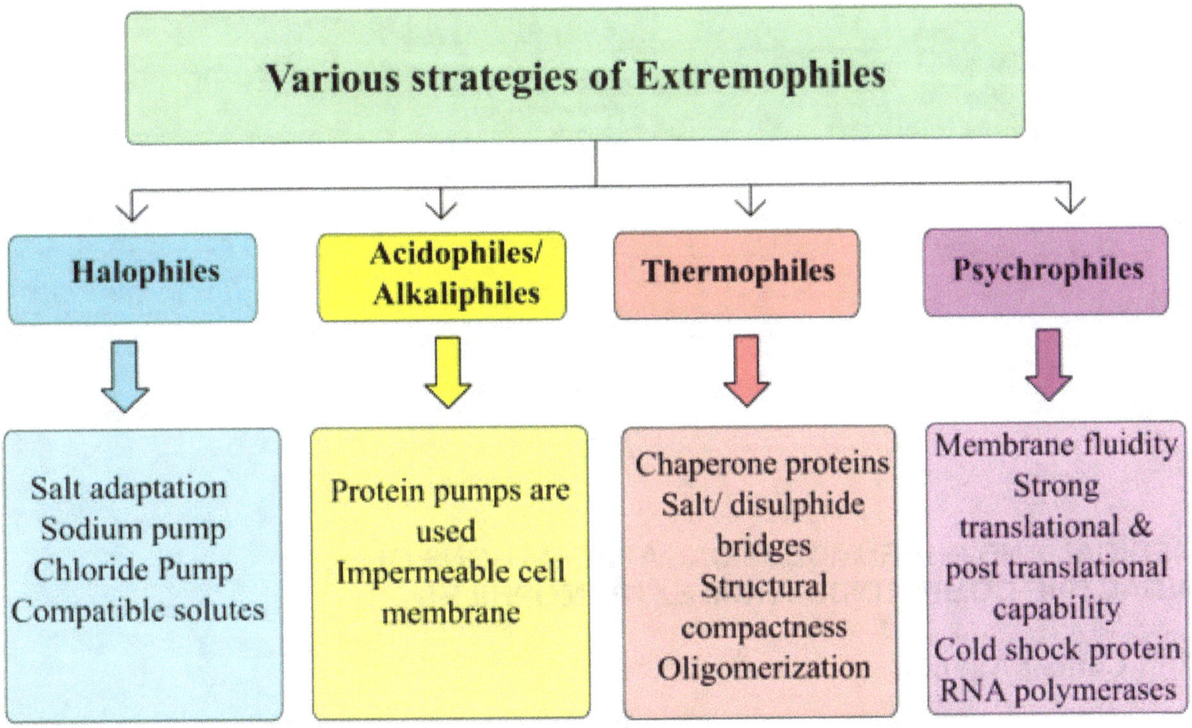

BIOACTIVE COMPOUNDS FROM EXTREMOPHILES

Extremophiles harbour bioactive compounds that range from enzymes, fatty acids, antibiotics, polymeric substances, bio-surfactant, halocins, and proteins, etc., which seems to have wide applicability in biotechnological processes. However, few bioactive compounds originated from extremophiles their peculiarity and properties are described in Table 1. The extremophilic system has acquired unique adaptive traits that are responsible for survival under extremities. The obstruction for biomolecules is its potential use less explored. There is wide scope exists for novel bioactive compounds from extremophiles that have potential applications for sustainable development.

Table 1. Bioactive compounds originated from extremophiles

Bioactive compounds	Microorganisms	Source	Peculiar extremity	Properties	References
Extremozymes	*Bacillus subtilis JK-79*	Marine soil collected from Parangipettai costal area	NI	NI	Kiruthika et al., 2018
	Vibrio alginolyticus 1374	Fresh water samples from coastal region of andhrapradesh	Therapeutic purpose most active and stable over a wide range of pH, salt and temperature	Anti-proliferative activity	Unissa et al., 2016
	Haloferax lucentensis GUBF-2 MG076078	NI	NI	Used as bio additive in detergent formulation	(Gaonkar and Furtado, 2021)
	Leucosporidium muscorum	Antarctic marine-sediment	Asparginase from cold adapted yeast that confers lower side effects of immunological reaction	Essential drug in the treatment of acute lymphoblastic leukemia (ALL)	(Freire et al., 2021)
	Bacillus cereus	Marine hydrothermal vent crabs	NI	Novel Keratin degrading subtilisin	(Gurunathan et al., 2021)
Exopolysaccharide	*Aeribacillus pallidus* 418	Hydrothermal springs in Bulgaria	Degradation temperature 176°C to 226°C	Pseudoplastic and rheological property	(Radchenkova et al., 2015)
	Geobacillus thermodenitrificans ArzA-6	Shallow marine vent	Stable at high temperature	NI	(Panosyan et al., 2018)
	Lactobacillus sakei TMW 1.411	Japanese beverage *Sake*	Tolerate high concentration of sucrose	NI	(Prechtl et al., 2018)
	Pseudoalteromonas sp. MER144	Terra nova bay, ross sea antarctic	Tolerate high concentration of mercury and cadmium	Heavy metal chelation, cryoprotection	(Caruso et al., 2018)
	Chromohalobacter canadensis 28	Pomorie salterns	Degradation temperature 250°C	Emulsifying and stabilizing property	(Radchenkova et al., 2018)
Polyhydroxyalkanoates	*Aneurinibacillus* sp.	Gudao oilfield, China	Stable at high temperature	Thermostability, water resistant capacity	(Xiao et al., 2015)
	Halomonas bluephagenesis	Salt-lake in Xinjiang province of China	Stable in high salt concentration	Biodegradability, flexibility	(Ye et al., 2018)
	Haloferax mediterranei	Rice-based ethanol manufacture at IFB Agro Industries, Noorpur (India)	Highly stable at high pH and salinity	Capacity to remove impurities from wastewater	(Bhattacharyya et al., 2014)
	Rubrobacter xylanophilus	German Collection of Microorganisms and Cell Cultures, Braunschweig, Germany	Multi-extremophilic	Enhance stress robustness of bacteria	(Kouřilová et al., 2021)
	Rubrobacter spartanus	German Collection of Microorganisms and Cell Cultures, Braunschweig, Germany	Multi-extremophilic	Enhance stress robustness of bacteria	(Kouřilová et al., 2021)
Halocins	*S8a Haloarchaea*	NI	First characterized microhalocin with 36 amino acids	Desalted easily and heat resistant	(Price and Shand, 2000)

MEDICAL AND PHARMACEUTICAL APPLICATIONS OF BIOACTIVE COMPOUNDS FROM EXTREMOPHILES

Anti-Cancer Agents

Anti-cancer compounds inhibit cancer proliferation; those sourced from extremophiles have received attention for intensive research due to their atypical applicability Figure 2. Extremophiles synthesize a variety of extremolytes that include a spectrum of antimicrobial and antitumor compounds (Singh et al., 2010). Recently, a halophilic extracellular polymeric substance from *Halomonas stenophila* was reported with anticancer, prevents cardiovascular disorders, and with antiaging properties. Blocking the synthetic lipids molecules lead to inhibit the growth and proliferation of tumours. (Amjres et al.,

2015). The extremozymes such as amylases, ligases, xylanases, α-glucosidases, endoglucanases, and Exopolymeric substances sourced from acidophiles are cited as a preventive measures for ulcer disease and various gastric cancer. (Babu et al., 2015). The extremolytes generally cold-active by-products from psychrophiles are used in food technology, molecular and genetics, pharmaceutical research, and also in medicinal purposes; on contrary, the cold stable by-products are used for the production of novel compounds viz. primary and secondary metabolites generally antibiotics used in cancer therapy and synthesis of nutraceutical. The Mycosporine-like amino acids (MAAs), scytonemin are the few bioactive compounds from extremophiles has been reported to have potential applications as preventive agents in cancers such as melanoma, which are induced by harmful UV radiation. Furthermore, the prospect for MAAs compounds may be associated with therapeutic implications (Abraham et al., 2021). For the advancement of novel pharmacophore, the most cited extremolyte is scytonemin found in sunscreens produces protein kinase inhibitors such as antiproliferative and anti-inflammatory drugs. The radio-resistant microbes that belong to *Halobacterium* and *Rubrobacter* sp. are reported to secrete bacterioruberin have been cited (Singh and Gabani, 2011) to have applicability in preventing human skin cancer due to restoring DNA impaired strands occurs by harmful UV radiation. The deinoxanthin obtained from the radioresistant bacterium *Deinococcus radiodurans* causes the cancer cell apoptosis as a chemopreventive agent. The extremolytes can inhibit protein aggregation or misfold thereby it seems to play a fascinating role in drug development for disease (Faria et al., 2019); for skin-related by-products, firoin and ectoine both have been reported to reduce signal-dependent pathways resulting from acquaintance towards carbon nanoparticles both *in-vivo* as well as *in-vitro* that broadens the applicability of compatible solutes. Many processes involve the activation of mitogen-activated protein kinases that upregulation apoptosis, proinflammatory cytokines, the proliferation of lung epithelium which leads to lung cancer, chronic pulmonary and fibrosis disease (Autengruber et al., 2014). The hyperthermophile i.e. aerobic archaeon *Aeropyrum pernix* K1 synthesizes thermostable nucleoside phosphorylase that has wide applicability for nucleoside analogues synthesis, which has antiviral properties (Zhu et al., 2013). Additionally, other extremolytes involve thermolysin from thermophiles to clean up DNA before PCR amplification and in various DNA processing enzymes. Liposome from *Halobacterium cutirubrum* synthesizes novel biomolecules that have useful applications in medicinal and pharmaceutical purposes particularly for the treatment of disease (Litchfield, 2011).

Drug Delivery, Immunosuppressant, Antibiotics

Extremophiles mainly involved in enzyme production and antibiotic preparation find their application in the medical field (Shukla et al., 2020). Some halophilic bacteria produce lipids primarily used as delivery vehicles for vaccines and drugs. Moreover, compounds like siderophores, found in halophilic archaea, are involved in the treatment of iron deficiency diseases or enhance antibiotic activity against bacteria. Halophiles are involved in nanoparticles synthesis that is the basis of gene therapy and drug delivery, *in-vitro* diagnosis, and *in-vivo* imaging (Ryan and Brayden, 2014). A halophilic archaeon, *Halococcus salifodinae* BK3 was employed to generate tellurium nanoparticles (Srivastava et al., 2015), and spherical intracellular selenium nanoparticles which were active against a range of Gram-positive and Gram-negative bacterial species. Halophilic archaea are widely exploited as carriers for topical drug delivery across the skin, due to the unique properties of their membrane lipids (Carrer et al., 2014). Extremophiles such as *Geobacillus thermodenitrificans* and *Bacillus licheniformis* produce EPS, which can act as strong stimulators for cell-mediated immunity. In the treatment of immunocompromised patients, these immunomodulatory agents can play a crucial role. *Halomonas* sp. produced PHAs which have multiple biomedical applications such as in tissue implants, comprising blood vessels, nerve conduits, artificial heart valves, tendons or cartilage, bone replacements, oesophagus replacements (Luo et al., 2019), porous micro spherical implant-scaffolds for microsurgery, surgical sutures, and hydrogels/organogels (Zhang et al., 2019). Additionally, PHA produced by extremophiles proves to demonstrate anti-osteoporosis effects (Cao et al., 2014). PHAs can also be converted into nanoparticles for controllable and targeted drug delivery to the appropriate sites (Li and Loh, 2017), for surface display of numerous target proteins and protein purification (Wong & Rehm, 2018), or as PHA Nano-vaccines.

Figure 2. Applications of bioactive compounds from Extremophiles

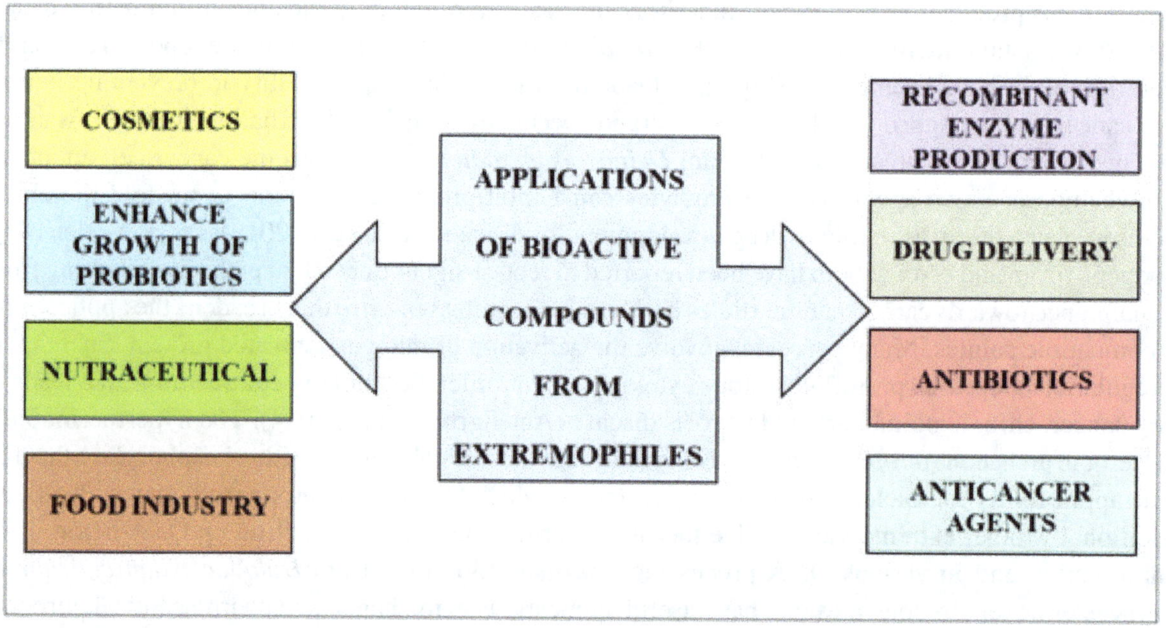

Recombinant Enzyme Production

The overall cost of intrinsic enzymes are high & obstructs its applicability for medicinal and pharmaceutical purpose so there is a constant upsurge for a recombinant form of the enzyme. The recombinant enzyme is produced generally by employing site-directed mutagenesis, genetic engineering tools in which genes can be modified by CRISPR and related technologies. The extremophiles-instigated microbial system is scarcely reported for recombinant enzyme synthesis. However, by employing molecular tools and various approaches the progression in enzyme technology can be possible. *Escherichia coli* is the foremost studied system for cloning and overexpression of genes of interest due to ease of handling and growth requirements (Nieuwkoop et al. 2019). The L-glutaminase gene sourced from halophilic bacteria *Micrococcus luteus* K3 was cloned in *E.coli* JM109 and allowed to express in vectors pUC19 and pKK2233 that leads to an increase in the expression level of the gene that has therapeutic applications, particularly in cancer therapy (Binod et al., 2017). Additionally, *Pseudomonas* sp. is reported for recombinant L-asparaginase production that has potential implications (Sindhu & Manonmani, 2018).

The alkaliphilic microorganisms *Thermoleophilum album,* and *Exiguobacterium oxidotolerans* produce recombinant catalase that can be used in food processing and genetic engineering processes. There are another amidohydrolase group of therapeutic enzymes that have potential applicability in the medicinal and pharmaceutical sectors. With the advent of the metagenomics technique, the high scale mining of data can be possible and results in novel gene sequences from various extreme surroundings (Tripathi et al., 2018). (In the present section, the information related to various omics approaches has been mentioned).

As a Potent Antioxidant Molecule

Halomonas nitroreducens WB1, a halothermophilic bacterium produced EPS with antioxidant properties against DPPH and hydroxyl radicals (Chikkanna et al., 2018). Halophilic EPS plays a crucial role as natural antioxidants for the inhibition of oxidative damage in humans. Thermophilic and halophilic EPS produced by *Geobacillus* sp. strain TS3-9 and *Halolactibacillus miurensis* respectively; both showed dose-dependent scavenging activity against hydroxyl, DPPH (2,2-diphenyl-1-picrylhydrazyl), and superoxide free radicals (Arun et al., 2017). A psychrophilic *Polaribacter* sp. SM1127 reported to produce a hyper-branched EPS showed significantly excessive antioxidant activity than that of hyaluronic acid, an industrial seizing agent for scavenging radicals. These extremophilic EPSs with outstanding antioxidative capacity may be efficient in the therapeutic purpose to treat neurodegenerative diseases. Extensive research inputs are required to determine the biofilm exopolysaccharides potential use (Baria et al., 2020).

Synthesis of Nutraceutical and Application in the Food Industry

Stephen DeFelice coined the word "nutraceutical" in 1989 by combining the terms "nutrition" and "pharmaceutical." Raw foods, fortified foods, and dietary supplements that include biologically active molecules that give health advantages beyond basic nutrition are referred to as bioactive molecules. Certain peptides, vitamins, fatty acids, polysaccharides, and phytochemicals are examples of bioactive substances that are naturally present in foods, may be added to foods to provide fortified or functional foods, or can be made as dietary supplements. These bioactive compounds can be extracted from natural sources or synthesized chemically and biotechnologically (Fernandes et al., 2019). Marine nutraceutical goods account for a significant percentage of the worldwide market and are generated from a variety of sources that include a wide spectrum of bioactive compounds.

Due to their unique activity under abnormal circumstances, biomolecules such as enzymes extracted from extremophiles can be extremely helpful in the food sector and is commonly acknowledged that extremophiles have a great potential for application in biotechnology. Biosynthesis of L-glutaminase, which has a variety of uses in the food industry, pharmaceutical manufacturing, and nutraceutical production. Theanine, for example, can enhance the growth of probiotic microorganisms. *Stenotrophomonas maltophilia* produces a salt-tolerant glutaminase, which is widely utilized in Japanese soy sauce production (Patel et al., 2021). Many commercially available aminopeptidases from microbial sources are used in the food industry, such as flavoenzyme from *Aspergillus oryzae*, neutralize from *A. oryzae* used in the dairy and fish industries, corolase from *Aspergillus sojae* used for hydrolysis of animal and vegetable protein, debitrase from *Lactococcus lactis*, and *A. oryzae* for bitterness prevention during fermentation (Nandan & Nampoothiri 2017).

Cold active enzymes produced by psychrophiles signify an effective tool in the food industry. There is an increment in industrial trend for food products treated under low temperature to prevent it from detrimental effects on texture, nutritional value, and taste as well as energy-efficient. Cold-active β-galactosidase is an essential enzyme in the dairy industry. It can break down lactose into glucose and galactose at freezing temperatures. The main application is for synthesizing lactose-free milk and other derived foodstuffs for the lactose intolerant, world population. Moreover, utilized to alter lactose in whey to D-tagatose, which is a natural sweetener with value-added application viz. low glycemic index and calories (Struvay & Feller, 2012). The cold-active pectinases showed significant applications in the fruit juice industry as they maintain the quality and nutritional properties of the fruit juices and allow various processing steps such as liquefaction, clarification, and extraction of juice; at low temperature (Adapa et al., 2014). Xylanase produced by *Pseudoalteromonas haloplanktis*, an Antarctic bacterium, efficiently enhanced the mechanical properties of dough and the final quality of bread with an increase in loaves. Cold-active xylanases hydrolyze β-1,4-xylan found in all flours, are the chief constituents of industrial dough conditioners utilized at low temperatures essential for dough resting to enhance bread quality (Struvay & Feller, 2012). Additionally, in the food industry extremolytes are the potent candidates for the production of food products, functional foods that supplement positive health benefits by improving short-range functioning ability or by the long-lasting improvement of particular diseases. For instance, ectoine accumulation (up to 89 mg/100 g of product) when cheeses are treated with *Brevibacterium linens* for surface ripening of the product (Giuffrida et al., 2020).

Thermozymes produced by thermophiles have several applications in the food industry, as their stability under high temperatures ensures that they resist degradation at high temperatures greater productivity, and lesser process time. Carbohydrate-degrading enzymes such as glycosyl hydrolases have abundant applications in the animal feed and food industries as well as in lignocellulose degradation (Aulitto et al., 2019). They comprised pullulanase, α- and β-amylases, α-glucosidases, and glucoamylases all of which are essential for the complete breakdown of starch. Xylanases and cellulases are widely used in food processing. Alcalase, a protease used to treat soy meal and thermolysin, synthesizes the artificial sweetener aspartame (Barzkar et al, 2018).

Cold-adapted proteases are implied for tenderisation, during the processing of fish and meat, and their application has been associated with enzymes from Atlantic cold-water fish (Parvizpour et al., 2021). Many of these enzymes are used for eggs processing, eliminating skin from squid mantle or fish fillets, or fish bones, or extracting fish membranes. Generally, the processing of seafood involved the use of several cold-adapted enzymes (Venugopal, 2016). Some cold-adapted microorganisms produce antifreeze small glycoproteins that inhibit the formation of ice crystals in body fluids and cells, first revealed in fish from polar waters. Ocean pout fish synthesize antifreeze protein called ice-structuring protein and is overexpressed in *Saccharomyces cerevisiae*. Unilever's ice cream prevent the recrystallization of ice after freeze-thaw cycles, which may affect the ice cream's texture and taste (Margesin and Feller, 2010).

Role of Probiotics

Probiotics are generally referred to as beneficial live microorganisms which confer many health benefits when administered in ample proportion, thereby the use of probiotic-rich food is very widely spread (Ouwehand, 2017). There are many probiotic strains reported but very few are extremophilic hence still there is a constant upsurge for an effective strain that can confer many health benefits particularly for therapeutic purposes. Various criteria that one needs to take into consideration while selecting potent probiotic strains include resistance towards enzymes and pH found to be present in the oral cavity, ability to resist bile and pancreatic juice present in small intestine and susceptibility to antibiotics, synthesis of antimicrobial compounds, colonization intestinal epithelial cells. Health benefits confer by LAB may involve therapeutic (synthesis of vitamins) / nutritional (immunomodulation, prevention of various types of cancers i.e. gastric and stomach ulcers, antimutagenic agent, etc.) (Śliżewska et al., 2021). Acidophilic organisms are common intestinal micro-flora, certain strains of *Lactobacillus acidophilus* prevent many diarrheal infections viz. traveller's diarrhoea, infectious diarrhoea, and antibiotic-associated diarrhoea (Ouwehand et al., 2014). The Commonest bowel disease associated with probiotics is Crohn's disease and ulcerative colitis, the most usual feature is chronic diarrhoea. The bacteria *Helicobacter pylori* was reported to cause severe ulcers in the stomach and duodenum. A thorough study of acidophiles and their stability concerning acid is possible through crystal structure analysis. The most prompting purpose of acidophiles is the formulation of a novel drug for treatments to those who are more prone to gastric cancers and stomach ulcers (Dos Reis et al., 2017). However, to regulate the pH of cells these bacteria usually pump out hydrogen ions inside the membrane to respond to change. The exact mechanism of acidophile and its regulatory machinery to retain the pH stability of cells is unknown. Extremozymes found in the genome of acidophilic bacteria are a major focus of the research area. Hence, the wide scope still exists for an upsurge of novel probiotics from extremophiles that has biotechnological implications (Lebeer et al., 2018).

Cosmetics

Psychrophilic bacteria exhibit promising applications in the cosmetics and pharmaceutical industries. Antarctic bacterium *Pseudoalteromonas* produced Antarticine-NF3, which is an antifreeze glycoprotein and highly efficient in scar treatment. It has been also used in cosmetic regeneration creams. Furthermore, *Candida antarctica* which is the polar yeast used to synthesize two cold-active lipases, A and B included in an enormous application associated with cosmetics (Guo et al., 2020).

Radiation-resistant microorganisms can produce various primary and secondary metabolites that defend them against high UV radiation, which destruct living cells. UV-A rays in sunlight can lead to skin ageing as well as produces ROS, which can damage DNA (Gabani & Singh, 2013), however, UV-B rays are the most mutagenic and cytotoxic for human skin, causing skin cancer. Photoprotective compounds such as mycosporine-like amino acids (MAAs) and scytonemin have chiefly been exploited in cosmetics and potential applications in the pharmaceutical sector. Some extremophilic cyanobacteria such as *Nostoc*, *Anabaena*, *Scytonema*, and *Lyngbya* generate a small molecule (Mr 544) called Scytonemin; synthesized during condensation of two cyclopentanone and indole rings. Its mode of action leads to decreasing the *in-vivo* production of ROS, in addition to thymine dimers in DNA (Rastogi et al., 2014). Scytonemin is a yellow-brown to the red pigment that absorbs blue and near UV radiation, but light with a higher wavelength used for photosynthesis is not absorbed by this molecule and can diminish inhibition of photosynthesis by UV-rays (Fuentes-Tristan et al., 2019). MAAs have a similar mode of action and function. In the cosmetic industry, MAAs are used as a preventive agent against UV rays-induced skin cancer. MAAs compounds are potent candidates in UV protective sunscreens (Shang et al., 2018).

Various other molecules are known, such as palythene, palythinol, porphyra-334, palythine, asterina, and shinorine. They absorb UV rays and prevent the creation of pyrimidine dimers in DNA (Singh & Gabani, 2011) but they allow the cell to disperse energy as heat, without the generation of ROS. These molecules offer defense against UV-B rays, according to their intracellular localization and structure. These compounds have anti-inflammatory properties in response to UV rays indicate that they may have potential skin anti-ageing activity, as shown by their ability to inhibit COX-1 mRNA production.

These compounds are widely used in sunscreens and other skincare products (Derikvand et al., 2017). *Chromohalobacter canadensis* 28, a halophilic bacterium synthesized exopolysaccharide; constitute Poly-gamma glutamic acid (γ-PGA) and carbohydrate with high lipophilic and hydrophilic properties are an obvious choice for ointments and skin creams (Radchenkova et al., 2018).

Case Study

The bioactive compounds sourced from extremophiles play a very crucial role in the medical field. However, the DNA polymerase obtained from the thermophilic bacterium *Thermus aquaticus* is known to play a critical role in molecular biology and now in molecular diagnostics as well. Several extremophiles synthesize antimicrobial peptides i.e. diketopiperazines are reported to possess antimicrobial, antifungal, antiviral, and antitumor properties in addition to the blood clotting phenomenon. Many other halocins from halophiles viz. *Naloterrigena hispanica* and *Natronococcus occultus* are cited to regulate the quorum-sensing pathways an essential one for the pathogenic bacteria such as *Pseudomonas aeruginosa*, a causative agent of pneumonia and an also infection found in patients with cystic fibrosis. Thus an ample scope exists in this field and it can give an altogether different dimension to the presently existing methods. This could be shaped as an alternative to combat multi drug-resistant bacterial infection. Haloarchaea have long been reported for the synthesis of gas vesicles and polar lipids capable of eliciting an immune response in mice and even is less toxic to cells (Coker, 2016). Recently the amidohydrolase group of enzymes from the terrestrial origin is studied for cancer treatment particularly a case study on acute lymphoblastic leukaemia in patients with Native Asparaginase from *E.coli* and *Erwinia sp.* is reported, (Nunes et al., 2020) and available commercially under the brand name of Oncaspar, Elspar, Leukanase, Novozymes and many more. Thus working in the same direction identical enzymes of the marine microbial system can be purified and employed (Patel et al., 2021). It is to the best of our knowledge that there is a scarce report on the therapeutic usage of bioactive compounds from extremophiles, which encourage researchers to work harder in this context. Though much scope of this field is still in its infancy and a mammoth task to be done for exploration of *in-vivo* use of bioactive molecules for medicinal purpose.

CONCLUSION AND FUTURE PROSPECTS

Extremophilic microorganisms synthesize various types of bioactive compounds that involve primary and secondary metabolites and have profound benefits for humans. The unusual surroundings are known to harbour the most potent extremophiles that serve as a prospect for various biotechnological purposes. Particularly by employing molecular tools, the enhancement of peculiar features of extremophiles may be possible through metabolic and genetic proficiencies. However, recently the bioactive compounds from extremophiles are seemed to have wide perseverance for medicinal and pharmaceutical purposes for sustainable development as reported from the literary works. Furthermore, the molecular biology and metagenomics approach has paved way for in-depth research on extremophilic metabolites many of which are reported novel and have plenty of scope in future.

ACKNOWLEDGMENT

Authors NYP, DMB and VHR acknowledge the infrastructural support provided by the Department of Microbiology & Biotechnology, Gujarat University, Ahmedabad. Authors SMY and VHR kindly acknowledge the infrastructural support from Christ College and Department of Biosciences, Saurashtra University, Rajkot respectively. VHR heartily acknowledge the junior & senior research fellowships by the Department of Biotechnology (DBT), Science and Engineering Research Board (SERB) Young Scientist Fellowship, and Universities Grant Commission (UGC)-Basic Science Research (BSR) Start-Up [Research Grant No.F.30-521/2020]. The Research fellowship [Reference No.202001380174] & [Reference No.201901380189] under the "SHODH - ScHeme of Developing High-Quality Research" by the Department of Higher Education, Government of Gujarat to NYP & DMB is also heartily acknowledged.

REFERENCES

Abraham, R. E., Alghazwi, M., Liang, Q., & Zhang, W. (2021). Advances on marine-derived natural radioprotection compounds: Historic development and future perspective. *Marine Life Science & Technology*, 1–14. doi:10.100742995-021-00095-x

Adapa, V., Ramya, L. N., Pulicherla, K. K., & Rao, K. S. (2014). Cold active pectinases: Advancing the food industry to the next generation. *Applied Biochemistry and Biotechnology*, *172*(5), 2324–2337. doi:10.100712010-013-0685-1 PMID:24390855

Adesioye, F. A., Makhalanyane, T. P., Vikram, S., Sewell, B. T., Schubert, W. D., & Cowan, D. A. (2018). Structural characterization and directed evolution of a novel acetyl xylan esterase reveal thermostability determinants of the carbohydrate esterase 7 family. *Applied and Environmental Microbiology*, *84*(8), e02695–e17. doi:10.1128/AEM.02695-17 PMID:29453256

Afra, S., Makhdoumi, A., Matin, M. M., & Feizy, J. (2017). A novel red pigment from marine Arthrobacter sp. G20 with specific anticancer activity. *Journal of Applied Microbiology*, *123*(5), 1228–1236. doi:10.1111/jam.13576 PMID:28862783

Amjres, H., Béjar, V., Quesada, E., Carranza, D., Abrini, J., Sinquin, C., Ratiskol, J., Colliec-Jouaultm, S., & Llamas, I. (2015). Characterization of halo glycan, an exopolysaccharide produced by *Halomonas stenophila* HK30. *International Journal of Biological Macromolecules*, *72*, 117–124. doi:10.1016/j.ijbiomac.2014.07.052 PMID:25109455

Amoozegar, M. A., Safarpour, A., Noghabi, K. A., Bakhtiary, T., & Ventosa, A. (2019). Halophiles and their vast potential in biofuel production. *Frontiers in Microbiology*, *10*, 1895. doi:10.3389/fmicb.2019.01895 PMID:31507545

Anderson, A. W. (1956). Studies on a radio resistant *Micrococcus*. 1. Isolation, morphology, cultural characteristics, and resistance to γ-radiation. *Food Technology*, *10*, 575–578.

Annamalai, N., Rajeswari, M. V., & Balasubramanian, T. (2016). Thermostable and alkaline cellulases from marine sources. In *New and future developments in microbial biotechnology and bioengineering* (pp. 91–98). Elsevier. doi:10.1016/B978-0-444-63507-5.00009-5

Arun, J., Selvakumar, S., Sathishkumar, R., Moovendhan, M., Ananthan, G., Maruthiah, T., & Palavesam, A. (2017). *In-vitro* antioxidant activities of an exopolysaccharide from a salt pan bacterium *Halolactibacillus miurensis. Carbohydrate Polymers, 155*, 400–406. doi:10.1016/j.carbpol.2016.08.085 PMID:27702528

Atalah, J., Cáceres-Moreno, P., Espina, G., & Blamey, J. M. (2019). Thermophiles and the applications of their enzymes as new biocatalysts. *Bioresource Technology, 280*, 478–488. doi:10.1016/j.biortech.2019.02.008 PMID:30826176

Aulitto, M., Fusco, S., Limauro, D., Fiorentino, G., Bartolucci, S., & Contursi, P. (2019). Galactomannan degradation by thermophilic enzymes: A hot topic for biotechnological applications. *World Journal of Microbiology & Biotechnology, 35*(2), 32. doi:10.100711274-019-2591-3 PMID:30701316

Autengruber, A., Sydlik, U., Kroker, M., Hornstein, T., Ale-Agha, N., Stöckmann, D., Bilstein, A., Albrecht, C., Paunel-Gorgulu, A., Suschek, C. V., Krutmann, J., & Unfried, K. (2014). Signalling-dependent adverse health effects of carbon nanoparticles are prevented by the compatible solute mannosylglycerate (firoin) *in-vitro* and *in-vivo*. *PLoS One, 9*(11), e111485. doi:10.1371/journal.pone.0111485 PMID:25415441

Babu, P., Chandel, A. K., & Singh, O. V. (2015). Therapeutic Implications of Extremophiles. In: Extremophiles and Their Applications in Medical Processes. Springer. doi:10.1007/978-3-319-12808-5_3

Baria, D. M., Patel, N. Y., Arya, P. S., Rajput, K. N., Panchal, R. R., & Raval, V. H. (2020). Exploration of marine bacteria for their Exopolysaccharide (EPS) producing ability along various regions of the Gujarat coast. *Bioscience Biotechnology Research Communications, 13*(1), 78-84. http://bbrc.in/bbrc/wp-content/uploads/2020/05/FINAL-PDF-Special-Issue-Vol-13-No-1-2020.pdf

Barzkar, N., Homaei, A., Hemmati, R., & Patel, S. (2018). Thermostable marine microbial proteases for industrial applications: Scopes and risks. *Extremophiles, 22*(3), 335–346. doi:10.100700792-018-1009-8 PMID:29442247

Besse, A., Vandervennet, M., Goulard, C., Peduzzi, J., Isaac, S., Rebuffat, S., & Carré-Mlouka, A. (2017). Halocin C8: an antimicrobial peptide distributed among four halophilic archaeal genera: *Natrinema, Haloterrigena, Haloferax*, and *Halobacterium. Extremophiles, 21*(3), 623–638. doi:10.100700792-017-0931-5 PMID:28386790

Bhattacharyya, A., Saha, J., Haldar, S., Bhowmic, A., Mukhopadhyay, U. K., & Mukherjee, J. (2014). Production of poly-3- (hydroxybutyrate-co-hydroxy valerate) by *Haloferax mediterranei* using rice-based ethanol stillage with simultaneous recovery and re-use of medium salts. *Extremophiles, 18*(2), 463–470. doi:10.100700792-013-0622-9 PMID:24442255

Binod, P., Sindhu, R., Madhavan, A., Abraham, A., Mathew, A. K., Beevi, U. S., Sukumaran, R. K., Singh, S. P., & Pandey, A. (2017). Recent developments in L-glutaminase production and applications–An overview. *Bioresource Technology, 245*, 1766–1774. doi:10.1016/j.biortech.2017.05.059 PMID:28549811

Cao, Q., Zhang, J., Liu, H., Wu, Q., Chen, J., & Chen, G. Q. (2014). The mechanism of anti-osteoporosis effects of 3-hydroxybutyrate and derivatives under simulated microgravity. *Biomaterials, 35*(28), 8273–8283. doi:10.1016/j.biomaterials.2014.06.020 PMID:24976243

Carrer, D. C., Higa, L. H., Tesoriero, M. V. D., Morilla, M. J., Roncaglia, D. I., & Romero, E. L. (2014). Structural features of ultra deformable archaeosomes for topical delivery of ovalbumin. *Colloids and Surfaces. B, Biointerfaces, 121*, 281–289. doi:10.1016/j.colsurfb.2014.05.015 PMID:24974012

Caruso, C., Rizzo, C., Mangano, S., Poli, A., Di Donato, P., Nicolaus, B., Di Marco, G., Michaud, L., & Giudice, A. L. (2018). Extracellular polymeric substances with metal adsorption capacity produced by *Pseudoalteromonas* sp. MER144 from Antarctic seawater. *Environmental Science and Pollution Research International, 25*(5), 4667–4677. doi:10.100711356-017-0851-z PMID:29197057

Casanueva, A., Tuffin, M., Cary, C., & Cowan, D. A. (2010). Molecular adaptations to psychrophily: The impact of 'omic' technologies. *Trends in Microbiology, 18*(8), 374–381. doi:10.1016/j.tim.2010.05.002 PMID:20646925

Chakravorty, D., Khan, M. F., & Patra, S. (2017). Multifactorial level of extremostability of proteins: Can they be exploited for protein engineering. *Extremophiles, 21*(3), 419–444. doi:10.100700792-016-0908-9 PMID:28283770

Chen, S., Sun, S., Korfanty, G. A., Liu, J., & Xiang, H. (2019). A halocin promotes DNA uptake in *Haloferax mediterranei. Frontiers in Microbiology, 10*, 1960. doi:10.3389/fmicb.2019.01960 PMID:31620096

Chen, X. (2021). *Thriving at Low pH: Adaptation Mechanisms of Acidophiles*. Academic Press.

Chikkanna, A., Ghosh, D., & Kishore, A. (2018). Expression and characterization of a potential exopolysaccharide from a newly isolated halophilic thermotolerant bacteria *Halomonas nitroreducens* strain WB1. *Peer-Reviewed Scientific Mega Journal, 6*, e4684. doi:10.7717/peerj.4684 PMID:29707437

Choi, B. H., Hwang, H. J., Lee, J. E., Oh, S. H., Hwang, J. S., Lee, B. Y., & Lee, P. C. (2020). Microbial Production of Retinyl Palmitate and its Application as a Cosmeceutical. *Antioxidants, 9*(11), 1130. doi:10.3390/antiox9111130 PMID:33202630

Chung, D., Sarai, N. S., Himmel, M. E., & Bomble, Y. J. (2020). Genetics of Unstudied Thermophiles for Industry. In *Metabolic Pathway Engineering* (pp. 5–19). Humana. doi:10.1007/978-1-0716-0195-2_2

Clarke, A. (2014). The thermal limits to life on Earth. *International Journal of Astrobiology, 13*(2), 141–154. doi:10.1017/S1473550413000438

Coker, J. A. (2016). Extremophiles and biotechnology: Current uses and prospects. *F1000 Research, 5*, 396. Advance online publication. doi:10.12688/f1000research.7432.1 PMID:27019700

Daoud, L., & Ali, M. B. (2020). Halophilic microorganisms: Interesting group of extremophiles with important applications in biotechnology and the environment. In *Physiological and Biotechnological Aspects of Extremophiles* (pp. 51–64). Academic Press. doi:10.1016/B978-0-12-818322-9.00005-8

DasSarma, S., & DasSarma, P. (2015). Halophiles and their enzymes: Negativity put to good use. *Current Opinion in Microbiology, 25*, 120–126. doi:10.1016/j.mib.2015.05.009 PMID:26066288

de Almeida Couto, C. R., Alvarez, V. M., Marques, J. M., de Azevedo Jurelevicius, D., & Seldin, L. (2015). Exploiting the aerobic endospore-forming bacterial diversity in saline and hypersaline environments for biosurfactant production. *BMC Microbiology, 15*(1), 1–17. doi:10.118612866-015-0575-5 PMID:26511622

De Maayer, P., Anderson, D., Cary, C., & Cowan, D. A. (2014). Some like it cold: Understanding the survival strategies of psychrophiles. *EMBO Reports*, *15*(5), 508–517. doi:10.1002/embr.201338170 PMID:24671034

de Menezes, G. C., Godinho, V. M., Porto, B. A., Gonçalves, V. N., & Rosa, L. H. (2017). *Antarctomyces pellizariae* sp. nov., a new, endemic, blue, snow resident psychrophilic ascomycete fungus from Antarctica. *Extremophiles*, *21*(2), 259–269. doi:10.100700792-016-0895-x PMID:27900476

Defelice, S. L. (1989). *The nutraceutical revolution: Fueling a powerful, new international market*. Harvard University Advanced Management Program in Biomedical Research and Development.

Derikvand, P., Llewellyn, C. A., & Purton, S. (2017). Cyanobacterial metabolites as a source of sunscreens and moisturizers: A comparison with current synthetic compounds. *European Journal of Phycology*, *52*(1), 43–56. doi:10.1080/09670262.2016.1214882

Di Donato, P., Buono, A., Poli, A., Finore, I., Abbamondi, G. R., Nicolaus, B., & Lama, L. (2018). Exploring marine environments for the identification of extremophiles and their enzymes for sustainable and green bioprocesses. *Sustainability*, *11*(1), 1–20. doi:10.3390u11010149

Dos Reis, S. A., da Conceição, L. L., Siqueira, N. P., Rosa, D. D., da Silva, L. L., & Maria do Carmo, G. P. (2017). Review of the mechanisms of probiotic actions in the prevention of colorectal cancer. *Nutrition Research (New York, N.Y.)*, *37*, 1–19. doi:10.1016/j.nutres.2016.11.009 PMID:28215310

Elazzazy, A. M., Abdelmoneim, T. S., & Almaghrabi, O. A. (2015). Isolation and characterization of biosurfactant production under extreme environmental conditions by alkali-halo-thermophilic bacteria from Saudi Arabia. *Saudi Journal of Biological Sciences*, *22*(4), 466–475. doi:10.1016/j.sjbs.2014.11.018 PMID:26150754

Faria, J., Ahmed, S., Gerritsen, K. G., Mihaila, S. M., & Masereeuw, R. (2019). Kidney-based *in-vitro* models for drug-induced toxicity testing. *Archives of Toxicology*, *93*(12), 3397–3418. doi:10.100700204-019-02598-0 PMID:31664498

Feng, S., Qiu, Y., Huang, Z., Yin, Y., Zhang, H., Zhu, D., Tong, Y., & Yang, H. (2021). The adaptation mechanisms of *Acidithiobacillus caldus* CCTCC M 2018054 to extreme acid stress: Bioleaching performance, physiology, and transcriptomics. *Environmental Research*, *199*, 111341. doi:10.1016/j.envres.2021.111341 PMID:34015291

Fernandes, S. S., Coelho, M. S., & de las Mercedes Salas-Mellado, M. (2019). Bioactive compounds as ingredients of functional foods: polyphenols, carotenoids, peptides from animal and plant sources new. In *Bioactive Compounds* (pp. 129–142). Woodhead Publishing. doi:10.1016/B978-0-12-814774-0.00007-4

Freire, R. K. B., Mendonça, C. M. N., Ferraro, R. B., Moguel, I. S., Tonso, A., Lourenco, F. R., Santos, J. H. P. M., Sette, L. D., & Pessoa, A. Jr. (2021). Glutaminase-free L-asparaginase production by *Leucosporidium muscorum* isolated from Antarctic marine-sediment. *Preparative Biochemistry & Biotechnology*, *51*(3), 277–288. doi:10.1080/10826068.2020.1815053 PMID:32921254

Fuentes-Tristan, S., Parra-Saldivar, R., Iqbal, H. M., & Carrillo-Nieves, D. (2019). Bioinspired biomolecules: Mycosporine-like amino acids and scytonemin from *Lyngbya* sp. with UV-protection potentialities. *Journal of Photochemistry and Photobiology. B, Biology*, *201*, 111684. doi:10.1016/j.jphotobiol.2019.111684 PMID:31733505

Gabani, P., & Singh, O. V. (2013). Radiation-resistant extremophiles and their potential in biotechnology and therapeutics. *Applied Microbiology and Biotechnology*, *97*(3), 993–1004. doi:10.100700253-012-4642-7 PMID:23271672

Gaonkar, S. K., & Furtado, I. J. (2021). Valorization of low-cost agro-wastes residues for the maximum production of protease and lipase haloextremozymes by *Haloferax lucentensis* GUBF-2 MG076078. *Process Biochemistry*, *101*, 72–88. doi:10.1016/j.procbio.2020.10.019

Ghanmi, F., Carré-Mlouka, A., Zarai, Z., Mejdoub, H., Peduzzi, J., Maalej, S., & Rebuffat, S. (2020). The extremely halophilic archaeon *Halobacterium salinarum* ETD5 from the solar saltern of Sfax (Tunisia) produces multiple halocins. *Research in Microbiology*, *171*(2), 80–90. doi:10.1016/j.resmic.2019.09.003 PMID:31560985

Gil, J. F., Mesa, V., Estrada-Ortiz, N., Lopez-Obando, M., Gómez, A., & Plácido, J. (2021). Viruses in extreme environments, current overview, and biotechnological potential. *Viruses*, *13*(1), 81. doi:10.3390/v13010081 PMID:33430116

Giuffrida, D., Monnet, C., Laurent, F., Cacciola, F., Oteri, M., Le Piver, M., Caro, Y., Donato, P., Mondello, L., Roueyre, D., & Dufossé, L. (2020). Carotenoids from the ripening bacterium *Brevibacterium linens* impart color to the rind of the French cheese, Fourme de Montbrison (PDO). *Natural Product Research*, *34*(1), 10–15. doi:10.1080/14786419.2019.1622107 PMID:31140308

Guan, Z., Tian, B., Perfumo, A., & Goldfine, H. (2013). The polar lipids of *Clostridium psychrophilum*, an anaerobic psychrophile. *Biochimica et Biophysica Acta (BBA)-. Molecular and Cell Biology of Lipids*, *1831*(6), 1108–1112. doi:10.1016/j.bbalip.2013.02.004 PMID:23454375

Guo, J., Sun, S., & Liu, J. (2020). Conversion of waste frying palm oil into biodiesel using free lipase A from *Candida antarctica* as a novel catalyst. *Fuel*, *267*, 117323. doi:10.1016/j.fuel.2020.117323

Gurunathan, R., Huang, B., Ponnusamy, V. K., Hwang, J. S., & Dahms, H. U. (2021). Novel recombinant keratin degrading subtilisin-like serine alkaline protease from *Bacillus cereus* isolated from marine hydrothermal vent crabs. *Scientific Reports*, *11*(1), 12007. doi:10.103841598-021-90375-4 PMID:34099743

Hagagy, N. I., Saddiq, A. A., Hamedo, H. A., & Selim, S. A. (2021). Extremophiles Inhabiting Unique Ecosystems in Egypt. In *Extreme Environments* (pp. 252–262). CRC Press. doi:10.1201/9780429343452-15

Hedrich, S., & Schippers, A. (2021). Distribution of acidophilic microorganisms in natural and man-made acidic environments. *Current Issues in Molecular Biology*, *40*(1), 25–48. doi:10.21775/cimb.040.025 PMID:32159522

Horikoshi, K. (2016). Alkaliphiles. In *Extremophiles* (pp. 53–78). Springer. doi:10.1007/978-4-431-55408-0_4

Jawahar, G., Rajasheker, G., Maheshwari, P., Punita, D. L., Jalaja, N., Kumari, P. H., Kumar, S. A., Afreen, R., Karumanchi, A. R., Rathnagiri, P., Sreenivasulu, N., & Kishor, P. B. K. (2019). Osmolyte diversity, distribution, and their biosynthetic pathways. In *Plant Signaling Molecules* (pp. 449–458). Woodhead Publishing. doi:10.1016/B978-0-12-816451-8.00028-9

Joyce, C. M. (2021). *Engineering thermostable regulators for inducible gene expression in thermophiles*. Academic Press.

Kaczmarek, D. K., Gwiazdowska, D., Juś, K., Klejdysz, T., Wojcieszak, M., Materna, K., & Pernak, J. (2021). Glycine betaine-based ionic liquids and their influence on bacteria, fungi, insects and plants. *New Journal of Chemistry*, *45*(14), 6344–6355. doi:10.1039/D1NJ00498K

Kateb, G. (2021). *The inner ocean*. Cornell University Press.

Kaushik, S., Alatawi, A., Djiwanti, S. R., Pande, A., Skotti, E., & Soni, V. (2021). Potential of Extremophiles for Bioremediation. In *Microbial Rejuvenation of Polluted Environment* (pp. 293–328). Springer., doi:10.1007/978-981-15-7447-4_12

Khemili-Talbi, S., Kebbouche-Gana, S., Akmoussi-Toumi, S., Angar, Y., & Gana, M. L. (2015). Isolation of an extremely halophilic archaeon *Natrialba* sp. C21 can degrade aromatic compounds and produce stable biosurfactants at high salinity. *Extremophiles*, *19*(6), 1109–1120. doi:10.100700792-015-0783-9 PMID:26334644

Kiruthika, J., Saraswathy, N., & Murugesan, S. (2018). Maximizing L-glutaminase production from marine *Bacillus subtilis* JK-79 under solid-state fermentation. *African Journal of Biotechnology*, *17*(9), 288–305. doi:10.5897/AJB2017.16015

Kouřilová, X., Schwarzerová, J., Pernicová, I., Sedlář, K., Mrázová, K., Krzyžánek, V., Nebesarova, J., & Obruča, S. (2021). The First Insight into Polyhydroxyalkanoates Accumulation in Multi-Extremophilic *Rubrobacter xylanophilus* and *Rubrobacter spartanus*. *Microorganisms*, *9*(5), 909. doi:10.3390/microorganisms9050909 PMID:33923216

Kushveer, J. S., Rashmi, M., & Sarma, V. V. (2021). Bioactive compounds from marine-derived fungi and their potential applications. In Fungi Bio-Prospects in Sustainable Agriculture, Environment and Nano-technology, (pp. 91-173). Academic Press. doi:10.1016/B978-0-12-821734-4.00014-9

Kushwaha, K., Saxena, J., Tripathi, B. K., & Agarwal, M. K. (2014). Detection of carotenoids in psychrotrophic bacteria by spectroscopic approach. *Journal of BioScience and Biotechnology*, *3*(3).

Lebeer, S., Bron, P. A., Marco, M. L., Van Pijkeren, J. P., Motherway, M. O. C., Hill, C., Pot, B., Roos, S., & Klaenhammer, T. (2018). Identification of probiotic effector molecules: Present state and future perspectives. *Current Opinion in Biotechnology*, *49*, 217–223. doi:10.1016/j.copbio.2017.10.007 PMID:29153882

Li, Z., & Loh, X. J. (2017). Recent advances of using polyhydroxyalkanoate-based nano vehicles as therapeutic delivery carriers. *Wiley Interdisciplinary Reviews. Nanomedicine and Nanobiotechnology*, *9*(3), e1429. doi:10.1002/wnan.1429 PMID:27595635

Litchfield, C. D. (2011). Potential for industrial products from the halophilic Archaea. *Journal of Industrial Microbiology & Biotechnology*, *38*(10), 1635–1647. doi:10.100710295-011-1021-9 PMID:21853327

Luo, Z., Wu, Y. L., Li, Z., & Loh, X. J. (2019). Recent Progress in Polyhydroxyalkanoates-Based Copolymers for Biomedical Applications. *Biotechnology Journal*, *14*(12), 1900283. doi:10.1002/biot.201900283 PMID:31469496

Margesin, R., & Feller, G. (2010). Biotechnological applications of psychrophiles. *Environmental Technology*, *31*(8-9), 835–844. doi:10.1080/09593331003663328 PMID:20662375

Martin, A., & McMinn, A. (2018). Sea ice, extremophiles, and life on extra-terrestrial ocean worlds. *International Journal of Astrobiology*, *17*(1), 1–16. doi:10.1017/S1473550416000483

Meknaci, R., Lopes, P., Servy, C., Le Caer, J. P., Andrieu, J. P., Hacene, H., & Ouazzani, J. (2014). Agar-supported cultivation of *Halorubrum sp.* SSR, and production of halocin C8 on the scale-up prototype Platotex. *Extremophiles*, *18*(6), 1049–1055. doi:10.100700792-014-0682-5 PMID:25138277

Mirete, S., Morgante, V., & González-Pastor, J. E. (2017). Acidophiles: diversity and mechanisms of adaptation to acidic environments. In *Adaption of microbial life to environmental extremes* (pp. 227–251). Springer. doi:10.1007/978-3-319-48327-6_9

Nandan, A., & Nampoothiri, K. M. (2017). Microbial aminopeptidases. In *Current Developments in Biotechnology and Bioengineering* (pp. 491–507). Elsevier. doi:10.1016/B978-0-444-63662-1.00021-X

Nieuwkoop, T., Claassens, N. J., & Van der Oost, J. (2019). Improved protein production and codon optimization analyses in *Escherichia coli* by bicistronic design. *Microbial Biotechnology*, *12*(1), 173–179. doi:10.1111/1751-7915.13332 PMID:30484964

Nunes, J. C., Cristóvão, R. O., Freire, M. G., Santos-Ebinuma, V. C., Faria, J. L., Silva, C. G., & Tavares, A. P. (2020). Recent strategies and applications for L-asparaginase confinement. *Molecules*, *25*(24), 5827. doi:10.3390/molecules25245827 PMID:33321857

Olaniyan, O. T., & Adetunji, C. O. (2021). Biological, Biochemical, and Biodiversity of Biomolecules from marine-based beneficial microorganisms: Industrial Perspective. Microorganisms for Sustainability, 27. doi:10.1007/978-981-15-7459-7_4

Ouwehand, A. C. (2017). A review of dose-responses of probiotics in human studies. *Beneficial Microbes*, *8*(2), 143–151. doi:10.3920/BM2016.0140 PMID:28008787

Ouwehand, A. C., DongLian, C., Weijian, X., Stewart, M., Ni, J., Stewart, T., & Miller, L. E. (2014). Probiotics reduce symptoms of antibiotic use in a hospital setting: A randomized dose-response study. *Vaccine*, *32*(4), 458–463. doi:10.1016/j.vaccine.2013.11.053 PMID:24291194

Pal, S., Sar, A., & Dam, B. (2019). Moderate halophilic bacteria, but not extreme halophilic archaea can alleviate the toxicity of short-alkyl side chain imidazolium-based ionic liquids. *Ecotoxicology and Environmental Safety*, *184*, 109634. doi:10.1016/j.ecoenv.2019.109634 PMID:31520950

Pandey, N., Jain, R., Pandey, A., & Tamta, S. (2018). Optimization and characterization of the orange pigment produced by a cold-adapted strain of *Penicillium* sp. (GBPI_P155) isolated from mountain ecosystem. *Mycology*, *9*(2), 81–92. doi:10.1080/21501203.2017.1423127 PMID:30123664

Panosyan, H., Di Donato, P., Poli, A., & Nicolaus, B. (2018). Production and characterization of exopolysaccharides by *Geobacillus thermodenitrificans* ArzA-6 and *Geobacillus toebii* ArzA-8 strains isolated from an Armenian geothermal spring. *Extremophiles, 22*(5), 725–737. doi:10.100700792-018-1032-9 PMID:29779131

Parvizpour, S., Hussin, N., Shamsir, M. S., & Razmara, J. (2021). Psychrophilic enzymes: Structural adaptation, pharmaceutical, and industrial applications. *Applied Microbiology and Biotechnology, 105*(3), 1–9. doi:10.100700253-020-11074-0 PMID:33427934

Patel, N. Y. (2020). *Exploring the potential of marine bacteria for L-glutaminase biosynthesis and its biochemical profiling* (M.Phil. Thesis). Gujarat University.

Patel, N. Y., Baria, D. M., Arya, P. S., Rajput, K. N., Panchal, R. R., & Raval, V. H. (2020). L-glutaminase biosynthesis from marine bacteria. *Bioscience Biotechnology Research Communications, 13*(1), 67-72. http://bbrc.in/bbrc/wp-content/uploads/2020/05/FINAL-PDF-Special-Issue-Vol-13-No-1-2020.pdf

Patel, N. Y., Baria, D. M., Yagnik, S. M., Rajput, K. N., Panchal, R. R., & Raval, V. H. (2021). Bio-prospecting the future in perspective of amidohydrolase L-glutaminase from marine habitats. *Applied Microbiology and Biotechnology, 105*(13), 5325–5340. doi:10.100700253-021-11416-6 PMID:34236482

Peng, S., Kai, M., Yang, X., Luo, Y., & Bai, L. (2020). Study on the osmoregulation of "*Halomonas socia*" NY-011 and the degradation of organic pollutants in the saline environment. *Extremophiles, 24*(6), 843–861. doi:10.100700792-020-01199-5 PMID:32930883

Postec, A., Quemeneur, M., Lecoeuvre, A., Chabert, N., Joseph, M., & Erauso, G. (2021). *Alkaliphilus serpentinus* sp. nov. and *Alkaliphilus pronyensis* sp. nov., two novel anaerobic alkaliphilic species isolated from the serpentinite-hosted Prony Bay Hydrothermal Field (New Caledonia). *Systematic and Applied Microbiology, 44*(2), 126175. doi:10.1016/j.syapm.2020.126175 PMID:33422701

Prechtl, R. M., Wefers, D., Jakob, F., & Vogel, R. F. (2018). Cold and salt stress modulates the amount, molecular and macromolecular structure of a *Lactobacillus sakei* dextran. *Food Hydrocolloids, 82*, 73–81. doi:10.1016/j.foodhyd.2018.04.003

Price, L. B., & Shand, R. F. (2000). Halocin S8: A 36-amino-acid microhalocin from the *haloarchaeal* strain S8a. *Journal of Bacteriology, 182*(17), 4951–4958. doi:10.1128/JB.182.17.4951-4958.2000 PMID:10940040

Purohit, M. K., Raval, V. H., & Singh, S. P. (2014). Haloalkaliphilic bacteria: molecular diversity and biotechnological applications. In *Geomicrobiology and Biogeochemistry* (pp. 61–79). Springer. doi:10.1007/978-3-642-41837-2_4

Radchenkova, N., Boyadzhieva, I., Atanasova, N., Poli, A., Finore, I., Di Donato, P., Nicolaus, B., Panchev, I., Kuncheva, M., & Kambourova, M. (2018). Extracellular polymer substance synthesized by a halophilic bacterium *Chromohalobacter canadensis* 28. *Applied Microbiology and Biotechnology, 102*(11), 4937–4949. doi:10.100700253-018-8901-0 PMID:29616312

Radchenkova, N., Panchev, I., Vassilev, S., Kuncheva, M., Dobreva, S., & Kambourova, M. (2015). Continuous cultivation of a thermophilic bacterium *Aeribacillus pallidus* 418 for production of an exopolysaccharide applicable in cosmetic creams. *Journal of Applied Microbiology, 119*(5), 1301–1309. doi:10.1111/jam.12944 PMID:26332083

Rajpurohit, Y. S., Sharma, D. K., & Misra, H. S. (2021). PprA protein inhibits DNA strand exchange and ATP hydrolysis of *Deinococcus* RecA and regulates the recombination in gamma-irradiated cells. *Frontiers in Cell and Developmental Biology, 9*, 636178. Advance online publication. doi:10.3389/fcell.2021.636178 PMID:33959605

Rastogi, R. P., Sonani, R. R., & Madamwar, D. (2014). The high-energy radiation protectant extracellular sheath pigment scytonemin and its reduced counterpart in the *cyanobacterium Scytonema* sp. R77DM. *Bioresource Technology, 171*, 396–400. doi:10.1016/j.biortech.2014.08.106 PMID:25226055

Rathinam, N. K., & Sani, R. K. (2018). Bioprospecting of extremophiles for biotechnology applications. In *Extremophilic microbial processing of lignocellulosic feedstocks to biofuels, value-added products, and usable power* (pp. 1–23). Springer. doi:10.1007/978-3-319-74459-9_1

Raval, V. H., Bhatt, H. B., & Singh, S. P. (2018). Adaptation Strategies in Halophilic Bacteria. In *Extremophiles* (pp. 137–164). CRC Press. doi:10.1201/9781315154695-7

Raval, V. H., Purohit, M. K., & Singh, S. P. (2013). Diversity, population dynamics, and biocatalytic potential of cultivable and non-cultivable bacterial communities of the saline ecosystems. In *Marine Enzymes for Biocatalysis* (pp. 165–189). Woodhead Publishing. doi:10.1533/9781908818355.2.165

Raval, V. H., Purohit, M. K., & Singh, S. P. (2015). Extracellular proteases from halophilic and haloalkaliphilic bacteria: Occurrence and biochemical properties. In *Halophiles* (pp. 421–449). Springer. doi:10.1007/978-3-319-14595-2_16

Ryan, S. M., & Brayden, D. J. (2014). Progress in the delivery of nanoparticle constructs: Towards clinical translation. *Current Opinion in Pharmacology, 18*, 120–128. doi:10.1016/j.coph.2014.09.019 PMID:25450066

Santra, H. K., & Banerjee, D. (2021). Extremophilic Microbes and their Extremozymes for Industry and Allied Sectors. In *Microbiomes of Extreme Environments* (pp. 1–42). CRC Press. doi:10.1201/9780429328633-1

Shang, J. L., Zhang, Z. C., Yin, X. Y., Chen, M., Hao, F. H., Wang, K., Feng, J. L., Xu, H. F., Yin, Y. C., Tang, H. R., & Qiu, B. S. (2018). UV-B induced biosynthesis of a novel sunscreen compound in solar radiation and desiccation-tolerant cyanobacteria. *Environmental Microbiology, 20*(1), 200–213. doi:10.1111/1462-2920.13972 PMID:29076601

Sharshar, M. M., Samak, N. A., Hao, X., Mu, T., Zhong, W., Yang, M., Peh, S., Ambreen, S., & Xing, J. (2019). Enhanced growth-driven stepwise inducible expression system development in haloalkaliphilic desulfurizing *Thioalkalivibrio versutus. Bioresource Technology, 288*, 121486. doi:10.1016/j.biortech.2019.121486 PMID:31128536

Shukla, P. J., Bhatt, V. D., Suriya, J., & Mootapally, C. (2020). Marine Extremophiles: Adaptations and Biotechnological Applications. Encyclopedia of Marine Biotechnology, 3, 1753-1771. doi:10.1002/9781119143802.ch74

Sindhu, R., & Manonmani, H. K. (2018). Expression and characterization of recombinant L-asparaginase from *Pseudomonas fluorescens. Protein Expression and Purification, 143*, 83–91. doi:10.1016/j.pep.2017.09.009 PMID:29079538

Singh, A. J., Xu, C. X., Xu, X., West, L. M., Wilmes, A., Chan, A., Hamel, E., Miller, J. H., Northcote, P. T., & Ghosh, A. K. (2010). Peloruside B, a potent antitumor macrolide from the New Zealand marine sponge *Mycale hentscheli*: Isolation, structure, total synthesis, and bioactivity. *The Journal of Organic Chemistry, 75*(1), 2–10. doi:10.1021/jo9021265 PMID:19957922

Singh, O. V., & Gabani, P. (2011). Extremophiles: Radiation resistance microbial reserves and therapeutic implications. *Journal of Applied Microbiology, 110*(4), 851–861. doi:10.1111/j.1365-2672.2011.04971.x PMID:21332616

Singh, S. P., Raval, V., & Purohit, M. K. (2013). Strategies for the salt tolerance in bacteria and archaea and its implications in developing crops for adverse conditions. In *Plant acclimation to environmental stress* (pp. 85–99). Springer. doi:10.1007/978-1-4614-5001-6_4

Śliżewska, K., Markowiak-Kopeć, P., & Śliżewska, W. (2021). The role of probiotics in cancer prevention. *Cancers (Basel), 13*(1), 20. doi:10.3390/cancers13010020 PMID:33374549

Srivastava, P., Nikhil, E. V., Bragança, J. M., & Kowshik, M. (2015). Anti-bacterial TeNPs biosynthesized by haloarcheaon *Halococcus salifodinae* BK3. *Extremophiles, 19*(4), 875–884. doi:10.100700792-015-0767-9 PMID:26085473

Struvay, C., & Feller, G. (2012). Optimization to low-temperature activity in psychrophilic enzymes. *International Journal of Molecular Sciences, 13*(9), 11643–11665. doi:10.3390/ijms130911643 PMID:23109875

Suyal, D. C., Soni, R., Yadav, A. N., & Goel, R. (2021). Cold Adapted Microorganisms: Survival Mechanisms and Applications. In Microbiomes of Extreme Environments (pp. 177-191). CRC Press.

Taylor, T. J., & Vaisman, I. I. (2010). Discrimination of thermophilic and mesophilic proteins. *BMC Structural Biology, 10*(1), 1–10. doi:10.1186/1472-6807-4-1 PMID:20487512

Tripathi, M., Singh, D. N., Vikram, S., Singh, V. S., & Kumar, S. (2018). Metagenomic approach towards bioprospection of novel biomolecule (s) and environmental bioremediation. *Annual Research & Review in Biology, 22*(2), 1–12. doi:10.9734/ARRB/2018/38385

Unissa, R., Sudhakar, M., & Reddy, A. S. K. (2016). Evaluation of *in-vitro* Anti-proliferative activity of L-arginine deiminase from Novel Marine Bacterial Isolate. *Microbiology Research Journal International, 13*(5), 1–10. doi:10.9734/BMRJ/2016/23592

Venugopal, V. (2016). Enzymes from seafood processing waste and their applications in seafood processing. *Advances in Food and Nutrition Research, 78*, 47–69. doi:10.1016/bs.afnr.2016.06.004 PMID:27452165

Villarreal, P., Carrasco, M., Barahona, S., Alcaíno, J., Cifuentes, V., & Baeza, M. (2016). Tolerance to ultraviolet radiation of psychrotolerant yeasts and analysis of their carotenoid, mycosporine, and ergosterol content. *Current Microbiology, 72*(1), 94–101. doi:10.100700284-015-0928-1 PMID:26483082

Wong, J. X., & Rehm, B. H. (2018). Design of modular polyhydroxyalkanoate scaffolds for protein immobilization by directed ligation. *Biomacromolecules, 19*(10), 4098–4112. doi:10.1021/acs.biomac.8b01093 PMID:30199627

Xiao, Z., Zhang, Y., Xi, L., Huo, F., Zhao, J. Y., & Li, J. (2015). Thermophilic production of polyhydroxyalkanoates by a novel *Aneurinibacillus* strain isolated from Gudao oilfield, China. *Journal of Basic Microbiology, 55*(9), 1125–1133. doi:10.1002/jobm.201400843 PMID:25832555

Yadav, A. N., Kaur, T., Devi, R., Kour, D., & Yadav, N. (2021). Biodiversity and Biotechnological Applications of Extremophilic Microbiomes: Current Research and Future Challenges. Microbiomes of Extreme Environments, 278-290.

Ye, J., Huang, W., Wang, D., Chen, F., Yin, J., Li, T., Zhang, H., & Chen, G. Q. (2018). Pilot scale-up of poly (3-hydroxybutyrateco-4-hydroxybutyrate) production by *Halomonas bluephagenesis* via cell growth adapted optimization process. *Biotechnology Journal, 13*(5), e1800074. doi:10.1002/biot.201800074 PMID:29578651

Zarinviarsagh, M., Ebrahimipour, G., & Sadeghi, H. (2017). Lipase and biosurfactant from *Ochrobactrum intermedium* strain MZV101 isolated by washing powder for detergent application. *Lipids in Health and Disease, 16*(1), 1–13. doi:10.118612944-017-0565-8 PMID:28923075

Zhang, X., Li, Z., Che, X., Yu, L., Jia, W., Shen, R., Chen, J., Ma, Y., & Chen, G. Q. (2019). Synthesis and characterization of polyhydroxyalkanoate organo/hydrogels. *Biomacromolecules, 20*(9), 3303-3312. .biomac.9b00479 doi:10.1021/acs

Zhu, D., Adebisi, W. A., Ahmad, F., Sethupathy, S., Danso, B., & Sun, J. (2020). The recent development of extremophilic bacteria and their application in biorefinery. *Frontiers in Bioengineering and Biotechnology, 8*, 483. doi:10.3389/fbioe.2020.00483 PMID:32596215

Zhu, S., Song, D., Gong, C., Tang, P., Li, X., Wang, J., & Zheng, G. (2013). Biosynthesis of nucleoside analogs via thermostable nucleoside phosphorylase. *Applied Microbiology and Biotechnology, 97*(15), 6769–6778. doi:10.100700253-012-4542-x PMID:23160980

Chapter 14
Extremophiles in Sustainable Bioenergy Production as Microbial Fuel Cells

Mukta Kothari
Modern College of Arts, Science, and Commerce (Autonomous), India

Leena Gaurav Kulkarni
Praj Matrix R&D Center, Praj Industries Ltd., India

Divita Gupta
Institut de Physique de Rennes, CNRS, University of Rennes, France

Rebecca Thombre
Blue Marble Space Institute of Science, USA

ABSTRACT

Microbial fuel cell (MFC) technology is considered one of the renewable sources of energy for the production of bioelectricity from waste. Due to the depletion of fossil fuels and environmental considerations, MFC haa garnered increasing importance as it is a sustainable and environmentally-friendly method of generation of bioenergy. In MFC, electroactive bacteria (EAB) and biofilms are harnessed to convert organic substances to electrical energy. Extremophiles survive in extreme environments, and they have demonstrated potential applications in microbial electrical systems (MES) and MFC technology. The key limitations of MFC are the low power output and engineering constraints of the fuel cell. Hence, it is imperative to understand the genetics, key metabolic pathways, and molecular mechanisms of the EAB for enhancing the power generation in MFC. This chapter gives a brief overview of the scope and applications of extremophiles in wastewater treatment, bioelectricity, and biohydrogen production using MFC, eventually enhancing the functional efficiency of MFC.

DOI: 10.4018/978-1-7998-9144-4.ch014

INTRODUCTION

Massive industrialization in the recent past has increased the energy needs and usage many fold. Excessive utilization of fossil fuels has not only led to depletion of natural reserves but also has posed major problems in terms of global warming, atmospheric changes and potential threat to natural ecosystems. Hence, alternative mechanisms are being sought for generation of sustainable renewable energy using environment friendly approaches namely wind energy, solar energy, tidal energy etc. Biohydrogen and bioelectricity production using Microbial fuel cells (MFCs) is also the one step in this direction. Journey of MFCs began in 1911 with glucose as a substrate and Platinum as electrode using *S. cerevisiae,* till the present time with several advancements in terms of usage of efficient microbial communities, electrodes, substrates and modes of operations leading to environment friendly mode of energy production (Anshu, 2021; Latika Bhatia et al., 2020). Still, even today the commercialization of MFC is a big challenge due to high cost of production and comparative low yield of electricity. Extremophiles being able to thrive in extreme physical and geothermal conditions can be used in MFCs. Thombre et al (2016a; 2016b) reported eco-friendly methods of usage of halophilic archaea and role of Bacteriorhodopsin in bioelectricity generation. Further, different approaches like de novo protein engineering and genetic engineering can be explored and could possibly make this energy option cost effective in near future. Present chapter explores the recent progress in metabolic and genetic engineering for manipulation of electroactive bacteria (EAB) and scope of extremophiles for enhanced production of bioenergy (Latika Bhatia et al., 2020; Jamile Mohammadi Moradian et al., 2021).

BACKGROUND

MFCs are bioelectrochemical devices which consist of anode and cathode chambers which are physically separated by a proton exchange membrane (PEM). Microbial cells work as biocatalysts where they oxidize the organic substrates to generate electrons and protons in anode chamber (Rahimnejad et al.,2015; Yibrah Tekle & Addisu Demeke, 2015). Microorganism transfers the electrons to anode through which electrons then travel to the cathode with the help of an external circuit to produce electricity. General reactions at anode and cathode are as follows: (Chang et al., 2006; Yibrah Tekle & Addisu Demeke, 2015)

Anode oxidation reaction:

$$(CH_2O)n + nH_2O \rightarrow nCO_2 + 4ne - +4nH^+ \quad (1)$$

Cathode reduction reaction:

$$4e - +4H^+ + O_2 \rightarrow 2H_2O \quad (2)$$

There are different variants of MFCs available based on its construction. Double-chamber MFC is the simplest of all types and consists of two separate chambers of anode and cathode; and chambers are compartmentalized with the help of membrane permeable for proton exchange. Single- chamber MFCs are made up of single chamber containing anode and cathode electrodes separated by proton exchange membrane (PEM). Up flow MFC is widely used in waste water and it is made up of a cylinder in which cathode and anode are separated by a layer of glass wool and glass beads. Stacked MFC is a system in which several MFCs are connected in series or in parallel to increase the total power output (Kumar et al., 2017). Different types of MFC technologies can be harnessed for environment friendly sustainable generation of biohydrogen and bioelectricity from waste water or organic waste (Venkata Mohan et al. 2007a; Hallenbeck et al., 2012; Chandra et al., 2015; Mark Dopson et al., 2015).

The bacteria used in MFCs and microbial electrosynthesis (MES) are electroactive bacteria (EAB) transfer the electrons from biological membranes to or from the environment. In MFCs bacteria utilize organic compounds to generate electrons while in MES bacteria utilize the electrons that are produced by external sources like electrodes (Logan et al., 2006). Interest in understanding the mechanism of electron transfer and the genes underlying the same are important from a fundamental point of view in order to harness the maximal electrogenic properties of the microbes as it serves as one of the renewable sources of energy. A crucial aspect in a microbial fuel cell is establishing the "link" for the electron transfer. As described by Schröder U. (2007), this linking species must be able to establish physical contact with the electrode surface, must be electrochemically active and its standard potential should be close to the redox potential of the primary substrate.

The first way in which the electron transfer can occur is when the microbes transfer electrons directly to the anode; referred to as Direct Electron Transfer (DET). It has been demonstrated in bacteria like *G. sulfurreducens* (Reguera et al., 2005), *Shewanella putrifaciens* (Kim B.H. et al., 1999b) and among many others. The second method of electron transfer involves mediators, which are either naturally present (Stams et al., 2006) or are produced by the microbes (Rabaey et al., 2005), which transport the electrons from the inside of the microbial cell to the electrode. An important point to be noted here is that the "mediator-less microbial fuel cells'' do not imply DET being the electron transfer mechanism. "Mediator-less"rather means that no artificial mediators have been added/inserted in these systems (Park & Zeikus, 2000; Prasad et al., 2007)

Although the DET method of transfer had been long debated since the living cells were generally assumed to be electrically non-conducting (Schröder U., 2007). Kim H.J. et al. (2002) at the end of the 20th century marked a remarkable breakthrough by establishing this mode of electron transfer. The direct contact is established through c-type cytochromes, while the long range contact is achieved via electrically conductive pili or nanowires (Reguera et al., 2005). In cases of direct contact, only one layer can take part in electron transfer, and the power densities understandably can be low. DET with conductive pili or nanowires turns out to be operationally more stable and yields much higher coulombic efficiency (Chang et al., 2006; Kumar et al., 2017).

Very few species using direct electron transfer have been found. Along with direct electron transfer, *Rhodoferox ferrireducens* can use complex substrates, like glucose (Chaudhuri & Lovley, 2003). Other species like *Geobacter* and *Shewanella* can utilise only low molecular organic alcohols and acids which limits their applications. The bacterial strain *Geobacter sulfurreducens* KN400 attracted a wide interest towards microbial fuel cells; as it was named as one of the 50 most important inventions by Time Magazine in 2009.

There has also been interest in genetically engineered microbial species which will further enhance the current output by reducing the electron transfer resistance. This effort, including adaptive selection, will depend on modifying the cytochromes and the pili structures as they play a significant role in deciding the microbe's ability to form biofilms and the carbon-based substrates it can metabolize. Mediated Electron Transfer (MET) is another method widely reported in the literature which involves organisms like *Klebsiella pneumoniae* which uses HNQ as mediator (Rhoads & Lewandowski, 2005), *Streptococcus lactis* which uses Ferric chelate complex as mediator (Vega & Fernandez, 1987) and *Proteus mirabilis* which uses Thionin as mediator (Choi et al., 2003). This method is more widely reported as the cell membranes in most bacteria consist of non-conductive lipids, peptidoglycans and lipopolysaccharides (Guo et al., 2012), rendering direct electron transfer impossible. The mediators can broadly be classified into two kinds: exogenous and endogenous. Exogenous mediators are the "electron shuttles" which are already present in the environment, whereas the endogenous mediators are produced by biocatalysts.

The exogenous mediators include compounds like phenazines, phenoxazines, phenothiazines and quinones which can facilitate the transfer of electrons from the bacterial cultures to the electrodes (Schröder U., 2007a; Bennetto et al., 1983; Stirling et al., 1983). The study on these systems from application point of view however, remains limited as they are expensive and their regular external addition is necessary, making these systems environment-unfriendly and commercially inconvenient. Moreover, the power densities achieved from the use of these are low compared to the systems operating on other kinds of mediators (Bullen et al., 2006). The different mesophilic organisms involved in generation of bioelectricity are enlisted in Table 1.

Table 1. Mesophiles used for producing bio-electricity in MFC

Microorganisms (Mediator required)	Microorganisms (Mediator less)
Erwinia dissolvens	*Shewanella putrefaciens IR-1*
Lactobacillus plantarum	*Geobacter sulfurreducens*
Proteus vulgaris	*Geobacter metallireducens*
Escherichia coli	*Desulfobulbus propionicus*
Gluconobacter oxydans	*Desulfovibrio desulfuricans*
Pseudomonas aeruginosa	*Aeromonas hydrophila*

The endogenous mediators can further be of two types: primary and secondary metabolites. The primary metabolites, in contrast to secondary metabolites, formed by anaerobic respiration and fermentation are more closely associated with the oxidative substrate degradation (Schröder U., 2007). Examples of these oxidizable metabolites include H_2 and H_2S, which were studied using *E. coli* (Hernandez & Newman, 2001; Schröder et al., 2003) and *Sulfurospirillum delayianum* (Niessen et al., 2004) respectively.

Secondar metabolites like pyocyanin and ACNQ (Straub & Schink, 2004) can play the role of electron shuttlers and are produced via metabolic pathways by the biocatalysts. The mediators of this type get reduced in the cytoplasm while the organic or carbon substrates get oxidized inside the bacterial cells. These reduced mediators then migrate to the anode where they release electrons and become oxidized. The released electrons can be absorbed by the oxidized mediators, hence repeating the cycle (Schröder U., 2007). The examples of these mediators remain limited as their identification is highly challenging and even their efficiency has been argued (Lovley, 2006) in case of open (flow) systems with their possible steady loss. Although the systems involving MET have lower Coulombic efficiency (compared to DET), the overall free energy in case of DET based systems is lower as they require high surface areas (Schröder U., 2007).

Metabolic Engineering Approaches for Improving MFC

For efficient performance of the Microbial fuel cell, the study of interaction between microorganisms and electrodes, related to low power densities which is considered to be a major drawback are essential. Another drawback of MFC is slow electron transfer at the anode and microorganism's interface (Wang et al., 2013). There is an urgent need to study all the factors contributing to the performance of Microbial fuel cells and modify the working in such a way that the obstacles are overcome. Microbial fuel cells are called so because the source of energy is micro-organisms. To increase the power densities, one needs to study the nature of the microorganisms.

Genetic and metabolic engineering involves modification or manipulation of a genetic makeup or a biochemical pathway to increase the concentration of a particular product respectively (Gonzalez, 2013). The biochemical pathways which are utilized for power generation by microorganisms can be modified at the genetic level to improve the performance of MFC by modifying the genes involved in the process. In this section, we are focusing on modification of biochemical pathways at the genetic level.

There are several studies carried out in which metabolic engineering techniques are utilized for improving the performance of micro-organisms in MFCs. The entire metabolic processes of *Escherichia coli* were extensively studied and experiments were performed to investigate the effect of modified the levels of NADH and $FADH_2$ on power generation of MFC. Other extensively studied microorganism is *Rhodopseudomonas palustris* (*R. palustris*) which utilizes ATP as a source of energy to reduce the downstream compounds and releases H_2 as a byproduct. H_2 is used for nitrogen fixation wherein a part of H_2 is transported to hydrogenase enzymes and the compounds are reduced. The hydrogen release was suppressed by genetic engineering technology wherein the genes of nitrogenase enzyme were deleted. The deletion of the genes for nitrogenase enzyme led to efficient production of electrons by the bacterium. An improved electron generation of the bacterium was followed by an increased number of reduced equivalents. This ultimately led to an increased amount of power generation which was calculated to be 18.3 mW cm^{-2} while the wild type strain could produce 11.7 mW cm^{-2} (Alfonta, 2010) This was a notable change in terms of power generation and of course, a very significant finding.

A study was carried out using uncommon strain *Pseudomonas aeruginosa*, commonly found in wastewater for power generation in MFC. It produces an organic compound Phenazine which is used in enhancing electron transport. The biosynthesis of Phenazines is regulated by the Quinolone system also known as PQS system (Qiao et al., 2017, Rabaey et al., 2005). However, utilization of *P. aeruginosa* can restrict power generation as the PQS system remains inactive for the majority of the time in the absence of oxygen. Also, the growth of the microorganism is inhibited in anaerobic conditions. In order to overcome these hindrances, metabolic engineering is implemented to seek an alternative method for power generation. The synthetic compound known as PqsE was found to be an alternative to the limitations. A PQS system deficient yet phenazine over-producing mutant of *P. aeruginosa* was modified by introducing a PqsE gene to the strain. The mutant, PqsC, was selected from a transposon mutagenesis library which was constructed using Mariner transposon vector pBT20. Out of the population, mutants carrying the transposon insertion were picked and cultured. The mutants producing low levels of pyocyanin were selected and the sequence flanking Mariner transposon was found out by performing Polymerase chain reaction. A PqsC mutant producing low levels of pyocyanin was identified using the PqsA-E operon in trans on plasmid pLG10. The sPqsE gene was introduced in PqsC mutant and PAO1 by electroporation and selected by growing them on LB agar containing 200 mg/l Carbencillin. The power generation using these mutants was investigated. After 3 days of operation of the mutants in Microbial fuel cells, a current of 0.5mA/cm^2 was achieved; this was five times higher than the rest of the strains (Wang et al., 2013). Hence, genetic engineering of the microorganisms by deleting the genes for the PQS system and developing a new strain of micro-organism containing a new gene altogether proved to be more efficient.

Genetic Engineering and Genome Microarray Analysis of EAB in MFC

Many bacteria are capable of producing bioelectricity but *Geobacter sulfurreducens* is a choice of micro-organism in MFCs and as a model organism to study extracellular respiration because it effectively produces large current, does not need mediator for transfer of electrons to anode and whole genome sequence is also available for it (Rollefson, 2009). *Geobacter sulfurreducens* show extracellular respiration with anode which uses c-type cytochrome complexes and nanopili for transfer of electrons directly to anode by formation of biofilm (Choudhary et al., 2017). Extensive study has been done on *G. sulfurreducens* and *Shewanella oneidensis.* A total of 111 and 42 genes for multiheme c-type cytochromes have been reported in *G. sulfurreducens* and *S. oneidensis.* These genes help in mediating the flow of electrons to the outer membrane. Whole genome microarray analysis of bioelectricity producing biofilms revealed that OmcZ is one of the most highly expressed genes in electrochemically active bacteria. Further genetic studies demonstrated the outer-membrane octa-heme c-type cytochrome, OmcZ is essential for high-density power production and is present at biofilm-anode interface.

A microarray analysis of *Geobacter* based MFCs gave insights into the genes of electrically conductive pili called nanopili or nanowires and various c-type cytochromes (Choudhary et al, 2016). In *G.sulfurreducens,* OmcF (GSU2432) a monohemec-type cytochrome is present in the outer membrane. Deletion of this gene results in less Fe (III) reduction and low current generation as compared to the wild type strain (Kim *et al.,* 2005; Choudhary et al., 2016). OmcF plays an important role in regulating other genes which are directly or indirectly involved in reduction and electricity production but the regulation mechanism is different for both. OmcB an outer membrane c-type cytochrome is actively involved in Fe (III) reduction but not needed for Mn (IV) reduction. OmcB is the favored route and not the only route for electron transfer to cell surface.

Recent studies suggest that hexa-heme c-type cytochrome OmcS (GSU2504) performs electron transfer to metal oxide and is present on type IV pili of *G. sulfurreducens.* It is proposed that electrons travel to conductive pili and then to OmcS which then transfers it to metal oxide. OmcS and OmcT (GSU2503) are homologues and are transcribed in the same operon. Further genetic studies have proved that OmcT is not essential for metal oxide reduction. OmcS and OmcE (GSU0618) outer membrane c-type cytochromes play an important role in electron transport to anode for electricity production and are thought to be regulated by OmcF. OmcF(GSU2432) is essential for optimum current generation as it plays an important role in proper positioning of OmcS and OmcE in the outer membrane (Kim et al., 2008).

The knocking out of OmcG (GSU2882) and OmcH (GSU2883) will show decrease in level of cytochrome OmcB and in Fe (III) reduction rates without affecting OmcB at transcriptional level. Study suggests OmcG and OmcH have roles in either stabilizing OmcB protein or efficient translation of OmcB transcript (Kim *et al.,* 2006). OmpB and OmpC, multicopper oxidase proteins are present in the outer membrane supposed to be involved in Fe (III) and/or Mn (IV) oxide reduction. Although OmpB is found to be essential for Fe (III) and/or Mn (IV) oxide reduction, OmpC has a less important role to play as compared to OmpB. Another outer membrane channel protein OmpJ (GSU3304) has an indirect role in regulation of electron transfer. OmpJ is thought to be involved in posttranslational modification of proteins which actively participate in electron transfer.

PpcA(GSU0612) a 9.6 kDa periplasmic c-type cytochrome is an intermediary electron carrier. A study demonstrated that PpcA and OmcB are possibly a part of the same route for electron transfer but has got alternatives in Fe (III) and Mn (IV) reduction pathways (Lloyd et al., 2003). PccH (GSU3274) is involved in assimilation of Fe (II) and further investigations on its role in Fe (III) reduction is needed. A putative periplasmic c-type cytochromes, CbcD (GSU0591) is encoded by an operon where b-type cytochrome CbcB (GSU0593), a membrane protein CbcE (GSU0590) and two other c-type cytochromes CbcA and CbcC (GSU0594, GSU0592), are a part of this operon. Together these five proteins that form menaquinol, a ferricytochrome c oxidoreductase are involved in electron transfers.

The b-type cytochrome CbcU (GSU0070) and iron–sulfur cluster-binding protein CbcT (GSU0069) work together along with c-type cytochrome CbcS (GSU0068) as another menaquinol: ferricytochrome c oxidoreductase complex named as Cbc4 complex. Gene GSU2210 encodes a large periplasmic c-type cytochrome with 27 putative heam-binding sites. Deletion mutants are not available to demonstrate its role in reduction of metal oxides and electron transfer. Other periplasmic c-type cytochrome PccF (GSU2201) with eight haem-binding sites is considered to be important for electron transport in the *Geobacteraceae* family as it is present in all *Geobacter* strains. In contrast, PccF mutant strain had no significant effect on metal oxide reduction or electron transfer.

Two genes cccA (GSU2811) and ccpA (GSU2813) code for periplasmic di-heam cytochrome c catalase and peroxidase proteins respectively. These genes protect the bacterial cell against damage caused by free radicals. In *Shewanella,* CcpA was present in abundance during dissimilatory Fe (III) reduction and was found to be important in growth under reducing conditions. In contrast, genetic analysis of CcpA mutant strain of *G. sulfurreducens* was found to be similar to the wild strain in terms of types of electron acceptors. Another peroxidase, macA (GSU0466) can transfer electrons to a c-type cytochrome, PpcA which plays a significant role in growth on Fe (III) citrate. Mutant analysis and gene expression patterns of MacA and OmcB are very similar, which concludes that both of them are a part of either same or similar routes of electron transfer.

The [NiFe]-hydrogenase complex called Hya is made up of three subunits hyaB (GSU0121), hyaS (GSU0123) and hyaL (GSU0122). These genes are thought to be involved in recycling of hydrogen, a byproduct of nitrogen fixation. This is assumed to be the similar case in Fe (III) oxide grown cells. Also, Hya catalyzes hydrogen dependent reduction of oxygen through menaquinone and participates in resistance mechanisms against oxidative stress. In *G. sulfurreducens,* ac-type cytochrome P460 is encoded by gene GSU0746. Similar cytochrome P460 present in methanotrophs oxidizes hydroxylamine to nitrite (Bergmann *et al.,* 1998). Expression of Hya and c-type cytochrome P460 indicates that though *G. sulfurreducens* is grown in anaerobic conditions, reactive oxygen and nitrogen species are formed during growth on insoluble metal oxides. OmcV (GSU2724) is a c-type cytochrome present in outer membrane of *G. sulfurreducens* and yedY (GSU2723) gene, a catalytic subunit of sulfooxidereductase present in periplasmic space have specific role in growth on Fe (III) oxide. Mutation in OmcV does not have any significant effect on growth on Fe (III) citrate, Mn (IV) oxide or fumarate but impairs the growth of *G. sulfurreducens* on Fe (III) oxide.

Nanowires (conductive pili) are present on many *Geobacter* species which are thought to be involved in long range electrons transport and the role of OmcS is restricted to facilitation of electrons from pili to Fe (III) oxide (Lovley, 2012). Thermodynamic analysis of half of the protein GSU1996 at the C- terminal end in *G. sulfurreducens* suggests that once entered, it acts as a nanowire through which electrons flow. Pili deficient mutants of G. *sulfurreducens* are unable to reduce Fe (III) but are capable of attaching themselves to anode (Reguera et al., 2005). Therefore, there is a possibility that along with pilisome other organelles like fimbriae help bacteria to colonize on heavy metals and metal oxides (Choudhary et al., 2016). A pilA expression is associated with pili structural proteins. PilA expression is regulated by a two-component system where pilR is an enhancer binding protein and acts as a transcriptional regulator. Gene knockout studies of pilR impairs both soluble and insoluble Fe (III) reduction (Juárez et al., 2009). PilA-N gene encodes the core domain 1 of geopilin, the structural protein of type IV pili, pilA-C (GSU1497) gene encodes a protein similar to variable domain 2 of geopilin. Many genes in the pilA-C adjacent to extracellular polysaccharide (xap) gene cluster together (GSU1498–GSU1508), for biofilm formation. Some other genes involved in the electron transfer are pilL (GSU2039), pilY1-2 (GSU2038), fimU (GSU2037), pilV-2 (GSU2036), pilM (GSU2032), pilP (GSU2029) and pilQ (GSU2028). In *G. sulfurreducens,* the upregulation of pili genes was observed and supports the fact of its involvement in electron transfer via OmcS to finally metal oxides

G. sulfurreducens form an extracellular matrix rich in cytochromes and polysaccharides. A gene cluster (GSU1498 to GSU1508) designated as *xap* is proved to be involved in the formation of extracellular anchoring polysaccharides. A xapD gene (GSU1501) is possibly involved in bacterial cell agglutination, attachment of bacterial cell to and surface for biofilm formation and localization of certain cytochromes in extracellular matrix (Aklujkar et al., 2013).

A study conducted on 46 different mutants of *Shewanella oneidensis* (*S. oneidensis*) for electricity production and metal reduction showed only 5 cytochrome deletions could produce electricity which was comparatively less than wild type. Other cytochrome deletions showed at least 20% higher current outputs as compared to wild type strain. This study concludes the highly complex mechanism of metal oxide reduction and electron flow through cytochromes (Choudhary et al., 2017). All the potential target genes for genetic engineering in EAB *Geobacter sulfurreducens* are enlisted in Table 2.

Metabolic Enzymes and Genes in Biohydrogen Production using MFC

MFC technology can be harnessed for the production of biohydrogen (Chandra et al., 2015). Dairy waste, waste water, effluents and organic waste can be utilized for generation of bio hydrogen using MFC (Hallenbeck *et al.,* 2012; Venkata Mohan et al., 2007a, 2007b). For Microbial fuel cells particularly, precious metals are not used at anode (Logan, 2009). Current wastewater treatment technology has drawbacks of high energy and cost input requirement as compared to the target recovery (He et al.2017). Therefore, a cost-effective wastewater treatment could be obtained by introduction of microbial fuel cells either by using consortium or pure cultures. In this, pure culture has got an edge superior due to ease of clarified electron transfer mechanisms and reducing the complexity of mixed cultures. The additional benefit of microbial fuel cells is that there is no any requirement of a mediator's addition to the system (Cao et.al 2019).

Clostridium strain DMHC-10^T (*aka Clostridium biohydrogenum* sp. nov) and strain MCMB-509^T (Kamalaskar et al., 2010, 2016) can be used to produce bio-electricity from both simple (glucose) and complex substrates (distillery effluent) in a microbial fuel cell.

In a recent study conducted by Kamalaskar et al. (2016); two dual chambered microbial fuel cells; one with glucose (1%) as substrate and second with distillery effluent (*ca.* 90 gCOD/L) were assembled with a working volume of 1 litre containing long cylindrical graphite electrodes (D = 1.5 cm and L = 11 cm) and connected through KCl salt bridge. Both the fuel cells were charged with 10% DMHC -10 (8.96 x 10^8cells/mL culture media) with initial pH 5.5 for glucose-based fuel cell and pH 5.3 for distillery effluent-based fuel cell. The fuel cells were kept at room temperature (32-35 °C) and the change in voltage and current were measured at regular intervals with a digital multimeter. Glucose based fuel cell showed a maximum voltage of 415 mV and power density, 34780 mA/m^2 at 24 h with a final pH of 4.29. In contrast, the distillery effluent-based fuel cell showed a voltage of 187 mV and power density, 10167 mA/m^{2}at 54 h with a final pH 5.27. The low bioelectricity generation potential from distillery effluent-based fuel cells might be due to the complex nature and high substrate load that resulted in its slow consumption by *Clostridium* strain DMHC-10. The results indicated considerable potential of the newly isolated *Clostridium* strain DMHC-10 to generate bioelectricity from glucose and high strength distillery effluent. It is interesting to note that in spite of the low-cost fuel cells (membrane-less and with graphite electrodes) used in this study, we could still observe the production of a substantial amount of bio-electricity. In accordance with these findings, Venkata Mohan *et. al.* (2007) have documented the feasibility of bioelectricity generation from anaerobic wastewater using a MFC fabricated with low-cost anode materials such as non-coated plain graphite electrodes without any toxic mediators and aerated cathode. Scott and Murano (2007) have documented an environment friendly and mediator-less microbial fuel cell (MFC), utilizing manure as a fuel, without any catalyst or a proton exchange membrane and is thus environmentally friendly (by using no toxic substances) in treating waste.

Table 2. Potential target genes for genetic engineering in EAB Geobacter sulfurreducens

Locus ID	Gene annotation	Gene	Subcellular location
GSU0466	Cytochrome c peroxidase	macA	Periplasm
GSU0612	Cytochrome c	ppcA	Periplasm
GSU0618	Cytochrome c	omcE	Outer membrane
GSU1394	Laccase family multicopper oxidase	ompB	Outer membrane
GSU1496	Geopilin domain 1 protein	pilA-N	Outer membrane
GSU1501	ABC transporter, ATP-binding protein	xapD	Cytoplasm
GSU2432	Lipoprotein cytochrome c	omcF	Outer membrane
GSU2503	Cytochrome c	omcT	Outer membrane
GSU2504	Cytochrome c	omcS	Outer membrane
GSU2657	Multicopper oxidase, manganese oxidase family	ompC	Outer membrane
GSU2737	Lipoprotein cytochrome c	omcB	Outer membrane
GSU2731	Lipoprotein cytochrome c	omcC	Outer membrane
GSU2882	Cytochrome c	omcG	Outer membrane
GSU2883	Cytochrome c	omcH	Outer membrane
GSU2884	Cytochrome c	omcA	Outer membrane
GSU3304	Outer membrane channel OmpJ	ompJ	Outer membrane
GSU2912	Cytochrome c	omcO	Outer membrane
GSU2898	Lipoprotein cytochrome c	omcN	Outer membrane
GSU2913	Cytochrome c	omcP	Outer membrane
GSU2723	Periplasmic sulfoxide reductase, catalytic subunit	yedY	Periplasm
GSU2501	Cytochrome c	omcU	Outer membrane

Source: Aklujkar et al., 2013

In MFCs, anodic chamber micro-organisms extract electrons and protons by the dissimilative oxidizing process of organic substrates. The electrons are absorbed by the anode and then flow through an electrical circuit with a load or resistor to the cathode. The potential difference (measured in Volt) across the anode and the cathode together with the flow of electrons (measured in Ampere) resulting in the generation of electricity (Watt). After crossing a PEM or a salt bridge, the protons enter the cathode chamber where they combine with oxygen to form water (Figure 1).

Figure 1. Construction and working of MFC using MCMB-509 and waste substrate or glucose

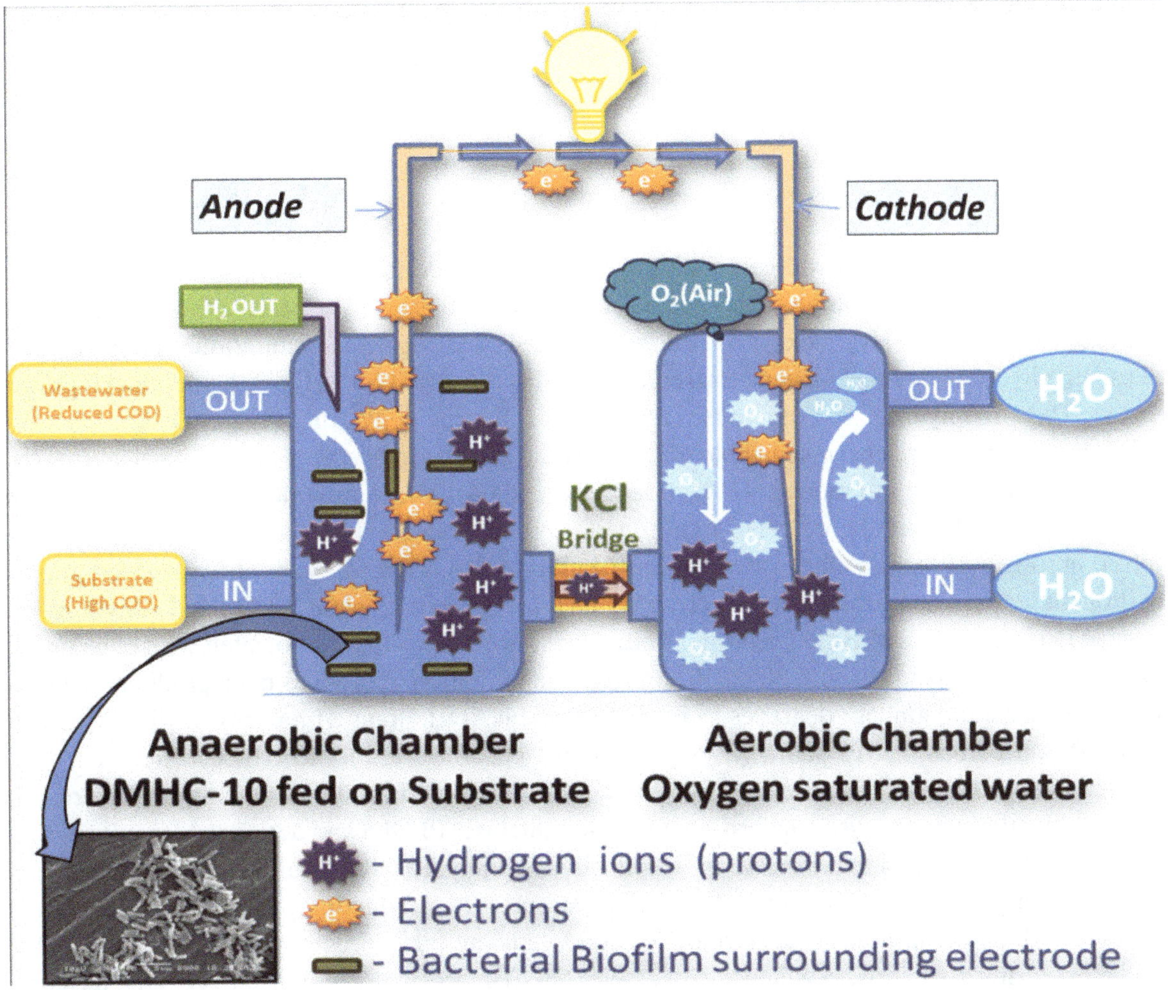

Typical electrode reactions are shown below using acetate as an example for VFAs.
Anodic reaction:

$$CH_3COO^- + 2H_2O \xrightarrow{microbes} 2CO_2 + 7H_2 + 8e^- \tag{3}$$

Cathode reaction:

$$O_2 + 4e^- + 4H^+ + \rightarrow 2H_2O \tag{4}$$

Due to high production of hydrogen and electricity, further investigation for molecular and genomic aspects to understand the function of enzymes in this bacterium was undertaken. Genome annotation by RAST revealed that MCM B-509 encodes genes that are involved in acidogenic phase that contribute to hydrogen production. The genes encoding hydrogen production such as periplasmic Fe hydrogenase, Fe-Fe-hydrogenase maturation protein, NAD dependent hydroxylbutyrate dehydrogenase and D-3 Phosphoglycerate dehydrogenase were detected in the genome (Kamalaskar et. al., 2016). These are the enzymes involved in generation of H^+ ions as well as production of biohydrogen (Das et al., 2006). These enzymes produce hydrogen by catalyzing a redox reaction:

$$2H^+ + 2e^- \xrightarrow{Hydrogenases} 4H_2 \tag{5}$$

To facilitate this reaction, hydrogen-evolving enzymes have complex metalloclusters as active sites and require special maturation proteins (Mathews & Wang, 2009). In MCM B-509 genome, hydrogenase maturation protein HyaD like 3 homologues of Fe-Fe hydrogenase maturation proteins HydE as well as NiFe hydrogenase metallocentre assembly proteins namely HypC, HypD, HypE and HypF were detected. The NiFe hydrogenase subunits referred to as "Energy-converting Hydrogenases'', not only produce hydrogen but are also responsible for the periplasmic H_2-uptake by utilizing the electrons from hydrogen and using them to reduce NADP (Hedderich, 2004). These subunits were found to be clustered together with the regulatory genes hypC, hypD and hypE and control their expression. The genes involved in solvent genesis: lactate dehydrogenase, butyryl COA dehydrogenase, 3-hydroxy butyrate are dehydrogenase, pyruvate formate lyase activating enzyme, glyceraldehyde 3-PO_4, and dehydrogenase were also detected in MCM B-509 genome. These genes are known to be associated with consumption of H+ ions as well as NADH, both of which contribute to inhibition of bio-hydrogen production. During glucose fermentation, glucose degrades to form ATP and NADH along with pyruvate via the Embden Meyerhoff and Parnas (EMP) pathway. Pyruvate then gets converted to acetyl CoA and CO_2, which generate a reduced ferredoxin molecule (Fd_{red}) that is further reoxidized by producing H_2. The reaction is catalysed by the enzyme pyruvate ferredoxin oxidoreductase (PFOR) and decreased ferredoxin transfers electrons to a hydrogenase: ferredoxin-dependent [FeFe] hydrogenase (Fd-[FeFe]), which is linked to PFOR driving hydrogen evolution, which in turn assures the production of two moles of hydrogen per mole of glucose consumed. Furthermore, second pathway for hydrogen production is via NADH reoxidation during glycolysis, in which the cytosolic (NiFe) hydrogenase, coupled to NADH: ferredoxin oxidoreductase (NFOR), wherein NADH acts as electron donor to generate hydrogen (Vardar-Schara *et al.*, 2008). The reoxidation of NADH produces additional hydrogen molecules via two other hydrogenases, i.e., NADH-dependent (NADH-[FeFe]) and NADH-Fd_{red} dependent hydrogenase (NADH-Fd_{red}-[FeFe]). Finally depending on the metabolic pathway, 2-4 mol of hydrogen per mole of glucose can be obtained. However, in practice, yields of 2 mol of hydrogen per mole of glucose was obtained as the NADH oxidation by NFOR is inhibited under operating conditions (Angenent et al., 2004; Kraemer & Bagley, 2007). As discussed earlier, Periplasmic Fe hydrogenase and maturation proteins for Fe-Fe- and

Ni-Fe-hydrogenases were detected in MCM B509, marking a potential role in the above-mentioned biochemical pathway leading to hydrogen production in MCM B-509 operation.

Extremophiles, Their Need and Scope in MFCs/ MESs

Extreme environments are not suitable for most of the life on earth but some special organisms like archaea and some bacteria that flourish in these sites are called extremophiles. Extremophiles could prevail either on natural sites such as salt lakes, volcanic eruption sites, marine sediments or could be obtained from acidic mine waters, nuclear waste sites etc. There are different types of extremophiles according to the physical or geochemical extremity in which they prevail such as Acidophiles (low pH environment), Alkaliphiles (high pH environment), Halophiles (high salinity condition), Thermophiles (high temperature environment), Psychrophiles (low temperature environment) and so on (Mark Dopson et al., 2015; Namita Shrestha et al., 2018).

Electrode potential, type of microorganism and its electron transport chain as well as its ability to deal with environmental conditions play a major role in the energy required for microbial activity in MFCs. Gibbs free energy calculated in theory differs from the actual extreme conditions developed as a result of change in temperature, pressure and pH of the electrolyte (Mark Dopson et al., 2015). Many studies conducted highlight an increase in bioelectricity production due to their potential to remain active under altered extreme conditions in MFCs. Different types of extremophiles best suited for this application are listed down in Table 3 and discussed in detail thereafter.

Table 3. Different classes of extremophiles used in MES and MFCs

Extremophile	Habitat	Example (used in MESs/MFCs)
Acidophiles	Optimum growth at pH<5	*Acidiphilium cryptum, Acidithiobacillus sp.*
Alkaliphiles	pH between 8.5 to 11	*Geoalkalibacter ferrihydriticus Desulfuromonas acetexigens*
Psychrophiles	optimum growth temperature <15oC and being able to replicate at <0oC	*Shewanella psychrophila Geobacter psychrophilus*
Thermophiles	Temperature between 41°C to 122°C	*Thermincola sp. strain JR Calditerrivibrio nitroreducens Yu37-1*

Acidophiles

Acidophiles flourish at pH<5 (low to high temperature adaptations) and contain bacteria, archaea and eukaryota. Acidophiles adjust with low pH conditions by influx of protons and are primarily involved in oxidation of ferrous iron, sulfur along with organic carbon as a substrate. Acidophiles are potential biocatalysts with a power of sustaining under extremely acidic conditions and high concentration of metal ions (Deochen Zhu et al., 2020). Abhijeet P. Borole et al. (2008), with two chambered bottle designs (graphite anode and platinum cathode), for the first time demonstrated the use of acidophile *Acidiphilium cryptum* in MFCs as an anode biocatalyst in the presence of an iron mediator at low pH (≤ 4) and further noted an increase in power generation to 12.7 mW m^{-2} on addition of nitrilotriacetic acid (NTA) as chelator and phenosafranin as secondary mediator. Study conducted used the mediator for electron transfer rather than direct electron transfer by microorganisms itself. Another strain *Acidiphilium sp. strain 3.2 sup 5* obtained from Rio Tinto, Spain demonstrated the direct electron transfer potentially via iron in c-cytochrome to anode at pH 2.5 (Mark Dopson et al., 2015). Chemolithotrophic acidophiles namely *Leptospirillum sp., Ferroplasma sp., Acidithiobacillus sp.* were successfully tested for pyrite enriched mining and processing catholyte with four-fold elevation in power generation. These cathode biocatalysts boosted redox processes with rise in reduction current (D.I. Stom et al., 2020). Considering the performances, acidophiles found to have huge potential in MESs and MFCs enhancement.

Alkaliphiles

Alkaliphiles can tolerate high pH from 8.5 to 11 as well as many times even tolerate the high salt concentrations called "haloalkaliphiles". Alkaliphiles employ a unique mechanism whereby they can maintain the pH ≥ t2 pH units compared to surrounding making use of sodium motive force, Mrp Na+/H+ antiporter, additional transport systems and a voltage-gated sodium channel and so on. Haloalkaliphiles are very promising in MESs which are involved in the treatment of alkali mediums or effluents. In working MFCs the anode becomes acidic while the cathode becomes more alkaline and as a consequence the cell voltage and power output decreases. Bioanode working in alkali condition has advantages such as reduced competition for substrate and increased cell voltage hence utilization of alkaliphiles capable of sustaining in a high pH environment are suitable for enhancing the power output of MFCs. *Bacillus sp.* working at a pH range between 9.5 to11 accompanying redox mediator was the very first alkaliphile to be tested for MFC applications. Further several different alkaliphiles such as *Corynebacterium strain MFC03* (glucose), *Geoalkalibacter ferrihydriticus* (acetate and yeast extract), *Corynebacterium humireducens MFC-5* (humic acid), *Bacillus pseudofirmus MC02* (humic acid), *Pseudomonas alcaliphila* (citrate), *Desulfuromonas acetexigens* (alkaline paper mill effluent) were also examined for their potential role in treatment of effluents and augmenting the power generation in MFCs (Mark Dopson et al., 2015). Also, study conducted on Lonar lake isolates *Oceanobacillus iheyensis BS1(2)* and *Bacillus alkalogaya BW2(1)* by Vishal Dhundale et al. (2018a, 2020b) supported this fact with increased power output of MFCs. Similar study related to Lonar lake isolates with dominance of *Geoalkalibacter sp.* in bioanode demonstrated increased power output (Sukrampal Yadav & Sunil Patil, 2020). Recently a study on Brazilian marine sediment enriched with alkaliphiles and acetogens namely *Desulfomicrobium, Advenella, Tindallia, Clostridium* & *Desulfuromonas* in treatment of synthetic wastewaters contaminated with iron and sulfate and acetate as carbon source has exhibited their potential in manipulating MFC bioanode positively (Lucca Bonjy Kikuti Mancílio et al., 2020).

Psychrophiles/ Psychrotolerant

Psychrophiles flourish at temperature < 15°C and replicate at < 0°C, whereas psychrotolerant grows at low temperatures with optimal temperature >15°C. Psychrophiles and psychrotolerant species have exploited several mechanisms for their survival at low temperature including high GC content in genome, expression of optimally flexible proteins, expression of cryoprotectant and antifreeze proteins, expression of cold shock proteins, expression of extracellular polysaccharides and so on. Temperature plays a crucial role in MFC working by directly affecting the Gibbs free energy of reactions and performance of biocatalysts. Rate of reaction escalates with rise in temperature and drop in temperature shows adverse impact on power output. Temperature change also influences the functioning of microbial enzymes, affecting microbial performance. Moreover, scaling up of MFCs used in wastewater treatment are expected to be tolerant to seasonal temperature fluctuations without compromising on energy efficiency. Iain S. Michie et al. (2011) demonstrated the stable acclimatization of MFC over the range of 8 to 35°C by using psychrotolerant species with higher COD removal rates at 35°C and higher coulombic efficiencies at low temperatures. Bioanode composed of psychrophilic bacteria *Shewanella psychrophila* and *Geobacter psychrophilus* and acetate as a substrate showed promising results with low anodic resistance and greater energy generation at two low temperatures 5°C & 10°C (Olga Tkach et al., 2017). Psychrophile *Pseudomonas fragi* DRR-2 isolated from goat rumen fluid used in MFC at four different temperatures (4°C, 10°C, 20°C and room temperature around 30°C) indicated maximum power output at 20°C (Deepika Jothinathan et al., 2017). Psychrophiles are equally beneficial in biohydrogen production, biodegradation of antibiotic Chloramphenicol at low temperatures. For biohydrogen production various substrates such as glucose, acetate, waste activated sludge, molasses wastewater etc. can be utilized. Trehalose protects the biocatalysts against the low temperature and hence addition of trehalose in waste-activated sludge and acetate-fed hydrogen-producing MEC working at 0oC showed higher coulombic efficiencies (Mark Dopson et al., 2015).

Thermophiles

Thermophiles are mostly the archaea capable of surviving over a temperature range of 41°C to 122°C as they have high GC content in their genome, more saturated fatty acids in membranes, heat stable proteins, and positively supercoiled DNA. Application of thermophiles in MFCs offer many advantages namely, increased microbial activity, increased solubility of substrate and less risk of contamination with more energy output. Along with advantages it also imposes potential limitation of excessive evaporation of electrolyte and could be overcomed by continuous addition of anolyte and catholyte during MFC operation (Deochen Zhu et al., 2020). MFC operated at 55°C fed with acetate as a substrate and methanogenic anaerobic digester derived microflora remained stable for over 100 days with coulombic efficiency as high as 89%. 16S rRNA microarray analysis of microflora revealed the dominance of the Firmicutes in energy production. Isolate of Firmicute genus *Thermincola sp. strain JR* employed direct electron transfer to anode in this MFC (Wrighton et al., 2008). *Thermincola ferriacetica* strain Z-0001 showed high coulombic efficiency in MFC operating in the presence of acetate further highlighted the extracellular electron transfer (EET) mechanism without any exogenous mediator (Marshall & May, 2009). Another Gram negative thermophilic *Calditerrivibrio nitroreducens Yu37-1* displayed stable electricity generation at 55°C by direct electron transfer to anode (Qian Fu et al., 2013). Further studies focused on possible mechanisms of direct electron transfer to anode led to the presence of many c type cytochromes in periplasm or cell surface of the thermophiles. Different strains of thermophiles were examined for different substrates other than acetate such as glucose and lactose (Mark Dopson et al., 2015). Thermophiles are highly beneficial in treatment of wastewater of thermophilic range with more efficient removal rate as compared to mesophiles corresponding to their fastidious growth at relatively high temperature (Bassam Abu Baker et al., 2020).

CONCLUSION

MFC is a promising technology for the generation of sustainable and environment friendly bioenergy. Most of the methods for enhancing the power generation in MFC are based on improving the technology, instruments or bioengineering. In the current chapter we have highlighted the understanding of the molecular mechanisms and the interplay of genes in generation of bioelectricity in MFC. Further highlighting extremophilic electroactive bacteria, their use in MFCs for stable high production of bioelectricity. Extremophiles have very high expectations in quality improvement and commercialization of MFCs. In the future, along with the use of extremophiles; application of systems biology, synthetic biology and CRISPR will create a paradigm shift in broadening the various methods of increasing the yield of MFC that have plausible use in sustainable bioenergy production to space applications as well.

REFERENCES

Abu Baker, B., Tawabini, B., Nazal, M., AlThaqfi, J., & Khalil, A. (2020). Efficiency of Thermophilic Bacteria in Wastewater Treatment. *Arabian Journal for Science and Engineering*, *46*(1), 123–128. doi:10.100713369-020-04830-x

Aklujkar, M., Coppi, M. V., Leang, C., Kim, B. C., Chavan, M. A., Perpetua, L. A., Giloteaux, L., Liu, A., & Holmes, D. E. (2013). Proteins involved in electron transfer to Fe(III) and Mn(IV) oxides by *Geobacter sulfurreducens* and *Geobacter uraniireducens*. *Microbiology*, *159*(Pt_3), 515–535. doi:10.1099/mic.0.064089-0 PMID:23306674

Alfonta, L. (2010). Genetically Engineered Microbial Fuel Cells. *Electroanalysis*, *22*, 822–831.

Alves, M. N., Fernandes, A. P., Salgueirob, C. A., & Paquete, C. M. (2015). Unraveling the electron transfer processes of a nanowire protein from Geobacter sulfurreducens. *Biochimica et Biophysica Acta*, *1857*(1), 7–13. PubMed

Angenent, L. T., Karim, K., Al-Dahhan, M. H., Wrenn, B. A., & Domíguez-Espinosa, R. (2004). Production of bioenergy and biochemicals from industrial and agricultural wastewater. *Trends in Biotechnology*, *22*(9), 477–485. doi:10.1016/j.tibtech.2004.07.001 PMID:15331229

Bennetto, H. P., Stirling, J. L., Tanaka, K., & Vega, C. A. (1983). Anodic reactions in microbial fuel cells. *Biotechnology and Bioengineering*, *25*(2), 559–568. doi:10.1002/bit.260250219 PMID:18548670

Bond, D. R., & Lovley, D. R. (2003). Electricity production by *Geobacter sulfurreducens* attached to electrodes. *Applied and Environmental Microbiology*, *69*(3), 1548–1555. doi:10.1128/AEM.69.3.1548-1555.2003 PMID:12620842

Borole, A. P., O'Neill, H., Tsouris, C., & Cesar, S. (2008). A microbial fuel cell operating at low pH using the acidophile *Acidiphilium cryptum*. *Biotechnology Letters*, *30*(8), 1367–1372. doi:10.100710529-008-9700-y PMID:18368296

Bullen, R. A., Arnot, T. C., Lakeman, J. B., & Walsh, F. C. (2006). Biofuel cells and their development. *Biosensors & Bioelectronics*, *21*(11), 2015–2045. doi:10.1016/j.bios.2006.01.030 PMID:16569499

Cao, Y., Zhang, R., Liu, W., Zhao, G., Niu, W., Guo, J., Xian, M., & Liu, H. (2019). Manipulation of the precursor supply for high-level production of longifolene by metabolically engineered *Escherichia coli*. *Scientific Reports*, *9*(1), 1–10. doi:10.103841598-018-36495-w PMID:30643175

Chandra, R., Nikhil, G. N., & Mohan, S. V. (2015). Single-Stage Operation of Hybrid Dark-Photo Fermentation to Enhance Biohydrogen Production through Regulation of System Redox Condition: Evaluation with Real-Field Wastewater. *International Journal of Molecular Sciences*, *16*(5), 9540–9556. doi:10.3390/ijms16059540 PMID:25927577

Chang, I. S., Moon, H., Bretschger, O., Jang, J. K., Park, H. I., Kenneth, H., Nealson, K. H., & Kim, B. H. (2006). Electrochemically Active Bacteria (EAB) and Mediator-Less Microbial Fuel Cells. *Journal of Microbiology and Biotechnology*, *16*(2), 163–177.

Chaudhuri, S. K., & Lovley, D. R. (2003). Electricity generation by direct oxidation of glucose in mediator less microbial fuel cells. *Nature Biotechnology*, *21*(10), 1229–1232. doi:10.1038/nbt867 PMID:12960964

Choi, Y., Kim, N., Kim, S., & Jung, S. (2003). Dynamic behaviors of redox mediators within the hydrophobic layers as an important factor for effective microbial fuel cell operation. *Bulletin of the Korean Chemical Society*, *24*(4), 437–440. doi:10.5012/bkcs.2003.24.4.437

Choudhary, P., Shankar, U. P., Bandopadhyay, T. K., Ray, R. N., & Bhunia, B. (2017). Performance improvement of microbial fuel cell (MFC) using suitable electrodes and Bioengineered organisms: A review. *Bioengineered*, *8*(5), 471–487. doi:10.1080/21655979.2016.1267883 PMID:28453385

Das, D., Dutta, T., Nath, K., Kotay, S. M., Das, A. K., & Veziroglu, T. N. (2006). Role of Fe hydrogenase in biological hydrogen production. *Current Science*, *90*(12), 1627–1637.

Dhundale, V., Hemke, V., Desai, D., Aher, G., Khemchandani, P., Dikonda, P., & Thosare, B. (2018). Improvement and Stable High Bioelectricity Generation Using Alkaliphilic *Oceanobacillus iheyensis* BS1(2) in Microbial Fuel Cells and Effect of Different Anodic Operating Conditions. *Annals of Applied Microbiology & Biotechnology Journal*, *2*(1), 1010. doi:10.36876/aamb.1010

Dhundale, V., Hemke, V., Desai, D., Khemchandani, P., Aher, G., & Dikonda, P. (2020). Evaluation of bioelectricity productivity using alkaliphilic *Bacillus alkalogaya* BW2(1) as a possible exoelectrogens for improvement of microbial fuel cell performance. *Journal of Applied Biology and Biotechnology*, *8*(01), 69–75. doi:10.7324/JABB.2020.80112

Dopson, Ni, & Sleutels. (2015). Possibilities for extremophilic microorganisms in microbial electrochemical systems. *Federation of European Microbiological Societies Microbiology Reviews*.

Fu, Q., Kobayashi, H., Kawaguchi, H., Wakayama, T., Maeda, H., & Sato, K. (2013). A thermophilic gram-negative nitrate-reducing bacterium, *Calditerrivibrio nitroreducens*, exhibiting electricity generation capability. *Environmental Science & Technology*, *47*(21), 12583–12590. doi:10.1021/es402749f PMID:24053548

Gonzalez, R. (2013). Metabolic engineering: Use of system-level approaches and application to fuel production in *Escherichia coli*. *Electronic Journal of Biotechnology*, *16*(3). Advance online publication. doi:10.2225/vol16-issue3-fulltext-14

Guo, K., Hassett, D. J., & Gu, T. (2012). *Microbial Fuel Cells: Electricity Generation from Organic Wastes by Microbes. In Microbial Biotechnology: Energy and Environment*. CAB International.

Hallenbeck, P. C., Abo-Hashesh, M., & Ghosh, D. (2012). Strategies for improving biological hydrogen production. *Bioresource Technology*, *110*, 1–9. doi:10.1016/j.biortech.2012.01.103 PMID:22342581

He, L., Du, P., Chen, Y., Lu, H., Cheng, X., Chang, B., & Wang, Z. (2017). Advances in microbial fuel cells for wastewater treatment. *Renewable & Sustainable Energy Reviews*, *71*, 388–403. doi:10.1016/j.rser.2016.12.069

Hedderich, R. (2004). Energy-converting [NiFe] hydrogenases from archaea and extremophiles: Ancestors of complex I. *Journal of Bioenergetics and Biomembranes*, *36*(1), 65–75. doi:10.1023/B:JOBB.0000019599.43969.33 PMID:15168611

Hernandez, M. E., & Newman, D. K. (2001). Extracellular electron transfer. *Cellular and Molecular Life Sciences CMLS*, *58*(11), 1562–1571. doi:10.1007/PL00000796 PMID:11706984

Holmes, D. E., Nicoll, J. S., Bond, D. R., & Lovley, D. R. (2004). Potential role of a novel psychrotolerant member of the family Geobacteraceae, *Geopsychrobacter electrodiphilus* gen. nov., sp. nov., in electricity production by a marine sediment fuel cell. *Applied and Environmental Microbiology*, *70*(10), 6023–6030. doi:10.1128/AEM.70.10.6023-6030.2004 PMID:15466546

Jothinathan, D., Sundaram, M., & Thilagaraj, R. (2017). The Optimization of Parameters for Increased Electricity Production by a Microbial Fuel Cell Using Rumen Fluid. *International Journal of Green Energy*, *12*(4), 333–338.

Juárez, K., Kim, B. C., Nevin, K., Olvera, L., Reguera, G., Lovley, D. R., & Methé, B. A. (2009). PilR, a Transcriptional Regulator for Pilin and Other Genes Required for Fe(III) Reduction in <i>Geobacter sulfurreducens</i>. *Journal of Molecular Microbiology and Biotechnology*, *16*(3-4), 146–158. doi:10.1159/000115849 PMID:18253022

Kamalaskar, L., Kapse, N., Pore, S., Dhakephalkar, A. P., Ranade, D. R., & Dhakephalkar, P. K. (2016). Genome sequence and gene expression studies reveal novel hydrogenases mediated hydrogen production by *Clostridium biohydrogenum* sp. nov., MCM B-509T. *International Journal of Hydrogen Energy*, *41*(28), 11990–11999. doi:10.1016/j.ijhydene.2016.05.101

Kamalaskar, L. B., Dhakephalkar, P. K., Meher, K. K., & Ranade, D. R. (2010). High biohydrogen yielding *Clostridium* sp. DMHC-10 isolated from sludge of distillery waste treatment plant. *International Journal of Hydrogen Energy*, *35*(19), 10639–10644. doi:10.1016/j.ijhydene.2010.05.020

Kelly, C. (2008). A novel ecological role of the Firmicutes identified in thermophilic microbial fuel cells. *International Society for Microbial Ecology*, *2*, 1146–1156.

Kim, B. C., Leang, C., Ding, Y. R., Glaven, R. H., Coppi, M. V., & Lovley, D. R. (2005). OmcF, a Putative *c*-Type Monoheme Outer Membrane Cytochrome required for the Expression of Other Outer Membrane Cytochromes in *Geobacter sulfurreducens. Journal of Bacteriology*, *187*(13), 4505–4513. doi:10.1128/JB.187.13.4505-4513.2005 PMID:15968061

Kim, B. C., Postier, B. L., DiDonato, R. J., Chaudhuri, S. K., Nevin, K. P., & Lovley, D. R. (2008). Insights into genes involved in electricity generation in *Geobacter sulfurreducens* via whole genome microarray analysis of the OmcF-deficient mutant. *Bioelectrochemistry (Amsterdam, Netherlands), 73*(1), 70–75. doi:10.1016/j.bioelechem.2008.04.023 PMID:18538641

Kim, B. C., Qian, X., Leang, C., Coppi, M. V., & Lovley, D. R. (2006). Two Putative *c*-Type Multiheme Cytochromes Required for the Expression of OmcB, an Outer Membrane Protein Essential for Optimal Fe(III) Reduction in *Geobacter sulfurreducens. Journal of Bacteriology, 188*(8), 3138–3142. doi:10.1128/JB.188.8.3138-3142.2006 PMID:16585776

Kim, B. H., Ikeda, T., Park, H., Kim, H., Hyun, M. S., Kano, K., Takagi, K., & Tatsumi, H. (1999a). Electrochemical activity of an Fe(III)-reducing bacterium, *Shewanella putrefaciens* IR-1, in the presence of alternative electron acceptors. *Biotechnology Techniques, 13*(7), 475–478. doi:10.1023/A:1008993029309

Kim, B. H., Kim, H. J., Hyun, M. S., & Park, D. H. (1999b). Direct electrode reaction of Fe (III)-reducing bacterium, *Shewanella putrefaciens. Journal of Microbiology and Biotechnology, 9*, 127–131.

Kim, H. J., Park, H. S., Hyun, M. S., Chang, I. S., Kim, M., & Kim, B. H. (2002). A mediator-less microbial fuel cell using a metal reducing bacterium, *Shewanella putrefaciens. Enzyme and Microbial Technology, 30*(2), 145–152. doi:10.1016/S0141-0229(01)00478-1

Kim, N., Choi, Y., Jung, S., & Kim, S. (1999). Development of microbial fuel cells using *Proteus vulgaris. Bulletin of the Korean Chemical Society, 21*(1), 44–48.

Kraemer, J. T., & Bagley, D. M. (2007). Improving the yield from fermentative hydrogen production. *Biotechnology Letters, 29*(5), 685–695. doi:10.100710529-006-9299-9 PMID:17279447

Kumar, R., Singh, L., & Zularisam, A. W. (2017). *Microbial Fuel Cells: Types and Applications: Waste biomass management- A Holistic Approach.* Springer.

Lloyd, J. R., Leang, C., Hodges Myerson, A. L., Coppi, M. V., Cuifo, S., Methe, B., Sandler, S. J., Logan, B. E., Hamelers, B., Rozendal, R., Schröder, U., Keller, J., Freguia, S., & Rabaey, K. (2006). Microbial Fuel Cells: Methodology and Technology. *Environmental Science & Technology, 40*(17), 5181–5192. doi:10.1021/es0605016 PMID:16999087

Logan, B. E. (2008). *Microbial fuel cells. John Wiley* & Sons.

Lovley, D. R. (2003). Biochemical and genetic characterization of PpcA, a periplasmic c-type cytochrome in *Geobacter sulfurreducens. The Biochemical Journal, 369*(Pt 1), 153–161. PMID:12356333

Lovley, D. R. (2006). Microbial fuel cells: Novel microbial physiologies and engineering approaches. *Current Opinion in Biotechnology, 17*(3), 327–332. doi:10.1016/j.copbio.2006.04.006 PMID:16679010

Lovley, D. R. (2012). Long-range electron transport to Fe(III) oxide via pili with metallic-like conductivity. *Biochemical Society Transactions, 40*(6), 1186–1190. doi:10.1042/BST20120131 PMID:23176452

Marshall, C. W., & May, H. D. (2009). Electrochemical evidence of direct electrode reduction by a thermophilic Gram-positive bacterium, *Thermincola ferriacetica. Energy & Environmental Science, 2*(6), 699–705. doi:10.1039/b823237g

Mathews, J., & Wang, G. (2009). Metabolic pathway engineering for enhanced biohydrogen production. *International Journal of Hydrogen Energy*, *34*(17), 7404–7416. doi:10.1016/j.ijhydene.2009.05.078

Michie, I. S., Kim, J. R., Dinsdale, R. M., Guwy, A. J., & Premier, G. C. (2011). The influence of psychrophilic and mesophilic start-up temperature on microbial fuel cell system performance. *Energy & Environmental Science*, *2011*(4), 1011–1019. doi:10.1039/c0ee00483a

Moradian, J. M., Fang, Z., & Yong, Y.-C. (2021). Recent advances on biomass-fueled microbial fuel cell. *Bioresources and Bioprocessing*, *8*(1), 14. doi:10.118640643-021-00365-7

Niessen, J., Schröder, U., & Scholz, F. (2004). Exploiting complex carbohydrates for microbial electricity generation–a bacterial fuel cell operating on starch. *Electrochemistry Communications*, *6*(9), 955–958. doi:10.1016/j.elecom.2004.07.010

Park, D. H., & Zeikus, J. G. (2000). Electricity generation in microbial fuel cells using neutral red as an electronophore. *Applied and Environmental Microbiology*, *66*(4), 1292–1297. doi:10.1128/AEM.66.4.1292-1297.2000 PMID:10742202

Prasad, D., Arun, S., Murugesan, M., Padmanaban, S., Satyanarayanan, R. S., Berchmans, S., & Yegnaraman, V. (2007). Direct electron transfer with yeast cells and construction of a mediatorless microbial fuel cell. *Biosensors & Bioelectronics*, *22*(11), 2604–2610. doi:10.1016/j.bios.2006.10.028 PMID:17129722

Qiao, Y.-J., Qiao, Y., Zou, L., Wu, X.-S., & Liu, J.-H. (2017). Biofilm promoted current generation of *Pseudomonas aeruginosa* microbial fuel cell via improving the interfacial redox reaction of phenazines. *Bioelectrochemistry (Amsterdam, Netherlands)*, *117*, 34–39. doi:10.1016/j.bioelechem.2017.04.003 PMID:28575838

Rabaey, K., Boon, N., Höfte, M., & Verstraete, W. (2005). Microbial phenazine production enhances electron transfer in biofuel cells. *Environmental Science & Technology*, *39*(9), 3401–3408. doi:10.1021/es048563o PMID:15926596

Rahimnejad, M., Adhami, A., Darvari, S., Zirepour, A., & Oh, S.-E. (2015). Microbial fuel cell as new technology for bioelectricity generation: A review. *Alexandria Engineering Journal*, *54*(3), 745–756. doi:10.1016/j.aej.2015.03.031

Reguera, G., Mccarthy, K. D., Mehta, T., Nicoll, J. S., Tuominen, M. T., & Lovley, D. R. (2005). Extracellular electron transfer via microbial nanowires. *Nature*, *435*(7045), 1098–1101. doi:10.1038/nature03661 PMID:15973408

Rhoads, A., Beyenal, H., & Lewandowski, Z. (2005). Microbial fuel cell using anaerobic respiration as an anodic reaction and biomineralized manganese as a cathodic reactant. *Environmental Science & Technology*, *39*(12), 4666–4671. doi:10.1021/es048386r PMID:16047807

Mancílio, Ribeiro, Lopes, Kishi, Martins-Santana, de Siqueira, Andrade, Guazzaroni, & Reginatto. (2020). Unusual microbial community and impact of iron and sulfate on microbial fuel cell ecology and performance. *Current Research in Biotechnology*, 64-73.

Rollefson, J. B., Levar, C. E., & Bond, D. R. (2009). Identification of Genes Involved in Biofilm Formation and Respiration via Mini-*Himar* Transposon Mutagenesis of *Geobacter sulfurreducens*. *Journal of Bacteriology, 191*(13), 4207–4217. doi:10.1128/JB.00057-09 PMID:19395486

Schröder, U. (2007). Anodic electron transfer mechanisms in microbial fuel cells and their energy efficiency. *Physical Chemistry Chemical Physics, 9*(21), 2619–2629. doi:10.1039/B703627M PMID:17627307

Schröder, U., Niessen, J., & Scholz, F. (2003). A generation of microbial fuel cells with current outputs boosted by more than one order of magnitude. *Angewandte Chemie International Edition in English, 42*(25), 2880–2883. doi:10.1002/anie.200350918 PMID:12833347

Scott, K., & Murano, C. (2007). Microbial fuel cells utilising carbohydrates. *Journal of Chemical Technology and Biotechnology (Oxford, Oxfordshire), 82*(1), 92–100. doi:10.1002/jctb.1641

Shrestha, N., Chilkoor, G., Vemuri, B., Rathinam, N., Sani, R. K., & Gadhamshetty, V. (2018). Extremophiles for microbial-electrochemistry applications: A critical review. *Bioresource Technology, 255*, 318–330. doi:10.1016/j.biortech.2018.01.151 PMID:29433771

Stams, A. J., De Bok, F. A., Plugge, C. M., Van Eekert, M. H., Dolfing, J., & Schraa, G. (2006). Exocellular electron transfer in anaerobic microbial communities. *Environmental Microbiology, 8*(3), 371–382. doi:10.1111/j.1462-2920.2006.00989.x PMID:16478444

Stirling, J. L., Bennetto, H. P., Delaney, G. M., Mason, J. R., Roller, S. D., Tanaka, K., & Thurston, C. F. (1983). Microbial fuel cells. *Biochemical Society Transactions, 11*(4), 451–453. doi:10.1042/bst0110451 PMID:6352363

Stom, D. I., Zhdanova, G. O., Kalashnikova, O. B., Bulaev, A. G., Kashevskii, A. V., Kupchinsky, A. B., Vardanyan, N. S., Ponamoreva, O. N., Alferov, S. V., Saksonov, M. N., Chesnokova, A. N., & Tolstoy, M. Yu. (2020). Acidophilic Microorganisms *Leptospirillum sp., Acidithiobacillus sp., Ferroplasma sp.* As a Cathodic Bioagents in a MFC. *Geomicrobiology, 38*(4), 340–346. doi:10.1080/01490451.2020.1856980

Straub, K. L., & Schink, B. (2004). Ferrihydrite-dependent growth of *Sulfurospirillum deleyianum* through electron transfer via sulfur cycling. *Applied and Environmental Microbiology, 70*(10), 5744–5749. doi:10.1128/AEM.70.10.5744-5749.2004 PMID:15466509

Tekle & Demeke. (2015). Review on microbial fuel cell. *Basic Research Journal of Microbiology*, 5-17.

The 50 Best Inventions of 2009. (n.d.). *Time*. Retrieved from https://www.time.com/time/specials/packages/article/0,28804,1934027_1934003_1933965,00.html

Thombre, R. S., Shinde, V. D., Oke, R. S., Dhar, S. K., & Shouche, Y. S. (2016a). Biology and survival of extremely halophilic archaeon *Haloarcula marismortui* RR12 isolated from Mumbai salterns, India in response to salinity stress. *Scientific Reports, 6*(1), 25642. doi:10.1038rep25642 PMID:27231230

Thombre, R. S., Shinde, V. D., Thaiparambil, E., Zende, S., & Mehta, S. (2016b). Antimicrobial activity and mechanism of inhibition of silver nanoparticles against extreme halophilic archaea. *Frontiers in Microbiology, 7*, 1424. doi:10.3389/fmicb.2016.01424 PMID:27679615

Tkach, O., Sangeetha, T., Maria, S., & Wang, A. (2017). Performance of low temperature Microbial Fuel Cells (MFCs) catalyzed by mixed bacterial consortia. *Journal of Environmental Sciences (China)*, *52*, 284–292. doi:10.1016/j.jes.2016.11.006 PMID:28254049

Vardar-Schara, G., Maeda, T., & Wood, T. K. (2008). Metabolically engineered bacteria for producing hydrogen via fermentation. *Microbial Biotechnology*, *1*(2), 107–125. doi:10.1111/j.1751-7915.2007.00009.x PMID:21261829

Vega, C. A., & Fernández, I. (1987). Mediating effect of ferric chelate compounds in microbial fuel cells with *Lactobacillus plantarum, Streptococcus lactis*, and *Erwinia dissolvens. Bioelectrochemistry and Bioenergetics*, *17*(2), 217–222. doi:10.1016/0302-4598(87)80026-0

Venkata Mohan, S., Lalit Babu, V., & Sarma, P. N. (2007a). Anaerobic biohydrogen production from dairy wastewater treatment in sequencing batch reactor (AnSBR): Effect of organic loading rate. *Enzyme and Microbial Technology*, *41*(4), 506–515. doi:10.1016/j.enzmictec.2007.04.007

Venkata Mohan, S., Raghavulu, S., Srikanth, S., & Sarma, P. (2007b). Bioelectricity production by mediatorless microbial fuel cell under acidophilic condition using wastewater as substrate: Influence of substrate loading rate. *Current Science*, *92*(12), 1720–1726.

Wang, W. B., Chua, S.-L., Cao, B., Seviour, T., Nesatyy, V. J., Marsili, E., Kjelleberg, S., Givskov, M., Tolker-Nielsen, T., Song, H., Loo, J. S. C., & Yang, L. (2013). Engineering PQS Biosynthesis Pathway for Enhancement of Bioelectricity Production in *Pseudomonas aeruginosa* Microbial Fuel Cells. *PLoS One*, *8*(5), e63129. doi:10.1371/journal.pone.0063129 PMID:23700414

Yadav & Patil. (2020). Microbial electroactive biofilms dominated by *Geoalkalibacter spp.* from a highly saline–alkaline environment. *NPJ Biofilms Microbiomes, 6*(1), 38.

Zhu, D., Adebisi, W. A., Ahmad, F., Sethupathy, S., Danso, B., & Sun, J. (2020). Recent development of extremophilic bacteria and their application in biorefinery. *Frontiers in Bioengineering and Biotechnology*, *8*, 483. doi:10.3389/fbioe.2020.00483 PMID:32596215

KEY TERMS AND DEFINITIONS

Bioelectricity: Electric potentials or currents generated by or present within living cells, tissues, or organisms.

Biofilm: Complex microbiome where different bacterial colonies, embedded in extracellular polymeric substances, stick to each other, or a surface.

Electroactive Bacteria (EAB): Bacterial species capable of transferring electrons from a microbial cell to an electrode.

Gene Knockout: Mutating a DNA/gene in a way that its expression is permanently prevented.

Genetic Engineering: Process of altering the genetic makeup of an organism by using biotechnology/recombinant DNA technology.

Metabolic Engineering: Process of modifying/optimizing the endogenous metabolic pathways in species.

Methanotrophs: Prokaryotic species, like bacteria or archaea, whose source of energy is via methane oxidation.

Mutant: Any species where the DNA sequence has been modified, either during cell division or exposure to external stimuli like ionizing radiation or chemicals.

Nanowire: Conductive appendages/structures, a few nanometers wide and deep but with a longer length, produced by a number of bacteria.

Oxidation: Addition of an oxygen atom, or more generally loss of electrons.

Polymerase Chain Reaction (PCR): Method used to make millions to billions copies of specific DNA sequence.

Reduction: Elimination of an oxygen atom, or more generally gain of electrons.

Transposon: DNA sequences capable of moving to different locations within a genome.

Chapter 15
Extremophiles as a Source of Biotechnological Products

Pushpa S. Murthy
Council of Scientific and Industrial Research, Central Food Technological Research Institute, India

Vedashree M.
Council of Scientific and Industrial Research, Central Food Technological Research Institute, India

Sneha H. P.
Council of Scientific and Industrial Research, Central Food Technological Research Institute, India

Inderjit Prakash
Council of Scientific and Industrial Research, Central Food Technological Research Institute, India

ABSTRACT

Extremophile and extremozyme capabilities to uphold catalytic actions under extreme situations open up a varied array of biotechnological applications. Extremophiles are a rich supply of biocatalysts used for innumerable purposes. Bioactive molecules and enzymes isolated from organisms inhabiting risky environments being used in biological innovation pipelines and pharmaceutical have positive claims. The species biodiversity has favourable reservoir of the unexploited amalgams with biotechnological significance. Prospective solicitations of extremozymes, chiefly as catalysis of multistep progressions, quorum sensing, bioremediation, biofuel, biodiversity and prospecting, biomining, and genetic technology are explored. To boost the biotechnological uses of extremozymes, research and development efforts are needed to address hurdles such as extremophile culture, gene expression in host cells, and extremozyme bioprocessing. Extremophiles can be a resource for innovative biotechnological comprising industrial biotechnology, agriculture, medical, food, and environmental biotechnology.

DOI: 10.4018/978-1-7998-9144-4.ch015

INTRODUCTION

The term extremophile was coined by Mac Elroy in 1974, more than a quarter-century ago. An extremophile is a relative term, as what is "extreme" for one entity may be "essential" for another's survival. Extremophiles are organisms that potentially grow in extreme environmental conditions. Some groups of organisms are competent to survive in multiple harsh conditions. The intense environmental condition can include temperature, pressure, radiation, salt conditions, pH, oxygen condition, redox potential (Rothschild & Mancinelli, 2001). Extremophiles are classified as thermophiles (Higher temperature 55-113°C), psychrophiles (Low and below 0°C), halophiles (High salinity), acidophiles. Alkalophilic (alkaline or acidic pH), piezophiles (higher pressure), saccharophiles (high sugar concentration), metallophiles (resistance for heavy metal), and polyextremophiles (Two or more extreme environments). Numerous taxonomic studies have revealed that these species are classified broadly into three groups: Archaea, Bacteria, and Eukarya. (Dalmaso, Ferreira, & Vermelho, 2015). Biotechnology has a crucial part in day-to-day lives from food and drinks like lactose-free milk (Coker & Brenchley, 2006), Bio-insecticide (Rubio-Infante & Moreno-Fierros, 2016), Enzymes like cellulase (Miettinen-Oinonen & Suominen, 2002), lipases (Joseph, Ramteke, & Thomas, 2008). Extremozymes are enzymes that function in extreme or unusual physicochemical conditions, such as extreme heat and cold, low pH and pressure, high salinity, low water activity, low oxygen, and so on. In many industries which utilize proteins, enzyme processing is carried out in extreme conditions, which include high or low temperature, pH, and salinity with an intervention of biotechnology and extremophiles, which can prevent the deproteination and loss of activity of enzymes. In this chapter, we initially discuss the extremozymes and their successful industrial and biotechnological applications, further followed by the application of extremophiles in biomining, biofuel production in industries where they can be the source of invaluable components through the application of biotechnology. Flow chart of outline extremophiles classification and application is described in figure 1. A number of reports and reviews highlight the advancement of methods for investigating extremophiles, including the use of proteomics for studying extremophiles and a multiplex approach for quantifying fine-scale temperature-induced proteome alterations. Modern biotechnological approaches can generate or activate the radiation-responsive metabolites, pigments, and enzymes they create to produce effective pharmaceuticals, including anti-cancer treatments, as well as antibiotics and commercially important agricultural goods. The benefits of extremozymes in the fields of therapeutics and biotechnology, on the other hand, have not been proven. The purpose of this article is to explain the tactics that bacteria adapt to flourish in radiation-rich settings, as well as their possible applications in biotechnology and therapeutics.

EXTREMOZYME

The extremozyme is the enzyme from thermophilic, psychrophilic, acidophilic, alkaliphilic, and halophilic microorganisms, which are resistant to extreme environmental conditions. Due to their biodegradability and exceptional stability, extremophilic microbes are a source of extremozymes with an extensive range of commercial uses. Cold-tolerant extremozymes, acid-tolerant extremozymes, alkali-tolerant extremozymes, and salt-tolerant extremozymes have all helped to generate a varied range of resistant biomolecules for industrial purposes. (Ricardo Cavicchioli, Siddiqui, Andrews, & Sowers, 2002).

Figure 1.

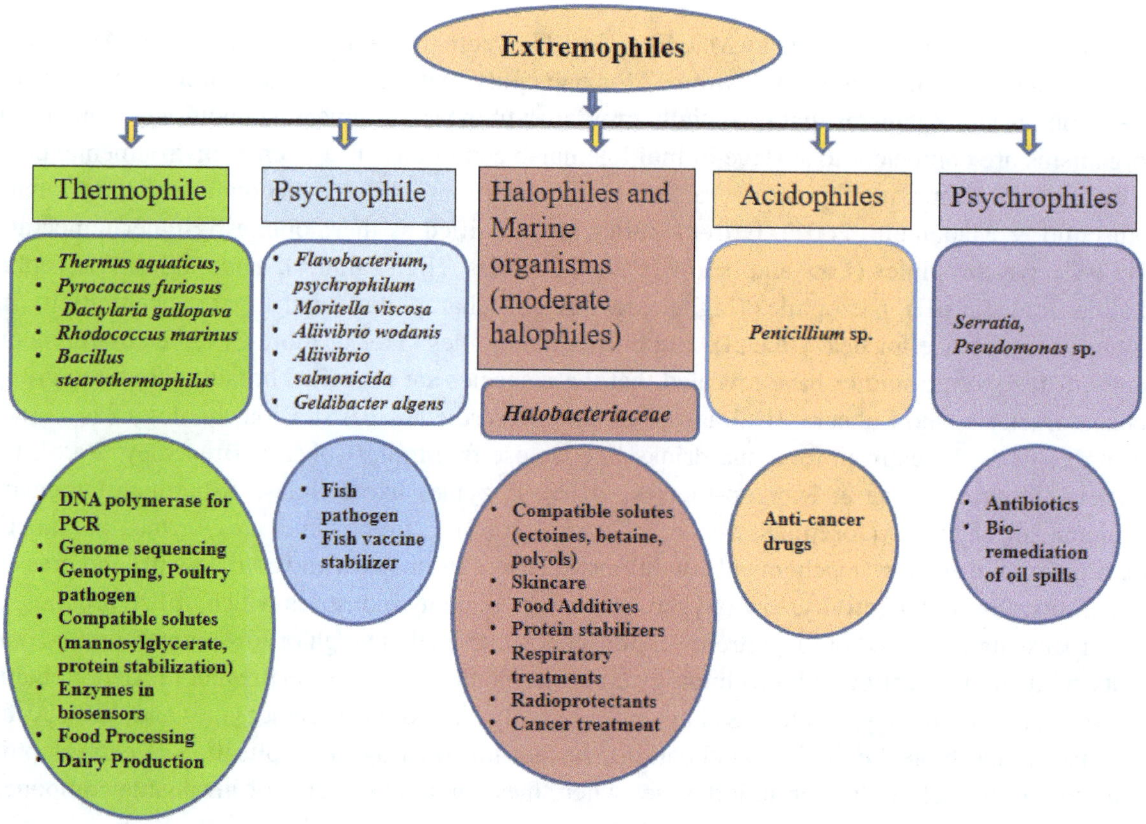

Thermozymes

Thermozymes are enzymes that are effective and functional under various temperatures, pH levels, substrate concentrations, and pressure and are extremely resilient to denaturants and organic solvents. Thermozymes are active at high temperatures by tighter packing in their hydrophic core, increased number of Coulomb interactions by replacing polar residues with charged residues, and a systematic shortening of surface-exposed loops. The advantages of employing thermozymes include lower contamination risk, decreased viscosity, and increased substrate solubility. The major drawback of utilizing enzyme from microbial sources is obtaining enzymes in pure form when cultured at an industrial level and with a lower yield. Through biotechnological approaches like cloning, many studies are conducted to obtain higher yields. Toplak et al. (2013) discovered a gene in the Gram-positive, anaerobic, thermophilic bacteria that codes for a subtilase called proteolysis. *Coprothermobacter proteolytic* is a kind of bacteria that expresses oneself functionally. The enzyme is made by inserting a gene encoding a putative serine protease into *E. coli*. It has been refined and found to be extremely thermostable in the presence of organic solvents and detergents, with a high concentration level of activity at elevated temperatures across a wide pH range (up to 80 °C), making it a good candidate for use in the breakdown of thermophilic organic solid waste. (Toplak, Wu, Fusetti, Quaedflieg, & Janssen, 2013). Thermozymes from *Pyrococcus furiosus,* such as amylase, have been used in mutational investigations. Maltoheptaose synthesis from -cycloamyloses was elevated due to a mutation in pancreatic fistula amylase. Maltoheptaose is a sugar carrier used in the food, cosmetics, and pharmaceutical industries. (Rosenbaum et al., 2012).

Psychrophilic Enzymes

Low-temperature enzymes (psychrophilic) offer various advantages for commercial applications and have been utilized in sectors as varied as food processing, molecular biology, besides precision chemical synthesis (Cavicchioli et al., 2011). Psychrophilic enzymes have maximized activity at low temperatures by their factors like substrate affinity, lowering the transition state's free energy barrier. These enzymes are also optimized by the instability of the structures containing the active site or by destabilization of the whole molecule paving way for a decrease in the quantity and intensity of all forms of weak interactions, as well as the elimination of stability factors, resulting in better dynamics of active site residues under cold conditions. (Merlino et al., 2010). The addition of β-glycanases to detergents allows efficient cold-water washing. These enzymes might potentially be used in the paper sector, such as in pulp manipulation or bioremediation initiatives (Cummings & Black, 1999). The β- galactosidase is used in the production of lactose-free milk, cheese, and the baking industry. Amylases, Proteases, Xylases, pectinase, protease enzymes are used in the food industry for flavor, texture development, and tenderization of meat (He, Chen, Li, Zhang, & Gao, 2004). Deep-sea psychrophiles are exploited for detergent, textile, and food industries. Jinag et al. report revealed a cold-adapted alpha-amylase from sea bacteria *Bacillus sp sh 19-1* that reaches its peak activity at 20°C and could be used in cosmetics, pharmaceuticals, and food (Cavicchioli et al., 2011).

Halophilic Extremozyme

Halophiles are microorganisms that can strive in high salt concentration (~ 1 M NaCl) like deep-sea hypersaline anoxic lakes, which produce enzymes that are stable and active in high salt concentration referred has holoenzymes or halophilic extremozymes (R. Cavicchioli et al., 2011). Halophiles produce enzymes stable in high salt, organic acids with certain adapting mechanisms and extremely high stability in the low water activity and other chemicals (Dumorné, Córdova, Astorga-Eló, & Renganathan, 2017). Different adaption strategies are used by halophilic proteins. Retainment of stability and active at high ionic strength, halophilic organisms' peptides have a skewed amino acid composition. On the surface of halophilic proteins, there is usually an abundance of acidic amino acids (glutamate and aspartate). They are negatively charged because halophilic proteins bind to large quantities of hydrated ions due to lower surface hydrophobicity and are less likely to assemble at high salt concentrations (Oren, 2002). Stabilizing compounds like betaine, ectoine, and polysaccharides, as well as secreted polymers for use in biodegradable plastics or halo-tolerant lipids from these strains, are the focus of research in applied halophilic archaeal groups (Madigan & Oren, 1999). The strong salt tolerance of haloarchaeal compounds like Retinal proteins similar to bacteriorhodopsin, for example, have found usage in holographic films and other light-sensitive or "bioelectric" applications (Rodriguez-Valera, 1992). Similarly, lipolytic deep-sea halophilic enzymes, such as esterases, offer a lot of promise in biodiesel, polyunsaturated fatty acids, and food production (Litchfield, 2011).

Piezophilic Extremozyme

Piezophilic extremozymes are proteins produced by piezophiles that thrive in high hydrostatic pressure like the deep sea, volcano areas. The organism requires physiological adaption, a combination of modification of gene structure and regulation to thrive in pressure conditions (Oger & Cario, 2014). Even though piezophilic enzymes offer a lot of promise for industrial applications, there hasn't been much study done on them. Abe and Horikoshi discovered that alpha-amylase from piezophilic proteins generates trisaccharide instead of maltobiose and tetrasaccharide when maltooligosaccharide is used as a substrate at high pressure and low energy. This response has a lot of industrial and biotech potential, especially in the food business (Abe & Horikoshi, 2001). Peptidase from Pyrococcous horikoshi shows high-pressure stability suggesting that it might be beneficial in food processing (Rosenbaum et al., 2012).

Acidophilic Extremozymes

Acidophilic extremozyme are enzymes produced by acidophiles that grow in optimum pH of 3-4 (Jaenicke, 1981). Acidophiles have adopted specific mechanisms for retaining normal internal pH to survive in acidic surroundings, which suggest that, a pH differential unit across the cell membrane (Baker-Austin & Dopson, 2007). The starch industries can benefit from acid-stable α-amylases produced by *A. acidocaldarius* and *Bacillus acidicola*. Glucoamylases are a different type of amylolytic enzyme that is employed in the starch business (Sharma, Kawarabayasi, & Satyanarayana, 2012; Bai et al., 2012). The use of acid-stable xylanases (pH 5.3) from the acidophilic fungus *Aspergillus foetidus* as a whole wheat bread additive was described. Walker and Phillips (2008) reported that *Alicyclobacillus* spores are acid and heat-resistant. They can withstand typical pasteurization processes (92°C for 10 seconds) and induce spoiling. Pasteurization for high acidity foods is designed using *A. acidoterrestris* as the benchmark organism (Walker & Phillips, 2007).

Radioresistent Extremozymes

Radiotolerant bacteria generate enzymes that have a wide range of applications. As an example, Lipases from radiation-tolerant bacteria were studied by Shao et al. (2013). *Deinococcus radiodurans* bacteria expressed in E. coli. They were shown to have a predilection for three of the short-chain esters, thermally stable in the presence of surfactants and organic compounds (Shao, Xu, & Yan, 2014). It contributes to mending broken DNA strands induced by ionizing UV radiation; it's been suggested that the bacterioruberin produced by radioresistant bacteria (*Halobacterium and Rubrobacter*) could be useful in avoiding human skin cancer (Singh and Gabani, 2011). Deinoxanthin derived from the radioresistant bacteria *D. radiodurans* induced cancer cells to die, indicating that this carotenoid could be employed as a chemopreventive medication, according to Choi et al. (2014). Thus, these enzymes application various industrial operations are listed in Table 1.

Table 1. Applications of Extremophiles in various industries

Extremophiles	Active Biomolecules	Applications
Thermophiles	Alpha-amylase DNA Polymerase, Proteases	Food processing, PCR, paper industry
Alkaliphiles	Proteases, Cellulases, amylases, lipase Antibiotics	Detergents-polymer breakdown Drug for infections and allergies
Psychrophiles	Amylases, lipases, Unsaturated fatty acids Dehydrogenases	Food additives, Biosensors
Halophiles	Solutes, glycerol, carotenoids, Carotene	Pharmaceuticals, Food adjuvants
Piezophilic extremozymes	Proteases and piezophilic α-amylase	Processing of food under high pressure
Radioresistant extremozymes	*Sporothrix schenckii, Cryptococcus neoformans, Pseudoterranova decipiens and Alexandrium tamarense*	Sunscreen, antioxidant activity Antiproliferative agent and immune-stimulatory activity

Biofuels

Extremophiles are a vigorous group of organisms capable of producing stable enzymes, which effectively tolerate conditions of environmental changes such as temperature and pH. The Role and applications of such organisms and their enzymes in biotechnology are vast, and one such utilization is in the production of biofuel. Fossil fuels are the major source of resource in energy arcades. Conventional fuels (natural gas, coal, oil) will continue to be used in the near future. The demand for natural gas and petroleum has augmented because of the high economic growth of developing countries. The finest alternative source for fuel is biofuels. They are fuel products acquired from biomass like sugarcane, wheat, corn, sunflower, soybean, palm, coconut, and from the biodegradable components of manufacturing and municipality waste. The first attempt to generate biofuels on a commercial scale dates back to 1975, but there has only been a revived interest in biofuels in the last five years (Luque et al., 2008). The mounting supply of crude oil, the demand for fossil fuels, and global warming result in increasing greenhouse gas emissions have catalyzed the interest of the development of biofuels as an alternative fuel; Chemical and thermos-chemical processes are thus the current biofuel production technologies. (Barnard, Casanueva, Tuffin, & Cowan, 2010, Luque et al., 2008).

Types of Biofuels

The initial generation of biofuels was made from commonly available crops like sugar, starch, and vegetable oils, which may be extracted using standard technologies like bioethanol, biodiesel, and biobutanol. Materials that are difficult to hydrolyze, such as lignocellulosic material, are used to make second-generation biofuels. Second-generation biofuels have a significant effect on the reduction of CO2, expensive due to technical problems such as converting recalcitrant lignocellulosic biomass into fermentable sugars and enzyme price (Sims, Mabee, Saddler, & Taylor, 2010; Naik, Goud, Rout, & Dalai, 2010). Currently, bioethanol is the most commonly used biofuel worldwide; bioethanol and biodiesel encompass 90% of the total biofuel market (Antoni, Zverlov, & Schwarz, 2007).

Bioethanol

The bioethanol visibility in the global matrix of fuels, linked to ecological appeal and the possibility of using new raw materials justifies the increasing investments for the development of new processes. The processes of bioethanol production from the utilization of starch, sugarcane, and microbial fermentation of sugar compounds considers as the main source to dispense with classical fuel which causes noticeable pollution for our planet. Researchers who work in the field of fermentation give more attention to lignocellulose. The bacterium *Zymomonas mobilis* can withstand up to 120 g/L of ethanol produced by *Saccharomyces cerevisiae* strains from sugarcane molasses or enzymatically degraded starch (Antoni et al., 2007). (Rogers, Lee, Skotnicki, & Tribe, 1982). lignocellulosic material becomes essential for the synthesize of bioethanol which involves the hydrolysis of cellulose, followed by hexose sugar and hemicellulose (Kumar, Singh, & Singh, 2008*). S. cerevisiae* and *Z. mobilis* can only use hexose sugar, whereas thermophiles can ferment both hexose and pentose sugar obtained from biomass and hydrolysates (Zaldivar, Nielsen, & Olsson, 2001), allowing for faster growth and metabolism on both cellulose and hemicellulose.

(a) Thermophilic Clostridia

Thermophilic clostridia are fermentative anaerobes that thrive at temperatures between 60 and 65 degrees Celsius. They have multiple cellulases and hemicellulases within the cellulosome and the capability to degrade lignin-containing materials, such as lignocellulosic waste .cellulosome (Demain, Newcomb, & Wu, 2005). Because *C. thermocellum* can ferment cellobiose and cellodextrins to ethanol, acetate, lactate, H2, and CO2, it is a good candidate for ethanol fermentation from cellulosic biomass. However, there are a number of drawbacks associated with its use in bioethanol production. The fact that most strains are susceptible to high ethanol concentrations is one among them (Antoni et al., 2007). Genetically engineered strains of *C. thermocellum* have been developed to generate up to 26-60 g/L of ethanol (Lee R Lynd, van Zyl, McBride, & Laser, 2005). The prevailing disadvantage is its ability to utilize hexose sugar from cellulose and not the pentose sugar derived from hemicellulose (Taylor et al., 2009a). The application of thermophilic anaerobic mixed cultures can reduce half the production cost (Lee Rybeck Lynd, n.d.). Five stages would be required to fully degrade and utilize lignocellulose: Initially, *C. thermocellum* produces cellulase and hemicellulase enzymes; hydrolysis of cellulose to cello-oligomers and cellobiose, and hemicellulose to xylans, xylobiose, and other pentose sugars; and sugar absorption; the fermentation of hexose carbohydrates into ethanol by C. thermocellum, and the fermentation of pentose sugars into ethanol by another thermophilic bacterium. In mixed fermentations using *Thermoanaerobacterium thermosaccharolyticum, C. thermocellum* has been employed, *Thermoanaerobacter thermohydrosulphuricum* (Mori, 1990), *Geobacillus stearothermophilus, Thermoanaerobacter brockii,* and *Thermoanaerobacterium saccharolyticum* (Demain, 2009).

(b) Thermoanaerobacterium Saccharolyticum

Thermoanaerobacterium is a thermophilic hemicellulolytic anaerobe. It produces ethanol and organic acid from pentose sugars like xylose (acetic acid is produced by pyruvate: ferredoxin oxidoreductase (POR), phosphate acetyltransferase (Pta), and acetate kinase (Ack), while lactic acid is produced by L-lactate dehydrogenase (Ldh) (Shaw et al., 2008). Metabolic engineering of the single knockout mutant for lactate dehydrogenase in *Thermoanaerobacterium saccharolyticum* helps in increasing the ethanol yields (Desai, Guerinot, & Lynd, 2004). knocking down the genes involved in lactate (ldh) and acetate (ack/PTA) production, the *Saccharolyticum strain ALK1* was designed to produce ethanol as the only organic product (Shaw et al., 2008).

(c) Thermoanaerobacter

Thermoanaerobacter species are classified as Clostridium species which are thermophilic anaerobes very similar to thermophilic clostridia. Lactic acid and ethanol are the main products of *Thermoanaerobacter* fermentation (Collins et al., 1994, Lamed & Zeikus, 1980). *Thermoanaerobacter ethanolicus* can ferment both D-glucose and D-xylose (Lacis & Lawford, 1991) and form ethanol, but their ethanol tolerance is only 4% (v/w) (Burdette, Jung, Shen, Hollingsworth, & Zeikus, 2002). *Thermoanaerobacter BG1L1,* a lactate dehydrogenase mutant lacking lactic acid production, has an 8.3% ethanol resistance level (equivalent to 65 g/L) after constant acquaintance with the product. The strain can digest corn stover and wheat straw hydrolysate (Georgieva & Ahring, 2007; Georgieva, Mikkelsen, & Ahring, 2007) with pre-treatment with dilute acid and also can utilize xylose from lignocellulosic hydrolysates (Georgieva et al., 2007). *T. thermohydrosulphuricus* and *T. brockii* are two more species that have been examined for ethanol production. (Georgieva et al., 2007; Mori & Inaba, 1990).

(d) Geobacillus Geobacillus

Geobacillus Geobacillus are thermophilic bacteria with high catabolic flexibility that can be metabolically engineered (Taylor, Esteban, & Leak, 2008). Few species can ferment sugars like D-glucose, D-xylose, and L-arabinose at temperatures ranging from 55 to 70 degrees Celsius, producing lactate, formate, acetate, and ethanol from glucose (San Martin, Bushell, Leak, & Hartley, 1992). Few Geobacillus strains digest complex polysaccharides like xylans, resulting in the presence of xylanases. (Wu, Liu, & Zhang, 2006). *Geobacillus thermoglucosidasius* strain can tolerate ethanol as high as 10% (v/v), *Geobacillus stearothermophilus* strains that are equivalent to *S. cerevisiae* generate ethanol at 70°C (Hartley & Payton, 1983). The pyruvate decarboxylase (PDC) and alcohol dehydrogenase (ADH) genes of *Saccharomyces cerevisiae* and Zymomonas mobilis create ethanol (Dawes, Ribbons, & Large, 1966) and have the ability to convert intercellular pools of pyruvate and NADH to ethanol, with alcohol dehydrogenase converting pyruvate-derived acetaldehyde to ethanol. The bacterial PDC genes are uncommon, but few with kinetic properties and thermal stability have been identified. Attempts have been undertaken to design this route in a variety of organisms, including Geobacillus species (Ingram et al., 1998; Talarico, Gil, Yomano, Ingram, & Maupin-Furlow, 2005).

Biodiesel

The generation of biodiesel is a well-established technology for use in compression-ignition (diesel) engines. However, the cost of production constitutes 70% of plant raw materials (Behzadi & Farid, 2007), involving the transformation of monoalkyl esters of the plant fatty acids from vegetable oils. Tactlessly, fatty acid production from these oleaginous plants accounts for only 5% of their total biomass. Therefore, a more effective innovative source of oil, such as microalgae or extremophilic organisms, must be researched for biodiesel to become a commercially viable resource.

(a) Photosynthetic Microalgae for the Synthesize of Biodiesel

Microalgae are eukaryotic photosynthetic microorganisms that produce algal biomass from sunlight, water, and CO2. These organisms synthesize fatty acids for esterification into glycerol-based membrane lipids, which can account for 5–20 percent of their dry cell weight under ideal growth conditions (Hu et al., 2008). Some microalgae, such as *Botryococcus braunii,* can create very long-chain hydrocarbons (C23 to C40), analogous to those found in petroleum, that can account for more than 80% of their dry cell weight under harsh environmental conditions (Banerjee, Sharma, Chisti, & Banerjee, 2002; Metzger & Largeau, 2005). There are several advantages to using microalgae for lipid production for biodiesel conversion: huge volumes of lipids and oils, quick growth within 24 hours (Chisti, 2008), and the ability to grow in saline environments and wastewaters without needing fresh water, Photobioreactors for microalgae growth can be set up in dry or semi-arid environments, with nutrients coming from waste sources such agricultural runoff, industrial or municipal wastewater, and animal feeds (Hu et al., 2008), After lipid extraction, the algal biomass can be anaerobically converted into biogas (Lantz, Svensson, Björnsson, & Börjesson, 2007). Contamination within photobioreactors is reduced by using extremophilic microalgae, which is an issue in outdoor culture. Despite the usage of microalgae in biodiesel production, the higher price of algal oil and low growth rates (Robles-Medina, González-Moreno, Esteban-Cerdán, & Molina-Grima, n.d.) are a hurdle. To attain better output under controlled growth rate settings genetic engineering of microalgae is required, and it is thought that this is the key to making this technology commercially feasible.

(b) Fermentative Use of Microalgae for the Synthesize of Biodiesel

Solazyme, a San Francisco-based firm, converts industrial and agricultural biomass directly into renewable oils using microalgae and ordinary commercial fermentation technology (www.solazyme.com). Microalgae are developed in light-protected fermenters converting carbohydrates to oil. Soladiesel® biodiesel is produced by Solazyme, which is based on the same technology. Solazyme is the only commercial biodiesel derived from microbes that can be produced in abundance. Natural and artificial strains of microalgae are being used, and extremophilic microalgae may be used in the future.

Biobutanol

Biobutanol manufacturing has recently sparked an interest. Clostridium acetone–butanol–ethanol (ABE) fermentation is still the most common method; however, the poor butanol production, as well as the organisms' limited product tolerance, is forcing researchers to consider other options, such as solvent-tolerant organisms. Clostridial species are widely known for their ability to create butanol via ABE fermentation, despite low yield due to the solvents' tendency to hinder the organism's growth. Metabolic engineering is progressing in two areas: the engineering of *C. acetobutylicum* to produce hyperbutanol-producing strains and the engineering of *C. acetobutylicum* to provide hyperbutanol-producing strains (Lee, Jang, Lee, Papoutsakis, & Lee, 2009), developing a number of additional clostridial strains capable of producing large butanol yields and hydrolyzing biomass (Berezina, Brandt, Yarotsky, Schwarz, & Zverlov, 2009). By enhancing the effectiveness of the ABE process, a very stable and robust mutant strain of *Clostridium beijerinckii* has been created (Barnard et al., 2010). When compared to the wildtype, this production strain exhibits higher butanol selectivity; lower product inhibition, and can efficiently employ low-cost cellulosic feedstocks. ABE fermentation has yet to be discovered in any *thermophilic clostridia* (Taylor et al., 2009b).

BIOMINING

Biomining is the process of using microorganisms (microbes) to extract metals of economic interest from rock ores or mine waste. Biomining Techniques may also be used to clean up sites that have been polluted with metals, it is a progressively functional biotechnological technique for the processing of ores in the mining industry. Currently, the manufacture of copper form low-grade ores is the utmost imperative industrial application and a major part of world copper production already originates form heap or dump/stockpile bioleaching. Bioleaching is a conversion of an insoluble valuable metal into a soluble form by means of microorganisms. The significant role of extremophiles and their enzymes are identified in the mining sector and is also an efficient and environmentally friendly technique using microorganisms in the elimination of metal oxides or sulfides than the traditional heap leaching (Podar & Reysenbach, 2006, Vera, Schippers, & Sand, 2013), where harmful chemicals like cyanide are used to bind or separate minerals/metals. Biomining accounted for only 10% of the in 1992, Commercially important metals like gold, silver, copper, zinc, nickel, and uranium are successfully biomined using acidophiles like *Acidithiobacillus* and *Ferroplasma*, and thermophilic strains like *Sulfolobus* and *Metallosphaera* are used based on the condition (Vera et al., 2013). Biomining is normally safe but can result in acid mine drainage (AMD) due to the leaching or flowering out of acidic water produced by the oxidation of sulfides during mining. Because the acidophiles engaged in biomining can survive in both acidic and heavy-metal settings, AMD thrives in a setting that is both acidic and heavy-metal-rich. AMD41 is most commonly found in copper, zinc, and nickel mines. Acidophiles that are mesophilic and occasionally psychrophilic, are the main causes of AMD41. When thermophiles participate in biomining, however, the chances of AMD are minimized, and costs associated with tank cooling are maintained to a minimum.

QUORUM SENSING

The control of gene expression in response to changes in cell-population density is known as quorum sensing and it can be used to combat (antibiotic-resistant) bacterial infections, by application synergistically with antibiotics. Quorum-sensing (QS) systems, which include the engineering of cell-cell communication, are required to establish synthetic consortia (Brenner, You, & Arnold, 2008). QS regulates the behavioral, metabolic, and structural dynamics of microbial communities via signaling molecules [e.g., the acyl-homoserine lactones family (AHL)] (Shong, Jimenez Diaz, & Collins, 2012). Auto inducer (AI-1) system, peptide-based system, AI-2, and AI-3 are the relevant QS systems in mesophiles. Biofilm generation by *Bifidobacterium longum* (Sun, He, Brancaccio, Yuan, & Riedel, 2014), *Staphylococcus epidermidis RP62A* (Xue, Ni, Shang, Chen, & Zhang, 2015), and pathogenicity in *E. coli 0152:H7* are examples of QS medicated regulation in mesophiles (Kim et al., 2016). Deinococcus biofilm generation and oxidative response radiodurans (radiodurans) were reported by Li et al., 2017. Similar studies are also carried out in halophiles and showed effectiveness in finding the synthesis of acylated homoserine lactones AHLs in culture and speculated in the creation of biofilms and the synthesis of exopolysaccharide (EPS). Which is known to protect cells from desiccation and increase communication by forming specialized channels (Chassaing, Convert, & Lavielle, 1986,Decho, 2000). Acidophiles were also investigated in QS studied. Examples include acidophile Leptospirillum ferrooxidans, which is employed in biomining, During biofilm formation, the genes ygiU and ygiT (homologues of E.coli's mqsR and mqsA) are activated. MqsR (motility and QS regulator) is known to be associated with biofilm formation. as well as the formation of persister cells (Kim & Wood, 2010). Reports of QS studies in thermophiles are available where *Thermotoga maritima* was cocultivated with *Methanococcus jannaschii*, the scientists saw an increase in cell density and exopolysaccharide synthesis. They also revealed that in the coculture, a gene encoding a polypeptide (TM0504) with a pattern similar to peptides implicated in QS processes was increased. Exopolysaccharide production was also enhanced at low cell density when a synthetic exogenous peptide (based on TM0504) was added to the culture (Johnson et al., 2004). Few studies of the radio-resistant organism are available but not completely understood. *Deinococcus gobiensis*, radioresistant bacteria, has exhibited several of molecular reactions in response to UV irradiation. One of these has yet to be investigated: the gidA gene, which encodes the glucose-inhibited cell division protein A that modulates the post-transcriptional regulation of quorum sensing genes in *Pseudomonas aeruginosa*. This is accomplished by RhlR-dependent and RhlR-independent pathways in *P. aeruginosa*, which are analogous to many quorum sensing soil symbionts (Yuan et al., 2012).

BIOREMEDIATION

Bioremediation is a process of neutralizing xenobiotic chemicals, which are chemical contaminants like Polychlorinated biphenyls, polyaromatic hydrocarbons (PAH), and other chemical substances that are poisonous, mutagenic, and carcinogenic which cannot be degraded naturally. Extremophiles have gained attention in this process because of their enzymatic solid and biocatalytic systems, making them ideal for removing toxins from contaminated environments. Metals and radionuclides (M&Rs) are pollutants produced by a variety of industrial processes, including nuclear power plants, electronic waste, and mining operations.

The exceptional resistance contrivances of extremophilic bacteria and archaea provide bioremediation for metals and radionuclides. The extremophilic bacteria/archaea lead M&R bioremediation via applications and waste management. Due to refined metal detoxification pathways, fast-adapting transcriptional and translational mechanisms that activate and/or inhibit much anti-oxidative stress, metal-binding, metal-transport, and membrane-permeability responses, acidophilic and/or metalophilic microorganisms, and thermophilic enzymes, such as the esterase EstATII, are critical for bioremediation. (Voica, Bartha, Banciu, & Oren, 2016; Mukherjee, Wheaton, Blum, & Kelly, 2012; Dekker, Arsène-Ploetze, & Santini, 2016). Extremophiles biosynthesize economically profitable inorganic Nanoparticles by transforming M&Rs intracellularly or extracellularly (Ulloa et al., 2016). Extremozymes can therefore be used to remove and detoxify a variety of contaminants. Bioengineering techniques such as in silico flux analysis, biostatistics, and multi-omics analysis will enable us to tap into the limitless potential of extremophilic bacteria for the management of harmful contaminants in the environment.

BIODIVERSITY AND PROSPECTING

Over the last few decades, the richness and dispersion of extremophilic microbiomes surviving in severe environments have been intensively investigated, focusing on culturable and metagenomic techniques (Kumar, Yadav, Saxena, Paul, & Tomar, 2021). Extremophilic microbiomes have diverse microbial biodiversity. Using different extremophilic microorganisms from a variety of extreme natural environments may be identified and counted. Methylotrophic bacteria, for example, require ammonium mineral salt, but Rhizobium requires Congo red yeast mannitol. DSMZ-97, DSMZ-823, DSMZ-1184, DSMZ-97, DSMZ-823, DSMZ-1184, DSMZ-97, DSMZ-823, DSMZ-1184, DSMZ-97, DSMZ-823, DSMZ-1184, DSMZ-97, DSMZ-823, DSMZ-1184, DSMZ Halophilic medium, chemically defined medium, complex media, and OS media with NaCl levels ranging from 10% to 25% (w/v) are used for halophilic archaea and bacteria. For example, Jensen's agar is used to isolate N2-fixing bacteria, King'B agar for Pseudomonads, nutritional agar for heterotrophic microorganisms, and nutrient agar with crystal violet for Gram-negative bacteria (Archna, Priyank, Nath, & Kumar, 2015; Sahay, Singh, Kaushik, Saxena, & Arora, 2011; Verma et al., 2016).

In all three realms of life extremophilic microbes have been discovered: archaea, bacteria, and eukarya from several phyla including *Actinobacteria, Deinococcus-Thermus, Bacteroidetes, Ascomycota, Crenarchaeota, Basidiomycota, Euryarchaeota, Proteobacteria, Proteobacteria, Proteobacteria, Firmicutes* of many genera, including *Alkalibacillus, Arthrobacter, Bacillus, Burkholderia, Desemzia, Exiguobacterium, Flavobacterium.* The dominating phylum was *Firmicutes,* preceded by *Proteobacteria* and *Actinobacteria*. The phylum *Deinococcus-Thermus* contains the smallest number of microorganisms, followed by Bacteroidetes. Based on a study of the literature, the dominating genera in all extremophilic habitats include Bacillus and Bacillus derived genera (BBDG), *Halomonas, Pseudomonas, and Staphylococcus* (Kour et al., 2019).

CAROTENOIDS

Carotenoids are natural pigments present in a variety of plants, animals, and bacteria that give them their yellow, orange, red, and purple colors. Plants, algae and microalgae, bacteria, fungi, and yeasts generate lipid-soluble, economically and biotechnologically essential pigments. Natural, nature-identical, synthetic, and inorganic colours are the four categories of pigments. Artificial colourants are the most commonly used in the food and pharmaceutical industries due to their wide colour range, low cost, resistance to oxygen destruction, and solubility. Microbes' ability to use a wide range of low-cost substrates, better cultivation control and, shorter production times are all benefits of microbial manufacturing. Carotenoids are commonly recovered after fermentation by cell rupture, solvent extraction, and concentration. A schematic representation of steps involved in carotenoids production is given below (Fig 2).

Extremophile microorganisms, such as microalgae, may grow in extreme-pH, high-salinity, or high-temperature conditions (López-Rodas, Marvá, Costas, & Flores-Moya, 2008). Furthermore, microalgae have long been known to produce essential biomolecules such as carotenoids (Olaizola, 2003). Isoprenoid pigments such as carotenoids are widely employed in the food, chemical, textile, pharmaceutical, and cosmetic industries. They have antioxidants, photoprotective, and antibacterial capabilities in addition to being dyes and provitamins. Carotenoids are the natural pigments commonly linked with halophilic archaea and algae in extremophiles. Most of the carotenoids cannot be synthesized/extracted at industrial levels from organisms; however, three exceptions exist bacteriorhodopsin, canthaxanthin, and β-carotene (Chandi & Gill, 2011).

Figure 2.

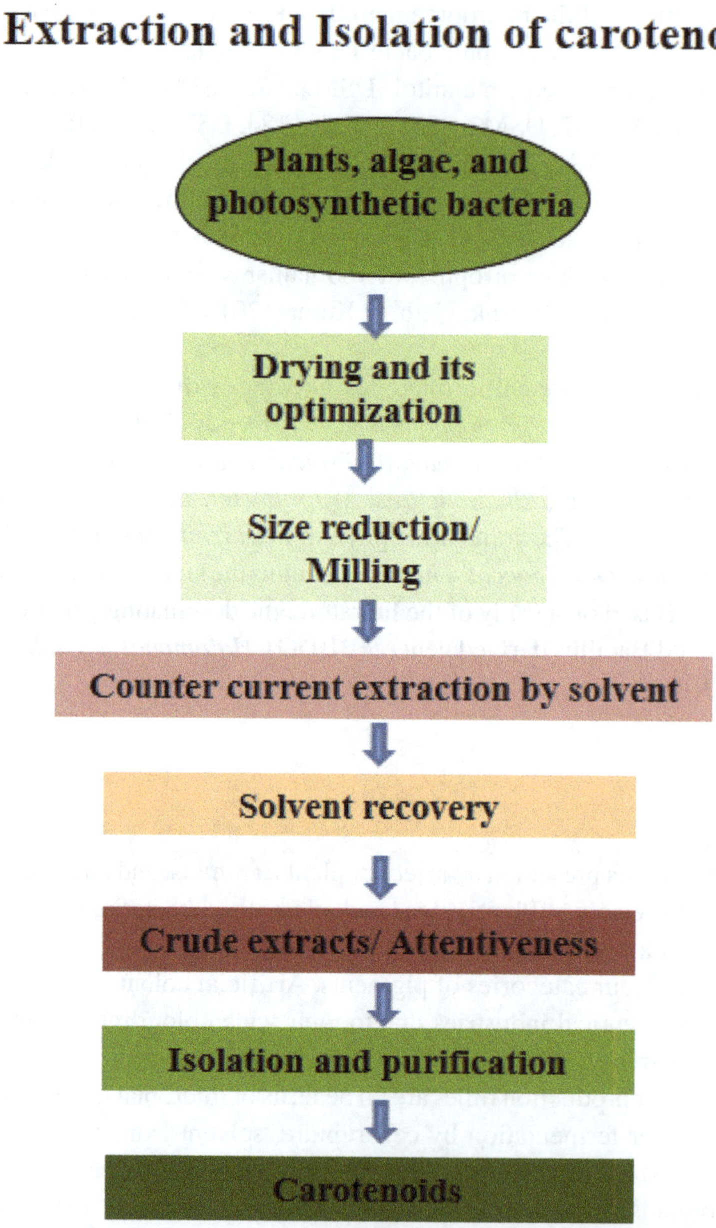

Bacteriorhodopsin (BR) is a transmembrane protein found in archaeobacteria, especially *Halobacterium salinarum,* that functions as a light-dependent proton pump (Soppa, 2006). This protein is located in the purple membrane (PM), a 2D crystal lattice found spontaneously in the plasma membrane of archaeobacteria. Undergrowth conditions that favour PM biogenesis, PM patches can cover more than half of the cell membrane (Henderson et al., 1990). This protein absorbs a lot of green light and then pumps protons outward from the plasma membrane, transforming light into an electrochemical gradient that can be utilized to make ATP (Koch & Oesterhelt, 2005). Optical appliances, electrical devices, chemical sensing, filter devices, therapeutic/diagnostic applications, and research all benefit from BR's ability to convert light energy into chemical energy (Puthenveetil & Vinogradova, 2013; Kawasaki et al., 2001; Hampp & Oesterhelt, 2005). Treatment of degenerative retinal blindness and eye disorders, therapeutic vaccination therapy, treatment of malignant tumours and other diseases, gene transcription regulation, drug delivery, transport and release of medicines, cell signalling, causing apoptosis or death of neoplastic cells, control of cell signalling, and creation of neuro-stimulation devices are just a few of the therapeutic applications of gene transcription regulation. (Koch & Oesterhelt, 2005 ; Grout et al., 2012 ; Kahya, Brown, & Schwille, 2005).

Canthaxanthin (Cx) is a diketocarotenoid (-carotene-4,4-dione) that promotes animal maturation and shields plant and animal tissues from oxidizing free radicals (Mller et al., 2000). Most cis-cx research has used the all-trans isomer, except studies on brine shrimp Artemia, where cis-cx was collected from the ovaries of female Artemia and assigned to its possible participation in reproduction and embryonic development (Nelis et al., 1984a). In poultry and aquaculture, Cx is commonly utilized as a feed supplements (Baker, 2001). Thus, chemical synthesis has been an alternate source of supplement for a long time. Since there are harmful repercussions of synthetic carotenoid by-products, interest in natural carotenoid sources has resurfaced, resulting in a renewed focus on microbial sources (Bhosale, 2004). Haloferax Alexandrinus TMT (Asker & Ohta, 2002) and Gordonia jacobaea MV-1 are some microorganisms that can synthesize cx on a commercial scale (Veiga-Crespo, Blasco, Rosa-Dos-Santos, Poza, & Villa, 2005). It's also utilized in cosmetics industries and tanning pills. These are preferably produced by halophilic archaea, such as bacteriorhodopsin, with Haloferax Alexandrinus (Asker & Ohta, 2002).

β-carotene, a red/orange pigment found in carrots, pumpkins, and halophilic bacteria, is the primary colorant in all three. Due to its chemical nature as a lipid/oil and water-soluble molecule, Dunaliella salina, a halophilic alga, is an essential source of -carotene (DasSarma, Coker, Huse, & DasSarma, 2010). Thus, it makes an addition for baking (e.g. food colouring), emulsions (e.g. confectionery and prepared foods), and as a food supplement (DasSarma et al., 2010).

MEDICAL APPLICATIONS

Microorganisms, particularly extremophiles, produce a wide range of antibiotics, antifungals, and anticancer chemicals (Littlechild, 2015). In addition to antibiotics produced by mesophilic microorganisms, extremophiles create antimicrobial peptides and diketopiperazines (Lechevalier, 1992; Waksman, Schatz, & Reynolds, 2010; Martins & Carvalho, 2007). Both *Halobacteriaceae* and *Sulfolobus* species have been found to contain antimicrobial peptides (phylogenetic family including all halophilic archaea). However, Halocins (halophilic archaea peptides) have a different activity range, and some are more effective than others against a more extensive range of bacteria (DasSarma et al., 2010). Although halocins have been identified to kill archaeal cells, there is no evidence that they kill bacteria harmful to people. However, they are used in the recovery of dogs after surgery (Shand & Leyva, 2008). Diketopiperazines (cyclic dipeptides) have been shown to have antibacterial, antifungal, antiviral, and anticancer properties and the ability to impact blood coagulation processes. Pathogens such as *Pseudomonas aeruginosa*, one of the leading causes of pneumonia and a common infection in cystic fibrosis patients, employ these pathways (Abed et al., 2013).

Finally, extremophiles' development of an alternative vaccine delivery method is a fascinating addition to medicine (Stuart, Morshed, Sremac, & DasSarma, 2001). Internal gas vesicles are small gas-filled proteinaceous structures produced by a variety of microorganisms. Halophilic archaea are the best studied. These structures were built in the Halobacterium species NRC-1 to create a recombinant form that makes simian immunodeficiency virus sections outside (Stuart, Morshed, Sremac, & DasSarma, 2004). When these recombinant vesicles were put into mice, they produced a strong antibody response and immunological memory. Adjuvants (e.g. cholera toxin B) are commonly used with recombinant vaccines to elicit a significant enough immune response (Stuart et al., 2004). Halobacterium species NRC-1 recombinant gas vesicles can be used as an adjuvant which is ether-linked rather than the more typical ester-linked molecules that provoke a robust immune response. NRC-1's polar lipids and recombinant gas vesicles as a nasal vaccine in mice were toxic.

GENETIC TECHNOLOGY

Extremozymes are enzymes produced by bacteria that live in extreme environments (Hough & Danson, 1999). Taq DNA polymerase is a prototype of these physiologically active proteins that has been widely used in diagnostic and molecular research laboratories worldwide. Thermus aquaticus was the first bacteria from which this enzyme was isolated (hence the name Taq). In 1969, it was discovered in Yellowstone National Park, Wyoming, in thermal springs (Brock and Freeze). Heat stability at 95°C makes this enzyme appropriate for polymerase chain reaction (PCR), which involves heating DNA to denature and separate the strands before polymerase amplification. Taq's heat stability eliminates the need for additional polymerase during the reaction, making such reactions efficient enough to be repeated regularly. Other thermostable polymerases have recently become available for specific PCR methods. Pfu polymerase (derived from Pyrococcus furiosus, a hyperthermophile) has greater replication fidelity than Taq. PCR techniques are constantly improving, and they're becoming more widespread in veterinary diagnostics and research. PCR can be utilized for a wide range of diagnostic applications. One of these is real-time quantification of feline herpesvirus 1 DNA in ocular fluid. Taqman PCR (Vögtlin et al., 2002) is a multiplex PCR approach for detecting *Brucella* and *Leptospira spp*., as well as diagnosing *Mycobacterium Bovis* Infection in calves sensitized by other mycobacteria (Amadori et al., 2002).

CONCLUSION AND FUTURE PROSPECT

The impact of biotechnology on our lives is inescapable. Extremozymes are found in organisms that thrive at extremes of temperature (122°C -12°C), pressure (as high as 1000 atm), salt (saturating levels), and pH (0 to 6 and 8 to 12).In recent years, many improvements have been made in the numerous technologies for biofuel production. However, no biofuel can replace petroleum, and high production costs still render biofuels unprofitable without subsidies. The cost of cellulosic ethanol will rise two to three folds than the present fare of gasoline depending on the source of feedstock. The biodiesel from microalgae is seven times higher than the present fare of diesel. The cost of feedstock sources, biomass treatment, enzyme manufacturing, and the absence of appropriate organisms for biofuel production all necessitate new insights for the commercialization of commercially viable biofuels. The benefits of high process temperatures, and advantages of using extremophilic organisms and their enzymes suggest that extremophiles have the potential for second-generation biofuel production. The discovery of new microorganisms, understanding of their metabolic pathways, one-step fermenters will provide new alternatives for improving biofuel production. Functional metagenomics is a powerful tool to screen libraries from extreme environments. These approaches should provide new biocatalysts as promising alternatives for future production processes, biofuels production process; mining and carotenoids used in the food and cosmetic industries are four success stories.

Breakthroughs have aided biotechnology and related commercial industries by providing new levels of understanding about biological adaptation mechanisms, informing us about the evolution of life on Earth, increasing the number of sites suitable for searching for extraterrestrial life, and increasing the number of sites suitable for searching for extraterrestrial life. The development of bioreactors made of materials that can endure the extremes of temperature, pH, and salinity required for extremophile culture is a considerable issue. Extremophiles may be able to improve metal bioremediation, recycling, and also as new comprehensive improvements, resources, and cost-effective techniques. The QS system explored has paved the way for biofilms, biomining, and exopolysaccharide production; further exploration can increase our further chances. Biodiversity and prospective study is very limited in extremophiles due to the extreme environmental changes. Further studies should be carried out. Synthetic biology tools and metabolic engineering on extremophiles can be engineered as new platforms (or chassis) that house stable and multiple intended functions and transcriptional and translational machinery to be active in extreme settings ranging from the lab to the field. To overcome constraints associated with uncultured extremophiles on a large scale, new approaches are required.

REFERENCES

Abe, F., & Horikoshi, K. (2001). The biotechnological potential of piezophiles. *Trends in Biotechnology*, *19*(3), 102–108. doi:10.1016/S0167-7799(00)01539-0 PMID:11179803

Abed, R. M. M., Dobretsov, S., Al-Fori, M., Gunasekera, S. P., Sudesh, K., & Paul, V. J. (2013). Quorum-sensing inhibitory compounds from extremophilic microorganisms isolated from a hypersaline cyanobacterial mat. *Journal of Industrial Microbiology & Biotechnology*, *40*(7), 759–772. doi:10.100710295-013-1276-4 PMID:23645384

Amadori, M., Tagliabue, S., Lauzi, S., Finazzi, G., Lombardi, G., Teló, P., Pacciarini, L., & Bonizzi, L. (2002). Diagnosis of Mycobacterium bovis infection in calves sensitized by mycobacteria of the avium/intracellulare group. *Journal of Veterinary Medicine. B, Infectious Diseases and Veterinary Public Health*, *49*(2), 89–96. doi:10.1046/j.1439-0450.2002.00513.x PMID:12002425

Antoni, D., Zverlov, V. V., & Schwarz, W. H. (2007). Biofuels from microbes. *Applied Microbiology and Biotechnology*, *77*(1), 23–35. doi:10.100700253-007-1163-x PMID:17891391

Archna, S., Priyank, V., Nath, Y. A., & Kumar, S. A. (2015). Bioprospecting for extracellular hydrolytic enzymes from culturable thermotolerant bacteria isolated from Manikaran thermal springs. *Research Journal of Biotechnology*, *10*(4), 33–42.

Asker, D., & Ohta, Y. (2002). Production of canthaxanthin by Haloferax alexandrinus under non-aseptic conditions and a simple, rapid method for its extraction. *Applied Microbiology and Biotechnology*, *58*(6), 743–750. doi:10.100700253-002-0967-y PMID:12021793

Auman, A. J., Breezee, J. L., Gosink, J. J., Kämpfer, P., & Staley, J. T. (2006). Psychromonas ingrahamii sp. nov., a novel gas vacuolate, psychrophilic bacterium isolated from Arctic polar sea ice. *International Journal of Systematic and Evolutionary Microbiology*, *56*(Pt 5), 1001–1007. doi:10.1099/ijs.0.64068-0 PMID:16627645

Bai, Y., Huang, H., Meng, K., Shi, P., Yang, P., Luo, H., Luo, C., Feng, Y., Zhang, W., & Yao, B. (2012). Identification of an acidic α-amylase from Alicyclobacillus sp. A4 and assessment of its application in the starch industry. *Food Chemistry*, *131*(4), 1473–1478. doi:10.1016/j.foodchem.2011.10.036

Baker, R. T. (2001). Canthaxanthin in aquafeed applications: Is there any risk? *Trends in Food Science & Technology*, *12*(7), 240–243. doi:10.1016/S0924-2244(01)00091-7

Baker-Austin, C., & Dopson, M. (2007). Life in acid: pH homeostasis in acidophiles. *Trends in Microbiology*, *15*(4), 165–171. doi:10.1016/j.tim.2007.02.005 PMID:17331729

Banerjee, A., Sharma, R., Chisti, Y., & Banerjee, U. C. (2002). Botryococcus braunii: A renewable source of hydrocarbons and other chemicals. *Critical Reviews in Biotechnology*, *22*(3), 245–279. doi:10.1080/07388550290789513 PMID:12405558

Barnard, D., Casanueva, A., Tuffin, M., & Cowan, D. (2010). Extremophiles in biofuel synthesis. *Environmental Technology*, *31*(8–9), 871–888. doi:10.1080/09593331003710236 PMID:20662378

Behzadi, S., & Farid, M. M. (2007). Review: Examining the use of different feedstock for the production of biodiesel. *Asia-Pacific Journal of Chemical Engineering*, *2*(5), 480–486. doi:10.1002/apj.85

Berezina, O. V., Brandt, A., Yarotsky, S., Schwarz, W. H., & Zverlov, V. V. (2009). Isolation of a new butanol-producing Clostridium strain: High level of hemicellulosic activity and structure of solventogenesis genes of a new Clostridium saccharobutylicum isolate. *Systematic and Applied Microbiology*, *32*(7), 449–459. doi:10.1016/j.syapm.2009.07.005 PMID:19674858

Bhosale, P. (2004). Environmental and cultural stimulants in the production of carotenoids from microorganisms. *Applied Microbiology and Biotechnology*, *63*(4), 351–361. doi:10.100700253-003-1441-1 PMID:14566431

Brenner, K., You, L., & Arnold, F. H. (2008). Engineering microbial consortia: A new frontier in synthetic biology. *Trends in Biotechnology*, *26*(9), 483–489. doi:10.1016/j.tibtech.2008.05.004 PMID:18675483

Burdette, D. S., Jung, S.-H., Shen, G.-J., Hollingsworth, R. I., & Zeikus, J. G. (2002). Physiological function of alcohol dehydrogenases and long-chain (C(30)) fatty acids in alcohol tolerance of Thermoanaerobacter ethanolicus. *Applied and Environmental Microbiology*, *68*(4), 1914–1918. doi:10.1128/AEM.68.4.1914-1918.2002 PMID:11916712

Carriquiry, M. A., Du, X., & Timilsina, G. R. (2011). Second generation biofuels: Economics and policies. *Energy Policy*, *39*(7), 4222–4234. doi:10.1016/j.enpol.2011.04.036

Cavicchioli, R., Charlton, T., Ertan, H., Omar, S. M., Siddiqui, K. S., & Williams, T. J. (2011). Biotechnological uses of enzymes from psychrophiles. *Microbial Biotechnology*, *4*(4), 449–460. doi:10.1111/j.1751-7915.2011.00258.x PMID:21733127

Cavicchioli, R., Siddiqui, K. S., Andrews, D., & Sowers, K. R. (2002). Low-temperature extremophiles and their applications. *Current Opinion in Biotechnology*, *13*(3), 253–261. doi:10.1016/S0958-1669(02)00317-8 PMID:12180102

Chandi, G. K., & Gill, B. S. (2011). Production and Characterization of Microbial Carotenoids as an Alternative to Synthetic Colors: A Review. *International Journal of Food Properties*, *14*(3), 503–513. doi:10.1080/10942910903256956

Chassaing, G., Convert, O., & Lavielle, S. (1986). Preferential conformation of substance P in solution. *European Journal of Biochemistry*, *154*(1), 77–85. doi:10.1111/j.1432-1033.1986.tb09361.x PMID:2417837

Chisti, Y. (2008). Biodiesel from microalgae beats bioethanol. *Trends in Biotechnology*, *26*(3), 126–131. doi:10.1016/j.tibtech.2007.12.002 PMID:18221809

Choi, Y.-J., Hur, J.-M., Lim, S., Jo, M., Kim, D. H., & Choi, J.-I. (2014). Induction of apoptosis by deinoxanthin in human cancer cells. *Anticancer Research*, *34*(4), 1829–1835. PMID:24692716

Coker, J. A., & Brenchley, J. E. (2006). Protein engineering of a cold-active β-galactosidase from Arthrobacter sp. SB to increase lactose hydrolysis reveals new sites affecting low temperature activity. *Extremophiles*, *10*(6), 515–524. doi:10.100700792-006-0526-z PMID:16736094

Collins, M. D., Lawson, P. A., Willems, A., Cordoba, J. J., Fernandez-Garayzabal, J., Garcia, P., Cai, J., Hippe, H., & Farrow, J. A. (1994). The phylogeny of the genus Clostridium: Proposal of five new genera and eleven new species combinations. *International Journal of Systematic Bacteriology*, *44*(4), 812–826. doi:10.1099/00207713-44-4-812 PMID:7981107

Cummings, S. P., & Black, G. W. (1999). Polymer hydrolysis in a cold climate. *Extremophiles: Life under Extreme Conditions*, *3*(2), 81–87. doi:10.1007/s007920050102

Dalmaso, G., Ferreira, D., & Vermelho, A. (2015). Marine Extremophiles: A Source of Hydrolases for Biotechnological Applications. *Marine Drugs*, *13*(4), 1925–1965. doi:10.3390/md13041925 PMID:25854643

DasSarma, P., Coker, J. A., Huse, V., & DasSarma, S. (2010). Halophiles, Industrial Applications. In *Encyclopedia of Industrial Biotechnology*. John Wiley & Sons, Inc. doi:10.1002/9780470054581.eib439

Dawes, E. A., Ribbons, D. W., & Large, P. J. (1966). The route of ethanol formation in Zymomonas mobilis. *The Biochemical Journal*, *98*(3), 795–803. doi:10.1042/bj0980795 PMID:4287842

Decho, A. W. (2000). Microbial biofilms in intertidal systems: An overview. *Continental Shelf Research*, *20*(10–11), 1257–1273. doi:10.1016/S0278-4343(00)00022-4

Dekker, L., Arsène-Ploetze, F., & Santini, J. M. (2016). Comparative proteomics of Acidithiobacillus ferrooxidans grown in the presence and absence of uranium. *Research in Microbiology*, *167*(3), 234–239. doi:10.1016/j.resmic.2016.01.007 PMID:26829305

Demain, A. L. (2009). Biosolutions to the energy problem. *Journal of Industrial Microbiology & Biotechnology, 36*(3), 319–332. doi:10.100710295-008-0521-8 PMID:19137336

Desai, S. G., Guerinot, M. L., & Lynd, L. R. (2004). Cloning of l-lactate dehydrogenase and elimination of lactic acid production via gene knockout in Thermoanaerobacterium saccharolyticum JW/SL-YS485. *Applied Microbiology and Biotechnology, 65*(5). Advance online publication. doi:10.100700253-004-1575-9 PMID:15007569

Dumorné, K., Córdova, D. C., Astorga-Eló, M., & Renganathan, P. (2017). Extremozymes: A Potential Source for Industrial Applications. *Journal of Microbiology and Biotechnology, 27*(4), 649–659. doi:10.4014/jmb.1611.11006 PMID:28104900

Elleuche, S., Schröder, C., Sahm, K., & Antranikian, G. (2014). Extremozymes—Biocatalysts with unique properties from extremophilic microorganisms. *Current Opinion in Biotechnology, 29*, 116–123. doi:10.1016/j.copbio.2014.04.003 PMID:24780224

Georgieva, T. I., & Ahring, B. K. (2007). Evaluation of continuous ethanol fermentation of dilute-acid corn stover hydrolysate using thermophilic anaerobic bacterium Thermoanaerobacter BG1L1. *Applied Microbiology and Biotechnology, 77*(1), 61–68. doi:10.100700253-007-1149-8 PMID:17899073

Georgieva, T. I., Mikkelsen, M. J., & Ahring, B. K. (2007). Ethanol Production from Wet-Exploded Wheat Straw Hydrolysate by Thermophilic Anaerobic Bacterium Thermoanaerobacter BG1L1 in a Continuous Immobilized Reactor. In *Biotechnology for Fuels and Chemicals* (pp. 99–110). Humana Press. doi:10.1007/978-1-60327-526-2_12

Gerday, C., & Glansdorff, N. (Eds.). (2007). *Physiology and Biochemistry of Extremophiles*. ASM Press. doi:10.1128/9781555815813

Grout, T., Hong, Y., Basara, J., Balasundaram, B., Kong, Z., & Bukkapatnam, S. T. S. (2012). Significant Winter Weather Events and Associated Socioeconomic Impacts (Federal Aid Expenditures) across Oklahoma: 2000–10. *Weather, Climate, and Society, 4*(1), 48–58. doi:10.1175/WCAS-D-11-00057.1

Hampp, N., & Oesterhelt, D. (2005). Bacteriorhodopsin and Its Potential in Technical Applications. In Nanobiotechnology (pp. 146–167). Wiley-VCH Verlag GmbH & Co. KGaA. doi:10.1002/3527602453.ch11

Hartley, B. S., & Payton, M. A. (1983). Industrial prospects for thermophiles and thermophilic enzymes. *Biochemical Society Symposium, 48*, 133–146. PMID:6400480

He, H., Chen, X., Li, J., Zhang, Y., & Gao, P. (2004). Taste improvement of refrigerated meat treated with cold-adapted Protease. *Food Chemistry, 84*(2), 307–311. doi:10.1016/S0308-8146(03)00242-5

Henderson, R., Baldwin, J. M., Ceska, T. A., Zemlin, F., Beckmann, E., & Downing, K. H. (1990). Model for the structure of bacteriorhodopsin based on high-resolution electron cryo-microscopy. *Journal of Molecular Biology, 213*(4), 899–929. doi:10.1016/S0022-2836(05)80271-2 PMID:2359127

Hough, D. W., & Danson, M. J. (1999). Extremozymes. *Current Opinion in Chemical Biology, 3*(1), 39–46. doi:10.1016/S1367-5931(99)80008-8 PMID:10021406

Hu, Q., Sommerfeld, M., Jarvis, E., Ghirardi, M., Posewitz, M., Seibert, M., & Darzins, A. (2008). Microalgal triacylglycerols as feedstocks for biofuel production: Perspectives and advances. *The Plant Journal : For Cell and Molecular Biology*, *54*(4), 621–639. doi:10.1111/j.1365-313X.2008.03492.x PMID:18476868

Ingram, G., Lai, M., & Wood, Y. (1998). Metabolic engineering of bacteria for ethanol production. *Biotechnology and Bioengineering*, *58*(2–3), 204–214. doi:10.1002/(SICI)1097-0290(19980420)58:2/3<204::AID-BIT13>3.0.CO;2-C PMID:10191391

Ishino, S., & Ishino, Y. (2014). DNA polymerases as useful reagents for biotechnology â€“ the history of developmental research in the field. *Frontiers in Microbiology*, *5*. Advance online publication. doi:10.3389/fmicb.2014.00465 PMID:25221550

Jaenicke, R. (1981). Enzymes Under Extremes of Physical Conditions. *Annual Review of Biophysics and Bioengineering*, *10*(1), 1–67. doi:10.1146/annurev.bb.10.060181.000245 PMID:7020570

Johnson, D. B. (2014). Biomining-biotechnologies for extracting and recovering metals from ores and waste materials. *Current Opinion in Biotechnology*, *30*, 24–31. doi:10.1016/j.copbio.2014.04.008 PMID:24794631

Johnson, M. R., Montero, C. I., Conners, S. B., Shockley, K. R., Bridger, S. L., & Kelly, R. M. (2004). Population density-dependent regulation of exopolysaccharide formation in the hyperthermophilic bacterium Thermotoga maritima. *Molecular Microbiology*, *55*(3), 664–674. doi:10.1111/j.1365-2958.2004.04419.x PMID:15660994

Joseph, B., Ramteke, P. W., & Thomas, G. (2008). Cold active microbial lipases: Some hot issues and recent developments. *Biotechnology Advances*, *26*(5), 457–470. doi:10.1016/j.biotechadv.2008.05.003 PMID:18571355

Kahya, N., Brown, D. A., & Schwille, P. (2005). Raft partitioning and dynamic behavior of human placental alkaline phosphatase in giant unilamellar vesicles. *Biochemistry*, *44*(20), 7479–7489. doi:10.1021/bi047429d PMID:15895991

Kato, C., Li, L., Nogi, Y., Nakamura, Y., Tamaoka, J., & Horikoshi, K. (1998). Extremely barophilic bacteria isolated from the Mariana Trench, Challenger Deep, at a depth of 11,000 meters. *Applied and Environmental Microbiology*, *64*(4), 1510–1513. doi:10.1128/AEM.64.4.1510-1513.1998 PMID:9546187

Kawasaki, S., Borchert, C., Deyholos, M., Wang, H., Brazille, S., Kawai, K., Galbraith, D., & Bohnert, H. J. (2001). Gene expression profiles during the initial phase of salt stress in rice. *The Plant Cell*, *13*(4), 889–905. doi:10.1105/tpc.13.4.889 PMID:11283343

Kim, Y., & Wood, T. K. (2010). Toxins Hha and CspD and small RNA regulator Hfq are involved in persister cell formation through MqsR in Escherichia coli. *Biochemical and Biophysical Research Communications*, *391*(1), 209–213. doi:10.1016/j.bbrc.2009.11.033 PMID:19909729

Kim, Y.-G., Lee, J.-H., Gwon, G., Kim, S.-I., Park, J. G., & Lee, J. (2016). Essential Oils and Eugenols Inhibit Biofilm Formation and the Virulence of Escherichia coli O157:H7. *Scientific Reports*, *6*(1), 36377. doi:10.1038rep36377 PMID:27808174

Koch, M. K., & Oesterhelt, D. (2005). MpcT is the transducer for membrane potential changes in Halobacterium salinarum. *Molecular Microbiology, 55*(6), 1681–1694. doi:10.1111/j.1365-2958.2005.04516.x PMID:15752193

Kour, D., Rana, K. L., Kaur, T., Singh, B., Chauhan, V. S., Kumar, A., ... Gupta, V. K. (2019). Extremophiles for Hydrolytic Enzymes Productions: Biodiversity and Potential Biotechnological Applications. In *Bioprocessing for Biomolecules Production* (pp. 321–372). John Wiley & Sons, Ltd. doi:10.1002/9781119434436.ch16

Kumar, M., Yadav, A. N., Saxena, R., Paul, D., & Tomar, R. S. (2021). Biodiversity of pesticides degrading microbial communities and their environmental impact. *Biocatalysis and Agricultural Biotechnology, 31*, 101883. doi:10.1016/j.bcab.2020.101883

Kumar, R., Singh, S., & Singh, O. V. (2008). Bioconversion of lignocellulosic biomass: Biochemical and molecular perspectives. *Journal of Industrial Microbiology & Biotechnology, 35*(5), 377–391. doi:10.100710295-008-0327-8 PMID:18338189

Lacis, L. S., & Lawford, H. G. (1991). Thermoanaerobacter ethanolicus Growth and Product Yield from Elevated Levels of Xylose or Glucose in Continuous Cultures. *Applied and Environmental Microbiology, 57*(2), 579–585. doi:10.1128/aem.57.2.579-585.1991 PMID:16348422

Lantz, M., Svensson, M., Björnsson, L., & Börjesson, P. (2007). The prospects for an expansion of biogas systems in Sweden—Incentives, barriers and potentials. *Energy Policy, 35*(3), 1830–1843. doi:10.1016/j.enpol.2006.05.017

Lechevalier, H. (1992). Actinomycetes and their products: A look at the future. *World Journal of Microbiology & Biotechnology, 8*(S1), 72–73. doi:10.1007/BF02421498 PMID:24425651

Lee, J. Y., Jang, Y.-S., Lee, J., Papoutsakis, E. T., & Lee, S. Y. (2009). Metabolic engineering of Clostridium acetobutylicum M5 for highly selective butanol production. *Biotechnology Journal, 4*(10), 1432–1440. doi:10.1002/biot.200900142 PMID:19830716

Li, T., Weng, Y., Ma, X., Tian, B., Dai, S., Jin, Y., Liu, M., Li, J., Yu, J., & Hua, Y. (2017). Deinococcus radiodurans Toxin–Antitoxin MazEF-dr Mediates Cell Death in Response to DNA Damage Stress. *Frontiers in Microbiology, 8*, 1427. Advance online publication. doi:10.3389/fmicb.2017.01427 PMID:28798741

Litchfield, C. D. (2011). Potential for industrial products from the halophilic Archaea. *Journal of Industrial Microbiology & Biotechnology, 38*(10), 1635–1647. doi:10.100710295-011-1021-9 PMID:21853327

Littlechild, J. A. (2015). Archaeal Enzymes and Applications in Industrial Biocatalysts. *Archaea (Vancouver, B.C.), 147671*, 1–10. Advance online publication. doi:10.1155/2015/147671 PMID:26494981

López-Rodas, V., Marvá, F., Costas, E., & Flores-Moya, A. (2008). Microalgal adaptation to a stressful environment (acidic, metal-rich mine waters) could be due to selection of pre-selective mutants originating in non-extreme environments. *Environmental and Experimental Botany, 64*(1), 43–48. doi:10.1016/j.envexpbot.2008.01.001

Luque, R., Herrero-Davila, L., Campelo, J. M., Clark, J. H., Hidalgo, J. M., Luna, D., Marinas, J. M., & Romero, A. A. (2008). Biofuels: A technological perspective. *Energy & Environmental Science*, *1*(5), 542. doi:10.1039/b807094f

Lynd, L. R. (n.d.). Production of ethanol from lignocellulosic materials using thermophilic bacteria: Critical evaluation of potential and review. In *Lignocellulosic Materials* (pp. 1–52). Springer-Verlag. doi:10.1007/BFb0007858

Lynd, L. R., van Zyl, W. H., McBride, J. E., & Laser, M. (2005). Consolidated bioprocessing of cellulosic biomass: An update. *Current Opinion in Biotechnology*, *16*(5), 577–583. doi:10.1016/j.copbio.2005.08.009 PMID:16154338

Madigan, M. T., & Oren, A. (1999). Thermophilic and halophilic extremophiles. *Current Opinion in Microbiology*, *2*(3), 265–269. doi:10.1016/S1369-5274(99)80046-0 PMID:10383863

Martins, M. B., & Carvalho, I. (2007). Diketopiperazines: Biological activity and synthesis. *Tetrahedron*, *63*(40), 9923–9932. doi:10.1016/j.tet.2007.04.105

Merlino, A., Krauss, I. R., Castellano, I., De Vendittis, E., Rossi, B., Conte, M., ... Sica, F. (2010). Structure and flexibility in cold-adapted iron superoxide dismutases: The case of the enzyme isolated from Pseudoalteromonas haloplanktis. *Journal of Structural Biology*, *172*(3), 343–352. doi:10.1016/j.jsb.2010.08.008 PMID:20732427

Metzger, P., & Largeau, C. (2005). Botryococcus braunii: A rich source for hydrocarbons and related ether lipids. *Applied Microbiology and Biotechnology*, *66*(5), 486–496. doi:10.100700253-004-1779-z PMID:15630516

Miettinen-Oinonen, A., & Suominen, P. (2002). Enhanced Production of Trichoderma reesei Endoglucanases and Use of the New Cellulase Preparations in Producing the Stonewashed Effect on Denim Fabric. *Applied and Environmental Microbiology*, *68*(8), 3956–3964. doi:10.1128/AEM.68.8.3956-3964.2002 PMID:12147496

Møller, A. P., Biard, C., Blount, J. D., Houston, D. C., Ninni, P., Saino, N., & Surai, P. F. (2000). Carotenoid-dependent Signals: Indicators of Foraging Efficiency, Immunocompetence or Detoxification Ability? *Avian and Poultry Biology Reviews*, *11*(3), 137–159.

Moran-Reyna, A., & Coker, J. A. (2014). The effects of extremes of pH on the growth and transcriptomic profiles of three haloarchaea. *F1000 Research*, *3*, 168. doi:10.12688/f1000research.4789.1 PMID:25285207

Mori, Y. (1990). Characterization of a Symbiotic Coculture of Clostridium thermohydrosulfuricum YM3 and Clostridium thermocellum YM4. *Applied and Environmental Microbiology*, *56*(1), 37–42. doi:10.1128/aem.56.1.37-42.1990 PMID:16348106

Mori, Y., & Inaba, T. (1990). Ethanol production from starch in a pervaporation membrane bioreactor using Clostridium thermohydrosulfuricum. *Biotechnology and Bioengineering*, *36*(8), 849–853. doi:10.1002/bit.260360815 PMID:18597284

Mukherjee, A., Wheaton, G. H., Blum, P. H., & Kelly, R. M. (2012). Uranium extremophily is an adaptive, rather than intrinsic, feature for extremely thermoacidophilic Metallosphaera species. *Proceedings of the National Academy of Sciences of the United States of America*, *109*(41), 16702–16707. doi:10.1073/pnas.1210904109 PMID:23010932

Naik, S. N., Goud, V. V., Rout, P. K., & Dalai, A. K. (2010). Production of first and second generation biofuels: A comprehensive review. *Renewable & Sustainable Energy Reviews*, *14*(2), 578–597. doi:10.1016/j.rser.2009.10.003

Nelis, H. J., Lavens, P., Moens, L., Sorgeloos, P., Jonckheere, J. A., Criel, G. R., & De Leenheer, A. P. (1984a). cis-Canthaxanthins. Unusual carotenoids in the eggs and the reproductive system of female brine shrimp artemia. *The Journal of Biological Chemistry*, *259*(10), 6063–6066. doi:10.1016/S0021-9258(20)82105-X PMID:6427206

Oger, P., & Cario, A. (2014). La vie sous pression des microorganismes piézophiles. *Biologie Aujourd'hui*, *208*(3), 193–206. doi:10.1051/jbio/2014023 PMID:25474000

Olaizola, M. (2003). Commercial development of microalgal biotechnology: From the test tube to the marketplace. *Biomolecular Engineering*, *20*(4–6), 459–466. doi:10.1016/S1389-0344(03)00076-5 PMID:12919832

Oren, A. (2002). Molecular ecology of extremely halophilic Archaea and Bacteria. *FEMS Microbiology Ecology*, *39*(1), 1–7. doi:10.1111/j.1574-6941.2002.tb00900.x PMID:19709178

Oren, A. (2010, July). Industrial and environmental applications of halophilic microorganisms. *Environmental Technology*, *31*(8–9), 825–834. doi:10.1080/09593330903370026 PMID:20662374

Podar, M., & Reysenbach, A.-L. (2006). New opportunities revealed by biotechnological explorations of extremophiles. *Current Opinion in Biotechnology*, *17*(3), 250–255. doi:10.1016/j.copbio.2006.05.002 PMID:16701993

Puthenveetil, R., & Vinogradova, O. (2013). Optimization of the design and preparation of nanoscale phospholipid bilayers for its application to solution NMR. *Proteins*, *81*(7), 1222–1231. doi:10.1002/prot.24271 PMID:23436707

Richtzenhain, L. J., Cortez, A., Heinemann, M. B., Soares, R. M., Sakamoto, S. M., Vasconcellos, S. A., Higa, Z. M. M., Scarcelli, E., & Genovez, M. E. (2002). A multiplex PCR for the detection of Brucella spp. and Leptospira spp. DNA from aborted bovine fetuses. *Veterinary Microbiology*, *87*(2), 139–147. doi:10.1016/S0378-1135(02)00049-4 PMID:12034542

Robles-Medina, A., González-Moreno, P. A., Esteban-Cerdán, L., & Molina-Grima, E. (2009, July). Biocatalysis: Towards ever greener biodiesel production. *Biotechnology Advances*, *27*(4), 398–408. doi:10.1016/j.biotechadv.2008.10.008 PMID:19268517

Rodriguez-Valera, F. (1992). Biotechnological potential of halobacteria. *Biochemical Society Symposium*, *58*, 135–147. PMID:1445403

Rogers, P. L., Lee, K. J., Skotnicki, M. L., & Tribe, D. E. (1982). *Ethanol production by Zymomonas mobilis*. doi:10.1007/3540116982_2

Rosenbaum, E., Gabel, F., Durá, M. A., Finet, S., Cléry-Barraud, C., Masson, P., & Franzetti, B. (2012). Effects of hydrostatic pressure on the quaternary structure and enzymatic activity of a large peptidase complex from Pyrococcus horikoshii. *Archives of Biochemistry and Biophysics, 517*(2), 104–110. doi:10.1016/j.abb.2011.07.017 PMID:21896270

Rothschild, L. J., & Mancinelli, R. L. (2001). Life in extreme environments. *Nature, 409*(6823), 1092–1101. doi:10.1038/35059215 PMID:11234023

Rubio-Infante, N., & Moreno-Fierros, L. (2016). An overview of the safety and biological effects of Bacillus thuringiensis Cry toxins in mammals. *Journal of Applied Toxicology, 36*(5), 630–648. doi:10.1002/jat.3252 PMID:26537666

Sahay, H., Singh, S., Kaushik, R., Saxena, A. K., & Arora, D. K. (2011). Characterization of halophilic bacteria from environmental samples from the brackish water of Pulicat Lake, India. *Biologia, 66*(5), 741–747. doi:10.247811756-011-0094-2

San Martin, R., Bushell, D., Leak, D. J., & Hartley, B. S. (1992). Development of a synthetic medium for continuous anaerobic growth and ethanol production with a lactate dehydrogenase mutant of Bacillus stearothermophilus. *Journal of General Microbiology, 138*(5), 987–996. doi:10.1099/00221287-138-5-987 PMID:1645133

Shah, A. R., Shah, R. K., & Madamwar, D. (2006). Improvement of the quality of whole wheat bread by supplementation of xylanase from Aspergillus foetidus. *Bioresource Technology, 97*(16), 2047–2053. doi:10.1016/j.biortech.2005.10.006 PMID:16307877

Shand, R. F., & Leyva, K. J. (2008). Archaeal Antimicrobials: An Undiscovered Country. In Archaea: New Models for Prokaryotic Biology (p. 248). Academic Press.

Shao, H., Xu, L., & Yan, Y. (2014). Thermostable lipases from extremely radioresistant bacterium Deinococcus radiodurans: Cloning, expression, and biochemical characterization. *Journal of Basic Microbiology, 54*(9), 984–995. doi:10.1002/jobm.201300434 PMID:24002979

Sharma, A., Kawarabayasi, Y., & Satyanarayana, T. (2012). Acidophilic bacteria and archaea: acid stable biocatalysts and their potential applications. *Extremophiles: Life under Extreme Conditions, 16*(1), 1–19. doi:10.1007/s00792-011-0402-3

Shaw, A. J., Podkaminer, K. K., Desai, S. G., Bardsley, J. S., Rogers, S. R., Thorne, P. G., Hogsett, D. A., & Lynd, L. R. (2008). Metabolic engineering of a thermophilic bacterium to produce ethanol at high yield. *Proceedings of the National Academy of Sciences of the United States of America, 105*(37), 13769–13774. doi:10.1073/pnas.0801266105 PMID:18779592

Shong, J., Jimenez Diaz, M. R., & Collins, C. H. (2012). Towards synthetic microbial consortia for bioprocessing. *Current Opinion in Biotechnology, 23*(5), 798–802. doi:10.1016/j.copbio.2012.02.001 PMID:22387100

Sims, R. E. H., Mabee, W., Saddler, J. N., & Taylor, M. (2010). An overview of second generation biofuel technologies. *Bioresource Technology, 101*(6), 1570–1580. doi:10.1016/j.biortech.2009.11.046 PMID:19963372

Soppa, J. (2006). From genomes to function: Haloarchaea as model organisms. *Microbiology (Reading, England)*, *152*(Pt 3), 585–590. doi:10.1099/mic.0.28504-0 PMID:16514139

Stuart, E. S., Morshed, F., Sremac, M., & DasSarma, S. (2001). Antigen presentation using novel particulate organelles from halophilic archaea. *Journal of Biotechnology*, *88*(2), 119–128. doi:10.1016/S0168-1656(01)00267-X PMID:11403846

Stuart, E. S., Morshed, F., Sremac, M., & DasSarma, S. (2004). Cassette-based presentation of SIV epitopes with recombinant gas vesicles from halophilic archaea. *Journal of Biotechnology*, *114*(3), 225–237. doi:10.1016/j.jbiotec.2004.01.005 PMID:15522433

Sun, Z., He, X., Brancaccio, V. F., Yuan, J., & Riedel, C. U. (2014). Bifidobacteria exhibit LuxS-dependent autoinducer 2 activity and biofilm formation. *PLoS One*, *9*(2), e88260. doi:10.1371/journal.pone.0088260 PMID:24505453

Talarico, L. A., Gil, M. A., Yomano, L. P., Ingram, L. O., & Maupin-Furlow, J. A. (2005). Construction and expression of an ethanol production operon in Gram-positive bacteria. *Microbiology*, *151*(12), 4023–4031. doi:10.1099/mic.0.28375-0 PMID:16339947

Taylor, M. P., Eley, K. L., Martin, S., Tuffin, M. I., Burton, S. G., & Cowan, D. A. (2009a). Thermophilic ethanologenesis: Future prospects for second-generation bioethanol production. *Trends in Biotechnology*, *27*(7), 398–405. doi:10.1016/j.tibtech.2009.03.006 PMID:19481826

Taylor, M. P., Eley, K. L., Martin, S., Tuffin, M. I., Burton, S. G., & Cowan, D. A. (2009b). Thermophilic ethanologenesis: Future prospects for second-generation bioethanol production. *Trends in Biotechnology*, *27*(7), 398–405. doi:10.1016/j.tibtech.2009.03.006 PMID:19481826

Taylor, M. P., Esteban, C. D., & Leak, D. J. (2008). Development of a versatile shuttle vector for gene expression in Geobacillus spp. *Plasmid*, *60*(1), 45–52. doi:10.1016/j.plasmid.2008.04.001 PMID:18501964

Toplak, A., Wu, B., Fusetti, F., Quaedflieg, P. J. L. M., & Janssen, D. B. (2013). Proteolysin, a novel highly thermostable and cosolvent-compatible protease from the thermophilic bacterium Coprothermobacter proteolyticus. *Applied and Environmental Microbiology*, *79*(18), 5625–5632. doi:10.1128/AEM.01479-13 PMID:23851086

Ulloa, G., Collao, B., Araneda, M., Escobar, B., Álvarez, S., Bravo, D., & Pérez-Donoso, J. M. (2016). Use of acidophilic bacteria of the genus Acidithiobacillus to biosynthesize CdS fluorescent nanoparticles (quantum dots) with high tolerance to acidic pH. *Enzyme and Microbial Technology*, *95*, 217–224. doi:10.1016/j.enzmictec.2016.09.005 PMID:27866618

Vane, L. M., & Alvarez, F. R. (2008). Membrane-assisted vapor stripping: Energy efficient hybrid distillation-vapor permeation process for alcohol-water separation. *Journal of Chemical Technology and Biotechnology (Oxford, Oxfordshire)*, *83*(9), 1275–1287. doi:10.1002/jctb.1941

Veiga-Crespo, P., Blasco, L., Rosa-Dos-Santos, F., Poza, M., & Villa, T. G. (2005). Influence of culture conditions of Gordonia jacobaea MV-26 on canthaxanthin production. *International Microbiology : The Official Journal of the Spanish Society for Microbiology*, *8*(1), 55–58. PMID:15906262

Vera, M., Schippers, A., & Sand, W. (2013). Progress in bioleaching: Fundamentals and mechanisms of bacterial metal sulfide oxidation—part A. *Applied Microbiology and Biotechnology, 97*(17), 7529–7541. doi:10.100700253-013-4954-2 PMID:23720034

Verma, P., Yadav, A. N., Khannam, K. S., Kumar, S., Saxena, A. K., & Suman, A. (2016). Molecular diversity and multifarious plant growth promoting attributes of Bacilli associated with wheat (Triticum aestivum L.) rhizosphere from six diverse agro-ecological zones of India. *Journal of Basic Microbiology, 56*(1), 44–58. doi:10.1002/jobm.201500459 PMID:26567901

Vögtlin, A., Fraefel, C., Albini, S., Leutenegger, C. M., Schraner, E., Spiess, B., Lutz, H., & Ackermann, M. (2002). Quantification of feline herpesvirus 1 DNA in ocular fluid samples of clinically diseased cats by real-time TaqMan PCR. *Journal of Clinical Microbiology, 40*(2), 519–523. doi:10.1128/JCM.40.2.519-523.2002 PMID:11825966

Voica, D. M., Bartha, L., Banciu, H. L., & Oren, A. (2016). Heavy metal resistance in halophilic Bacteria and Archaea. *FEMS Microbiology Letters, 363*(14), fnw146. Advance online publication. doi:10.1093/femsle/fnw146 PMID:27279625

Waksman, S. A., Schatz, A., & Reynolds, D. M. (2010). Production of antibiotic substances by actinomycetes. *Annals of the New York Academy of Sciences, 1213*(1), 112–124. doi:10.1111/j.1749-6632.2010.05861.x PMID:21175680

Walker, M., & Phillips, C. A. (2007). Alicyclobacillus acidoterrestris: An increasing threat to the fruit juice industry? *International Journal of Food Science & Technology, 43*(2), 250–260. doi:10.1111/j.1365-2621.2006.01427.x

Wu, S., Liu, B., & Zhang, X. (2006). Characterization of a recombinant thermostable xylanase from deep-sea thermophilic Geobacillus sp. MT-1 in East Pacific. *Applied Microbiology and Biotechnology, 72*(6), 1210–1216. doi:10.100700253-006-0416-4 PMID:16607523

Xue, T., Ni, J., Shang, F., Chen, X., & Zhang, M. (2015). Autoinducer-2 increases biofilm formation via an ica- and bhp-dependent manner in Staphylococcus epidermidis RP62A. *Microbes and Infection, 17*(5), 345–352. doi:10.1016/j.micinf.2015.01.003 PMID:25637952

Yuan, M., Chen, M., Zhang, W., Lu, W., Wang, J., Yang, M., Zhao, P., Tang, R., Li, X., Hao, Y., Zhou, Z., Zhan, Y., Yu, H., Teng, C., Yan, Y., Ping, S., Wang, Y., & Lin, M. (2012). Genome sequence and transcriptome analysis of the radioresistant bacterium Deinococcus gobiensis: Insights into the extreme environmental adaptations. *PLoS One, 7*(3), e34458. doi:10.1371/journal.pone.0034458 PMID:22470573

Zaldivar, J., Nielsen, J., & Olsson, L. (2001). Fuel ethanol production from lignocellulose: A challenge for metabolic engineering and process integration. *Applied Microbiology and Biotechnology, 56*(1–2), 17–34. doi:10.1007002530100624 PMID:11499926

Chapter 16
Thermophilic Bacterial Exopolysaccharides:
From Bio-Physicochemical Characterization to Biotechnological Applications

Rakesh Goswami
Raiganj University, India

Bidyut Bandyopadhyay
Oriental Institute of Science and Technology, India

Sanjoy Sadhukhan
https://orcid.org/0000-0002-2619-8700
Raiganj University, India

ABSTRACT

Bacterial exopolysaccharides have enormous diversity with valuable characteristics, synthesized by various pathways in extreme conditions like salinity, geothermal springs, or hydrothermal vents. Due to extreme environments, these microorganisms have various adaption principles (e.g., low pH, high temperature, high saltation, and high radiation). Exopolysaccharide is an organic compound produced by most bacteria during fermentation using various carbon sources, resulting in a jelly-like or mass network structure outside the cell wall. This biopolymer has an adherent cohesive layer throughout the cell layer. Hot spring bacterial polysaccharides contain diverse extracellular polymeric substances. With a gain in popularity in applications of thermophilic microbial polysaccharides and its demand in diverse value-added industrial products, this chapter aims to provide valuable information on the physicochemical function and biotechnological applications in the field of food, medical imaging, nano-drugs, bioremediation, cancer, anti-bacterial, tissue engineering, etc.

DOI: 10.4018/978-1-7998-9144-4.ch016

1. INTRODUCTION

Over the last few decades, vast numbers of extracellular polysaccharides or exopolysaccharides from extremophiles have been extensively studied all over the world. In the year 1972 Sutherland, proposed the word 'exopolysaccharide' (EPS) which stands for a variety of bacterial and microalgal heterogeneous long-chain polysaccharides that are synthesized and released outside the cell wall into their ambience during growth. Hot spring microorganisms produce a large number of structurally diverse extracellular polymeric substances i.e. EPSs. Diverse macromolecules e.g. peptidoglycan, lipopolysaccharide, and exopolysaccharide are the most important component of bacterial polysaccharides and these polysaccharides make the structure of their cell wall (e.g. Peptidoglycan) and Poly N-acetyl glucosamine plays a major role in bacteria to survive in the unfavourable environments (Kazak et al., 2010; Nichols et al., 2005; Ruffing & Chen, 2006). The thermophilic bacterium *Geobacillus* sp. (WSUCF1) has a great production rate in the production of two exopolysaccharides with prominent quantities. Following purification of these two exopolysaccharides, it was discovered that EPS-1 is composed of glucomannan with a 1:0.21 molar ratio of mannose and glucose, whereas EPS-2 was made up entirely of mannan. Both EPSs have molecular weights of around 1000 kDa, and their FTIR and NMR spectra revealed the existence of α-type glycosidic linkages in a linear structure. XRD examination revealed a low degree of crystallinity i.e. for EPS-1 and EPS-2 is 0.11 and 0.27 respectively (Wang et al., 2021). Most of the exopolysaccharides have their applications in food, pharmaceutical, and other industries. Exopolysaccharides produced by thermophiles have been used for various biotechnological processes like fermentation and food emulsification (Ruffing & Chen, 2006). A novel exopolysaccharide (EPS-B3-15) isolated from marine thermo-tolerant *Bacillus licheniformis* strain B3-15 showed high-temperature stability up to 80°C. Depending on this property of thermal stability it can be used for nano-medicine development (Caccamo et al., 2020). High-temperature-loving (thermophilic) bacteria from different classes of Archaea and Bacteria have been discovered from a variety of thermal habitats, including both deep and shallow marine hot springs, as well as terrestrial hot springs, which have provided the genesis for the separation of microbial EPS producers. Some thermophilic bacteria e.g. *Archaeoglobus fulgidus*, *Thermococcus litoralis*, *Pseudomonas aeruginosa* are good EPSs producers (Lapaglia & Hartzell, 1997; Nicolaus et al., 1993).

Presently, bioremediation technology is a new and challenging field of research on environmental issues. EPSs are ubiquitous and low-cost chemicals that have been utilised to adsorb oil and those are biodegradable. EPSs possess a high number of negatively charged functional groups, therefore they can remove a variety of heavy metals and organic pollutants successfully (Lakzian et al., 2008). EPSs of *Ensifer meliloti* have potent adsorption potential for Lead, Nickel and Zinc from industrial waste (Lakzian et al., 2008). EPS acts as an excellent biosorbent material for arsenic bioremediation. Recently, novel strains of *Exiguobacterium profundum* PT2 and *Ochrobactrum ciceri* SW1 have been isolated which can produce biopolymers and possess a large number of polyanionic functional groups on their surface and can sequester arsenic (a potent carcinogen), through electrostatic or covalent interactions (Saba et al., 2019). Researches on bioremediation reveal that the dead biomass of bacteria with exopolysaccharides is an important substrate for heavy metal sequestration. Dead biomass of *Ochrobactrum anthropi* removes cadmium ions along with other toxic metals under specific pH and initial metal ion concentration. The dead biomass has high heavy metal tolerance activity (up to 30 mg/L of cadmium ion concentration) (Ozdemir et al., 2003). Numerous studies have established that EPSs from *Lactobacillus* sp. can be utilized in diverse health benefits like anticancer, antiulcer, anti-viral properties and also act as immune-modulator, antiviral, antioxidant, bifidogenic, and cholesterol-lowering properties etc. (Freitas et al., 2011). In 1986 Louis Pasteur first isolated dextran; it is also the first commercial exopolysaccharide, which is composed of simple glucose derivatives. Currently, dextran hydrogel micelles are used for nano-drug delivery systems or nanocarrier systems (Banerjee & Bandopadhyay, 2016). Exopolysaccharides can be used to degrade industrial textile dye, an industrial pollutant (Pathak & Navneet, 2017). Polysaccharides obtained from *Leuconostoc lactis* can be used in silver nanoparticles formation, which has been used for the degradation of industrial textile dyes like methyl red and congo red (Saravanan et al., 2017). TGA-TGD analysis showed that EPS based silver nanoparticles have thermo stability at 437.1°C, therefore, they can be used to treat textile effluent at a higher temperature. The research by Saravanan et al. (2017) reveals that exopolysaccharides based nanoparticles should be used as an environment friendly with a very low-cost production strategy for the degradation of harmful dyes.

2. STRUCTURAL AND FUNCTIONAL PROPERTIES OF BACTERIAL EXOPOLYSACCHARIDES

Exopolysaccharides are high molecular weight polysaccharides having various structures that are being secreted into the adjacent environment by diverse groups of microorganisms (Sutherland, 1972). Numerous studies on the primary conformation of exopolysaccharides have deciphered the structure and composition of the bacterial polysaccharides. The cellulosic backbone of the EPS xanthan obtained from *Xanthomonas campestris* revealed that it has β (1, 4) - or β (1, 3) - linkages, thus accounting for rigidity of the backbone structure (Jansson et al., 1975). Whereas, other studies showed that flexible polysaccharides with α (1, 2) or α (1, 6) linkages are present in several dextrans (Sutherland, 1994). Several dominant forces like hydrogen bond, electrostatic forces are involved in polysaccharides backbone, even ionic interaction are also involved which help to form a fine chain-chain complex and this complex forms a gaggle networks which are insoluble in solvent (Sutherland, 2001). The chemical and physical properties of polysaccharides can be exploited in EPSs producing bacteria in industrial, food and medical applications (Zannini et al., 2016). There are various techniques involved for purification and structural characterization of EPSs, like two-dimensional NMR using, ^{1}H and ^{13}C and some enzymatic reactions detected by MS study. NMR is used to find out the repeating unit of the glucose polymer, methylation and sulfated groups (Vincent et al., 2001). High-performance anion-exchange chromatography and methylation studies in *Halomonas* sp. revealed the repeating units of the polysaccharides, which are composed of (2, 6)-D-fructofuranosyl residues, and NMR studies showed the repeating units of the polysaccharides, which were composed of (2, 6)-D-fructofuranosyl residues. The presence of fructose residue in polysaccharide composition was verified using GC-MS and acid hydrolysis methods (Nicolaus et al., 2010). A similar study showed that *Lactobacillus* sp. produced heteropolysaccharides which are composed of 1, 6-α and 1, 4-α linkages (Nicolaus et al., 2010). On the other hand, Raman Microspectroscopy and Atomic Force Microscopy, have been used to study the structural cellular surface of bacterial biopolymer (McEwen et al., 2010). Extracellular biopolymer binds biofilm-bacteria together and protects them from harsh environments, including contaminated environments. As a result, the development of biofilms is an important process for bacteria (van Hullebusch et al., 2003). *Anoxybacillus gonensis* YK25 is a thermophilic bacterium that has been isolated from hot springs in Erzurum province (Turkey) produces potent exopolysaccharides under optimized culture conditions (initial pH 7.0, temperature 55°C, molasses concentration 100 mL/L, peptone concentration 6 g/L and incubation time 96 h). These EPSs are composed of heteropolysaccharide, with 91.4% carbohydrate, 4.1% protein, xylose 59%, sucrose 18%, glucose 16% and galactose 5%. FTIR examination confirmed the presence of hydroxyl, carboxyl, amide and sulphate groups. Due to this chemical structure, these EPSs showed moderate solubility in DMSO and high solubility in water (Karadayi et al., 2021). *Anoxybacillus pushchinoensis* G11 (MN720646), a thermophilic bacterium, has been discovered to produce heteropolysaccharide with arabinose (57%), fructose (26%), glucose (12%), and galactose (5%), 93% carbohydrates and 1.08% protein. FTIR analysis of the EPSs revealed the presence of sulfate ester (band at 1217 cm^{-1}) in the EPSs (Genc et al., 2021).

2.1. Bacterial Capsule

The bacterial capsule is the outermost covering of some bacterial cells which serves in the protection of cells from their surrounding environments and against desiccation. Exopolysaccharides forms fine strands of carbohydrates that are linked to the bacterial cell wall producing a complex network around the cell surface (Figure 1). These capsules are made up of various polysaccharides structures and are known as capsular polysaccharides. The bacterial capsule is joined to the cell surface of the bacterium via a covalent bond with phospholipid or lipid-A molecules (Reckseidler-Zenteno, 2012). Homo or heteropolysaccharides are present in the capsular polysaccharides that are composed of repeating monosaccharides joined by glycosidic linkages (Roberts, 1996). The bacterial capsule has much structural diversity e.g. eighty different types of diversity have been shown in *Escherichia coli* alone and some different taxa have the same structural properties like *Escherichia coli* and *Neisseria meningitidis* capsules containing sialic acid (Ghuysen, 1968). Based on various bacterial capsular structures, new findings have opened in the field of capsules' evolution. The bacterial capsules may produce a hydrated gel around the surface of the bacterial cell, which may protect the bacteria from the harmful effects of the extreme environment. By using bacterial capsules each bacteria can form a bacterial colony through the surface of the cell that formed a bacterial biofilm (Costerton et al., 1987).

2.2. Exopolysaccharides

According to the composition and synthesis mechanisms, exopolysaccharides can be divided into heteropolysaccharides and homopolysaccharides. Heteropolysaccharides are integrated by more than one monosaccharide molecule and are synthesized intracellularly, but homopolysaccharides are composed of only a single type of monosaccharide, which has been produced externally to the cell by an enzyme secreted by the bacterium (Badel et al., 2011).

Most of the heteropolysaccharides have a simple sugar polymer chain and contain D-glucose, D-galactose and L-rhamnose. Mesophilic and thermophilic bacteria can produce heteropolysaccharides by activation of intracellular glycosyltransferases with sugar molecules (Vieille & Zeikus, 2001). These sugar molecules formed linear or branched structures with high molecular weight up to 10^6 Da. Three to eight monosaccharides can be joined together through different linkage patterns. Bacterial monosaccharide's mostly formed by β-(1, 4) or β-(1, 3) linkages which provide flexibility (Werning et al., 2012). The heteropolysaccharide backbone is composed of diverse monosaccharides e.g. pentoses (d-arabinose, d-ribose, d-xylose), hexoses (d-glucose, d-galactose, d-mannose, d-allose, l-rhamnose, l-fucose), amino sugars (d-glucosamine and d-galactosamine), and the modified monosaccharides uronic acids such as d-glucuronic acids, d-galacturonic acids (Davey & Amos, 2002). In the backbone formation, some inorganic and organic substances are also involved such as sulfate, phosphate, acetate, succinate, and pyruvate etc. The most important natural heteropolysaccharides are - gellan gum, xanthan gum, and kefiran etc. (Davey & Amos, 2002; Rinker & Kelly, 2000).

Figure 1. Bacterial cell wall showing the different surface polysaccharides

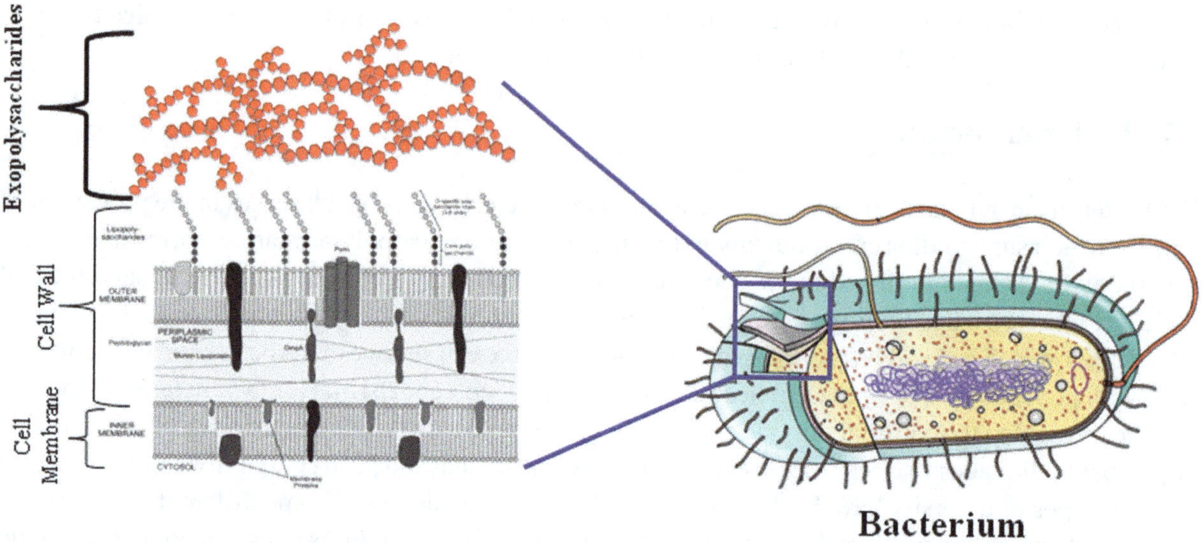

Homopolysaccharides are composed of a single unit of monosaccharide either glucose or fructose. Glycansucrase is the major enzyme that helps to break extracellular sucrose units (Monsan et al., 2001). By glycoside hydrolases, glycansucrases that synthesize glucans and fructans and this enzyme are classified into two groups (GH70 and GH68). Various *Lactobacillus* species along with *Streptococcus, Leuconostoc, Oenococcus, Weissella* are responsible for the secretion of homopolysaccharides in high quantity (Dimopoulou et al., 2016; van Hijum et al., 2006). Glucosidic or fructosidic is the major unit to form homopolysaccharides polymer backbone. The major homopolysaccharides are dextran, reuteran etc. Dextran contains predominantly α-(1→6) linkages between the glucosyl units and reuteran contains predominantly α-(1→4) linkages (Meng et al., 2016). Whereas inulin and levan consist of β-(2→1) and β-(2→6) linkages but levan and inulin are fructan type homopolysaccharides (Monsan et al., 2001).

2.3. Important Polysaccharides from Thermophilic Bacteria

Thermophilic bacteria are extremophiles, growing at high temperature that is between, 45°C to 125°C. Thermophilic (heat-loving) bacteria can be categorised based on their optimal growth temperatures into simple (50–64°C), extreme (65–79°C), and hyperthermophiles (80°C and beyond). Among the various thermophilic sources like- marine hot springs, terrestrial hot springs have various novel bacterial strains, which can produce many essential exopolysaccharides. Marine shallow thermophilic bacteria like - *Bacillus thermoantarcticus, Geobacillus thermodenitrificans* and *B. licheniformis* can produce exopolysaccharides in a large amount (Manca et al., 1996). Bulgarian Hot Spring research opined that thermostable gellan lyase activity is too high for mesophilic enzymes. On the other hand, thermostable enzymes xylanase was able to degrade more than 60% xylan of beechwood. From this hot spring, various novel enzymes and exopolysaccharides have been isolated for broad-spectrum application (Kambourova, 2018). *Pseudomonas elodea* is non-pathogenic bacteria, produce gellan, and mesophilic bacteria *Alcaligenes faecalis* produce curdlan. This gellan has the unique property of forming thermo-stable gels, therefore, have commercial value in food and pharmaceutical applications on large scales (Fialho et al., 2008). The production-based research study revealed that thermophilic strains like *Lactobacillus delbrueckii, L. delbrueckii, Streptococcus macedonicus and Streptococcus thermophilus*, were able to produce more EPSs than mesophilic bacteria such as *Enterococcus faecalis, Enterococcus faecium, Lactobacillus casei, Lactobacillus paracasei* (Mozzi et al., 2006). *Geobacillus* sp. strain 4004 has been isolated from sediment in marine hot springs in Italy; which produces pentasaccharide EPSs structure with repeating units and these EPSs are used in pharmaceuticals industries (Nicolaus et al., 2002). Both *Bacillus thermodenitrificans* strain B3-72 and *Bacillus licheniformis* strain B3-15 have been isolated from a shallow marine hot spring, Italy, produced high quantity exopolysaccharides with trisaccharide repeating unit and a manno-pyranosidic configuration; this configuration based EPSs has potent medicinal activity such as Immunomodulatory, antibacterial and antiviral properties (Poli et al., 2010). *Streptococci thermophilus* produces hyaluronic acid which is an important material for medical, cosmetic and food applications especially in dairy industries (Vaningelgem et al., 2004). This hyaluronic acid improves applications of EPS for better consumer satisfaction for appealing taste and healthier dairy products (Izawa et al., 2009). Exopolysaccharide supernatants, which has been isolated from *Geobacillus stearothermophilus* bacterium can be used to treat immunological disorders associated with functions of IFN-α, IL-12, IFN-γ, TNF-α, and IL-18 (Arena et al., 2009). This bacteria also generates gellan exopolysaccharide, which has recently been discovered and is widely utilised in the food and pharmaceutical sectors (Novik et al., 2018).

2.4. Exopolysaccharides in Bacterial Biofilm

In the natural environment, few micro-organisms especially bacteria do not live singly but always form polymicrobial aggregates such as films, mats, flocks, sludge or so-called 'biofilms'. This biofilm consists of an agglomerate of different types of biopolymers such as extracellular polymeric substances, which formed a three-dimensional structure of the biofilm and is responsible for adhesion to surfaces (Flemming & Wingender, 2010). Biofilm formation depends upon the favourable environment or suitable conditions in which the biofilms are formed (Donlan, 2002). An extracellular matrix of the biofilms holds the cells together (Branda et al., 2005). Globally, bacteria secreting exopolysaccharides, illustrate a prominent source of the reduced-carbon reservoir in soils and sediments and suspended aggregates in oceans and freshwater. Exopolysaccharides have gained a wide range of matrix biopolymer, therefore it is very difficult to analyze their structures, that is why EPS have been called "the dark matters of biofilm" (Flemming & Wingender, 2010). The matrix biopolymer also contains proteins, nucleic acids, lipids and other biopolymers such as humic substances (Allison et al., 2003). Numerous microorganisms are enclosed within exopolysaccharides, which help them to survive in adverse conditions. EPS has a specific attachment role more closely with the cell surfaces than other substances. Thus microbial biofilm formation always depends upon the production rate and quantity of major exopolysaccharides. The dense area, pores, channels and as well as three-dimensional structure of the biofilm provides more stability than others. Morphologically, some mushroom-like macrocolonies and rough, fluffy, smooth like surfaces have been also seen in the biofilm structures. These characteristics resulted in stationary biofilm cells that have a wide range of effects on environments while also favouring biodiversity under certain ecological circumstances. The interaction of anionic exopolysaccharides, which contain carboxylic groups in a complex network, can have a big impact on biofilm models (Flemming & Wingender, 2010; Körstgens et al., 2001). Exopolysaccharide from thermophilic *Bacillus licheniformis* T14 (Eolian Island, Italy) was found to have an antibiofilm impact against various multi-resistant clinical strains. Microtiter plate assays and confocal laser scanning microscopic images revealed that 2% concentration of the cell-free supernatants (CFS) showed a strong antibiofilm activity (C50% of biofilm reduction) only against *E. coli*. At higher concentrations i.e. 54 - 69% CFC reduces the formation of biofilm against *K. pneumoniae* and *S. aureus*. Clinical studies revealed that EPS1-T14 has a dose-dependent (25 to 400 μg mL^{-1}) inhibitory effect on biofilm formation by *E. coli* and *S. aureus*. This novel T14 exopolysaccharides (EPS1-T14) is a highly water-soluble molecule in which non-cytotoxic exopolymeric characters prevent the formation of biofilm. Due to this character of EPS1-T14 may be used as a promising therapeutic agent for combating bacterial biofilm-associated infections (Spanò et al., 2016). Based on several research works, exopolysaccharides have several functions associated with the biofilm, which are categorized in Table 1.

Table 1. Basic relationship of exopolysaccharides in Biofilm formation

Major Functions	Relationship of the Biofilms	Major components of EPS	References
Cohesion of biofilms	The network of biofilm matrix i.e. the hydrated polymer is composed of both neutral and charged exopolysaccharides provides mechanical stability to biofilms along with multivalent cations that regulate biofilm framework and permit intercellular crosstalk	Neutral and charged polysaccharides, proteins and DNA	(Decho & Gutierrez, 2017; Donlan, 2002)
Adhesion	For colonization, exopolysaccharides make provision for the initial steps for the long-term attachment of biofilms	Polysaccharides, proteins, DNA and amphiphilic molecules	(Allison & Sutherland, 1987; Dertli et al., 2015)
Bacterial cell grouping	Bacterial cells built a bridging between cells are occupied by exopolysaccharides.	Polysaccharides, proteins and DNA	(Costa et al., 2018)
Water Capacity	Exopolysaccharides also make a water intake capacity in the highly hydrated microenvironment around biofilm organisms, resulting in survival in water-deficient environments.	Hydrophilic polysaccharides and, possibly, proteins.	(Bogino et al., 2013; Costa et al., 2018)
Tenacity of biofilms	A hydrated polymer network gives the biofilm matrix strong mechanical stability.	Neutral and charged polysaccharides, proteins and DNA	(Stewart, 2014)
Nutrient source	Exopolysaccharides contained carbon, nitrogen and phosphorus-containing compounds, which make a biofilm community.	Potentially all EPS components	(Costa et al., 2018; Decho & Gutierrez, 2017)
Exchange of genetic information	Horizontal gene transfer happening between biofilm cells.	DNA	(Madsen et al., 2012)
Storage of excess energy	Exopolysaccharides store excess carbon under unbalanced carbon to nitrogen ratios.	Organic compounds.	(Nwodo et al., 2012)
Enzymes binding properties.	Polysaccharides interaction makes the accumulation, retention and stabilization of enzymes.	Polysaccharides and protein	(Flemming, 2016)
Electron donor or acceptor	Redox activity in the biofilm matrix	Pili and nanowires, some proteins.	(Arnaouteli et al., 2017; Kataky & Knowles, 2018)

3. VARIOUS BIOTECHNOLOGICAL USES OF EXOPOLYSACCHARIDES

Exopolysaccharides, particularly those from thermophiles, are useful because of their thermo stability and low viscosity at high temperatures. The significance of understanding the functional properties of thermophilic bacterial exopolysaccharides, as well as their applications in the medical, cosmetics, and food industry sectors, as well as in heavy metal bioremediation, is rapidly increasing. The most advantages of using exopolysaccharides are as follows:

Figure 2. Schematic diagram of potential applications of Exopolysaccharides in food, tissue engineering, medical sciences and bioremediation

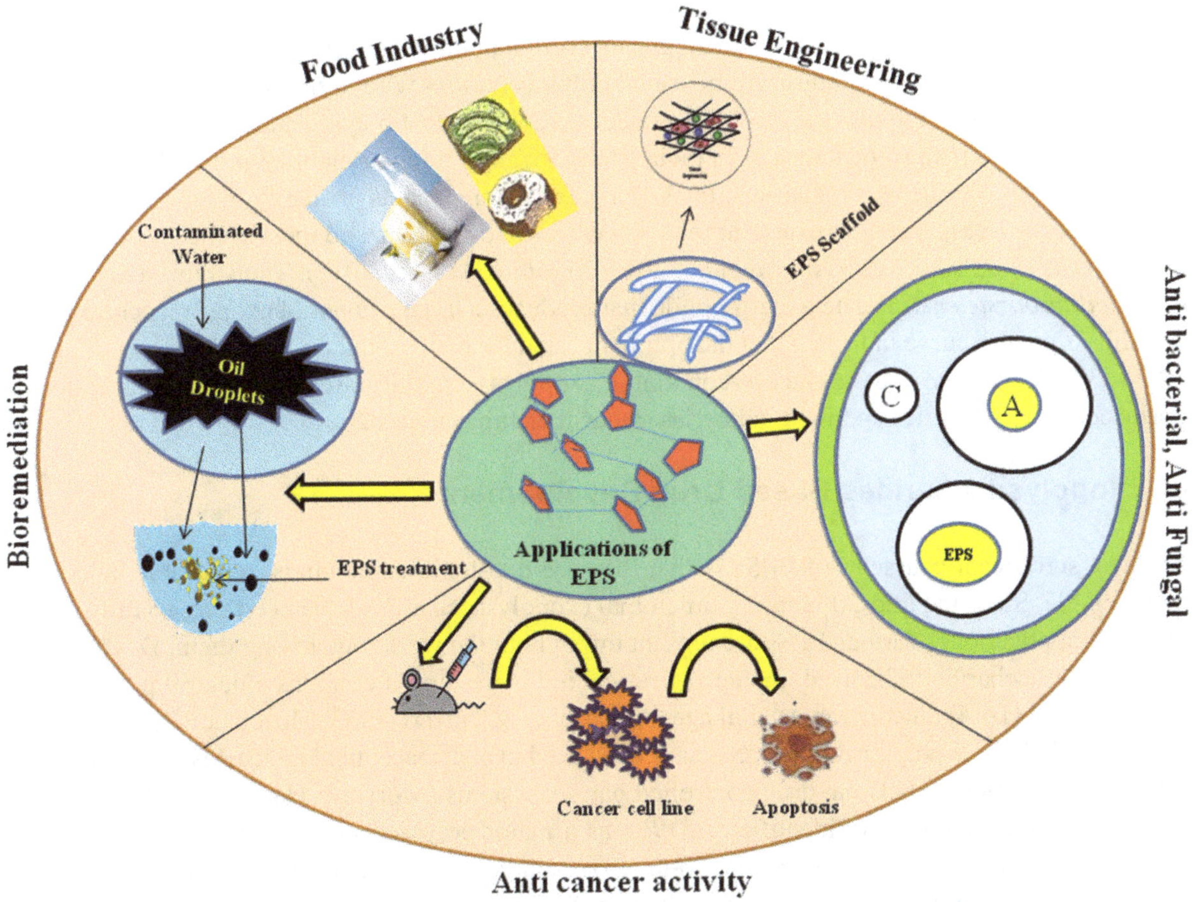

3.1 Role of Exopolysaccharides in the Food Industry

Exopolysaccharides are employed as viscosifying agents, stabilisers, emulsifiers, gelling agents, and water-binding agents in the food industry. EPS of microbial sources have exclusive rheological properties due to their ability to form excessively viscous solutions at low concentrations and their pseudoplastic nature (Becker et al., 1998). Two important commercially available microbial-derived EPSs composed of xanthan and gellan, which are synthesized by the bacteria *Xanthomonas campestris* and *Pseudomonas paucimobilis* respectively (Becker et al., 1998; Patel & Prajapat, 2013).

In terms of its relevance in both culinary as well as non-food uses, xanthan has been termed as a "gold standard" product (Sutherland, 1998). Dairy, soft beverages, confectionery, dressing, bread items, syrups, and dog food, as well as the oil, pharmaceutical, cosmetic, paper, paint, and textile sectors, utilise xanthan gum (Cho & Yoo, 2015; Palaniraj & Jayaraman, 2011; Patel & Prajapat, 2013). Gellan gum is becoming more popular in the food industry, because of its functional properties, which include great heat and acid resistance, variable gel elasticity and stiffness, high transparency, and good flavour. Gellan gum is mainly used as a stabilizer, suspending agent, structuring and versatile gelling agent in a wide variety of applications in food products that include bakery fillings, confections, dairy products, dessert gels, frostings, icings and glazes, jams and jellies, low-fat spreads, microwavable foods, puddings, sauces, structured foods, and toppings (Prajapati et al., 2013).

Bacterial EPSs such as dextran, xanthan, gellan, curdlan, pullulan, acetan, and levan are used commercially and are well known industrial microbial polysaccharides with numerous applications and has a considerable market (Freitas et al., 2011; Kazak et al., 2010). Most polysaccharides are hydrocolloids that are widely used in the food industry for their extensive range of functionalities and applications. They can be utilized as stabilizers, emulsifiers, thickeners, and gelling agents primarily in food products such as bread, sauces, syrup, ice cream, instant food, beverages and ketchup, and most of these hydrocolloids are termed as 'food additives' (Ahmad et al., 2015; Milani & Maleki, 2012). The strain MK878423 of *Brevibacillus borstelensis* is a new thermophile that produces a large amount of exopolysaccharide. The bacterium's growth curve indicated that the optimal period for EPS production is 40 hours, with highest yield at 30 g/L glucose. These glucose-containing exopolysaccharides are suitable candidates for the pharmaceutical, cosmetics and food industries (Dhagat & Jujjavarapu, 2021).

3.2 Exopolysaccharides Based Drug Development

The most significant character of EPSs is that they are non-toxic compounds, renewable, and easily water-soluble. Since the mid-20th century, microbial exopolysaccharides have been successfully used in medicinal applications. Various EPSs are present throughout the world but among them, Dextran is the first exopolysaccharide to be used in pharmaceutical industries and it contains a neutral polymer with α-(1$\rightarrow$6) and α-(1$\rightarrow$4) glucopyranosyl linkage (Amspacher & Curreri, 1953; Moscovici, 2015). Xanthan is a branched-chain of polysaccharides that have been used in pharmaceutical industries for cream-based ointment preparation and also in drug controlled barrier systems (Morris & Harding, 2009). Alginates extracted from seaweeds were identified in 1964 as a major polysaccharide, and these EPSs include acetyl groups within a linear arrangement of β-(1$\rightarrow$4) mannuronic acid and α-(1$\rightarrow$4) guluronic acid. This character gives a prominent disintegrating agent for tablet formation (Cyber Colloids Ltd., McHugh, 1987). The technique of microencapsulation gradually increasing in many pharmaceutical industries, in this field, alginate is a good transporter for the nano-drug delivery system (Mukherjee & Atala, 2005; Nwodo et al., 2012). Gellan exopolysaccharides are reliable agents for many drug development systems e.g. nasal drugs, various ophthalmic drugs, tablet formation (Felt et al., 2001; Hägerström, 2003). Leaven is yet another notable exopolysaccharide for anti-tumour drug preparation; it has the potency to control cholesterol levels in the human blood system. Leaven posses various derivatives such as leaven sulfates, phosphates, and acetates which have been used in pharmaceutical industries (Roberts & Garegg, 1998; Yamamoto et al., 1999). *Agrobacterium* produces an exopolysaccharide i.e. curdlan which is composed of triple-helical conformation with β (1, 3)- linked D-glucose (El-Naggar et al., 2020). Curdlan has anti-tumour and anti-HIV activity (Osawa et al., 1993; Yoshida et al., 1990). Magnetic Resonance Imaging (MRI) is one of the most important medical-technique for human disease identification. In this technique, some nanoprobe materials are used. Fucopol is one of the exopolysaccharide that has been isolated from *Enterobacter* A47 DSM 23139 using glycerol as a carbon source (Torres et al., 2012). It is used for the coating of iron oxide magnetic nanoparticles (MNP), which is applicable for better imaging in MRI applications. Covalent bonding happens between EPSs and DMSA - stabilized nanoparticles which formed a hybrid magnetic-biopolymeric nanosystem (MNPDMSA-EPS) with a hydrodynamic size of 170 nm, negative surface charge at physiological conditions and transverse to longitudinal relaxivities ratio, r2/r1, of 148. This MNP-DMSA-EPS *in vitro* experiment showed excellent performance on colorectal carcinoma - HCT116 - and neural stem/progenitor cells - ReNcell VM. It can be opined that bacterial exopolysaccharides would bring a new area for high-quality MRI techniques (Palma et al., 2015). Hyaluronic acids, which include N-acetyl-D-glucosamine and glucuronic acid, are linear disaccharides. *Streptococcus equisimilis*, *Bacillus subtilis*, and other bacteria produce hyaluronic acid which has a high viscosity and molecular mass due to its polymeric and polyelectrolyte characteristics. Hyaluronic acid can be used as a diagnostic marker for many diseases including cancer, rheumatoid arthritis and liver diseases, and it can be used for supplementation of impaired synovial fluid in arthritic patients (Kogan et al., 2007).

3.3 Role of Exopolysaccharides in 3-D Printing

3D printing technology is the most demanding and promising component of computational biology in current medical biology. Bacterial exopolysaccharides were shown to serve a potential function in the production of scaffolds in 3D technology. Bacterial exopolysaccharides exhibit three-dimensional networks. Hyaluronic acid (HA) is an abundant biopolymer that is secreted by *Streptococcus equisimilis, Bacillus subtilis etc.* Tissue engineering research demonstrates that hyaluronic based scaffolds can be successfully used in 3D printing. The HA-mediated scaffold has some excellent properties, it can bind to proteins and cells through cell surface receptors such as CD44 (Knudson et al., 2002). Intercellular adhesion molecules (ICAM) can connect with HA-scaffold molecules in endothelial and macrophage cells, and this binding may help to regulate ICAM-1-mediated inflammatory activation (Chen & Abatangelo, 1999). Hydrogels are a major component in tissue engineering since hydrogel has high mechanical strength. In this regard, double strength hydrogels were formed by using gelatin methacrylamide for the soft and ductile network (Shin et al., 2012). Immobilization is another technique in tissue grafting, the dextran-coated scaffold is a better adhesive peptide and it can promote endothelial cell adhesion as well as cell proliferation. Dextran coating peptides has low-fouling properties and was efficiently modified with biomolecules. Most of the peptides like - RGD, YIGSR, and REDV have potent cell adhesiveness and proliferation activity (Noel et al., 2016). In rabbits, it was revealed that alginate gel-based hydrogel produces new autologous cartilage at the subcutaneous dorsal side of rabbits. After 9 weeks of alginate hydrogel treatment, immunodeficient mice exhibited a substantial improvement. A rapid-curing alginate gel system has been developed for articular cartilage tissue engineering applications (Stevens et al., 2004). The thermal technique for generating highly osteoconductive α-(1, 3)-glucan/HA scaffolds for bone tissue engineering is a novel field. Curdlan gelation allows the structural modification of α-(1, 3)-glucan/HA scaffold which is more thermostable and promotes osteoblast growth and proliferation as well as increases bone alkaline phosphatase level thereby enhancing cell differentiation (Klimek et al., 2016). The combined structure of polycaprolactone, gelatin and levan is a potent formula of 3D printing. *In vitro* cell culture assay using these three combined structures proved human osteoblast cells have increased biocompatibility of the printing materials with increasing leaven content. So it can be opined that levan may be used as a good component for 3D printing (Duymaz et al., 2019). Co-axial and Single-Needle Techniques has been used for levan based fibrous Scaffolds formation. This levan based scaffold technology may be used as scaffold grafts to reduce the risk of restenosis and thrombosis and prolong graft life (Avsar et al., 2018). Chondroitin Sulfate is another important biopolymer used in the tissue grafting of bone cells. It is composed of two units i.e. N-acetyl-D-galactosamine and D-glucuronic acid. These features lead to be proven as excellent bio-characteristics including the binding and modulation of certain growth factors on bone cells. Chondroitin (CS) sulfate scaffold is a mimic of the natural cartilage system. Also, the bilayer of gelatin-CS-hyaluronan biopolymer has potent wound treatment properties (Han et al., 2014; Van Vlierberghe et al., 2011). Cellulose is the best and easily available bacterial polysaccharides that have been used successfully in tissue engineering technology. It has two types of cellulose-based scaffolds, 6-carboxy cellulose with 2.1 or 6.6 wt% of –COOH groups (Dutta et al., 2019). The unique surface chemistry and exclusive properties of bacterial cellulose make user-friendly scaffolds than other synthetic scaffolds in tissue engineering. Bacterial cellulose has good mechanical strength, microporosity and biodegradability which lead to a unique role in 3D printing technology (Rajwade et al., 2015). In bone, tissue-engineering biodegradation is a very critical issue. That's why bacterial cellulose could not degrade rapidly upon implant which is a good characteristic to hold scaffold in place for timely Osseo integration (Hickey & Pelling, 2019). Besides, different strategies such as chemical, enzymatic and genetic engineering could be adopted to enhance the degradability of bacterial cellulose in various conditions without altering its unique features for bone regeneration (Torgbo & Sukyai, 2018).

3.4 Role of Exopolysaccharides in Antimicrobial and Antifungal Activity

Multidrug-resistant microorganisms are currently posing a new hazard to public health. Exopolysaccharides based on nanoparticles have emerged as a promising antibacterial therapy in this area. Exopolysaccharides obtained from *Bacillus subtilis* were used to make the silver nanoparticles, which showed good antibacterial action against *Pseudomonas aeroginosa* and *Staphylococcus aureus* (Selvakumar et al., 2014). Using a green synthesis technique, exopolysaccharides capping based Zn and Ni metal nanoparticles were generated from *Rhodotorula mucilaginosa* UANL-001L. The average size of the Zn and Ni nanoparticles are 8 and 26 nm, respectively. With extremely low doses i.e. 3 and 2 mg/mL of Ni-EPS exhibits significant antibacterial and antibiofilm action against resistant strains of *Staphylococcus aureus* and *Pseudomonas aeruginosa*, respectively. On the other hand, 1mg/ml Zn-EPS was formulated against resistant strains and gave potent activity against *Staphylococcus aureus* (Garza-Cervantes et al., 2019). Exopolysaccharides from both *Bifidobacterium bifidum* and *Lactobacillus plantarum* were found to have potent antibacterial activity against *Listeria monocytogenes* CMCC54007, *Staphylococcus aureus* CGMCC26003, *Bacillus cereus* ATCC14579, and Salmonella *typhimurium* ATCC13311 but inhibition zones were lower than those obtained with 50 µg/mL ampicillin. Exopolysaccharides of *Lactobacillus plantarum* have an inhibition zone of (7.83 ± 1.04 and 10.67 ±0.29 mm) which is significantly greater than *Bifidobacterium bifidum* EPS against *Cronobacter sakazakii* ATCC29544 and *Shigella sonnei* ATCC25931. EPS of *Lactobacillus plantarum* prevented the growth of *Candida albicans* which is similar to ampicillin dosages (Li et al., 2014). *Mesoflavibacter zeaxanthinifaciens* based silver nanomaterial has a significant antibacterial and antibiofilm effect against pathogenic bacteria e.g. - *Bacillus subtilis* and methicillin-resistant *Staphylococcus aureus* (Oves et al., 2019). The dose-dependence of silver nanomaterial revealed a maximum of zone inhibition of about 22 and 18 mm against *B. subtilis* and *S. aureus* at a dose of 50 µg/well and a minimum inhibitory concentration of 8 and 10 µg/ml, respectively (Oves et al., 2019). Dextran is also used as a prominent agent for capping and reduction in the silver nanoparticle synthesis process. Dextran based silver nanoparticles size in the range of 10–60 nm and concentration of 0.2 mg/ml showed significant antibacterial properties against *Bacillus subtilis, Bacillus cereus, Escherichia coli, Staphylococcus aureus and Pseudomonas aeruginosa* (Bankura et al., 2012). Lactic acid bacteria are also able to synthesize silver nanoparticles of 0.2 nm–10 nm and 0.0–10 nm particles size. These nanoparticles have antibacterial activity against both Gram-positive and Gram-negative bacteria and the zone of inhibition ranged between 12-26 mm. Yoghurt starter based polysaccharides have potent antibacterial features against *Escherichia coli, Staphylococcus aureus* and the yeast of *Candida albicans* by disc diffusion method and inhibitions zones ranging from 9 to 13 mm (Adebayo-Tayo & Popoola, 2017; Ghalem, 2017). Silver nanoparticles composed of bacterial exopolysaccharide have both anti-fungal and anti-bacterial properties. The Agar well diffusion method indicates that it has potent inhibition properties against the pathogenic *E. coli*, the food-borne pathogen *L. monocytogenes*, and the multidrug-resistant pathogens *K. pneumonia* and *P. aeruginosa*. Antifungal activity was analyzed against food-borne fungal pathogens e.g. - *Aspergillus* spp., *Penicillium* spp., which were inhibited by the concentration-dependent manner with silver nanoparticle concentration ranging from 0.2 mg to 2 mg/ml (Kanmani & Lim, 2013). KPEP1, KPEP2, KPEP3, KPEP4, bacterial strains were too good for exopolysaccharide production; all bacteria were isolated from the sugar cane field. The ability of different EPSs against test organisms, EPSs of KPEP1 and KPEP3 were found to inhibit the growth of *E. coli* and *Bacillus cereus* at the MIC values of 8.7 and 9.4 mg/ml respectively, KPEP4 was found to inhibit the growth of *Bacillus subtilis* at MIC of 6.8 mg/ml as compared to other EPSs and 7.6 mg/ml was found to inhibit the growth of *Vibrio cholerae* by KPEP 2 (Patel et al., 2018).

3.5 Role of Exopolysaccharides in Bioremediation

Heavy metals are discharged into the environment on a daily basis through diverse anthropogenic activities, resulting in harmful contamination of the air, water, and atmosphere. These have encouraged numerous researchers to convert heavy metals from hazardous form to safe form by utilizing diverse engineering and biological methods. Since most exopolysaccharides are acidic, therefore they can withstand heavy metal stress hence they are preferred in heavy metal removal (Bhunia et al., 2018). Therefore, bacterial exopolysaccharides have ushered in a new era of fruitful study in this sector. Many countries use uranium as the major fuel in nuclear reactors. *Deinococcus radiodurans* is a well-known microbe that is used in the bioremediation of heavy metals, especially radioactive ones. *Deinococcus radiodurans* strain (DR1-bf) can form a biofilm with uranium for its remediation. Arsenazo III dye method and Plasma-Atomic Emission Spectroscopy (ICP-AES) method reveals that the DR1-bf biofilm can remove ~75 ± 2% of 1000 mg/L uranium from uranyl nitrate aqueous solution after 30 minutes of treatment. An engineered *D. radiodurans* has been utilized for detoxification of mercury; degrade toluene and chromium reduction in polluted nature (Brim et al., 2000; Brim et al., 2006; Manobala et al., 2019). Some bacterial strains were isolated from the plant species *Tagetes minuta* found near soil surrounding an automobile workshop and these bacterial EPSs have good bioabsorption capacity for heavy metals e.g. KA24, KA18 and KA25 able to accumulate 118.2, 121.87 and 90 mg chromium ion per gram of biomass respectively. They are even able to accumulate Ni and cadmium ions. Strontium is a nonradioactive metal that causes lung cancer in humans. EPS of *Cupriavidus metallidurans* CH34 is the most potent candidate in strontium removal from infected sites at pH 7.5 (Salem et al., 2013). The bacterium *Rhizobium radiobacter* VBCK1062 isolated from the rhizosphere of *Vigna radiata* is effective in arsenic removal. *Rhizobium radiobacter* VBCK1062 bacterial biomass can uptake 0.068 mg of arsenate per gram of biomass (Deepika et al., 2016). *Azotobacter* is a well-known bacterium for plant growth-promoting the EPS obtained from *Azotobacter* can remove mercury (Hindersah et al., 2017). Chromium is a well known mutagenic and carcinogenic agent, which reduces the growth of mustard plants. *Pseudomonas aeruginosa* OSG41 isolated from the mustard field contains Cr (VI) at a concentration of 1800 mg/ml. This strain can reduce the maximum Cr (VI) level at pH 6–8 temperature 30–40°C (Oves et al., 2013). Heavy metal contamination of water is a serious concern all over the world. Fluoride and arsenic compounds in water are serious pollution problems in all corners of the planet. The raw biomass of *Providencia vermicola* (KX926492) is an effective absorbent in removing fluoride from contaminated water (Mukherjee et al., 2017). Large scale study revealed that two-Gram positive bacteria e.g. *Rhodococcus opacus* and *Rhodococcus rhodochrous* were able to form EPS in high amounts and are also efficient in the adsorption of Cadmium, Lead, Nickel, Cobalt, and Chromium. The pH, time and temperature play a vital role. The pH of 2-7.5 is crucial for different heavy metal's adsorption by the bacterial EPS. These bacterial EPS are efficient in adsorbing chromium ions at 25°C and Nickel ions at 35°C (Dobrowolski et al., 2017). Rivers and lakes are polluted by mercury with a concentration range of 0.5-3 mg/l while it is too high in coastal seawater with a range of 2-15 mg/l. EPS obtained from *Bacillus lichenifonnis* can remove 70% Hg when pH is 7 (Upadhyay et al., 2017). Marine bacteria *Alteromonas sp.* is a good producer of EPS, which shows that it has an enormous capacity to remove heavy metals. At pH 5.0, the EPS reveals the highest bioabsorption capacity for Cu^{2+} and Ni^{2+} i.e. 140.8 ± 8.2 mg/g and 226.3 ± 3.3 mg/g, respectively; whereas for Cr^{6+} the bioabsorption capacity is 215.2 ± 5.1 mg/g (Zhang et al., 2017). EPS obtained from *Alteromonas* sp. reveals that C=O and C-O-C are the major functional groups involved in the adsorption of heavy metals (Zhang et al., 2017). A high molecular-weight glycoprotein with an unusually high content of xylose has been isolated from the marine bacterium *Pseudoalteromona* sp. This glycoprotein has good emulsifying properties and metal ions binding capacity from sea sediments (Gutierrez et al., 2008). The EPS of *Pseudoalteromona* sp. isolated from Antarctic seawater exhibits a mechanism for rapid removal of heavy metals like as cadmium and mercury (Caruso et al., 2018). Lactic acid-based bacterial nonglucan exopolysaccharide i.e. EPS-605 can modify itself to form spherical nano-sized particles of 88 nm. EPS-605 is composed of mannose, glucose and galactose with several modifications including acylation, phosphorylation, sulfation, and carboxylation, and posses a highly negative charge. In comparison to other reported EPSs, EPS-605 has strong bioabsorption ability against Pb^{2+}, Cu^{2+}, Cd^{2+} and methylene blue (Mb). The characteristics of EPS-605 as a biosorbent and nanosorbent, are determined by a number of variables, including pH, temperature, initial adsorbate concentration, contact time, and the presence of background electrolytes (Li et al., 2017). The exopolysaccharides of *Pseudomonas* sp. W6 has a unique component for lead removal from wastewater. A comparative biosorption study among different bacterial strains i.e. *P. aeruginosa* MTCC2474, *P. alcaligenes* MJ7 from forest soil, and *P. ficuserectae* PKRS11 from uranium-rich soil reveal the remediation capacity of 65% and 61.2% of Pb from the Synthetic Bangladesh Ground Water medium in batch culture and column respectively (Kalita & Joshi, 2017).

3.6 Role of Exopolysaccharides in Anti-Cancer and Anti-Tumour Activity

Cancer is a very widespread disease in today's world, and there is no effective treatment or medicine available to help cancer sufferers. The use of bacterial exopolysaccharides is a relatively recent field in cancer treatment. Exopolysaccharides from bacteria have anti-tumour and anti-cancer effects. *Bacillus amyloliquefaciens*, an entophytic bacterium, has an inhibitory effect on cancer cells. The research opined that the exopolysaccharide of *Bacillus amyloliquefaciens* has antitumor activity against gastric carcinoma cell lines (MC-4 and SGC-7901). MTT assay, cell viability, and microscopy were employed in this study to evaluate changes in cell effects in response to exopolysaccharide concentrations of 14, 22, and 30 g/L, respectively. The exopolysaccharide shows concentration-dependent inhibitory effects against the MC-4 and SGC-7901 cells lines having IC_{50} values of 19.7 and 26.8 g/L, respectively (Chen et al., 2013). Recently, Sulfur and uronic acids containing exopolysaccharides have gained new attention for the treatment of cancer. Marine bacterial samples containing sulfur and uronic acids in their exopolysaccharides has the highest cytotoxicity, IC_{50} 218 µg mL^{-1} (Abdelnasser et al., 2017). EPS11 an exopolysaccharide obtained from marine bacteria can induce apoptosis of A549 cells by stimulating III-tubulin associated anoikis and decreases the phosphorylation of protein kinase B. Microscopic and western blot investigations revealed that the filiform structure of EPS kills cancer cells over time, and the cell adhesion-related proteins are down-regulated following treatment with EPS11 (Cao et al., 2018). *In vitro* study of a gastric cancer cell line treated with exopolysaccharides of *Bifidobacterium bifidum*, shows promising results in the field of gastric cancer. This experiment confirmed that *Bifidobacterium bifidum* EPS inhibits the growth of gastric cancer cell BGC-823 through telomerase reverse transcriptase activity. RT-PCR result showed that after treatment with the exopolysaccharide expression level of telomerase rate-limiting factor hTERT's mRNA gradually decreases in the cell, which is directly dependent on dosages of exopolysaccharides (Chen et al., 2009). In lung cancer treatment numerous bacterial exopolysaccharides e.g. *Lactobacillus acidophilus, L. plantarum, and L. lactis* have proved their efficacy. These EPSs can protect DNA from oxidation through scavenging free radicals activity, by producing different antioxidant enzymes which can suppress the growth of the tumour and also decrease the expression level of the p53 gene (Deepak et al., 2016). Osteosarcoma is a common cancer disease among teenagers. In this field, marine bacterial oversulfated exopolysaccharides of *Alteromonas infernus* showed a potent inhibition zone of migration, invasion and also the proliferation of cells in human and murine osteosarcoma cells line without any side effects (Heymann et al., 2016). The halophilic bacterium, *Halomonas maura* produces mauran exopolysaccharides. It has a high uronic acid containing sulfated with a high molecular mass, visco-elasticity; pseudo-plasticity; have thixotropic properties; and is resistant to high pH, temperature, and salt content. This high sulfate property is the major key to success to prevent the growth and progression level of the breast cancer cell. Mauran based nano-drug has anti-proliferative activity and also it can kill about 80% of a human breast adenocarcinoma cell line synergistically when loaded with 5-fluorouracil drug (Sivakumar et al., 2014). Exopolysaccharides of *Aphanothece halophytica* GR02 (EPSAH) have a promising effect against cervical cancer cells. The EPASH triggers programmed cell death by affecting the protein Grp78, a key regulator of the unfolded protein response (UPR), thereby activates mitochondrial apoptosis and the p53 survivin pathway; which ultimately helps to activate the caspase-3 pathways leading to programmed cell death (Ou et al., 2014). The halophilic bacterium *Halomonas stenophila* strain B100 produces heteropolysaccharides that are over-sulfated, that induces the T cell line through the caspase-dependent pathway (Ruiz-Ruiz et al., 2011). A new strain belonging to the probiotic bacteria *Bifidobacterium breve* lw01 produce exopolysaccharides, which are comprised of rhamnose, arabinose, galactose, glucose, and mannose with a molar ratio of 0.35:0.44:1.38:0.67:1.65 respectively. The exopolysaccharide demonstrates remarkable anticancerous properties against head and neck squamous cell carcinoma cell line, by controlling cell cycle inhibition and promoting programmed cell death (Wang et al., 2019). *Bifidobacterium bifidum* BGN4 bacterial strain has been isolated from human faeces. This bacterial strain produces a unique composition of exopolysaccharides; composed of chiroinositol, rhamnose, glucose, galactose, and ribose. This chiroinositol-containing polysaccharide has potent anticancer activity against fHT-29 and HCT-116 cancer cells (You et al., 2004). Lactic acid bacteria *Bifidobacteria* spp. is known as ambivalent type; four different strains have been isolated from the gastrointestinal tract of a human. Among them, the EPS from the SK2 strain is a promising agent for cancer treatment (Prosekov et al., 2015). *Anoxybacillus gonensis* YK25 is a thermophilic bacterium that has been isolated from hot springs in Erzurum province (Turkey) which produces potent exopolysaccharides containing hydroxyl, carboxyl, amide and sulphate groups. These EPSs showed significant anticancer activity against SH-SY5Y, HT29, DU145 and A549 cell lines in a dose-dependent manner and the IC_{50} values of EPS were examined as 5.5, 9.61, 15.62 and 17.35 mg/mL respectively (Karadayi et al., 2021). Exopolysaccharides obtained from *Anoxybacillus pushchinoensis* contained sulfate ester which has significant antitumor [lung (A-549) and colon (Caco-2 and HT-29) cancer] activities (Genc et al., 2021).

4. DISCUSSION AND FUTURE ASPECTS

Exopolysaccharides have a diverse range of uses in the fields of medicine, wastewater management, cosmetics, and food industry etc., but still, many more research works are wanted to fully understand their benefits. In the field of drug development such as drug-targeting and carriers, EPS nanoparticulate properties open a wide area against the synthetic drug development, such as biocompatibility and with very low toxicity. Exopolysaccharides are also involved in medical accessories development like fucopol have been used in MRI for better imaging which leads to better diagnosis of the diseases. On the other hand, hyaluronic acid is used as a potent biomarker for detecting cancer and other cancerous tumour cell growth. Wound healing, skin repairing based diagnosis is also done by various exopolysaccharides therapy. Hyaluronic based scaffold is successfully used in tissue grafting with a high success rate. Not only hyaluronic acid but also dextran and curdlan based polymer are used in tissue engineering for the growth of fractured osteoblast cells. The liposome-based drug is targeting by EPS coated drugs already approved for clinical trials. Further research is required in drug development on the topic of stroke, tumours, Alzheimer's and oral cancer treatment. In the field of antimicrobial and antifungal research, exopolysaccharides exhibited prominent data, such as nano-based exopolysaccharides showed antibacterial role against *Pseudomonas aeroginosa* and *Staphylococcus aureus*. Zn and Ni metal-based nanoparticles (3 and 2 mg/mL) provides a prominent result against antibacterial activity without harmful effects in the human cell. Gellan based silver nanoparticles have potent antifungal activity against *Candida albicans* (9 to 13 mm inhibition zone). But in the field of multidrug-resistant, bacteria-based anti-bacterial nanocapsules exopolysaccharides have not been examined in detail. *In vitro* study of Curdlan shows an anti-HIV effect. But there is a big gap in the field of virological treatment with exopolysaccharides. Researches on life-threatening viruses such as COVID-19, HIV, and Dengue using EPSs can be opened with good hope and researchers must also investigate numerous new exopolysaccharides for therapy. In the present scenario, the environment is getting polluted day by day with various pollutants like- heavy metals (Lead, Fluoride, Arsenic, Mercury etc.). *Deinococcus radiodurans* is well-known exopolysaccharide producing bacteria; this polysaccharide is also used for the degradation of uranium. Most of the stains of *Rhizobium radiobacter* have been used for the bioabsorption of arsenic. *Pseudoalteromona* sp. collected from the Arctic region has high removal mechanism rate of heavy metals like- cadmium and mercury at higher concentrations. *Pseudomonas* sp. strain W6 (North East hot spring) exopolysaccharides have a unique component for lead removal from wastewater. Oversulfated exopolysaccharides of *Alteromonas infernus* has potent anti-cancerous property against osteocarcinoma cell. Because exopolysaccharides contain high uronic acid-containing sulfated EPS which has a high molecular weight, viscoelasticity; pseudo-plasticity; thixotropic properties; and resistance to high pH, temperature, and salt content, this composition lead to good anti-cancerous activity. *Aphanothece halaphytica* and *Halomonas stenophila* formed heteropolysaccharides which induce the T cell line through a caspase-dependent pathway. Exopolysaccharide produced by *Bifidobacterium* sp., which is composed up of rhamnose, arabinose, galactose, and glucose, reveals significant anticancerous properties when tested against a head and neck squamous cell carcinoma cell line mainly controlling cell cycle inhibition and promoting cell death. In the field of cancer treatment, various polysaccharides have been explored but till now there are no prominent principles that show a strong review against cancer. Exopolysaccharides are simple monomer-based polysaccharides that have been effectively employed in the food and cosmetic sectors. In the food industry, thickening agents, stabilisers, and emulsifiers are all required for food processing and EPS include all of these necessary ingredients. Therefore, exopolysaccharides are very promising agents in food industries. Throughout the world, food products have always been in high demand and simultaneously additives are increasing day by day, in this situation exopolysaccharides is reliable additive components without any toxicity. Exopolysaccharides have recently been utilised as a texture enhancer in food as non-fat alternatives. They are typically helpful at low concentrations and also have health benefits for obese consumers. Presently, the use of different microbial polysaccharides is increasing in food technology having various anti-oxidant properties, resulting in better human health.

5. CONCLUSION

Microbial exopolysaccharides, particularly thermophilic-based exopolysaccharides, are the industrially important microbial product that is utilized as the raw materials in sectors of food, pharmaceuticals, health care, tissue engineering, 3-D printing and bioremediation. They have been implicated as a beneficial agent in various aspects of human existence. The physicochemical characteristics and biological activities of EPS are determined by their structural complexity that defines their functional use. In addition to structural alterations, EPS can be modified with natural or synthetic polymers to expand its range of applications. Researchers are constantly searching and exploring unique EPS having multifunctional characters with novel and beneficial functions, which can provide industrially value-added products.

Recent research has provided insight into the roles of EPS in health-promoting effects like antitumor, antiulcer, immunomodulatory, antiviral and cholesterol-lowering activities. EPSs are in high demand as industrial due to their outstanding rheological, emulsifying, water-retention and biocompatibility characteristics. Furthermore, the thermophilic EPSs are stable across a broad range of temperature and pH range, thereby increasing their industrial demand. More researches are expected in future by utilizing structural modifications of natural EPS or synthetic polymers in the diverse field of food, pharmaceutical, tissue engineering, agriculture, bioremediation, etc. for the benefit of human life.

REFERENCES

Abdelnasser, S. M., Yahya, S. M., Mohamed, W. F., Asker, M. M., Shady, H. M. A., Mahmoud, M. G., & Gadallah, M. A. (2017). Antitumor exopolysaccharides derived from novel marine bacillus: Isolation, characterization aspect and biological activity. *Asian Pacific Journal of Cancer Prevention, 18*(7), 1847. PMID:28749119

Adebayo-Tayo, B. C., & Popoola, A. O. (2017). Biogenic synthesis and antimicrobial activity of silver nanoparticle using exopolysaccharides from lactic acid bacteria. *International Journal of Nanodimension, 8*(1), 61–69.

Ahmad, N. H., Mustafa, S., & Che Man, Y. B. (2015). Microbial polysaccharides and their modification approaches: A review. *International Journal of Food Properties, 18*(2), 332–347. doi:10.1080/10942912.2012.693561

Allison, D., Sutherland, I., & Neu, T. (2003). EPS: What's an acronym. *Biofilm communities: Order from chaos*, 381-387.

Allison, D. G., & Sutherland, I. W. (1987). The role of exopolysaccharides in adhesion of freshwater bacteria. *Microbiology, 133*(5), 1319–1327. doi:10.1099/00221287-133-5-1319

Amspacher, W. H., & Curreri, A. R. (1953). Use of dextran in control of shock resulting from war wounds. *A.M.A. Archives of Surgery, 66*(6), 730–740. doi:10.1001/archsurg.1953.01260030750004 PMID:13050239

Arena, A., Gugliandolo, C., Stassi, G., Pavone, B., Iannello, D., Bisignano, G., & Maugeri, T. L. (2009). An exopolysaccharide produced by *Geobacillus thermodenitrificans* strain B3-72: Antiviral activity on immunocompetent cells. *Immunology Letters, 123*(2), 132–137. doi:10.1016/j.imlet.2009.03.001 PMID:19428560

Arnaouteli, S., Ferreira, A. S., Schor, M., Morris, R. J., Bromley, K. M., Jo, J., ... Dietrich, L. E. (2017). Bifunctionality of a biofilm matrix protein controlled by redox state. *Proceedings of the National Academy of Sciences of the United States of America, 114*(30), E6184–E6191. doi:10.1073/pnas.1707687114 PMID:28698374

Avsar, G., Agirbasli, D., Agirbasli, M. A., Gunduz, O., & Oner, E. T. (2018). Levan based fibrous scaffolds electrospun via co-axial and single-needle techniques for tissue engineering applications. *Carbohydrate Polymers, 193*, 316–325. doi:10.1016/j.carbpol.2018.03.075 PMID:29773387

Badel, S., Bernardi, T., & Michaud, P. (2011). New perspectives for *Lactobacilli* exopolysaccharides. *Biotechnology Advances, 29*(1), 54–66. doi:10.1016/j.biotechadv.2010.08.011 PMID:20807563

Banerjee, A., & Bandopadhyay, R. (2016). Use of dextran nanoparticle: A paradigm shift in bacterial exopolysaccharide based biomedical applications. *International Journal of Biological Macromolecules, 87*, 295–301. doi:10.1016/j.ijbiomac.2016.02.059 PMID:26927936

Bankura, K., Maity, D., Mollick, M. M., Mondal, D., Bhowmick, B., Bain, M., Chakraborty, A., Sarkar, J., Acharya, K., & Chattopadhyay, D. (2012). Synthesis, characterization and antimicrobial activity of dextran stabilized silver nanoparticles in aqueous medium. *Carbohydrate Polymers*, *89*(4), 1159–1165. doi:10.1016/j.carbpol.2012.03.089 PMID:24750927

Becker, A., Katzen, F., Pühler, A., & Ielpi, L. (1998). Xanthan gum biosynthesis and application: A biochemical/genetic perspective. *Applied Microbiology and Biotechnology*, *50*(2), 145–152. doi:10.1007002530051269 PMID:9763683

Bhunia, B., Uday, U. S. P., Oinam, G., Mondal, A., Bandyopadhyay, T. K., & Tiwari, O. N. (2018). Characterization, genetic regulation and production of cyanobacterial exopolysaccharides and its applicability for heavy metal removal. *Carbohydrate Polymers*, *179*, 228–243. doi:10.1016/j.carbpol.2017.09.091 PMID:29111047

Bogino, P., Abod, A., Nievas, F., & Giordano, W. (2013). Water-limiting conditions alter the structure and biofilm-forming ability of bacterial multispecies communities in the alfalfa rhizosphere. *PLoS One*, *8*(11), e79614. doi:10.1371/journal.pone.0079614 PMID:24223979

Branda, S. S., Vik, S., Friedman, L., & Kolter, R. (2005). Biofilms: The matrix revisited. *Trends in Microbiology*, *13*(1), 20–26. doi:10.1016/j.tim.2004.11.006 PMID:15639628

Brim, H., McFarlan, S. C., Fredrickson, J. K., Minton, K. W., Zhai, M., Wackett, L. P., & Daly, M. J. (2000). Engineering *Deinococcus radiodurans* for metal remediation in radioactive mixed waste environments. *Nature Biotechnology*, *18*(1), 85–90. doi:10.1038/71986 PMID:10625398

Brim, H., Osborne, J. P., Kostandarithes, H. M., Fredrickson, J. K., Wackett, L. P., & Daly, M. J. (2006). *Deinococcus radiodurans* engineered for complete toluene degradation facilitates Cr (VI) reduction. *Microbiology*, *152*(8), 2469–2477. doi:10.1099/mic.0.29009-0 PMID:16849809

Caccamo, M. T., Gugliandolo, C., Zammuto, V., & Magazù, S. (2020). Thermal properties of an exopolysaccharide produced by a marine thermotolerant *Bacillus licheniformis* by ATR-FTIR spectroscopy. *International Journal of Biological Macromolecules*, *145*, 77–83. doi:10.1016/j.ijbiomac.2019.12.163 PMID:31870877

Cao, R., Jin, W., Shan, Y., Wang, J., Liu, G., Kuang, S., & Sun, C. (2018). Marine bacterial polysaccharide EPS11 inhibits cancer cell growth via blocking cell adhesion and stimulating anoikis. *Marine Drugs*, *16*(3), 85. doi:10.3390/md16030085 PMID:29518055

Caruso, C., Rizzo, C., Mangano, S., Poli, A., Di Donato, P., Nicolaus, B., ... Giudice, A. L. (2018). Extracellular polymeric substances with metal adsorption capacity produced by *Pseudoalteromonas* sp. MER144 from Antarctic seawater. *Environmental Science and Pollution Research International*, *25*(5), 4667–4677. doi:10.100711356-017-0851-z PMID:29197057

Chen, W. Y., & Abatangelo, G. (1999). Functions of hyaluronan in wound repair. *Wound Repair and Regeneration*, *7*(2), 79–89. doi:10.1046/j.1524-475X.1999.00079.x PMID:10231509

Chen, X., Jiang, H., Yang, Y., & Liu, N. (2009). Effect of exopolysaccharide from *Bifidobacterium bifidum* on cell of gastric cancer and human telomerase reverse transcriptase. *Wei sheng wu xue bao=. Acta Microbiologica Sinica*, *49*(1), 117–122. PMID:19388274

Chen, Y.-T., Yuan, Q., Shan, L.-T., Lin, M.-A., Cheng, D.-Q., & Li, C.-Y. (2013). Antitumor activity of bacterial exopolysaccharides from the endophyte *Bacillus amyloliquefaciens* sp. isolated from *Ophiopogon japonicus*. *Oncology Letters*, *5*(6), 1787–1792. doi:10.3892/ol.2013.1284 PMID:23833642

Cho, H. M., & Yoo, B. (2015). Rheological characteristics of cold thickened beverages containing xanthan gum–based food thickeners used for dysphagia diets. *Journal of the Academy of Nutrition and Dietetics*, *115*(1), 106–111. doi:10.1016/j.jand.2014.08.028 PMID:25441963

Costa, O. Y., Raaijmakers, J. M., & Kuramae, E. E. (2018). Microbial extracellular polymeric substances: Ecological function and impact on soil aggregation. *Frontiers in Microbiology*, *9*, 1636. doi:10.3389/fmicb.2018.01636 PMID:30083145

Costerton, J. W., Cheng, K., Geesey, G. G., Ladd, T. I., Nickel, J. C., Dasgupta, M., & Marrie, T. J. (1987). Bacterial biofilms in nature and disease. *Annual Review of Microbiology*, *41*(1), 435–464. doi:10.1146/annurev.mi.41.100187.002251 PMID:3318676

Davey, K. R., & Amos, S. A. (2002). Study of the effects of temperature, pH and yeast extract on growth and exopolysaccharides production by *Propionibacterium acidi-propionici* on milk microfiltrate using a response surface methodology. *Journal of Applied Microbiology*, *92*(3), 583–587. doi:10.1046/j.1365-2672.2002.1617a.x PMID:11922033

Decho, A. W., & Gutierrez, T. (2017). Microbial extracellular polymeric substances (EPSs) in ocean systems. *Frontiers in Microbiology*, *8*, 922. doi:10.3389/fmicb.2017.00922 PMID:28603518

Deepak, V., Ramachandran, S., Balahmar, R. M., Pandian, S. R. K., Sivasubramaniam, S. D., Nellaiah, H., & Sundar, K. (2016). *In vitro* evaluation of anticancer properties of exopolysaccharides from *Lactobacillus acidophilus* in colon cancer cell lines. *In Vitro Cellular & Developmental Biology. Animal*, *52*(2), 163–173. doi:10.100711626-015-9970-3 PMID:26659393

Deepika, K., Raghuram, M., Kariali, E., & Bramhachari, P. (2016). Biological responses of symbiotic *Rhizobium radiobacter* strain VBCK1062 to the arsenic contaminated rhizosphere soils of mung bean. *Ecotoxicology and Environmental Safety*, *134*, 1–10. doi:10.1016/j.ecoenv.2016.08.008 PMID:27566287

Dertli, E., Mayer, M. J., & Narbad, A. (2015). Impact of the exopolysaccharide layer on biofilms, adhesion and resistance to stress in *Lactobacillus johnsonii* FI9785. *BMC Microbiology*, *15*(1), 1–9. doi:10.118612866-015-0347-2 PMID:25648083

Dhagat, S., & Jujjavarapu, S. E. (2021). Biorefinery system for production of thermostable exopolysaccharide by a novel thermophile *Brevibacillus borstelensis* MK878423 and its study on impact of glucose utilization. *Biomass Conversion and Biorefinery*, 1–11.

Dimopoulou, M., Bardeau, T., Ramonet, P.-Y., Miot-Certier, C., Claisse, O., Doco, T., Petrel, M., Lucas, P., & Dols-Lafargue, M. (2016). Exopolysaccharides produced by *Oenococcus oeni*: From genomic and phenotypic analysis to technological valorization. *Food Microbiology*, *53*, 10–17. doi:10.1016/j.fm.2015.07.011 PMID:26611165

Dobrowolski, R., Szcześ, A., Czemierska, M., & Jarosz-Wikołazka, A. (2017). Studies of cadmium (II), lead (II), nickel (II), cobalt (II) and chromium (VI) sorption on extracellular polymeric substances produced by *Rhodococcus opacus* and *Rhodococcus rhodochrous. Bioresource Technology, 225*, 113–120. doi:10.1016/j.biortech.2016.11.040 PMID:27888727

Donlan, R. M. (2002). Biofilms: Microbial life on surfaces. *Emerging Infectious Diseases, 8*(9), 881–890. doi:10.3201/eid0809.020063 PMID:12194761

Dutta, S. D., Patel, D. K., & Lim, K.-T. (2019). Functional cellulose-based hydrogels as extracellular matrices for tissue engineering. *Journal of Biological Engineering, 13*(1), 1–19. doi:10.118613036-019-0177-0 PMID:31249615

Duymaz, B. T., Erdiler, F. B., Alan, T., Aydogdu, M. O., Inan, A. T., Ekren, N., ... Oktar, F. N. (2019). 3D bio-printing of levan/polycaprolactone/gelatin blends for bone tissue engineering: Characterization of the cellular behavior. *European Polymer Journal, 119*, 426–437. doi:10.1016/j.eurpolymj.2019.08.015

El-Naggar, M. E., Othman, S. I., Allam, A. A., & Morsy, O. M. (2020). Synthesis, drying process and medical application of polysaccharide-based aerogels. *International Journal of Biological Macromolecules, 145*, 1115–1128. doi:10.1016/j.ijbiomac.2019.10.037 PMID:31678101

Felt, O., Einmahl, S., Furrer, P., Baeyens, V., & Gurny, R. (2001). Polymeric systems for ophthalmic drug delivery. *Polymeric Biomaterials*, 377-421.

Fialho, A. M., Moreira, L. M., Granja, A. T., Popescu, A. O., Hoffmann, K., & Sá-Correia, I. (2008). Occurrence, production, and applications of gellan: Current state and perspectives. *Applied Microbiology and Biotechnology, 79*(6), 889–900. doi:10.100700253-008-1496-0 PMID:18506441

Flemming, H.-C. (2016). EPS—Then and now. *Microorganisms, 4*(4), 41. doi:10.3390/microorganisms4040041 PMID:27869702

Flemming, H.-C., & Wingender, J. (2010). The biofilm matrix. *Nature Reviews. Microbiology, 8*(9), 623–633. doi:10.1038/nrmicro2415 PMID:20676145

Freitas, F., Alves, V. D., & Reis, M. A. (2011). Advances in bacterial exopolysaccharides: From production to biotechnological applications. *Trends in Biotechnology, 29*(8), 388–398. doi:10.1016/j.tibtech.2011.03.008 PMID:21561675

Garza-Cervantes, J. A., Escárcega-González, C. E., Castro, E. D. B., Mendiola-Garza, G., Marichal-Cancino, B. A., López-Vázquez, M. A., & Morones-Ramirez, J. R. (2019). Antimicrobial and antibiofilm activity of biopolymer-Ni, Zn nanoparticle biocomposites synthesized using *R. mucilaginosa* UANL-001L exopolysaccharide as a capping agent. *International Journal of Nanomedicine, 14*, 2557–2571. doi:10.2147/IJN.S196470 PMID:31118605

Genc, B., Taskin, M., & Adiguzel, A. (2021). Exopolysaccharide of *Anoxybacillus pushchinoensis* G11 has antitumor and antibiofilm activities. *Archives of Microbiology*, 1–18. PMID:33604750

Ghalem, B. R. (2017). Antioxidant and antimicrobial activities of exopolysaccharides from yoghurt starter. *American Journal of Chemical and Biochemical Engineering, 3*(3), 35–39.

Ghuysen, J.-M. (1968). Use of bacteriolytic enzymes in determination of wall structure and their role in cell metabolism. *Bacteriological Reviews*, *32*(4 Pt 2), 425–464. doi:10.1128/br.32.4_pt_2.425-464.1968 PMID:4884715

Gutierrez, T., Shimmield, T., Haidon, C., Black, K., & Green, D. H. (2008). Emulsifying and metal ion binding activity of a glycoprotein exopolymer produced by *Pseudoalteromonas* sp. strain TG12. *Applied and Environmental Microbiology*, *74*(15), 4867–4876. doi:10.1128/AEM.00316-08 PMID:18552188

Hägerström, H. (2003). *Polymer gels as pharmaceutical dosage forms: rheological performance and physicochemical interactions at the gel-mucus interface for formulations intended for mucosal drug delivery* Acta Universitatis Upsaliensis.

Han, W., Wang, W., Zhao, M., Sugahara, K., & Li, F. (2014). A novel eliminase from a marine bacterium that degrades hyaluronan and chondroitin sulfate. *The Journal of Biological Chemistry*, *289*(40), 27886–27898. doi:10.1074/jbc.M114.590752 PMID:25122756

Heymann, D., Ruiz-Velasco, C., Chesneau, J., Ratiskol, J., Sinquin, C., & Colliec-Jouault, S. (2016). Anti-metastatic properties of a marine bacterial exopolysaccharide-based derivative designed to mimic glycosaminoglycans. *Molecules (Basel, Switzerland)*, *21*(3), 309. doi:10.3390/molecules21030309 PMID:26959001

Hickey, R. J., & Pelling, A. E. (2019). Cellulose biomaterials for tissue engineering. *Frontiers in Bioengineering and Biotechnology*, *7*, 45. doi:10.3389/fbioe.2019.00045 PMID:30968018

Hindersah, R., Mulyani, O., & Osok, R. (2017). Proliferation and exopolysaccharide production ofAzotobacter in the presence of mercury. *Biodiversity Journal*, *8*(1), 21–26.

Izawa, N., Hanamizu, T., Iizuka, R., Sone, T., Mizukoshi, H., Kimura, K., & Chiba, K. (2009). *Streptococcus thermophilus* produces exopolysaccharides including hyaluronic acid. *Journal of Bioscience and Bioengineering*, *107*(2), 119–123. doi:10.1016/j.jbiosc.2008.11.007 PMID:19217548

Jansson, P.-E., Kenne, L., & Lindberg, B. (1975). Structure of the extracellular polysaccharide from *Xanthomonas campestris*. *Carbohydrate Research*, *45*(1), 275–282. doi:10.1016/S0008-6215(00)85885-1 PMID:1212669

Kalita, D., & Joshi, S. (2017). Study on bioremediation of Lead by exopolysaccharide producing metallophilic bacterium isolated from extreme habitat. *Biotechnology Reports (Amsterdam, Netherlands)*, *16*, 48–57. doi:10.1016/j.btre.2017.11.003 PMID:29167759

Kambourova, M. (2018). Thermostable enzymes and polysaccharides produced by thermophilic bacteria isolated from Bulgarian hot springs. *Engineering in Life Sciences*, *18*(11), 758–767. doi:10.1002/elsc.201800022 PMID:32624870

Kanmani, P., & Lim, S. T. (2013). Synthesis and structural characterization of silver nanoparticles using bacterial exopolysaccharide and its antimicrobial activity against food and multidrug resistant pathogens. *Process Biochemistry*, *48*(7), 1099–1106. doi:10.1016/j.procbio.2013.05.011

Karadayi, Y. I., Aykutoglu, G., Arslan, N. P., Baltaci, M. O., Adiguzel, A., & Taskin, M. (2021). Production of water-soluble sulfated exopolysaccharide with anticancer activity from *Anoxybacillus gonensis* YK25. *Journal of Chemical Technology and Biotechnology (Oxford, Oxfordshire)*, *96*(5), 1258–1266. doi:10.1002/jctb.6638

Kataky, R., & Knowles, E. (2018). Biofilm formation on abiotic surfaces and their redox activity. *Current Opinion in Electrochemistry*, *12*, 121–128. doi:10.1016/j.coelec.2018.07.007

Kazak, H., Oner, E., & Dekker, R. F. (2010). Extremophiles as sources of exopolysaccharides. *Carbohydrate Polymers: Development, Properties and Applications*, 605-619.

Klimek, K., Przekora, A., Pałka, K., & Ginalska, G. (2016). New method for the fabrication of highly osteoconductive β-1, 3-glucan/HA scaffold for bone tissue engineering: Structural, mechanical, and biological characterization. *Journal of Biomedical Materials Research. Part A*, *104*(10), 2528–2536. doi:10.1002/jbm.a.35798 PMID:27239050

Knudson, W., Chow, G., & Knudson, C. B. (2002). CD44-mediated uptake and degradation of hyaluronan. *Matrix Biology*, *21*(1), 15–23. doi:10.1016/S0945-053X(01)00186-X PMID:11827788

Kogan, G., Šoltés, L., Stern, R., & Gemeiner, P. (2007). Hyaluronic acid: A natural biopolymer with a broad range of biomedical and industrial applications. *Biotechnology Letters*, *29*(1), 17–25. doi:10.100710529-006-9219-z PMID:17091377

Körstgens, V., Flemming, H.-C., Wingender, J., & Borchard, W. (2001). Influence of calcium ions on the mechanical properties of a model biofilm of mucoid *Pseudomonas aeruginosa. Water Science and Technology*, *43*(6), 49–57. doi:10.2166/wst.2001.0338 PMID:11381972

Lakzian, A., Berenji, A. R., Karimi, E., & Razavi, S. (2008). Adsorption capability of lead, nickel and zinc by exopolysaccharide and dried cell of *Ensifer meliloti. Asian Journal of Chemistry*, *20*(8), 6075.

Lapaglia, C., & Hartzell, P. L. (1997). Stress-induced production of biofilm in the hyperthermophile *Archaeoglobus fulgidus. Applied and Environmental Microbiology*, *63*(8), 3158–3163. doi:10.1128/aem.63.8.3158-3163.1997 PMID:16535671

Li, C., Zhou, L., Yang, H., Lv, R., Tian, P., Li, X., Zhang, Y., Chen, Z., & Lin, F. (2017). Self-assembled exopolysaccharide nanoparticles for bioremediation and green synthesis of noble metal nanoparticles. *ACS Applied Materials & Interfaces*, *9*(27), 22808–22818. doi:10.1021/acsami.7b02908 PMID:28613815

Li, S., Huang, R., Shah, N. P., Tao, X., Xiong, Y., & Wei, H. (2014). Antioxidant and antibacterial activities of exopolysaccharides from *Bifidobacterium bifidum* WBIN03 and Lactobacillus plantarum R315. *Journal of Dairy Science*, *97*(12), 7334–7343. doi:10.3168/jds.2014-7912 PMID:25282420

Madsen, J. S., Burmølle, M., Hansen, L. H., & Sørensen, S. J. (2012). The interconnection between biofilm formation and horizontal gene transfer. *FEMS Immunology and Medical Microbiology*, *65*(2), 183–195. doi:10.1111/j.1574-695X.2012.00960.x PMID:22444301

Manca, M. C., Lama, L., Improta, R., Esposito, E., Gambacorta, A., & Nicolaus, B. (1996). Chemical composition of two exopolysaccharides from *Bacillus thermoantarcticus. Applied and Environmental Microbiology*, *62*(9), 3265–3269. doi:10.1128/aem.62.9.3265-3269.1996 PMID:16535400

Manobala, T., Shukla, S. K., Rao, T. S., & Kumar, M. D. (2019). A new uranium bioremediation approach using radio-tolerant *Deinococcus radiodurans* biofilm. *Journal of Biosciences, 44*(5), 1–9. doi:10.100712038-019-9942-y PMID:31719231

McEwen, G. D., Wu, Y., & Zhou, A. (2010). Probing nanostructures of bacterial extracellular polymeric substances versus culture time by Raman microspectroscopy and atomic force microscopy. *Biopolymers: Original Research on Biomolecules, 93*(2), 171–177. doi:10.1002/bip.21315 PMID:19787720

Meng, X., Pijning, T., Dobruchowska, J. M., Yin, H., Gerwig, G. J., & Dijkhuizen, L. (2016). Structural determinants of alternating (α1→ 4) and (α1→ 6) linkage specificity in reuteransucrase of *Lactobacillus reuteri. Scientific Reports, 6*(1), 1–14. doi:10.1038rep35261 PMID:27748434

Milani, J., & Maleki, G. (2012). *Hydrocolloids in Food Industry. Food Industrial Processes-Methods and Equipment*. Tech.

Monsan, P., Bozonnet, S., Albenne, C., Joucla, G., Willemot, R.-M., & Remaud-Siméon, M. (2001). Homopolysaccharides from lactic acid bacteria. *International Dairy Journal, 11*(9), 675–685. doi:10.1016/S0958-6946(01)00113-3

Morris, G., & Harding, S. (2009). Polysaccharides, microbial. In *Encyclopedia of microbiology* (pp. 482–494). Elsevier Inc. doi:10.1016/B978-012373944-5.00135-8

Moscovici, M. (2015). Present and future medical applications of microbial exopolysaccharides. *Frontiers in Microbiology, 6*, 1012. doi:10.3389/fmicb.2015.01012 PMID:26483763

Mozzi, F., Vaningelgem, F., Hébert, E. M., Van der Meulen, R., Moreno, M. R. F., de Valdez, G. F., & De Vuyst, L. (2006). Diversity of heteropolysaccharide-producing lactic acid bacterium strains and their biopolymers. *Applied and Environmental Microbiology, 72*(6), 4431–4435. doi:10.1128/AEM.02780-05 PMID:16751563

Mukherjee, P., & Atala, A. (2005). Biomaterials for genitourinary tissue engineering. *Scaffolding in Tissue Engineering*, 357.

Mukherjee, S., Sahu, P., & Halder, G. (2017). Microbial remediation of fluoride-contaminated water via a novel bacterium *Providencia vermicola* (KX926492). *Journal of Environmental Management, 204*, 413–423. doi:10.1016/j.jenvman.2017.08.051 PMID:28915476

Nichols, C. M., Guezennec, J., & Bowman, J. (2005). Bacterial exopolysaccharides from extreme marine environments with special consideration of the southern ocean, sea ice, and deep-sea hydrothermal vents: A review. *Marine Biotechnology (New York, N.Y.), 7*(4), 253–271. doi:10.100710126-004-5118-2 PMID:16075348

Nicolaus, B., Kambourova, M., & Oner, E. T. (2010). Exopolysaccharides from extremophiles: From fundamentals to biotechnology. *Environmental Technology, 31*(10), 1145–1158. doi:10.1080/09593330903552094 PMID:20718297

Nicolaus, B., Lama, L., Panico, A., Moriello, V. S., Romano, I., & Gambacorta, A. (2002). Production and characterization of exopolysaccharides excreted by thermophilic bacteria from shallow, marine hydrothermal vents of Flegrean Ares (Italy). *Systematic and Applied Microbiology, 25*(3), 319–325. doi:10.1078/0723-2020-00128 PMID:12421069

Nicolaus, B., Manca, M. C., Romano, I., & Lama, L. (1993). Production of an exopolysaccharide from two thermophilic archaea belonging to the genus *Sulfolobus. FEMS Microbiology Letters, 109*(2-3), 203–206. doi:10.1111/j.1574-6968.1993.tb06168.x

Noel, S., Fortier, C., Murschel, F., Belzil, A., Gaudet, G., Jolicoeur, M., & De Crescenzo, G. (2016). Co-immobilization of adhesive peptides and VEGF within a dextran-based coating for vascular applications. *Acta Biomaterialia, 37*, 69–82. doi:10.1016/j.actbio.2016.03.043 PMID:27039978

Novik, G., Savich, V., & Meerovskaya, O. (2018). Geobacillus Bacteria: potential commercial applications in industry, bioremediation, and bioenergy production. In Growing and Handling of Bacterial Cultures. IntechOpen.

Nwodo, U. U., Green, E., & Okoh, A. I. (2012). Bacterial exopolysaccharides: Functionality and prospects. *International Journal of Molecular Sciences, 13*(11), 14002–14015. doi:10.3390/ijms131114002 PMID:23203046

Osawa, Z., Morota, T., Hatanaka, K., Akaike, T., Matsuzaki, K., Nakashima, H., ... Mimura, T. (1993). Synthesis of sulfated derivatives of curdlan and their anti-HIV activity. *Carbohydrate Polymers, 21*(4), 283–288. doi:10.1016/0144-8617(93)90060-H

Ou, Y., Xu, S., Zhu, D., & Yang, X. (2014). Molecular mechanisms of exopolysaccharide from *Aphanothece halaphytica* (EPSAH) induced apoptosis in HeLa cells. *PLoS One, 9*(1), e87223. doi:10.1371/journal.pone.0087223 PMID:24466342

Oves, M., Khan, M. S., & Zaidi, A. (2013). Chromium reducing and plant growth promoting novel strain *Pseudomonas aeruginosa* OSG41 enhance chickpea growth in chromium amended soils. *European Journal of Soil Biology, 56*, 72–83. doi:10.1016/j.ejsobi.2013.02.002

Oves, M., Rauf, M. A., Hussain, A., Qari, H. A., Khan, A. A. P., Muhammad, P., Rehman, M. T., Alajmi, M. F., & Ismail, I. I. (2019). Antibacterial silver nanomaterial synthesis from *Mesoflavibacter zeaxanthinifaciens* and targeting biofilm formation. *Frontiers in Pharmacology, 10*, 801. doi:10.3389/fphar.2019.00801 PMID:31427961

Ozdemir, G., Ozturk, T., Ceyhan, N., Isler, R., & Cosar, T. (2003). Heavy metal biosorption by biomass of *Ochrobactrum anthropi* producing exopolysaccharide in activated sludge. *Bioresource Technology, 90*(1), 71–74. doi:10.1016/S0960-8524(03)00088-9 PMID:12835060

Palaniraj, A., & Jayaraman, V. (2011). Production, recovery and applications of xanthan gum by *Xanthomonas campestris. Journal of Food Engineering, 106*(1), 1–12. doi:10.1016/j.jfoodeng.2011.03.035

Palma, S. I., Rodrigues, C. A., Carvalho, A., del Puerto Morales, M., Freitas, F., Fernandes, A. R., ... Roque, A. C. (2015). A value-added exopolysaccharide as a coating agent for MRI nanoprobes. *Nanoscale, 7*(34), 14272–14283. doi:10.1039/C5NR01979F PMID:26186402

Patel, A., & Prajapat, J. (2013). Food and health applications of exopolysaccharides produced by lactic acid bacteria. *Journal of Advances in Dairy Research*, 1–8.

Patel, K. K., Shah, F., & Dwivedi, M. (2018). Exopolysaccharide (EPS) Producing Isolates from Sugarcane Field Soil and Antibacterial Activity of Extracted EPSs. *Acta Scientific Microbiology*, *1*(4), 6–13. doi:10.31080/ASMI.2018.01.0032

Pathak, V. M., & Navneet. (2017). Review on the current status of polymer degradation: a microbial approach. *Bioresources and Bioprocessing, 4*, 15.

Poli, A., Anzelmo, G., & Nicolaus, B. (2010). Bacterial exopolysaccharides from extreme marine habitats: Production, characterization and biological activities. *Marine Drugs*, *8*(6), 1779–1802. doi:10.3390/md8061779 PMID:20631870

Prajapati, V. D., Jani, G. K., Zala, B. S., & Khutliwala, T. A. (2013). An insight into the emerging exopolysaccharide gellan gum as a novel polymer. *Carbohydrate Polymers*, *93*(2), 670–678. doi:10.1016/j.carbpol.2013.01.030 PMID:23499110

Prosekov, A., Dyshlyuk, L., Milentyeva, I., Sukhih, S., Babich, O., Ivanova, S., ... Matskova, L. (2015). Antioxidant, antimicrobial and antitumor activity of bacteria of the genus bifidobacterium, selected from the gastrointestinal tract of human. *Integrative Molecular Medicine*, *2*(5), 295–303.

Rajwade, J., Paknikar, K., & Kumbhar, J. (2015). Applications of bacterial cellulose and its composites in biomedicine. *Applied Microbiology and Biotechnology*, *99*(6), 2491–2511. doi:10.100700253-015-6426-3 PMID:25666681

Reckseidler-Zenteno, S. L. (2012). Capsular polysaccharides produced by the bacterial pathogen *Burkholderia pseudomallei. The Complex World of Polysaccharides*, 128-152.

Rinker, K. D., & Kelly, R. M. (2000). Effect of carbon and nitrogen sources on growth dynamics and exopolysaccharide production for the hyperthermophilic archaeon *Thermococcus litoralis* and bacterium *Thermotoga maritima. Biotechnology and Bioengineering*, *69*(5), 537–547. doi:10.1002/1097-0290(20000905)69:5<537::AID-BIT8>3.0.CO;2-7 PMID:10898863

Roberts, E., & Garegg, P. (1998). Levan derivatives, their preparation, composition and applications including medical and food applications. *World Intellectual Property Organization Patent Application No. WO, 98*, 03184.

Roberts, I. S. (1996). The biochemistry and genetics of capsular polysaccharide production in bacteria. *Annual Review of Microbiology*, *50*(1), 285–315. doi:10.1146/annurev.micro.50.1.285 PMID:8905082

Ruffing, A., & Chen, R. R. (2006). Metabolic engineering of microbes for oligosaccharide and polysaccharide synthesis. *Microbial Cell Factories*, *5*(1), 1–9. doi:10.1186/1475-2859-5-25 PMID:16859553

Ruiz-Ruiz, C., Srivastava, G. K., Carranza, D., Mata, J. A., Llamas, I., Santamaría, M., Quesada, E., & Molina, I. J. (2011). An exopolysaccharide produced by the novel halophilic bacterium *Halomonas stenophila* strain B100 selectively induces apoptosis in human T leukaemia cells. *Applied Microbiology and Biotechnology*, *89*(2), 345–355. doi:10.100700253-010-2886-7 PMID:20890756

Saba, R., Rehman, Y., Ahmed, M., & Sabri, A. N. (2019). Potential role of bacterial extracellular polymeric substances as biosorbent material for arsenic bioremediation. *Bioremediation Journal, 23*(2), 72–81. doi:10.1080/10889868.2019.1602107

Salem, I. B., Sghaier, H., Monsieurs, P., Moors, H., Van Houdt, R., Fattouch, S., ... Leys, N. (2013). Strontium-induced genomic responses of *Cupriavidus metallidurans* and strontium bioprecipitation as strontium carbonate. *Annals of Microbiology, 63*(3), 833–844. doi:10.100713213-012-0462-3

Saravanan, C., Rajesh, R., Kaviarasan, T., Muthukumar, K., Kavitake, D., & Shetty, P. H. (2017). Synthesis of silver nanoparticles using bacterial exopolysaccharide and its application for degradation of azo-dyes. *Biotechnology Reports (Amsterdam, Netherlands), 15*, 33–40. doi:10.1016/j.btre.2017.02.006 PMID:28664148

Selvakumar, R., Aravindh, S., Ashok, A. M., & Balachandran, Y. L. (2014). A facile synthesis of silver nanoparticle with SERS and antimicrobial activity using *Bacillus subtilis* exopolysaccharides. *Journal of Experimental Nanoscience, 9*(10), 1075–1087. doi:10.1080/17458080.2013.778425

Shin, H., Olsen, B. D., & Khademhosseini, A. (2012). The mechanical properties and cytotoxicity of cell-laden double-network hydrogels based on photocrosslinkable gelatin and gellan gum biomacromolecules. *Biomaterials, 33*(11), 3143–3152. doi:10.1016/j.biomaterials.2011.12.050 PMID:22265786

Sivakumar, B., Aswathy, R. G., Sreejith, R., Nagaoka, Y., Iwai, S., Suzuki, M., ... Maekawa, T. (2014). Bacterial exopolysaccharide based magnetic nanoparticles: A versatile nanotool for cancer cell imaging, targeted drug delivery and synergistic effect of drug and hyperthermia mediated cancer therapy. *Journal of Biomedical Nanotechnology, 10*(6), 885–899. doi:10.1166/jbn.2014.1820 PMID:24749386

Spanò, A., Laganà, P., Visalli, G., Maugeri, T. L., & Gugliandolo, C. (2016). In vitro antibiofilm activity of an exopolysaccharide from the marine thermophilic *Bacillus licheniformis* T14. *Current Microbiology, 72*(5), 518–528. doi:10.100700284-015-0981-9 PMID:26750122

Stevens, M. M., Qanadilo, H. F., Langer, R., & Shastri, V. P. (2004). A rapid-curing alginate gel system: Utility in periosteum-derived cartilage tissue engineering. *Biomaterials, 25*(5), 887–894. doi:10.1016/j.biomaterials.2003.07.002 PMID:14609677

Stewart, P. S. (2014). Biophysics of biofilm infection. *Pathogens and Disease, 70*(3), 212–218. doi:10.1111/2049-632X.12118 PMID:24376149

Sutherland, I. (1972). Bacterial exopolysaccharides. *Advances in Microbial Physiology, 8*, 143–213. doi:10.1016/S0065-2911(08)60190-3 PMID:4581483

Sutherland, I. W. (1994). Structure-function relationships in microbial exopolysaccharides. *Biotechnology Advances, 12*(2), 393–448. doi:10.1016/0734-9750(94)90018-3 PMID:14545899

Sutherland, I. W. (1998). Novel and established applications of microbial polysaccharides. *Trends in Biotechnology, 16*(1), 41–46. doi:10.1016/S0167-7799(97)01139-6 PMID:9470230

Sutherland, I. W. (2001). Biofilm exopolysaccharides: A strong and sticky framework. *Microbiology, 147*(1), 3–9. doi:10.1099/00221287-147-1-3 PMID:11160795

Torgbo, S., & Sukyai, P. (2018). Bacterial cellulose-based scaffold materials for bone tissue engineering. *Applied Materials Today*, *11*, 34–49. doi:10.1016/j.apmt.2018.01.004

Torres, C. A., Antunes, S., Ricardo, A. R., Grandfils, C., Alves, V. D., Freitas, F., & Reis, M. A. (2012). Study of the interactive effect of temperature and pH on exopolysaccharide production by *Enterobacter* A47 using multivariate statistical analysis. *Bioresource Technology*, *119*, 148–156. doi:10.1016/j.biortech.2012.05.106 PMID:22728195

Upadhyay, K. H., Vaishnav, A. M., Tipre, D. R., Patel, B. C., & Dave, S. R. (2017). Kinetics and mechanisms of mercury biosorption by an exopolysaccharide producing marine isolate *Bacillus licheniformis*. *Biotech, 7*(5), 1-10.

van Hijum, S. A., Kralj, S., Ozimek, L. K., Dijkhuizen, L., & van Geel-Schutten, I. G. (2006). Structure-function relationships of glucansucrase and fructansucrase enzymes from lactic acid bacteria. *Microbiology and Molecular Biology Reviews*, *70*(1), 157–176. doi:10.1128/MMBR.70.1.157-176.2006 PMID:16524921

van Hullebusch, E. D., Zandvoort, M. H., & Lens, P. N. (2003). Metal immobilisation by biofilms: Mechanisms and analytical tools. *Reviews in Environmental Science and Biotechnology*, *2*(1), 9–33. doi:10.1023/B:RESB.0000022995.48330.55

Van Vlierberghe, S., Dubruel, P., & Schacht, E. (2011). Biopolymer-based hydrogels as scaffolds for tissue engineering applications: A review. *Biomacromolecules*, *12*(5), 1387–1408. doi:10.1021/bm200083n PMID:21388145

Vaningelgem, F., Zamfir, M., Mozzi, F., Adriany, T., Vancanneyt, M., Swings, J., & De Vuyst, L. (2004). Biodiversity of exopolysaccharides produced by *Streptococcus thermophilus* strains is reflected in their production and their molecular and functional characteristics. *Applied and Environmental Microbiology*, *70*(2), 900–912. doi:10.1128/AEM.70.2.900-912.2004 PMID:14766570

Vieille, C., & Zeikus, G. J. (2001). Hyperthermophilic enzymes: Sources, uses, and molecular mechanisms for thermostability. *Microbiology and Molecular Biology Reviews*, *65*(1), 1–43. doi:10.1128/MMBR.65.1.1-43.2001 PMID:11238984

Vincent, S. J., Faber, E. J., Neeser, J.-R., Stingele, F., & Kamerling, J. P. (2001). Structure and properties of the exopolysaccharide produced by Streptococcus macedonicus Sc136. *Glycobiology*, *11*(2), 131–139. doi:10.1093/glycob/11.2.131 PMID:11287400

Wang, J., Salem, D. R., & Sani, R. K. (2021). Two new exopolysaccharides from a thermophilic bacterium *Geobacillus* sp. WSUCF1: Characterization and bioactivities. *New Biotechnology*, *61*, 29–39. doi:10.1016/j.nbt.2020.11.004 PMID:33188978

Wang, L., Wang, Y., Li, Q., Tian, K., Xu, L., Liu, G., & Guo, C. (2019). Exopolysaccharide, Isolated From a Novel Strain *Bifidobacterium breve* lw01 Possess an Anticancer Effect on Head and Neck Cancer–Genetic and Biochemical Evidences. *Frontiers in Microbiology*, *10*, 1044. doi:10.3389/fmicb.2019.01044 PMID:31143171

Werning, M. L., Notararigo, S., Nácher, M., Fernández de Palencia, P., Aznar, R., & López, P. (2012). Biosynthesis, purification and biotechnological use of exopolysaccharides produced by lactic acid bacteria. *Food Additives*, 83-114.

Yamamoto, Y., Takahashi, Y., Kawano, M., Iizuka, M., Matsumoto, T., Saeki, S., & Yamaguchi, H. (1999). *In vitro* digestibility and fermentability of levan and its hypocholesterolemic effects in rats. *The Journal of Nutritional Biochemistry*, *10*(1), 13–18. doi:10.1016/S0955-2863(98)00077-1 PMID:15539245

Yoshida, T., Hatanaka, K., Uryu, T., Kaneko, Y., Suzuki, E., Miyano, H., Mimura, T., Yoshida, O., & Yamamoto, N. (1990). Synthesis and structural analysis of curdlan sulfate with a potent inhibitory effect in vitro of AIDS virus infection. *Macromolecules*, *23*(16), 3717–3722. doi:10.1021/ma00218a001

You, H. J., Oh, D.-K., & Ji, G. E. (2004). Anticancerogenic effect of a novel chiroinositol-containing polysaccharide from *Bifidobacterium bifidum* BGN4. *FEMS Microbiology Letters*, *240*(2), 131–136. doi:10.1016/j.femsle.2004.09.020 PMID:15522499

Zannini, E., Waters, D. M., Coffey, A., & Arendt, E. K. (2016). Production, properties, and industrial food application of lactic acid bacteria-derived exopolysaccharides. *Applied Microbiology and Biotechnology*, *100*(3), 1121–1135. doi:10.100700253-015-7172-2 PMID:26621802

Zhang, Z., Cai, R., Zhang, W., Fu, Y., & Jiao, N. (2017). A novel exopolysaccharide with metal adsorption capacity produced by a marine bacterium *Alteromonas* sp. JL2810. *Marine Drugs*, *15*(6), 175. doi:10.3390/md15060175 PMID:28604644

Compilation of References

Pire, C., Marhuenda-egea, F. C., Esclapez, J., Alcaraz, L., Ferrer, J., & José Bonete, M. (2004). Stability and enzymatic studies of glucose dehydrogenase from the archaeon Haloferax mediterranei in reverse micelles. *Biocatalysis and Biotransformation*, *22*(1), 17–23. doi:10.1080/10242420310001643179

Kletzin, A. (2007). General characteristics and important model organisms. *Archaea: molecular and cellular biology*, 14-92.

Schleper, C., Jurgens, G., & Jonuscheit, M. (2005). Genomic studies of uncultivated archaea. *Nature Reviews. Microbiology*, *3*(6), 479–488. doi:10.1038/nrmicro1159 PMID:15931166

Slade, D., & Radman, M. (2011). Oxidative stress resistance in Deinococcus radiodurans. *Microbiology and Molecular Biology Reviews*, *75*(1), 133–191. doi:10.1128/MMBR.00015-10 PMID:21372322

Pavlopoulou, A., Savva, G. D., Louka, M., Bagos, P. G., Vorgias, C. E., Michalopoulos, I., & Georgakilas, A. G. (2016). Unraveling the mechanisms of extreme radioresistance in prokaryotes: Lessons from nature. *Mutation Research/Reviews in Mutation Research*, *767*, 92–107. doi:10.1016/j.mrrev.2015.10.001 PMID:27036069

Cox, M. M., Keck, J. L., & Battista, J. R. (2010). Rising from the Ashes: DNA Repair in *Deinococcus radiodurans. PLOS Genetics*, *6*(1), e1000815. doi:10.1371/journal.pgen.1000815 PMID:20084105

Lockhart, J. S., & DeVeaux, L. C. (2013). The essential role of the Deinococcus radiodurans ssb gene in cell survival and radiation tolerance. *PLoS One*, *8*(8), e71651. doi:10.1371/journal.pone.0071651 PMID:23951213

DeVeaux, L. C., Müller, J. A., Smith, J., Petrisko, J., Wells, D. P., & DasSarma, S. (2007). Extremely radiation-resistant mutants of a halophilic archaeon with increased single-stranded DNA-binding protein (RPA) gene expression. *Radiation Research*, *168*(4), 507–514. doi:10.1667/RR0935.1 PMID:17903038

Minsky, A., Shimoni, E., & Englander, J. (2006). Ring-like nucleoids and DNA repair through error-free nonhomologous end joining in Deinococcus radiodurans. *Journal of Bacteriology*, *188*(17), 6047–6051. doi:10.1128/JB.01951-05 PMID:16923869

Daly, M. J., Gaidamakova, E. K., Matrosova, V. Y., Vasilenko, A., Zhai, M., Leapman, R. D., Lai, B., Ravel, B., Li, S.-M. W., Kemner, K. M., & Fredrickson, J. K. (2007). Protein oxidation implicated as the primary determinant of bacterial radioresistance. *PLoS Biology*, *5*(4), e92. doi:10.1371/journal.pbio.0050092 PMID:17373858

Terpe, K. (2013). Overview of thermostable DNA polymerases for classical PCR applications: From molecular and biochemical fundamentals to commercial systems. *Applied Microbiology and Biotechnology*, *97*(24), 10243–10254. doi:10.100700253-013-5290-2 PMID:24177730

Jones, D. L., & Baxter, B. K. (2017). DNA repair and photoprotection: Mechanisms of overcoming environmental ultraviolet radiation exposure in halophilic archaea. *Frontiers in Microbiology*, *8*, 1882. doi:10.3389/fmicb.2017.01882 PMID:29033920

Quintana-Cabrera, R., Fernandez-Fernandez, S., Bobo-Jimenez, V., Escobar, J., Sastre, J., Almeida, A., & Bolaños, J. P. (2012). γ-Glutamylcysteine detoxifies reactive oxygen species by acting as glutathione peroxidase-1 cofactor. *Nature Communications*, *3*(1), 1–8. doi:10.1038/ncomms1722 PMID:22395609

Saiki, R. K., Gelfand, D. H., Stoffel, S., Scharf, S. J., Higuchi, R., Horn, G. T., Mullis, K. B., & Erlich, H. A. (1988). Primer-directed enzymatic amplification of DNA with a thermostable DNA polymerase. *Science*, *239*(4839), 487–491. doi:10.1126cience.239.4839.487 PMID:2448875

Elleuche, S., Schäfers, C., Blank, S., Schröder, C., & Antranikian, G. (2015). Exploration of extremophiles for high temperature biotechnological processes. *Current Opinion in Microbiology*, *25*, 113–119. doi:10.1016/j.mib.2015.05.011 PMID:26066287

Tomazic, S. J., & Klibanov, A. M. (1988). Mechanisms of irreversible thermal inactivation of Bacillus alpha-amylases. *The Journal of Biological Chemistry*, *263*(7), 3086–3091. doi:10.1016/S0021-9258(18)69038-6 PMID:3257756

Champdore, M. D., Staiano, M., Rossi, M., & D'Auria, S. (2007). Proteins from extremophiles as stable tools for advanced biotechnological applications of high social interest. *Journal of the Royal Society, Interface*, *4*(13), 183–191. doi:10.1098/rsif.2006.0174 PMID:17251151

Mevarech, M., Frolow, F., & Gloss, L. M. (2000). Halophilic enzymes: Proteins with a grain of salt. *Biophysical Chemistry*, *86*(2-3), 155–164. doi:10.1016/S0301-4622(00)00126-5 PMID:11026680

Karan, R., Capes, M. D., & DasSarma, S. (2012). Function and biotechnology of extremophilic enzymes in low water activity. *Aquatic Biosystems*, *8*(1), 1–15. doi:10.1186/2046-9063-8-4 PMID:22480329

Zhang, G., & Ge, H. (2013). Protein hypersaline adaptation: Insight from amino acids with machine learning algorithms. *The Protein Journal*, *32*(4), 239–245. doi:10.100710930-013-9484-3 PMID:23592219

Ito, S., Kobayashi, T., Ara, K., Ozaki, K., Kawai, S., & Hatada, Y. (1998). Alkaline detergent enzymes from alkaliphiles: Enzymatic properties, genetics, and structures. *Extremophiles*, *2*(3), 185–190. doi:10.1007007920050059 PMID:9783164

Vannier, P., Michoud, G., Oger, P., Marteinsson, V., & Jebbar, M. (2015). Genome expression of *Thermococcus barophilus* and *Thermococcus kodakarensis* in response to different hydrostatic pressure conditions. *Research in Microbiology*, *166*(9), 717–725. doi:10.1016/j.resmic.2015.07.006 PMID:26239966

Amrani, A., Bergon, A., Holota, H., Tamburini, C., Garel, M., Ollivier, B., Imbert, J., Dolla, A., & Pradel, N. (2014). Transcriptomics reveal several gene expression patterns in the piezophile *Desulfovibrio hydrothermalis* in response to hydrostatic pressure. *PLoS One*, *9*(9), e106831. doi:10.1371/journal.pone.0106831 PMID:25215865

D'Auria, S., Moracci, M., Febbraio, F., Tanfani, F., Nucci, R., & Rossi, M. (1998). Structure-function studies on β-glycosidase from *Sulfolobus solfataricus*. Molecular bases of thermostability. *Biochimie*, *80*(11), 949–957. doi:10.1016/S0300-9084(00)88892-6 PMID:9893955

D'Auria, S., Herman, P., Rossi, M., & Lakowicz, J. R. (1999). The fluorescence emission of the apo-glucose oxidase from Aspergillus niger as probe to estimate glucose concentrations. *Biochemical and Biophysical Research Communications*, *263*(2), 550–553. doi:10.1006/bbrc.1999.1330 PMID:10491329

D'Auria, S., Di Cesare, N., Gryczynski, Z., Gryczynski, I., Rossi, M., & Lakowicz, J. R. (2000). A thermophilic apoglucose dehydrogenase as nonconsuming glucose sensor. *Biochemical and Biophysical Research Communications*, *274*(3), 727–731. doi:10.1006/bbrc.2000.3172 PMID:10924344

Pettit, R. K. (2011). Culturability and secondary metabolite diversity of extreme microbes: Expanding contribution of deep sea and deep-sea vent microbes to natural product discovery. *Marine Biotechnology (New York, N.Y.)*, *13*(1), 1–11. doi:10.100710126-010-9294-y PMID:20437069

Zhang, Y., Li, X., Bartlett, D. H., & Xiao, X. (2015). Current developments in marine microbiology: High-pressure biotechnology and the genetic engineering of piezophiles. *Current Opinion in Biotechnology*, *33*, 157–164. doi:10.1016/j.copbio.2015.02.013 PMID:25776196

Rosenbaum, E., Gabel, F., Durá, M. A., Finet, S., Cléry-Barraud, C., Masson, P., & Franzetti, B. (2012). Effects of hydrostatic pressure on the quaternary structure and enzymatic activity of a large peptidase complex from *Pyrococcus horikoshii*. *Archives of Biochemistry and Biophysics*, *517*(2), 104–110. doi:10.1016/j.abb.2011.07.017 PMID:21896270

Hay, S., Evans, R. M., Levy, C., Loveridge, E. J., Wang, X., Leys, D., Allemann, R. K., & Scrutton, N. S. (2009). Are the catalytic properties of enzymes from piezophilic organisms pressure adapted? *ChemBioChem*, *10*(14), 2348–2353. doi:10.1002/cbic.200900367 PMID:19681091

Reed, C. J., Lewis, H., Trejo, E., Winston, V., & Evilia, C. (2013). Protein adaptations in archaeal extremophiles. *Archaea*. PMID:24151449

Zhuang, X., Han, Z., Bai, Z., Zhuang, G., & Shim, H. (2010). Progress in decontamination by halophilic microorganisms in saline wastewater and soil. *Environmental Pollution*, *158*(5), 1119–1126. doi:10.1016/j.envpol.2010.01.007 PMID:20163899

Usui, K., Hiraki, T., Kawamoto, J., Kurihara, T., Nogi, Y., Kato, C., & Abe, F. (2012). Eicosapentaenoic acid plays a role in stabilizing dynamic membrane structure in the deep-sea piezophile Shewanella violacea: A study employing high-pressure time-resolved fluorescence anisotropy measurement. *Biochimica et Biophysica Acta (BBA)- Biomembranes*, *1818*(3), 574–583. doi:10.1016/j.bbamem.2011.10.010 PMID:22037146

Jebbar, M., Franzetti, B., Girard, E., & Oger, P. (2015). Microbial diversity and adaptation to high hydrostatic pressure in deep-sea hydrothermal vents prokaryotes. *Extremophiles*, *19*(4), 721–740. doi:10.100700792-015-0760-3 PMID:26101015

Simonato, F., Campanaro, S., Lauro, F. M., Vezzi, A., D'Angelo, M., Vitulo, N., Valle, G., & Bartlett, D. H. (2006). Piezophilic adaptation: A genomic point of view. *Journal of Biotechnology*, *126*(1), 11–25. doi:10.1016/j.jbiotec.2006.03.038 PMID:16780980

Wright, P. C., Westacott, R. E., & Burja, A. M. (2003). Piezotolerance as a metabolic engineering tool for the biosynthesis of natural products. *Biomolecular Engineering*, *20*(4-6), 325–331. doi:10.1016/S1389-0344(03)00042-X PMID:12919816

Baker-Austin, C., & Dopson, M. (2007). Life in acid: pH homeostasis in acidophiles. *Trends in Microbiology*, *15*(4), 165–171. doi:10.1016/j.tim.2007.02.005 PMID:17331729

Dopson, M., Baker-Austin, C., Koppineedi, P. R., & Bond, P. L. (2003). Growth in sulfidic mineral environments: Metal resistance mechanisms in acidophilic micro-organisms. *Microbiology*, *149*(8), 1959–1970. doi:10.1099/mic.0.26296-0 PMID:12904536

Gemmell, R. T., & Knowles, C. J. (2000). Utilisation of aliphatic compounds by acidophilic heterotrophic bacteria. The potential for bioremediation of acidic wastewaters contaminated with toxic organic compounds and heavy metals. *FEMS Microbiology Letters*, *192*(2), 185–190. doi:10.1111/j.1574-6968.2000.tb09380.x PMID:11064193

Rani, A., Souche, Y. S., & Goel, R. (2009). Comparative assessment of in situ bioremediation potential of cadmium resistant acidophilic Pseudomonas putida 62BN and alkalophilic Pseudomonas monteilli 97AN strains on soybean. *International Biodeterioration & Biodegradation*, *63*(1), 62–66. doi:10.1016/j.ibiod.2008.07.002

Tindall, B. J., Ross, H. N. M., & Grant, W. D. (1984). Natronobacterium gen. nov. and Natronococcus gen. nov., two new genera of haloalkaliphilic archaebacteria. *Systematic and Applied Microbiology*, *5*(1), 41–57. doi:10.1016/S0723-2020(84)80050-8

Hough, D. W., & Danson, M. J. (1999). Extremozymes. *Current Opinion in Chemical Biology*, *3*(1), 39–46. doi:10.1016/S1367-5931(99)80008-8 PMID:10021406

Krulwich, T. A. (1995). Alkaliphiles: Basicmolecular problems of pH tolerance and bioenergetics. *Molecular Microbiology*, *15*(3), 403–410. doi:10.1111/j.1365-2958.1995.tb02253.x PMID:7783613

Horikoshi, K. (1999). Alkaliphiles: Some applications of their products for biotechnology. *Microbiology and Molecular Biology Reviews*, *63*(4), 735–750. doi:10.1128/MMBR.63.4.735-750.1999 PMID:10585964

Fujinami, S., & Fujisawa, M. (2010). Industrial applications of alkaliphiles and their enzymes–past, present and future. *Environmental Technology*, *31*(8-9), 845–856. doi:10.1080/09593331003762807 PMID:20662376

Gomes, J., & Steiner, W. (2004). The biocatalytic potential of extremophiles and extremozymes. *Food Technology and Biotechnology*, *42*(4), 223–225.

Kaul, P., & Asano, Y. (2012). Strategies for discovery and improvement of enzyme function: State of the art and opportunities. *Microbial Biotechnology*, *5*(1), 18–33. doi:10.1111/j.1751-7915.2011.00280.x PMID:21883976

Adams, M. W., Perler, F. B., & Kelly, R. M. (1995). Extremozymes: Expanding the limits of biocatalysis. *Bio/technology*, *13*(7), 662–668. PMID:9634802

DasSarma, P., Coker, J. A., Huse, V., & DasSarma, S. (2010). Halophiles, biotechnology. Encyclopedia of industrial biotechnology, bioprocess, bioseparation, and cell technology, 2769-2777.

Bowers, K. J., Mesbah, N. M., & Wiegel, J. (2009). Biodiversity of poly-extremophilic Bacteria: Does combining the extremes of high salt, alkaline pH and elevated temperature approach a physico-chemical boundary for life? *Saline Systems*, *5*(1), 1–8. doi:10.1186/1746-1448-5-9 PMID:19930649

Aanniz, T., Ouadghiri, M., & Melloul, M. (2015). Thermophilic bacteria in Moroccan hot springs, salt marshes and desert soils. *Brazilian Journal of Microbiology*, *46*(2), 443–453. doi:10.1590/S1517-838246220140219

Abdel-Mageed, W. M., Milne, B. F., Wagner, M., Schumacher, M., Sandor, P., Pathom-aree, W., Goodfellow, M., Bull, A. T., Horikoshi, K., Ebel, R., Diederich, M., Fiedler, H.-P., & Jaspars, M. (2010). Dermacozines, a new phenazine family from deep-sea dermacocci isolated from a Mariana Trench sediment. *Organic & Biomolecular Chemistry*, *8*(10), 2352–2362. doi:10.1039/c001445a PMID:20448892

Abdelnasser, S. M., Yahya, S. M., Mohamed, W. F., Asker, M. M., Shady, H. M. A., Mahmoud, M. G., & Gadallah, M. A. (2017). Antitumor exopolysaccharides derived from novel marine bacillus: Isolation, characterization aspect and biological activity. *Asian Pacific Journal of Cancer Prevention*, *18*(7), 1847. PMID:28749119

Abdel-Shafy, H. I., & Mansour, M. S. M. (2018). Solid waste issue: Sources, composition, disposal, recycling, and valorization. *Egyptian Journal of Petroleum*, *27*(4), 1275–1290. doi:10.1016/j.ejpe.2018.07.003

Abdullah, M. A., Usman, S. M., Shah, A. A., & El-Sayed, H. (2015). Algal biotechnology for bioenergy, environmental remediation and high-value biochemical. In D. Thangadurai & J. Sangeetha (Eds.), *Biotechnology and Bioinformatics* (pp. 301–344). CRC Press.

Abed, R. M. M., Dobretsov, S., Al-Fori, M., Gunasekera, S. P., Sudesh, K., & Paul, V. J. (2013). Quorum-sensing inhibitory compounds from extremophilic microorganisms isolated from a hypersaline cyanobacterial mat. *Journal of Industrial Microbiology & Biotechnology*, *40*(7), 759–772. doi:10.100710295-013-1276-4 PMID:23645384

Abe, F. (2004). Piezophysiology of yeast: Occurrence and significance. *Cellular and Molecular Biology*, *50*(4), 437–445. PMID:15529753

Abe, F., & Horikoshi, K. (2001). The biotechnological potential of piezophiles. *Trends in Biotechnology*, *19*(3), 102–108. doi:10.1016/S0167-7799(00)01539-0 PMID:11179803

Abomohra, A. F., El-Sheekh, M., & Hanelt, D. (2017). Screening of marine microalgae isolated from the hyper saline Bardawil lagoon for biodiesel feedstock. *Renewable Energy*, *101*(C), 1266–1272. doi:10.1016/j.renene.2016.10.015

Abraham, R. E., Alghazwi, M., Liang, Q., & Zhang, W. (2021). Advances on marine-derived natural radioprotection compounds: Historic development and future perspective. *Marine Life Science & Technology*, 1–14. doi:10.100742995-021-00095-x

Abu Baker, B., Tawabini, B., Nazal, M., AlThaqfi, J., & Khalil, A. (2020). Efficiency of Thermophilic Bacteria in Wastewater Treatment. *Arabian Journal for Science and Engineering*, *46*(1), 123–128. doi:10.100713369-020-04830-x

Acuña-Rodríguez, I. S., Hansen, H., Gallardo-Cerda, J., Atala, C., & Molina-Montenegro, M. A. (2019). Antarctic Extremophiles: Biotechnological Alternative to Crop Productivity in Saline Soils. *Frontiers in Bioengineering and Biotechnology, 7*. doi:10.3389/fbioe.2019.00022

Adalsteinsson, B. T., Kristjansdottir, T., Merre, W., Helleux, A., Dusaucy, J., Tourigny, M., Fridjonsson, O., & Hreggvidsson, G. O. (2021). Efficient genome editing of an extreme thermophile, *Thermus thermophilus,* using a thermostable Cas9 variant. *Scientific Reports*, *11*(1), 9586. doi:10.103841598-021-89029-2 PMID:33953310

Adams, H. E., Crump, B. C., & Kling, G. W. (2010). Temperature controls on aquatic bacterial production and community dynamics in arctic lakes and streams. *Environmental Microbiology*, *12*(5), 1319–1333. doi:10.1111/j.1462-2920.2010.02176.x PMID:20192972

Adapa, V., Ramya, L. N., Pulicherla, K. K., & Rao, K. S. (2014). Cold active pectinases: Advancing the food industry to the next generation. *Applied Biochemistry and Biotechnology*, *172*(5), 2324–2337. doi:10.100712010-013-0685-1 PMID:24390855

Adebayo-Tayo, B. C., & Popoola, A. O. (2017). Biogenic synthesis and antimicrobial activity of silver nanoparticle using exopolysaccharides from lactic acid bacteria. *International Journal of Nanodimension*, *8*(1), 61–69.

Adesioye, F. A., Makhalanyane, T. P., Vikram, S., Sewell, B. T., Schubert, W. D., & Cowan, D. A. (2018). Structural characterization and directed evolution of a novel acetyl xylan esterase reveal thermostability determinants of the carbohydrate esterase 7 family. *Applied and Environmental Microbiology*, *84*(8), e02695–e17. doi:10.1128/AEM.02695-17 PMID:29453256

Adiego-Perez, B., Randazzo, P., Daran, J. M., Verwaal, R., Roubos, J., Daran-Lapujade, P., & van der Oost, J. (2019). Multiplex genome editing of microorganisms using CRISPR-Cas. *FEMS Microbiology Letters*, *366*(8), fnz086. doi:10.1093/femsle/fnz086 PMID:31087001

Afinah, S., Yazid, A. M., Anis Shobirin, M. H., & Shuhaimi, M. (2010). Phytase: Application in food industry. *International Food Research Journal*, *17*(1), 13–21.

Afra, S., Makhdoumi, A., Matin, M. M., & Feizy, J. (2017). A novel red pigment from marine *Arthrobacter* sp. G20 with specific anticancer activity. *Journal of Applied Microbiology*, *123*(5), 1228–1236. doi:10.1111/jam.13576 PMID:28862783

Aghajari, N., van Petegem, F., Villeret, V., Chessa, J.-P., Gerday, C., Haser, R., & van Beeumen, J. (2003). Crystal structures of a psychrophilic metalloprotease reveal new insights into catalysis by cold-adapted proteases. *Proteins*, *50*(4), 636–647. doi:10.1002/prot.10264 PMID:12577270

Agnew, M. D., Koval, S. F., & Jarrell, K. F. (1995). Isolation and Characterization of Novel Alkaliphiles from Bauxite-Processing Waste and Description of *Bacillus vedderi* sp. nov., a New Obligate Alkaliphile. *Systematic and Applied Microbiology*, *18*(2), 221–230. doi:10.1016/S0723-2020(11)80392-9

Ahern, H. E., Walsh, K. A., Hill, T. C. J., & Moffett, B. F. (2007). Fluorescent pseudomonads isolated from Hebridean cloud and rain water produce biosurfactants but do not cause ice nucleation. *Biogeosciences*, *4*(1), 115–124. doi:10.5194/bg-4-115-2007

Ahmad, N. H., Mustafa, S., & Che Man, Y. B. (2015). Microbial polysaccharides and their modification approaches: A review. *International Journal of Food Properties*, *18*(2), 332–347. doi:10.1080/10942912.2012.693561

Aino, K., Hirota, K., Matsuno, T., Morita, N., Nodasaka, Y., Fujiwara, T., Matsuyama, H., Yoshimune, K., & Yumoto, I. (2008). *Bacillus polygoni* sp. nov., a moderately halophilic, non-motile obligate alkaliphile isolated from indigo balls. *International Journal of Systematic and Evolutionary Microbiology*, *58*(1), 120–124. doi:10.1099/ijs.0.65193-0 PMID:18175695

Akal, A. L., Karan, R., Hohl, A., Alam, I., Vogler, M., Grötzinger, S. W., Eppinger, J., & Rueping, M. (2019). A poly-extremophilic alcohol dehydrogenase from the Atlantis II Deep Red Sea brine pool. *FEBS Open Bio*, *9*(2), 194–205. doi:10.1002/2211-5463.12557 PMID:30761247

Akinola, S. A., & Babalola, O. O. (2020). The importance of adverse soil microbiomes in the light of omics: Implications for food safety. *Plant Soil and Environment, 66*(9), 421–430. doi:10.17221/118/2020-PSE

Aklujkar, M., Coppi, M. V., Leang, C., Kim, B. C., Chavan, M. A., Perpetua, L. A., Giloteaux, L., Liu, A., & Holmes, D. E. (2013). Proteins involved in electron transfer to Fe(III) and Mn(IV) oxides by *Geobacter sulfurreducens* and *Geobacter uraniireducens*. *Microbiology*, *159*(Pt_3), 515–535. doi:10.1099/mic.0.064089-0 PMID:23306674

Alam, S. I., Dube, S., Reddy, G. S. N., Bhattacharya, B. K., Shivaji, S., & Singh, L. (2005). Purification and characterisation of extracellular protease produced by *Clostridium* sp. from Schirmacher oasis, Antarctica. *Enzyme and Microbial Technology*, *36*(5-6), 824–831. doi:10.1016/j.enzmictec.2005.01.011

Alavi, S., Rafieyan, S., Yavari-Bafghi, M., & Amoozegar, M. A. (2020). Extremophiles: A Powerful Choice for Bioremediation of Toxic Oxyanions. *Microbial Bioremediation & Biodegradation*, 203–249. doi:10.1007/978-981-15-1812-6_9

Alfonta, L. (2010). Genetically Engineered Microbial Fuel Cells. *Electroanalysis*, *22*, 822–831.

Alford, I., Pilon-Smits, E. A. H., & Paschke, M. W. (2010). Metallophytesa view from the rhizosphere. *Plant and Soil, 337*(1-2), 33–50. doi:10.1007/s11104-010-0482-3

Alfreider, A., Pernthaler, J., Amann, R., Sattler, B., Glockner, F., Wille, A., & Psenner, R. (1996). Community analysis of the bacterial assemblages in the winter cover and pelagic layers of a high mountain lake by in situ hybridization. *Applied and Environmental Microbiology*, *62*(6), 2138–2144. doi:10.1128/aem.62.6.2138-2144.1996 PMID:16535341

Allan, R. N., Lebbe, L., Heyrman, J., De Vos, P., Buchanan, C. J., & Logan, N. A. (2005). Brevibacillus levickii sp. nov. and Aneurinibacillus terranovensis sp. nov., two novel thermoacidophiles isolated from geothermal soils of northern Victoria Land, Antarctica. *International Journal of Systematic and Evolutionary Microbiology*, *55*(3), 1039–1050. doi:10.1099/ijs.0.63397-0 PMID:15879231

Allison, D., Sutherland, I., & Neu, T. (2003). EPS: What's an acronym. *Biofilm communities: Order from chaos*, 381-387.

Allison, D. G., & Sutherland, I. W. (1987). The role of exopolysaccharides in adhesion of freshwater bacteria. *Microbiology*, *133*(5), 1319–1327. doi:10.1099/00221287-133-5-1319

Alsharif, W., Saad, M. M., & Hirt, H. (2020). Desert Microbes for Boosting Sustainable Agriculture in Extreme Environments. *Frontiers in Microbiology, 11*. doi:10.3389/fmicb.2020.01666

Alves, M. N., Fernandes, A. P., Salgueirob, C. A., & Paquete, C. M. (2015). Unraveling the electron transfer processes of a nanowire protein from Geobacter sulfurreducens. *Biochimica et Biophysica Acta, 1857*(1), 7–13. PubMed

Amadori, M., Tagliabue, S., Lauzi, S., Finazzi, G., Lombardi, G., Teló, P., Pacciarini, L., & Bonizzi, L. (2002). Diagnosis of Mycobacterium bovis infection in calves sensitized by mycobacteria of the avium/intracellulare group. *Journal of Veterinary Medicine. B, Infectious Diseases and Veterinary Public Health, 49*(2), 89–96. doi:10.1046/j.1439-0450.2002.00513.x PMID:12002425

Amasha, R. H. (2018). Use of NMR to determine compatible solutes in halophilic bacteria isolated from highly saline areas. *Applied Microbiology: Open Access, 4*. Advance online publication. doi:10.4172/2471-9315-C3-017

Amato, P., & Christner, B. C. (2009). Energy metabolism response to low-temperature and frozen conditions in Psychrobacter cryohalolentis. *Applied and Environmental Microbiology, 75*(3), 711–718. doi:10.1128/AEM.02193-08 PMID:19060163

Amato, P., Ménager, M., Sancelme, M., Laj, P., Mailhot, G., & Delort, A. M. (2005). Microbial population in cloud water at the Puy de Dôme: Implications for the chemistry of clouds. *Atmospheric Environment, 39*(22), 4143–4153. doi:10.1016/j.atmosenv.2005.04.002

Amato, P., Parazols, M., Sancelme, M., Mailhot, G., Laj, P., & Delort, A. M. (2007b). An important oceanic source of micro-organisms for cloud water at the Puy de Dôme (France). *Atmospheric Environment, 41*(37), 8253–8263. doi:10.1016/j.atmosenv.2007.06.022

Amend, J. P., & Shock, E. L. (2001). Energetics of overall metabolic reactions of thermophilic and hyperthermophilic Archaea and Bacteria. *FEMS Microbiology Reviews, 25*(2), 175–243. doi:10.1111/j.1574-6976.2001.tb00576.x PMID:11250035

Amjres, H., Béjar, V., Quesada, E., Carranza, D., Abrini, J., Sinquin, C., Ratiskol, J., Colliec-Jouaultm, S., & Llamas, I. (2015). Characterization of halo glycan, an exopolysaccharide produced by *Halomonas stenophila* HK30. *International Journal of Biological Macromolecules, 72*, 117–124. doi:10.1016/j.ijbiomac.2014.07.052 PMID:25109455

Amoozegar, M. A., Safarpour, A., Noghabi, K. A., Bakhtiary, T., & Ventosa, A. (2019). Halophiles and their vast potential in biofuel production. *Frontiers in Microbiology, 10*, 1–17. doi:10.3389/fmicb.2019.01895 PMID:31507545

Amoozegar, M. A., Salehghamari, E., Khajeh, K., Kabiri, M., & Naddaf, S. (2008). Production of an extracellular thermohalophilic lipase from a moderately halophilic bacterium, Salinivibrio sp. strain SA-2. *Journal of Basic Microbiology, 48*(3), 160–167. doi:10.1002/jobm.200700361 PMID:18506896

Amspacher, W. H., & Curreri, A. R. (1953). Use of dextran in control of shock resulting from war wounds. *A.M.A. Archives of Surgery, 66*(6), 730–740. doi:10.1001/archsurg.1953.01260030750004 PMID:13050239

Anders, C., Niewoehner, O., Duerst, A., & Jinek, M. (2014). Structural basis of PAM-dependent target DNA recognition by the Cas9 endonuclease. *Nature, 513*(7519), 569–573. doi:10.1038/nature13579 PMID:25079318

Anderson, A. W. (1956). Studies on a radio resistant *Micrococcus*. 1. Isolation, morphology, cultural characteristics, and resistance to γ-radiation. *Food Technology, 10*, 575–578.

Ando, N., Barquera, B., Bartlett, D. H., Boyd, E., Burnim, A. A., Byer, A. S., Colman, D., Gillilan, R. E., Gruebele, M., Makhatadze, G., Royer, C. A., Shock, E., Wand, A. J., & Watkins, M. B. (2021). The Molecular Basis for Life in Extreme Environments. *Annual Review of Biophysics, 50*(1), 343–372. doi:10.1146/annurev-biophys-100120-072804 PMID:33637008

Anesio, A. M., Mindl, B., Laybourn-Parry, J., Hodson, A. J., & Sattler, B. (2007). Viral dynamics in cryoconite holes on a high Arctic glacier (Svalbard). *Journal of Geophysical Research, 112*(G4), G04S31. Advance online publication. doi:10.1029/2006JG000350

Angenent, L. T., Karim, K., Al-Dahhan, M. H., Wrenn, B. A., & Domíguez-Espinosa, R. (2004). Production of bioenergy and biochemicals from industrial and agricultural wastewater. *Trends in Biotechnology, 22*(9), 477–485. doi:10.1016/j.tibtech.2004.07.001 PMID:15331229

Annamalai, N., Rajeswari, M. V., & Balasubramanian, T. (2016). Thermostable and alkaline cellulases from marine sources. In *New and future developments in microbial biotechnology and bioengineering* (pp. 91–98). Elsevier. doi:10.1016/B978-0-444-63507-5.00009-5

Antoni, D., Zverlov, V. V., & Schwarz, W. H. (2007). Biofuels from microbes. *Applied Microbiology and Biotechnology, 77*(1), 23–35. doi:10.100700253-007-1163-x PMID:17891391

Anwar, U. B., Zwar, I. P., & de Souza, A. O. (2020). Biomolecules produced by extremophiles microorganisms and recent discoveries. In *New and Future Developments in Microbial Biotechnology and Bioengineering* (pp. 247–270). Elsevier. doi:10.1016/B978-0-444-64301-8.00012-3

Aparicio, T., Jensen, S. I., Nielsen, A. T., de Lorenzo, V., & Martínez-García, E. (2016). The Ssr protein (T1E_1405) from *Pseudomonas putida* DOT-T1E enables oligonucleotide-based recombineering in platform strain *P. putida* EM42. *Biotechnology Journal, 11*(10), 1309–1319. doi:10.1002/biot.201600317 PMID:27367544

Ara, K., Igarashi, K., Saeki, K., Kawai, S., & Ito, S. (1992). Purification and some properties of an alkaline pullulanase from alkalophilic *Bacillus* sp. KSM-1876. *Bioscience, Biotechnology, and Biochemistry, 56*(1), 62–65. doi:10.1271/bbb.56.62

Arasu, M. V., Esmail, G. A., & Abdullah Al-Dhabi, N. (2016). Hyper saline actinomycetes and their biological applications. In D. Dhanasekaran & J. Yi (Eds.), *Actinobacteria - Basics and Biotechnological Applications.* IntechOpen. doi:10.5772/61065

Arayes, M. A., Mabrouk, M. E. M., Sabry, S. A., & Abdella, B. (2021). Diversity and characterization of culturable halo-alkaliphilic bacteria from two distinct hypersaline lakes in northern Egypt. *Biologia, 76*(2), 751–761. doi:10.247811756-020-00609-5

Archna, S., Priyank, V., Nath, Y. A., & Kumar, S. A. (2015). Bioprospecting for extracellular hydrolytic enzymes from culturable thermotolerant bacteria isolated from Manikaran thermal springs. *Research Journal of Biotechnology, 10*(4), 33–42.

Arena, A., Gugliandolo, C., Stassi, G., Pavone, B., Iannello, D., Bisignano, G., & Maugeri, T. L. (2009). An exopolysaccharide produced by *Geobacillus thermodenitrificans* strain B3-72: Antiviral activity on immunocompetent cells. *Immunology Letters, 123*(2), 132–137. doi:10.1016/j.imlet.2009.03.001 PMID:19428560

Arnaouteli, S., Ferreira, A. S., Schor, M., Morris, R. J., Bromley, K. M., Jo, J., ... Dietrich, L. E. (2017). Bifunctionality of a biofilm matrix protein controlled by redox state. *Proceedings of the National Academy of Sciences of the United States of America, 114*(30), E6184–E6191. doi:10.1073/pnas.1707687114 PMID:28698374

Arras, W., Hussain, A., Hausler, R., & Guiot, S. R. (2019). Mesophilic, thermophilic and hyperthermophilic acidogenic fermentation of food waste in batch: Effect of inoculum source. *Waste Management (New York, N.Y.), 87*, 279–287. doi:10.1016/j.wasman.2019.02.011 PMID:31109527

Arsiccio, A., Ganguly, P., la Cortiglia, L., Shea, J.-E., & Pisano, R. (2020). ADD Force Field for Sugars and Polyols: Predicting the Additivity of Protein–Osmolyte Interaction. *The Journal of Physical Chemistry B, 124*(36), 7779–7790. doi:10.1021/acs.jpcb.0c05345 PMID:32790371

Arti, D., Park, J., Jung, T. Y., Song, H., Jang, M., & Han, N. S. (2012). Structural analysis of α-L-arabinofuranosidase from *Thermotoga maritima* reveals characteristics for thermostability and substrate specificity. *Journal of Microbiology and Biotechnology*, *22*(12), 1724–1730. doi:10.4014/jmb.1208.08043 PMID:23221536

Arulazhagan, P., Mnif, S., Rajesh Banu, J. R., Huda, Q., & Jalal, M. A. B. (2017). HC-0B-01: Biodegradation of hydrocarbons by extremophiles. In *Environmental Footprints and Eco-Design of Products and Processes* (pp. 137–162). Springer.

Arun, J., Selvakumar, S., Sathishkumar, R., Moovendhan, M., Ananthan, G., Maruthiah, T., & Palavesam, A. (2017). *In-vitro* antioxidant activities of an exopolysaccharide from a salt pan bacterium *Halolactibacillus miurensis. Carbohydrate Polymers*, *155*, 400–406. doi:10.1016/j.carbpol.2016.08.085 PMID:27702528

Arya, P., Trivedi, M., Ramani, N., Patel, K., & Rajput, K. N. (2020a). Isolation, Production and Applications of Alkaline Protease from Hot Springs Bacterial Isolates. *Bioscience Biotechnology Research Communications.*

Asao, M., Pinkart, H. C., & Madigan, M. T. (2011). Diversity of extremophilic purple phototrophic bacteria in Soap Lake, a Central Washington (USA) Soda Lake. *Environmental Microbiology*, *13*(8), 2146–2157. doi:10.1111/j.1462-2920.2011.02449.x PMID:21410624

Ashwini, R., Vijayanand, S., & Hemapriya, J. (2017). Photonic potential of Haloarchaeal pigment bacteriorhodopsin for future electronics. *Current Microbiology*, *74*(8), 996–1002. doi:10.100700284-017-1271-5 PMID:28573340

Asker, D., & Ohta, Y. (2002). Production of canthaxanthin by Haloferax alexandrinus under non-aseptic conditions and a simple, rapid method for its extraction. *Applied Microbiology and Biotechnology*, *58*(6), 743–750. doi:10.100700253-002-0967-y PMID:12021793

Asoodeh, A., Chamani, J., & Lagzian, M. (2010). A novel thermostable, acidophilic α-amylase from a new thermophilic "*Bacillus* sp. *Ferdowsicous*" isolated from Ferdows hot mineral spring in Iran: Purification and biochemical characterization. *International Journal of Biological Macromolecules*, *46*(3), 289–297. doi:10.1016/j.ijbiomac.2010.01.013 PMID:20109486

Atalah, J., Cáceres-Moreno, P., Espina, G., & Blamey, J. M. (2019). Thermophiles and the applications of their enzymes as new biocatalysts. *Bioresource Technology*, *280*, 478–488. doi:10.1016/j.biortech.2019.02.008 PMID:30826176

Atalo, K., & Gashe, B. A. (1993). Protease production by a thermophilic *Bacillus* species (P-001A) which degrades various kinds of fibrous proteins. *Biotechnology Letters*, *15*(11), 1151–1156. doi:10.1007/BF00131207

Atkinson, D. E. (1968). Energy charge of the adenylate pool as a regulatory parameter. Interaction with feedback modifiers. *Biochemistry*, *7*(11), 4030–4034. doi:10.1021/bi00851a033 PMID:4972613

Auernik, K. S., Maezato, Y., Blum, P. H., & Kelly, R. M. (2008). The genome sequence of the metal-mobilizing, extremely thermophilic archaeon Metallosphaera sedula provides insight into bioleaching-associated metabolism. *Applied and Environmental Microbiology*, *74*(3), 682–692. doi:10.1128/AEM.02019-07 PMID:18083856

Aulitto, M., Fusco, S., Limauro, D., Fiorentino, G., Bartolucci, S., & Contursi, P. (2019). Galactomannan degradation by thermophilic enzymes: A hot topic for biotechnological applications. *World Journal of Microbiology & Biotechnology*, *35*(2), 32. doi:10.100711274-019-2591-3 PMID:30701316

Auman, A. J., Breezee, J. L., Gosink, J. J., Kämpfer, P., & Staley, J. T. (2006). Psychromonas ingrahamii sp. nov., a novel gas vacuolate, psychrophilic bacterium isolated from Arctic polar sea ice. *International Journal of Systematic and Evolutionary Microbiology*, *56*(Pt 5), 1001–1007. doi:10.1099/ijs.0.64068-0 PMID:16627645

Autengruber, A., Sydlik, U., Kroker, M., Hornstein, T., Ale-Agha, N., Stöckmann, D., Bilstein, A., Albrecht, C., Paunel-Gorgulu, A., Suschek, C. V., Krutmann, J., & Unfried, K. (2014). Signalling-dependent adverse health effects of carbon nanoparticles are prevented by the compatible solute mannosylglycerate (firoin) *in-vitro* and *in-vivo*. *PLoS One*, *9*(11), e111485. doi:10.1371/journal.pone.0111485 PMID:25415441

Avsar, G., Agirbasli, D., Agirbasli, M. A., Gunduz, O., & Oner, E. T. (2018). Levan based fibrous scaffolds electrospun via co-axial and single-needle techniques for tissue engineering applications. *Carbohydrate Polymers*, *193*, 316–325. doi:10.1016/j.carbpol.2018.03.075 PMID:29773387

Awasthi, M. K., Wong, J. W. C., Kumar, S., Awasthi, S. K., Wang, Q., Wang, M., Ren, X., Zhao, J., Chen, H., & Zhang, Z. (2018). Biodegradation of food waste using microbial cultures producing thermostable A-amylase and cellulase under different pH and temperature. *Bioresource Technology*, *248*, 160–170. doi:10.1016/j.biortech.2017.06.160 PMID:28709882

Azubuike, C. C., Chikere, C. B., & Okpokwasili, G. C. (2016). Bioremediation techniques classification based on site of application: Principles, advantages, limitations and prospects. *World Journal of Microbiology & Biotechnology*, *32*(11), 180. doi:10.100711274-016-2137-x PMID:27638318

Babu, P., Chandel, A. K., & Singh, O. V. (2015). Therapeutic Implications of Extremophiles. In: Extremophiles and Their Applications in Medical Processes. Springer. doi:10.1007/978-3-319-12808-5_3

Badel, S., Bernardi, T., & Michaud, P. (2011). New perspectives for *Lactobacilli* exopolysaccharides. *Biotechnology Advances*, *29*(1), 54–66. doi:10.1016/j.biotechadv.2010.08.011 PMID:20807563

Bae, H. S., Moe, W. M., Yan, J., Tiago, I., da Costa, M. S., & Rainey, F. A. (2006). Propionicicella superfundia gen. nov., sp. nov, a color solvent-tolerant propionate forming, facultative anaerobic bacterium isolated from contaminated groundwater. *Systematic and Applied Microbiology*, *29*(5), 404–413. doi:10.1016/j.syapm.2005.11.004 PMID:16338112

Bagshaw, E. A., Tranter, M., Fountain, A. G., Welch, K. A., Basagic, H., & Lyons, W. B. (2007). Biogeochemical evolution of cryoconite holes on Canada glacier, Taylor Valley, Antarctica. *Journal of Geophysical Research. Biogeosciences*, *112*(G4).

Baharaeen, S., & Vishniac, H. S. (1982). Cryptococcus lupi sp. nov., an Antarctic basidioblastomycete. *International Journal of Systematic and Evolutionary Microbiology*, *32*(2), 229-232.

Bai, W., Cao, Y., Liu, J., Wang, Q., & Jia, Z. (2016). Improvement of alkalophilicity of an alkaline xylanase Xyn11A-LC from *Bacillus* sp. SN5 by random mutation and Glu135 saturation mutagenesis. *BMC Biotechnology*, *16*(1), 77. doi:10.118612896-016-0310-9 PMID:27825339

Bai, Y., Huang, H., Meng, K., Shi, P., Yang, P., Luo, H., Luo, C., Feng, Y., Zhang, W., & Yao, B. (2012). Identification of an acidic α-amylase from Alicyclobacillus sp. A4 and assessment of its application in the starch industry. *Food Chemistry*, *131*(4), 1473–1478. doi:10.1016/j.foodchem.2011.10.036

Baker-Austin, C., Dopson, M., Wexler, M., Sawers, R. G., & Bond, P. L. (2005). Molecular insight into extreme copper resistance in the extremophilic archaeon 'Ferroplasma acidarmanus' Fer1. *Microbiology*, *151*(8), 2637–2646. doi:10.1099/mic.0.28076-0 PMID:16079342

Bakermans, C., Tollaksen, S. L., Giometti, C. S., Wilkerson, C., Tiedje, J. M., & Thomashow, M. F. (2007). Proteomic analysis of Psychrobacter cryohalolentis K5 during growth at subzero temperatures. *Extremophiles*, *11*(2), 343–354. doi:10.100700792-006-0042-1 PMID:17123128

Baker, R. T. (2001). Canthaxanthin in aquafeed applications: Is there any risk? *Trends in Food Science & Technology*, *12*(7), 240–243. doi:10.1016/S0924-2244(01)00091-7

Banciu, H.L., Sorokin, D.Y., Tourova, T.P., Galinski, E.A., Muntyan, M.S., Kuenen, J.G., & Muyzer, G. (2008). Influence of salts and pH on growth and activity of a novel facultatively alkaliphilic, extremely salt-tolerant, obligately chemolithoautotrophic sufur-oxidizing Gammaproteobacterium *Thioalkalibacter halophilus* gen. nov., sp. nov. from South-Western Sibe. *Extremophiles: Life Under Extreme Conditions, 12*, 391–404. doi:10.1007/s00792-008-0142-1

Banciu, H., Sorokin, D. Y., Galinski, E. A., Muyzer, G., Kleerebezem, R., & Kuenen, J. G. (2004). Thialkalivibrio halophilus sp. nov., a novel obligately chemolithoautotrophic, facultatively alkaliphilic, and extremely salt-tolerant, sulfur-oxidizing bacterium from a hypersaline alkaline lake. *Extremophiles, 8*(4), 325–334. doi:10.100700792-004-0391-6 PMID:15309564

Banerjee, A., & Bandopadhyay, R. (2016). Use of dextran nanoparticle: A paradigm shift in bacterial exopolysaccharide based biomedical applications. *International Journal of Biological Macromolecules, 87*, 295–301. doi:10.1016/j.ijbiomac.2016.02.059 PMID:26927936

Banerjee, A., Sharma, R., Chisti, Y., & Banerjee, U. C. (2002). Botryococcus braunii: A renewable source of hydrocarbons and other chemicals. *Critical Reviews in Biotechnology, 22*(3), 245–279. doi:10.1080/07388550290789513 PMID:12405558

Bankura, K., Maity, D., Mollick, M. M., Mondal, D., Bhowmick, B., Bain, M., Chakraborty, A., Sarkar, J., Acharya, K., & Chattopadhyay, D. (2012). Synthesis, characterization and antimicrobial activity of dextran stabilized silver nanoparticles in aqueous medium. *Carbohydrate Polymers, 89*(4), 1159–1165. doi:10.1016/j.carbpol.2012.03.089 PMID:24750927

Ban, S., & Trites, A. W. (2007). Quantification of terrestrial haul-out and rookery characteristics of Steller sea lions. *Marine Mammal Science, 23*(3), 496–507. doi:10.1111/j.1748-7692.2007.00130.x

Baria, D. M., Patel, N. Y., Arya, P. S., Rajput, K. N., Panchal, R. R., & Raval, V. H. (2020). Exploration of marine bacteria for their Exopolysaccharide (EPS) producing ability along various regions of the Gujarat coast. *Bioscience Biotechnology Research Communications, 13*(1), 78-84. http://bbrc.in/bbrc/wp-content/uploads/2020/05/FINAL-PDF-Special-Issue-Vol-13-No-1-2020.pdf

Barkay, T., Miller, S. M., & Summers, A. O. (2003). Bacterial mercury resistance from atoms to ecosystems. *FEMS Microbiology Reviews, 27*(2–3), 355–384. doi:10.1016/S0168-6445(03)00046-9 PMID:12829275

Barnard, D., Casanueva, A., Tuffin, M., & Cowan, D. (2010). Extremophiles in biofuel synthesis. *Environmental Technology, 31*(8–9), 871–888. doi:10.1080/09593331003710236 PMID:20662378

Barrangou, R., Fremaux, C., Deveau, H., Richards, M., Boyaval, P., Moineau, S., Romero, D., & Philippe Horvath, P. (2007). CRISPR provides acquired resistance against viruses in prokaryotes. *Science, 315*(5819), 1709–1712. doi:10.1126cience.1138140 PMID:17379808

Bartley, M. L., Boeing, W. J., Corcoran, A. A., Holguin, F. O., & Schaub, T. (2013). Effects of salinity on growth and lipid accumulation of biofuel microalga *Nannochloropsis salina* and invading organisms. *Biomass and Bioenergy, 54*, 83–88. doi:10.1016/j.biombioe.2013.03.026

Barzkar, N., Homaei, A., Hemmati, R., & Patel, S. (2018). Thermostable marine microbial proteases for industrial applications: Scopes and risks. *Extremophiles, 22*(3), 335–346. doi:10.100700792-018-1009-8 PMID:29442247

Bass-Becking, L. M. (1931). Historical notes on salt and salt-manufacture. *The Scientific Monthly, 32*(5), 434–446.

Bauer, H., Kasper-Giebl, A., Löflund, M., Giebl, H., Hitzenberger, R., Zibuschka, F., & Puxbaum, H. (2002). The contribution of bacteria and fungal spores to the organic carbon content of cloud water, precipitation and aerosols. *Atmospheric Research, 64*(1-4), 109–119. doi:10.1016/S0169-8095(02)00084-4

Becker, A., Katzen, F., Pühler, A., & Ielpi, L. (1998). Xanthan gum biosynthesis and application: A biochemical/genetic perspective. *Applied Microbiology and Biotechnology*, *50*(2), 145–152. doi:10.1007002530051269 PMID:9763683

Beeby, M., O'Connor, B. D., Ryttersgaard, C., Boutz, D. R., Perry, L. J., & Yeates, T. O. (2005). The genomics of disulfide bonding and protein stabilization in thermophiles. *PLoS Biology*, *3*(9), 1549–1558. doi:10.1371/journal.pbio.0030309 PMID:16111437

Behzadi, S., & Farid, M. M. (2007). Review: Examining the use of different feedstock for the production of biodiesel. *Asia-Pacific Journal of Chemical Engineering*, *2*(5), 480–486. doi:10.1002/apj.85

Bell, E. (2012). *Life at extremes: Environments, organisms, and strategies for survival. CABI Invasive Species 1*. CABI. doi:10.1079/9781845938147.0000

Bennetto, H. P., Stirling, J. L., Tanaka, K., & Vega, C. A. (1983). Anodic reactions in microbial fuel cells. *Biotechnology and Bioengineering*, *25*(2), 559–568. doi:10.1002/bit.260250219 PMID:18548670

Benson, C. A., Bizzoco, R. W., Lipson, D. A., & Kelley, S. T. (2011). Microbial diversity in nonsulfur, sulfur and iron geothermal steam vents. *FEMS Microbiology Ecology*, *76*(1), 74–88. doi:10.1111/j.1574-6941.2011.01047.x PMID:21223339

Berben, T., Overmars, L., Sorokin, D. Y., & Muyzer, G. (2017). Comparative Genome Analysis of Three Thiocyanate Oxidizing Thioalkalivibrio Species Isolated from Soda Lakes. *Frontiers in Microbiology*, *0*(FEB), 254. doi:10.3389/fmicb.2017.00254 PMID:28293216

Berezina, O. V., Brandt, A., Yarotsky, S., Schwarz, W. H., & Zverlov, V. V. (2009). Isolation of a new butanol-producing Clostridium strain: High level of hemicellulosic activity and structure of solventogenesis genes of a new Clostridium saccharobutylicum isolate. *Systematic and Applied Microbiology*, *32*(7), 449–459. doi:10.1016/j.syapm.2009.07.005 PMID:19674858

Berezovsky, I. N., & Shakhnovich, E. I. (2005). Physics and evolution of thermophilic adaptation. *Proceedings of the National Academy of Sciences of the United States of America*, *102*(36), 12742–12747. doi:10.1073/pnas.0503890102 PMID:16120678

Berg, K., Leiros, I., & Williamson, A. (2019). Temperature adaptation of DNA ligases from psychrophilic organisms. *Extremophiles, 23*(3), 305–317. doi:10.1007/s00792-019-01082-y

Berger, F., Morellet, N., Menu, F., & Potier, P. (1996). Cold shock and cold acclimation proteins in the psychrotrophic bacterium Arthrobacter globiformis SI55. *Journal of Bacteriology*, *178*(11), 2999–3007. doi:10.1128/jb.178.11.2999-3007.1996 PMID:8655472

Bernhart Owen, R., Renaut, R. W., Muiruri, V. M., Rabideaux, N. M., Lowenstein, T. K., McNulty, E. P., Leet, K., Deocampo, D., Luo, S., Deino, A. L., Cohen, A., Sier, M. J., Campisano, C., Shen, C. C., Billingsley, A., Mbuthia, A., & Stockhecke, M. (2019). Quaternary history of the Lake Magadi Basin, southern Kenya Rift: Tectonic and climatic controls. *Palaeogeography, Palaeoclimatology, Palaeoecology*, *518*, 97–118. doi:10.1016/j.palaeo.2019.01.017

Bertoldo, C., & Antranikian, G. (2006). The order Thermococcales. In M. Dworkin, S. Falkow, E. Rosenberg, K. Schleifer, & E. Stackebrandt (Eds.), *The Prokaryotes* (pp. 69–81). Springer. doi:10.1007/0-387-30743-5_5

Bertoldo, C., Dock, C., & Antranikian, G. (2004). Thermoacidophilic microorganisms and their novel biocatalysts. *Engineering in Life Sciences*, *4*(6), 521–532. doi:10.1002/elsc.200402155

Besse, A., Peduzzi, J., Rebuffat, S., & Carre-Mlouka, A. (2015). Antimicrobial peptides and proteins in the face of extremes: Lessons from archaeocins. *Biochimie*, *118*, 344–355. doi:10.1016/j.biochi.2015.06.004 PMID:26092421

Besse, A., Vandervennet, M., Goulard, C., Peduzzi, J., Isaac, S., Rebuffat, S., & Carré-Mlouka, A. (2017). Halocin C8: an antimicrobial peptide distributed among four halophilic archaeal genera: *Natrinema, Haloterrigena, Haloferax*, and *Halobacterium. Extremophiles, 21*(3), 623–638. doi:10.100700792-017-0931-5 PMID:28386790

Bezsudnova, E. Y., Boyko, K. M., Polyakov, K. M., Dorovatovskiy, P. V., Stekhanova, T. N., Gumerov, V. M., Ravin, N. V., Skryabin, K. G., Kovalchuk, M. V., & Popov, V. O. (2012). Structural insight into the molecular basis of poly-extremophilicity of short-chain alcohol dehydrogenase from the hyperthermophilic archaeon *Thermococcus sibiricus. Biochimie, 94*(12), 2628–2638. doi:10.1016/j.biochi.2012.07.024 PMID:22885278

Bhalla, A., Bansal, N., Kumar, S., Bischoff, K. M., & Sani, R. K. (2013). Improved lignocellulose conversion to biofuels with thermophilic bacteria and thermostable enzymes. *Bioresource Technology, 128*, 751–759. doi:10.1016/j.biortech.2012.10.145 PMID:23246299

Bhattacharya, A., Goyal, N., & Gupta, A. (2017). Degradation of azo dye methyl red by alkaliphilic, halotolerant *Nesterenkonia lacusekhoensis* EMLA3: Application in alkaline and salt-rich dyeing effluent treatment. *Extremophiles, 21*(3), 479–490. doi:10.100700792-017-0918-2 PMID:28255636

Bhattacharyya, A., Saha, J., Haldar, S., Bhowmic, A., Mukhopadhyay, U. K., & Mukherjee, J. (2014). Production of poly-3- (hydroxybutyrate-co-hydroxy valerate) by *Haloferax mediterranei* using rice-based ethanol stillage with simultaneous recovery and re-use of medium salts. *Extremophiles, 18*(2), 463–470. doi:10.100700792-013-0622-9 PMID:24442255

Bhatt, H. B., Begum, M. A., Chintalapati, S., Chintalapati, V. R., & Singh, S. P. (2017). Desertibacillus haloalkaliphilus gen. nov., sp. nov., isolated from a saline desert. *International Journal of Systematic and Evolutionary Microbiology, 67*(11), 4435–4442. doi:10.1099/ijsem.0.002310 PMID:28920841

Bhaya, D., Davison, M., & Barrangou, R. (2011). CRISPR-Cas systems in bacteria and archaea: Versatile small RNAs for adaptive defense and regulation. *Annual Review of Genetics, 45*(1), 273–297. doi:10.1146/annurev-genet-110410-132430 PMID:22060043

Bhosale, H., Shaheen, U., & Kadam, T. (2016). Characterization of a Hyperthermostable Alkaline Lipase from *Bacillus sonorensis* 4R. *Enzyme Research, 2016*, 1–11. Advance online publication. doi:10.1155/2016/4170684 PMID:26904276

Bhosale, P. (2004). Environmental and cultural stimulants in the production of carotenoids from microorganisms. *Applied Microbiology and Biotechnology, 63*(4), 351–361. doi:10.100700253-003-1441-1 PMID:14566431

Bhunia, B., Uday, U. S. P., Oinam, G., Mondal, A., Bandyopadhyay, T. K., & Tiwari, O. N. (2018). Characterization, genetic regulation and production of cyanobacterial exopolysaccharides and its applicability for heavy metal removal. *Carbohydrate Polymers, 179*, 228–243. doi:10.1016/j.carbpol.2017.09.091 PMID:29111047

Bikard, D., Euler, C. W., Jiang, W., Nussenzweig, P. M., Goldberg, G. W., Duportet, X., Fischetti, V. A., & Marraffini, L. A. (2014). Exploiting CRISPR-Cas nucleases to produce sequence-specific antimicrobials. *Nature Biotechnology, 32*(11), 1146–1150. doi:10.1038/nbt.3043 PMID:25282355

Billi, D., Friedmann, E. I., Hofer, K. G., Caiola, M. G., & Ocampo-Friedmann, R. (2000). Ionizing-radiation resistance in the desiccation-tolerant cyanobacterium *Chroococcidiopsis. Applied and Environmental Microbiology, 66*(4), 1489–1492. doi:10.1128/AEM.66.4.1489-1492.2000 PMID:10742231

Binod, P., Sindhu, R., Madhavan, A., Abraham, A., Mathew, A. K., Beevi, U. S., Sukumaran, R. K., Singh, S. P., & Pandey, A. (2017). Recent developments in L-glutaminase production and applications–An overview. *Bioresource Technology, 245*, 1766–1774. doi:10.1016/j.biortech.2017.05.059 PMID:28549811

BioRender. (n.d.). *Reprinted from Antibiotic resistance mechanisms.* https://app.biorender.com/illustrations/60e56625 46850f00acfe0e3a

Bocchini, D. A., Damiano, V. B., Gomes, E., & Da Silva, R. (2003). Effect of *Bacillus circulans* D1 thermostable xylanase on biobleaching of eucalyptus kraft pulp. *Applied Biochemistry and Biotechnology, 105–108*(1-3), 393–402. doi:10.1385/ABAB:106:1-3:393 PMID:12721462

Bogino, P., Abod, A., Nievas, F., & Giordano, W. (2013). Water-limiting conditions alter the structure and biofilm-forming ability of bacterial multispecies communities in the alfalfa rhizosphere. *PLoS One, 8*(11), e79614. doi:10.1371/journal.pone.0079614 PMID:24223979

Bokhari, A., Essack, M., Lafi, F. F., Andres-Barrao, C., Jalal, R., Alamoudi, S., & Razali, R. (2019). Bioprospecting desert plant Bacillus endophytic strains for their potential to enhance plant stress tolerance. *Scientific Reports, 9*(1). doi:10.1038/s41598-019-54685-y

Bolotin, A., Quinquis, B., Sorokin, A., & Ehrlich, S. D. (2005). Clustered regularly interspaced short palindrome repeats (CRISPRs) have spacers of extrachromosomal origin. *Microbiology (Reading), 151*(Pt 8), 2551–2561. doi:10.1099/mic.0.28048-0 PMID:16079334

Bond, D. R., & Lovley, D. R. (2003). Electricity production by *Geobacter sulfurreducens* attached to electrodes. *Applied and Environmental Microbiology, 69*(3), 1548–1555. doi:10.1128/AEM.69.3.1548-1555.2003 PMID:12620842

Bonilla, S., Villeneuve, V., & Vincent, W. F. (2005). Benthic And Planktonic Algal Communities In A High Arctic Lake: Pigment Structure And Contrasting Responses To Nutrient Enrichment 1. *Journal of Phycology, 41*(6), 1120–1130. doi:10.1111/j.1529-8817.2005.00154.x

Borges, N., Ramos, A., Raven, N. D., Sharp, R. J., & Santos, H. (2002). Comparative study of the thermostabilizing properties of mannosylglycerate and other compatible solutes on model enzymes. *Extremophiles, 6*(3), 209–216. doi:10.1007/s007920100236

Borjian, F., Han, J., Hou, J., Xiang, H., & Berg, I. A. (2016). The methyl aspartate cycle in *Haloarchaea* and its possible role in carbon metabolism. *The ISME Journal, 10*(3), 546–557. doi:10.1038/ismej.2015.132 PMID:26241502

Borole, A. P., O'Neill, H., Tsouris, C., & Cesar, S. (2008). A microbial fuel cell operating at low pH using the acidophile *Acidiphilium cryptum. Biotechnology Letters, 30*(8), 1367–1372. doi:10.100710529-008-9700-y PMID:18368296

Bosley, K. S., Botchan, M., Bredenoord, A. L., Carroll, D., Charo, R. A., Charpentier, E., Cohen, R., Corn, J., Doudna, J., Feng, G., Greely, H. T., Isasi, R., Ji, W., Kim, J. S., Knoppers, B., Lanphier, E., Li, J., Lovell-Badge, R., Martin, G. S., ... Zhou, Q. (2015). CRISPR germline engineering—The community speaks. *Nature Biotechnology, 33*(5), 478–486. doi:10.1038/nbt.3227 PMID:25965754

Bosma, E. F., van de Weijer, A. H. P., Daas, M. J. A., van der Oost, J., de Vos, W. M., & van Kranenburg, R. (2015). Isolation and screening of thermophilic *Bacilli* from compost for electrotransformation and fermentation: Characterization of *Bacillus smithii* ET 138 as a new biocatalyst. *Applied and Environmental Microbiology, 81*(5), 1874–1883. doi:10.1128/AEM.03640-14 PMID:25556192

Boucher, N., & Noll, K. M. (2011). Ligands of thermophilic ABC transporters encoded in a newly sequenced genomic region of Thermotoga maritima MSB8 screened by differential scanning fluorimetry. *Applied and Environmental Microbiology, 77*(18), 6395–6399. doi:10.1128/AEM.05418-11 PMID:21764944

Bowman, J. P. (2008). *Genomic analysis of psychrophilic prokaryotes.* In R. Margesin, F. Schinner, J. C. Marx, & C. Gerday (Eds.), *Psychrophiles: from biodiversity to biotechnology* (pp. 265–284). Springer. doi:10.1007/978-3-540-74335-4_16

Bowman, J. P., Cavanagh, J., Austin, J. J., & Sanderson, K. (1996). Novel Psychrobacter species from Antarctic ornithogenic soils. *International Journal of Systematic and Evolutionary Microbiology, 46*(4), 841–848. PMID:8863407

Bowman, J. P., & Nichols, D. S. (2002). Aequorivita gen. nov., a member of the family Flavobacteriaceae isolated from terrestrial and marine Antarctic habitats. *International Journal of Systematic and Evolutionary Microbiology, 52*(5), 1533–1541. PMID:12361255

Bowman, J. S., & Deming, J. W. (2010). Elevated bacterial abundance and exopolymers in saline frost flowers and implications for atmospheric chemistry and microbial dispersal. *Geophysical Research Letters, 37*(13), n/a. doi:10.1029/2010GL043020

Boyer, E. W., & Ingle, M. B. (1972). Extracellular alkaline amylase from a *Bacillus* species. *Journal of Bacteriology, 110*(3), 992–1000. doi:10.1128/jb.110.3.992-1000.1972 PMID:4624112

Bozal, N., Montes, M. J., Tudela, E., Jiménez, F., & Guinea, J. (2002). Shewanella frigidimarina and Shewanella livingstonensis sp. nov. isolated from Antarctic coastal areas. *International Journal of Systematic and Evolutionary Microbiology, 52*(1), 195–205. doi:10.1099/00207713-52-1-195 PMID:11837303

Bozal, N., Tudela, E., Rossello-Mora, R., Lalucat, J., & Guinea, J. (1997). Pseudoalteromonas antarctica sp. nov., isolated from an Antarctic coastal environment. *International Journal of Systematic and Evolutionary Microbiology, 47*(2), 345–351. PMID:9103620

Branda, S. S., Vik, S., Friedman, L., & Kolter, R. (2005). Biofilms: The matrix revisited. *Trends in Microbiology, 13*(1), 20–26. doi:10.1016/j.tim.2004.11.006 PMID:15639628

Bräsen, C., Urbanke, C., & Schönheit, P. (2005). A novel octameric AMP-forming acetyl-CoA synthetase from the hyperthermophilic crenarchaeon Pyrobaculum aerophilum. *FEBS Letters, 579*(2), 477–482. doi:10.1016/j.febslet.2004.12.016 PMID:15642362

Brenner, K., You, L., & Arnold, F. H. (2008). Engineering microbial consortia: A new frontier in synthetic biology. *Trends in Biotechnology, 26*(9), 483–489. doi:10.1016/j.tibtech.2008.05.004 PMID:18675483

Brim, H., McFarlan, S. C., Fredrickson, J. K., Minton, K. W., Zhai, M., Wackett, L. P., & Daly, M. J. (2000). Engineering Deinococcus radiodurans for metal remediation in radioactive mixed waste environments. *Nature Biotechnology, 18*(1), 85–90. doi:10.1038/71986 PMID:10625398

Brim, H., Osborne, J. P., Kostandarithes, H. M., Fredrickson, J. K., Wackett, L. P., & Daly, M. J. (2006). *Deinococcus radiodurans* engineered for complete toluene degradation facilitates Cr (VI) reduction. *Microbiology, 152*(8), 2469–2477. doi:10.1099/mic.0.29009-0 PMID:16849809

Brininger, C., Spradlin, S., Cobani, L., & Evilia, C. (2018). The more adaptive to change, the more likely you are to survive: Protein adaptation in extremophiles. *Seminars in Cell & Developmental Biology, 84*, 158–169. doi:10.1016/j.semcdb.2017.12.016 PMID:29288800

Brittanica. (2014). *Chott El-Jarid | lake, Tunisia.* Author.

Britton, K. L., Baker, P. J., Fisher, M., Ruzheinikov, S., Gilmour, D. J., Bonete, M.-J., Ferrer, J., Pire, C., Esclapez, J., & Rice, D. W. (2006). Analysis of protein solvent interactions in glucose dehydrogenase from the extreme halophile <em>Haloferax mediterranei</em>. *Proceedings of the National Academy of Sciences of the United States of America, 103*(13), 4846–4851. doi:10.1073/pnas.0508854103 PMID:16551747

Broady, P. A. (1981b). The ecology of sublithic terrestrial algae at the Vestfold Hills, Antarctica. *British Phycological Journal, 16*(3), 231–240. doi:10.1080/00071618100650241

Brock, T. D. (1998). Early days in Yellowstone microbiology. *ASM News, 64*, 137–140.

Brock, T. D., & Freeze, H. (1969). *Thermus aquaticus* gen. n. and sp. n., a nonsporulating extreme thermophile. *Journal of Bacteriology, 98*(1), 289–297. doi:10.1128/jb.98.1.289-297.1969 PMID:5781580

Bronwyn, Barnard, Tuffin, & Cowan. (2011). Ecological Distributopn of Microorgnaisms in Terrestrial, Psychrophilic Habitats. In Extremophiles Handbook. Springer. DOI doi:10.1007/978-4-431-53898-1_6.5

Brouns, S. J., Jore, M. M., Lundgren, M., Westra, E. R., Slijkhuis, R. J., Snijders, A. P., Dickman, M. J., Makarova, K. S., Koonin, E. V., & van der Oost, J. (2008). Small CRISPR RNAs guide antiviral defense in prokaryotes. *Science*, *321*(5891), 960–964. doi:10.1126cience.1159689 PMID:18703739

Bruins, M. E., Janssen, A. E. M., & Boom, R. M. (2001). Thermozymes and their applications. *Applied Biochemistry and Biotechnology*, *90*(2), 155–186. doi:10.1385/ABAB:90:2:155

Brunner, N. A., Siebers, B., & Hensel, R. (2001). Role of two different glyceraldehyde-3-phosphate dehydrogenases in controlling the reversible Embden-Meyerhof-Parnas pathway in *Thermoproteus tenax*: Regulation on protein and transcript level. *Extremophiles*, *5*(2), 101–109. doi:10.1007007920100181 PMID:11354453

Buchner, J. (1996). Supervising the fold: Functional principles of molecular chaperones. *Federation of American Societies for Experimental Biology Journal*, *10*(1), 10–19. doi:10.1096/fasebj.10.1.8566529 PMID:8566529

Bulat, S. A., Alekhina, I. A., Blot, M., Petit, J. R., Vasileva, L. P., de Angelis, M., Wagenbach, D., Lipenkov, V. Y., Vasilyeva, L. P., Wloch, D., Raynaud, D., & Lukin, V. V. (2004). DNA signature of thermophilic bacteria from the aged accretion ice of Lake Vostok, Antarctica: Implications for searching for life in extreme icy environments. *International Journal of Astrobiology*, *3*(1), 1–12. doi:10.1017/S1473550404001879

Bull, A. T., & Asenjo, J. A. (2013). Microbiology of hyper-arid environments: Recent insights from the Atacama Desert, Chile. *Antonie van Leeuwenhoek*, *103*(6), 1173–1179. doi:10.100710482-013-9911-7 PMID:23564162

Bullen, R. A., Arnot, T. C., Lakeman, J. B., & Walsh, F. C. (2006). Biofuel cells and their development. *Biosensors & Bioelectronics*, *21*(11), 2015–2045. doi:10.1016/j.bios.2006.01.030 PMID:16569499

Burdette, D. S., Jung, S.-H., Shen, G.-J., Hollingsworth, R. I., & Zeikus, J. G. (2002). Physiological function of alcohol dehydrogenases and long-chain (C(30)) fatty acids in alcohol tolerance of Thermoanaerobacter ethanolicus. *Applied and Environmental Microbiology*, *68*(4), 1914–1918. doi:10.1128/AEM.68.4.1914-1918.2002 PMID:11916712

Burrows, S. M., Elbert, W., Lawrence, M. G., & Pöschl, U. (2009). Bacteria in the global atmosphere–Part 1: Review and synthesis of literature data for different ecosystems. *Atmospheric Chemistry and Physics*, *9*(23), 9263–9280. doi:10.5194/acp-9-9263-2009

Burstein, D., Harrington, L. B., Strutt, S. C., Probst, A. J., Anantharaman, K., Thomas, B. C., Doudna, J. A., & Banfield, J. F. (2017). New CRISPR-Cas systems from uncultivated microbes. *Nature*, *542*(7640), 237–241. doi:10.1038/nature21059 PMID:28005056

Bursy, J., Pierik, A. J., Pica, N., & Bremer, E. (2007). Osmotically induced synthesis of the compatible solute hydroxyectoine is mediated by an evolutionarily conserved ectoine hydroxylase. *The Journal of Biological Chemistry*, *282*(43), 31147–31155. doi:10.1074/jbc.M704023200 PMID:17636255

Butt, M. S., Tahir-Nadeem, M., Ahmad, Z., & Sultan, M. T. (2008). Xylanases and their applications in baking industry. *Food Technology and Biotechnology*, *46*(1), 22–31.

Caccamo, M. T., Gugliandolo, C., Zammuto, V., & Magazù, S. (2020). Thermal properties of an exopolysaccharide produced by a marine thermotolerant *Bacillus licheniformis* by ATR-FTIR spectroscopy. *International Journal of Biological Macromolecules*, *145*, 77–83. doi:10.1016/j.ijbiomac.2019.12.163 PMID:31870877

Callaghan, R., Luk, F., & Bebawy, M. (2014). Inhibition of the multidrug resistance P-glycoprotein: Time for a change of strategy? *Drug Metabolism and Disposition: the Biological Fate of Chemicals*, *42*(4), 623–631. doi:10.1124/dmd.113.056176 PMID:24492893

Cameron, R. E. (1969). Cold desert characteristics and problems relevant to other arid lands. *Arid Lands in Perspective.*

Cameron, R. E. (1972). Microbial and ecologic investigations in Victoria valley, southern Victoria Land, Antarctica. *Antarctic Terrestrial Biology*, *20*, 195-260.

Cao, Q., Zhang, J., Liu, H., Wu, Q., Chen, J., & Chen, G. Q. (2014). The mechanism of anti-osteoporosis effects of 3-hydroxybutyrate and derivatives under simulated microgravity. *Biomaterials*, *35*(28), 8273–8283. doi:10.1016/j.biomaterials.2014.06.020 PMID:24976243

Cao, R., Jin, W., Shan, Y., Wang, J., Liu, G., Kuang, S., & Sun, C. (2018). Marine bacterial polysaccharide EPS11 inhibits cancer cell growth via blocking cell adhesion and stimulating anoikis. *Marine Drugs*, *16*(3), 85. doi:10.3390/md16030085 PMID:29518055

Cao, Y., Zhang, R., Liu, W., Zhao, G., Niu, W., Guo, J., Xian, M., & Liu, H. (2019). Manipulation of the precursor supply for high-level production of longifolene by metabolically engineered *Escherichia coli. Scientific Reports*, *9*(1), 1–10. doi:10.103841598-018-36495-w PMID:30643175

Carpenter, E. J., Lin, S., & Capone, D. G. (2000). Bacterial activity in South Pole snow. *Applied and Environmental Microbiology*, *66*(10), 4514–4517. doi:10.1128/AEM.66.10.4514-4517.2000 PMID:11010907

Carrer, D. C., Higa, L. H., Tesoriero, M. V. D., Morilla, M. J., Roncaglia, D. I., & Romero, E. L. (2014). Structural features of ultra deformable archaeosomes for topical delivery of ovalbumin. *Colloids and Surfaces. B, Biointerfaces*, *121*, 281–289. doi:10.1016/j.colsurfb.2014.05.015 PMID:24974012

Carriquiry, M. A., Du, X., & Timilsina, G. R. (2011). Second generation biofuels: Economics and policies. *Energy Policy*, *39*(7), 4222–4234. doi:10.1016/j.enpol.2011.04.036

Caruso, C., Rizzo, C., Mangano, S., Poli, A., Di Donato, P., Nicolaus, B., Di Marco, G., Michaud, L., & Giudice, A. L. (2018). Extracellular polymeric substances with metal adsorption capacity produced by *Pseudoalteromonas* sp. MER144 from Antarctic seawater. *Environmental Science and Pollution Research International*, *25*(5), 4667–4677. doi:10.100711356-017-0851-z PMID:29197057

Casanueva, A., Tuffin, M., Cary, C., & Cowan, D. A. (2010). Molecular adaptations to psychrophily: The impact of 'omic' technologies. *Trends in Microbiology*, *18*(8), 374–381. doi:10.1016/j.tim.2010.05.002 PMID:20646925

Castenholz, R. W. (1969). The Thermophilic cyanophytes of Iceland and the upper temperature limit. *Journal of Phycology*, *5*(4), 360–368. doi:10.1111/j.1529-8817.1969.tb02626.x PMID:27096455

Castenholz, R. W., & Pierson, B. K. (1995). Ecology of Thermophilic Anoxygenic Phototrophs. *Anoxygenic Photosynthetic Bacteria*, *2*, 87–103. doi:10.1007/0-306-47954-0_5

Castilla, A., Panizza, P., Rodríguez, D., Bonino, L., Díaz, P., Irazoqui, G., & Giordano, S. R. (2017). A novel thermophilic and halophilic esterase from Janibacter sp. R02, the first member of a new lipase family (Family XVII). *Enzyme and Microbial Technology*, *98*, 86–95. doi:10.1016/j.enzmictec.2016.12.010 PMID:28110668

Castro, D., Torres, M., Sampedro, I., Martínez-Checa, F., Torres, B., & Béjar, V. (2020). Biological Control of Verticillium Wilt on Olive Trees by the Salt-Tolerant Strain Bacillus velezensis XT1. *Microorganisms, 8*(7), 1080. doi:10.3390/microorganisms8071080

Caumette, P. (1993). Ecology and physiology of phototrophic bacteria and sulfate-reducing bacteria in marine salterns. *Experientia, 49*(6), 473–481. doi:10.1007/BF01955148

Cavicchioli, R., Thomas, T., & Curmi, P. M. G. (2000). Cold stress response in Archaea. *Extremophiles, 4*(6), 321–331. doi:10.1007/s007920070001

Cavicchioli, R., Charlton, T., Ertan, H., Omar, S. M., Siddiqui, K. S., & Williams, T. J. (2011). Biotechnological uses of enzymes from psychrophiles. *Microbial Biotechnology, 4*(4), 449–460. doi:10.1111/j.1751-7915.2011.00258.x PMID:21733127

Cavicchioli, R., Siddiqui, K. S., Andrews, D., & Sowers, K. R. (2002). Low-temperature extremophiles and their applications. *Current Opinion in Biotechnology, 13*(3), 253–261. doi:10.1016/S0958-1669(02)00317-8 PMID:12180102

Cha, H. J., Yoon, H. G., Kim, Y. W., Lee, H. S., Kim, J. W., Kweon, K. S., Oh, B. H., & Park, K. H. (1998). Molecular and enzymatic characterization of a maltogenic amylase that hydrolyzes and transglycosylates acarbose. *European Journal of Biochemistry, 253*(1), 251–262. doi:10.1046/j.1432-1327.1998.2530251.x PMID:9578484

Chakravorty, D., Khan, M. F., & Patra, S. (2017). Multifactorial level of extremostability of proteins: Can they be exploited for protein engineering. *Extremophiles, 21*(3), 419–444. doi:10.100700792-016-0908-9 PMID:28283770

Chandi, G. K., & Gill, B. S. (2011). Production and Characterization of Microbial Carotenoids as an Alternative to Synthetic Colors: A Review. *International Journal of Food Properties, 14*(3), 503–513. doi:10.1080/10942910903256956

Chandra, R., Nikhil, G. N., & Mohan, S. V. (2015). Single-Stage Operation of Hybrid Dark-Photo Fermentation to Enhance Biohydrogen Production through Regulation of System Redox Condition: Evaluation with Real-Field Wastewater. *International Journal of Molecular Sciences, 16*(5), 9540–9556. doi:10.3390/ijms16059540 PMID:25927577

Chang, I. S., Moon, H., Bretschger, O., Jang, J. K., Park, H. I., Kenneth, H., Nealson, K. H., & Kim, B. H. (2006). Electrochemically Active Bacteria (EAB) and Mediator-Less Microbial Fuel Cells. *Journal of Microbiology and Biotechnology, 16*(2), 163–177.

Charpentier, E., Richter, H., van der Oost, J., & White, M. F. (2015). Biogenesis pathways of RNA guides in archaeal and bacterial CRISPR-Cas adaptive immunity. *FEMS Microbiology Reviews, 39*(3), 428–441. doi:10.1093/femsre/fuv023 PMID:25994611

Charron, C., Talfournier, F., Isupov, M. N., Littlechild, J. A., Branlant, G., Vitoux, B., & Aubry, A. (2000). The crystal structure of D-glyceraldehyde-3-phosphate dehydrogenase from the hyperthermophilic archaeon *Methanothermus fervidus* in the presence of NADP(+) at 2.1 A resolution. *Journal of Molecular Biology, 297*(2), 481–500. doi:10.1006/jmbi.2000.3565 PMID:10715215

Chassaing, G., Convert, O., & Lavielle, S. (1986). Preferential conformation of substance P in solution. *European Journal of Biochemistry, 154*(1), 77–85. doi:10.1111/j.1432-1033.1986.tb09361.x PMID:2417837

Chaudhuri, S. K., & Lovley, D. R. (2003). Electricity generation by direct oxidation of glucose in mediator less microbial fuel cells. *Nature Biotechnology, 21*(10), 1229–1232. doi:10.1038/nbt867 PMID:12960964

Chela-Flores, J. (2013). Polyextremophiles. *Summary and Conclusions*, 611–615. Advance online publication. doi:10.1007/978-94-007-6488-0_29

Chellapandi, P., & Ranjani, J. (2015). Knowledge-based discovery for designing CRISPRCAS systems against invading mobilomes in thermophiles. *Systems and Synthetic Biology, 9*(3), 97–106. doi:10.100711693-015-9176-8 PMID:26279704

Chen, X. (2021). *Thriving at Low pH: Adaptation Mechanisms of Acidophiles*. Academic Press.

Chen, B., Gilbert, L. A., Cimini, B. A., Schnitzbauer, J., Zhang, W., Li, G. W., Park, J., Blackburn, E. H., Weissman, J. S., Qi, L. S., & Huang, B. (2013). Dynamic imaging of genomic loci in living human cells by an optimized CRISPR/Cas system. *Cell, 155*(7), 1479–1491. doi:10.1016/j.cell.2013.12.001 PMID:24360272

Chen, C. S. (1987). Relationship between water activity and freezing point depression of food systems. *Journal of Food Science, 52*(2), 433–435. doi:10.1111/j.1365-2621.1987.tb06632.x

Chen, F., Castranova, V., & Shi, X. (2001). New insights into the role of nuclear factor-κB in cell growth regulation. *American Journal of Pathology, 159*(2), 387–397. doi:10.1016/S0002-9440(10)61708-7 PMID:11485895

Cheng, H., Sun, Y., Chang, H., Cui, F., Xue, H., Shen, Y., Wang, M., & Luo, J. (2020). Compatible solutes adaptive alterations in Arthrobacter simplex during exposure to ethanol, and the effect of trehalose on the stress resistance and biotransformation performance. *Bioprocess and Biosystems Engineering, 43*(5), 895–908. doi:10.1007/s00449-020-02286-9

Chen, G. Q., & Jiang, X. R. (2018). Next generation industrial biotechnology based on extremophilic bacteria. *Current Opinion in Biotechnology, 50*, 94–100. doi:10.1016/j.copbio.2017.11.016 PMID:29223022

Chen, S., Su, L., Billig, S., Zimmermann, W., Chen, J., & Wu, J. (2010). Biochemical characterization of the cutinases from Thermobifida fusca. *Journal of Molecular Catalysis. B, Enzymatic, 63*(3-4), 121–127. doi:10.1016/j.molcatb.2010.01.001

Chen, S., Sun, S., Korfanty, G. A., Liu, J., & Xiang, H. (2019). A halocin promotes DNA uptake in *Haloferax mediterranei. Frontiers in Microbiology, 10*, 1960. doi:10.3389/fmicb.2019.01960 PMID:31620096

Chen, W. C., Hsu, C. C., Lan, J. C. W., Chang, Y. K., Wang, L. F., & Wei, Y. H. (2018). Production and characterization of ectoine using a moderately halophilic strain Halomonas salina BCRC17875. *Journal of Bioscience and Bioengineering, 125*(5), 578–584. doi:10.1016/j.jbiosc.2017.12.011 PMID:29331525

Chen, W. Y., & Abatangelo, G. (1999). Functions of hyaluronan in wound repair. *Wound Repair and Regeneration, 7*(2), 79–89. doi:10.1046/j.1524-475X.1999.00079.x PMID:10231509

Chen, X., Bhandari, B., & Zhou, P. (2019). Insight into the effect of glycerol on stability of globular proteins in high protein model system. *Food Chemistry, 278*, 780–785. doi:10.1016/j.foodchem.2018.11.117 PMID:30583443

Chen, X., Jiang, H., Yang, Y., & Liu, N. (2009). Effect of exopolysaccharide from *Bifidobacterium bifidum* on cell of gastric cancer and human telomerase reverse transcriptase. *Wei sheng wu xue bao=. Acta Microbiologica Sinica, 49*(1), 117–122. PMID:19388274

Chen, X., Yu, L., Qiao, G., & Chen, G. (2018). Reprogramming *Halomonas* for industrial production of chemicals. *Journal of Industrial Microbiology & Biotechnology, 45*(7), 545–554. doi:10.100710295-018-2055-z PMID:29948194

Chen, Y.-G., Cui, X.-L., Wang, Y.-X., Zhang, Y.-Q., Tang, S.-K., Li, W.-J., Liu, Z.-X., Wen, M.-L., & Peng, Q. (2009). *Virgibacillus sediminis* sp. nov., a moderately halophilic bacterium isolated from a salt lake in China. *International Journal of Systematic and Evolutionary Microbiology, 59*(8), 2058–2063. doi:10.1099/ijs.0.003624-0 PMID:19605714

Chen, Y.-T., Yuan, Q., Shan, L.-T., Lin, M.-A., Cheng, D.-Q., & Li, C.-Y. (2013). Antitumor activity of bacterial exopolysaccharides from the endophyte *Bacillus amyloliquefaciens* sp. isolated from *Ophiopogon japonicus. Oncology Letters, 5*(6), 1787–1792. doi:10.3892/ol.2013.1284 PMID:23833642

Chen, Z., Wang, Q., Lin, L., Tang, Q., Edwards, J. L., Li, S., & Liu, S. (2012). Comparative evaluation of two isobaric labeling tags, DiART and iTRAQ. *Analytical Chemistry, 84*(6), 2908–2915. doi:10.1021/ac203467q PMID:22404494

Chester, F. D. (1897). Report of mycologist: Bacteriological work. *Delaware Agricultural Experiment Station Bulletin, 9*, 38–145.

Chhabra, S. R., Shockley, K. R., Conners, S. B., Scott, K. L., Wolfinger, R. D., & Kelly, R. M. (2003). Carbohydrate-induced differential gene expression patterns in the hyperthermophilic bacterium *Thermotoga maritima. The Journal of Biological Chemistry*, *278*(9), 7540–7552. doi:10.1074/jbc.M211748200 PMID:12475972

Chhabra, S. R., Shockley, K. R., Ward, D. E., & Kelly, R. M. (2002). Regulation of endo-acting glycosyl hydrolases in the hyperthermophilic bacterium *Thermotoga maritima* grown on glucan- and mannan-based polysaccharides. *Applied and Environmental Microbiology*, *68*(2), 545–554. doi:10.1128/AEM.68.2.545-554.2002 PMID:11823189

Chikkanna, A., Ghosh, D., & Kishore, A. (2018). Expression and characterization of a potential exopolysaccharide from a newly isolated halophilic thermotolerant bacteria *Halomonas nitroreducens* strain WB1. *Peer-Reviewed Scientific Mega Journal*, *6*, e4684. doi:10.7717/peerj.4684 PMID:29707437

Childers, S. E., Madeline, V., & Noll, K. M. (1992). Improved methods for cultivation of the extremely thermophilic bacterium *Thermotoga neapolitana. Applied and Environmental Microbiology*, *58*(12), 3949–3953. doi:10.1128/aem.58.12.3949-3953.1992 PMID:16348824

Chinnathambi, A. (2015). Industrial Important Enzymes from Alkaliphiles – An overview. *Biosciences Biotechnology Research Asia*, *12*(3), 2007–2016. doi:10.13005/bbra/1868

Chisti, Y. (2008). Biodiesel from microalgae beats bioethanol. *Trends in Biotechnology*, *26*(3), 126–131. doi:10.1016/j.tibtech.2007.12.002 PMID:18221809

Chneiweiss, H., Hirsch, F., Montoliu, L., Müller, A. M., Fenet, S., Abecassis, M., Merchant, J., Baertschi, B., Botbol-Baum, M., Houghton, J. A., Kritikos, M., Mifsud, J., Bartnik, E., Rath, J., Druml, C., Friedrich, B., Carvalho, A. S., Lanzerath, D., & Raymond, A. S. (2017). Fostering responsible research with genome editing technologies: A European perspective. *Transgenic Research*, *26*(5), 709–713. doi:10.100711248-017-0028-z PMID:28730514

Cho, H. M., & Yoo, B. (2015). Rheological characteristics of cold thickened beverages containing xanthan gum–based food thickeners used for dysphagia diets. *Journal of the Academy of Nutrition and Dietetics*, *115*(1), 106–111. doi:10.1016/j.jand.2014.08.028 PMID:25441963

Choi, B. H., Hwang, H. J., Lee, J. E., Oh, S. H., Hwang, J. S., Lee, B. Y., & Lee, P. C. (2020). Microbial Production of Retinyl Palmitate and its Application as a Cosmeceutical. *Antioxidants*, *9*(11), 1130. doi:10.3390/antiox9111130 PMID:33202630

Choi, Y.-J., Hur, J.-M., Lim, S., Jo, M., Kim, D. H., & Choi, J.-I. (2014). Induction of apoptosis by deinoxanthin in human cancer cells. *Anticancer Research*, *34*(4), 1829–1835. PMID:24692716

Choi, Y., Kim, N., Kim, S., & Jung, S. (2003). Dynamic behaviors of redox mediators within the hydrophobic layers as an important factor for effective microbial fuel cell operation. *Bulletin of the Korean Chemical Society*, *24*(4), 437–440. doi:10.5012/bkcs.2003.24.4.437

Cho, J. S., Choi, K. R., Prabowo, C. P. S., Shin, J. H., Yang, D., Jang, J., & Lee, S. Y. (2017). CRISPR/Cas9-coupled recombineering for metabolic engineering of *Corynebacterium glutamicum. Metabolic Engineering*, *42*, 157–167. doi:10.1016/j.ymben.2017.06.010 PMID:28649005

Choudhary, P., Shankar, U. P., Bandopadhyay, T. K., Ray, R. N., & Bhunia, B. (2017). Performance improvement of microbial fuel cell (MFC) using suitable electrodes and Bioengineered organisms: A review. *Bioengineered*, *8*(5), 471–487. doi:10.1080/21655979.2016.1267883 PMID:28453385

Christner, B. C. (2002). Incorporation of DNA and Protein Precursors into Macromolecules by Bacteria at– 15oC. *Applied and Environmental Microbiology*, *68*(12), 6435–6438. doi:10.1128/AEM.68.12.6435-6438.2002 PMID:12450874

Christner, B. C. (2010). Bioprospecting for microbial products that affect ice crystal formation and growth. *Applied Microbiology and Biotechnology*, *85*(3), 481–489. doi:10.100700253-009-2291-2 PMID:19841917

Christner, B. C., Kvitko, B. H. II, & Reeve, J. N. (2003a). Molecular identification of bacteria and eukarya inhabiting an Antarctic cryoconite hole. *Extremophiles*, *7*(3), 177–183. doi:10.100700792-002-0309-0 PMID:12768448

Christner, B. C., Skidmore, M. L., Priscu, J. C., Tranter, M., & Foreman, C. M. (2008b). Bacteria in subglacial environments. In *Psychrophiles: from biodiversity to biotechnology* (pp. 51–71). Springer. doi:10.1007/978-3-540-74335-4_4

Chung, D., Sarai, N. S., Himmel, M. E., & Bomble, Y. J. (2020). Genetics of Unstudied Thermophiles for Industry. In *Metabolic Pathway Engineering* (pp. 5–19). Humana. doi:10.1007/978-1-0716-0195-2_2

Cihan, A. C., Cokmus, C., Koc, M., & Ozcan, B. (2014). *Anoxybacillus calidus* sp. nov., a thermophilic bacterium isolated from soil near a thermal power plant. *International Journal of Systematic and Evolutionary Microbiology*, *64*(1), 211–219. doi:10.1099/ijs.0.056549-0 PMID:24052627

Citorik, R. J., Mimee, M., & Lu, T. K. (2014). Sequence-specific antimicrobials using efficiently delivered RNA-guided nucleases. *Nature Biotechnology*, *32*(11), 1141–1145. doi:10.1038/nbt.3011 PMID:25240928

Clarke, A. (2014). The thermal limits to life on Earth. *International Journal of Astrobiology*, *13*(2), 141–154. doi:10.1017/S1473550413000438

Cleland, D., Krader, P., McCree, C., Tang, J., & Emerson, D. (2004). Glycine betaine as a cryoprotectant for prokaryotes. *Journal of Microbiological Methods*, *58*(1), 31–38. doi:10.1016/j.mimet.2004.02.015 PMID:15177901

Cobucci-Ponzano, B., Zorzetti, C., Strazzulli, A., Carillo, S., Bedini, E., Corsaro, M. M., Comfort, D. A., Kelly, R. M., Rossi, M., & Moracci, M. (2011). A novel α-D-galactosynthase from *Thermotoga maritima* converts β-D-galactopyranosyl azide to α-galacto-oligosaccharides. *Glycobiology*, *21*(4), 448–456. doi:10.1093/glycob/cwq177 PMID:21084405

Cockell, C. S., & Córdoba-Jabonero, C. (2004). Coupling of climate change and biotic UV exposure through changing snow-ice covers in terrestrial habitats. *Photochemistry and Photobiology*, *79*(1), 26–31. PMID:14974712

Cockell, C. S., Kelly, L. C., & Marteinsson, V. (2013). Actinobacteria—An Ancient Phylum Active in Volcanic Rock Weathering. *Geomicrobiology Journal*, *30*(8), 706–720. doi:10.1080/01490451.2012.758196

Cockell, C. S., & Stokes, M. D. (2004). Widespread colonization by polar hypoliths. *Nature*, *431*(7007), 414–414. doi:10.1038/431414a PMID:15386002

Cockell, C., Rettberg, P., Horneck, G., Scherer, K., & Stokes, D. M. (2003). Measurements of microbial protection from ultraviolet radiation in polar terrestrial microhabitats. *Polar Biology*, *26*(1), 62–69. doi:10.100700300-002-0438-z

Coker, J. A. (2019). Recent advances in understanding extremophiles. *F1000Research*, *8*, 1917. doi:10.12688/f1000research.20765.1

Coker, J. (2016). "Sohail Siddiqui of New South Wales, K." Extremophiles and Biotechnology: Current Uses and Prospects. *F1000 Research*, *5*, 396. doi:10.12688/f1000research.7432.1

Coker, J. A., & Brenchley, J. E. (2006). Protein engineering of a cold-active β-galactosidase from Arthrobacter sp. SB to increase lactose hydrolysis reveals new sites affecting low temperature activity. *Extremophiles*, *10*(6), 515–524. doi:10.100700792-006-0526-z PMID:16736094

Collins, T., & Margesin, R. (2019). Psychrophilic lifestyles: mechanisms of adaptation and biotechnological tools. *Applied Microbiology and Biotechnology*, *103*(7), 2857–2871. doi:10.1007/s00253-019-09659-5

Collins, T., D'Amico, S., Marx, J.-C., Feller, G., & Gerday, C. (2014). Cold-Adapted Enzymes. *Physiology and Biochemistry of Extremophiles*, 165–179. doi:10.1128/9781555815813.ch13

Collins, M. D., Lawson, P. A., Willems, A., Cordoba, J. J., Fernandez-Garayzabal, J., Garcia, P., Cai, J., Hippe, H., & Farrow, J. A. (1994). The phylogeny of the genus Clostridium: Proposal of five new genera and eleven new species combinations. *International Journal of Systematic Bacteriology*, *44*(4), 812–826. doi:10.1099/00207713-44-4-812 PMID:7981107

Collins, M. D., Lund, B. M., Farrow, J. A. E., & Schleifer, K. H. (1983). Chemotaxonomic Study of an Alkalophilic Bacterium, *Exiguobacterium aurantiacum* gen. nov., sp. nov. *Microbiology*, *129*(7), 2037–2042. doi:10.1099/00221287-129-7-2037

Collins, R. E., Rocap, G., & Deming, J. W. (2010). Persistence of bacterial and archaeal communities in sea ice through an Arctic winter. *Environmental Microbiology*, *12*(7), 1828–1841. doi:10.1111/j.1462-2920.2010.02179.x PMID:20192970

Comfort, D. A., Bobrov, K. S., Ivanen, D. R., Shabalin, K. A., Harris, J. M., Kulminskaya, A. A., Brumer, H., & Kelly, R. M. (2007). Biochemical analysis of *Thermotoga maritima* GH36 α -galactosidase (TmGalA) confirms the mechanistic commonality of clan GH-D glycoside hydrolases. *Biochemistry*, *46*(11), 3319–3330. doi:10.1021/bi061521n PMID:17323919

Connell, L. B., Redman, R., Rodriguez, R., Barrett, A., Iszard, M., & Fonseca, A. (2010). Dioszegia antarctica sp. nov. and Dioszegia cryoxerica sp. nov., psychrophilic basidiomycetous yeasts from polar desert soils in Antarctica. *International Journal of Systematic and Evolutionary Microbiology*, *60*(6), 1466–1472. doi:10.1099/ijs.0.015412-0 PMID:19671723

Conners, S. B., Mongodin, E. F., Johnson, M. R., Montero, C. I., Nelson, K. E., & Kelly, R. M. (2006). Microbial biochemistry, physiology, and biotechnology of hyperthermophilic *Thermotoga* species. *FEMS Microbiology Reviews*, *30*(6), 872–905. doi:10.1111/j.1574-6976.2006.00039.x PMID:17064285

Coradin, J. H., Braun, A., Viccini, G., de Lima Luz, L. F., Krieger, N., & Mitchell, D. A. (2011). A three-dimensional discrete lattice-based system for modeling the growth of aerial hyphae of filamentous fungi on solid surfaces: A tool for investigating micro-scale phenomena in solid-state fermentation. *Biochemical Engineering Journal*, *54*(3), 164–171. doi:10.1016/j.bej.2011.02.012

Corral, P., Amoozegar, M. A., & Ventosa, A. (2020). Halophiles and their biomolecules: Recent advances and future applications in biomedicine. *Marine Drugs*, *18*(1), 33. doi:10.3390/md18010033 PMID:31906001

Costa, O. Y., Raaijmakers, J. M., & Kuramae, E. E. (2018). Microbial extracellular polymeric substances: Ecological function and impact on soil aggregation. *Frontiers in Microbiology*, *9*, 1636. doi:10.3389/fmicb.2018.01636 PMID:30083145

Costerton, J. W., Cheng, K., Geesey, G. G., Ladd, T. I., Nickel, J. C., Dasgupta, M., & Marrie, T. J. (1987). Bacterial biofilms in nature and disease. *Annual Review of Microbiology*, *41*(1), 435–464. doi:10.1146/annurev.mi.41.100187.002251 PMID:3318676

Cowan, D. A., Ramond, J. B., Makhalanyane, T. P., & De Maayer, P. (2015). Metagenomics of extreme environments. *Current Opinion in Microbiology*, *25*, 97–102. doi:10.1016/j.mib.2015.05.005 PMID:26048196

Cowan, D. A., Russell, N. J., Mamais, A., & Sheppard, D. M. (2002). Antarctic Dry Valley mineral soils contain unexpectedly high levels of microbial biomass. *Extremophiles*, *6*(5), 431–436. doi:10.100700792-002-0276-5 PMID:12382121

Cramp, R. A., & Cowan, D. A. (1999). Molecular characterisation of a novel thermophilic nitrile hydratase. *Biochimica et Biophysica Acta (BBA)-. Protein Structure and Molecular Enzymology*, *1431*(1), 249–260. doi:10.1016/S0167-4838(99)00010-2 PMID:10209297

Crowley, E. (2017). Compatible Solute Ectoine Review: Protection Mechanisms and Production Methods. [*Journal of Undergraduate Studies at Trent*, *5*(1), 32–39. https://ojs.trentu.ca/ojs/index.php/just/article/view/79

Cui, Y., Li, Y., Gorge, O., Platonov, M. E., Yan, Y., Guo, Z., Pourcel, C., Dentovskaya, S. V., Balakhonov, S. V., Wang, X., Song, Y., Anisimov, A. P., Vergnaud, G., & Yang, R. (2008). Insight into microevolution of *Yersinia pestis* by clustered regularly interspaced short palindromic repeats. *PLoS One*, *3*(7), e2652. doi:10.1371/journal.pone.0002652 PMID:18612419

Cullen, J. J., & MacIntyre, H. L. (2016). On the use of the serial dilution culture method to enumerate viable phytoplankton in natural communities of plankton subjected to ballast water treatment. *Journal of Applied Phycology*, *28*(1), 279–298. doi:10.100710811-015-0601-x PMID:26893536

Cummings, S. P., & Black, G. W. (1999). Polymer hydrolysis in a cold climate. *Extremophiles: Life under Extreme Conditions*, *3*(2), 81–87. doi:10.1007/s007920050102

Cuneo, M. J., Beese, L. S., & Hellinga, H. W. (2009). Structural analysis of semi-specific oligosaccharide recognition by a cellulose-binding protein of *Thermotoga maritima* reveals adaptations for functional diversification of the oligopeptide periplasmic binding protein fold. *The Journal of Biological Chemistry*, *284*(48), 33217–33223. doi:10.1074/jbc.M109.041624 PMID:19801540

Cycil, L. M., DasSarma, S., Pecher, W., & McDonald, R. (1567). AbdulSalam, M., & Hasan, F. (2020). Metagenomic Insights Into the Diversity of Halophilic Microorganisms Indigenous to the Karak Salt Mine, Pakistan. *Frontiers in Microbiology*, *0*. Advance online publication. doi:10.3389/fmicb.2020.01567 PMID:32793134

D'Amico, S., Collins, T., Marx, J. C., Feller, G., Gerday, C., & Gerday, C. (2006). Psychrophilic microorganisms: Challenges for life. *EMBO Reports*, *7*(4), 385–389. doi:10.1038j.embor.7400662 PMID:16585939

D'Amico, S., Gerday, C., & Feller, G. (2001). Structural Determinants of Cold Adaptation and Stability in a Large Protein. *The Journal of Biological Chemistry*, *276*(28), 25791–25796. doi:10.1074/jbc.M102741200 PMID:11325973

D'Amico, S., Marx, J. C., Gerday, C., & Feller, G. (2003). Activity-stability relationships in extremophilic enzymes. *The Journal of Biological Chemistry*, *278*(10), 7891–7896. doi:10.1074/jbc.M212508200 PMID:12511577

D'Amico, S., Sohier, J. S., & Feller, G. (2006). Kinetics and Energetics of Ligand Binding Determined by Microcalorimetry: Insights into Active Site Mobility in a Psychrophilic α-Amylase. *Journal of Molecular Biology*, *358*(5), 1296–1304. doi:10.1016/j.jmb.2006.03.004 PMID:16580683

D'Elia, T., Veerapaneni, R., & Rogers, S. O. (2008). Isolation of microbes from Lake Vostok accretion ice. *Applied and Environmental Microbiology*, *74*(15), 4962–4965. doi:10.1128/AEM.02501-07 PMID:18552196

D'Ippolito, G., Dipasquale, L., & Fontana, A. (2014). Recycling of carbon dioxide and acetate as lactic acid by the hydrogen-producing bacterium *Thermotoga neapolitana*. *ChemSusChem*, *7*(9), 2678–2683. doi:10.1002/cssc.201402155 PMID:25116280

Dalmaso, G. Z. L., Ferreira, D., & Vermelho, A. B. (2015). Marine extremophiles: A source of hydrolases for biotechnological applications. *Marine Drugs*, *13*(4), 1925–1965. doi:10.3390/md13041925 PMID:25854643

Damiano, V. B., Ward, R., Gomes, E., Alves-Prado, H. F., & Da Silva, R. (2006). Purification and characterization of two xylanases from alkalophilic and thermophilic *Bacillus licheniformis* 77-2. *Applied Biochemistry and Biotechnology*, *129–132*(1-3), 289–302. doi:10.1385/ABAB:129:1:289 PMID:16915648

Dang, H., Zhu, H., Wang, J., & Li, T. (2009). Extracellular hydrolytic enzyme screening of culturable heterotrophic bacteria from deep-sea sediments of the Southern Okinawa Trough. *World Journal of Microbiology & Biotechnology*, *25*(1), 71–79. doi:10.100711274-008-9865-5

Daoud, L., & Ali, M. B. (2020). Halophilic microorganisms: Interesting group of extremophiles with important applications in biotechnology and the environment. In *Physiological and Biotechnological Aspects of Extremophiles* (pp. 51–64). Academic Press. doi:10.1016/B978-0-12-818322-9.00005-8

Daranas, N., Roselló, G., Cabrefiga, J., Donati, I., Francés, J., Badosa, E., & Spinelli, F. (2018). Biological control of bacterial plant diseases withLactobacillus plantarumstrains selected for their broad-spectrum activity. *Annals of Applied Biology*, *174*(1), 92–105. doi:10.1111/aab.12476

Darwesh, O. M., El-Hawary, A. S., El Kelany, U. S., & El-Sherbiny, G. M. (2019). Nematicidal activity of thermostable alkaline protease produced by *Saccharomonospora viridis* strain Hw G550. *Biotechnology Reports (Amsterdam, Netherlands)*, *24*, e00386. doi:10.1016/j.btre.2019.e00386 PMID:31763199

Das, D., Dutta, T., Nath, K., Kotay, S. M., Das, A. K., & Veziroglu, T. N. (2006). Role of Fe hydrogenase in biological hydrogen production. *Current Science*, *90*(12), 1627–1637.

Das, R., & Kazy, S. K. (2014). Microbial diversity, community composition and metabolic potential in hydrocarbon contaminated oily sludge: Prospects for in situ bioremediation. *Environmental Science and Pollution Research International*, *21*(12), 7369–7389. doi:10.100711356-014-2640-2 PMID:24682711

DasSarma & Arora. (2001). *Halophiles*. e LS.

DasSarma, P., Coker, J. A., Huse, V., & DasSarma, S. (2009). *Halophiles, industrial applications. Encyclopedia of industrial biotechnology: bioprocess, bioseparation, and cell technology*, 1-43.

DasSarma, P., Coker, J. A., Huse, V., & DasSarma, S. (2010). Halophiles, Industrial Applications. In *Encyclopedia of Industrial Biotechnology*. John Wiley & Sons, Inc. doi:10.1002/9780470054581.eib439

DasSarma, S., Capes, M. D., Karan, R., & DasSarma, P. (2013). Amino Acid Substitutions in Cold-Adapted Proteins from Halorubrum lacusprofundi, an Extremely Halophilic Microbe from Antarctica. *PLoS One*, *8*(3), e58587. doi:10.1371/journal.pone.0058587 PMID:23536799

DasSarma, S., & DasSarma, P. (2015). Halophiles and their enzymes: Negativity put to good use. *Current Opinion in Microbiology*, *25*, 120–126. doi:10.1016/j.mib.2015.05.009 PMID:26066288

Davey, K. R., & Amos, S. A. (2002). Study of the effects of temperature, pH and yeast extract on growth and exopolysaccharides production by *Propionibacterium acidi-propionici* on milk microfiltrate using a response surface methodology. *Journal of Applied Microbiology*, *92*(3), 583–587. doi:10.1046/j.1365-2672.2002.1617a.x PMID:11922033

Dawes, E. A., Ribbons, D. W., & Large, P. J. (1966). The route of ethanol formation in Zymomonas mobilis. *The Biochemical Journal*, *98*(3), 795–803. doi:10.1042/bj0980795 PMID:4287842

Dawson, K. S., Strąpoć, D., Huizinga, B., Lidstrom, U., Ashby, M., & Macalady, J. L. (2012). Quantitative fluorescence in situ hybridization analysis of microbial consortia from a biogenic gas field in Alaska's Cook Inlet basin. *Applied and Environmental Microbiology*, *78*(10), 3599–3605. doi:10.1128/AEM.07122-11 PMID:22427501

Dayma, P. B., Mangrola, A. V., Suriyaraj, S. P., Dudhagara, P., & Patel, R. K. (2019). Synthesis of bio-silver nanoparticles using desert isolated *Streptomyces intermedius* and its antimicrobial activity. *J. Pharm. Chem. Biol. Sci*, *7*, 94–101.

de Almeida Couto, C. R., Alvarez, V. M., Marques, J. M., de Azevedo Jurelevicius, D., & Seldin, L. (2015). Exploiting the aerobic endospore-forming bacterial diversity in saline and hypersaline environments for biosurfactant production. *BMC Microbiology*, *15*(1), 1–17. doi:10.118612866-015-0575-5 PMID:26511622

de Kato, C. (2011). High pressure and prokaryotes. In Extremophiles Handbook. Academic Press.

De Maayer, P., Anderson, D., Cary, C., & Cowan, D. A. (2014). Some like it cold: Understanding the survival strategies of psychrophiles. *EMBO Reports*, *15*(5), 508–517. doi:10.1002/embr.201338170 PMID:24671034

de Menezes, G. C., Godinho, V. M., Porto, B. A., Gonçalves, V. N., & Rosa, L. H. (2017). *Antarctomyces pellizariae* sp. nov., a new, endemic, blue, snow resident psychrophilic ascomycete fungus from Antarctica. *Extremophiles*, *21*(2), 259–269. doi:10.100700792-016-0895-x PMID:27900476

de Siqueira, F. G., de Siqueira, E. G., Jaramillo, P. M. D., Silveira, M. H. L., Andreaus, J., Couto, F. A., Batista, L. R., & Ferreira Filho, E. X. (2010). The potential of agro-industrial residues for production of holocellulase from filamentous fungi. *International Biodeterioration & Biodegradation*, *64*(1), 20–26. doi:10.1016/j.ibiod.2009.10.002

de Vos, W. M., Kengen, S. W., Voorhorst, W. G., & van der Oost, J. (1998). Sugar utilization and its control in hyperthermophiles. *Extremophiles*, *2*(3), 201–205. doi:10.1007007920050061 PMID:9783166

De Vrije, T., De Haas, G., Tan, G., Keijsers, E., & Claassen, P. (2002). Pretreatment of Miscanthus for hydrogen production by Thermotoga elfii. *International Journal of Hydrogen Energy*, *27*(11-12), 1381–1390. doi:10.1016/S0360-3199(02)00124-6

De Vrije, T., Mars, A., Budde, M., Lai, M., Dijkema, C., De Waard, P., & Claassen, P. (2007). Glycolytic pathway and hydrogen yield studies of the extreme thermophile *Caldicellulosiruptor saccharolyticus. Applied Microbiology and Biotechnology*, *74*(6), 1358–1367. doi:10.100700253-006-0783-x PMID:17216445

Dean-Ross, D., Moody, J., & Cerniglia, C. E. (2002). Utilization of mixtures of polycyclic aromatic hydrocarbons by bacteria isolated from contaminated sediment. *FEMS Microbiology Ecology*, *41*(1), 1–7. doi:10.1111/j.1574-6941.2002.tb00960.x PMID:19709233

DeBoy, R. T., Mongodin, E. F., Emerson, J. B., & Nelson, K. E. (2006). Chromosome evolution in the Thermotogales: Large-scale inversions and strain diversification of CRISPR sequences. *Journal of Bacteriology*, *188*(7), 2364–2374. doi:10.1128/JB.188.7.2364-2374.2006 PMID:16547022

Decho, A. W. (2000). Microbial biofilms in intertidal systems: An overview. *Continental Shelf Research*, *20*(10–11), 1257–1273. doi:10.1016/S0278-4343(00)00022-4

Decho, A. W., & Gutierrez, T. (2017). Microbial extracellular polymeric substances (EPSs) in ocean systems. *Frontiers in Microbiology*, *8*, 922. doi:10.3389/fmicb.2017.00922 PMID:28603518

Deepak, V., Ramachandran, S., Balahmar, R. M., Pandian, S. R. K., Sivasubramaniam, S. D., Nellaiah, H., & Sundar, K. (2016). *In vitro* evaluation of anticancer properties of exopolysaccharides from *Lactobacillus acidophilus* in colon cancer cell lines. *In Vitro Cellular & Developmental Biology. Animal*, *52*(2), 163–173. doi:10.100711626-015-9970-3 PMID:26659393

Deepika, K., Raghuram, M., Kariali, E., & Bramhachari, P. (2016). Biological responses of symbiotic *Rhizobium radiobacter* strain VBCK1062 to the arsenic contaminated rhizosphere soils of mung bean. *Ecotoxicology and Environmental Safety*, *134*, 1–10. doi:10.1016/j.ecoenv.2016.08.008 PMID:27566287

Defelice, S. L. (1989). *The nutraceutical revolution: Fueling a powerful, new international market*. Harvard University Advanced Management Program in Biomedical Research and Development.

Deguillaume, L., Leriche, M., Amato, P., Ariya, P. A., Delort, A.-M., Po¨schl, U., Chaumerliac, N., Bauer, H., Flossmann, A. I., & Morris, C. E. (2008). Microbiology and atmospheric processes: Chemical interactions of primary biological aerosols. *Biogeosciences*, *5*(4), 1073–1084. doi:10.5194/bg-5-1073-2008

Dehghanikhah, F., Shakarami, J., & Asoodeh, A. (2020). Purification and Biochemical Characterization of Alkalophilic Cellulase from the Symbiotic *Bacillus subtilis* BC1 of the Leopard Moth, Zeuzera pyrina (L.) (Lepidoptera: Cossidae). *Current Microbiology*, *77*(7), 1254–1261. doi:10.100700284-020-01938-z PMID:32125446

Dekker, L., Arsène-Ploetze, F., & Santini, J. M. (2016). Comparative proteomics of Acidithiobacillus ferrooxidans grown in the presence and absence of uranium. *Research in Microbiology*, *167*(3), 234–239. doi:10.1016/j.resmic.2016.01.007 PMID:26829305

Deltcheva, E., Chylinski, K., Sharma, C., Gonzales, K., Chao, Y., Pirzada, Z., Eckert, M., Vogel, J., & Charpentier, E. (2011). CRISPR RNA maturation by *trans*-encoded small RNA and host factor RNase III. *Nature*, *471*(7340), 602–607. doi:10.1038/nature09886 PMID:21455174

Demain, A. L. (2009). Biosolutions to the energy problem. *Journal of Industrial Microbiology & Biotechnology*, *36*(3), 319–332. doi:10.100710295-008-0521-8 PMID:19137336

Deming, J. W. (2009b). Extremophiles: Cold Environments. In M. Schaechter (Ed.), *Encyclopedia of Microbiology* (pp. 147–158). Elsevier. doi:10.1016/B978-012373944-5.00280-7

Demirel, B., & Scherer, P. (2008). The roles of acetotrophic and hydrogenotrophic methanogens during anaerobic conversion of biomass to methane: a review. *Reviews in Environmental Science and Bio/Technology*, *7*(2), 173–190. doi:10.1007/s11157-008-9131-1

Denizci, A. A., Kazan, D., & Erarslan, A. (2010). *Bacillus marmarensis* sp. nov., an alkaliphilic, protease-producing bacterium isolated from mushroom compost. *International Journal of Systematic and Evolutionary Microbiology*, *60*(7), 1590–1594. Advance online publication. doi:10.1099/ijs.0.012369-0 PMID:19700450

Dennis, P. P. (2020). The Molecular Biology of Halophilic Archaebacteria. The Biology of Halophilic Bacteria, 255–287. doi:10.1201/9781003069140-9/MOLECULAR-BIOLOGY-HALOPHILIC-ARCHAEBACTERIA-PATRICK-DENNIS

Deole, R., & Hoff, W. D. (2020). A potassium chloride to glycine betaine osmoprotectant switch in the extreme halophile Halorhodospira halophila. *Scientific Reports*, *10*(1), 1–10. doi:10.1038/s41598-020-59231-9

Derikvand, P., Llewellyn, C. A., & Purton, S. (2017). Cyanobacterial metabolites as a source of sunscreens and moisturizers: A comparison with current synthetic compounds. *European Journal of Phycology*, *52*(1), 43–56. doi:10.1080/09670262.2016.1214882

Dertli, E., Mayer, M. J., & Narbad, A. (2015). Impact of the exopolysaccharide layer on biofilms, adhesion and resistance to stress in *Lactobacillus johnsonii* FI9785. *BMC Microbiology*, *15*(1), 1–9. doi:10.118612866-015-0347-2 PMID:25648083

Desai, S. G., Guerinot, M. L., & Lynd, L. R. (2004). Cloning of l-lactate dehydrogenase and elimination of lactic acid production via gene knockout in Thermoanaerobacterium saccharolyticum JW/SL-YS485. *Applied Microbiology and Biotechnology*, *65*(5). Advance online publication. doi:10.100700253-004-1575-9 PMID:15007569

Desmarais, D., Jablonski, P. E., Fedarko, N. S., & Roberts, M. F. (1997). 2-sulfotrehalose, a novel osmolyte in haloalkaliphilic archaea. *Journal of Bacteriology*, *179*(10), 3146–3153. doi:10.1128/jb.179.10.3146-3153.1997 PMID:9150208

Deveau, H., Barrangou, R., Garneau, J. E., Labonté, J., Fremaux, C., Boyaval, P., Romero, D. A., Horvath, P., & Moineau, S. (2008). Phage response to CRISPR-encoded resistance in Streptococcus thermophilus. *Journal of Bacteriology*, *190*(4), 1390–1400. doi:10.1128/JB.01412-07 PMID:18065545

Dewan, S. S. (2014). *Global markets for enzymes in industrial applications.* BCC Research.

Dhagat, S., & Jujjavarapu, S. E. (2021). Biorefinery system for production of thermostable exopolysaccharide by a novel thermophile *Brevibacillus borstelensis* MK878423 and its study on impact of glucose utilization. *Biomass Conversion and Biorefinery*, 1–11.

Dhakar, K., & Pandey, A. (2016). Wide pH range tolerance in extremophiles: Towards understanding an important phenomenon for future biotechnology. *Applied Microbiology and Biotechnology*, *100*(6), 2499–2510. doi:10.100700253-016-7285-2 PMID:26780356

Dhundale, V., Hemke, V., Desai, D., Aher, G., Khemchandani, P., Dikonda, P., & Thosare, B. (2018). Improvement and Stable High Bioelectricity Generation Using Alkaliphilic *Oceanobacillus iheyensis* BS1(2) in Microbial Fuel Cells and Effect of Different Anodic Operating Conditions. *Annals of Applied Microbiology & Biotechnology Journal*, *2*(1), 1010. doi:10.36876/aamb.1010

Dhundale, V., Hemke, V., Desai, D., Khemchandani, P., Aher, G., & Dikonda, P. (2020). Evaluation of bioelectricity productivity using alkaliphilic *Bacillus alkalogaya* BW2(1) as a possible exoelectrogens for improvement of microbial fuel cell performance. *Journal of Applied Biology and Biotechnology*, *8*(01), 69–75. doi:10.7324/JABB.2020.80112

Di Donato, P., Buono, A., Poli, A., Finore, I., Abbamondi, G. R., Nicolaus, B., & Lama, L. (2018). Exploring marine environments for the identification of extremophiles and their enzymes for sustainable and green bioprocesses. *Sustainability*, *11*(1), 1–20. doi:10.3390u11010149

Di Giulio, M. (2005). A comparison of proteins from Pyrococcus furiosus and Pyrococcus abyssi: Barophily in the physicochemical properties of amino acids and in the genetic code. *Gene*, *346*, 1–6. doi:10.1016/j.gene.2004.10.008 PMID:15716096

Dikilitas, M., Karakas, S., Simsek, E., & Yadav, A. N. (2021). Microbes from Cold Deserts and Their Applications in Mitigation of Cold Stress in Plants. In *Microbiomes of Extreme Environments* (pp. 126–152). CRC Press. doi:10.1201/9780429328633-7

Dimopoulou, M., Bardeau, T., Ramonet, P.-Y., Miot-Certier, C., Claisse, O., Doco, T., Petrel, M., Lucas, P., & Dols-Lafargue, M. (2016). Exopolysaccharides produced by *Oenococcus oeni*: From genomic and phenotypic analysis to technological valorization. *Food Microbiology*, *53*, 10–17. doi:10.1016/j.fm.2015.07.011 PMID:26611165

Ding, Z. G., Ji, Y., Yin, M., Zhao, Y. R., Feng, Y. Z., Chunyu, W. X., & Tang, S. K. (2019). *Phytoactinopolyspora halophila* sp. nov., a halophilic actinomycete isolated from a saline soil. *International Journal of Systematic and Evolutionary Microbiology*, *69*(2), 384–389. doi:10.1099/ijsem.0.003153 PMID:30543502

Dipippo, J. L., Nesbø, C. L., Dahle, H., Doolittle, W. F., Birkland, N. K., & Noll, K. M. (2009). Kosmotogaolearia gen. nov., sp. nov., a thermophilic, anaerobic heterotroph isolated from an oil production fluid. *International Journal of Systematic and Evolutionary Microbiology*, *59*(12), 2991–3000. doi:10.1099/ijs.0.008045-0 PMID:19643902

Dirmeier, R., Keller, M., Hafenbradl, D., Braun, F.-J., Rachel, R., Burggraf, S., & Stetter, K.O. (1998). *Thermococcus acidaminovorans sp. nov., a new hyperthermophilic alkalophilic archaeon growing on amino acids.* Academic Press.

Dobrowolski, R., Szcześ, A., Czemierska, M., & Jarosz-Wikołazka, A. (2017). Studies of cadmium (II), lead (II), nickel (II), cobalt (II) and chromium (VI) sorption on extracellular polymeric substances produced by *Rhodococcus opacus* and *Rhodococcus rhodochrous*. *Bioresource Technology*, *225*, 113–120. doi:10.1016/j.biortech.2016.11.040 PMID:27888727

Dolev, M. B., Braslavsky, I., & Davies, P. L. (2016). *Ice-Binding Proteins and Their Function.* doi:10.1146/annurev-biochem-060815-014546

Dolfing, J. (2014). Thermodynamic constraints on syntrophic acetate oxidation. *Applied and Environmental Microbiology, 80*(4), 1539–1541. doi:10.1128/AEM.03312-13 PMID:24480834

Dong, X., & Chen, Z. (2012). Psychrotolerant methanogenic archaea: Diversity and cold adaptation mechanisms. *Science China Life Sciences, 55*(5), 415–421. doi:10.1007/s11427-012-4320-0

Donlan, R. M. (2002). Biofilms: Microbial life on surfaces. *Emerging Infectious Diseases, 8*(9), 881–890. doi:10.3201/eid0809.020063 PMID:12194761

Don, T. M., Chen, C. W., & Chan, T. H. (2006). Preparation and characterization of poly (hydroxyalkanoate) from the fermentation of *Haloferax mediterranei. Journal of Biomaterials Science. Polymer Edition, 17*(12), 1425–1438. doi:10.1163/156856206778937208 PMID:17260512

Dopson, Ni, & Sleutels. (2015). Possibilities for extremophilic microorganisms in microbial electrochemical systems. *Federation of European Microbiological Societies Microbiology Reviews.*

Dörr, C., Zaparty, M., Tjaden, B., Brinkmann, H., & Siebers, B. (2003). The Hexokinase of the Hyperthermophile *Thermoproteus tenax* ATP-Dependent Hexokinases and ADP-Dependent glucokinases, two alternatives for glucose phosphorylation in archaea. *The Journal of Biological Chemistry, 278*(21), 18744–18753. PMID:12626506

Dos Reis, S. A., da Conceição, L. L., Siqueira, N. P., Rosa, D. D., da Silva, L. L., & Maria do Carmo, G. P. (2017). Review of the mechanisms of probiotic actions in the prevention of colorectal cancer. *Nutrition Research (New York, N.Y.), 37*, 1–19. doi:10.1016/j.nutres.2016.11.009 PMID:28215310

Dua, A., Joshi, S., & Satyanarayana, T. (2016). Recombinant exochitinase of the thermophilic mould Myceliopthora thermophila BJA: Characteristics and utility in generating N-acetyl glucosamine and in biocontrol of phytopathogenic fungi. *Biotechnology Progress, 33*(1), 70–80. doi:10.1002/btpr.2370

Dua, M., Singh, A., Sethunathan, N., & Johri, A. K. (2002). Biotechnology and bioremediation: Successes and limitations. *Applied Microbiology and Biotechnology, 59*(2–3), 143–152. PMID:12111139

Dumorné, K., Córdova, D. C., Astorga-Eló, M., & Renganathan, P. (2017). *Extremozymes: A potential source for industrial applications.* Academic Press.

Dumorné, K., Córdova, D. C., Astorga-Eló, M., & Renganathan, P. (2017). Extremozymes: A potential source for industrial applications. *Journal of Microbiology and Biotechnology, 27*(4), 649–659. doi:10.4014/jmb.1611.11006 PMID:28104900

Dutta, S. D., Patel, D. K., & Lim, K.-T. (2019). Functional cellulose-based hydrogels as extracellular matrices for tissue engineering. *Journal of Biological Engineering, 13*(1), 1–19. doi:10.118613036-019-0177-0 PMID:31249615

Duymaz, B. T., Erdiler, F. B., Alan, T., Aydogdu, M. O., Inan, A. T., Ekren, N., ... Oktar, F. N. (2019). 3D bio-printing of levan/polycaprolactone/gelatin blends for bone tissue engineering: Characterization of the cellular behavior. *European Polymer Journal, 119*, 426–437. doi:10.1016/j.eurpolymj.2019.08.015

Ece, F., Kaya, A., Avci, F. G., Sayar, N. A., Kazan, D., Altinisik Kaya, F. E., Avci, F. G., Sayar, N. A., Kazan, D., Sayar, A. A., & Sariyar Akbulut, B. (2018). What Are the Multi-Omics Mechanisms for Adaptation by Microorganisms to High Alkalinity? A Transcriptomic and Proteomic Study of a Bacillus Strain with Industrial Potential. *OMICS: A Journal of Integrative Biology, 22*(11), 717–732. doi:10.1089/omi.2018.0127 PMID:30457468

Edbeib, M. F., Wahab, R. A., & Huyop, F. (2016). Halophiles: Biology, adaptation, and their role in decontamination of hyper saline environments. *World Journal of Microbiology & Biotechnology, 32*(8), 1–23. doi:10.100711274-016-2081-9 PMID:27344438

Eddehech, A., Zied, Z., Aloui, F., Smichi, N., Noiriel, A., Abousalham, A., & Gargouri, Y. (2019). Production, purification and biochemical characterization of a thermoactive, alkaline lipase from a newly isolated *Serratia* sp. W3 Tunisian strain. *International Journal of Biological Macromolecules, 123*, 792–800. doi:10.1016/j.ijbiomac.2018.11.050 PMID:30439433

Eichler, J. (2001). Biotechnological uses of archaeal extremozymes. *Biotechnology Advances, 19*(4), 261–278. doi:10.1016/S0734-9750(01)00061-1 PMID:14538076

Ekström, S., Nozière, B., Hultberg, M., Alsberg, T., Magnér, J., Nilsson, E. D., & Artaxo, P. (2010). A possible role of ground-based microorganisms on cloud formation in the atmosphere. *Biogeosciences, 7*(1), 387–394. doi:10.5194/bg-7-387-2010

Elazzazy, A. M., Abdelmoneim, T. S., & Almaghrabi, O. A. (2015). Isolation and characterization of biosurfactant production under extreme environmental conditions by alkali-halo-thermophilic bacteria from Saudi Arabia. *Saudi Journal of Biological Sciences, 22*(4), 466–475. doi:10.1016/j.sjbs.2014.11.018 PMID:26150754

Elleuche, S., Schroeder, C., Sahm, K., & Antranikian, G. (2014). Extremozymes—Biocatalysts with unique properties from extremophilic microorganisms. *Current Opinion in Biotechnology, 29*, 116–123. doi:10.1016/j.copbio.2014.04.003 PMID:24780224

Elmahdi, S., DaSilva, L. V., & Parveen, S. (2016). Antibiotic resistance of *Vibrio parahaemolyticus* and *Vibrio vulnificus* in various countries: A review. *Food Microbiology, 57*, 128–134. doi:10.1016/j.fm.2016.02.008 PMID:27052711

Elmore, J. R., Yokooji, Y., Sato, T., Olson, S., Claiborne, V. C., Glover, I. III, Graveley, B. R., Atomi, H., Terns, R. M., & Terns, M. P. (2013). Programmable plasmid interference by the CRISPR-Cas system in *Thermococcus kodakarensis*. *RNA Biology, 10*(5), 828–840. doi:10.4161/rna.24084 PMID:23535213

El-Naggar, M. E., Othman, S. I., Allam, A. A., & Morsy, O. M. (2020). Synthesis, drying process and medical application of polysaccharide-based aerogels. *International Journal of Biological Macromolecules, 145*, 1115–1128. doi:10.1016/j.ijbiomac.2019.10.037 PMID:31678101

Elumalai, P., Parthipan, P., Karthikeyan, O. P., & Rajasekar, A. (2017). *Enzyme-mediated biodegradation of long-chain n-alkanes (C32 and C40) by thermophilic bacteria. Biotech, 7(2).*

Empadinhas, N., & da Costa, M. S. (2011). Diversity, biological roles and biosynthetic pathways for sugar-glycerate containing compatible solutes in bacteria and archaea. *Environmental Microbiology, 13*(8), 2056–2077. doi:10.1111/j.1462-2920.2010.02390.x PMID:21176052

Esakkiraj, P., Prabakaran, G., Maruthiah, T., Immanuel, G., & Palavesam, A. (2016). Purification and Characterization of Halophilic Alkaline Lipase from *Halobacillus* sp. *Proceedings of the National Academy of Sciences. India. Section B, Biological Sciences, 86*(2), 309–314. doi:10.100740011-014-0437-1

Evilia, C., & Hou, Y.-M. (2006). Acquisition of an Insertion Peptide for Efficient Aminoacylation by a Halophile tRNA Synthetase. *Biochemistry, 45*(22), 6835–6845. doi:10.1021/bi0521386 PMID:16734420

Ezeokoli, O. T., Bezuidenhout, C. C., Maboeta, M. S., Khasa, D. P., & Adeleke, R. A. (2020). Structural and functional differentiation of bacterial communities in post-coal mining reclamation soils of South Africa: bioindicators of soil ecosystem restoration. *Scientific Reports, 10*(1). doi:10.1038/s41598-020-58576-5

Fan, G., Wu, Q., Li, Q., Sun, B., Ma, Y., Wu, K., Wang, C., Teng, C., Yang, R., & Li, X. (2020). Impact of the disulfide bond on hydrolytic characteristics of a xylanase from Talaromyces thermophiles F1208. *International Journal of Biological Macromolecules, 164*, 1748–1757. doi:10.1016/j.ijbiomac.2020.07.270 PMID:32777426

Fang, J., Zhang, L., & Bazylinski, D. A. (2010). Deep-sea piezosphere and piezophiles: Geomicrobiology and biogeochemistry. *Trends in Microbiology*, *18*(9), 413–422. doi:10.1016/j.tim.2010.06.006 PMID:20663673

Fang, Z. M., Li, T. L., Chang, F., Zhou, P., Fang, W., Hong, Y. Z., Zhang, X. C., Peng, H., & Xiao, Y. Z. (2012). A new marine bacterial laccase with chloride-enhancing, alkaline-dependent activity and dye decolorization ability. *Bioresource Technology*, *111*, 36–41. doi:10.1016/j.biortech.2012.01.172 PMID:22377476

Faria, J., Ahmed, S., Gerritsen, K. G., Mihaila, S. M., & Masereeuw, R. (2019). Kidney-based *in-vitro* models for drug-induced toxicity testing. *Archives of Toxicology*, *93*(12), 3397–3418. doi:10.100700204-019-02598-0 PMID:31664498

Farraj, D. A. al, Hadibarata, T., Yuniarto, A., Alkufeidy, R. M., Alshammari, M. K., & Syafiuddin, A. (2020). Exploring the potential of halotolerant bacteria for biodegradation of polycyclic aromatic hydrocarbon. *Bioprocess and Biosystems Engineering, 43*(12), 2305–2314. doi:10.1007/s00449-020-02415-4

Feder, M. E., & Hofmann, G. E. (1999). Heat-shock proteins, molecular chaperones, and the stress response: Evolutionary and ecological physiology. *Annual Review of Physiology*, *61*(1), 243–282. doi:10.1146/annurev.physiol.61.1.243 PMID:10099689

Felföldi, T., Somogyi, B., Márialigeti, K., & Vörös, L. (2009). Characterization of photoautotrophic picoplankton assemblages in turbid, alkaline lakes of the Carpathian Basin (Central Europe). *Journal of Limnology*, *68*(2), 385–395. doi:10.4081/jlimnol.2009.385

Felip, M., Wille, A., Sattler, B., & Psenner, R. (2002). Microbial communities in the winter cover and the water column of an alpine lake: System connectivity and uncoupling. *Aquatic Microbial Ecology*, *29*(2), 123–134. doi:10.3354/ame029123

Feller, G. (2007). Life at low temperatures: Is disorder the driving force? *Extremophiles*, *11*(2), 211–216. doi:10.100700792-006-0050-1 PMID:17160345

Feller, G. (2010). Protein stability and enzyme activity at extreme biological temperatures. *Journal of Physics Condensed Matter*, *22*(32), 323101. doi:10.1088/0953-8984/22/32/323101 PMID:21386475

Feller, G. (2013). Psychrophilic Enzymes: From Folding to Function and Biotechnology. *Scientifica*, *2013*, 1–28. doi:10.1155/2013/512840 PMID:24278781

Felt, O., Einmahl, S., Furrer, P., Baeyens, V., & Gurny, R. (2001). Polymeric systems for ophthalmic drug delivery. *Polymeric Biomaterials*, 377-421.

Fendrihan, S., Legat, A., Pfaffenhuemer, M., Gruber, C., Weidler, G., Gerbl, F., & Stan-Lotter, H. (2006). Extremely halophilic archaea and the issue of long-term microbial survival. *Life in Extreme Environments*, (2-3), 125–140. doi:10.100711157-006-0007-y PMID:21984879

Feng, S., Qiu, Y., Huang, Z., Yin, Y., Zhang, H., Zhu, D., Tong, Y., & Yang, H. (2021). The adaptation mechanisms of *Acidithiobacillus caldus* CCTCC M 2018054 to extreme acid stress: Bioleaching performance, physiology, and transcriptomics. *Environmental Research*, *199*, 111341. doi:10.1016/j.envres.2021.111341 PMID:34015291

Feng, X. M., Zhai, X. H., Liu, S. W., Wang, C. M., Sun, C. H., Zhou, X., Zhang, Y. J., & Li, X. J. (2020). *Phytoactinopolyspora mesophila* sp. Nov., isolated from a saline–alkaline soil. *International Journal of Systematic and Evolutionary Microbiology*, *70*(7), 4179–4185. doi:10.1099/ijsem.0.004274 PMID:32558636

Feniouk, B. A., Suzuki, T., & Yoshida, M. (2007). Regulatory interplay between proton motive force, ADP, phosphate, and subunit ε in bacterial ATP synthase. *The Journal of Biological Chemistry*, *282*(1), 764–772. doi:10.1074/jbc.M606321200 PMID:17092944

Fernandes, S. S., Coelho, M. S., & de las Mercedes Salas-Mellado, M. (2019). Bioactive compounds as ingredients of functional foods: polyphenols, carotenoids, peptides from animal and plant sources new. In *Bioactive Compounds* (pp. 129–142). Woodhead Publishing. doi:10.1016/B978-0-12-814774-0.00007-4

Fialho, A. M., Moreira, L. M., Granja, A. T., Popescu, A. O., Hoffmann, K., & Sá-Correia, I. (2008). Occurrence, production, and applications of gellan: Current state and perspectives. *Applied Microbiology and Biotechnology*, *79*(6), 889–900. doi:10.100700253-008-1496-0 PMID:18506441

Fierer, N., & Jackson, R. B. (2006). The diversity and biogeography of soil bacterial communities. *Proceedings of the National Academy of Sciences of the United States of America*, *103*(3), 626–631. doi:10.1073/pnas.0507535103 PMID:16407148

Fischel, U., & Oren, A. (1993). Fate of compatible solutes during dilution stress in Ectothiorhodospira marismortui. *FEMS Microbiology Letters*, *113*(1), 113–118. doi:10.1111/j.1574-6968.1993.tb06497.x

Flemming, H.-C. (2016). EPS—Then and now. *Microorganisms*, *4*(4), 41. doi:10.3390/microorganisms4040041 PMID:27869702

Flemming, H.-C., & Wingender, J. (2010). The biofilm matrix. *Nature Reviews. Microbiology*, *8*(9), 623–633. doi:10.1038/nrmicro2415 PMID:20676145

Foreman, C. M., Sattler, B., Mikucki, J. A., Porazinska, D. L., & Priscu, J. C. (2007). Metabolic activity and diversity of cryoconites in the Taylor Valley, Antarctica. *Journal of Geophysical Research. Biogeosciences*, *112*(G4).

Forró, L., Nédli, J., Csata, E., Krízsik, V., Balogh, C., & Tóth, L. G. (2017). Morphometric characteristics and coi haplotype diversity of *Arctodiaptomus spinosus* (copedoda) populations in Soda Pans in Hungary. *Acta Biologica Hungarica*, *68*, 279–289. doi:10.1556/018.68.2017.3.5 PMID:28901808

Forster, J. (1887). Ueber einige Eigenschaften leuchtender Bakterien. *Centr. Bakteriol. Parasitenk.*, *2*, 337–340.

Fraissinet-Tachet, L., & Fevre, M. (1996). Regulation by galacturonic acid of pectinolytic enzyme production by *Sclerotinia sclerotiorum*. *Current Microbiology*, *33*(1), 49–53. doi:10.1007002849900073 PMID:8661689

Francis, C. A., Roberts, K. J., Beman, J. M., Santoro, A. E., & Oakley, B. B. (2005). Ubiquity and diversity of ammonia-oxidizing archaea in water columns and sediments of the ocean. *Proceedings of the National Academy of Sciences of the United States of America*, *102*(41), 14683–14688. doi:10.1073/pnas.0506625102 PMID:16186488

Freire, R. K. B., Mendonça, C. M. N., Ferraro, R. B., Moguel, I. S., Tonso, A., Lourenco, F. R., Santos, J. H. P. M., Sette, L. D., & Pessoa, A. Jr. (2021). Glutaminase-free L-asparaginase production by *Leucosporidium muscorum* isolated from Antarctic marine-sediment. *Preparative Biochemistry & Biotechnology*, *51*(3), 277–288. doi:10.1080/10826068.2020.1815053 PMID:32921254

Freitas, F., Alves, V. D., & Reis, M. A. (2011). Advances in bacterial exopolysaccharides: From production to biotechnological applications. *Trends in Biotechnology*, *29*(8), 388–398. doi:10.1016/j.tibtech.2011.03.008 PMID:21561675

Friedmann, E. I., & Ocampo, R. (1976). Endolithic blue-green algae in the dry valleys: Primary producers in the Antarctic desert ecosystem. *Science*, *193*(4259), 1247–1249. doi:10.1126cience.193.4259.1247 PMID:17837022

Frock, A. D., Gray, S. R., & Kelly, R. M. (2012). Hyperthermophilic Thermotoga species differ with respect to specific carbohydrate transporters and glycoside hydrolases. *Applied and Environmental Microbiology*, *78*(6), 1978–1986. doi:10.1128/AEM.07069-11 PMID:22247137

Fuentes-Tristan, S., Parra-Saldivar, R., Iqbal, H. M., & Carrillo-Nieves, D. (2019). Bioinspired biomolecules: Mycosporine-like amino acids and scytonemin from *Lyngbya* sp. with UV-protection potentialities. *Journal of Photochemistry and Photobiology. B, Biology*, *201*, 111684. doi:10.1016/j.jphotobiol.2019.111684 PMID:31733505

Fujii, M., Takano, Y., Kojima, H., Hoshino, T., Tanaka, R., & Fukui, M. (2010). Microbial community structure, pigment composition, and nitrogen source of red snow in Antarctica. *Microbial Ecology*, *59*(3), 466–475. doi:10.100700248-009-9594-9 PMID:19847476

Fujimoto, K., & Morita, T. (2006). Aerobic removal of technetium by a marine Halomonas strain. *Applied and Environmental Microbiology*, *72*(12), 7922–7924. doi:10.1128/AEM.00819-06 PMID:17056690

Fujisawa, M., Fackelmayer, O. J., Liu, J., Krulwich, T. A., & Hicks, D. B. (2010). The ATP synthase a-subunit of extreme alkaliphiles is a distinct variant: Mutations in the critical alkaliphile-specific residue Lys-180 and other residues that support alkaliphile oxidative phosphorylation. *The Journal of Biological Chemistry*, *285*(42), 32105–32115. doi:10.1074/jbc.M110.165084 PMID:20716528

Fujiwara, N., Masui, A., & Imanaka, T. (1993). Purification and properties of the highly thermostable alkaline protease from an alkaliphilic and thermophilic *Bacillus* sp. *Journal of Biotechnology*, *30*(2), 245–256. doi:10.1016/0168-1656(93)90117-6 PMID:7764036

Fukuchi, S., & Nishikawa, K. (2001). Protein surface amino acid compositions distinctively differ between thermophilic and mesophilic bacteria. *Journal of Molecular Biology*, *309*(4), 835–843. doi:10.1006/jmbi.2001.4718 PMID:11399062

Fu, Q., Kobayashi, H., Kawaguchi, H., Wakayama, T., Maeda, H., & Sato, K. (2013). A thermophilic gram-negative nitrate-reducing bacterium, *Calditerrivibrio nitroreducens*, exhibiting electricity generation capability. *Environmental Science & Technology*, *47*(21), 12583–12590. doi:10.1021/es402749f PMID:24053548

Gabani, P., & Singh, O. V. (2013). Radiation-resistant extremophiles and their potential in biotechnology and therapeutics. *Applied Microbiology and Biotechnology*, *97*(3), 993–1004. doi:10.100700253-012-4642-7 PMID:23271672

Gaisin, V. A., Kalashnikov, A. M., Grouzdev, D. S., Sukhacheva, M. V., Kuznetsov, B. B., & Gorlenko, V. M. (2017). *Chloroflexus islandicus* sp. nov., a thermophilic filamentous anoxygenic phototrophic bacterium from a geyser. *International Journal of Systematic and Evolutionary Microbiology*, *67*(5), 1381–1386. doi:10.1099/ijsem.0.001820 PMID:28126046

Galinski, E. A. (1993). Compatible solutes of halophilic eubacteria: molecular principles, water-solute interaction, stress protection. *Experientia, 49*(6), 487–496. doi:10.1007/BF01955150

Galinski, E. A., & Herzog, R. M. (1990). The role of trehalose as a substitute for nitrogen-containing compatible solutes (Ectothiorhodospira halochloris). *Archives of Microbiology*, *153*(6), 607–613. doi:10.1007/BF00245273

Gallardo-Cerda, J., Levihuan, J., Lavín, P., Oses, R., Atala, C., Torres-Díaz, C., & Cuba-Díaz, M. (2018). Antarctic rhizobacteria improve salt tolerance and physiological performance of the Antarctic vascular plants. *Polar Biology, 41*(10), 1973–1982. doi:10.1007/s00300-018-2336-z

Ganguly, J., Martin-Pascual, M., & van Kranenburg, R. (2020). CRISPR interference (CRISPRi) as transcriptional repression tool for *Hungateiclostridium thermocellum* DSM 1313. *Microbial Biotechnology*, *13*(2), 339–349. doi:10.1111/1751-7915.13516 PMID:31802632

Gaonkar, S. K., & Furtado, I. J. (2021). Valorization of low-cost agro-wastes residues for the maximum production of protease and lipase haloextremozymes by *Haloferax lucentensis* GUBF-2 MG076078. *Process Biochemistry*, *101*, 72–88. doi:10.1016/j.procbio.2020.10.019

Gao, Y. C., Wang, J. N., Guo, S. H., Hu, Y. L., Li, T. T., Mao, R., & Zeng, D. H. (2015). Effects of salinization and crude oil contamination on soil bacterial community structure in the Yellow River Delta region, China. *Applied Soil Ecology*, *86*, 165–173. doi:10.1016/j.apsoil.2014.10.011

Garneau, J. E., Dupuis, M. È., Villion, M., Romero, D. A., Barrangou, R., Boyaval, P., Fremaux, C., Horvath, P., Magadán, A. H., & Moineau, S. (2010). The CRISPR/Cas bacterial immune system cleaves bacteriophage and plasmid DNA. *Nature*, *468*(7320), 67–71. doi:10.1038/nature09523 PMID:21048762

Garnier, F., Couturier, M., Débat, H., & Nadal, M. (2021). Archaea: A Gold Mine for Topoisomerase Diversity. *Frontiers in Microbiology*, *0*, 1083. doi:10.3389/fmicb.2021.661411 PMID:34113328

Garza-Cervantes, J. A., Escárcega-González, C. E., Castro, E. D. B., Mendiola-Garza, G., Marichal-Cancino, B. A., López-Vázquez, M. A., & Morones-Ramirez, J. R. (2019). Antimicrobial and antibiofilm activity of biopolymer-Ni, Zn nanoparticle biocomposites synthesized using *R. mucilaginosa* UANL-001L exopolysaccharide as a capping agent. *International Journal of Nanomedicine*, *14*, 2557–2571. doi:10.2147/IJN.S196470 PMID:31118605

Gasiunas, G., Barrangou, R., Horvath, P., & Siksnys, V. (2012). Cas9-crRNA ribonucleoprotein complex mediates specific DNA cleavage for adaptive immunity in bacteria. *Proceedings of the National Academy of Sciences of the United States of America*, *109*(39), E2579–E2586. doi:10.1073/pnas.1208507109 PMID:22949671

Gatti-Lafranconi, P., Natalello, A., Rehm, S., Doglia, S. M., Pleiss, J., & Lotti, M. (2010). Evolution of stability in a cold-active enzyme elicits specificity relaxation and highlights substrate-related effects on temperature adaptation. *Journal of Molecular Biology*, *395*(1), 155–166. doi:10.1016/j.jmb.2009.10.026 PMID:19850050

Gaur, R., & Tiwari, S. (2015). Isolation, production, purification and characterization of an organic-solvent-thermostable alkalophilic cellulase from *Bacillus vallismortis* RG-07. *BMC Biotechnology*, *15*(1), 19. doi:10.118612896-015-0129-9 PMID:25886936

Geissler, S., Siegers, K., & Schiebel, E. (1998). A novel protein complex promoting formation of functional alpha and gamma-tubulin. *The EMBO Journal*, *17*(4), 952–966. doi:10.1093/emboj/17.4.952 PMID:9463374

Genc, B., Taskin, M., & Adiguzel, A. (2021). Exopolysaccharide of *Anoxybacillus pushchinoensis* G11 has antitumor and antibiofilm activities. *Archives of Microbiology*, 1–18. PMID:33604750

Georgieva, T. I., & Ahring, B. K. (2007). Evaluation of continuous ethanol fermentation of dilute-acid corn stover hydrolysate using thermophilic anaerobic bacterium Thermoanaerobacter BG1L1. *Applied Microbiology and Biotechnology*, *77*(1), 61–68. doi:10.100700253-007-1149-8 PMID:17899073

Georgieva, T. I., Mikkelsen, M. J., & Ahring, B. K. (2007). Ethanol Production from Wet-Exploded Wheat Straw Hydrolysate by Thermophilic Anaerobic Bacterium Thermoanaerobacter BG1L1 in a Continuous Immobilized Reactor. In *Biotechnology for Fuels and Chemicals* (pp. 99–110). Humana Press. doi:10.1007/978-1-60327-526-2_12

Georlette, D., Damien, B., Blaise, V., Depiereux, E., Uversky, V. N., Gerday, C., & Feller, G. (2003). Structural and functional adaptations to extreme temperatures in psychrophilic, mesophilic, and thermophilic DNA ligases. *The Journal of Biological Chemistry*, *278*(39), 37015–37023. doi:10.1074/jbc.M305142200 PMID:12857762

Gerday, C., & Glansdorff, N. (Eds.). (2007). *Physiology and Biochemistry of Extremophiles*. ASM Press. doi:10.1128/9781555815813

Ghalem, B. R. (2017). Antioxidant and antimicrobial activities of exopolysaccharides from yoghurt starter. *American Journal of Chemical and Biochemical Engineering*, *3*(3), 35–39.

Ghanmi, F., Carré-Mlouka, A., Zarai, Z., Mejdoub, H., Peduzzi, J., Maalej, S., & Rebuffat, S. (2020). The extremely halophilic archaeon *Halobacterium salinarum* ETD5 from the solar saltern of Sfax (Tunisia) produces multiple halocins. *Research in Microbiology*, *171*(2), 80–90. doi:10.1016/j.resmic.2019.09.003 PMID:31560985

Ghelani, A., Patel, R., Mangrola, A., & Dudhagara, P. (2014). Cultivation-independent comprehensive survey of bacterial diversity in Tulsi Shyam Hot Springs, India. *Genomics Data*, *4*, 54–56. doi:10.1016/j.gdata.2015.03.003 PMID:26484176

Ghimire-Rijal, S., Lu, X., Myles, D. A., & Cuneo, M. J. (2014). Duplication of genes in an ATP-binding cassette transport system increases dynamic range while maintaining ligand specificity. *The Journal of Biological Chemistry*, *289*(43), 30090–30100. doi:10.1074/jbc.M114.590992 PMID:25210043

Ghorbel-Bellaaj, O., Hayet, B. K., Bayoudh, A., Younes, I., Hmidet, N., Jellouli, K., & Nasri, M. (2012). *Pseudomonas aeruginosa* A2 elastase: Purification, characterization and biotechnological applications. *International Journal of Biological Macromolecules*, *50*(3), 679–686. doi:10.1016/j.ijbiomac.2012.01.038 PMID:22326423

Ghosh, R., Chattopadhyay, P., Chattopadhyay, B., & Pal, D. (2010). Antibiotic resistance profile of halophilic microorganisms isolated from tannery effluent. *Indian Journal of Biotechnology*, *9*(1), 80–86.

Ghuysen, J.-M. (1968). Use of bacteriolytic enzymes in determination of wall structure and their role in cell metabolism. *Bacteriological Reviews*, *32*(4 Pt 2), 425–464. doi:10.1128/br.32.4_pt_2.425-464.1968 PMID:4884715

Giaquinto, L., Curmi, P. M. G., Siddiqui, K. S., Poljak, A., DeLong, E., DasSarma, S., & Cavicchioli, R. (2007). Structure and function of cold shock proteins in archaea. *Journal of Bacteriology*, *189*(15), 5738–5748. doi:10.1128/JB.00395-07 PMID:17545280

Giavasis, I., Harvey, L. M., & McNeil, B. (2000). Gellan Gum. *Critical Reviews in Biotechnology*, *20*(3), 177–211. doi:10.1080/07388550008984169 PMID:11039329

Gilbert, J. A., Hill, P. J., Dodd, C. E., & Laybourn-Parry, J. (2004). Demonstration of antifreeze protein activity in Antarctic lake bacteria. *Microbiology*, *150*(1), 171–180. doi:10.1099/mic.0.26610-0 PMID:14702410

Gilbert, L. A., Larson, M. H., Morsut, L., Liu, Z., Brar, G. A., Torres, S. E., Stern-Ginossar, N., Brandman, O., Whitehead, E. H., Doudna, J. A., Lim, W. A., Weissman, J. S., & Qi, L. S. (2013). CRISPR-mediated modular RNA-guided regulation of transcription in eukaryotes. *Cell*, *154*(2), 442–451. doi:10.1016/j.cell.2013.06.044 PMID:23849981

Gilichinsky, D., Vishnivetskaya, T., Petrova, M., Spirina, E., Mamykin, V., & Rivkina, E. (2008). Bacteria in permafrost. In *Psychrophiles: from biodiversity to biotechnology* (pp. 83–102). Springer. doi:10.1007/978-3-540-74335-4_6

Gil, J. F., Mesa, V., Estrada-Ortiz, N., Lopez-Obando, M., Gómez, A., & Plácido, J. (2021). Viruses in extreme environments, current overview, and biotechnological potential. *Viruses*, *13*(1), 81. doi:10.3390/v13010081 PMID:33430116

Gimza, B. D., Larias, M. I., Budny, B. G., & Shaw, L. N. (2019). Mapping the Global Network of Extracellular Protease Regulation in *Staphylococcus aureus*. *MSphere*, *4*(5). Advance online publication. doi:10.1128/mSphere.00676-19 PMID:31645429

Giuffrida, D., Monnet, C., Laurent, F., Cacciola, F., Oteri, M., Le Piver, M., Caro, Y., Donato, P., Mondello, L., Roueyre, D., & Dufossé, L. (2020). Carotenoids from the ripening bacterium *Brevibacterium linens* impart color to the rind of the French cheese, Fourme de Montbrison (PDO). *Natural Product Research*, *34*(1), 10–15. doi:10.1080/14786419.2019.1622107 PMID:31140308

Godon, J. J., Galès, A., Latrille, E., Ouichanpagdee, P., & Seyer, J.-P. (2020). An "overlooked" habitat for thermophilic bacteria: The phyllosphere. *BioDiscovery*, *23*, 1–14. doi:10.3897/biodiscovery.23.e47033

Gonzalez, R. (2013). Metabolic engineering: Use of system-level approaches and application to fuel production in *Escherichia coli. Electronic Journal of Biotechnology, 16*(3). Advance online publication. doi:10.2225/vol16-issue3-fulltext-14

Goodchild, A., Saunders, N. F., Ertan, H., Raftery, M., Guilhaus, M., Curmi, P. M., & Cavicchioli, R. (2004). A proteomic determination of cold adaptation in the Antarctic archaeon, Methanococcoides burtonii. *Molecular Microbiology, 53*(1), 309–321. doi:10.1111/j.1365-2958.2004.04130.x PMID:15225324

Gophna, U., Allers, T., & Marchfelder, A. (2017). Finally, archaea get their CRISPR-Cas toolbox. *Trends in Microbiology, 25*(6), 430–432. doi:10.1016/j.tim.2017.03.009 PMID:28391963

Gosink, J. J., Irgens, R. L., & Staley, J. T. (1993). Vertical distribution of bacteria in arctic sea ice. *FEMS Microbiology Ecology, 11*(2), 85–90. doi:10.1111/j.1574-6968.1993.tb05799.x

Gostinčar, C., Grube, M., De Hoog, S., Zalar, P., & Gunde-Cimerman, N. (2009). Extremotolerance in fungi: Evolution on the edge. *FEMS Microbiology Ecology, 71*(1), 2–11. doi:10.1111/j.1574-6941.2009.00794.x PMID:19878320

Goswami, D., Dhandhukia, P., Patel, P., & Thakker, J. N. (2014). Screening of PGPR from saline desert of Kutch: Growth promotion in Arachis hypogea by *Bacillus licheniformis* A2. *Microbiological Research, 169*(1), 66–75. doi:10.1016/j.micres.2013.07.004 PMID:23896166

Gouda, H. A., Moubasher, A. H., Ismail, M. A., & Hussein, N. A. (2020). Alkaliphilic and alkalitolerant fungi in mud and salt crusts collected from Wadi El-Natrun lakes. Journal of Agriculture. *Food and Environment, 01*(02), 29–34. doi:10.47440/JAFE.2020.1205

Goude, R., Renaud, S., Bonnassie, S., Bernard, T., & Blanco, C. (2004). Glutamine, glutamate, and α-glucosylglycerate are the major osmotic solutes accumulated by Erwinia chrysanthemi strain 3937. *Applied and Environmental Microbiology, 70*(11), 6535–6541. doi:10.1128/AEM.70.11.6535-6541.2004 PMID:15528516

Goyer, R. A. (1997). Toxic and essential metal interactions. *Annual Review of Nutrition, 17*(1), 37–50. doi:10.1146/annurev.nutr.17.1.37 PMID:9240918

Graham, D., Pereira, R., Barfield, D., & Cowan, D. (2000). Nitrile biotransformations using free and immobilized cells of a thermophilic *Bacillus* spp. *Enzyme and Microbial Technology, 26*(5-6), 368–373. doi:10.1016/S0141-0229(99)00169-6 PMID:10713209

Grant, W. D., & Heaphy, S. (2010). Metagenomics and recovery of enzyme genes from alkaline saline environments. *Environmental Technology, 31*(10), 1135–1143. doi:10.1080/09593331003646661 PMID:20718296

Grant, W. D., & Sorokin, D. Y. (2011). *Distribution and Diversity of Soda Lake Alkaliphiles*. Extremophiles Handbook. doi:10.1007/978-4-431-53898-1_3

Green, J. L., Bohannan, B. J., & Whitaker, R. J. (2008). Microbial biogeography: from taxonomy to traits. *Science, 320*(5879), 1039-1043.

Grkovic, T., Abdelmohsen, U. R., Othman, E. M., Stopper, H., Edrada-Ebel, R., Hentschel, U., & Quinn, R. J. (2014). Two new antioxidant actinosporin analogues from the calcium alginate beads culture of sponge-associated *Actinokineospora* sp. strain EG49. *Bioorganic & Medicinal Chemistry Letters, 24*(21), 5089–5092. doi:10.1016/j.bmcl.2014.08.068 PMID:25266784

Grout, T., Hong, Y., Basara, J., Balasundaram, B., Kong, Z., & Bukkapatnam, S. T. S. (2012). Significant Winter Weather Events and Associated Socioeconomic Impacts (Federal Aid Expenditures) across Oklahoma: 2000–10. *Weather, Climate, and Society, 4*(1), 48–58. doi:10.1175/WCAS-D-11-00057.1

Grum-Grzhimaylo, A. A., Debets, A. J. M., van Diepeningen, A. D., Georgieva, M. L., & Bilanenko, E. N. (2013). *Sodiomyces alkalinus*, a new holomorphic alkaliphilic ascomycete within the Plectosphaerellaceae. Persoonia. *Molecular Phylogeny and Evolution of Fungi*, *31*(1), 147–158. doi:10.3767/003158513X673080 PMID:24761040

Guan, Z., Tian, B., Perfumo, A., & Goldfine, H. (2013). The polar lipids of *Clostridium psychrophilum*, an anaerobic psychrophile. *Biochimica et Biophysica Acta (BBA)-. Molecular and Cell Biology of Lipids*, *1831*(6), 1108–1112. doi:10.1016/j.bbalip.2013.02.004 PMID:23454375

Gu, J., & Hilser, V. J. (2009). Sequence-based analysis of protein energy landscapes reveals nonuniform thermal adaptation within the proteome. *Molecular Biology and Evolution*, *26*(10), 2217–2227. doi:10.1093/molbev/msp140 PMID:19592668

Gundala, P., & Chinthala, P. (2017). Extremophilic pectinases. In R. K. Sani & R. N. Krishnaraj (Eds.), *Extremophilic enzymatic processing of lignocellulosic feedstocks to bioenergy* (p. 155). Springer. doi:10.1007/978-3-319-54684-1_9

Gunde-Cimerman, N., Plemenitaš, A., & Oren, A. (2018). Strategies of adaptation of microorganisms of the three domains of life to high salt concentrations. *FEMS Microbiology Reviews*, *42*(3), 353–375. doi:10.1093/femsre/fuy009 PMID:29529204

Gunde-Cimerman, N., Zalar, P., de Hoog, S., & Plemenitaš, A. (2000). Hypersaline waters in salterns–natural ecological niches for halophilic black yeasts. *FEMS Microbiology Ecology*, *32*(3), 235–240. PMID:10858582

GunjalAparna, B., Waghmode Meghmala, S., & Patil Neha, N. (2021). Role of extremozymes in bioremediation. *Research Journal of Biotechnology*, *16*, 3.

Guo, J., Sun, S., & Liu, J. (2020). Conversion of waste frying palm oil into biodiesel using free lipase A from *Candida antarctica* as a novel catalyst. *Fuel*, *267*, 117323. doi:10.1016/j.fuel.2020.117323

Guo, K., Hassett, D. J., & Gu, T. (2012). *Microbial Fuel Cells: Electricity Generation from Organic Wastes by Microbes. In Microbial Biotechnology: Energy and Environment*. CAB International.

Guo, L., Brugger, K., Liu, C., Shah, S. A., Zheng, H., Zhu, Y., Wang, S., Lillestol, R. K., Chen, L., Frank, J., Prangishvili, D., Paulin, L., She, Q., Huang, L., & Garrett, R. A. (2011). Genome analyses of Icelandic strains of Sulfolobus islandicus, model organisms for genetic and virus-host interaction studies. *Journal of Bacteriology*, *193*(7), 1672–1680. doi:10.1128/JB.01487-10 PMID:21278296

Gupta, G., Srivastava, S., Khare, S. K., & Prakash, V. (2014). Extremophiles: An overview of microorganism from extreme environment. *International Journal of Agriculture Environment and Biotechnology*, *7*(2), 371–380. doi:10.5958/2230-732X.2014.00258.7

Gupta, M., Aggarwal, S., Navani, N. K., & Choudhury, B. (2015). Isolation and characterization of a protease-producing novel haloalkaliphilic bacterium *Halobiforma* sp. strain BNMIITR from Sambhar lake in Rajasthan, India. *Annals of Microbiology*, *65*(2), 677–686. doi:10.100713213-014-0906-z

Gupta, N., Reddy, V. S., Maiti, S., & Ghosh, A. (2000). Cloning, expression, and sequence analysis of the gene encoding the alkali-stable, thermostable endoxylanase from alkalophilic, mesophilic *Bacillus* sp. strain NG-27. *Applied and Environmental Microbiology*, *66*(6), 2631–2635. doi:10.1128/AEM.66.6.2631-2635.2000 PMID:10831448

Gurumurthy, D. M., & Neelagund, S. E. (2012). Molecular characterization of industrially viable extreme thermostable novel alpha-amylase of *Geobacillus* sp Iso5 Isolated from geothermal spring. *Journal of Pure & Applied Microbiology*, *6*, 1759–1773.

Gurunathan, R., Huang, B., Ponnusamy, V. K., Hwang, J. S., & Dahms, H. U. (2021). Novel recombinant keratin degrading subtilisin-like serine alkaline protease from *Bacillus cereus* isolated from marine hydrothermal vent crabs. *Scientific Reports, 11*(1), 12007. doi:10.103841598-021-90375-4 PMID:34099743

Gutierrez, T., Shimmield, T., Haidon, C., Black, K., & Green, D. H. (2008). Emulsifying and metal ion binding activity of a glycoprotein exopolymer produced by *Pseudoalteromonas* sp. strain TG12. *Applied and Environmental Microbiology, 74*(15), 4867–4876. doi:10.1128/AEM.00316-08 PMID:18552188

Haft, D. H., Selengut, J., Mongodin, E. F., & Nelson, K. E. (2005). A guild of 45 CRISPR-associated (Cas) protein families and multiple CRISPR/Cas subtypes exist in prokaryotic genomes. *PLoS Computational Biology, 1*(6), e60. doi:10.1371/journal.pcbi.0010060 PMID:16292354

Hagagy, N. I., Saddiq, A. A., Hamedo, H. A., & Selim, S. A. (2021). Extremophiles Inhabiting Unique Ecosystems in Egypt. In *Extreme Environments* (pp. 252–262). CRC Press. doi:10.1201/9780429343452-15

Hägerström, H. (2003). *Polymer gels as pharmaceutical dosage forms: rheological performance and physicochemical interactions at the gel-mucus interface for formulations intended for mucosal drug delivery* Acta Universitatis Upsaliensis.

Hãggblom, M., & Margesin, R. (2005). Microbial life in cold ecosystems. FEMS Microbiology Ecology, 53(1), 1–2. doi:10.1016/j.femsec.2005.03.006

Hakamada, Y., Koike, K., Yoshimatsu, T., Mori, H., Kobayashi, T., & Ito, S. (1997). Thermostable alkaline cellulase from an alkaliphilic isolate, *Bacillus* sp. KSM-S237. *Extremophiles, 1*(3), 151–156. doi:10.1007007920050028 PMID:9680321

Hallenbeck, P. C., Abo-Hashesh, M., & Ghosh, D. (2012). Strategies for improving biological hydrogen production. *Bioresource Technology, 110*, 1–9. doi:10.1016/j.biortech.2012.01.103 PMID:22342581

Hamdan, A. (2018). Psychrophiles: Ecological significance and potential industrial application. *South African Journal of Science, 114*(5-6), 1–6. doi:10.17159ajs.2018/20170254

Hamed, A., Khedr, M., & Abdelraof, M. (2019). Molecular characterization of alkaline protease-coding gene from *Bacillus licheniformis* MK90 mutants with biofilm inhibitory activity. *Egyptian Pharmaceutical Journal, 18*(4), 419. doi:10.4103/epj.epj_47_19

Hamer, R. J. (1995). Enzymes in the baking industry. In *Enzymes in food processing* (pp. 190–222). Springer. doi:10.1007/978-1-4615-2147-1_6

Hampp, N., & Oesterhelt, D. (2005). Bacteriorhodopsin and Its Potential in Technical Applications. In Nanobiotechnology (pp. 146–167). Wiley-VCH Verlag GmbH & Co. KGaA. doi:10.1002/3527602453.ch11

Han, W., Wang, W., Zhao, M., Sugahara, K., & Li, F. (2014). A novel eliminase from a marine bacterium that degrades hyaluronan and chondroitin sulfate. *The Journal of Biological Chemistry, 289*(40), 27886–27898. doi:10.1074/jbc.M114.590752 PMID:25122756

Harbak, L., & Thygesen, H. V. (2002). Safety evaluation of a xylanase expressed in *Bacillus subtilis. Food and Chemical Toxicology, 40*(1), 1–8. doi:10.1016/S0278-6915(01)00092-8 PMID:11731030

Harishchandra, R. K., Sachan, A. K., Kerth, A., Lentzen, G., Neuhaus, T., & Galla, H.-J. (2011). Compatible solutes: Ectoine and hydroxyectoine improve functional nanostructures in artificial lung surfactants. *Biochimica et Biophysica Acta (BBA) - Biomembranes, 1808*(12), 2830–2840. doi:10.1016/j.bbamem.2011.08.022 PMID:21889490

Harishchandra, R. K., Wulff, S., Lentzen, G., Neuhaus, T., & Galla, H. J. (2010). The effect of compatible solute ectoines on the structural organization of lipid monolayer and bilayer membranes. *Biophysical Chemistry, 150*(1–3), 37–46. doi:10.1016/j.bpc.2010.02.007 PMID:20206435

Harnpicharnchai, P., Pinngoen, W., Teanngam, W., Sornlake, W., Sae-Tang, K., Manitchotpisit, P., & Tanapongpipat, S. (2016). Production of high activity Aspergillus niger BCC4525 β-mannanase in Pichia pastoris and its application for mannooligosaccharides production from biomass hydrolysis. *Bioscience, Biotechnology, and Biochemistry*, *80*(12), 2298–2305. doi:10.1080/09168451.2016.1230003 PMID:27648762

Harrison, J. P., Gheeraert, N., Tsigelnitskiy, D., & Cockell, C. S. (2013). The limits for life under multiple extremes. *Trends in Microbiology*, *21*(4), 204–212. doi:10.1016/j.tim.2013.01.006 PMID:23453124

Hartley, B. S., & Payton, M. A. (1983). Industrial prospects for thermophiles and thermophilic enzymes. *Biochemical Society Symposium*, *48*, 133–146. PMID:6400480

Hayden, M. S., & Ghosh, S. (2008). Shared Principles in NF-κB Signaling. *Cell*, *132*(3), 344–362. doi:10.1016/j.cell.2008.01.020 PMID:18267068

Hedderich, R. (2004). Energy-converting [NiFe] hydrogenases from archaea and extremophiles: Ancestors of complex I. *Journal of Bioenergetics and Biomembranes*, *36*(1), 65–75. doi:10.1023/B:JOBB.0000019599.43969.33 PMID:15168611

Hedlund, B. P., Dodsworth, J. A., Murugapiran, S. K., Rinke, C., & Woyke, T. (2014). Impact of single-cell genomics and metagenomics on the emerging view of extremophile "microbial dark matter.". *Extremophiles*, *18*(5), 865–875. doi:10.100700792-014-0664-7 PMID:25113821

Hedrich, S., & Schippers, A. (2021). Distribution of acidophilic microorganisms in natural and man-made acidic environments. *Current Issues in Molecular Biology*, *40*(1), 25–48. doi:10.21775/cimb.040.025 PMID:32159522

He, H., Chen, X., Li, J., Zhang, Y., & Gao, P. (2004). Taste improvement of refrigerated meat treated with cold-adapted protease. *Food Chemistry*, *84*(2), 307–311. doi:10.1016/S0308-8146(03)00242-5

He, L., Du, P., Chen, Y., Lu, H., Cheng, X., Chang, B., & Wang, Z. (2017). Advances in microbial fuel cells for wastewater treatment. *Renewable & Sustainable Energy Reviews*, *71*, 388–403. doi:10.1016/j.rser.2016.12.069

Helmann, J., Wu, M. F. W., Kobel, P. A., Gamo, F.-J., Wilson, M., Morshedi, M. M., Navre, M., & Paddon, C. (2001). Global transcriptional response of *Bacillus subtilis* to heat shock. *Journal of Bacteriology*, *183*(24), 7318–7328. doi:10.1128/JB.183.24.7318-7328.2001 PMID:11717291

Henderson, R., Baldwin, J. M., Ceska, T. A., Zemlin, F., Beckmann, E., & Downing, K. H. (1990). Model for the structure of bacteriorhodopsin based on high-resolution electron cryo-microscopy. *Journal of Molecular Biology*, *213*(4), 899–929. doi:10.1016/S0022-2836(05)80271-2 PMID:2359127

Hendrick, J. P., & Hartl, F. (1993). Molecular chaperone functions of heat-shock proteins. *Annual Review of Biochemistry*, *62*(1), 349–384. doi:10.1146/annurev.bi.62.070193.002025 PMID:8102520

Hermans, S. M., Buckley, H. L., Case, B. S., Curran-Cournane, F., Taylor, M., & Lear, G. (2017). Bacteria as Emerging Indicators of Soil Condition. Applied and Environmental Microbiology, 83(1). doi:10.1128/AEM.02826-16

Hernandez, M. E., & Newman, D. K. (2001). Extracellular electron transfer. *Cellular and Molecular Life Sciences CMLS*, *58*(11), 1562–1571. doi:10.1007/PL00000796 PMID:11706984

Heymann, D., Ruiz-Velasco, C., Chesneau, J., Ratiskol, J., Sinquin, C., & Colliec-Jouault, S. (2016). Anti-metastatic properties of a marine bacterial exopolysaccharide-based derivative designed to mimic glycosaminoglycans. *Molecules (Basel, Switzerland)*, *21*(3), 309. doi:10.3390/molecules21030309 PMID:26959001

Hickey, R. J., & Pelling, A. E. (2019). Cellulose biomaterials for tissue engineering. *Frontiers in Bioengineering and Biotechnology*, *7*, 45. doi:10.3389/fbioe.2019.00045 PMID:30968018

Hill, R. (1970). *The chemistry of life: eight lectures on the history of biochemistry*. CUP Archive.

Hindersah, R., Mulyani, O., & Osok, R. (2017). Proliferation and exopolysaccharide production ofAzotobacter in the presence of mercury. *Biodiversity Journal, 8*(1), 21–26.

Hippchen, B., Röll, A., & Poralla, K. (1981). Occurrence in soil of thermo-acidophilic bacilli possessing ω-cyclohexane fatty acids and hopanoids. *Archives of Microbiology, 129*(1), 53–55. doi:10.1007/BF00417180

Hirota, K., Aino, K., Nodasaka, Y., & Yumoto, I. (2013). *Oceanobacillus indicireducens* sp. nov., a facultative alkaliphile that reduces an indigo dye. *International Journal of Systematic and Evolutionary Microbiology, 63*(Pt_4), 1437–1442. doi:10.1099/ijs.0.034579-0 PMID:22843722

Hirsch, P., Mevs, U., Kroppenstedt, R. M., Schumann, P., & Stackebrandt, E. (2004). Cryptoendolithic actinomycetes from Antarctic sandstone rock samples: Micromonospora endolithica sp. nov. and two isolates related to Micromonospora coerulea Jensen 1932. *Systematic and Applied Microbiology, 27*(2), 166–174. doi:10.1078/072320204322881781 PMID:15046305

Hmidet, N., Bayoudh, A., Berrin, J. G., Kanoun, S., Juge, N., & Nasri, M. (2008). Purification and biochemical characterization of a novel α-amylase from *Bacillus licheniformis* NH1: Cloning, nucleotide sequence and expression of amyN gene in Escherichia coli. *Process Biochemistry, 43*(5), 499–510. doi:10.1016/j.procbio.2008.01.017

Hodson, A., Anesio, A. M., Tranter, M., Fountain, A., Osborn, M., Priscu, J., Laybourn-Parry, J., & Sattler, B. (2008). Glacial ecosystems. *Ecological Monographs, 78*(1), 41–67. doi:10.1890/07-0187.1

Hoham, R. W., & Duval, B. (2001). Microbial Ecology of Snow and Freshwater Ice with Emphasis on Snow Algae. In H. G. Jones, J. W. Pomeroy, D. A. Walker, & R. W. Hoham (Eds.), *Snow Ecology* (pp. 168–228). Cambridge University Press.

Holden, J. F. (2012). Hot Environments. In M. Schaechter (Ed.), *Schmidt TM. Topics in ecological and environmental microbiology* (pp. 313–332). Elsevier Inc.

Holmes, D. E., Nicoll, J. S., Bond, D. R., & Lovley, D. R. (2004). Potential role of a novel psychrotolerant member of the family Geobacteraceae, *Geopsychrobacter electrodiphilus* gen. nov., sp. nov., in electricity production by a marine sediment fuel cell. *Applied and Environmental Microbiology, 70*(10), 6023–6030. doi:10.1128/AEM.70.10.6023-6030.2004 PMID:15466546

Holwerda, E. K., Olson, D. G., Ruppertsberger, N. M., Stevenson, D. M., Murphy, S. J. L., Maloney, M. I., Lanahan, A. A., Amador-Noguez, D., & Lynd, L. R. (2020). Metabolic and evolutionary responses of *Clostridium thermocellum* to genetic interventions aimed at improving ethanol production. *Biotechnology for Biofuels, 13*(1), 1-20. doi:10.1186/s13068-020-01680-5

Hongo, K., Hirai, H., Uemura, C., Ono, S., Tsunemi, J., Higurashi, T., Mizobata, T., & Kawata, Y. (2006). A novel ATP/ADP hydrolysis activity of hyperthermostable group II chaperonin in the presence of cobalt or manganese ion. *FEBS Letters, 580*(1), 34–40. doi:10.1016/j.febslet.2005.11.043 PMID:16343486

Hoondal, G., Tiwari, R., Tewari, R., Dahiya, N. B. Q. K., & Beg, Q. (2002). Microbial alkaline pectinases and their industrial applications: A review. *Applied Microbiology and Biotechnology, 59*(4), 409–418. PMID:12172603

Horikoshi, K. (1971). Production of Alkaline Enzymes by Alkalophilic Microorganisms Part II. Alkaline Amylase Produced by *Bacillus* No. A-40-2. *Agricultural and Biological Chemistry, 35*, 1783–1791.

Horikoshi, K. (1973). *Characters of Bacillus No . C-59-2 The isolated bacteria The charac was that the Table I summarizes the of carbonate an important 37*. Academic Press.

Horikoshi, K. (2007). Past present and future of extremophiles. *Extremophiles, 12*(1), 1–2. doi:10.1007/s00792-007-0127-5

Horikoshi, K. (1971). Production of Alkaline Enzymes by Alkalophilic Microorganisms: Part I. Alkaline Protease Produced by *Bacillus* No. 221. *Agricultural and Biological Chemistry, 35*(9), 1407–1414. doi:10.1080/00021369.1971.10860094

Horikoshi, K. (1998). Barophiles: Deep-sea microorganisms adapted to an extreme environment. *Current Opinion in Microbiology, 1*(3), 291–295. doi:10.1016/S1369-5274(98)80032-5 PMID:10066496

Horikoshi, K. (2004). Alkaliphiles. *Proceedings of the Japan Academy. Series B, Physical and Biological Sciences, 80*(4), 166–178. doi:10.2183/pjab.80.166

Horikoshi, K. (2011). *Extremophiles Handbook*. Extremophiles Handbook. doi:10.1007/978-4-431-53898-1

Horikoshi, K. (2016). Alkaliphiles. In *Extremophiles* (pp. 53–78). Springer. doi:10.1007/978-4-431-55408-0_4

Horneck, G., Klaus, D. M., & Mancinelli, R. L. (2010). Space microbiology. *Microbiology and Molecular Biology Reviews, 74*(1), 121–156. doi:10.1128/MMBR.00016-09 PMID:20197502

Horne, I., Sutherland, T. D., Harcourt, R. L., Russell, R. J., & Oakeshott, J. G. (2002). Identification of an opd (organophosphate degradation) gene in an Agrobacterium isolate. *Applied and Environmental Microbiology, 68*(7), 3371–3376. doi:10.1128/AEM.68.7.3371-3376.2002 PMID:12089017

Horvath, P., & Barrangou, R. (2010). CRISPR/Cas, the immune system of bacteria and archaea. *Science, 327*(5962), 167–170. doi:10.1126cience.1179555 PMID:20056882

Horvath, P., Romero, D. A., Coûté-Monvoisin, A. C., Richards, M., Deveau, H., Moineau, S., Boyaval, P., Fremaux, C., & Barrangou, R. (2008). Diversity, activity, and evolution of CRISPR loci in Streptococcus thermophilus. *Journal of Bacteriology, 190*(4), 1401–1412. doi:10.1128/JB.01415-07 PMID:18065539

Hottenstein, J. D., Neilson, J. W., Gil-Loaiza, J., Root, R. A., White, S. A., Chorover, J., & Maier, R. M. (2019). Soil Microbiome Dynamics During Pyritic Mine Tailing Phytostabilization: Understanding Microbial Bioindicators of Soil Acidification. *Frontiers in Microbiology, 10*. doi:10.3389/fmicb.2019.01211

Hozzein, W. N., Li, W. J., Ali, M. I. A., Hammouda, O., Mousa, A. S., Xu, L. H., & Jiang, C. L. (2004). *Nocardiopsis alkaliphila* sp. nov., a novel alkaliphilic actinomycete isolated from desert soil in Egypt. *International Journal of Systematic and Evolutionary Microbiology, 54*(1), 247–252. doi:10.1099/ijs.0.02832-0 PMID:14742488

Hsu, P. D., Lander, E. S., & Zhang, F. (2014). Development and applications of CRISPR-Cas9 for genome engineering. *Cell, 157*(6), 1262–1278. doi:10.1016/j.cell.2014.05.010 PMID:24906146

Huber, R., & Hannig, M. (2006). Thermotogales. In The Prokaryotes. Springer.

Huber, G., & Stetter, K. O. (1991). *Sulfolobus metallicus*, sp. nov., a novel strictly chemolithoautotrophic thermophilic archaeal species of metal-mobilizers. *Systematic and Applied Microbiology, 14*(4), 372–378. doi:10.1016/S0723-2020(11)80312-7

Huber, R., Wilharm, T., Huber, D., Trincone, A., Burggraf, S., König, H., Reinhard, R., Rockinger, I., Fricke, H., & Stetter, K. O. (1992). *Aquifex pyrophilus* gen. nov. sp. nov., represents a novel group of marine hyperthermophilic hydrogen-oxidizing bacteria. *Systematic and Applied Microbiology, 15*(3), 340–351. doi:10.1016/S0723-2020(11)80206-7

Hu, J., Cai, W., Wang, C., Du, X., Lin, J., & Cai, J. (2018). Purification and characterization of alkaline lipase production by *Pseudomonas aeruginosa* HFE733 and application for biodegradation in food wastewater treatment. *Biotechnology, Biotechnological Equipment, 32*(3), 583–590. doi:10.1080/13102818.2018.1446764

Hu, Q., Sommerfeld, M., Jarvis, E., Ghirardi, M., Posewitz, M., Seibert, M., & Darzins, A. (2008). Microalgal triacylglycerols as feedstocks for biofuel production: Perspectives and advances. *The Plant Journal : For Cell and Molecular Biology, 54*(4), 621–639. doi:10.1111/j.1365-313X.2008.03492.x PMID:18476868

Hu, Q., Zhou, N., Gong, K., Liu, H., Liu, Q., Sun, D., Wang, Q., Shao, Q., Liu, H., Qiu, B., & Guo, Z. (2019). Intracellular Polymer Substances Induced Conductive Polyaniline for Improved Methane Production from Anaerobic Wastewater Treatment. *ACS Sustainable Chemistry & Engineering, 7*(6), 5912–5920. doi:10.1021/acssuschemeng.8b05847

Hussain, A., Ahmad, M., Nafees, M., Iqbal, Z., Luqman, M., Jamil, M., & Maqsood, A. (2020). Plant-growth-promoting Bacillus and Paenibacillus species improve the nutritional status of Triticum aestivum L. PLOS ONE, 15(12), e0241130. doi:10.1371/journal.pone.0241130

Igiehon, N. O., Babalola, O. O., & Aremu, B. R. (2019). Genomic insights into plant growth promoting rhizobia capable of enhancing soybean germination under drought stress. *BMC Microbiology, 19*(1). doi:10.1186/s12866-019-1536-1

Iizuka, R., Sugano, Y., Ide, N., Ohtaki, A., Yoshida, T., Fujiwara, S., Imanaka, T., & Yohda, M. (2008). Functional characterization of recombinant prefoldin complexes from a hyperthermophilic archaeon, *Thermococcus* sp. strain KS-1. *Journal of Molecular Biology, 377*(3), 972–983. doi:10.1016/j.jmb.2008.01.070 PMID:18295793

Ikeda, M., & Clark, D. S. (1998). Molecular cloning of extremely thermostable esterase gene from hyperthermophilic archaeon *Pyrococcus furiosus* in Escherichia coli. *Biotechnology and Bioengineering, 57*(5), 624–629. doi:10.1002/(SICI)1097-0290(19980305)57:5<624::AID-BIT15>3.0.CO;2-B PMID:10099242

Ikehata, K., Buchannan, I., Pickard, M., & Smith, D. (2005). Purification, characterization and evaluation of extracellular peroxidase from two species for aqueous phenol treatment. *Bioresource Technology, 96*(16), 1758–1770. doi:10.1016/j.biortech.2005.01.019 PMID:16051082

Imanaka, H., Yamatsu, A., Fukui, T., Atomi, H., & Imanaka, T. (2006). Phosphoenolpyruvate synthase plays an essential role for glycolysis in the modified Embden-Meyerhof pathway in *Thermococcus kodakarensis. Molecular Microbiology, 61*(4), 898–909. doi:10.1111/j.1365-2958.2006.05287.x PMID:16879645

Imhoff, J. F. (2020). Osmotic Adaptation in Halophilic and Halotolerant Microorganisms. *The Biology of Halophilic Bacteria*, 211–253. doi:10.1201/9781003069140-8

Imhoff, J. F., & Rodriguez-Valera, F. (1984). Betaine is the main compatible solute of halophilic eubacteria. *Journal of Bacteriology, 160*(1), 478–479. doi:10.1128/jb.160.1.478-479.1984 PMID:6148337

Ingram, G., Lai, M., & Wood, Y. (1998). Metabolic engineering of bacteria for ethanol production. *Biotechnology and Bioengineering, 58*(2–3), 204–214. doi:10.1002/(SICI)1097-0290(19980420)58:2/3<204::AID-BIT13>3.0.CO;2-C PMID:10191391

Irwin, J. A. (2020). Overview of extremophiles and their food and medical applications. Physiological and Biotechnological Aspects of Extremophiles, 65-87.

Irwin, J. A. (2020). Overview of extremophiles and their food and medical applications. In *Physiological and Biotechnological Aspects of Extremophiles* (pp. 65–87). Academic Press. doi:10.1016/B978-0-12-818322-9.00006-X

Ishino, S., & Ishino, Y. (2014). DNA polymerases as useful reagents for biotechnology â€" the history of developmental research in the field. *Frontiers in Microbiology, 5*. Advance online publication. doi:10.3389/fmicb.2014.00465 PMID:25221550

Ishino, Y., Krupovic, M., & Forterre, P. (2018). History of CRISPR-Cas from Encounter with a Mysterious Repeated Sequence to Genome Editing Technology. *Journal of Bacteriology, 200*(7), 5429. doi:10.1128/JB.00580-17 PMID:29358495

Ishino, Y., Shinagawa, H., Makino, K., Amemura, M., & Nakata, A. (1987). Nucleotide sequence of the iap gene, responsible for alkaline phosphatase isozyme conversion in *Escherichia coli,* and identification of the gene product. *Journal of Bacteriology, 169*(12), 5429–5433. doi:10.1128/jb.169.12.5429-5433.1987 PMID:3316184

Isken, S., & de Bont, J. A. M. (1998). Bacteria tolerant to organic solvents. *Extremophiles, 2*(3), 229–238. doi:10.1007/s007920050065

Ito, S., Shdcata, S., Ozaki, K., Kawai, S., Okamoto, K., Inoue, S., Takei, A., Ohta, Y., & Satoh, T. (1989). *Alkaline Cellulase for Laundry Detergents: Production by Bacillus sp. KSM-635 and Enzymatic Properties*. Academic Press.

Itoh, T., Yamaguchi, T., Zhou, P., & Takashina, T. (2005). *Natronolimnobius baerhuensis* gen. nov., sp. nov. and *Natronolimnobius innermongolicus* sp. nov., novel haloalkaliphilic archaea isolated from soda lakes in Inner Mongolia, China. *Extremophiles, 9*(2), 111–116. doi:10.100700792-004-0426-z PMID:15841343

Izawa, N., Hanamizu, T., Iizuka, R., Sone, T., Mizukoshi, H., Kimura, K., & Chiba, K. (2009). *Streptococcus thermophilus* produces exopolysaccharides including hyaluronic acid. *Journal of Bioscience and Bioengineering, 107*(2), 119–123. doi:10.1016/j.jbiosc.2008.11.007 PMID:19217548

Jadhav, K., Kushwah, B., & Jadhav, I. (2018). Insight into Compatible Solutes from Halophiles: Exploring Significant Applications in Biotechnology. *Microbial Bioprospecting for Sustainable Development*, 291–307. doi:10.1007/978-981-13-0053-0_16

Jaenicke, R. (1981). Enzymes Under Extremes of Physical Conditions. *Annual Review of Biophysics and Bioengineering, 10*(1), 1–67. doi:10.1146/annurev.bb.10.060181.000245 PMID:7020570

Jaenicke, R., & Bohm, G. (1998). The stability of proteins in extreme environments. *Current Opinion in Structural Biology, 8*(6), 738–748. doi:10.1016/S0959-440X(98)80094-8 PMID:9914256

Jansen, R., Embden, J. D., Gaastra, W., & Schouls, L. M. (2002). Identification of genes that are associated with DNA repeats in prokaryotes. *Molecular Microbiology, 43*(6), 1565–1575. doi:10.1046/j.1365-2958.2002.02839.x PMID:11952905

Jansson, P.-E., Kenne, L., & Lindberg, B. (1975). Structure of the extracellular polysaccharide from *Xanthomonas campestris. Carbohydrate Research, 45*(1), 275–282. doi:10.1016/S0008-6215(00)85885-1 PMID:1212669

Jaouadi, B., Abdelmalek, B., Zarai, N., & Bejar, S. 2011. The Bioengineering and Industrial Applications of Bacterial Alkaline Proteases: The Case of SAPB and KERAB. In Progress in Molecular and Environmental Bioengineering - From Analysis and Modeling to Technology Applications. InTech. doi:10.5772/23850

Jaradat, Z., Al Aboudi, A., Shatnawi, M., & Ababneh, Q. (2013). Partial purification and characterization of alkalophilic protease from *Pseudomonas aeruginosa. Journal of Microbiology, Biotechnology and Food Sciences, 2*, 2455–2461.

Jawahar, G., Rajasheker, G., Maheshwari, P., Punita, D. L., Jalaja, N., Kumari, P. H., Kumar, S. A., Afreen, R., Karumanchi, A. R., Rathnagiri, P., Sreenivasulu, N., & Kishor, P. B. K. (2019). Osmolyte diversity, distribution, and their biosynthetic pathways. In *Plant Signaling Molecules* (pp. 449–458). Woodhead Publishing. doi:10.1016/B978-0-12-816451-8.00028-9

Jayakumar, J., & Seenivasan, N. (2019). Evaluation of Biocontrol Agents and Organic Amendments for the Management of Root Knot Nematode and Spiral Nematode in Banana. *International Journal of Current Microbiology and Applied Sciences, 8*(5), 613–621. doi:10.20546/ijcmas.2019.805.071

Jeffrey, J. (2007). Human Telomerase RNA Pseudoknot and Hairpin Thermal Stability with Glycine Betaine and Urea: Preferential Interactions with RNA Secondary and Tertiary Structures†. *Biochemistry, 46*(31), 9068–9079. doi:10.1021/bi602637v PMID:17630773

Jiang, Y., Chen, B., Duan, C., Sun, B., Yang, J., & Yang, S. (2015). Multigene editing in the *Escherichia coli* genome via the CRISPR-Cas9 system. *Applied and Environmental Microbiology, 81*(7), 2506–2514. doi:10.1128/AEM.04023-14 PMID:25636838

Jiang, Z., Li, X., Yang, S., Li, L., & Tan, S. (2005). Improvement of the breadmaking quality of wheat flour by the hyperthermophilic xylanase B from Thermotoga maritima. *Food Research International, 38*(1), 37–43. doi:10.1016/j.foodres.2004.07.007

Jinek, M., Chylinski, K., Fonfara, I., Hauer, M., Doudna, J. A., & Charpentier, E. (2012). A programmable dual-RNA-guided DNA endonuclease in adaptive bacterial immunity. *Science, 337*(6096), 816–821. doi:10.1126cience.1225829 PMID:22745249

Jin, M., Gai, Y., Guo, X., Hou, Y., & Zeng, R. (2019). Properties and applications of extremozymes from deep-sea extremophilic microorganisms: A mini review. *Marine Drugs, 17*(12), 656. doi:10.3390/md17120656 PMID:31766541

Johnson, D. B. (2014). Biomining-biotechnologies for extracting and recovering metals from ores and waste materials. *Current Opinion in Biotechnology, 30*, 24–31. doi:10.1016/j.copbio.2014.04.008 PMID:24794631

Johnson, K., Jiang, Y., Kleerebezem, R., Muyzer, G., & Van Loosdrecht, M. C. M. (2009). Enrichment of a mixed bacterial culture with a high polyhydroxyalkanoate storage capacity. *Biomacromolecules, 10*(4), 670–676. doi:10.1021/bm8013796 PMID:19193058

Johnson, M. R., Montero, C. I., Conners, S. B., Shockley, K. R., Bridger, S. L., & Kelly, R. M. (2004). Population density-dependent regulation of exopolysaccharide formation in the hyperthermophilic bacterium Thermotoga maritima. *Molecular Microbiology, 55*(3), 664–674. doi:10.1111/j.1365-2958.2004.04419.x PMID:15660994

Jones, H. G. (1999). The ecology of snow-covered systems: A brief overview of nutrient cycling and life in the cold. *Hydrological Processes, 13*(14-15), 2135–2147. doi:10.1002/(SICI)1099-1085(199910)13:14/15<2135::AID-HYP862>3.0.CO;2-Y

Jorda, J., & Yeates, T. O. (2011). Widespread disulfide bonding in proteins from thermophilic archaea. *Archaea (Vancouver, B.C.), 2011*, 1–9. doi:10.1155/2011/409156 PMID:21941460

Joseph, B., Kumar, V., & Ramteke, P. W. (2019). Psychrophilic Enzymes: Potential Biocatalysts for Food Processing. In Enzymes in Food Biotechnology (pp. 817-825). Academic Press.

Joseph, B., Ramteke, P. W., & Thomas, G. (2008). Cold active microbial lipases: Some hot issues and recent developments. *Biotechnology Advances, 26*(5), 457–470. doi:10.1016/j.biotechadv.2008.05.003 PMID:18571355

Joshi, S., & Satyanarayana, T. (2013). Characteristics and applications of a recombinant alkaline serine protease from a novel bacterium *Bacillus lehensis*. *Bioresource Technology, 131*, 76–85. doi:10.1016/j.biortech.2012.12.124 PMID:23340105

Jothinathan, D., Sundaram, M., & Thilagaraj, R. (2017). The Optimization of Parameters for Increased Electricity Production by a Microbial Fuel Cell Using Rumen Fluid. *International Journal of Green Energy, 12*(4), 333–338.

Joyce, C. M. (2021). *Engineering thermostable regulators for inducible gene expression in thermophiles*. Academic Press.

Juárez, K., Kim, B. C., Nevin, K., Olvera, L., Reguera, G., Lovley, D. R., & Methé, B. A. (2009). PilR, a Transcriptional Regulator for Pilin and Other Genes Required for Fe(III) Reduction in <i>Geobacter sulfurreducens</i>. *Journal of Molecular Microbiology and Biotechnology, 16*(3-4), 146–158. doi:10.1159/000115849 PMID:18253022

Jungblut, A. D., Lovejoy, C., & Vincent, W. F. (2010). Global distribution of cyanobacterial ecotypes in the cold biosphere. *The ISME Journal, 4*(2), 191–202. doi:10.1038/ismej.2009.113 PMID:19890368

Junge, K., Eicken, H., & Deming, J. W. (2004). Bacterial activity at− 2 to− 20 C in Arctic wintertime sea ice. *Applied and Environmental Microbiology*, *70*(1), 550–557. doi:10.1128/AEM.70.1.550-557.2004 PMID:14711687

Junge, K., Eicken, H., Swanson, B. D., & Deming, J. W. (2006). Bacterial incorporation of leucine into protein down to− 20 C with evidence for potential activity in sub-eutectic saline ice formations. *Cryobiology*, *52*(3), 417–429. doi:10.1016/j.cryobiol.2006.03.002 PMID:16647051

Junge, K., & Swanson, B. D. (2008). High-resolution ice nucleation spectra of sea-ice bacteria: Implications for cloud formation and life in frozen environments. *Biogeosciences*, *5*(3), 865–873. doi:10.5194/bg-5-865-2008

Juturu, V., & Wu, J. C. (2012). Microbial xylanases: Engineering, production and industrial applications. *Biotechnology Advances*, *30*(6), 1219–1227. doi:10.1016/j.biotechadv.2011.11.006 PMID:22138412

Kaczmarek, D. K., Gwiazdowska, D., Juś, K., Klejdysz, T., Wojcieszak, M., Materna, K., & Pernak, J. (2021). Glycine betaine-based ionic liquids and their influence on bacteria, fungi, insects and plants. *New Journal of Chemistry*, *45*(14), 6344–6355. doi:10.1039/D1NJ00498K

Kahya, N., Brown, D. A., & Schwille, P. (2005). Raft partitioning and dynamic behavior of human placental alkaline phosphatase in giant unilamellar vesicles. *Biochemistry*, *44*(20), 7479–7489. doi:10.1021/bi047429d PMID:15895991

Kakkar, P., & Wadhwa, N. (2021). Extremozymes used in textile industry. *Journal of the Textile Institute*, 1–9. doi:10.1080/00405000.2021.1948251

Kalanetra, K. M., Bano, N., & Hollibaugh, J. T. (2009). Ammonia-oxidizing Archaea in the Arctic Ocean and Antarctic coastal waters. *Environmental Microbiology*, *11*(9), 2434–2445. doi:10.1111/j.1462-2920.2009.01974.x PMID:19601959

Kalita, D., & Joshi, S. (2017). Study on bioremediation of Lead by exopolysaccharide producing metallophilic bacterium isolated from extreme habitat. *Biotechnology Reports (Amsterdam, Netherlands)*, *16*, 48–57. doi:10.1016/j.btre.2017.11.003 PMID:29167759

Kalwasińska, A., Felföldi, T., Szabó, A., Deja-Sikora, E., Kosobucki, P., & Walczak, M. (2017). Microbial communities associated with the anthropogenic, highly alkaline environment of a saline soda lime, Poland. *Antonie van Leeuwenhoek*, *110*(7), 945–962. doi:10.100710482-017-0866-y PMID:28382378

Kamalaskar, L. B., Dhakephalkar, P. K., Meher, K. K., & Ranade, D. R. (2010). High biohydrogen yielding *Clostridium* sp. DMHC-10 isolated from sludge of distillery waste treatment plant. *International Journal of Hydrogen Energy*, *35*(19), 10639–10644. doi:10.1016/j.ijhydene.2010.05.020

Kamalaskar, L., Kapse, N., Pore, S., Dhakephalkar, A. P., Ranade, D. R., & Dhakephalkar, P. K. (2016). Genome sequence and gene expression studies reveal novel hydrogenases mediated hydrogen production by *Clostridium biohydrogenum* sp. nov., MCM B-509T. *International Journal of Hydrogen Energy*, *41*(28), 11990–11999. doi:10.1016/j.ijhydene.2016.05.101

Kambam, P. K. R., & Henson, M. A. (2010). Engineering bacterial processes for cellulosic ethanol production. *Biofuels*, *1*(5), 729–743. doi:10.4155/bfs.10.46

Kambourova, M. (2018). Thermostable enzymes and polysaccharides produced by thermophilic bacteria isolated from Bulgarian hot springs. *Engineering in Life Sciences*, *18*(11), 758–767. doi:10.1002/elsc.201800022 PMID:32624870

Kambura, A. K. (2016). *Metagenomic and metatranscriptomic analysis of bacterial, Archaeal and Fungal communities within the hot springs of Lake Magadi in Kenya* (PhD Thesis). Jomo Kenyatta University of Agrioculture and Technology.

Kamerbeek, J., Schouls, L., Kolk, A., van Agterveld, M., van Soolingen, D., Kuijper, S., Bunschoten, A., Molhuizen, H., Shaw, R., Goyal, M., & van Embden, J. (1997). Simultaneous detection and strain differentiation of *Mycobacterium tuberculosis* for diagnosis and epidemiology. *Journal of Clinical Microbiology, 35*(4), 907–914. doi:10.1128/jcm.35.4.907-914.1997 PMID:9157152

Kanekar, P., Kanekar, S., Kelkar, A., & Dhakephalkar, P. (2012). Halophiles - Taxonomy, diversity, physiology and applications. In *Microorganisms in Environmental Management: Microbes and Environment* (pp. 1–34). Springer, Netherlands. doi:10.1007/978-94-007-2229-3_1

Kaneko, R., Hayashi, T., Tanahashi, M., & Naganuma, T. (2007). Phylogenetic diversity and distribution of dissimilatory sulfite reductase genes from deep-sea sediment cores. *Marine Biotechnology (New York, N.Y.), 9*(4), 429–436. doi:10.100710126-007-9003-7 PMID:17497195

Kanmani, P., & Lim, S. T. (2013). Synthesis and structural characterization of silver nanoparticles using bacterial exopolysaccharide and its antimicrobial activity against food and multidrug resistant pathogens. *Process Biochemistry, 48*(7), 1099–1106. doi:10.1016/j.procbio.2013.05.011

Karadayi, Y. I., Aykutoglu, G., Arslan, N. P., Baltaci, M. O., Adiguzel, A., & Taskin, M. (2021). Production of water-soluble sulfated exopolysaccharide with anticancer activity from *Anoxybacillus gonensis* YK25. *Journal of Chemical Technology and Biotechnology (Oxford, Oxfordshire), 96*(5), 1258–1266. doi:10.1002/jctb.6638

Karan, R., Capes, M. D., DasSarma, P., & DasSarma, S. (2013). Cloning, overexpression, purification, and characterization of a polyextremophilic β-galactosidase from the Antarctic haloarchaeon Halorubrum lacusprofundi. *BMC Biotechnology, 13*(1), 1–11. doi:10.1186/1472-6750-13-3

Karan, R., Kumar, S., Sinha, R., & Khare, S. K. (2012b). Halophilic microorganisms as sources of novel enzymes. In *Microorganisms in sustainable agriculture and biotechnology* (pp. 555–579). Springer. doi:10.1007/978-94-007-2214-9_25

Karataş, H., Uyar, F., Tolan, V., & Baysal, Z. (2013). Optimization and enhanced production of α-amylase and protease by a newly isolated *Bacillus licheniformis* ZB-05 under solid-state fermentation. *Annals of Microbiology, 63*(1), 45–52. doi:10.100713213-012-0443-6

Karigar, C. S., & Rao, S. S. (2011). Role of microbial enzymes in the bioremediation of pollutants: A review. *Enzyme Research, 2011*, 805187. doi:10.4061/2011/805187 PMID:21912739

Kasting, J. (1993). Earth's early atmosphere. *Science, 259*(5097), 920–926. doi:10.1126cience.11536547 PMID:11536547

Kästner, M., Breuer-Jammali, M., & Mahro, B. (1994). Enumeration and characterization of the soil microflora from hydrocarbon-contaminated soil sites able to mineralize polycyclic aromatic hydrocarbons (PAH). *Applied Microbiology and Biotechnology, 41*(2), 267–273. doi:10.1007/BF00186971

Kastritis, P. L., Papandreou, N. C., & Hamodrakas, S. J. (2007). Haloadaptation: Insights from comparative modeling studies of halophilic archaeal DHFRs. *International Journal of Biological Macromolecules, 41*(4), 447–453. doi:10.1016/j.ijbiomac.2007.06.005 PMID:17675150

Kataky, R., & Knowles, E. (2018). Biofilm formation on abiotic surfaces and their redox activity. *Current Opinion in Electrochemistry, 12*, 121–128. doi:10.1016/j.coelec.2018.07.007

Kateb, G. (2021). *The inner ocean*. Cornell University Press.

Kato, C., Li, L., Nogi, Y., Nakamura, Y., Tamaoka, J., & Horikoshi, K. (1998). Extremely barophilic bacteria isolated from the Mariana Trench, Challenger Deep, at a depth of 11,000 meters. *Applied and Environmental Microbiology, 64*(4), 1510–1513. doi:10.1128/AEM.64.4.1510-1513.1998 PMID:9546187

Kato, C., & Nogi, Y. (2001). Correlation between phylogenetic structure and function: Examples from deep-sea Shewanella. *FEMS Microbiology Ecology, 35*(3), 223–230. doi:10.1111/j.1574-6941.2001.tb00807.x PMID:11311432

Kaur, R., Macleod, J., Foley, W., & Nayudu, M. (2006). Gluconic acid: An antifungal agent produced by Pseudomonas species in biological control of take-all. *Phytochemistry, 67*(6), 595–604. doi:10.1016/j.phytochem.2005.12.011

Kaur, S., Dhanjal, D. S., Singh, S., Chopra, C., & Chopra, R. S. (2020). Isolation and biochemical characterization of an alkalophilic, cold-active cellulase from bacteria isolated from Boniyar region of Kashmir. *Plant Archives, 20*, 2548–2553.

Kaushik, S., Alatawi, A., Djiwanti, S. R., Pande, A., Skotti, E., & Soni, V. (2021). Potential of Extremophiles for Bioremediation. Microbial Rejuvenation of Polluted Environment. Springer Singapore. 10.1007%2F978-981-15-7447-4_12

Kaushik, S., Alatawi, A., Djiwanti, S. R., Pande, A., Skotti, E., & Soni, V. (2021). Potential of Extremophiles for Bioremediation. In *Microbial Rejuvenation of Polluted Environment* (pp. 293–328). Springer., doi:10.1007/978-981-15-7447-4_12

Kawakami, Y., Hayashi, N., Ema, M., & Nakayama, M. (2007). Effects of divalent cations on Halobacterium salinarum cell aggregation. *Journal of Bioscience and Bioengineering, 104*(1), 42–46. doi:10.1263/jbb.104.42 PMID:17697982

Kawamori, M., Takayama, K., & Takasawa, S. (1987). Production of Cellulases by a Thermophilic Fungus, *Thermoascus aurantiacus* A-131. *Agricultural and Biological Chemistry, 51*(3), 647–654. doi:10.1080/00021369.1987.10868111

Kawasaki, S., Borchert, C., Deyholos, M., Wang, H., Brazille, S., Kawai, K., Galbraith, D., & Bohnert, H. J. (2001). Gene expression profiles during the initial phase of salt stress in rice. *The Plant Cell, 13*(4), 889–905. doi:10.1105/tpc.13.4.889 PMID:11283343

Kazak, H., Oner, E., & Dekker, R. F. (2010). Extremophiles as sources of exopolysaccharides. *Carbohydrate Polymers: Development, Properties and Applications*, 605-619.

Keasling, J. D., Van Dien, S. J., Trelstad, P., Renninger, N., & McMahon, K. (2000). Application of polyphosphate metabolism to environmental and biotechnological problems. Biochemistry. *Biokhimiia (Moscow, Russia), 65*(3), 324–331. PMID:10739475

Kelly, C. (2008). A novel ecological role of the Firmicutes identified in thermophilic microbial fuel cells. *International Society for Microbial Ecology, 2*, 1146–1156.

Kempf, B., & Bremer, E. (1998). Uptake and synthesis of compatible solutes as microbial stress responses to high-osmolality environments. *Archives of Microbiology, 170*(5), 319–330. doi:10.1007002030050649 PMID:9818351

Kengen, S., De Bok, F., Van Loo, N., Dijkema, C., Stams, A., & De Vos, W. (1994). Evidence for the operation of a novel Embden-Meyerhof pathway that involves ADP-dependent kinases during sugar fermentation by *Pyrococcus furiosus. The Journal of Biological Chemistry, 269*(26), 17537–17541. doi:10.1016/S0021-9258(17)32474-2 PMID:8021261

Kengen, S., Stams, A. J., & Vos, W. M. (1996). Sugar metabolism of hyperthermophiles. *FEMS Microbiology Reviews, 18*(2-3), 119–137. doi:10.1111/j.1574-6976.1996.tb00231.x

Kent, A. D., & Triplett, E. W. (2002). Microbial communities and their interactions in soil and rhizosphere ecosystems. *Annual Review of Microbiology, 56*(1), 211–236. doi:10.1146/annurev.micro.56.012302.161120 PMID:12142496

Khan, M., & Sathya, T. A. (2018). Extremozymes from metagenome: Potential applications in food processing. *Critical Reviews in Food Science and Nutrition, 58*(12), 2017–2025. doi:10.1080/10408398.2017.1296408 PMID:28605203

Khatoon, H., Rai, J. P. N., & Jillani, A. (2021). Role of fungi in bioremediation of contaminated soil. In *Fungi Bio-Prospects in Sustainable Agriculture Environment and Nano-technology* (pp. 121–156). Elsevier. doi:10.1016/B978-0-12-821925-6.00007-1

Khelaifia, S., Fardeau, M.-L., Pradel, N., Aussignargues, C., Garel, M., Tamburini, C., Cayol, J.-L., Gaudron, S., Gaill, F., & Ollivier, B. (2011). Desulfovibrio piezophilus sp. nov., a piezophilic, sulfate-reducing bacterium isolated from wood falls in the Mediterranean Sea. *International Journal of Systematic and Evolutionary Microbiology*, *61*(11), 2706–2711. doi:10.1099/ijs.0.028670-0 PMID:21169465

Khemili-Talbi, S., Kebbouche-Gana, S., Akmoussi-Toumi, S., Angar, Y., & Gana, M. L. (2015). Isolation of an extremely halophilic archaeon *Natrialba* sp. C21 can degrade aromatic compounds and produce stable biosurfactants at high salinity. *Extremophiles*, *19*(6), 1109–1120. doi:10.100700792-015-0783-9 PMID:26334644

Kim, B. C., Leang, C., Ding, Y. R., Glaven, R. H., Coppi, M. V., & Lovley, D. R. (2005). OmcF, a Putative *c*-Type Monoheme Outer Membrane Cytochrome required for the Expression of Other Outer Membrane Cytochromes in *Geobacter sulfurreducens*. *Journal of Bacteriology*, *187*(13), 4505–4513. doi:10.1128/JB.187.13.4505-4513.2005 PMID:15968061

Kim, B. C., Postier, B. L., DiDonato, R. J., Chaudhuri, S. K., Nevin, K. P., & Lovley, D. R. (2008). Insights into genes involved in electricity generation in *Geobacter sulfurreducens* via whole genome microarray analysis of the OmcF-deficient mutant. *Bioelectrochemistry (Amsterdam, Netherlands)*, *73*(1), 70–75. doi:10.1016/j.bioelechem.2008.04.023 PMID:18538641

Kim, B. C., Qian, X., Leang, C., Coppi, M. V., & Lovley, D. R. (2006). Two Putative *c*-Type Multiheme Cytochromes Required for the Expression of OmcB, an Outer Membrane Protein Essential for Optimal Fe(III) Reduction in *Geobacter sulfurreducens*. *Journal of Bacteriology*, *188*(8), 3138–3142. doi:10.1128/JB.188.8.3138-3142.2006 PMID:16585776

Kim, B. H., Ikeda, T., Park, H., Kim, H., Hyun, M. S., Kano, K., Takagi, K., & Tatsumi, H. (1999a). Electrochemical activity of an Fe(III)-reducing bacterium, *Shewanella putrefaciens* IR-1, in the presence of alternative electron acceptors. *Biotechnology Techniques*, *13*(7), 475–478. doi:10.1023/A:1008993029309

Kim, B. H., Kim, H. J., Hyun, M. S., & Park, D. H. (1999b). Direct electrode reaction of Fe (III)-reducing bacterium, *Shewanella putrefaciens*. *Journal of Microbiology and Biotechnology*, *9*, 127–131.

Kim, H. J., Park, H. S., Hyun, M. S., Chang, I. S., Kim, M., & Kim, B. H. (2002). A mediator-less microbial fuel cell using a metal reducing bacterium, *Shewanella putrefaciens*. *Enzyme and Microbial Technology*, *30*(2), 145–152. doi:10.1016/S0141-0229(01)00478-1

Kim, H. S., Kim, Y. J., & Seo, Y. R. (2015). An overview of carcinogenic heavy metal: Molecular toxicity mechanism and prevention. *Journal of Cancer Prevention*, *20*(4), 232–240. doi:10.15430/JCP.2015.20.4.232 PMID:26734585

Kim, J.-Y., Hur, S.-H., & Hong, J.-H. (2005). Purification and characterization of an alkaline cellulase from a newly isolated alkalophilic *Bacillus* sp. HSH-810. *Biotechnology Letters*, *27*(5), 313–316. doi:10.100710529-005-0685-5 PMID:15834791

Kim, N., Choi, Y., Jung, S., & Kim, S. (1999). Development of microbial fuel cells using *Proteus vulgaris*. *Bulletin of the Korean Chemical Society*, *21*(1), 44–48.

Kim, S. J., Koh, D. C., Park, S. J., Cha, I. T., Park, J. W., Na, J. H., Roh, Y., Ko, K. S., Kim, K., & Rhee, S. K. (2012). Molecular analysis of spatial variation of iron-reducing bacteria in riverine alluvial aquifers of the Mankyeong River. *Journal of Microbiology (Seoul, Korea)*, *50*(2), 207–217. doi:10.100712275-012-1342-z PMID:22538648

Kim, Y.-G., Lee, J.-H., Gwon, G., Kim, S.-I., Park, J. G., & Lee, J. (2016). Essential Oils and Eugenols Inhibit Biofilm Formation and the Virulence of Escherichia coli O157:H7. *Scientific Reports*, *6*(1), 36377. doi:10.1038rep36377 PMID:27808174

Kim, Y., & Wood, T. K. (2010). Toxins Hha and CspD and small RNA regulator Hfq are involved in persister cell formation through MqsR in Escherichia coli. *Biochemical and Biophysical Research Communications*, *391*(1), 209–213. doi:10.1016/j.bbrc.2009.11.033 PMID:19909729

Kiruthika, J., Saraswathy, N., & Murugesan, S. (2018). Maximizing L-glutaminase production from marine *Bacillus subtilis* JK-79 under solid-state fermentation. *African Journal of Biotechnology*, *17*(9), 288–305. doi:10.5897/AJB2017.16015

Kleine, J., & Liebl, W. (2006). Comparative characterization of deletion derivatives of the modular xylanase XynA of *Thermotoga maritima. Extremophiles*, *10*(5), 373–381. doi:10.100700792-006-0509-0 PMID:16550304

Kletzin, A., Mukund, S., Kelley-Crouse, T. L., Chan, M. K., Rees, D. C., & Adams, M. (1995). Molecular characterization of the genes encoding the tungsten-containing aldehyde ferredoxin oxidoreductase from Pyrococcus furiosus and formaldehyde ferredoxin oxidoreductase from *Thermococcus litoralis. Journal of Bacteriology*, *177*(16), 4817–4819. doi:10.1128/jb.177.16.4817-4819.1995 PMID:7642512

Klimek, K., Przekora, A., Pałka, K., & Ginalska, G. (2016). New method for the fabrication of highly osteoconductive β-1, 3-glucan/HA scaffold for bone tissue engineering: Structural, mechanical, and biological characterization. *Journal of Biomedical Materials Research. Part A*, *104*(10), 2528–2536. doi:10.1002/jbm.a.35798 PMID:27239050

Knief, C., Delmotte, N., Chaffron, S., Stark, M., Innerebner, G., Wassmann, R., von Mering, C., & Vorholt, J. (2012). Metaproteogenomic analysis of microbial communities in the phyllosphere and rhizosphere of rice. *The ISME Journal*, *6*(7), 1378–1390. doi:10.1038/ismej.2011.192 PMID:22189496

Knudson, W., Chow, G., & Knudson, C. B. (2002). CD44-mediated uptake and degradation of hyaluronan. *Matrix Biology*, *21*(1), 15–23. doi:10.1016/S0945-053X(01)00186-X PMID:11827788

Koch, R., Spreinat, A., Lemke, K., & Antranikian, G. (1991). Purification and properties of a hyperthermoactive α-amylase from the archaeobacterium Pyrococcus woesei. *Archives of Microbiology*, *155*(6), 572–578. doi:10.1007/BF00245352

Koch, M. K., & Oesterhelt, D. (2005). MpcT is the transducer for membrane potential changes in Halobacterium salinarum. *Molecular Microbiology*, *55*(6), 1681–1694. doi:10.1111/j.1365-2958.2005.04516.x PMID:15752193

Koendjbiharie, J. G., Wiersma, K., & van Kranenburg, R. (2018). Investigating the central metabolism of *Clostridium thermosuccinogenes. Applied and Environmental Microbiology*, *84*(13). Advance online publication. doi:10.1128/AEM.00363-18 PMID:29678919

Kogan, G., Šoltés, L., Stern, R., & Gemeiner, P. (2007). Hyaluronic acid: A natural biopolymer with a broad range of biomedical and industrial applications. *Biotechnology Letters*, *29*(1), 17–25. doi:10.100710529-006-9219-z PMID:17091377

Koga, S., Yoshioka, I., Sakuraba, H., Takahashi, M., Sakasegawa, S., Shimizu, S., & Ohshima, T. (2000). Biochemical characterization, cloning, and sequencing of ADP-dependent (AMP-forming) glucokinase from two hyperthermophilic archaea, *Pyrococcus furiosus* and *Thermococcus litoralis. Journal of Biochemistry*, *128*(6), 1079–1085. doi:10.1093/oxfordjournals.jbchem.a022836 PMID:11098152

Köhl, J., Kolnaar, R., & Ravensberg, W. J. (2019). Mode of Action of Microbial Biological Control Agents Against Plant Diseases: Relevance Beyond Efficacy. *Frontiers in Plant Science*, *10*. doi:10.3389/fpls.2019.00845

Kohli, I., Joshi, N. C., Mohapatra, S., & Varma, A. (2020). Extremophile An Adaptive Strategy for Extreme Conditions and Applications. *Current Genomics*, *21*(2), 96–110. doi:10.2174/1389202921666200401105908

Koide, T., Vêncio, R. Z. N., & Gomes, S. L. (2006). Global gene expression analysis of the heat shock response in the phytopathogen *Xylella fastidiosa. Journal of Bacteriology*, *188*(16), 5821–5830. doi:10.1128/JB.00182-06 PMID:16885450

Kojima, M., Kanai, M., Tominaga, M., Kitazume, S., Inoue, A., & Horikoshi, K. (2006). Isolation and characterization of a feather-degrading enzyme from *Bacillus pseudofirmus* FA30-01. *Extremophiles, 10*(3), 229–235. doi:10.100700792-005-0491-y PMID:16489414

Koonin, E. V., Makarova, K. S., & Zhang, F. (2017). Diversity, classification and evolution of CRISPR-Cas systems. *Current Opinion in Microbiology, 37*, 67–78. doi:10.1016/j.mib.2017.05.008 PMID:28605718

Ko, R., Smith, L. T., & Smith, G. M. (1994). Glycine betaine confers enhanced osmotolerance and cryotolerance on Listeria monocytogenes. *Journal of Bacteriology, 176*(2), 426–431. doi:10.1128/jb.176.2.426-431.1994 PMID:8288538

Korashy, H. M., Attafi, I. M., Famulski, K. S., Bakheet, S. A., Hafez, M. M., Alsaad, A. M. S., & Al-Ghadeer, A. R. M. (2017). Gene expression profiling to identify the toxicities and potentially relevant human disease outcomes associated with environmental heavy metals exposure. *Environmental Pollution, 221*, 64–74. doi:10.1016/j.envpol.2016.10.058 PMID:27916491

Körstgens, V., Flemming, H.-C., Wingender, J., & Borchard, W. (2001). Influence of calcium ions on the mechanical properties of a model biofilm of mucoid *Pseudomonas aeruginosa. Water Science and Technology, 43*(6), 49–57. doi:10.2166/wst.2001.0338 PMID:11381972

Kosicki, M., Tomberg, K., & Bradley, A. (2018). Repair of double-strand breaks induced by CRISPR-Cas9 leads to large deletions and complex rearrangements. *Nature Biotechnology, 36*(8), 765–771. doi:10.1038/nbt.4192 PMID:30010673

Kottwitz, B., Upadek, H., & Carrer, G. (1994). Application and benefits of enzymes in detergents. *Chimica Oggi, 12*(11-12), 21–24.

Koul, B., Chaudhary, R., & Taak, P. (2021). Extremophilic microbes and their application in bioremediation of environmental contaminants. In *Microbe Mediated Remediation of Environmental Contaminants* (pp. 115–128). Elsevier. doi:10.1016/B978-0-12-821199-1.00012-2

Kour, D., Rana, K. L., Kaur, T., Singh, B., Chauhan, V. S., Kumar, A., Rastegari, A. A., Yadav, N., Yadav, A. N., & Gupta, V. K. (2019). Extremophiles for Hydrolytic Enzymes Productions: Biodiversity and Potential Biotechnological Applications. *Bioprocessing for Biomolecules Production*, 321–372. doi:10.1002/9781119434436.ch16

Kour, D., Rana, K. L., Kaur, T., Singh, B., Chauhan, V. S., Kumar, A., Rastegari, A.A., Yadav, N., Yadav, A.N. & Gupta, V. K. (2019). Extremophiles for hydrolytic enzymes productions: biodiversity and potential biotechnological applications. *Bioprocessing for biomolecules production*, 321-372.

Kouřilová, X., Schwarzerová, J., Pernicová, I., Sedlář, K., Mrázová, K., Krzyžánek, V., Nebesarova, J., & Obruča, S. (2021). The First Insight into Polyhydroxyalkanoates Accumulation in Multi-Extremophilic *Rubrobacter xylanophilus* and *Rubrobacter spartanus. Microorganisms, 9*(5), 909. doi:10.3390/microorganisms9050909 PMID:33923216

Kraemer, J. T., & Bagley, D. M. (2007). Improving the yield from fermentative hydrogen production. *Biotechnology Letters, 29*(5), 685–695. doi:10.100710529-006-9299-9 PMID:17279447

Krembs, C., Eicken, H., Junge, K., & Deming, J. W. (2002). High concentrations of exopolymeric substances in Arctic winter sea ice: Implications for the polar ocean carbon cycle and cryoprotection of diatoms. *Deep-sea Research. Part I, Oceanographic Research Papers, 49*(12), 2163–2181. doi:10.1016/S0967-0637(02)00122-X

Kryachko, Y., Dong, X., Sensen, C. W., & Voordouw, G. (2012). Compositions of microbial communities associated with oil and water in a mesothermic oil field. *Antonie van Leeuwenhoek, 101*(3), 493–506. doi:10.100710482-011-9658-y PMID:22038128

Kulkarni, S., Dhakar, K., & Joshi, A. (2019). Alkaliphiles. In *Microbial Diversity in the Genomic Era* (pp. 239–263). Elsevier. doi:10.1016/B978-0-12-814849-5.00015-0

Kulshreshtha, N. M., Kumar, R., Begum, Z., Shivaji, S., & Kumar, A. (2013). *Exiguobacterium alkaliphilum* sp. nov. isolated from alkaline wastewater drained sludge of a beverage factory. *International Journal of Systematic and Evolutionary Microbiology, 63*(Pt_12), 4374–4379. doi:10.1099/ijs.0.039123-0 PMID:23838447

Kumar, L., Kumar, D., Nagar, S., Gupta, R., Garg, N., Kuhad, R.C., & Gupta, V.K. (2014). Modulation of xylanase production from alkaliphilic *Bacillus pumilus* VLK-1 through process optimization and temperature shift operation. *Biotech, 3*(4), 345–356. doi:10.1007/s13205-013-0160-2

Kumar, D. J. M., Rajan, R., Lawrence, L., Priyadarshini, S., Chittybabu, S., & Kalaichelvan, P. T. (2012). Destaining and dehairing capability of partially purified *Bacillus subtilis* protease from optimized fermentation medium. *Asian Journal of Experimental Biological Sciences, 3*, 613–620.

Kumar, L., Awasthi, G., & Singh, B. (2011). Extremophiles: A novel source of industrially important enzymes. *Biotechnology, 10*(2), 121–135. doi:10.3923/biotech.2011.121.135

Kumar, M., Yadav, A. N., Saxena, R., Paul, D., & Tomar, R. S. (2021). Biodiversity of pesticides degrading microbial communities and their environmental impact. *Biocatalysis and Agricultural Biotechnology, 31*, 101883. doi:10.1016/j.bcab.2020.101883

Kumar, M., Yadav, A. N., Tiwari, R., Prasanna, R., & Saxena, A. K. (2014). Deciphering the diversity of culturable thermotolerant bacteria from Manikaran hot springs. *Annals of Microbiology, 64*(2), 741–751. doi:10.100713213-013-0709-7

Kumar, N., & Hovik, A. (2019). Extremophiles : Applications and roles in environmental sustainability. *Environmental Sustainability, 2*(3), 217–218. doi:10.100742398-019-00082-0

Kumar, R., Singh, L., & Zularisam, A. W. (2017). *Microbial Fuel Cells: Types and Applications: Waste biomass management- A Holistic Approach*. Springer.

Kumar, R., Singh, S., & Singh, O. V. (2008). Bioconversion of lignocellulosic biomass: Biochemical and molecular perspectives. *Journal of Industrial Microbiology & Biotechnology, 35*(5), 377–391. doi:10.100710295-008-0327-8 PMID:18338189

Kumar, S., & Nussinov, R. (2001). How do thermophilic proteins deal with heat? *Cellular and Molecular Life Sciences, 58*(9), 1216–1233. doi:10.1007/PL00000935 PMID:11577980

Kumar, V., Saxena, J., & Tiwari, S. K. (2016). Description of a halocin-producing *Haloferax larsenii* HA1 isolated from Pachpadra salt lake in Rajasthan. *Archives of Microbiology, 198*(2), 181–192. doi:10.100700203-015-1175-3 PMID:26659359

Kurihara, T., & Esaki, N. (2008). Proteomic studies of psychrophilic microorganisms. In *Psychrophiles: from biodiversity to biotechnology* (pp. 333–344). Springer. doi:10.1007/978-3-540-74335-4_19

Kushveer, J. S., Rashmi, M., & Sarma, V. V. (2021). Bioactive compounds from marine-derived fungi and their potential applications. In Fungi Bio-Prospects in Sustainable Agriculture, Environment and Nano-technology, (pp. 91-173). Academic Press. doi:10.1016/B978-0-12-821734-4.00014-9

Kushwaha, K., Saxena, J., Tripathi, B. K., & Agarwal, M. K. (2014). Detection of carotenoids in psychrotrophic bacteria by spectroscopic approach. *Journal of BioScience and Biotechnology, 3*(3).

Kutty, S. N., & Philip, R. (2008). Marine yeasts—A review. *Yeast (Chichester, England), 25*(7), 465–483. doi:10.1002/yea.1599 PMID:18615863

Kuvarina, A. E., Gavryushina, I. A., Kulko, A. B., Ivanov, I. A., Rogozhin, E. A., Georgieva, M. L., & Sadykova, V. S. (2021). The emericellipsins a-e from an alkalophilic fungus emericellopsis alkalina show potent activity against multidrug-resistant pathogenic fungi. *Journal of Fungi (Basel, Switzerland)*, *7*(2), 1–17. doi:10.3390/jof7020153 PMID:33669976

Kwatra, B., Yadav, P., Balasubramaniam, S., Garg, S., & Koyande, T. (2021). Reviewing major microbes as a single cell protein. *International Journal of Medical and Biomedical Studies*, *5*(3). doi:10.32553/ijmbs.v5i3.1842

Labbé, D., Margesin, R., Schinner, F., Whyte, L. G., & Greer, C. W. (2007). Comparative phylogenetic analysis of microbial communities in pristine and hydrocarbon-contaminated Alpine soils. *FEMS Microbiology Ecology*, *59*(2), 466–475. doi:10.1111/j.1574-6941.2006.00250.x PMID:17313586

Labes, A., & Schönheit, P. (2001). Sugar utilization in the hyperthermophilic, sulfate-reducing archaeon *Archaeoglobus fulgidus* strain 7324: Starch degradation to acetate and CO_2 via a modified Embden-Meyerhof pathway and acetyl-CoA synthetase (ADP-forming). *Archives of Microbiology*, *176*(5), 329–338. doi:10.1007002030100330 PMID:11702074

Lacis, L. S., & Lawford, H. G. (1991). Thermoanaerobacter ethanolicus Growth and Product Yield from Elevated Levels of Xylose or Glucose in Continuous Cultures. *Applied and Environmental Microbiology*, *57*(2), 579–585. doi:10.1128/aem.57.2.579-585.1991 PMID:16348422

Ladenstein, R., & Antranikian, G. (1998). (1098). Proteins from hyperthermophiles: Stability and enzymatic catalysis close to the boiling point of water. *Advances in Biochemical Engineering/Biotechnology*, *61*, 37–85. doi:10.1007/BFb0102289 PMID:9670797

Laksanalamai, P., & Robb, F. T. (2004). Small heat shock proteins from extremophiles: A review. *Extremophiles*, *8*(1), 1–11. doi:10.100700792-003-0362-3 PMID:15064984

Lakzian, A., Berenji, A. R., Karimi, E., & Razavi, S. (2008). Adsorption capability of lead, nickel and zinc by exopolysaccharide and dried cell of *Ensifer meliloti*. *Asian Journal of Chemistry*, *20*(8), 6075.

Lambert, D., & Draper, D. E. (2007). Effects of Osmolytes on RNA Secondary and Tertiary Structure Stabilities and RNA-Mg2+ Interactions. *Journal of Molecular Biology*, *370*(5), 993–1005. doi:10.1016/j.jmb.2007.03.080 PMID:17555763

Lantz, M., Svensson, M., Björnsson, L., & Börjesson, P. (2007). The prospects for an expansion of biogas systems in Sweden—Incentives, barriers and potentials. *Energy Policy*, *35*(3), 1830–1843. doi:10.1016/j.enpol.2006.05.017

Lanzén, A., Simachew, A., Gessesse, A., Chmolowska, D., Jonassen, I., & Øvreås, L. (2013). Surprising Prokaryotic and Eukaryotic Diversity, Community Structure and Biogeography of Ethiopian Soda Lakes. *PLoS One*, *8*(8), e72577. Advance online publication. doi:10.1371/journal.pone.0072577 PMID:24023625

Lapaglia, C., & Hartzell, P. L. (1997). Stress-induced production of biofilm in the hyperthermophile *Archaeoglobus fulgidus*. *Applied and Environmental Microbiology*, *63*(8), 3158–3163. doi:10.1128/aem.63.8.3158-3163.1997 PMID:16535671

Lauro, F. M., & Bartlett, D. H. (2008). Prokaryotic lifestyles in deep sea habitats. *Extremophiles*, *12*(1), 15–25. doi:10.100700792-006-0059-5 PMID:17225926

Lawson, C. E., & Lücker, S. (2018). Complete ammonia oxidation: An important control on nitrification in engineered ecosystems? *Current Opinion in Biotechnology*, *50*, 158–165. doi:10.1016/j.copbio.2018.01.015 PMID:29414055

Laybourn-Parry, J., Marshall, W. A., & Madan, N. J. (2007). Viral dynamics and patterns of lysogeny in saline Antarctic lakes. *Polar Biology*, *30*(3), 351–358. doi:10.100700300-006-0191-9

Laybourn-Parry, J., & Pearce, D. A. (2007). The biodiversity and ecology of Antarctic lakes: Models for evolution. *Philosophical Transactions of the Royal Society of London. Series B, Biological Sciences*, *362*(1488), 2273–2289. doi:10.1098/rstb.2006.1945 PMID:17553775

Le Roes-Hill, M., Rohland, J., Meyers, P. R., Cowan, D. A., & Burton, S. G. (2009). Streptomyces hypolithicus sp. nov., isolated from an Antarctic hypolith community. *International Journal of Systematic and Evolutionary Microbiology, 59*(8), 2032–2035. doi:10.1099/ijs.0.007971-0 PMID:19567563

Lebeer, S., Bron, P. A., Marco, M. L., Van Pijkeren, J. P., Motherway, M. O. C., Hill, C., Pot, B., Roos, S., & Klaenhammer, T. (2018). Identification of probiotic effector molecules: Present state and future perspectives. *Current Opinion in Biotechnology*, *49*, 217–223. doi:10.1016/j.copbio.2017.10.007 PMID:29153882

Lebre, P. H., De Maayer, P., & Cowan, D. A. (2017). Xerotolerant bacteria: Surviving through a dry spell. *Nature Reviews. Microbiology*, *15*(5), 285–296. doi:10.1038/nrmicro.2017.16 PMID:28316329

Lechevalier, H. (1992). Actinomycetes and their products: A look at the future. *World Journal of Microbiology & Biotechnology*, *8*(S1), 72–73. doi:10.1007/BF02421498 PMID:24425651

Lee, D., Koh, Y., Kim, K., & B., K.-F. M. (1999). *Isolation and characterization of a thermophilic lipase from* Bacillus thermoleovorans *ID-1*. Academic.oup.com

Lee, C.-F., Makhatadze, G. I., & Wong, K.-B. (2005). Effects of Charge-to-Alanine Substitutions on the Stability of Ribosomal Protein L30e from Thermococcus celer†. *Biochemistry*, *44*(51), 16817–16825. doi:10.1021/bi0519654 PMID:16363795

Lee, C., Kang, H. J., von Ballmoos, C., Newstead, S., Uzdavinys, P., Dotson, D. L., Iwata, S., Beckstein, O., Cameron, A. D., & Drew, D. (2013). A two-domain elevator mechanism for sodium/proton antiport. *Nature*, *501*(7468), 573–577. doi:10.1038/nature12484 PMID:23995679

Lee, J. Y., Jang, Y.-S., Lee, J., Papoutsakis, E. T., & Lee, S. Y. (2009). Metabolic engineering of Clostridium acetobutylicum M5 for highly selective butanol production. *Biotechnology Journal*, *4*(10), 1432–1440. doi:10.1002/biot.200900142 PMID:19830716

Lentzen, G., & Schwarz, T. (2006). Extremolytes: Natural compounds from extremophiles for versatile applications. *Applied Microbiology and Biotechnology*, *72*(4), 623–634. doi:10.100700253-006-0553-9 PMID:16957893

Leroux, M. R., Fandrich, M., Klunker, D., Siegers, K., Lupas, A. N., Brown, J. R., Schiebel, E., Dobson, C. M., & Hartl, F. U. (1999). MtGimC, a novel archaeal chaperone related to the eukaryotic chaperonin cofactor GimC/ prefoldin. *The EMBO Journal*, *18*(23), 6730–6743. doi:10.1093/emboj/18.23.6730 PMID:10581246

Letchumanan, V., Pusparajah, P., Tan, L. H., Yin, W. F., Lee, L. H., & Chan, K. G. (2015). Occurrence and antibiotic resistance of *Vibrio parahaemolyticus* from shellfish in Selangor, Malaysia. *Frontiers in Microbiology*, *6*, 1417. doi:10.3389/fmicb.2015.01417 PMID:26697003

Le, Y., Fu, Y., & Sun, J. (2021). Genome editing of the anaerobic thermophile Thermoanaerobacter ethanolicus using thermostable Cas9. *Applied and Environmental Microbiology*, *87*, e01773–e20. PMID:33067194

Li, W. F., Zhou, X. X., & Lu, P. (2005). Structural features of thermozymes. *Biotechnology Advances, 23*(4), 271-281. doi:.biotechadv.2005.01.002 doi:10.1016/j

Liao, H., McKenzie, T., & Hageman, R. (1986). Isolation of a thermostable enzyme variant by cloning and selection in a thermophile. *Proceedings of the National Academy of Sciences of the United States of America*, *83*(3), 576–580. doi:10.1073/pnas.83.3.576 PMID:3003740

Li, C., Zhou, L., Yang, H., Lv, R., Tian, P., Li, X., Zhang, Y., Chen, Z., & Lin, F. (2017). Self-assembled exopolysaccharide nanoparticles for bioremediation and green synthesis of noble metal nanoparticles. *ACS Applied Materials & Interfaces*, *9*(27), 22808–22818. doi:10.1021/acsami.7b02908 PMID:28613815

Li, H., Liao, Z., Sun, J., Jiang, B., Wang, J., & Yang, Y. (2019). Modelling and simulation of two-bed PSA process for separating H2 from methane steam reforming. *Chinese Journal of Chemical Engineering, 27*(8), 1870–1878. doi:10.1016/j.cjche.2018.11.022

Lin, X., Yang, B., Shen, J., & Du, N. (2009). Biodegradation of Crude Oil by an Arctic Psychrotrophic Bacterium Pseudoalteromomas sp. P29. *Current Microbiology, 59*(3), 341–345. doi:10.1007/s00284-009-9440-9

Ling, H. L., Rahmat, Z., Bakar, F. D. A., Murad, A. M. A., & Illias, R. M. (2018). Secretome analysis of alkaliphilic bacterium *Bacillus lehensis* G1 in response to pH changes. *Microbiological Research, 215*, 46–54. doi:10.1016/j.micres.2018.06.006 PMID:30172308

Lino, C. A., Harper, J. C., Carney, J. P., & Timlin, J. A. (2018). Delivering CRISPR: A review of the challenges and approaches. *Drug Delivery, 25*(1), 1234–1257. doi:10.1080/10717544.2018.1474964 PMID:29801422

Lipson, D. A. (2007). Relationships between temperature responses and bacterial community structure along seasonal and altitudinal gradients. *FEMS Microbiology Ecology, 59*(2), 418–427. doi:10.1111/j.1574-6941.2006.00240.x PMID:17328121

Li, S. J., Hua, Z.-S., Huang, L., Li, J., Shi, S.-H., Chen, L.-X., Kuang, J.-L., Liu, J., Hu, M., & Shu, W.-S. (2014). Microbial communities evolve faster in extreme environments. *Scientific Reports, 4*(6205), 1–9. doi:10.1038rep06205 PMID:25158668

Li, S., Huang, R., Shah, N. P., Tao, X., Xiong, Y., & Wei, H. (2014). Antioxidant and antibacterial activities of exopolysaccharides from *Bifidobacterium bifidum* WBIN03 and Lactobacillus plantarum R315. *Journal of Dairy Science, 97*(12), 7334–7343. doi:10.3168/jds.2014-7912 PMID:25282420

Li, T., Weng, Y., Ma, X., Tian, B., Dai, S., Jin, Y., Liu, M., Li, J., Yu, J., & Hua, Y. (2017). Deinococcus radiodurans Toxin–Antitoxin MazEF-dr Mediates Cell Death in Response to DNA Damage Stress. *Frontiers in Microbiology, 8*, 1427. Advance online publication. doi:10.3389/fmicb.2017.01427 PMID:28798741

Litchfield, C. D. (2011). Potential for industrial products from the halophilic Archaea. *Journal of Industrial Microbiology & Biotechnology, 38*(10), 1635–1647. doi:10.100710295-011-1021-9 PMID:21853327

Littlechild, J. A. (2015). Archaeal Enzymes and Applications in Industrial Biocatalysts. *Archaea (Vancouver, B.C.), 147671*, 1–10. Advance online publication. doi:10.1155/2015/147671 PMID:26494981

Littlechild, J. A. (2015). Enzymes from extreme environments and their industrial applications. *Frontiers in Bioengineering and Biotechnology, 3*, 161. doi:10.3389/fbioe.2015.00161 PMID:26528475

Liu, F., Kariyawasam, S., Jayarao, B. M., Barrangou, R., Gerner-Smidt, P., Ribot, E. M., Knabel, S. J., & Dudley, E. G. (2011). Subtyping *Salmonella enterica* serovar enteritidis isolates from different sources by using sequence typing based on virulence genes and clustered regularly interspaced short palindromic repeats (CRISPRs). *Applied and Environmental Microbiology, 77*(13), 4520–4526. doi:10.1128/AEM.00468-11 PMID:21571881

Liu, H., Zhou, Y., Liu, R., Zhang, K.-Y., & Lai, R. (2009). Bacillus solisalsi sp. nov., a halotolerant, alkaliphilic bacterium isolated from soil around a salt lake. *International Journal of Systematic and Evolutionary Microbiology, 59*(6), 1460–1464. doi:10.1099/ijs.0.000653-0 PMID:19502335

Liu, T. Y., Iavarone, A. T., & Doudna, J. A. (2017). RNA and DNA targeting by a reconstituted *Thermus thermophiles* Type III-A CRISPR-Cas system. *PLoS One, 12*(1), 1–20. doi:10.1371/journal.pone.0170552

Liu, Y. Q., Yao, T. D., Jiao, N. Z., Kang, S. C., Xu, B. Q., Zeng, Y. H., Huang, S. J., & Liu, X. B. (2009b). Bacterial diversity in the snow over Tibetan Plateau Glaciers. *Extremophiles*, *13*(3), 411–423. doi:10.100700792-009-0227-5 PMID:19159068

Liu, Y., Yao, T., Kang, S., Jiao, N., Zeng, Y., Shi, Y., ... Huang, S. (2006). Seasonal variation of snow microbial community structure in the East Rongbuk glacier, Mt. Everest. *Chinese Science Bulletin*, *51*(12), 1476–1486.

Liu, Z. Y., Jia, D. C., di Zhang, K., Zhu, H. F., Zhang, Q., Jiang, W. H., Gu, Y., & Li, F. L. (2020). Ethanol metabolism dynamics in *Clostridium ljungdahlii* grown on carbon monoxide. *Applied and Environmental Microbiology*, *86*(14). Advance online publication. doi:10.1128/AEM.00730-20 PMID:32414802

Li, X., & Yu, H. Y. (2013). Halostable cellulase with organic solvent tolerance from *Haloarcula* sp. LLSG7 and its application in bioethanol fermentation using agricultural wastes. *Journal of Industrial Microbiology & Biotechnology*, *40*(12), 1357–1365. doi:10.100710295-013-1340-0 PMID:24037323

Lixia, Z. (2007). Soil microbial characteristics as bioindicators of soil health. Biodiversity Science, 15(2), 162. doi:10.1360/biodiv.060290

Li, Y., Engle, M., Weiss, N., Mandelco, L., & Wiegel, J. (1994). *Clostridium thermoalcaliphilum* sp. nov., an Anaerobic and Thermotolerant Facultative Alkaliphile. *International Journal of Systematic Bacteriology*, *44*(1), 111–118. doi:10.1099/00207713-44-1-111 PMID:8123553

Li, Y., Pan, S., Zhang, Y., Ren, M., Feng, M., Peng, N., Chen, L., Liang, Y. X., & She, Q. (2016). Harnessing Type I and Type III CRISPR-Cas systems for genome editing. *Nucleic Acids Research*, *44*(4), e34. doi:10.1093/nar/gkv1044 PMID:26467477

Li, Z., Kawamura, Y., Shida, O., Yamagata, S., Deguchi, T., & Ezaki, T. (2002). *Bacillus okuhidensis* sp. nov., isolated from the Okuhida spa area of Japan. *International Journal of Systematic and Evolutionary Microbiology*, *52*(4), 1205–1209. doi:10.1099/00207713-52-4-1205 PMID:12148629

Li, Z., & Loh, X. J. (2017). Recent advances of using polyhydroxyalkanoate-based nano vehicles as therapeutic delivery carriers. *Wiley Interdisciplinary Reviews. Nanomedicine and Nanobiotechnology*, *9*(3), e1429. doi:10.1002/wnan.1429 PMID:27595635

Lloyd, J. R., Leang, C., Hodges Myerson, A. L., Coppi, M. V., Cuifo, S., Methe, B., Sandler, S. J., & Lovley, D. R. (2003). Biochemical and genetic characterization of PpcA, a periplasmic c-type cytochrome in Geobacter sulfurreducens. *The Biochemical Journal*, *369*(1), 153–161. doi:10.1042/bj20020597 PMID:12356333

Logan, B. E. (2008). *Microbial fuel cells. John Wiley* & Sons.

Logan, B. E., Hamelers, B., Rozendal, R., Schröder, U., Keller, J., Freguia, S., Aelterman, P., Verstraete, W., & Rabaey, K. (2006). Microbial Fuel Cells: Methodology and Technology. *Environmental Science & Technology*, *40*(17), 5181–5192. doi:10.1021/es0605016 PMID:16999087

Longo, L. M., Lee, J., & Blaber, M. (2013). Simplified protein design biased for prebiotic amino acids yields a foldable, halophilic protein. *Proceedings of the National Academy of Sciences of the United States of America*, *110*(6), 2135–2139. doi:10.1073/pnas.1219530110 PMID:23341608

López-Rodas, V., Marvá, F., Costas, E., & Flores-Moya, A. (2008). Microalgal adaptation to a stressful environment (acidic, metal-rich mine waters) could be due to selection of pre-selective mutants originating in non-extreme environments. *Environmental and Experimental Botany*, *64*(1), 43–48. doi:10.1016/j.envexpbot.2008.01.001

Loukas, A., Kappas, I., & Abatzopoulos, T. J. (2018). HaloDom: A new database of halophiles across all life domains. *Journal of Biological Research (Greece), 25*(1), 2. Advance online publication. doi:10.118640709-017-0072-0 PMID:29796383

Lovley, D. R. (2006). Microbial fuel cells: Novel microbial physiologies and engineering approaches. *Current Opinion in Biotechnology, 17*(3), 327–332. doi:10.1016/j.copbio.2006.04.006 PMID:16679010

Lovley, D. R. (2012). Long-range electron transport to Fe(III) oxide via pili with metallic-like conductivity. *Biochemical Society Transactions, 40*(6), 1186–1190. doi:10.1042/BST20120131 PMID:23176452

Luca, D., Denier van der Gon, A. W., Anita, V., Ponjee, M. W. G., Brongersma, H. H., & Popa, G. (2001). Surface nitridation processes and non-linear behaviour of the reactive magnetron discharge with titanium target. *Vacuum, 61*(2–4), 163–167. doi:10.1016/S0042-207X(00)00474-7

Luo, Z., Wu, Y. L., Li, Z., & Loh, X. J. (2019). Recent Progress in Polyhydroxyalkanoates-Based Copolymers for Biomedical Applications. *Biotechnology Journal, 14*(12), 1900283. doi:10.1002/biot.201900283 PMID:31469496

Luque, R., Bejar, V., Quesada, E., & Llamas, I. (2014). Diversity of halophilic bacteria isolated from Rambla Salada, Murcia (Spain). *Canadian Journal of Microbiology, 60*(12), 839–846. doi:10.1139/cjm-2014-0342 PMID:25403824

Luque, R., Herrero-Davila, L., Campelo, J. M., Clark, J. H., Hidalgo, J. M., Luna, D., Marinas, J. M., & Romero, A. A. (2008). Biofuels: A technological perspective. *Energy & Environmental Science, 1*(5), 542. doi:10.1039/b807094f

Lynd, L. R. (n.d.). Production of ethanol from lignocellulosic materials using thermophilic bacteria: Critical evaluation of potential and review. In *Lignocellulosic Materials* (pp. 1–52). Springer-Verlag. doi:10.1007/BFb0007858

Lynd, L. R., van Zyl, W. H., McBride, J. E., & Laser, M. (2005). Consolidated bioprocessing of cellulosic biomass: An update. *Current Opinion in Biotechnology, 16*(5), 577–583. doi:10.1016/j.copbio.2005.08.009 PMID:16154338

Lyu, Z., Shao, N., Akinyemi, T., & Whitman, W. B. (2018). Methanogenesis. *Current Biology, 28*(13), R727–R732. doi:10.1016/j.cub.2018.05.021 PMID:29990451

MacElroy, R. D. (1974). Some comments on the evolution of extremophiles. *Bio Systems, 6*(1), 74–75. doi:10.1016/0303-2647(74)90026-4

Mader, H. M., Pettitt, M. E., Wadham, J. L., Wolff, E. W., & Parkes, R. J. (2006). Subsurface ice as a microbial habitat. *Geology, 34*(3), 169–172. doi:10.1130/G22096.1

Madern, D., Ebel, C., & Zaccai, G. (2000). Halophilic adaptation of enzymes. *Extremophiles, 4*(2), 91–98. doi:10.1007/s007920050142

Madigan, M. T., & Oren, A. (1999). Thermophilic and halophilic extremophiles. *Current Opinion in Microbiology, 2*(3), 265–269. doi:10.1016/S1369-5274(99)80046-0 PMID:10383863

Madigan, M. T., Takigiku, R., Lee, R. G., Gest, H., & Hayes, J. M. (1989). Carbon isotope fractionation by thermophilic phototrophic sulfur bacteria: Evidence for autotrophic growth in natural populations. *Applied and Environmental Microbiology, 55*(3), 639–644. doi:10.1128/aem.55.3.639-644.1989 PMID:11536609

Madsen, J. S., Burmølle, M., Hansen, L. H., & Sørensen, S. J. (2012). The interconnection between biofilm formation and horizontal gene transfer. *FEMS Immunology and Medical Microbiology, 65*(2), 183–195. doi:10.1111/j.1574-695X.2012.00960.x PMID:22444301

Maier, L. K., Fischer, S., Stoll, B., Brendel, J., Pfeiffer, F., Dyall-Smith, M., & Marchfelder, A. (2012). The immune system of halophilic archaea. *Mobile Genetic Elements, 2*(5), 228–232. doi:10.4161/mge.22530 PMID:23446883

Ma, K., & Adams, M. W. (1999). An unusual oxygen-sensitive, iron-and zinc-containing alcohol dehydrogenase from the hyperthermophilic archaeon *Pyrococcus furiosus. Journal of Bacteriology, 181*(4), 1163–1170. doi:10.1128/JB.181.4.1163-1170.1999 PMID:9973342

Ma, K., Maeda, T., You, H., & Shirai, Y. (2014). Open fermentative production of l-lactic acid with high optical purity by thermophilic Bacillus coagulans using excess sludge as nutrient. *Bioresource Technology, 151*, 28–35. doi:10.1016/j.biortech.2013.10.022 PMID:24201025

Makarova, K. S., Grishin, N. V., Shabalina, S. A., Wolf, Y. I., & Koonin, E. V. (2006). A putative RNA-interference-based immune system in prokaryotes: Computational analysis of the predicted enzymatic machinery, functional analogies with eukaryotic RNAi, and hypothetical mechanisms of action. *Biology Direct, 1*(1), 7. doi:10.1186/1745-6150-1-7 PMID:16545108

Makarova, K. S., Haft, D. H., Barrangou, R., Brouns, S. J., Charpentier, E., Horvath, P., Moineau, S., Mojica, F. J., Wolf, Y. I., Yakunin, A. F., van der Oost, J., & Koonin, E. V. (2011b). Evolution and classification of the CRISPR-Cas systems. *Nature Reviews. Microbiology, 9*(6), 467–477. doi:10.1038/nrmicro2577 PMID:21552286

Makarova, K. S., Wolf, Y. I., Alkhnbashi, O. S., Costa, F., Shah, S. A., Saunders, S. J., Barrangou, R., Brouns, S. J., Charpentier, E., Haft, D. H., Horvath, P., Moineau, S., Mojica, F. J., Terns, R. M., Terns, M. P., White, M. F., Yakunin, A. F., Garrett, R. A., van der Oost, J., ... Koonin, E. V. (2015). An updated evolutionary classification of CRISPR-Cas systems. *Nature Reviews. Microbiology, 13*(11), 722–736. doi:10.1038/nrmicro3569 PMID:26411297

Makarova, K. S., Wolf, Y. I., Forterre, P., Prangishvili, D., Krupovic, M., & Koonin, E. V. (2014). Dark matter in archaeal genomes: A rich source of novel mobile elements, defense systems and secretory complexes. *Extremophiles, 18*(5), 877–893. doi:10.100700792-014-0672-7 PMID:25113822

Makarova, K., Aravind, L., Wolf, Y., & Koonin, E. (2011). Unification of Cas protein families and a simple scenario for the origin and evolution of CRISPR-Cas systems. *Biology Direct, 6*(1), 38. doi:10.1186/1745-6150-6-38 PMID:21756346

Malashenko, Y. R., Romanovskaya, V. A., Bogachenko, V. N., & Shved, A. D. (1975). Thermophilic and thermotolerant methane-assimilating bacteria. *Mikrobiologiya, 44*, 855–862. PMID:1207503

Mallick, P., Boutz, D. R., Eisenberg, D., & Yeates, T. O. (2002). Genomic evidence that the intracellular proteins of archaeal microbes contain disulfide bonds. *Proceedings of the National Academy of Sciences of the United States of America, 99*(15), 9679–9684. doi:10.1073/pnas.142310499 PMID:12107280

Mamo, G. (2019). Challenges and Adaptations of Life in Alkaline Habitats. *Advances in Biochemical Engineering/Biotechnology, 172*, 85–133. doi:10.1007/10_2019_97 PMID:31273417

Mamo, G., & Mattiasson, B. (2016). Alkaliphilic Microorganisms in Biotechnology. In *Biotechnology of Extremophiles* (pp. 243–272). Springer International Publishing. doi:10.1007/978-3-319-13521-2_8

Manca, M. C., Lama, L., Improta, R., Esposito, E., Gambacorta, A., & Nicolaus, B. (1996). Chemical composition of two exopolysaccharides from *Bacillus thermoantarcticus. Applied and Environmental Microbiology, 62*(9), 3265–3269. doi:10.1128/aem.62.9.3265-3269.1996 PMID:16535400

Mancílio, Ribeiro, Lopes, Kishi, Martins-Santana, de Siqueira, Andrade, Guazzaroni, & Reginatto. (2020). Unusual microbial community and impact of iron and sulfate on microbial fuel cell ecology and performance. *Current Research in Biotechnology*, 64-73.

Mangrola, A. V., Dudhagara, P. R., Koringa, P. G., Joshi, C. G., & Patel, R. K. (2018). Metagenomic microbial community profiling of Unnai hot spring by Ion-Torrent based shotgun sequencing. *Microbiology, 87*(1), 143–146. doi:10.1134/S0026261718010113

Mangrola, A. V., Dudhagara, P., Koringa, P., Joshi, C. G., & Patel, R. K. (2015). Shotgun metagenomic sequencing based microbial diversity assessment of Lasundra hot spring, India. *Genomics Data*, *4*, 73–75. doi:10.1016/j.gdata.2015.03.005 PMID:26484181

Mangrola, A., Dudhagara, P., Koringa, P., Joshi, C. G., Parmar, M., & Patel, R. (2015). Deciphering the microbiota of Tuwa hot spring, India using shotgun metagenomic sequencing approach. *Genomics Data*, *4*, 153–155. doi:10.1016/j.gdata.2015.04.014 PMID:26484204

Manjula, D., Jeevitha, P., Ramya, I., Hemapriya, J., & Ravi, A. (2018). Biotechnological applications of halophilic pigments - An overview. *International Journal of Current Microbiology and Applied Sciences*, *7*(7), 4392–4398. doi:10.20546/ijcmas.2018.707.512

Männistö, M. K., & Häggblom, M. M. (2006). Characterization of psychrotolerant heterotrophic bacteria from Finnish Lapland. *Systematic and Applied Microbiology*, *29*(3), 229–243. doi:10.1016/j.syapm.2005.09.001 PMID:16564959

Männistö, M. K., Tiirola, M., & Häggblom, M. M. (2007). Bacterial communities in Arctic fjelds of Finnish Lapland are stable but highly pH-dependent. *FEMS Microbiology Ecology*, *59*(2), 452–465. doi:10.1111/j.1574-6941.2006.00232.x PMID:17328122

Manobala, T., Shukla, S. K., Rao, T. S., & Kumar, M. D. (2019). A new uranium bioremediation approach using radio-tolerant *Deinococcus radiodurans* biofilm. *Journal of Biosciences*, *44*(5), 1–9. doi:10.100712038-019-9942-y PMID:31719231

Marg, B.-L., Schweimer, K., Sticht, H., & Oesterhelt, D. (2004). A Two-α-Helix Extra Domain Mediates the Halophilic Character of a Plant-Type Ferredoxin from Halophilic Archaea†,‡. *Biochemistry*, *44*(1), 29–39. doi:10.1021/bi0485169 PMID:15628843

Margesin, R., Dieplinger, H., Hofmann, J., Sarg, B., & Lindner, H. (2005). A cold-active extracellular metalloprotease from Pedobacter cryoconitis—Production and properties. *Research in Microbiology*, *156*(4), 499–505. doi:10.1016/j.resmic.2004.12.008 PMID:15862448

Margesin, R., & Feller, G. (2010). Biotechnological applications of psychrophiles. *Environmental Technology*, *31*(8-9), 835–844. doi:10.1080/09593331003663328 PMID:20662375

Margesin, R., & Fell, J. W. (2008). Mrakiella cryoconiti gen. nov., sp. nov., a psychrophilic, anamorphic, basidiomycetous yeast from alpine and arctic habitats. *International Journal of Systematic and Evolutionary Microbiology*, *58*(12), 2977–2982. doi:10.1099/ijs.0.2008/000836-0 PMID:19060092

Margesin, R., & Miteva, V. (2011). Diversity and ecology of psychrophilic microorganisms. *Research in Microbiology*, *162*(3), 346–361. doi:10.1016/j.resmic.2010.12.004 PMID:21187146

Margesin, R., Neuner, G., & Storey, K. B. (2007). Cold-loving microbes, plants, and animals—Fundamental and applied aspects. *Naturwissenschaften*, *94*(2), 77–99. doi:10.100700114-006-0162-6 PMID:17039344

Margesin, R., Spröer, C., Schumann, P., & Schinner, F. (2003). Pedobacter cryoconitis sp. nov., a facultative psychrophile from alpine glacier cryoconite. *International Journal of Systematic and Evolutionary Microbiology*, *53*(5), 1291–1296. doi:10.1099/ijs.0.02436-0 PMID:13130009

Margesin, R., Zacke, G., & Schinner, F. (2002). Characterization of heterotrophic microorganisms in alpine glacier cryoconite. *Arctic, Antarctic, and Alpine Research*, *34*(1), 88–93. doi:10.1080/15230430.2002.12003472

Marraffini, L. A., & Sontheimer, E. J. (2008). CRISPR interference limits horizontal gene transfer in staphylococci by targeting DNA. *Science*, *322*(5909), 1843–1845. doi:10.1126cience.1165771 PMID:19095942

Marshall, C. W., & May, H. D. (2009). Electrochemical evidence of direct electrode reduction by a thermophilic Gram-positive bacterium, *Thermincola ferriacetica. Energy & Environmental Science*, *2*(6), 699–705. doi:10.1039/b823237g

Martin, A., & McMinn, A. (2018). Sea ice, extremophiles, and life on extra-terrestrial ocean worlds. *International Journal of Astrobiology*, *17*(1), 1–16. doi:10.1017/S1473550416000483

Martinez-Espinosa, R. M., Richardson, D. J., & Bonete, M. J. (2015). Characterisation of chlorate reduction in the haloarchaeon *Haloferax mediterranei. Biochimica et Biophysica Acta (BBA)- General Subjects*, *1850*(4), 587–594. doi:10.1016/j.bbagen.2014.12.011 PMID:25512066

Martins, M. B., & Carvalho, I. (2007). Diketopiperazines: Biological activity and synthesis. *Tetrahedron*, *63*(40), 9923–9932. doi:10.1016/j.tet.2007.04.105

Maru, B. T., Bielen, A. A. M., Kengen, S. W. M., Constantí, M., & Medina, F. (2012). Biohydrogen Production from Glycerol using *Thermotoga* spp. *Energy Procedia*, *29*, 300–307. doi:10.1016/j.egypro.2012.09.036

Maruyama, T., Dougan, S. K., Truttmann, M. C., Bilate, A. M., Ingram, J. R., & Ploegh, H. L. (2015). Increasing the efficiency of precise genome editing with CRISPR-Cas9 by inhibition of nonhomologous end joining. *Nature Biotechnology*, *33*(5), 538–542. doi:10.1038/nbt.3190 PMID:25798939

Mathews, D. J., Chan, S., Donovan, P. J., Douglas, T., Gyngell, C., Harris, J., Regenberg, A., & Lovell-Badge, R. (2015). CRISPR: A path through the thicket. *Nature*, *527*(7577), 159–161. doi:10.1038/527159a PMID:26560284

Mathews, J., & Wang, G. (2009). Metabolic pathway engineering for enhanced biohydrogen production. *International Journal of Hydrogen Energy*, *34*(17), 7404–7416. doi:10.1016/j.ijhydene.2009.05.078

Matte-Tailliez, O. M., Brochier, C., Forterre, P., & Philippe, H. (2002). Archaeal phylogeny based on ribosomal proteins. *Molecular Biology and Evolution*, *19*(5), 631–639. doi:10.1093/oxfordjournals.molbev.a004122 PMID:11961097

Mayer, F., Küper, U., Meyer, C., Daxer, S., Müller, V., Rachel, R., & Huber, H. (2012). AMP-forming acetyl coenzyme a synthetase in the outermost membrane of the hyperthermophilic crenarchaeon ignicoccus hospitalis. *Journal of Bacteriology*, *194*(6), 1572–1581. doi:10.1128/JB.06130-11 PMID:22247508

McBeath, J. B. (2002). Snow mold-plant-antagonist interactions: Survival of the fittest under the snow. *The Plant Health Instructor. Science and Society*. Advance online publication. 10.1094%2Fphi-i-2002-1010-01

McEwen, G. D., Wu, Y., & Zhou, A. (2010). Probing nanostructures of bacterial extracellular polymeric substances versus culture time by Raman microspectroscopy and atomic force microscopy. *Biopolymers: Original Research on Biomolecules*, *93*(2), 171–177. doi:10.1002/bip.21315 PMID:19787720

McMillan, D. G. G., Velasquez, I., Nunn, B. L., Goodlett, D. R., Hunter, K. A., Lamont, I., Sander, S. G., & Cook, G. M. (2010). Acquisition of iron by alkaliphilic *Bacillus* species. *Applied and Environmental Microbiology*, *76*(20), 6955–6961. doi:10.1128/AEM.01393-10 PMID:20802068

Medigue, C., Krin, E., Pascal, G., Barbe, V., Bernsel, A., Bertin, P. N., Cheung, F., Cruveiller, S., D'Amico, S., Duilio, A., Fang, G., Feller, G., Ho, C., Mangenot, S., Marino, G., Nilsson, J., Parrilli, E., Rocha, E. P. C., Rouy, Z., ... Danchin, A. (2005). Coping with cold: The genome of the versatile marine Antarctica bacterium Pseudoalteromonas haloplanktis TAC125. *Genome Research*, *15*(10), 1325–1335. doi:10.1101/gr.4126905 PMID:16169927

Medina, A., Schmidt-Heydt, M., Rodriguez, A., Parra, R., Geisen, R., & Magan, N. (2015). Impacts of environmental stress on growth, secondary metabolite biosynthetic gene clusters and metabolite production of xerotolerant/xerophilic fungi. *Current Genetics*, *61*(3), 325–334. doi:10.100700294-014-0455-9 PMID:25381155

Medrano-Santillana, M., Souza-Brito, E. M., Duran, R., Gutierrez-Corona, F., & Reyna-Lopez, G. E. (2017). Bacterial diversity in fumarole environments of the Paricutín volcano, Michoacán (Mexico). *Extremophiles, 21*(3), 499–511. doi:10.100700792-017-0920-8 PMID:28289847

Meena, K. K., Kumar, M., Kalyuzhnaya, M. G., Yandigeri, M. S., Singh, D. P., Saxena, A. K., & Arora, D. K. (2011). Epiphytic pink-pigmented methylotrophic bacteria enhance germination and seedling growth of wheat (Triticum aestivum) by producing phytohormone. *Antonie van Leeuwenhoek, 101*(4), 777–786. doi:10.1007/s10482-011-9692-9

Mehrshad, M., Amoozegar, M. A., Makhdoumi, A., Rasooli, M., Asadi, B., Schumann, P., & Ventosa, A. (2015a). *Halovarius luteusgen* nov., sp. nov., a novel extremely halophilic archaeon from Urmia salt lake. *International Journal of Systematic and Evolutionary Microbiology, 65*(Pt_8), 2420–2425. doi:10.1099/ijs.0.000279 PMID:25899505

Meijer, W. G., Nienhuis-Kuiper, M. E., & Hansen, T. A. (1999). Fermentative bacteria from estuarine mud: Phylogenetic position of Acidaminobacter hydrogenoformans and description of a new type of gram-negative, propionigenic bacterium as Propionibacter pelophilus gen. nov., sp. nov. *International Journal of Systematic Bacteriology, 49*(3), 1039–1044. PMID:10425761

Meknaci, R., Lopes, P., Servy, C., Le Caer, J. P., Andrieu, J. P., Hacene, H., & Ouazzani, J. (2014). Agar-supported cultivation of *Halorubrum sp.* SSR, and production of halocin C8 on the scale-up prototype Platotex. *Extremophiles, 18*(6), 1049–1055. doi:10.100700792-014-0682-5 PMID:25138277

Meng, X., Pijning, T., Dobruchowska, J. M., Yin, H., Gerwig, G. J., & Dijkhuizen, L. (2016). Structural determinants of alternating (α1→ 4) and (α1→ 6) linkage specificity in reuteransucrase of *Lactobacillus reuteri. Scientific Reports, 6*(1), 1–14. doi:10.1038rep35261 PMID:27748434

Merino, N., Aronson, H. S., Bojanova, D. P., Feyhl-Buska, J., Wong, M. L., Zhang, S., & Giovannelli, D. (2019). Living at the Extremes: Extremophiles and the Limits of Life in a Planetary Context. *Frontiers in Microbiology, 10*, 780. Advance online publication. doi:10.3389/fmicb.2019.00780 PMID:31037068

Merlino, A., Krauss, I. R., Castellano, I., De Vendittis, E., Rossi, B., Conte, M., ... Sica, F. (2010). Structure and flexibility in cold-adapted iron superoxide dismutases: The case of the enzyme isolated from Pseudoalteromonas haloplanktis. *Journal of Structural Biology, 172*(3), 343–352. doi:10.1016/j.jsb.2010.08.008 PMID:20732427

Merone, L., Mandrich, L., Rossi, M., & Manco, G. (2005). A thermostable phosphotriesterase from the archaeon Sulfolobus solfataricus: Cloning, overexpression and properties. *Extremophiles: Life under Extreme Conditions, 9*(4), 297–305.

Mesbah, N., & Wiegel, J. (2018). Improvement of Activity and Thermostability of Agar-Entrapped, Thermophilic, Haloalkaliphilic Amylase AmyD8. *Catalysis Letters, 148*(9), 2665–2674. doi:10.100710562-018-2493-2

Methe´, B. A., Nelson, K. E., Deming, J. W., Momen, B., Melamud, E., Zhang, X., Moult, J., Madupa, R., Nelson, W. C., Dodson, R. J., Brinkac, L. M., Daugherty, S. C., Durkin, A. S., DeBoy, R. T., Kolonay, J. F., Sullivan, S. A., Zhou, L., Davidsen, T. M., Wu, M., ... Fraser, C. M. (2005). The psychrophilic lifestyle as revealed by the genome sequence of Colwellia psychrerythraea 34H through genomic and proteomic analyses. *Proceedings of the National Academy of Sciences of the United States of America, 102*(31), 10913–10918. doi:10.1073/pnas.0504766102 PMID:16043709

Metzger, P., & Largeau, C. (2005). Botryococcus braunii: A rich source for hydrocarbons and related ether lipids. *Applied Microbiology and Biotechnology, 66*(5), 486–496. doi:10.100700253-004-1779-z PMID:15630516

Mevada, V., Patel, S., Pandya, J., Joshi, H., & Patel, R. (2017). Whole genome sequencing and annotation of halophilic *Salinicoccus sp.* BAB 3246 isolated from the coastal region of Gujarat. *Genomics Data, 13*, 30–34. doi:10.1016/j.gdata.2017.06.006 PMID:28702355

Mevs, U., Stackebrandt, E., Schumann, P., Gallikowski, C. A., & Hirsch, P. (2000). Modestobacter multiseptatus gen. nov., sp. nov., a budding actinomycete from soils of the Asgard Range (Transantarctic Mountains). *International Journal of Systematic and Evolutionary Microbiology, 50*(1), 337–346. doi:10.1099/00207713-50-1-337 PMID:10826821

Miao, L.-L., Hou, Y.-J., Fan, H.-X., Qu, J., Qi, C., Liu, Y., Li, D.-F., & Liu, Z.-P. (2016). Molecular structural basis for the cold adaptedness of the psychrophilic β-glucosidase BglU in Micrococcus antarcticus. *Applied and Environmental Microbiology, 82*(7), 2021–2030. doi:10.1128/AEM.03158-15 PMID:26801571

Michie, I. S., Kim, J. R., Dinsdale, R. M., Guwy, A. J., & Premier, G. C. (2011). The influence of psychrophilic and mesophilic start-up temperature on microbial fuel cell system performance. *Energy & Environmental Science, 2011*(4), 1011–1019. doi:10.1039/c0ee00483a

Michoud, G., & Jebbar, M. (2016). High hydrostatic pressure adaptive strategies in an obligate piezophile Pyrococcus yayanosii. *Scientific Reports, 6*(1), 1–10. doi:10.1038rep27289 PMID:27250364

Midhat, L., Ouazzani, N., Hejjaj, A., Ouhammou, A., & Mandi, L. (2019). Accumulation of heavy metals in metallophytes from three mining sites (Southern Centre Morocco) and evaluation of their phytoremediation potential. *Ecotoxicology and Environmental Safety, 169*, 150–160. doi:10.1016/j.ecoenv.2018.11.009

Miettinen-Oinonen, A., & Suominen, P. (2002). Enhanced Production of Trichoderma reesei Endoglucanases and Use of the New Cellulase Preparations in Producing the Stonewashed Effect on Denim Fabric. *Applied and Environmental Microbiology, 68*(8), 3956–3964. doi:10.1128/AEM.68.8.3956-3964.2002 PMID:12147496

Milani, J., & Maleki, G. (2012). *Hydrocolloids in Food Industry. Food Industrial Processes-Methods and Equipment.* Tech.

Milford, A. D., Achenbach, L. A., Jung, D. O., & Madigan, M. T. (2000). *Rhodobaca bogoriensis* gen. nov. and sp. nov., an alkaliphilic purple nonsulfur bacterium from African Rift Valley soda lakes. *Archives of Microbiology, 174*(1-2), 18–27. doi:10.1007002030000166 PMID:10985738

Mir, M. Y., Hamid, S., Rohela, G. K., Parray, J. A., & Kamili, A. N. (2021, March). *Composting and Bioremediation Potential of Thermophiles.* Wiley. doi:10.1002/9781119547976.ch7

Mirete, S., Morgante, V., & González-Pastor, J. E. (2017). Acidophiles: diversity and mechanisms of adaptation to acidic environments. In *Adaption of microbial life to environmental extremes* (pp. 227–251). Springer. doi:10.1007/978-3-319-48327-6_9

Miriam, L. M., Raj, R. E., Kings, A. J., & Visvanathan, M. A. (2017). Identification and characterization of a novel biodiesel producing halophilic *Aphanothece halophytica* and its growth and lipid optimization in various media. *Energy Conversion and Management, 141*, 91–100. doi:10.1016/j.enconman.2016.05.041

Miroshnichenko, M. L., & Bonch-Osmolovskaya, E. A. (2006). Recent developments in the thermophilic microbiology of deep-sea hydrothermal vents. *Extremophiles, 10*(2), 85–96. doi:10.100700792-005-0489-5 PMID:16418793

Mi, S., Song, J., Lin, J., Che, Y., Zheng, H., & Lin, J. (2011). Complete genome of Leptospirillum ferriphilum ML-04 provides insight into its physiology and environmental adaptation. *Journal of Microbiology (Seoul, Korea), 49*(6), 890–901. doi:10.100712275-011-1099-9 PMID:22203551

Mishra, P. K., Joshi, P., Bisht, S. C., Bisht, J. K., & Selvakumar, G. (2010). Cold-Tolerant Agriculturally Important Microorganisms. In *Plant Growth and Health Promoting Bacteria* (pp. 273–296). Springer Berlin Heidelberg. doi:10.1007/978-3-642-13612-2_12

Miteva, V. (2008). Bacteria in snow and glacier ice. In *Psychrophiles: from biodiversity to biotechnology* (pp. 31–50). Springer. doi:10.1007/978-3-540-74335-4_3

Miteva, V., Sowers, T., & Brenchley, J. (2007). Production of N 2 O by ammonia oxidizing bacteria at subfreezing temperatures as a model for assessing the N 2 O anomalies in the Vostok ice core. *Geomicrobiology Journal, 24*(5), 451–459. doi:10.1080/01490450701437693

Moghaddam, J. A., Boehringer, N., Burdziak, A., Kunte, H. J., Galinski, E. A., & Schaberle, T. F. (2016). Different strategies of osmoadaptation in the closely related marine Myxobacteria *Enhygromyxa salina* SWB007 and *Plesiocystis pacifica* SIR-1. *Microbiology, 162*(4), 651–661. doi:10.1099/mic.0.000250 PMID:26842314

Mohammadipanah, F., Hamedi, J., & Dehhaghi, M. (2015). Halophilic bacteria: Potentials and applications in biotechnology. In D. K. Maheshwari & M. Saraf (Eds.), *Halophiles - Biodiversity and Sustainable Exploitation* (pp. 277–321). Springer Publisher. doi:10.1007/978-3-319-14595-2_11

Mohanraju, P., Mougiakos, I., Albers, J., Mabuchi, M., Fuchs, R., Curcuru, J., van Kranenburg, R., Brett Robb, G., & van der Oost, J. (2021). Development of a Cas12a-Based Genome Editing Tool for Moderate Thermophiles. *The CRISPR Journal, 4*(1), 82–91. doi:10.1089/crispr.2020.0086 PMID:33538626

Möhler, O., DeMott, P. J., Vali, G., & Levin, Z. (2007). Microbiology and atmospheric processes: The role of biological particles in cloud physics. *Biogeosciences, 4*(6), 1059–1071. doi:10.5194/bg-4-1059-2007

Mojica, F. J. M., Díez-Villaseñor, C., García-Martínez, J., & Almendros, C. (2009). Short motif sequences determine the targets of the prokaryotic CRISPR defence system. *Microbiology (Reading), 155*(Pt 3), 733–740. doi:10.1099/mic.0.023960-0 PMID:19246744

Mojica, F. J., Díez-Villaseñor, C., Soria, E., & Juez, G. (2000). Biological significance of a family of regularly spaced repeats in the genomes of Archaea, Bacteria and mitochondria. *Molecular Microbiology, 36*(1), 244–246. doi:10.1046/j.1365-2958.2000.01838.x PMID:10760181

Mojica, M. J., Juez, G., & Rodríguez-Valera, F. (1993). Transcription at different salinities of *Haloferax mediterranei* sequences adjacent to partially modified PstI sites. *Molecular Microbiology, 9*(3), 613–621. doi:10.1111/j.1365-2958.1993.tb01721.x PMID:8412707

Mojsov, K., Andronikov, D., Janevski, A., Jordeva, S., Kertakova, M., Golomeova, S., Gaber, S., & Ignjatov, I. (2018). Production and application of α-amylase enzyme in textile industry. *Tekstilna industrija, 66*, 23–28.

Mojsov, K. (2012). Microbial cellulases and their applications in textile processing. *International Journal of Marketing and Technology, 2*(11), 12–29.

Mokrousov, I., Narvskaya, O., Limeschenko, E., & Vyazovaya, A. (2005). Efficient discrimination within a *Corynebacterium diphtheriae* epidemic clonal group by a novel macroarray-based method. *Journal of Clinical Microbiology, 43*(4), 1662–1668. doi:10.1128/JCM.43.4.1662-1668.2005 PMID:15814981

Møller, A. P., Biard, C., Blount, J. D., Houston, D. C., Ninni, P., Saino, N., & Surai, P. F. (2000). Carotenoid-dependent Signals: Indicators of Foraging Efficiency, Immunocompetence or Detoxification Ability? *Avian and Poultry Biology Reviews, 11*(3), 137–159.

Mongodin, E. F., Hance, I. R., DeBoy, R. T., Gill, S. R., Daugherty, S., Huber, R., Fraser, C. M., Stetter, K., & Nelson, K. E. (2005). Gene transfer and genome plasticity in *Thermotoga maritima*, a model hyperthermophilic species. *Journal of Bacteriology, 187*(14), 4935–4944. doi:10.1128/JB.187.14.4935-4944.2005 PMID:15995209

Monsan, P., Bozonnet, S., Albenne, C., Joucla, G., Willemot, R.-M., & Remaud-Siméon, M. (2001). Homopolysaccharides from lactic acid bacteria. *International Dairy Journal, 11*(9), 675–685. doi:10.1016/S0958-6946(01)00113-3

Moon, S., Kim, J., Koo, J., & Bae, E. (2019). Structural and mutational analyses of psychrophilic and mesophilic adenylate kinases highlight the role of hydrophobic interactions in protein thermal stability. *Structure and Dynamics*, *6*(2), 024702. doi:10.1063/1.5089707 PMID:31111079

Moradian, J. M., Fang, Z., & Yong, Y.-C. (2021). Recent advances on biomass-fueled microbial fuel cell. *Bioresources and Bioprocessing*, *8*(1), 14. doi:10.118640643-021-00365-7

Moran-Reyna, A., & Coker, J. A. (2014). The effects of extremes of pH on the growth and transcriptomic profiles of three haloarchaea. *F1000 Research*, *3*, 168. doi:10.12688/f1000research.4789.1 PMID:25285207

Moreno, R., & Rojo, F. (2014). Features of pseudomonads growing at low temperatures: another facet of their versatility. *Environmental Microbiology Reports, 6*(5), 417–426. doi:10.1111/1758-2229.12150

Mori, Y. (1990). Characterization of a Symbiotic Coculture of Clostridium thermohydrosulfuricum YM3 and Clostridium thermocellum YM4. *Applied and Environmental Microbiology*, *56*(1), 37–42. doi:10.1128/aem.56.1.37-42.1990 PMID:16348106

Mori, Y., & Inaba, T. (1990). Ethanol production from starch in a pervaporation membrane bioreactor using Clostridium thermohydrosulfuricum. *Biotechnology and Bioengineering*, *36*(8), 849–853. doi:10.1002/bit.260360815 PMID:18597284

Morris, C. E., Sands, D. C., Bardin, M., Jaenicke, R., Vogel, B., Leyronas, C., Ariya, P. A., & Psenner, R. (2008). Microbiology and atmospheric processes: An upcoming era of research on bio-meteorology. *Biogeosciences Discussions*, *5*(1), 191–212.

Morris, G., & Harding, S. (2009). Polysaccharides, microbial. In *Encyclopedia of microbiology* (pp. 482–494). Elsevier Inc. doi:10.1016/B978-012373944-5.00135-8

Moscovici, M. (2015). Present and future medical applications of microbial exopolysaccharides. *Frontiers in Microbiology*, *6*, 1012. doi:10.3389/fmicb.2015.01012 PMID:26483763

Mosier, A. C., Murray, A. E., & Fritsen, C. H. (2007). Microbiota within the perennial ice cover of Lake Vida, Antarctica. *FEMS Microbiology Ecology*, *59*(2), 274–288. doi:10.1111/j.1574-6941.2006.00220.x PMID:17092309

Mougiakos, I., Bosma, E. F., de Vos, W. M., van Kranenburg, R., & van der Oost, J. (2016). Next generation prokaryotic engineering: The CRISPR-Cas toolkit. *Trends in Biotechnology*, *34*(7), 575–587. doi:10.1016/j.tibtech.2016.02.004 PMID:26944793

Mougiakos, I., Mohanraju, P., Bosma, E., Vrouwe, V., Finger, M., Naduthodi, M., Gussak, A., Brinkman, R., van Kranenburg, R., & Oost, J. (2017). Characterizing a thermostable Cas9 for bacterial genome editing and silencing. *Nature Communications*, *8*(1), 1647. doi:10.103841467-017-01591-4 PMID:29162801

Mozzi, F., Vaningelgem, F., Hébert, E. M., Van der Meulen, R., Moreno, M. R. F., de Valdez, G. F., & De Vuyst, L. (2006). Diversity of heteropolysaccharide-producing lactic acid bacterium strains and their biopolymers. *Applied and Environmental Microbiology*, *72*(6), 4431–4435. doi:10.1128/AEM.02780-05 PMID:16751563

Mukherjee, P., & Atala, A. (2005). Biomaterials for genitourinary tissue engineering. *Scaffolding in Tissue Engineering*, 357.

Mukherjee, A., Wheaton, G. H., Blum, P. H., & Kelly, R. M. (2012). Uranium extremophily is an adaptive, rather than intrinsic, feature for extremely thermoacidophilic Metallosphaera species. *Proceedings of the National Academy of Sciences of the United States of America*, *109*(41), 16702–16707. doi:10.1073/pnas.1210904109 PMID:23010932

Mukherjee, S., Sahu, P., & Halder, G. (2017). Microbial remediation of fluoride-contaminated water via a novel bacterium *Providencia vermicola* (KX926492). *Journal of Environmental Management, 204*, 413–423. doi:10.1016/j.jenvman.2017.08.051 PMID:28915476

Mukhopadhyay, A., Dasgupta, A. K., & Chakrabarti, K. (2015). Enhanced functionality and stabilization of a cold active laccase using nanotechnology based activation-immobilization. *Bioresource Technology, 179*, 573–584. doi:10.1016/j.biortech.2014.12.070 PMID:25590281

Mukund, S., & Adams, M. (1991). The novel tungsten-iron-sulfur protein of the hyperthermophilic archaebacterium, *Pyrococcus furiosus*, is an aldehyde ferredoxin oxidoreductase. Evidence for its participation in a unique glycolytic pathway. *The Journal of Biological Chemistry, 266*(22), 14208–14216. doi:10.1016/S0021-9258(18)98669-2 PMID:1907273

Mukund, S., & Adams, M. (1993). Characterization of a novel tungsten-containing formaldehyde ferredoxin oxidoreductase from the hyperthermophilic archaeon, *Thermococcus litoralis*. A role for tungsten in peptide catabolism. *The Journal of Biological Chemistry, 268*(18), 13592–13600. doi:10.1016/S0021-9258(19)38690-9 PMID:8390467

Mukund, S., & Adams, M. W. (1995). Glyceraldehyde-3-phosphate ferredoxin oxidoreductase, a novel tungsten-containing enzyme with a potential glycolytic role in the hyperthermophilic archaeon *Pyrococcus furiosus*. *The Journal of Biological Chemistry, 270*(15), 8389–8392. doi:10.1074/jbc.270.15.8389 PMID:7721730

Müller-Santos, M., de Souza, E. M., Pedrosa, F. de O., Mitchell, D. A., Longhi, S., Carrière, F., Canaan, S., & Krieger, N. (2009). First evidence for the salt-dependent folding and activity of an esterase from the halophilic archaea Haloarcula marismortui. *Biochimica et Biophysica Acta (BBA) -. Molecular and Cell Biology of Lipids, 1791*(8), 719–729. doi:10.1016/j.bbalip.2009.03.006 PMID:19303051

Muñoz, P. A., Flores, P. A., Boehmwald, F. A., & Blamey, J. M. (2011). Thermophilic bacteria present in a sample from Fumarole Bay, Deception Island. *Antarctic Science, 23*(6), 549–555. doi:10.1017/S0954102011000393

Mutlu, M. B., & Guven, K. (2015). Bacterial diversity in Camalti Saltern, Turkey. *Polish Journal of Microbiology, 64*(1), 37–45. doi:10.33073/pjm-2015-005 PMID:26094314

Mwatha, V.E. (1990). *Microbial Ecology of Kenyan soda lakes*. doi:10.1016/B978-0-12-363522-8.X5001-9

Nafis, A., Raklami, A., Bechtaoui, N., Khalloufi, F. E., Alaoui, A. E., Glick, B. R., & Hafidi, M. (2019). Actinobacteria from Extreme Niches in Morocco and Their Plant Growth-Promoting Potentials. *Diversity, 11*(8), 139. doi:10.3390/d11080139

Naik, S. N., Goud, V. V., Rout, P. K., & Dalai, A. K. (2010). Production of first and second generation biofuels: A comprehensive review. *Renewable & Sustainable Energy Reviews, 14*(2), 578–597. doi:10.1016/j.rser.2009.10.003

Naik, S., & Furtado, I. (2014). Equilibrium and kinetics of adsorption of Mn+2 by haloarchaeon Halobacterium sp. GUSF (MTCC3265). *Geomicrobiology Journal, 31*(8), 708–715. doi:10.1080/01490451.2013.859769

Nakagawa, T., Mori, K., Kato, C., Takahashi, R., & Tokuyama, T. (2007). Distribution of cold-adapted ammonia-oxidizing microorganisms in the deep-ocean of the northeastern Japan Sea. *Microbes and Environments, 22*(4), 365–372. doi:10.1264/jsme2.22.365

Nakajima, K., Hirota, K., Nodasaka, Y., & Yumoto, I. (2005). *Alkalibacterium iburiense* sp. nov., an obligate alkaliphile that reduces an indigo dye. *International Journal of Systematic and Evolutionary Microbiology, 55*(4), 1525–1530. doi:10.1099/ijs.0.63487-0 PMID:16014476

Nanavati, D. M., Nguyen, T. N., & Noll, K. M. (2005). Substrate specificities and expression patterns reflect the evolutionary divergence of maltose ABC transporters in Thermotoga maritima. *Journal of Bacteriology, 187*(6), 2002–2009. doi:10.1128/JB.187.6.2002-2009.2005 PMID:15743948

Nanavati, D. M., Thirangoon, K., & Noll, K. M. (2006). Several archaeal homologs of putative oligopeptide-binding proteins encoded by Thermotoga maritima bind sugars. *Applied and Environmental Microbiology, 72*(2), 1336–1345. doi:10.1128/AEM.72.2.1336-1345.2006 PMID:16461685

Nandal, M., Solanki, P., Rastogi, M., & Hooda, R. (2015). Bioremediation: A sustainable tool for environmental management of oily sludge. *Nature Environment and Pollution Technology, 14*(1), 181.

Nandan, A., & Nampoothiri, K. M. (2017). Microbial aminopeptidases. In *Current Developments in Biotechnology and Bioengineering* (pp. 491–507). Elsevier. doi:10.1016/B978-0-444-63662-1.00021-X

Napolitano, M. J., & Shain, D. H. (2004). Four kingdoms on glacier ice: Convergent energetic processes boost energy levels as temperatures fall. *Proceedings of the Royal Society of London. Series B, Biological Sciences, 271*(suppl_5), S273–S276. PMID:15503992

Napolitano, M. J., & Shain, D. H. (2005). Distinctions in adenylate metabolism among organisms inhabiting temperature extremes. *Extremophiles, 9*(2), 93–98. doi:10.100700792-004-0424-1 PMID:15490283

Narsing Rao, M. P., Dong, Z. Y., Luo, Z. H., Li, M. M., Liu, B. B., Guo, S. X., Hozzein, W. N., Xiao, M., & Li, W. J. (2021). Physicochemical and Microbial Diversity Analyses of Indian Hot Springs. *Frontiers in Microbiology, 12*, 368. doi:10.3389/fmicb.2021.627200 PMID:33763045

Nath, A. (2016). Insights into the sequence parameters for halophilic adaptation. *Amino Acids, 48*(3), 751–762. doi:10.100700726-015-2123-x PMID:26520112

Navarro, F., Forján, E., Vázquez, M., Montero, Z., Bermejo, E., Castaño, M. Á., & Toimil, A. (2016). Microalgae as a safe food source for animals: nutritional characteristics of the acidophilic microalgaCoccomyxa onubensis. *Food & Nutrition Research, 60*(1), 30472. doi:10.3402/fnr.v60.30472

Nayak, D. D., & Metcalf, W. W. (2017). Cas9-mediated genome editing in the methanogenic archaeon *Methanosarcina acetivorans. Proceedings of the National Academy of Sciences of the United States of America, 114*(11), 2976–2981. doi:10.1073/pnas.1618596114 PMID:28265068

Nduka, J. K., Umeh, L. N., & Okerulu, I. O. (2012). Utilization of Different Microbes in Bioremediation of Hydrocarbon Contaminated Soils Stimulated With Inorganic and Organic Fertilizers. *Journal of Petroleum & Environmental Biotechnology, 3*(2). doi:10.4172/2157-7463.1000116

Nejad, A. S., & Zahedi, A. R. (2018). Optimization of biodiesel production as a clean fuel for thermal power plants using renewable energy source. *Renewable Energy, 119*, 365–374. doi:10.1016/j.renene.2017.11.087

Nelis, H. J., Lavens, P., Moens, L., Sorgeloos, P., Jonckheere, J. A., Criel, G. R., & De Leenheer, A. P. (1984a). cis-Canthaxanthins. Unusual carotenoids in the eggs and the reproductive system of female brine shrimp artemia. *The Journal of Biological Chemistry, 259*(10), 6063–6066. doi:10.1016/S0021-9258(20)82105-X PMID:6427206

Nelson, K. E., Clayton, R. A., Gill, S. R., Gwinn, M. L., Dodson, R. J., Haft, D. H., Hickey, E. K., Peterson, J. D., Nelson, W. C., Ketchum, K. A., McDonald, L., Utterback, T. R., Malek, J. A., Linher, K. D., Garrett, M. M., Stewart, A. M., Cotton, M. D., Pratt, M. S., Phillips, C. A., ... Fraser, C. M. (1999). Evidence for lateral gene transfer between archaea and bacteria from genome sequence of Thermotoga maritima. *Nature, 399*(6734), 323–329. doi:10.1038/20601 PMID:10360571

Nesbo, C. L., & Doolittle, W. F. (2003). Targeting clusters of transferred genes in *Thermotoga maritima. Environmental Microbiology*, *5*(11), 1144–1154. doi:10.1046/j.1462-2920.2003.00515.x PMID:14641594

Ng, C., De Maere, M. Z., Williams, T. J., Lauro, F. M., Raftery, M., Gibson, J. A. E., Andrews-Pfannkoch, C., Lewis, M., Hoffman, J. M., Thomas, T., & Cavicchioli, R. (2010). Metaproteogenomic analysis of a dominant green sulfur bacterium from Ace Lake, Antarctica. *The ISME Journal*, *4*(8), 1002–1019. doi:10.1038/ismej.2010.28 PMID:20237513

Nguyen, T. A. D., Kim, J. P., Kim, M. S., Oh, Y. K., & Sim, S. J. (2008). Optimization of hydrogen production by hyperthermophilic eubacteria, *Thermotoga maritima* and *Thermotoga neapolitana* in batch fermentation. *International Journal of Hydrogen Energy*, *33*(5), 1483–1488. doi:10.1016/j.ijhydene.2007.09.033

Ng, W. V., Kennedy, S. P., Mahairas, G. G., Berquist, B., Pan, M., Shukla, H. D., Lasky, S. R., Baliga, N. S., Thorsson, V., Sbrogna, J., Swartzell, S., Weir, D., Hall, J., Dahl, T. A., Welti, R., Goo, Y. A., Leithauser, B., Keller, K., Cruz, R., ... DasSarma, S. (2000). Genome sequence of *Halobacterium* species NRC-1. *Proceedings of the National Academy of Sciences of the United States of America*, *97*(22), 12176–12181. doi:10.1073/pnas.190337797 PMID:11016950

Nichols, C. M., Guezennec, J., & Bowman, J. (2005). Bacterial exopolysaccharides from extreme marine environments with special consideration of the southern ocean, sea ice, and deep-sea hydrothermal vents: A review. *Marine Biotechnology (New York, N.Y.)*, *7*(4), 253–271. doi:10.100710126-004-5118-2 PMID:16075348

Nicolaus, B., Kambourova, M., & Oner, E. T. (2010). Exopolysaccharides from extremophiles: From fundamentals to biotechnology. *Environmental Technology*, *31*(10), 1145–1158. doi:10.1080/09593330903552094 PMID:20718297

Nicolaus, B., Lama, L., Panico, A., Moriello, V. S., Romano, I., & Gambacorta, A. (2002). Production and characterization of exopolysaccharides excreted by thermophilic bacteria from shallow, marine hydrothermal vents of Flegrean Ares (Italy). *Systematic and Applied Microbiology*, *25*(3), 319–325. doi:10.1078/0723-2020-00128 PMID:12421069

Nicolaus, B., Manca, M. C., Romano, I., & Lama, L. (1993). Production of an exopolysaccharide from two thermophilic archaea belonging to the genus *Sulfolobus. FEMS Microbiology Letters*, *109*(2-3), 203–206. doi:10.1111/j.1574-6968.1993.tb06168.x

Nielsen, P., Fritze, D., & Priest, F. G. (1995). Phenetic diversity of alkaliphilic Bacillus strains: Proposal for nine new species. *Microbiology*, *141*(7), 1745–1761. doi:10.1099/13500872-141-7-1745

Nienow, J. A., & Friedmann, E. I. (1993) Terrestrial lithophytic (rock) communities. In Antarctic Microbiology. Wiley.

Nies, D. H. (2000). Heavy metal-resistant bacteria as extremophiles: Molecular physiology and biotechnological use of Ralstonia sp. CH34. *Extremophiles*, *4*(2), 77–82. doi:10.1007007920050140 PMID:10805561

Niessen, J., Schröder, U., & Scholz, F. (2004). Exploiting complex carbohydrates for microbial electricity generation–a bacterial fuel cell operating on starch. *Electrochemistry Communications*, *6*(9), 955–958. doi:10.1016/j.elecom.2004.07.010

Nieuwkoop, T., Claassens, N. J., & Van der Oost, J. (2019). Improved protein production and codon optimization analyses in *Escherichia coli* by bicistronic design. *Microbial Biotechnology*, *12*(1), 173–179. doi:10.1111/1751-7915.13332 PMID:30484964

Nishimasu, H., Cong, L., Yan, W. X., Ran, F. A., Zetsche, B., Li, Y., Kurabayashi, A., Ishitani, R., Zhang, F., & Nureki, O. (2015). Crystal Structure of *Staphylococcus aureus* Cas9. *Cell*, *162*(5), 1113–1126. doi:10.1016/j.cell.2015.08.007 PMID:26317473

Niyonzima, F. N. (2019). Detergent-compatible bacterial cellulases. *Journal of Basic Microbiology*, *59*(2), 134–147. doi:10.1002/jobm.201800436 PMID:30421443

Noel, S., Fortier, C., Murschel, F., Belzil, A., Gaudet, G., Jolicoeur, M., & De Crescenzo, G. (2016). Co-immobilization of adhesive peptides and VEGF within a dextran-based coating for vascular applications. *Acta Biomaterialia*, *37*, 69–82. doi:10.1016/j.actbio.2016.03.043 PMID:27039978

Nogales, J., Gudmundsson, S., & Thiele, I. (2012). An in silico re-design of the metabolism in Thermotoga maritima for increased biohydrogen production. *International Journal of Hydrogen Energy*, *37*(17), 12205–12218. doi:10.1016/j.ijhydene.2012.06.032

Nogi, Y. (2008). Bacteria in the deep sea: psychropiezophiles. In R. Margesin, F. Schinner, J. C. Marx, & C. Gerday (Eds.), *Psychrophiles: from Biodiversity to Biotechnology* (pp. 73–82). Springer. doi:10.1007/978-3-540-74335-4_5

Norton, C. F., & Grant, W. (1988). Survival of halobacteria within fluid inclusions in salt crystals. *Microbiology*, *134*(5), 1365–1373. doi:10.1099/00221287-134-5-1365

Novik, G., Savich, V., & Meerovskaya, O. (2018). Geobacillus Bacteria: potential commercial applications in industry, bioremediation, and bioenergy production. In Growing and Handling of Bacterial Cultures. IntechOpen.

Nowlan, B., Dodia, M. S., Singh, S. P., & Patel, B. K. C. (2006). Bacillus okhensis sp. nov., a halotolerant and alkali-tolerant bacterium from an Indian saltpan. *International Journal of Systematic and Evolutionary Microbiology*, *56*(5), 1073–1077. doi:10.1099/ijs.0.63861-0 PMID:16627657

Ntougias, S., & Russell, N. J. (2001). *Alkalibacterium olivoapovliticus* gen. nov., sp. nov., a new obligately alkaliphilic bacterium isolated from edible-olive wash-waters. *International Journal of Systematic and Evolutionary Microbiology*, *51*(3), 1161–1170. doi:10.1099/00207713-51-3-1161 PMID:11414268

Nunes, J. C., Cristóvão, R. O., Freire, M. G., Santos-Ebinuma, V. C., Faria, J. L., Silva, C. G., & Tavares, A. P. (2020). Recent strategies and applications for L-asparaginase confinement. *Molecules*, *25*(24), 5827. doi:10.3390/molecules25245827 PMID:33321857

Nuñez, J. K., Kranzusch, P. J., Noeske, J., Wright, A. V., Davies, C. W., & Doudna, J. A. (2014). Cas1-Cas2 complex formation mediates spacer acquisition during CRISPR-Cas adaptive immunity. *Nature Structural & Molecular Biology*, *21*(6), 528–534. doi:10.1038/nsmb.2820 PMID:24793649

Nwodo, U. U., Green, E., & Okoh, A. I. (2012). Bacterial exopolysaccharides: Functionality and prospects. *International Journal of Molecular Sciences*, *13*(11), 14002–14015. doi:10.3390/ijms131114002 PMID:23203046

Ogata-Gutiérrez, K., Alvarado, D., Chumpitaz-Segovia, C., & Zúñiga-Dávila, D. (2016). Characterization of Plant Growth Promoting Rhizobacteria Isolated from the Rhizosphere of Peruvian Highlands Native Crops. *International Journal of Plant & Soil Science*, *11*(1), 1–8. doi:10.9734/IJPSS/2016/24573

Oger, P., & Cario, A. (2014). The high pressure life of piezophiles. *Biologie Aujourd'hui*, *208*(3), 193–206. doi:10.1051/jbio/2014023 PMID:25474000

Ohtaki, A., Kida, H., Miyata, Y., Ide, N., Yonezawa, A., Arakawa, T., Iizuka, R., Noguchi, K., Kita, A., Odaka, M., Miki, K., & Yohda, M. (2008). Structure and molecular dynamics simulation of archaeal prefoldin: The molecular mechanism for binding and recognition of nonnative substrate proteins. *Journal of Molecular Biology*, *376*(4), 1130–1141. doi:10.1016/j.jmb.2007.12.010 PMID:18201719

Okongo, R. N., Puri, A. K., Wang, Z., Singh, S., & Permaul, K. (2019). Comparative biocontrol ability of chitinases from bacteria and recombinant chitinases from the thermophilic fungus Thermomyces lanuginosus. *Journal of Bioscience and Bioengineering*, *127*(6), 663–671. doi:10.1016/j.jbiosc.2018.11.007

Okubo, Y., Yokoigawa, K., Esaki, N., Soda, K., & Kawai, H. (1999). Characterization of Psychrophilic Alanine Racemase from *Bacillus psychrosaccharolyticus. Biochemical and Biophysical Research Communications, 256*(2), 333–340. doi:10.1006/bbrc.1999.0324 PMID:10080917

Oladipo, Awotoye, Olayinka, Bezuidenhout, & Maboeta. (2018). Heavy metal tolerance traits of filamentous fungi isolated from gold and gemstone mining sites. *Brazilian Journal of Microbiology, 49*, 29-37.

Olaizola, M. (2003). Commercial development of microalgal biotechnology: From the test tube to the marketplace. *Biomolecular Engineering, 20*(4–6), 459–466. doi:10.1016/S1389-0344(03)00076-5 PMID:12919832

Olaniyan, O. T., & Adetunji, C. O. (2021). Biological, Biochemical, and Biodiversity of Biomolecules from marine-based beneficial microorganisms: Industrial Perspective. Microorganisms for Sustainability, 27. doi:10.1007/978-981-15-7459-7_4

Ollivier, B., & Cayol, J. (2005). Fermentative, iron-reducing, and nitrate-reducing microorganisms. In B. Ollivier & M. Magot (Eds.), *Petroleum Microbiology* (pp. 71–88). ASM Press. doi:10.1128/9781555817589

Olsen, H. S., & Falholt, P. (1998). The role of enzymes in modern detergency. *Journal of Surfactants and Detergents, 1*(4), 555–567. doi:10.100711743-998-0058-7

Olson, D. G., Sparling, R., & Lynd, L. R. (2015). Ethanol production by engineered thermophiles. *Current Opinion in Biotechnology, 33*, 130–141. doi:10.1016/j.copbio.2015.02.006 PMID:25745810

Oren, A. (2006). Life at high salt concentrations. *The Prokaryotes, 3*, 263-282.

Oren, A. (2001). The order Haloanaerobiales. In M. Dworkin, S. Falkow, E. Rosenberg, K. H. Schleifer, & E. Stackebrandt (Eds.), *The Prokaryotes. A Handbook on the biology of bacteria: Eco physiology, isolation, identification, applications* (3rd ed.). Springer-Verlag. doi:10.1007/978-1-4757-2191-1

Oren, A. (2002). Diversity of halophilic microorganisms: Environment, phylogeny, physiology, and applications. *Journal of Industrial Microbiology & Biotechnology, 28*(1), 56–63. doi:10.1038j/jim/7000176 PMID:11938472

Oren, A. (2002). Molecular ecology of extremely halophilic Archaea and Bacteria. *FEMS Microbiology Ecology, 39*(1), 1–7. doi:10.1111/j.1574-6941.2002.tb00900.x PMID:19709178

Oren, A. (2010). Industrial and environmental applications of halophilic microorganisms. *Environmental Technology, 31*(8-9), 825–834. doi:10.1080/09593330903370026 PMID:20662374

Oren, A. (2013). Life at high salt concentrations, intracellular KCl concentrations, and acidic proteomes. *Frontiers in Microbiology, 4*(NOV). Advance online publication. doi:10.3389/fmicb.2013.00315 PMID:24204364

Oren, A. (2014). Taxonomy of halophilic Archaea: Current status and future challenges. *Extremophiles, 18*(5), 825–834. doi:10.100700792-014-0654-9 PMID:25102811

Oren, A. (2015). Halophilic microbial communities and their environments. *Current Opinion in Biotechnology, 33*, 119–124. doi:10.1016/j.copbio.2015.02.005 PMID:25727188

Oren, A. (2017). Glycerol metabolism in hyper saline environments. *Environmental Microbiology, 19*(3), 851–863. doi:10.1111/1462-2920.13493 PMID:27511298

Ortiz-Lemus, J. F., Campoy, S., & Martín, J. F. (2021). Biological control of mites by xerophile Eurotium species isolated from the surface of dry cured ham and dry beef cecina. *Journal of Applied Microbiology, 130*(3), 665–676. doi:10.1111/jam.14839 PMID:32869458

Ortiz-Ojeda, Ogata-Gutiérrez, & Zúñiga-Dávila. (2017). Evaluation of plant growth promoting activity and heavy metal tolerance of psychrotrophic bacteria associated with maca (emLepidium meyenii/em Walp.) rhizosphere. *AIMS Microbiology, 3*(2), 279–292. doi:10.3934/microbiol.2017.2.279

Osawa, Z., Morota, T., Hatanaka, K., Akaike, T., Matsuzaki, K., Nakashima, H., ... Mimura, T. (1993). Synthesis of sulfated derivatives of curdlan and their anti-HIV activity. *Carbohydrate Polymers, 21*(4), 283–288. doi:10.1016/0144-8617(93)90060-H

Oshima, T., & Moriya, T. (2008). A preliminary analysis of microbial and biochemical properties of high-temperature compost. *Annals of the New York Academy of Sciences, 1125*(1), 338–344. doi:10.1196/annals.1419.012 PMID:18378603

Ouwehand, A. C. (2017). A review of dose-responses of probiotics in human studies. *Beneficial Microbes, 8*(2), 143–151. doi:10.3920/BM2016.0140 PMID:28008787

Ouwehand, A. C., DongLian, C., Weijian, X., Stewart, M., Ni, J., Stewart, T., & Miller, L. E. (2014). Probiotics reduce symptoms of antibiotic use in a hospital setting: A randomized dose-response study. *Vaccine, 32*(4), 458–463. doi:10.1016/j.vaccine.2013.11.053 PMID:24291194

Ou, Y., Xu, S., Zhu, D., & Yang, X. (2014). Molecular mechanisms of exopolysaccharide from *Aphanothece halaphytica* (EPSAH) induced apoptosis in HeLa cells. *PLoS One, 9*(1), e87223. doi:10.1371/journal.pone.0087223 PMID:24466342

Ouyang, J., Ma, R., Zheng, Z., Cai, C., Zhang, M., & Jiang, T. (2013). Open fermentative production of L-lactic acid by Bacillus sp. strain NL01 using lignocellulosic hydrolyzates as low-cost raw material. *Bioresource Technology, 135*, 475–480. doi:10.1016/j.biortech.2012.09.096 PMID:23127843

Oves, M., Khan, M. S., & Zaidi, A. (2013). Chromium reducing and plant growth promoting novel strain *Pseudomonas aeruginosa* OSG41 enhance chickpea growth in chromium amended soils. *European Journal of Soil Biology, 56*, 72–83. doi:10.1016/j.ejsobi.2013.02.002

Oves, M., Rauf, M. A., Hussain, A., Qari, H. A., Khan, A. A. P., Muhammad, P., Rehman, M. T., Alajmi, M. F., & Ismail, I. I. (2019). Antibacterial silver nanomaterial synthesis from *Mesoflavibacter zeaxanthinifaciens* and targeting biofilm formation. *Frontiers in Pharmacology, 10*, 801. doi:10.3389/fphar.2019.00801 PMID:31427961

Oves, M., Saghir Khan, M., & Huda Qari, A., NadeenFelemban, M., & Almeelbi, T. (2016). Heavy metals: Biological importance and detoxification strategies. *Journal of Bioremediation & Biodegradation, 7*(2), 1–15.

Ozawa, T., Endo, K., Igarashi, K., Kitayama, K., Hayashi, Y., Hagihara, H., Kawai, S., Ito, S., & Ozaki, K. (2007). Improvement of the Thermal Stability of a Calcium-free, Alkaline. ALPHA.-Amylase by Site-directed Mutagenesis. *Journal of Applied Glycoscience, 54*(2), 77–83. doi:10.5458/jag.54.77

Ozdemir, G., Ozturk, T., Ceyhan, N., Isler, R., & Cosar, T. (2003). Heavy metal biosorption by biomass of *Ochrobactrum anthropi* producing exopolysaccharide in activated sludge. *Bioresource Technology, 90*(1), 71–74. doi:10.1016/S0960-8524(03)00088-9 PMID:12835060

Paar, A., Costa, S., Tzanov, T., Gudelj, M., Robra, K.-H., Cavaco-Paulo, A., & Gübitz, G. (2001). Thermo-alkali-stable catalases from newly isolated Bacillus sp. for the treatment and recycling of textile bleaching effluents. *Journal of Biotechnology, 89*(2-3), 147–153. doi:10.1016/S0168-1656(01)00305-4 PMID:11500208

Padan, E. (2014). Functional and structural dynamics of NhaA, a prototype for Na+ and H+ antiporters, which are responsible for Na+ and H+ homeostasis in cells. Biochimica et Biophysica Acta (BBA) -. *Bioenergetics, 1837*(7), 1047–1062. doi:10.1016/j.bbabio.2013.12.007 PMID:24361841

Padan, E., Venturi, M., Gerchman, Y., & Dover, N. (2001). Na+/H+ antiporters. Biochimica et Biophysica Acta (BBA) -. *Bioenergetics, 1505*(1), 144–157. doi:10.1016/S0005-2728(00)00284-X

Palaniraj, A., & Jayaraman, V. (2011). Production, recovery and applications of xanthan gum by *Xanthomonas campestris. Journal of Food Engineering, 106*(1), 1–12. doi:10.1016/j.jfoodeng.2011.03.035

Palanivel, P., Ashokkumar, L., & Balagurunathan, R. (2013). Production, purification and fibrinolytic characterization of alkaline protease from extremophilic soil fungi. *Int J Pharm Bio Sci, 4*, 101–110.

Palaniyandi, S. A., Yang, S. H., Zhang, L., & Suh, J.-W. (2013). Effects of actinobacteria on plant disease suppression and growth promotion. *Applied Microbiology and Biotechnology, 97*(22), 9621–9636. doi:10.100700253-013-5206-1 PMID:24092003

Palma, S. I., Rodrigues, C. A., Carvalho, A., del Puerto Morales, M., Freitas, F., Fernandes, A. R., ... Roque, A. C. (2015). A value-added exopolysaccharide as a coating agent for MRI nanoprobes. *Nanoscale, 7*(34), 14272–14283. doi:10.1039/C5NR01979F PMID:26186402

Pal, S., Sar, A., & Dam, B. (2019). Moderate halophilic bacteria, but not extreme halophilic archaea can alleviate the toxicity of short-alkyl side chain imidazolium-based ionic liquids. *Ecotoxicology and Environmental Safety, 184*, 109634. doi:10.1016/j.ecoenv.2019.109634 PMID:31520950

Panda, A. N., Ray, L., Mishra, S. R., & Raina, V. (2020). *Rheinheimera pleomorphica* sp. nov., a Novel Alkali-Tolerant Bacteria Isolated from Chilika Lake, India. *Current Microbiology, 77*(1), 158–165. doi:10.100700284-019-01802-9 PMID:31705390

Panda, M. K., Sahu, M. K., & Tayung, K. (2013). Isolation and characterization of a thermophilic *Bacillus* sp. with protease activity isolated from hot spring of Tarabalo, Odisha, India. *Iranian Journal of Microbiology, 5*(2), 159–165. PMID:23825735

Pandey, A., & Yarzábal, L. A. (2018). Bioprospecting cold-adapted plant growth promoting microorganisms from mountain environments. *Applied Microbiology and Biotechnology, 103*(2), 643–657. doi:10.1007/s00253-018-9515-2

Pandey, A., Dhakar, K., Sharma, A., Priti, P., Sati, P., & Kumar, B. (2015). Thermophilic bacteria that tolerate a wide temperature and pH range colonize the Soldhar (95°C) and Ringigad (80°C) hot springs of Uttarakhand, India. *Annals of Microbiology, 65*(2), 809–816. doi:10.100713213-014-0921-0

Pandey, A., Nigam, P., Soccol, C. R., Soccol, V. T., Singh, D., & Mohan, R. (2000). Advances in microbial amylases. *Biotechnology and Applied Biochemistry, 31*(2), 135–152. doi:10.1042/BA19990073 PMID:10744959

Pandey, N., Jain, R., Pandey, A., & Tamta, S. (2018). Optimization and characterization of the orange pigment produced by a cold-adapted strain of *Penicillium* sp. (GBPI_P155) isolated from mountain ecosystem. *Mycology, 9*(2), 81–92. doi:10.1080/21501203.2017.1423127 PMID:30123664

Pandey, S., Rakholiya, K. D., Raval, V. H., & Singh, S. P. (2012). Catalysis and stability of an alkaline protease from a haloalkaliphilic bacterium under non-aqueous conditions as a function of pH, salt and temperature. *Journal of Bioscience and Bioengineering, 114*(3), 251–256. doi:10.1016/j.jbiosc.2012.03.003 PMID:22727739

Pang, K. L., Chow, R. K., Chan, C. W., & Vrijmoed, L. P. (2011). Diversity and physiology of marine lignicolous fungi in Arctic waters: A preliminary account. *Polar Research, 30*(1), 5859–5865. doi:10.3402/polar.v30i0.5859

Panosyan, H., Di Donato, P., Poli, A., & Nicolaus, B. (2018). Production and characterization of exopolysaccharides by *Geobacillus thermodenitrificans* ArzA-6 and *Geobacillus toebii* ArzA-8 strains isolated from an Armenian geothermal spring. *Extremophiles, 22*(5), 725–737. doi:10.100700792-018-1032-9 PMID:29779131

Papke, R. T., Ramsing, N. B., Bateson, M. M., & Ward, D. M. (2003). Geographical isolation in hot spring cyanobacteria. *Environmental Microbiology*, *5*(8), 650–659. doi:10.1046/j.1462-2920.2003.00460.x PMID:12871232

Park, C., & Park, W. (2018). Survival and energy producing strategies of alkane degraders under extreme conditions and their biotechnological potential. *Frontiers in Microbiology*, *9*, 1081. doi:10.3389/fmicb.2018.01081 PMID:29910779

Park, D. H., & Zeikus, J. G. (2000). Electricity generation in microbial fuel cells using neutral red as an electronophore. *Applied and Environmental Microbiology*, *66*(4), 1292–1297. doi:10.1128/AEM.66.4.1292-1297.2000 PMID:10742202

Park, J. T., Song, H. N., Jung, T. Y., Lee, M. H., Park, S. G., Woo, E. J., & Park, K. H. (2013). A novel domain arrangement in a monomeric cyclodextrin-hydrolyzing enzyme from the hyperthermophile Pyrococcus furiosus. *Biochimica et Biophysica Acta (BBA) - Proteins and Proteomics*, *1834*(1), 380–386. doi:10.1016/j.bbapap.2012.08.001 PMID:22902546

Parvizpour, S., Hussin, N., Shamsir, M. S., & Razmara, J. (2021). Psychrophilic enzymes: Structural adaptation, pharmaceutical, and industrial applications. *Applied Microbiology and Biotechnology*, *105*(3), 1–9. doi:10.100700253-020-11074-0 PMID:33427934

Pastor, J., Salvador, M., Argandona, M., Bernal, V., Reina-Bueno, M., Csonka, L., Iborra, J. L., Vargas, C., Nieto, J. J., & Cánovas, M. (2010). Ectoines in cell stress protection: Uses and biotechnological production. *Biotechnology Advances*, *28*(6), 782–801. doi:10.1016/j.biotechadv.2010.06.005 PMID:20600783

Patel, N. Y. (2020). *Exploring the potential of marine bacteria for L-glutaminase biosynthesis and its biochemical profiling* (M.Phil. Thesis). Gujarat University.

Patel, N. Y., Baria, D. M., Arya, P. S., Rajput, K. N., Panchal, R. R., & Raval, V. H. (2020). L-glutaminase biosynthesis from marine bacteria. *Bioscience Biotechnology Research Communications, 13*(1), 67-72. http://bbrc.in/bbrc/wp-content/uploads/2020/05/FINAL-PDF-Special-Issue-Vol-13-No-1-2020.pdf

Patel, A., & Prajapat, J. (2013). Food and health applications of exopolysaccharides produced by lactic acid bacteria. *Journal of Advances in Dairy Research*, 1–8.

Patel, K. K., Shah, F., & Dwivedi, M. (2018). Exopolysaccharide (EPS) Producing Isolates from Sugarcane Field Soil and Antibacterial Activity of Extracted EPSs. *Acta Scientific Microbiology*, *1*(4), 6–13. doi:10.31080/ASMI.2018.01.0032

Patel, N. Y., Baria, D. M., Yagnik, S. M., Rajput, K. N., Panchal, R. R., & Raval, V. H. (2021). Bio-prospecting the future in perspective of amidohydrolase L-glutaminase from marine habitats. *Applied Microbiology and Biotechnology*, *105*(13), 5325–5340. Advance online publication. doi:10.100700253-021-11416-6 PMID:34236482

Patel, R., Mevada, V., Prajapati, D., Dudhagara, P., Koringa, P., & Joshi, C. G. (2015). Metagenomic sequence of saline desert microbiota from wild ass sanctuary, Little Rann of Kutch, Gujarat, India. *Genomics Data*, *3*, 137–139. doi:10.1016/j.gdata.2015.01.003 PMID:26484162

Pathak, V. M., & Navneet. (2017). Review on the current status of polymer degradation: a microbial approach. *Bioresources and Bioprocessing, 4*, 15.

Pearce, D. A., Bridge, P. D., Hughes, K. A., Sattler, B., Psenner, R., & Russell, N. J. (2009). Microorganisms in the atmosphere over Antarctica. *FEMS Microbiology Ecology*, *69*(2), 143–157. doi:10.1111/j.1574-6941.2009.00706.x PMID:19527292

Pearce, D. A., & Galand, P. E. (2008). *Microbial biodiversity and biogeography. In Polar lakes and rivers, limnology of Arctic and Antarctic aquatic ecosystems*. Oxford University Press.

Pearce, K. E. (2011). Convergence through mobile peer-to-peer file sharing in the Republic of Armenia. *International Journal of Communication*, *5*, 511–528.

Pedone, E., Fiorentino, G., Bartolucci, S., & Limauro, D. (2020). Enzymatic Antioxidant Signatures in Hyperthermophilic Archaea. *Antioxidants, 9*(8), 703. doi:10.3390/antiox9080703

Pedone, E., Ren, B., Ladenstein, R., Rossi, M., & Bartolucci, S. (2004). Functional properties of the protein disulfide oxidoreductase from the archaeon *Pyrococcus furiosus*: A member of a novel protein family related to protein disulfide-isomerase. *European Journal of Biochemistry, 271*(16), 3437–3448. doi:10.1111/j.0014-2956.2004.04282.x PMID:15291821

Peeples, T. L. (2014). Bioremediation using extremophiles. In *Microbial biodegradation and bioremediation* (pp. 251–268). Elsevier. doi:10.1016/B978-0-12-800021-2.00010-8

Peng, S., Kai, M., Yang, X., Luo, Y., & Bai, L. (2020). Study on the osmoregulation of "Halomonas socia" NY-011 and the degradation of organic pollutants in the saline environment. *Extremophiles, 24*(6), 843–861. doi:10.1007/s00792-020-01199-5

Peng, N., Han, W., Li, Y., Liang, Y., & She, Q. (2017). Genetic technologies for extremely thermophilic microorganisms of Sulfolobus, the only genetically tractable genus of crenarchaea. *Science China. Life Sciences, 60*(4), 370–385. doi:10.100711427-016-0355-8 PMID:28251462

Pentecost, A. (2009). The Marl Lakes of the British Isles. *Freshwater Reviews, 2*(2), 167–197. doi:10.1608/FRJ-2.2.4

Pérez, M. T., & Sommaruga, R. (2007). Interactive effects of solar radiation and dissolved organic matter on bacterial activity and community structure. *Environmental Microbiology, 9*(9), 2200–2210. doi:10.1111/j.1462-2920.2007.01334.x PMID:17686018

Perfumo, A., Banat, I. M., & Marchant, R. (2006). The use of thermophilic bacteria in accelerated hydrocarbon bioremediation. In *Environmental Problems in Coastal Regions VI.* WIT Press. doi:10.2495/CENV060071

Pernthaler, J., Glöckner, F. O., Unterholzner, S., Alfreider, A., Psenner, R., & Amann, R. (1998). Seasonal community and population dynamics of pelagic bacteria and archaea in a high mountain lake. *Applied and Environmental Microbiology, 64*(11), 4299–4306. doi:10.1128/AEM.64.11.4299-4306.1998 PMID:9797280

Pierson, B. K., & Castenholz, R. W. (1974). A phototrophic gliding filamentous bacterium of hot springs, *Chloroflexus aurantiacus*, gen. and sp. nov. *Archives of Microbiology, 100*(1), 5–24. doi:10.1007/BF00446302 PMID:4374148

Pikuta, E. V., Hoover, R. B., & Tang, J. (2007). Microbial Extremophiles at the Limits of Life. *Critical Reviews in Microbiology, 33*(3), 183–209. doi:10.1080/10408410701451948 PMID:17653987

Pinar, G., Krakova, L., Pangallo, D., Piombino-Mascali, D., Maixner, F., Zink, A., & Sterflinger, K. (2014). Halophilic bacteria are colonizing the exhibition areas of the Capuchin Catacombs in Palermo, Italy. *Extremophiles, 18*(4), 677–691. doi:10.100700792-014-0649-6 PMID:24863363

Pinto, E., Sigaud-Kutner, T. C. S., Leitao, M. A. S., Okamoto, O. K., Morse, D., & Colepicolo, P. (2003). Heavy metal–induced oxidative stress in algae 1. *Journal of Phycology, 39*(6), 1008–1018. doi:10.1111/j.0022-3646.2003.02-193.x

Piubeli, F., de Lourdes Moreno, M., Kishi, L. T., Henrique-Silva, F., García, M. T., & Mellado, E. (2015). Phylogenetic profiling and diversity of bacterial communities in the Death Valley, an extreme habitat in the Atacama Desert. *Indian Journal of Microbiology, 55*(4), 392–399. doi:10.100712088-015-0539-3 PMID:26543264

Plumb, J. J., Haddad, C. M., Gibson, J. A., & Franzmann, P. D. (2007). Acidianus sulfidivorans sp. nov., an extremely acidophilic, thermophilic archaeon isolated from a solfatara on Lihir Island, Papua New Guinea, and emendation of the genus description. *International Journal of Systematic and Evolutionary Microbiology, 57*(7), 1418–1423. doi:10.1099/ijs.0.64846-0 PMID:17625168

Podar, M., & Reysenbach, A.-L. (2006). New opportunities revealed by biotechnological explorations of extremophiles. *Current Opinion in Biotechnology, 17*(3), 250–255. doi:10.1016/j.copbio.2006.05.002 PMID:16701993

Podar, P. T., Yang, Z., Björnsdóttir, S. H., & Podar, M. (2020). Comparative Analysis of Microbial Diversity Across Temperature Gradients in Hot Springs From Yellowstone and Iceland. *Frontiers in Microbiology, 0*, 1625. doi:10.3389/fmicb.2020.01625 PMID:32760379

Pointing, S. B., Chan, Y., Lacap, D. C., Lau, M. C., Jurgens, J. A., & Farrell, R. L. (2009). Highly specialized microbial diversity in hyper-arid polar desert. *Proceedings of the National Academy of Sciences of the United States of America, 106*(47), 19964–19969. doi:10.1073/pnas.0908274106 PMID:19850879

Poli, A., Anzelmo, G., & Nicolaus, B. (2010). Bacterial exopolysaccharides from extreme marine habitats: Production, characterization and biological activities. *Marine Drugs, 8*(6), 1779–1802. doi:10.3390/md8061779 PMID:20631870

Poli, A., Esposito, E., Orlando, P., Lama, L., Giordano, A., de Appolonia, F., Nicolaus, B., & Gambacorta, A. (2007). *Halomonas alkaliantarctica* sp. nov., isolated from saline lake Cape Russell in Antarctica, an alkalophilic moderately halophilic, exopolysaccharide-producing bacterium. *Systematic and Applied Microbiology, 30*(1), 31–38. doi:10.1016/j.syapm.2006.03.003 PMID:16621401

Poli, A., Finore, I., Romano, I., Gioiello, A., Lama, L., & Nicolaus, B. (2017). Microbial diversity in extreme marine habitats and their biomolecules. *Microorganisms, 5*(2), 25. doi:10.3390/microorganisms5020025 PMID:28509857

Popa, A., Israel-Roming, F., & Cornea, C.P. (2016). *Microbial xylanase: A review SEE PROFILE XX*. Academic Press.

Popescu, G., & Dumitru, L. (2009). Biosorption of some heavy metals from media with high salt concentrations by halohilicarchaea. *Biotechnology and Biotechnological Equipment, 23*(sup1), 791–795.

Postec, A., Quemeneur, M., Lecoeuvre, A., Chabert, N., Joseph, M., & Erauso, G. (2021). *Alkaliphilus serpentinus* sp. nov. and *Alkaliphilus pronyensis* sp. nov., two novel anaerobic alkaliphilic species isolated from the serpentinite-hosted Prony Bay Hydrothermal Field (New Caledonia). *Systematic and Applied Microbiology, 44*(2), 126175. doi:10.1016/j.syapm.2020.126175 PMID:33422701

Pouliot, J., Galand, P. E., Lovejoy, C., & Vincent, W. F. (2009). Vertical structure of archaeal communities and the distribution of ammonia monooxygenase A gene variants in two meromictic High Arctic lakes. *Environmental Microbiology, 11*(3), 687–699. doi:10.1111/j.1462-2920.2008.01846.x PMID:19207564

Prajapati, V. D., Jani, G. K., Zala, B. S., & Khutliwala, T. A. (2013). An insight into the emerging exopolysaccharide gellan gum as a novel polymer. *Carbohydrate Polymers, 93*(2), 670–678. doi:10.1016/j.carbpol.2013.01.030 PMID:23499110

Prasad, A. A., & Rao, K. B. (2013). Aerobic biodegradation of Azo dye by *Bacillus cohnii* MTCC 3616; an obligately alkaliphilic bacterium and toxicity evaluation of metabolites by different bioassay systems. *Applied Microbiology and Biotechnology, 97*(16), 7469–7481. doi:10.100700253-012-4492-3 PMID:23070653

Prasad, D., Arun, S., Murugesan, M., Padmanaban, S., Satyanarayanan, R. S., Berchmans, S., & Yegnaraman, V. (2007). Direct electron transfer with yeast cells and construction of a mediatorless microbial fuel cell. *Biosensors & Bioelectronics, 22*(11), 2604–2610. doi:10.1016/j.bios.2006.10.028 PMID:17129722

Prechtl, R. M., Wefers, D., Jakob, F., & Vogel, R. F. (2018). Cold and salt stress modulates the amount, molecular and macromolecular structure of a *Lactobacillus sakei* dextran. *Food Hydrocolloids, 82*, 73–81. doi:10.1016/j.foodhyd.2018.04.003

Prenafeta-Boldú, F. X., Guivernau, M., Gallastegui, G., Viñas, M., de Hoog, G. S., & Elías, A. (2012). Fungal/bacterial interactions during the biodegradation of TEX hydrocarbons (toluene, ethylbenzene and p-xylene) in gas biofilters operated under xerophilic conditions. *FEMS Microbiology Ecology*, *80*(3), 722–734. doi:10.1111/j.1574-6941.2012.01344.x PMID:22380456

Price, L. B., & Shand, R. F. (2000). Halocin S8: A 36-amino-acid microhalocin from the *haloarchaeal* strain S8a. *Journal of Bacteriology*, *182*(17), 4951–4958. doi:10.1128/JB.182.17.4951-4958.2000 PMID:10940040

Price, P. B. (2000). A habitat for psychrophiles in deep Antarctic ice. *Proceedings of the National Academy of Sciences of the United States of America*, *97*(3), 1247–1251. doi:10.1073/pnas.97.3.1247 PMID:10655516

Price, S., Kuzhiumparambil, U., Pernice, M., & Ralph, P. J. (2020). Cyanobacterial polyhydroxybutyrate for sustainable bioplastic production: Critical review and perspectives. *Journal of Environmental Chemical Engineering*, *8*(4), 104007. doi:10.1016/j.jece.2020.104007

Prince, R. C. (1993). Petroleum spill bioremediation in marine environments. *Critical Reviews in Microbiology*, *19*(4), 217–242. doi:10.3109/10408419309113530 PMID:8305136

Prince, R. C., & Butler, J. D. (2014). A protocol for assessing the effectiveness of oil spill dispersants in stimulating the biodegradation of oil. *Environmental Science and Pollution Research International*, *21*(16), 9506–9510. doi:10.100711356-013-2053-7 PMID:23943003

Priscu, J. C., & Christner, B. C. (2003). Earth's icy biosphere. In Microbial diversity and bioprospecting. Washington, DC: ASM Press.

Priscu, J. C., & Foreman, C. M. (2009). Article. In G. E. Likens (Ed.), Encyclopedia of Inland Waters. Lakes of Antarctica (Vol. 2, pp. 555–566). Elsevier Press. doi:10.1016/B978-012370626-3.00038-7

Priscu, J. C., Christner, B. C., Foreman, C. M., & Royston-Bishop, G. (2007). Biological material in ice cores. Encyclopedia of Quaternary Sciences, 2, 1156-1166.

Priscu, J. C., Tulaczyk, S., Studinger, M., Kennicutt, M. C., Christner, B. C., & Foreman, C. M. (2008). *Antarctic subglacial water: origin, evolution and ecology. In Polar lakes and rivers: limnology of Arctic and Antarctic aquatic ecosystems*. Oxford University Press.

Priyadarshini, S., Pradhan, S. K., & Ray, P. (2020). Production, characterization and application of thermostable, alkaline α-amylase (AA11) from *Bacillus cereus* strain SP-CH11 isolated from Chilika Lake. *International Journal of Biological Macromolecules*, *145*, 804–812. doi:10.1016/j.ijbiomac.2019.11.149 PMID:31758985

Prosekov, A., Dyshlyuk, L., Milentyeva, I., Sukhih, S., Babich, O., Ivanova, S., ... Matskova, L. (2015). Antioxidant, antimicrobial and antitumor activity of bacteria of the genus bifidobacterium, selected from the gastrointestinal tract of human. *Integrative Molecular Medicine*, *2*(5), 295–303.

Psenner, R., Sattler, B., Wille, A., Fritsen, C. H., Priscu, J. C., Felip, M., & Catalan, J. (1999). Lake ice microbial communities in alpine and antarctic lakes. In R. Margeisn & F. Schinner (Eds.), *Cold-Adapted Organisms* (pp. 17–31). Springer. doi:10.1007/978-3-662-06285-2_2

Puchta, H. (2017). Applying CRISPR/Cas for genome engineering in plants: The best is yet to come. *Current Opinion in Plant Biology*, *36*, 1–8. doi:10.1016/j.pbi.2016.11.011 PMID:27914284

Purcell, D., Sompong, U., Lau, C. Y., Barraclough, T. G., Peerapornpisal, Y., & Pointing, S. B. (2007). The effects of temperature, pH and sulphide on community structure of hyperthermophilic streamers in hot springs of northern Thailand. *FEMS Microbiology Ecology*, *60*(3), 456–466. doi:10.1111/j.1574-6941.2007.00302.x PMID:17386034

Purich, D.L. (2010). Factors Influencing Enzyme Activity. *Enzyme Kinetics: Catalysis & Control,* 379–484. doi:10.1016/B978-0-12-380924-7.10007-9

Purohit, M. K., Raval, V. H., & Singh, S. P. (2014). Haloalkaliphilic bacteria: molecular diversity and biotechnological applications. In Geomicrobiology and Biogeochemistry (pp. 61-79). Springer.

Purohit, M. K., Raval, V. H., & Singh, S. P. (2014). Haloalkaliphilic bacteria: molecular diversity and biotechnological applications. In *Geomicrobiology and Biogeochemistry* (pp. 61–79). Springer. doi:10.1007/978-3-642-41837-2_4

Puthenveetil, R., & Vinogradova, O. (2013). Optimization of the design and preparation of nanoscale phospholipid bilayers for its application to solution NMR. *Proteins, 81*(7), 1222–1231. doi:10.1002/prot.24271 PMID:23436707

Qiao, Y.-J., Qiao, Y., Zou, L., Wu, X.-S., & Liu, J.-H. (2017). Biofilm promoted current generation of *Pseudomonas aeruginosa* microbial fuel cell via improving the interfacial redox reaction of phenazines. *Bioelectrochemistry (Amsterdam, Netherlands), 117*, 34–39. doi:10.1016/j.bioelechem.2017.04.003 PMID:28575838

Qu, D., Show, P.-L., & Miao, X. (2021). Transcription Factor ChbZIP1 from Alkaliphilic Microalgae *Chlorella* sp. BLD Enhancing Alkaline Tolerance in Transgenic *Arabidopsis thaliana. International Journal of Molecular Sciences, 22*(5), 2387. doi:10.3390/ijms22052387 PMID:33673599

Quillaguamán, J., Guzmán, H., Van-Thuoc, D., & Hatti-Kaul, R. (2010). Synthesis and production of polyhydroxyalkanoates by halophiles: Current potential and future prospects. *Applied Microbiology and Biotechnology, 85*(6), 1687–1696. doi:10.100700253-009-2397-6 PMID:20024541

Qvist, J., Ortega, G., Tadeo, X., Millet, O., & Halle, B. (2012). Hydration Dynamics of a Halophilic Protein in Folded and Unfolded States. *The Journal of Physical Chemistry B, 116*(10), 3436–3444. doi:10.1021/jp3000569 PMID:22329545

Rabaey, K., Boon, N., Höfte, M., & Verstraete, W. (2005). Microbial phenazine production enhances electron transfer in biofuel cells. *Environmental Science & Technology, 39*(9), 3401–3408. doi:10.1021/es048563o PMID:15926596

Rabus, R., Ruepp, A., Frickey, T., Rattei, T., Fartmann, B., Stark, M., Bauer, A., Zibat, A., Lombardot, T., Becker, I., Amann, J., Gellner, K., Teeling, H., Leuschner, W. D., Glockner, F.-O., Lupas, A. N., Amann, R., & Klenk, H.-P. (2004). The genome of Desulfotalea psychrophila, a sulfate-reducing bacterium from permanently cold Arctic sediments. *Environmental Microbiology, 6*(9), 887–902. doi:10.1111/j.1462-2920.2004.00665.x PMID:15305914

Radchenkova, N., Boyadzhieva, I., Atanasova, N., Poli, A., Finore, I., Di Donato, P., Nicolaus, B., Panchev, I., Kuncheva, M., & Kambourova, M. (2018). Extracellular polymer substance synthesized by a halophilic bacterium *Chromohalobacter canadensis* 28. *Applied Microbiology and Biotechnology, 102*(11), 4937–4949. doi:10.100700253-018-8901-0 PMID:29616312

Radchenkova, N., Panchev, I., Vassilev, S., Kuncheva, M., Dobreva, S., & Kambourova, M. (2015). Continuous cultivation of a thermophilic bacterium *Aeribacillus pallidus* 418 for production of an exopolysaccharide applicable in cosmetic creams. *Journal of Applied Microbiology, 119*(5), 1301–1309. doi:10.1111/jam.12944 PMID:26332083

Raddadi, N., Cherif, A., Daffonchio, D., & Fava, F. (2013). Halo-alkalitolerant and thermostable cellulases with improved tolerance to ionic liquids and organic solvents from Paenibacillus tarimensis isolated from the Chott El Fejej, Sahara desert, Tunisia. *Bioresource Technology, 150*, 121–128. doi:10.1016/j.biortech.2013.09.089 PMID:24161550

Raddadi, N., Cherif, A., Daffonchio, D., Neifar, M., & Fava, F. (2015). Biotechnological applications of extremophiles, extremozymes and extremolytes. *Applied Microbiology and Biotechnology, 99*(19), 7907–7913. doi:10.100700253-015-6874-9 PMID:26272092

Rahimnejad, M., Adhami, A., Darvari, S., Zirepour, A., & Oh, S.-E. (2015). Microbial fuel cell as new technology for bioelectricity generation: A review. *Alexandria Engineering Journal, 54*(3), 745–756. doi:10.1016/j.aej.2015.03.031

Raichand, R., Pareek, S., Singh, N. K., & Mayilraj, S. (2012). *Exiguobacterium aquaticum* sp. nov., a member of the genus exiguobacterium. *International Journal of Systematic and Evolutionary Microbiology, 62*(Pt_9), 2150–2155. doi:10.1099/ijs.0.035790-0 PMID:22058319

Raizel, R., Leite, J. S., Hypolito, T. M., Coqueiro, A. Y., Newsholme, P., Cruzat, V. F., & Tirapegui, J. (2016). Determination of the anti-inflammatory and cytoprotective effects of L-glutamine and L-alanine, or dipeptide, supplementation in rats submitted to resistance exercise. *British Journal of Nutrition, 116*(3), 470–479. doi:10.1017/S0007114516001999 PMID:27215379

Rajashree, R. P., & S., G. B. (2018). Isolation and characterization of thermophilic bacteria from different habitats and their assessment for antagonism against soil-borne fungal plant pathogens. *African Journal of Microbiology Research, 12*(24), 556–566. doi:10.5897/AJMR2015.7511

Rajpurohit, Y. S., Sharma, D. K., & Misra, H. S. (2021). PprA protein inhibits DNA strand exchange and ATP hydrolysis of *Deinococcus* RecA and regulates the recombination in gamma-irradiated cells. *Frontiers in Cell and Developmental Biology, 9*, 636178. Advance online publication. doi:10.3389/fcell.2021.636178 PMID:33959605

Rajwade, J., Paknikar, K., & Kumbhar, J. (2015). Applications of bacterial cellulose and its composites in biomedicine. *Applied Microbiology and Biotechnology, 99*(6), 2491–2511. doi:10.100700253-015-6426-3 PMID:25666681

Ramette, A., & Tiedje, J. M. (2007). Biogeography: An emerging cornerstone for understanding prokaryotic diversity, ecology, and evolution. *Microbial Ecology, 53*(2), 197–207. doi:10.100700248-005-5010-2 PMID:17106806

Rampelotto, P. H. (2013). Extremophiles and Extreme Environments. *Life 2013, 3*(3), 482–485. doi:10.3390/life3030482

Ranawat, P., & Rawat, S. (2017). Radiation resistance in thermophiles: Mechanisms and applications. *World Journal of Microbiology & Biotechnology, 33*(6), 112. doi:10.100711274-017-2279-5 PMID:28470425

Rao, M. A., Scelza, R., Acevedo, F., Diez, M. C., & Gianfreda, L. (2014). Enzymes as useful tools for environmental purposes. *Chemosphere, 107*, 145–162. doi:10.1016/j.chemosphere.2013.12.059

Rao, M. B., Tanksale, A. M., Ghatge, M. S., & Deshpande, V. V. (1998). Molecular and biotechnological aspects of microbial proteases. *Microbiology and Molecular Biology Reviews, 62*(3), 597–635. doi:10.1128/MMBR.62.3.597-635.1998 PMID:9729602

Rastädter, K., Wurm, D. J., Spadiut, O., & Quehenberger, J. (2021). Physiological Characterization of Sulfolobus acidocaldarius in a Controlled Bioreactor Environment. *International Journal of Environmental Research and Public Health*, 18. PMID:34064179

Rastogi, R. P., Sonani, R. R., & Madamwar, D. (2014). The high-energy radiation protectant extracellular sheath pigment scytonemin and its reduced counterpart in the *cyanobacterium Scytonema* sp. R77DM. *Bioresource Technology, 171*, 396–400. doi:10.1016/j.biortech.2014.08.106 PMID:25226055

Rath, K. M., Fierer, N., Murphy, D. V., & Rousk, J. (2018). Linking bacterial community composition to soil salinity along environmental gradients. *The ISME Journal, 13*(3), 836–846. doi:10.1038/s41396-018-0313-8

Rathinam, N. K., & Sani, R. K. (2018). Bioprospecting of extremophiles for biotechnology applications. In *Extremophilic microbial processing of lignocellulosic feedstocks to biofuels, value-added products, and usable power* (pp. 1–23). Springer. doi:10.1007/978-3-319-74459-9_1

Raval, V. H., Bhatt, H. B., & Singh, S. P. (2018). Adaptation strategies in halophilic bacteria. In *Extremophiles* (pp. 137–164). CRC Press. doi:10.1201/9781315154695-7

Raval, V. H., Pillai, S., Rawal, C. M., & Singh, S. P. (2014). Biochemical and structural characterization of a detergent-stable serine alkaline protease from seawater haloalkaliphilic bacteria. *Process Biochemistry*, *49*(6), 955–962. doi:10.1016/j.procbio.2014.03.014

Raval, V. H., Purohit, M. K., & Singh, S. P. (2013). Diversity, population dynamics and biocatalytic potential of cultivable and non-cultivable bacterial communities of the saline ecosystems. In *Marine Enzymes for Biocatalysis* (pp. 165–189). Woodhead Publishing. doi:10.1533/9781908818355.2.165

Raval, V. H., Purohit, M. K., & Singh, S. P. (2015). Extracellular proteases from halophilic and haloalkaliphilic bacteria: Occurrence and biochemical properties. In *Halophiles* (pp. 421–449). Springer. doi:10.1007/978-3-319-14595-2_16

Ray, A. (2012). Application of lipase in industry. *Asian. The Journal of Pharmacy Technology*, *2*(2), 33–37.

Raymond-Bouchard, I., & Whyte, L. G. (2017). From Transcriptomes to Metatranscriptomes: Cold Adaptation and Active Metabolisms of Psychrophiles from Cold Environments. In Psychrophiles: From Biodiversity to Biotechnology. Springer International Publishing. 10.1007%2F978-3-319-57057-0_18

Raymond, J. A., Christner, B. C., & Schuster, S. C. (2008). A bacterial ice-binding protein from the Vostok ice core. *Extremophiles*, *12*(5), 713–717. doi:10.100700792-008-0178-2 PMID:18622572

Razzaq, A., Shamsi, S., Ali, A., Ali, Q., Sajjad, M., Malik, A., & Ashraf, M. (2019). Microbial Proteases Applications. *Frontiers in Bioengineering and Biotechnology*, *7*, 110. Advance online publication. doi:10.3389/fbioe.2019.00110 PMID:31263696

Reckseidler-Zenteno, S. L. (2012). Capsular polysaccharides produced by the bacterial pathogen *Burkholderia pseudo-mallei. The Complex World of Polysaccharides*, 128-152.

Reed, C. J., Bushnell, S., & Evilia, C. (2014). Circular Dichroism and Fluorescence Spectroscopy of Cysteinyl-tRNA Synthetase from Halobacterium salinarum ssp. NRC-1 Demonstrates that Group I Cations Are Particularly Effective in Providing Structure and Stability to This Halophilic Protein. *PLoS One*, *9*(3), e89452. doi:10.1371/journal.pone.0089452 PMID:24594651

Reed, C. J., Lewis, H., Trejo, E., Winston, V., & Evilia, C. (2013). Protein adaptations in archaeal extremophiles. *Archaea (Vancouver, B.C.)*, *373275*, 1–14. Advance online publication. doi:10.1155/2013/373275 PMID:24151449

Reeve, J. N.JN. (1992). Molecular biology of methanogens. *Annual Review of Microbiology*, *46*(1), 165–191. doi:10.1146/annurev.mi.46.100192.001121 PMID:1444253

Reguera, G., Mccarthy, K. D., Mehta, T., Nicoll, J. S., Tuominen, M. T., & Lovley, D. R. (2005). Extracellular electron transfer via microbial nanowires. *Nature*, *435*(7045), 1098–1101. doi:10.1038/nature03661 PMID:15973408

Rezaei, S., Shahverdi, A. R., & Faramarzi, M. A. (2017). Isolation, one-step affinity purification, and characterization of a polyextremotolerant laccase from the halophilic bacterium *Aquisalibacillus elongatus* and its application in the delignification of sugar beet pulp. *Bioresource Technology*, *230*, 67–75. doi:10.1016/j.biortech.2017.01.036 PMID:28161622

Rhaisa, A. (2016). *Studies of Debaryomyces hansenii killer toxin and its effect on pathogenic bloodstream Candida isolates* (Master thesis). University of Nebraska, Lincoln, NE.

Rhoads, A., Beyenal, H., & Lewandowski, Z. (2005). Microbial fuel cell using anaerobic respiration as an anodic reaction and biomineralized manganese as a cathodic reactant. *Environmental Science & Technology*, *39*(12), 4666–4671. doi:10.1021/es048386r PMID:16047807

Ricard, M., & Reid, I. D. (2004). Purified pectinase lowers cationic demand in peroxide-bleached mechanical pulp. *Enzyme and Microbial Technology, 34*(5), 499–504. doi:10.1016/j.enzmictec.2003.12.005

Rice, T., Zannini, E., Arendt, E. K., & Coffey, A. (2019). *A review of polyols – biotechnological production, food applications, regulation, labeling and health effects*. doi:10.1080/10408398.2019.1625859

Richtzenhain, L. J., Cortez, A., Heinemann, M. B., Soares, R. M., Sakamoto, S. M., Vasconcellos, S. A., Higa, Z. M. M., Scarcelli, E., & Genovez, M. E. (2002). A multiplex PCR for the detection of Brucella spp. and Leptospira spp. DNA from aborted bovine fetuses. *Veterinary Microbiology, 87*(2), 139–147. doi:10.1016/S0378-1135(02)00049-4 PMID:12034542

Riley, M., Staley, J. T., Danchin, A., Wang, T. Z., Brettin, T. S., Hauser, L. J., Land, M. L., & Thompson, L. S. (2008). Genomics of an extreme psychrophile, Psychromonas ingrahamii. *BMC Genomics, 9*(1), 1–19. doi:10.1186/1471-2164-9-210 PMID:18460197

Rinker, K. D., & Kelly, R. M. (2000). Effect of carbon and nitrogen sources on growth dynamics and exopolysaccharide production for the hyperthermophilic archaeon *Thermococcus litoralis* and bacterium *Thermotoga maritima. Biotechnology and Bioengineering, 69*(5), 537–547. doi:10.1002/1097-0290(20000905)69:5<537::AID-BIT8>3.0.CO;2-7 PMID:10898863

Ritala, A., Häkkinen, S. T., Toivari, M., & Wiebe, M. G. (2017). Single Cell ProteinState-of-the-Art Industrial Landscape and Patents 20012016. *Frontiers in Microbiology, 8*. doi:10.3389/fmicb.2017.02009

Robb, F., Antranikian, G., Grogan, D., & Driessen, A. (2008). *Thermophiles, biology and technology at high temperatures*. CRC Press.

Roberts, E., & Garegg, P. (1998). Levan derivatives, their preparation, composition and applications including medical and food applications. *World Intellectual Property Organization Patent Application No. WO, 98*, 03184.

Roberts, I. S. (1996). The biochemistry and genetics of capsular polysaccharide production in bacteria. *Annual Review of Microbiology, 50*(1), 285–315. doi:10.1146/annurev.micro.50.1.285 PMID:8905082

Robles-Medina, A., González-Moreno, P. A., Esteban-Cerdán, L., & Molina-Grima, E. (2009, July). Biocatalysis: Towards ever greener biodiesel production. *Biotechnology Advances, 27*(4), 398–408. doi:10.1016/j.biotechadv.2008.10.008 PMID:19268517

Rodionov, D. A., Rodionova, I. A., Li, X., Ravcheev, D. A., Tarasova, Y., Portnoy, V. A., et al. (2013). Transcriptional regulation of the carbohydrate utilization network in *Thermotoga maritima. Front. Microbiol., 4*, 244. doi:.00244 doi:10.3389/fmicb.2013

Rodriguez-Valera, F. (1992). Biotechnological potential of halobacteria. *Biochemical Society Symposium, 58*, 135–147. PMID:1445403

Rogers, P. L., Lee, K. J., Skotnicki, M. L., & Tribe, D. E. (1982). *Ethanol production by Zymomonas mobilis*. doi:10.1007/3540116982_2

Roh, Y., Chon, C.-M., & Moon, J.-W. (2007). Metal reduction and biomineralization by an alkaliphilic metal-reducing bacterium, Alkaliphilus metalliredigens (QYMF). *Geosciences Journal, 11*(4), 415–423. doi:10.1007/BF02857056

Rollefson, J. B., Levar, C. E., & Bond, D. R. (2009). Identification of Genes Involved in Biofilm Formation and Respiration via Mini-*Himar* Transposon Mutagenesis of *Geobacter sulfurreducens. Journal of Bacteriology, 191*(13), 4207–4217. doi:10.1128/JB.00057-09 PMID:19395486

Ronimus, R. S., Koning, J., & Morgan, H. (1999). Purification and characterization of an ADP-dependent phosphofructokinase from *Thermococcus zilligii. Extremophiles, 3*(2), 121–129. doi:10.1007007920050107 PMID:10356998

Rose, K. C., Williamson, C. E., Saros, J. E., Sommaruga, R., & Fischer, J. M. (2009). Differences in UV transparency and thermal structure between alpine and subalpine lakes: Implications for organisms. *Photochemical & Photobiological Sciences, 8*(9), 1244–1256. doi:10.1039/b905616e PMID:19707613

Rothschild, L. J., & Mancinelli, R. L. (2001). Life in extreme environments. *Nature, 409*(6823), 1092–1101. doi:10.1038/35059215 PMID:11234023

Roy, A., Dutta, A., Pal, S., Gupta, A., Sarkar, J., Chatterjee, A., Saha, A., Sarkar, P., Sar, P., & Kazy, S. K. (2018). Biostimulation and bioaugmentation of native microbial community accelerated bioremediation of oil refinery sludge. *Bioresource Technology, 253*, 22–32. doi:10.1016/j.biortech.2018.01.004 PMID:29328931

Rubio-Infante, N., & Moreno-Fierros, L. (2016). An overview of the safety and biological effects of Bacillus thuringiensis Cry toxins in mammals. *Journal of Applied Toxicology, 36*(5), 630–648. doi:10.1002/jat.3252 PMID:26537666

Ruffing, A., & Chen, R. R. (2006). Metabolic engineering of microbes for oligosaccharide and polysaccharide synthesis. *Microbial Cell Factories, 5*(1), 1–9. doi:10.1186/1475-2859-5-25 PMID:16859553

Ruginescu, R., Gomoiu, I., Popescu, O., Cojoc, R., Neagu, S., Lucaci, I., Batrinescu-Moteau, C., & Enache, M. (2020). Bioprospecting for Novel Halophilic and Halotolerant Sources of Hydrolytic Enzymes in Brackish, Saline and Hypersaline Lakes of Romania. *Microorganisms, 8*(12), 1903. doi:10.3390/microorganisms8121903

Ruiz-Ruiz, C., Srivastava, G. K., Carranza, D., Mata, J. A., Llamas, I., Santamaría, M., Quesada, E., & Molina, I. J. (2011). An exopolysaccharide produced by the novel halophilic bacterium *Halomonas stenophila* strain B100 selectively induces apoptosis in human T leukaemia cells. *Applied Microbiology and Biotechnology, 89*(2), 345–355. doi:10.100700253-010-2886-7 PMID:20890756

Ruparelia, J., Saraf, M., & Jha. (2020). Screening and optimization for cellulase production by soil bacterial isolates JRC1 and JRC2. *BBRC, 13*, 159-164.

Russell, N. J. (2008). Membrane components and cold sensing. In *Psychrophiles: from biodiversity to biotechnology* (pp. 177–190). Springer. doi:10.1007/978-3-540-74335-4_11

Russell, R. J. M., Gerike, U., Danson, M. J., Hough, D. W., & Taylor, G. L. (1998). Structural adaptations of the cold-active citrate synthase from an Antarctic bacterium. *Structure (London, England), 6*(3), 351–361. doi:10.1016/S0969-2126(98)00037-9 PMID:9551556

Ryan, S. M., & Brayden, D. J. (2014). Progress in the delivery of nanoparticle constructs: Towards clinical translation. *Current Opinion in Pharmacology, 18*, 120–128. doi:10.1016/j.coph.2014.09.019 PMID:25450066

Ryanzhin, S. V., Subetto, D. A., Kochkov, N. V., Akhmetova, N. S., & Veinmeister, N. V. (2010). Polar lakes of the World: Current data and status of investigations. *Water Resources, 37*(4), 427–436. doi:10.1134/S0097807810040019

Saba, R., Rehman, Y., Ahmed, M., & Sabri, A. N. (2019). Potential role of bacterial extracellular polymeric substances as biosorbent material for arsenic bioremediation. *Bioremediation Journal, 23*(2), 72–81. doi:10.1080/10889868.2019.1602107

Sabath, N., Ferrada, E., Barve, A., & Wagner, A. (2013). Growth Temperature and Genome Size in Bacteria Are Negatively Correlated, Suggesting Genomic Streamlining During Thermal Adaptation. *Genome Biology and Evolution, 5*(5), 966–977. doi:10.1093/gbe/evt050 PMID:23563968

Sagot, B., Gaysinski, M., Mehiri, M., Guigonis, J.-M., le Rudulier, D., & Alloing, G. (2010). Osmotically induced synthesis of the dipeptide N-acetylglutaminylglutamine amide is mediated by a new pathway conserved among bacteria. *Proceedings of the National Academy of Sciences of the United States of America, 107*(28), 12652–12657. doi:10.1073/pnas.1003063107 PMID:20571117

Sahay, H., Singh, S., Kaushik, R., Saxena, A. K., & Arora, D. K. (2011). Characterization of halophilic bacteria from environmental samples from the brackish water of Pulicat Lake, India. *Biologia, 66*(5), 741–747. doi:10.247811756-011-0094-2

Sahlan, M., & Yohda, M. (2013). Molecular Chaperones in Thermophilic Eubacteria and Archaea. In Thermophilic Microbes in Environmental and Industrial Biotechnology: Biotechnology of Thermophiles, (pp. 375-394). Springer Science Business Media. doi:10.1007/978-94-007-5899-5

Sahoo, R. K., Sahu, A., & Subudhi, E. (2020). Bioremediation of Hydrocarbon Using Bacterial Lipase from Waste Biomass. Iranian Journal of Science and Technology, Transactions A. *Science, 44*(5), 1287–1293. doi:10.100740995-020-00935-y

Saini, R., Singh Saini, H., Dahiya, A., & Harnek Singh Saini, C. (2017). Amylases: Characteristics and industrial applications. *Journal of Pharmacognosy and Phytochemistry, 6*, 1865–1871.

Saito, K., Nishio, Z., Takata, K., Kuwabara, T., Yokota, A., Yamauchi, H., & Oda, Y. (2003). Bacterial Adjunct to Improve the Quality of Alkaline Noodles. Food Science and Technology Research -. *Food Science and Technology Research, 9*(1), 40–44. doi:10.3136/fstr.9.40

Sakuraba, H., Goda, S., & Ohshima, T. (2004). Unique sugar metabolism and novel enzymes of hyperthermophilic archaea. *Chemical Record (New York, N.Y.), 3*(5), 281–287. doi:10.1002/tcr.10066 PMID:14762828

Sakuraba, H., & Ohshima, T. (2002). Novel energy metabolism in anaerobic hyperthermophilic archaea: A modified Embden-Meyerhof pathway. *Journal of Bioscience and Bioengineering, 93*(5), 441–448. doi:10.1016/S1389-1723(02)80090-1 PMID:16233230

Sakuraba, H., Yoshioka, I., Koga, S., Takahashi, M., Kitahama, Y., Satomura, T., Kawakami, R., & Ohshima, T. (2002). ADP-dependent glucokinase/phosphofructokinase, a novel bifunctional enzyme from the hyperthermophilic archaeon Methanococcus jannaschii. *The Journal of Biological Chemistry, 277*(15), 12495–12498. doi:10.1074/jbc.C200059200 PMID:11856730

Salem, I. B., Sghaier, H., Monsieurs, P., Moors, H., Van Houdt, R., Fattouch, S., ... Leys, N. (2013). Strontium-induced genomic responses of *Cupriavidus metallidurans* and strontium bioprecipitation as strontium carbonate. *Annals of Microbiology, 63*(3), 833–844. doi:10.100713213-012-0462-3

Salwan, R., & Sharma, V. (2020). Physiology of extremophiles. Physiological and Biotechnological Aspects of Extremophiles. Elsevier. 10.1016%2Fb978-0-12-818322-9.00002-2

San Martin, R., Bushell, D., Leak, D. J., & Hartley, B. S. (1992). Development of a synthetic medium for continuous anaerobic growth and ethanol production with a lactate dehydrogenase mutant of Bacillus stearothermophilus. *Journal of General Microbiology, 138*(5), 987–996. doi:10.1099/00221287-138-5-987 PMID:1645133

Sander, D., Yu, Y., Sukul, P., Schäkermann, S., Bandow, J. E., Mukherjee, T., Mukhopadhyay, S. K., & Leichert, L. I. (2021). Metaproteomic Discovery and Characterization of a Novel Lipolytic Enzyme From an Indian Hot Spring. *Frontiers in Microbiology, 12*, 672727. Advance online publication. doi:10.3389/fmicb.2021.672727 PMID:34149658

Santos, R., Fernandes, J., Fernandes, N., Oliveira, F., & Cadete, M. (2007). Mycobacterium parascrofulaceum in Acidic Hot Springs in Yellowstone National Park. *Applied and Environmental Microbiology, 73*(15), 5071–5073. doi:10.1128/AEM.00353-07

Santra, H. K., & Banerjee, D. (2021). Extremophilic Microbes and their Extremozymes for Industry and Allied Sectors. In *Microbiomes of Extreme Environments* (pp. 1–42). CRC Press. doi:10.1201/9780429328633-1

Sanz-Montero, M. E., Cabestrero, Ó., & Sánchez-Román, M. (2019). Microbial Mg-rich carbonates in an extreme alkaline lake (Las Eras, Central Spain). *Frontiers in Microbiology, 10*, 148. Advance online publication. doi:10.3389/fmicb.2019.00148 PMID:30800103

Sapranauskas, R., Gasiunas, G., Fremaux, C., Barrangou, R., Horvath, P., & Siksnys, V. (2011). The *Streptococcus thermophilus* CRISPR/Cas system provides immunity in *Escherichia coli. Nucleic Acids Research, 39*(21), 9275–9282. doi:10.1093/nar/gkr606 PMID:21813460

Saravanan, P., Ilavarasan, N., Karthikeyan, A., & Padmanaban, B. (2020). Management of Root Knot Nematode Meloidogyne incognita in Banana (Cv. Robusta) through Biocontrol Agents and Neem Cake. *Indian Journal of Agricultural Research.* doi:10.18805/IJARe.A-5507

Saravanan, C., Rajesh, R., Kaviarasan, T., Muthukumar, K., Kavitake, D., & Shetty, P. H. (2017). Synthesis of silver nanoparticles using bacterial exopolysaccharide and its application for degradation of azo-dyes. *Biotechnology Reports (Amsterdam, Netherlands), 15*, 33–40. doi:10.1016/j.btre.2017.02.006 PMID:28664148

Sarethy, I. P., Saxena, Y., Kapoor, A., Sharma, M., Sharma, S. K., Gupta, V., & Gupta, S. (2011). Alkaliphilic bacteria: Applications in industrial biotechnology. *Journal of Industrial Microbiology & Biotechnology, 38*(7), 769–790. doi:10.100710295-011-0968-x PMID:21479938

Sarkar, G. K. S., & K, S. (2020). Extraction and characterization of alkaline protease from *Streptomyces* sp. GS-1 and its application as dehairing agent. *Biocatalysis and Agricultural Biotechnology, 25*, 101590. doi:10.1016/j.bcab.2020.101590

Sarkar, J., Kazy, S. K., Gupta, A., Dutta, A., Mohapatra, B., Roy, A., Bera, P., Mitra, A., & Sar, P. (2016). Biostimulation of indigenous microbial community for bioremediation of petroleum refinery sludge. *Frontiers in Microbiology, 7*, 1407. doi:10.3389/fmicb.2016.01407 PMID:27708623

Sarkar, J., Roy, A., Sar, P., & Kazy, S. K. (2020). Accelerated bioremediation of petroleum refinery sludge through biostimulation and bioaugmentation of native microbiome. In *Emerging technologies in environmental bioremediation* (pp. 23–65). Elsevier. doi:10.1016/B978-0-12-819860-5.00002-X

Sarkar, S., Banerjee, R., Chanda, S., Das, P., Ganguly, S., & Pal, S. (2010). Effectiveness of inoculation with isolated *Geobacillus* strains in the thermophilic stage of vegetable waste composting. *Bioresource Technology, 101*(8), 2892–2895. doi:10.1016/j.biortech.2009.11.095 PMID:20036532

Sattler, B., Puxbaum, H., & Psenner, R. (2001). Bacterial growth in supercooled cloud droplets. *Geophysical Research Letters, 28*(2), 239–242. doi:10.1029/2000GL011684

Sattler, B., & Storrie-Lombardi, M. C. (2009). LIFE in Antarctic lakes. In A. K. Bej, J. Aislabie, & R. M. Atlas (Eds.), *Polar Microbiology: The Ecology, Biodiversity, and Bioremediation Potential of Microorganisms in Extremely Cold Environments* (pp. 95–114). Taylor and Francis. doi:10.1201/9781420083880-c4

Satyanarayana, T., Raghukumar, C., & Shivaji, S. (2005). Extremophilic microbes: Diversity and perspectives. *Current Science*, 78–90.

Saunders, N. F. W., Thomas, T., Curmi, P. M. G., Mattick, J. S., Kuczek, E., Slade, R., Davis, J., Franzmann, P. D., Boone, D., Rusterholtz, K., Feldman, R., Gates, C., Bench, S., Sowers, K., Kadner, K., Aerts, A., Dehal, P., Detter, C., Glavina, T., ... Cavicchioli, R. (2003). Mechanisms of Thermal Adaptation Revealed From the Genomes of the Antarctic Archaea Methanogenium frigidum and Methanococcoides burtonii. *Genome Research, 13*(7), 1580–1588. doi:10.1101/gr.1180903 PMID:12805271

Säwström, C., Lisle, J., Anesio, A. M., Priscu, J. C., & Laybourn-Parry, J. (2008). Bacteriophage in polar inland waters. *Extremophiles, 12*(2), 167–175. doi:10.100700792-007-0134-6 PMID:18188502

Schagerl, M. (2016). *Bacteria*. Archaea and Viruses of Soda Lakes, Soda Lakes of East Africa. doi:10.1007/978-3-319-28622-8

Schelert, J., Rudrappa, D., Johnson, T., & Blum, P. (2013). Role of MerH in mercury resistance in the archaeon Sulfolobus solfataricus. *Microbiology, 159*(6), 1198–1208. doi:10.1099/mic.0.065854-0 PMID:23619003

Schiraldi, C., & De Rosa, M. (2002). The production of biocatalysts and biomolecules from extremophiles. *Trends in Biotechnology, 20*(12), 515–521. doi:10.1016/S0167-7799(02)02073-5 PMID:12443873

Schiraldi, C., Giuliano, M., & De Rosa, M. (2002). Perspectives on biotechnological applications of archaea. *Archaea (Vancouver, B.C.), 1*(2), 75–86. doi:10.1155/2002/436561 PMID:15803645

Schmidt, S. T., Yu, F. B., Blainey, P. C., May, A. P., & Quake, S. R. (2019). Nucleic acid cleavage with a hyperthermophilic Cas9 from an uncultured Ignavibacterium. *Proceedings of the National Academy of Sciences of the United States of America, 116*(46), 23100–23105. doi:10.1073/pnas.1904273116 PMID:31659048

Schnoor, M., Voß, P., Cullen, P., Böking, T., Galla, H. J., Galinski, E. A., & Lorkowski, S. (2004). Characterization of the synthetic compatible solute homoectoine as a potent PCR enhancer. *Biochemical and Biophysical Research Communications, 322*(3), 867–872. doi:10.1016/j.bbrc.2004.07.200 PMID:15336543

Schönheit, P., & Schäfer, T. (1995). Metabolism of hyperthermophiles. *World Journal of Microbiology & Biotechnology, 11*(1), 26–57. doi:10.1007/BF00339135 PMID:24414410

Schröder, C., Burkhardt, C., & Antranikian, G. (2020). What we learn from extremophiles. *Chemtexts, 6*(1), 8. doi:10.100740828-020-0103-6

Schröder, C., Selig, M., & Schönheit, P. (1994). Glucose fermentation to acetate, CO_2 and H2 in the anaerobic hyperthermophilic eubacterium *Thermotoga maritima*: Involvement of the Embden-Meyerhof pathway. *Archives of Microbiology, 161*(6), 460–470.

Schröder, U. (2007). Anodic electron transfer mechanisms in microbial fuel cells and their energy efficiency. *Physical Chemistry Chemical Physics, 9*(21), 2619–2629. doi:10.1039/B703627M PMID:17627307

Schröder, U., Niessen, J., & Scholz, F. (2003). A generation of microbial fuel cells with current outputs boosted by more than one order of magnitude. *Angewandte Chemie International Edition in English, 42*(25), 2880–2883. doi:10.1002/anie.200350918 PMID:12833347

Schultze-Lam, S., Ferris, F. G., Sherwood-Lollar, B., & Gerits, J. P. (1996). Ultrastructure and seasonal growth patterns of microbial mats in a temperate climate saline-alkaline lake: Goodenough Lake, British Columbia, Canada. *Canadian Journal of Microbiology, 42*(2), 147–161. doi:10.1139/m96-023

Schulz, A., Stöveken, N., Binzen, I. M., Hoffmann, T., Heider, J., & Bremer, E. (2017). Feeding on compatible solutes: A substrate-induced pathway for uptake and catabolism of ectoines and its genetic control by EnuR. *Environmental Microbiology, 19*(3), 926–946. doi:10.1111/1462-2920.13414 PMID:27318028

Schumann, P., Prauser, H., Rainey, F. A., Stackebrandt, E., & Hirsch, P. (1997). Friedmanniella antarctica gen. nov., sp. nov., an LL-diaminopimelic acid-containing actinomycete from Antarctic sandstone. *International Journal of Systematic and Evolutionary Microbiology, 47*(2), 278–283. PMID:9103610

Scott, K., & Murano, C. (2007). Microbial fuel cells utilising carbohydrates. *Journal of Chemical Technology and Biotechnology (Oxford, Oxfordshire), 82*(1), 92–100. doi:10.1002/jctb.1641

Seckbach, J., Oren, A., & Stan-Lotter, H. (Eds.). (2013). *Polyextremophiles. Cellular Origin, Life in Extreme Habitats and Astrobiology*. Springer.

Segawa, T., Miyamoto, K., Ushida, K., Agata, K., Okada, N., & Kohshima, S. (2005). Seasonal change in bacterial flora and biomass in mountain snow from the Tateyama Mountains, Japan, analyzed by 16S rRNA gene sequencing and real-time PCR. *Applied and Environmental Microbiology*, *71*(1), 123–130. doi:10.1128/AEM.71.1.123-130.2005 PMID:15640179

Sehar, S., & Hameed, A. (2011). Extracellular Alkaline Protease by a Newly Isolated *Halophilic Bacillus* sp. *Global Journal of Biotechnology & Biochemistry*, *6*, 142–148.

Selig, M., Xavier, K. B., Santos, H., & Schonheit, P. (1997). Comparative analysis of Embden-Meyerhof and Entner-Doudoroff glycolytic pathways in hyperthermophilic archaea and the bacterium *Thermotoga*. *Archives of Microbiology*, *167*(4), 217–232. doi:10.1007/BF03356097 PMID:9075622

Selvakumar, R., Aravindh, S., Ashok, A. M., & Balachandran, Y. L. (2014). A facile synthesis of silver nanoparticle with SERS and antimicrobial activity using *Bacillus subtilis* exopolysaccharides. *Journal of Experimental Nanoscience*, *9*(10), 1075–1087. doi:10.1080/17458080.2013.778425

Sevinc, N., & Demirkan, E. (2011). Production of Protease by *Bacillus* sp. N-40 Isolated from Soil and Its Enzymatic Properties. *Journal of Biological and Environmental Sciences*, *5*, 95–103.

Shah, A. R., Shah, R. K., & Madamwar, D. (2006). Improvement of the quality of whole wheat bread by supplementation of xylanase from Aspergillus foetidus. *Bioresource Technology*, *97*(16), 2047–2053. doi:10.1016/j.biortech.2005.10.006 PMID:16307877

Shah, S. R. (2013). Chemistry and Applications of Cellulase in Textile Wet Processing. *Research Journal of Engineering Sciences*, *2*, 1–5.

Shand, R. F., & Leyva, K. J. (2008). Archaeal Antimicrobials: An Undiscovered Country. In Archaea: New Models for Prokaryotic Biology (p. 248). Academic Press.

Shang, J. L., Zhang, Z. C., Yin, X. Y., Chen, M., Hao, F. H., Wang, K., Feng, J. L., Xu, H. F., Yin, Y. C., Tang, H. R., & Qiu, B. S. (2018). UV-B induced biosynthesis of a novel sunscreen compound in solar radiation and desiccation-tolerant cyanobacteria. *Environmental Microbiology*, *20*(1), 200–213. doi:10.1111/1462-2920.13972 PMID:29076601

Shanmugam, V., Chugh, P., & Sharma, P. (2014). Cold-tolerant Trichoderma species for the managementof Fusarium wilt of tomato plants. *Annals of Microbiology*, *65*(1), 543–551. doi:10.1007/s13213-014-0890-3

Shao, H., Xu, L., & Yan, Y. (2014). Thermostable lipases from extremely radioresistant bacterium Deinococcus radiodurans: Cloning, expression, and biochemical characterization. *Journal of Basic Microbiology*, *54*(9), 984–995. doi:10.1002/jobm.201300434 PMID:24002979

Shapovalova, A. A., Khijniak, T. V., Tourova, T. P., & Sorokin, D. Y. (2009). *Halomonas chromatireducens* sp. nov., a new denitrifying facultatively haloalkaliphilic bacterium from solonchak soil capable of aerobic chromate reduction. *Microbiology*, *78*(1), 102–111. doi:10.1134/S0026261709010135 PMID:19334604

Sharma, A., Parashar, D., & Satyanarayana, T. (2016). Acidophilic Microbes: Biology and Applications. In Biotechnology of Extremophiles (pp. 215–241). Springer International Publishing. doi:10.1007/978-3-319-13521-2_7

Sharma, Parashar, & *Satyanarayana. (2016). Acidophilic microbes: biology and applications. Biotechnology* of Extremophiles, 215-241.

Sharma, A., Kawarabayasi, Y., & Satyanarayana, T. (2012). Acidophilic bacteria and archaea: Acid stable biocatalysts and their potential applications. *Extremophiles*, *16*(1), 1–19. doi:10.100700792-011-0402-3 PMID:22080280

Sharma, A., & Satyanarayana, T. (2012). Cloning and expression of acidstable, high maltose-forming, Ca 2+-independent α-amylase from an acidophile Bacillus acidicola and its applicability in starch hydrolysis. *Extremophiles*, *16*(3), 515–522. doi:10.100700792-012-0451-2 PMID:22527045

Sharma, A., Singh, S. B., Sharma, R., Chaudhary, P., Pandey, A. K., Ansari, R., Vasudevan, V., Arora, A., Singh, S., Saha, S., & Nain, L. (2016). Enhanced biodegradation of PAHs by microbial consortium with different amendment and their fate in-situ condition. *Journal of Environmental Management*, *181*, 728–736. doi:10.1016/j.jenvman.2016.08.024 PMID:27558829

Sharshar, M. M., Samak, N. A., Hao, X., Mu, T., Zhong, W., Yang, M., Peh, S., Ambreen, S., & Xing, J. (2019). Enhanced growth-driven stepwise inducible expression system development in haloalkaliphilic desulfurizing *Thioalkalivibrio versutus*. *Bioresource Technology*, *288*, 121486. doi:10.1016/j.biortech.2019.121486 PMID:31128536

Shaw, A. J., Podkaminer, K. K., Desai, S. G., Bardsley, J. S., Rogers, S. R., Thorne, P. G., Hogsett, D. A., & Lynd, L. R. (2008). Metabolic engineering of a thermophilic bacterium to produce ethanol at high yield. *Proceedings of the National Academy of Sciences of the United States of America*, *105*(37), 13769–13774. doi:10.1073/pnas.0801266105 PMID:18779592

Shim, J., Soum, V., Kang, K.-W., Chuon, K., Cho, S.-G., Kim, J.-H., Meas, S., Pushkarev, A., Shin, K., & Jung, K.-H. (2021). Discovery of a microbial rhodopsin that is the most stable in extreme environments. *iScience*, *24*(6), 102620. Advance online publication. doi:10.1016/j.isci.2021.102620 PMID:34151231

Shin, B., Park, C., Lee, B. H., Lee, K. E., & Park, W. (2020). *Bacillus miscanthi* sp. Nov., a alkaliphilic bacterium from the rhizosphere of *Miscanthus sacchariflorus*. *International Journal of Systematic and Evolutionary Microbiology*, *70*(3), 1843–1849. doi:10.1099/ijsem.0.003982 PMID:31951191

Shinde, V., & Thombre, R. (2016). Antibiotic resistance profiling of marine halophilic bacteria and haloarchaea. *Journal of Applied Pharmaceutical Science*, *6*(10), 132–137. doi:10.7324/JAPS.2016.601018

Shin, H., Olsen, B. D., & Khademhosseini, A. (2012). The mechanical properties and cytotoxicity of cell-laden double-network hydrogels based on photocrosslinkable gelatin and gellan gum biomacromolecules. *Biomaterials*, *33*(11), 3143–3152. doi:10.1016/j.biomaterials.2011.12.050 PMID:22265786

Shivaji, S., & Prakash, J. S. (2010). How do bacteria sense and respond to low temperature? *Archives of Microbiology*, *192*(2), 85–95. doi:10.100700203-009-0539-y PMID:20049417

Shivaji, S., Rao, N. S., Saisree, L., Sheth, V., Reddy, G. S., & Bhargava, P. M. (1989). Isolation and identification of Pseudomonas spp. from Schirmacher Oasis, Antarctica. *Applied and Environmental Microbiology*, *55*(3), 767–770. doi:10.1128/aem.55.3.767-770.1989 PMID:2930174

Shivlata, L., & Satyanarayana, T. (2015). Thermophilic and alkaliphilic Actinobacteria: Biology and potential applications. *Frontiers in Microbiology*, *6*, 1014. doi:10.3389/fmicb.2015.01014 PMID:26441937

Shmakov, S., Smargon, A., Scott, D., Cox, D., Pyzocha, N., Yan, W., Abudayyeh, O. O., Gootenberg, J. S., Makarova, K. S., Wolf, Y. I., Severinov, K., Zhang, F., & Koonin, E. V. (2017). Diversity and evolution of class 2 CRISPR-Cas systems. *Nature Reviews. Microbiology*, *15*(3), 169–182. doi:10.1038/nrmicro.2016.184 PMID:28111461

Shockley, K. R., Ward, D. E., Chhabra, S. R., Conners, S. B., Montero, C. I., & Kelly, R. M. (2003). Heat shock response by the hyperthermophilic archaeon *Pyrococcus furiosus*. *Applied and Environmental Microbiology*, *69*(4), 2365–2371. doi:10.1128/AEM.69.4.2365-2371.2003 PMID:12676722

Shong, J., Jimenez Diaz, M. R., & Collins, C. H. (2012). Towards synthetic microbial consortia for bioprocessing. *Current Opinion in Biotechnology*, *23*(5), 798–802. doi:10.1016/j.copbio.2012.02.001 PMID:22387100

Showalter, A. R., Szymanowski, J. E. S., Fein, J. B., & Bunker, B. A. (2016). An x-ray absorption spectroscopy study of Cd binding onto a halophilic archaeon. *Journal of Physics: Conference Series*, *712*(1), 012079. doi:10.1088/1742-6596/712/1/012079

Shrestha, B. K., Karki, H. S., Groth, D. E., Jungkhun, N., & Ham, J. H. (2016). Biological Control Activities of Rice-Associated Bacillus sp. Strains against Sheath Blight and Bacterial Panicle Blight of Rice. PLOS ONE, 11(1), e0146764. doi:10.1371/journal.pone.0146764

Shrestha, N., Chilkoor, G., Vemuri, B., Rathinam, N., Sani, R. K., & Gadhamshetty, V. (2018). Extremophiles for microbial-electrochemistry applications: A critical review. *Bioresource Technology*, *255*, 318–330. doi:10.1016/j.biortech.2018.01.151 PMID:29433771

Shukla, A. K., & Singh, A. K. (2020). Exploitation of Potential Extremophiles for Bioremediation of Xenobiotics Compounds: A Biotechnological Approach. *Current Genomics*, *21*(3), 161–167. doi:10.2174/1389202921999200422122253

Shukla, P. J., Bhatt, V. D., Suriya, J., & Mootapally, C. (2020, August). *Marine Extremophiles.* Wiley. doi:10.1002/9781119143802.ch74

Shukla, R. J., & Singh, S. P. (2015). Production optimization, purification and characterization of α-amylase from thermophilic Bacillus licheniformis TSI-14. *Starch*, *67*(7-8), 629–639. doi:10.1002tar.201500046

Siddiqui, K. S., & Cavicchioli, R. (2006). Cold-adapted enzymes. *Annual Review of Biochemistry*, *75*(1), 403–433. doi:10.1146/annurev.biochem.75.103004.142723 PMID:16756497

Siddiqui, K. S., Williams, T. J., Wilkins, D., Yau, S., Allen, M. A., Brown, M. V., Lauro, F. M., & Cavicchioli, R. (2013). Psychrophiles. *Annual Review of Earth and Planetary Sciences*, *41*(1), 87–115. doi:10.1146/annurev-earth-040610-133514

Siebers, B., & Schönheit, P. (2005). Unusual pathways and enzymes of central carbohydrate metabolism in *Archaea. Current Opinion in Microbiology*, *8*(6), 695–705. doi:10.1016/j.mib.2005.10.014 PMID:16256419

Siebert, J., & Hirsch, P. (1988). Characterization of 15 selected coccal bacteria isolated from Antarctic rock and soil samples from the McMurdo-Dry Valleys (South-Victoria Land). *Polar Biology*, *9*(1), 37–44. doi:10.1007/BF00441762 PMID:11538335

Siegert, R., Leroux, M. R., Scheufler, C., Hartl, F. U., & Moarefi, I. (2000). Structure of the molecular chaperone prefoldin: Unique interaction of multiple coiled coil tentacles with unfolded proteins. *Cell*, *103*(4), 621–632. doi:10.1016/S0092-8674(00)00165-3 PMID:11106732

Siglioccolo, A., Paiardini, A., Piscitelli, M., & Pascarella, S. (2011). Structural adaptation of extreme halophilic proteins through decrease of conserved hydrophobic contact surface. *BMC Structural Biology*, *11*(1), 1–12. doi:10.1186/1472-6807-11-50

Siliakus, M. F., van der Oost, J., & Kengen, S. W. M. (2017). Adaptations of archaeal and bacterial membranes to variations in temperature, pH and pressure. *Extremophiles*, *21*(4), 651–670. doi:10.100700792-017-0939-x PMID:28508135

Simon, C., Wiezer, A., Strittmatter, A. W., & Daniel, R. (2009). Phylogenetic diversity and metabolic potential revealed in a glacier ice metagenome. *Applied and Environmental Microbiology*, *75*(23), 7519–7526. doi:10.1128/AEM.00946-09 PMID:19801459

Sims, R. E. H., Mabee, W., Saddler, J. N., & Taylor, M. (2010). An overview of second generation biofuel technologies. *Bioresource Technology*, *101*(6), 1570–1580. doi:10.1016/j.biortech.2009.11.046 PMID:19963372

Sindhu, R., & Manonmani, H. K. (2018). Expression and characterization of recombinant L-asparaginase from *Pseudomonas fluorescens. Protein Expression and Purification*, *143*, 83–91. doi:10.1016/j.pep.2017.09.009 PMID:29079538

Singh, L., Sharma, G., Awasthi, G., & Kumar, L. (2021). *The Impact of extremophilic Enzyme: Widely Application in Industrial uses*. Academic Press.

Singh, A. J., Xu, C. X., Xu, X., West, L. M., Wilmes, A., Chan, A., Hamel, E., Miller, J. H., Northcote, P. T., & Ghosh, A. K. (2010). Peloruside B, a potent antitumor macrolide from the New Zealand marine sponge *Mycale hentscheli*: Isolation, structure, total synthesis, and bioactivity. *The Journal of Organic Chemistry*, *75*(1), 2–10. doi:10.1021/jo9021265 PMID:19957922

Singh, G., Jain, V. K., & Singh, A. (2017). Effect of Temperature and other factors on Anaerobic Digestion Process, responsible for Bio Gas Production. *International Journal of Theoretical and Applied Mechanics.*, *12*(3), 637–657.

Singh, H., Kaur, M., Kaur, L., Sharma, S., Mishra, S., Tanuku, N. R. S., & Pinnaka, A. K. (2018). *Bacillus lacus* sp. nov., isolated from a water sample of a salt lake in India. *International Journal of Systematic and Evolutionary Microbiology*, *68*(3), 801–809. doi:10.1099/ijsem.0.002588 PMID:29458476

Singh, O. V., & Gabani, P. (2011). Extremophiles: Radiation resistance microbial reserves and therapeutic implications. *Journal of Applied Microbiology*, *110*(4), 851–861. doi:10.1111/j.1365-2672.2011.04971.x PMID:21332616

Singh, P., Hamid, B., Lone, M. A., Ranjan, K., Khan, A., Chaurse, V. K., & Sahay, S. (2012). Evaluation of pectinase activity from the psychrophilic fungal strain Truncatella angustata-BPF5 for use in wine industry. *Journal of Endocytobiosis and Cell Research*, *22*, 57–61.

Singh, S. K., Singh, S. K., Tripathi, V. R., Khare, S. K., & Garg, S. K. (2011). A novel psychrotrophic, solvent tolerant Pseudomonas putida SKG-1 and solvent stability of its psychro-thermoalkalistable protease. *Process Biochemistry*, *46*(7), 1430–1435. doi:10.1016/j.procbio.2011.03.012

Singh, S. P., Raval, V., & Purohit, M. K. (2013). Strategies for the salt tolerance in bacteria and archaea and its implications in developing crops for adverse conditions. In *Plant acclimation to environmental stress* (pp. 85–99). Springer. doi:10.1007/978-1-4614-5001-6_4

Singh, S. P., Thumar, J. T., Gohel, S. D., Kikani, B., Shukla, R., Sharma, A., & Dangar, K. (2013). Actinomycetes from marine habitats and their enzymatic potential. In *Marine Enzymes for Biocatalysis* (pp. 191–214). Woodhead Publishing. doi:10.1533/9781908818355.2.191

Sirevåg, R. (2021). Thermophilic Phototrophs. *Thermophilic Bacteria*, 195-219. doi:10.1201/9781003068334-8

Siroosi, M., Amoozegar, M. A., Khajeh, K., & Dabirmanesh, B. (2018). Decolourization of dyes by a novel sodium azide-resistant spore laccase from a halo tolerant bacterium, *Bacillus safensis* sp. strain S31. *Water Science and Technology*, *77*(12), 2867–2875. doi:10.2166/wst.2018.281 PMID:30065139

Sivakumar, B., Aswathy, R. G., Sreejith, R., Nagaoka, Y., Iwai, S., Suzuki, M., ... Maekawa, T. (2014). Bacterial exopolysaccharide based magnetic nanoparticles: A versatile nanotool for cancer cell imaging, targeted drug delivery and synergistic effect of drug and hyperthermia mediated cancer therapy. *Journal of Biomedical Nanotechnology*, *10*(6), 885–899. doi:10.1166/jbn.2014.1820 PMID:24749386

Śliżewska, K., Markowiak-Kopeć, P., & Śliżewska, W. (2021). The role of probiotics in cancer prevention. *Cancers (Basel)*, *13*(1), 20. doi:10.3390/cancers13010020 PMID:33374549

Smith, M. C., Bowman, J. P., Scott, F. J., & Line, M. A. (2000). Sublithic bacteria associated with Antarctic quartz stones. *Antarctic Science*, *12*(2), 177–184. doi:10.1017/S0954102000000237

Snyder, B. (2007). Why the NSF biology budget should be doubled. *Bioscience*, *57*(9), 727–728. doi:10.1641/B570902

Sogin, M. L., Morrison, H. G., Huber, J. A., Welch, D. M., Huse, S. M., Neal, P. R., Arrieta, J. M., & Herndl, G. J. (2006). Microbial diversity in the deep sea and the underexplored "rare biosphere". *Proceedings of the National Academy of Sciences of the United States of America, 103*(32), 12115–12120. doi:10.1073/pnas.0605127103 PMID:16880384

Soppa, J. (2006). From genomes to function: Haloarchaea as model organisms. *Microbiology (Reading, England), 152*(Pt 3), 585–590. doi:10.1099/mic.0.28504-0 PMID:16514139

Soppa, J. (2013). Evolutionary advantages of polyploidy in halophilic archaea. *Biochemical Society Transactions, 41*(1), 339–343. doi:10.1042/BST20120315 PMID:23356308

Sorokin, D. Y., Abbas, B., Erik, V. Z., & Muyzer, G. (2014). Isolation and characterization of an obligate chemolitho-autotrophic *Halothiobacillus* strain capable of growth on thiocyanate as an energy source. *FEMS Microbiology Letters, 354*(1), 69–74. doi:10.1111/1574-6968.12432 PMID:24673239

Sorokin, D. Y., Khijniak, T. V., Zakharycheva, A. P., Elcheninov, A. G., Hahnke, R. L., Boueva, O. V., Ariskina, E. V., Bunk, B., Kublanov, I. V., & Evtushenko, L. I. (2021). *Natronoglycomyces albus* gen. Nov., sp. nov, a haloalkaliphilic actinobacterium from a soda solonchak soil. *International Journal of Systematic and Evolutionary Microbiology, 71*(5), 004804. doi:10.1099/ijsem.0.004804 PMID:33999794

Sorokin, I. D., Kravchenko, I. K., Doroshenko, E. V., Boulygina, E. S., Zadorina, E. V., Tourova, T. P., & Sorokin, D. Y. (2008). *Haloalkaliphilic diazotroph*s in soda solonchak soils. *FEMS Microbiology Ecology, 65*(3), 425–433. doi:10.1111/j.1574-6941.2008.00542.x PMID:18616587

Sou, T., Meeusen, E. N., de Veer, M., Morton, D. A. V., Kaminskas, L. M., & McIntosh, M. P. (2011). New developments in dry powder pulmonary vaccine delivery. *Trends in Biotechnology, 29*(4), 191–198. doi:10.1016/j.tibtech.2010.12.009 PMID:21255854

Souza, E. C., Vessoni-Penna, T. C., & de Souza Oliveira, R. P. (2014). Biosurfactant-enhanced hydrocarbon bioremediation: An overview. *International Biodeterioration & Biodegradation, 89*, 88–94. doi:10.1016/j.ibiod.2014.01.007

Souza, K. A., & Deal, P. H. (1977). Characteristics of a novel extremely alkalophilic bacterium. *Journal of General Microbiology, 101*(1), 103–109. doi:10.1099/00221287-101-1-103

Souza, V., Eguiarte, L. E., Siefert, J., & Elser, J. J. (2008). Microbial endemism: Does phosphorus limitation enhance speciation. *Nature Reviews. Microbiology, 6*(7), 559–564. doi:10.1038/nrmicro1917 PMID:18521074

Spanò, A., Laganà, P., Visalli, G., Maugeri, T. L., & Gugliandolo, C. (2016). In vitro antibiofilm activity of an exopolysaccharide from the marine thermophilic *Bacillus licheniformis* T14. *Current Microbiology, 72*(5), 518–528. doi:10.100700284-015-0981-9 PMID:26750122

Srivastava, P., Nikhil, E. V., Bragança, J. M., & Kowshik, M. (2015). Anti-bacterial TeNPs biosynthesized by haloarcheaon *Halococcus salifodinae* BK3. *Extremophiles, 19*(4), 875–884. doi:10.100700792-015-0767-9 PMID:26085473

Staley, J. T., & Gosink, J. J. (1999). Poles apart: Biodiversity and biogeography of sea ice bacteria. *Annual Review of Microbiology, 53*(1), 189–215. doi:10.1146/annurev.micro.53.1.189 PMID:10547690

Stams, A. J., De Bok, F. A., Plugge, C. M., Van Eekert, M. H., Dolfing, J., & Schraa, G. (2006). Exocellular electron transfer in anaerobic microbial communities. *Environmental Microbiology, 8*(3), 371–382. doi:10.1111/j.1462-2920.2006.00989.x PMID:16478444

Stan-Lotter, H., & Fendrihan, S. (2012). *Adaption of microbial life to environmental extremes*. Springer. doi:10.1007/978-3-211-99691-1

Stapper, N. J., & John, V. (2015). Monitoring climate change with lichens as bioindicators. Pollution atmosphérique (N226). INIST-CNRS. 10.4267%2Fpollution-atmospherique.4936

Stathopoulou, P. M., Savvides, A. L., Karagouni, A. D., & Hatzinikolaou, D. G. (2013). Unraveling the lipolytic activity of thermophilic bacteria isolated from a volcanic environment. *BioMed Research International*, *2013*, 1–13. Advance online publication. doi:10.1155/2013/703130 PMID:23738330

Sterner, R., & Liebl, W. (2001). Thermophilic adaptation of proteins. *Critical Reviews in Biochemistry and Molecular Biology*, *36*(1), 39–106. doi:10.1080/20014091074174 PMID:11256505

Stetter, K. O. (1982). Ultrathin mycelia-forming organisms from submarine volcanic areas having an optimum growth temperature of 105 °C. *Nature*, *300*(5889), 258–260. doi:10.1038/300258a0

Stevens, M. M., Qanadilo, H. F., Langer, R., & Shastri, V. P. (2004). A rapid-curing alginate gel system: Utility in periosteum-derived cartilage tissue engineering. *Biomaterials*, *25*(5), 887–894. doi:10.1016/j.biomaterials.2003.07.002 PMID:14609677

Stewart, P. S. (2014). Biophysics of biofilm infection. *Pathogens and Disease*, *70*(3), 212–218. doi:10.1111/2049-632X.12118 PMID:24376149

Stibal, M., Elster, J., Sabacka, M., & Kastovska, K. (2007). Seasonal and diel changes in photosynthetic activity of the snow alga Chlamydomonas nivalis (Chlorophyceae) from Svalbard determined by pulse amplitude modulation fluorometry. *FEMS Microbiology Ecology*, *59*(2), 265–273. doi:10.1111/j.1574-6941.2006.00264.x PMID:17313577

Stingl, U., Cho, J. C., Foo, W., Vergin, K. L., Lanoil, B., & Giovannoni, S. J. (2008). Dilution-to-extinction culturing of psychrotolerant planktonic bacteria from permanently ice-covered lakes in the McMurdo Dry Valleys, Antarctica. *Microbial Ecology*, *55*(3), 395–405. doi:10.100700248-007-9284-4 PMID:17623231

Stirling, J. L., Bennetto, H. P., Delaney, G. M., Mason, J. R., Roller, S. D., Tanaka, K., & Thurston, C. F. (1983). Microbial fuel cells. *Biochemical Society Transactions*, *11*(4), 451–453. doi:10.1042/bst0110451 PMID:6352363

Stirling, P. C., Lundin, V. F., & Leroux, M. R. (2003). Getting a grip on non-native proteins. *EMBO Reports*, *4*(6), 565–570. doi:10.1038j.embor.embor869 PMID:12776175

Stom, D. I., Zhdanova, G. O., Kalashnikova, O. B., Bulaev, A. G., Kashevskii, A. V., Kupchinsky, A. B., Vardanyan, N. S., Ponamoreva, O. N., Alferov, S. V., Saksonov, M. N., Chesnokova, A. N., & Tolstoy, M. Yu. (2020). Acidophilic Microorganisms *Leptospirillum sp., Acidithiobacillus sp., Ferroplasma sp.* As a Cathodic Bioagents in a MFC. *Geomicrobiology*, *38*(4), 340–346. doi:10.1080/01490451.2020.1856980

Straub, K. L., & Schink, B. (2004). Ferrihydrite-dependent growth of *Sulfurospirillum deleyianum* through electron transfer via sulfur cycling. *Applied and Environmental Microbiology*, *70*(10), 5744–5749. doi:10.1128/AEM.70.10.5744-5749.2004 PMID:15466509

Strazzulli, A., Cobucci-Ponzano, B., Iacono, R., Giglio, R., Maurelli, L., Curci, N., Schiano-di-Cola, C., Santangelo, A., Contursi, P., Lombard, V., Henrissat, B., Lauro, F. M., Fontes, C. M. G. A., & Moracci, M. (2020). Discovery of hyperstable carbohydrate-active enzymes through metagenomics of extreme environments. *The FEBS Journal*, *287*(6), 1116–1137. doi:10.1111/febs.15080 PMID:31595646

Street, T. O., Bolen, D. W., & Rose, G. D. (2006). A molecular mechanism for osmolyte-induced protein stability. *Proceedings of the National Academy of Sciences of the United States of America*, *103*(38), 13997–14002. doi:10.1073/pnas.0606236103 PMID:16968772

Struvay, C., & Feller, G. (2012). Optimization to low temperature activity in psychrophilic enzymes. *International Journal of Molecular Sciences*, *13*(9), 11643–11665. doi:10.3390/ijms130911643 PMID:23109875

Stuart, E. S., Morshed, F., Sremac, M., & DasSarma, S. (2001). Antigen presentation using novel particulate organelles from halophilic archaea. *Journal of Biotechnology*, *88*(2), 119–128. doi:10.1016/S0168-1656(01)00267-X PMID:11403846

Stuart, E. S., Morshed, F., Sremac, M., & DasSarma, S. (2004). Cassette-based presentation of SIV epitopes with recombinant gas vesicles from halophilic archaea. *Journal of Biotechnology*, *114*(3), 225–237. doi:10.1016/j.jbiotec.2004.01.005 PMID:15522433

Sui, X., Wen, C., Yang, J., Guo, H., Zhao, W., Li, Q., Zhang, J., Zhu, Y., & Zhang, L. (2018). Betaine Combined with Membrane Stabilizers Enables Solvent-Free Whole Blood Cryopreservation and One-Step Cryoprotectant Removal. *ACS Biomaterials Science & Engineering*, *5*(2), 1083–1091. doi:10.1021/acsbiomaterials.8b01286 PMID:33405798

Suka, A., Oki, H., Kato, Y., Kawahara, K., Ohkubo, T., Maruno, T., Kobayashi, Y., Fujii, S., Wakai, S., Lisdiana, L., & Sambongi, Y. (2019). Stability of cytochromes c′ from psychrophilic and piezophilic Shewanella species: implications for complex multiple adaptation to low temperature and high hydrostatic pressure. *Extremophiles*, *23*(2), 239–248. doi:10.1007/s00792-019-01077-9

Sukumaran, R. K., Singhania, R. R., Mathew, G. M., & Pandey, A. (2009). Cellulase production using biomass feed stock and its application in lignocellulose saccharification for bio-ethanol production. *Renewable Energy*, *34*(2), 421–424. doi:10.1016/j.renene.2008.05.008

Sullivan, C. W., & Palmisano, A. C. (1984). Sea ice microbial communities: Distribution, abundance, and diversity of ice bacteria in McMurdo Sound, Antarctica, in 1980. *Applied and Environmental Microbiology*, *47*(4), 788–795. doi:10.1128/aem.47.4.788-795.1984 PMID:16346517

Sullivan, W. T. III, & Baross, J. (Eds.). (2007). *Planets and life: the emerging science of astrobiology*. Cambridge University Press. doi:10.1017/CBO9780511812958

Sun, M. M., Tolliday, N., Vetriani, C., Robb, F. T., & Clark, D. S. (1999). Pressure-induced thermostabilization of glutamate dehydrogenase from the hyperthermophile *Pyrococcus furiosus*. *Protein Science*, *8*(5), 1056–1063. doi:10.1110/ps.8.5.1056 PMID:10338016

Sun, Z., He, X., Brancaccio, V. F., Yuan, J., & Riedel, C. U. (2014). Bifidobacteria exhibit LuxS-dependent autoinducer 2 activity and biofilm formation. *PLoS One*, *9*(2), e88260. doi:10.1371/journal.pone.0088260 PMID:24505453

Survival of *Halobacteria* within fluid inclusions in salt crystals. (n.d.). *Microbiology Society*. Retrieved July 7, 2021, from https://www.microbiologyresearch.org/content/journal/micro/10.1099/00221287-134-5-1365?crawler=true

Sutherland, I. (1972). Bacterial exopolysaccharides. *Advances in Microbial Physiology*, *8*, 143–213. doi:10.1016/S0065-2911(08)60190-3 PMID:4581483

Sutherland, I. W. (1994). Structure-function relationships in microbial exopolysaccharides. *Biotechnology Advances*, *12*(2), 393–448. doi:10.1016/0734-9750(94)90018-3 PMID:14545899

Sutherland, I. W. (1998). Novel and established applications of microbial polysaccharides. *Trends in Biotechnology*, *16*(1), 41–46. doi:10.1016/S0167-7799(97)01139-6 PMID:9470230

Sutherland, I. W. (2001). Biofilm exopolysaccharides: A strong and sticky framework. *Microbiology*, *147*(1), 3–9. doi:10.1099/00221287-147-1-3 PMID:11160795

Sutton, N. B., Maphosa, F., Morillo, J. A., Abu Al-Soud, W., Langenhoff, A. A., Grotenhuis, T., Rijnaarts, H. H., & Smidt, H. (2013). Impact of long-term diesel contamination on soil microbial community structure. *Applied and Environmental Microbiology*, *79*(2), 619–630. doi:10.1128/AEM.02747-12 PMID:23144139

Suyal, D. C., Soni, R., Yadav, A. N., & Goel, R. (2021). Cold Adapted Microorganisms: Survival Mechanisms and Applications. In Microbiomes of Extreme Environments (pp. 177-191). CRC Press.

Suzuki, K. I., Sasaki, J., Uramoto, M., Nakase, T., & Komagata, K. (1997). Cryobacterium psychrophilum gen. nov., sp. nov., nom. rev., comb. nov., an obligately psychrophilic actinomycete to accommodate "Curtobacterium psychrophilum" Inoue and Komagata 1976. *International Journal of Systematic and Evolutionary Microbiology*, *47*(2), 474–478. PMID:9103638

Suzuki, Y., Haruki, M., Takano, K., Morikawa, M., & Kanaya, S. (2004). Possible involvement of an FKBP family member protein from a psychrotrophic bacterium *Shewanella* sp. SIB1 in cold adaptation. *European Journal of Biochemistry*, *271*(7), 1372–1381. doi:10.1111/j.1432-1033.2004.04049.x PMID:15030488

Suzuki, Y., Tsujimoto, Y., Matsui, H., & Watanabe, K. (2006). Decomposition of extremely hard-to-degrade animal proteins by thermophilic bacteria. *Journal of Bioscience and Bioengineering*, *102*(2), 73–81. doi:10.1263/jbb.102.73 PMID:17027867

Sysoev, M., Grötzinger, S. W., Renn, D., Eppinger, J., Rueping, M., & Karan, R. (2021). Bioprospecting of Novel Extremozymes From Prokaryotes—The Advent of Culture-Independent Methods. *Frontiers in Microbiology*, 12. PMID:33643258

Szczelkun, M. D., Tikhomirova, M. S., Sinkunas, T., Gasiunas, G., Karvelis, T., Pschera, P., Siksnys, V., & Seidel, R. (2014). Direct observation of R-loop formation by single RNA-guided Cas9 and Cascade e_ector complexes. *Proceedings of the National Academy of Sciences of the United States of America*, *111*(27), 9798–9803. doi:10.1073/pnas.1402597111 PMID:24912165

Taguchi, H., Konishi, J., Ishii, N., & Yoshida, M. A. (1991). Chaperonin from a thermophilic bacterium, *Thermus thermophilus*, that controls refolding of several thermophilic enzymes. *The Journal of Biological Chemistry*, *266*(33), 22411–22418. doi:10.1016/S0021-9258(18)54588-9 PMID:1682319

Takami, H., Nakamura, S., Aono, R., & Horikoshi, K. (1992). Degradation of Human Hair by a Thermostable Alkaline Protease from Alkaliphilic *Bacillus* sp. No. AH-101. *Bioscience, Biotechnology, and Biochemistry*, *56*(10), 1667–1669. doi:10.1271/bbb.56.1667

Talarico, L. A., Gil, M. A., Yomano, L. P., Ingram, L. O., & Maupin-Furlow, J. A. (2005). Construction and expression of an ethanol production operon in Gram-positive bacteria. *Microbiology*, *151*(12), 4023–4031. doi:10.1099/mic.0.28375-0 PMID:16339947

Tam, L. T., Antelmann, H., Eymann, C., Albrecht, D., Bernhardt, J., & Hecker, M.Tam le. (2006). Proteome signatures for stress and starvation in *Bacillus subtilis* as revealed by a 2-D gel image color coding approach. *Proteomics*, *6*(16), 4565–4585. doi:10.1002/pmic.200600100 PMID:16847875

Tamulaitis, G., Venclovas, C., & Siksnys, V. (2017). Type III CRISPR-Cas immunity: Major differences brushed aside. *Trends in Microbiology*, *25*(1), 49–61. doi:10.1016/j.tim.2016.09.012 PMID:27773522

Tanabe, H., Yoshihara, K., Tamura, K., Kobayashi, Y., Akamatsu, I., Niyomwan, N., & Footrakul, P. (1987). Pretreatment of pectic wastewater from orange canning process by an alkalophilic Bacillus sp. *Journal of Fermentation Technology*, *65*(2), 243–246. doi:10.1016/0385-6380(87)90173-7

Tan, C. W., Rukayadi, Y., Hasan, H., Thung, T. Y., Lee, E., Rollon, W. D., Hara, H., Kayali, A. Y., Nishibuchi, M., & Radu, S. (2020). Prevalence and antibiotic resistance patterns of *Vibrio parahaemolyticus* isolated from different types of seafood in Selangor, Malaysia. *Saudi Journal of Biological Sciences*, *27*(6), 1602–1608. doi:10.1016/j.sjbs.2020.01.002 PMID:32489301

Tan, D., Xue, Y. S., Aibaidula, G., & Chen, G. Q. (2011). Unsterile and continuous production of polyhydroxybutyrate by *Halomonas* TD01. *Bioresource Technology*, *102*(17), 8130–8136. doi:10.1016/j.biortech.2011.05.068 PMID:21680179

Tang, E. P., Tremblay, R., & Vincent, W. F. (1997). Cyanobacterial Dominance Of Polar Freshwater Ecosystems: Are High-Latitude Mat-Formers Adapted To Low Temperature? 1. *Journal of Phycology*, *33*(2), 171–181. doi:10.1111/j.0022-3646.1997.00171.x

Tang, X., Lowder, L. G., Zhang, T., Malzahn, A. A., Zheng, X., Voytas, D. F., Zhong, Z., Chen, Y., Ren, Q., Li, Q., Kirkland, E. R., Zhang, Y., & Qi, Y. (2017). A CRISPR-Cpf1 system for efficient genome editing and transcriptional repression in plants. *Nature Plants*, *3*(3), 17018. doi:10.1038/nplants.2017.18 PMID:28211909

Tang, Y. J., Sapra, R., Joyner, D., Hazen, T. C., Myers, S., Reichmuth, D., Blanch, H., & Keasling, J. D. (2009). Analysis of metabolic pathways and fluxes in a newly discovered thermophilic and ethanoltolerant *Geobacillus* strain. *Biotechnology and Bioengineering*, *102*(5), 1377–1386. doi:10.1002/bit.22181 PMID:19016470

Tanimura, K., Matsumoto, T., Nakayama, H., Tanaka, T., & Kondo, A. (2016). Improvement of ectoine productivity by using sugar transporter-overexpressing *Halomonas elongate*. *Enzyme and Microbial Technology*, *89*, 63–68. doi:10.1016/j.enzmictec.2016.03.006 PMID:27233128

Tapia-Vázquez, I., Sánchez-Cruz, R., Arroyo-Domínguez, M., Lira-Ruan, V., Sánchez-Reyes, A., Rayo Sánchez-Carbente, M. del, & Padilla-Chacón, D. (2020). Isolation and characterization of psychrophilic and psychrotolerant plant-growth promoting microorganisms from a high-altitude volcano crater in Mexico. *Microbiological Research, 232*, 126394. doi:10.1016/j.micres.2019.126394

Taylor, M. P., Eley, K. L., Martin, S., Tuffin, M. I., Burton, S. G., & Cowan, D. A. (2009a). Thermophilic ethanologenesis: Future prospects for second-generation bioethanol production. *Trends in Biotechnology*, *27*(7), 398–405. doi:10.1016/j.tibtech.2009.03.006 PMID:19481826

Taylor, M. P., Esteban, C. D., & Leak, D. J. (2008). Development of a versatile shuttle vector for gene expression in Geobacillus spp. *Plasmid*, *60*(1), 45–52. doi:10.1016/j.plasmid.2008.04.001 PMID:18501964

Taylor, M. P., van Zyl, L., Tuffin, I. M., Leak, D. J., & Cowan, D. A. (2011). Genetic tool development underpins recent advances in thermophilic whole-cell biocatalysts. *Microbial Biotechnology*, *4*(4), 438–448. doi:10.1111/j.1751-7915.2010.00246.x PMID:21310009

Taylor, T. J., & Vaisman, I. I. (2010). Discrimination of thermophilic and mesophilic proteins. *BMC Structural Biology*, *10*(1), 1–10. doi:10.1186/1472-6807-4-1 PMID:20487512

Tayyem, N. E. S. A., Elhello, K. T. M., & Elmanama, A. A. (2021). Isolation, identification and growth conditions of calcite producing bacteria from urea-rich soil. *African Journal of Microbiological Research*, *15*(1), 37–46. doi:10.5897/AJMR2020.9445

Tazi, L., Breakwell, D. P., Harker, A. R., & Crandall, K. A. (2014). Life in extreme environments: Microbial diversity in Great Salt Lake, Utah. *Extremophiles*, *18*(3), 525–535. doi:10.100700792-014-0637-x PMID:24682608

Tekle & Demeke. (2015). Review on microbial fuel cell. *Basic Research Journal of Microbiology*, 5-17.

Terra, L., Dyson, P. J., Hitchings, M. D., Thomas, L., Abdelhameed, A., Banat, I. M., Gazze, S. A., Vujaklija, D., Facey, P. D., Francis, L. W., & Quinn, G. A. (2018). A Novel Alkaliphilic *Streptomyces* Inhibits ESKAPE Pathogens. *Frontiers in Microbiology*, *9*, 2458. doi:10.3389/fmicb.2018.02458 PMID:30459722

Tetas, M., & Lowenstein, J. M. (1963). the effect of bivalent metal ions on the hydrolysis of adenosine Di-and triphosphate. *Biochemistry*, *2*(2), 350–357. doi:10.1021/bi00902a030 PMID:13980762

The 50 Best Inventions of 2009. (n.d.). *Time*. Retrieved from https://www.time.com/time/specials/packages/article/0,28804,1934027_1934003_1933965,00.html

Thomas, D. N. (2005). Photosynthetic microbes in freezing deserts. *Trends in Microbiology*, *13*(3), 87–88. doi:10.1016/j.tim.2004.11.002 PMID:15737723

Thomas-Hall, S., Watson, K., & Scorzetti, G. (2002). Cryptococcus statzelliae sp. nov. and three novel strains of Cryptococcus victoriae, yeasts isolated from Antarctic soils. *International Journal of Systematic and Evolutionary Microbiology*, *52*(6), 2303–2308. PMID:12508901

Thomas, L., Joseph, A., Singhania, R. R., Patel, A. K., & Pandey, A. (2017). Industrial enzymes: xylanases. In *Current developments in biotechnology and bioengineering* (pp. 127–148). Elsevier. doi:10.1016/B978-0-444-63662-1.00006-3

Thomas, T., & Cavicchioli, R. (1998). Archaeal cold-adapted proteins: Structural and evolutionary analysis of the elongation factor 2 proteins from psychrophilic, mesophilic and thermophilic methanogens. *FEBS Letters*, *439*(3), 281–286. doi:10.1016/S0014-5793(98)01375-1 PMID:9845338

Thombre, R. S., Bhalerao, A. R., Shinde, V. D., Dhar, S. K., & Shouche, Y. S. (2017). Response of Haloalkaliphilic Archaeon *Natronococcus Jeotgali* RR17 to Hypergravity. *Microgravity Science and Technology*, *29*(3), 191–200. doi:10.100712217-017-9538-9

Thombre, R. S., Shinde, V. D., Oke, R. S., Dhar, S. K., & Shouche, Y. S. (2016a). Biology and survival of extremely halophilic archaeon *Haloarcula marismortui* RR12 isolated from Mumbai salterns, India in response to salinity stress. *Scientific Reports*, *6*(1), 25642. doi:10.1038rep25642 PMID:27231230

Thombre, R. S., Shinde, V. D., Thaiparambil, E., Zende, S., & Mehta, S. (2016b). Antimicrobial activity and mechanism of inhibition of silver nanoparticles against extreme halophilic archaea. *Frontiers in Microbiology*, *7*, 1424. doi:10.3389/fmicb.2016.01424 PMID:27679615

Thongthai, C., McGenity, T., Suntinanalert, P., & Grant, W. (1992). Isolation and characterization of an extremely halophilic archaeobacterium from traditionally fermented Thai fish sauce (nam pla). *Letters in Applied Microbiology*, *14*(3), 111–114. doi:10.1111/j.1472-765X.1992.tb00661.x

Timms, B. V. (2021). Drivers restricting biodiversity in Australian saline lakes: A review. *Marine and Freshwater Research*, *72*(4), 462–468. doi:10.1071/MF20205

Tkach, O., Sangeetha, T., Maria, S., & Wang, A. (2017). Performance of low temperature Microbial Fuel Cells (MFCs) catalyzed by mixed bacterial consortia. *Journal of Environmental Sciences (China)*, *52*, 284–292. doi:10.1016/j.jes.2016.11.006 PMID:28254049

Toplak, A., Wu, B., Fusetti, F., Quaedflieg, P. J. L. M., & Janssen, D. B. (2013). Proteolysin, a novel highly thermostable and cosolvent-compatible protease from the thermophilic bacterium Coprothermobacter proteolyticus. *Applied and Environmental Microbiology*, *79*(18), 5625–5632. doi:10.1128/AEM.01479-13 PMID:23851086

Toral, L., Rodríguez, M., Béjar, V., & Sampedro, I. (2018). Antifungal Activity of Lipopeptides From Bacillus XT1 CECT 8661 Against Botrytis cinerea. *Frontiers in Microbiology, 9*. doi:10.3389/fmicb.2018.01315

Torgbo, S., & Sukyai, P. (2018). Bacterial cellulose-based scaffold materials for bone tissue engineering. *Applied Materials Today*, *11*, 34–49. doi:10.1016/j.apmt.2018.01.004

Torracchi, J. E., Morel, M. A., Tapia-Vázquez, I., Castro-Sowinski, S., Batista-García, R. A., & R., L. A. Y. (2020). Fighting plant pathogens with cold-active microorganisms: biopesticide development and agriculture intensification in cold climates. *Applied Microbiology and Biotechnology, 104*(19), 8243–8256. doi:10.1007/s00253-020-10812-8

Torres, M., Llamas, I., Torres, B., Toral, L., Sampedro, I., & Béjar, V. (2020). Growth promotion on horticultural crops and antifungal activity of Bacillus velezensis XT1. *Applied Soil Ecology, 150*, 103453. doi:10.1016/j.apsoil.2019.103453

Torres, C. A., Antunes, S., Ricardo, A. R., Grandfils, C., Alves, V. D., Freitas, F., & Reis, M. A. (2012). Study of the interactive effect of temperature and pH on exopolysaccharide production by *Enterobacter* A47 using multivariate statistical analysis. *Bioresource Technology, 119*, 148–156. doi:10.1016/j.biortech.2012.05.106 PMID:22728195

Torres-Díaz, C., Gallardo-Cerda, J., Lavin, P., Oses, R., Carrasco-Urra, F., Atala, C., & Acuña-Rodríguez, I. S. (2016). Biological Interactions and Simulated Climate Change Modulates the Ecophysiological Performance of Colobanthus quitensis in the Antarctic Ecosystem. PLOS ONE, 11(10), e0164844. doi:10.1371/journal.pone.0164844

Trabelsi, S., Ben Mabrouk, S., Kriaa, M., Ameri, R., Sahnoun, M., Mezghani, M., & Bejar, S. (2019). The optimized production, purification, characterization, and application in the bread making industry of three acid-stable alpha-amylases isoforms from a new isolated *Bacillus subtilis* strain US586. *Journal of Food Biochemistry, 43*(5), 1–13. doi:10.1111/jfbc.12826 PMID:31353531

Trent, J. D. (1996). A review of acquired thermotolerance, heat-shock proteins, and molecular chaperones in archaea. *FEMS Microbiology Reviews, 18*(2-3), 249–258. doi:10.1111/j.1574-6976.1996.tb00241.x

Tripathi, M., Singh, D. N., Vikram, S., Singh, V. S., & Kumar, S. (2018). Metagenomic approach towards bioprospection of novel biomolecule (s) and environmental bioremediation. *Annual Research & Review in Biology, 22*(2), 1–12. doi:10.9734/ARRB/2018/38385

Trivedi, S., RamaRao, S., & Gehlot, H.S. (2005). Nucleic Acid stability in thermophilic prokaryotes: A review. *Journal of Cell & Molecular Biology, 4*, 61–69.

Tse, C., & Ma, K. (2016). Growth and Metabolism of Extremophilic Microorganisms. In Biotechnology of Extremophiles, Grand Challenges in Biology and Biotechnology. Springer. doi:10.1007/978-3-319-13521-2_1

Turner, P., Mamo, G., & Karlsson, E. N. (2007). Potential and utilization of thermophiles and thermostable enzymes in biorefining. *Microbial Cell Factories, 6*(1), 1–23. doi:10.1186/1475-2859-6-9 PMID:17359551

Tyagi, B., & Kumar, N. (2021). Bioremediation: Principles and applications in environmental management. In Bioremediation for environmental sustainability (pp. 3–28). Elsevier.

Ulloa, G., Collao, B., Araneda, M., Escobar, B., Álvarez, S., Bravo, D., & Pérez-Donoso, J. M. (2016). Use of acidophilic bacteria of the genus Acidithiobacillus to biosynthesize CdS fluorescent nanoparticles (quantum dots) with high tolerance to acidic pH. *Enzyme and Microbial Technology, 95*, 217–224. doi:10.1016/j.enzmictec.2016.09.005 PMID:27866618

Unissa, R., Sudhakar, M., & Reddy, A. S. K. (2016). Evaluation of *in-vitro* Anti-proliferative activity of L-arginine deiminase from Novel Marine Bacterial Isolate. *Microbiology Research Journal International, 13*(5), 1–10. doi:10.9734/BMRJ/2016/23592

Upadhyay, K. H., Vaishnav, A. M., Tipre, D. R., Patel, B. C., & Dave, S. R. (2017). Kinetics and mechanisms of mercury biosorption by an exopolysaccharide producing marine isolate *Bacillus licheniformis. Biotech, 7*(5), 1-10.

US6855734B2 - Glycine betaine and its use - Google Patents. (n.d.). Retrieved August 15, 2021, from https://patents.google.com/patent/US6855734B2/en

Ushasree, M. V., Vidya, J., & Pandey, A. I. (2017). Other enzymes: phytases. In *Current developments in biotechnology and bioengineering* (pp. 309–333). Elsevier. doi:10.1016/B978-0-444-63662-1.00014-2

Uthandi, S., Saad, B., Humbard, M. A., & Maupin-Furlow, J. A. (2010). LccA, an archaeal laccase secreted as a highly stable glycoprotein into the extracellular medium by *Haloferax volcanii*. *Applied and Environmental Microbiology*, *76*(3), 733–743. doi:10.1128/AEM.01757-09 PMID:19966030

Utobo, E. B., & Tewari, L. T. L. (2015). Soil enzymes as bioindicators of soil ecosystem status. *Applied Ecology and Environmental Research, 13*(1). Advance online publication. 10.15666%2Faeer%2F1301_147169

VA, S., RB, P., PV, P. (2017). Comparative Study of Antimicrobial Potentials of Phospholipid Compound Produced by Halophilic and Alkaliphiles *Bacillus subtilis* isolated from Alkaline Meteorite Crater. *International Journal of Life Sciences (Amravati)*, *5*, 420–424.

Vainberg, I. E., Lewis, S. A., Rommelaere, H., Ampe, C., Vandekerckhove, J., Klein, H. L., & Cowan, N. J. (1998). Prefoldin, a chaperone that delivers unfolded proteins to cytosolic chaperonin. *Cell*, *93*(5), 863–873. doi:10.1016/S0092-8674(00)81446-4 PMID:9630229

Valones, M. A. A., Guimarães, R. L., Brandão, L. A. C., Souza, P. R. E., Carvalho, A. A. T., & Crovela, S. (2009). Principles and applications of polymerase chain reaction in medical diagnostic fields: A review. *Brazilian Journal of Microbiology*, *40*(1), 1–11. doi:10.1590/S1517-83822009000100001 PMID:24031310

Valverde, A., Tuffin, M., & Cowan, D. A. (2012). Biogeography of bacterial communities in hot springs: A focus on the actinobacteria. *Extremophiles*, *16*(4), 669–679. doi:10.100700792-012-0465-9 PMID:22711422

Van De Voorde, I., Goiris, K., Syryn, E., Van den Bussche, C., & Aerts, G. (2014). Evaluation of the cold-active *Pseudoalteromonas haloplanktis* β-galactosidase enzyme for lactose hydrolysis in whey permeate as primary step of d-tagatose production. *Process Biochemistry*, *49*(12), 2134–2140. doi:10.1016/j.procbio.2014.09.010

van der Oost, J., & Siebers, B. (2007). The glycolytic pathways of Archaea: evolution by tinkering. *Archaea: Evolution, Physiology and Molecular Biology, 22*, 247-260.

van der Oost, J., Schut, G., Kengen, S. M., Hagen, W. R., Thomm, M., & de Vos, W. M. (1998). The ferredoxin-dependent conversion of glyceraldehyde-3-phosphate in the hyperthermophilic archaeon *Pyrococcus furiosus* represents a novel site of glycolytic regulation. *The Journal of Biological Chemistry*, *273*(43), 28149–28154. doi:10.1074/jbc.273.43.28149 PMID:9774434

van Hijum, S. A., Kralj, S., Ozimek, L. K., Dijkhuizen, L., & van Geel-Schutten, I. G. (2006). Structure-function relationships of glucansucrase and fructansucrase enzymes from lactic acid bacteria. *Microbiology and Molecular Biology Reviews*, *70*(1), 157–176. doi:10.1128/MMBR.70.1.157-176.2006 PMID:16524921

van Hullebusch, E. D., Peerbolte, A., Zandvoort, M. H., & Lens, P. N. (2005). Sorption of cobalt and nickel on anaerobic granular sludges: Isotherms and sequential extraction. *Chemosphere*, *58*(4), 493–505. doi:10.1016/j.chemosphere.2004.09.017 PMID:15620741

van Hullebusch, E. D., Zandvoort, M. H., & Lens, P. N. (2003). Metal immobilisation by biofilms: Mechanisms and analytical tools. *Reviews in Environmental Science and Biotechnology*, *2*(1), 9–33. doi:10.1023/B:RESB.0000022995.48330.55

Van Vlierberghe, S., Dubruel, P., & Schacht, E. (2011). Biopolymer-based hydrogels as scaffolds for tissue engineering applications: A review. *Biomacromolecules*, *12*(5), 1387–1408. doi:10.1021/bm200083n PMID:21388145

Vane, L. M., & Alvarez, F. R. (2008). Membrane-assisted vapor stripping: Energy efficient hybrid distillation-vapor permeation process for alcohol-water separation. *Journal of Chemical Technology and Biotechnology (Oxford, Oxfordshire), 83*(9), 1275–1287. doi:10.1002/jctb.1941

Vaningelgem, F., Zamfir, M., Mozzi, F., Adriany, T., Vancanneyt, M., Swings, J., & De Vuyst, L. (2004). Biodiversity of exopolysaccharides produced by *Streptococcus thermophilus* strains is reflected in their production and their molecular and functional characteristics. *Applied and Environmental Microbiology, 70*(2), 900–912. doi:10.1128/AEM.70.2.900-912.2004 PMID:14766570

Van-Thuoc, D., Hashim, S. O., Hatti-Kaul, R., & Mamo, G. (2013). Ectoine-mediated protection of enzyme from the effect of pH and temperature stress: A study using *Bacillus halodurans* xylanase as a model. *Applied Microbiology and Biotechnology, 97*(14), 6271–6278. doi:10.100700253-012-4528-8 PMID:23132342

Vardar-Schara, G., Maeda, T., & Wood, T. K. (2008). Metabolically engineered bacteria for producing hydrogen via fermentation. *Microbial Biotechnology, 1*(2), 107–125. doi:10.1111/j.1751-7915.2007.00009.x PMID:21261829

Vargas, C., Jebbar, M., Carrasco, R., Blanco, C., Calderón, M. I., Iglesias-Guerra, F., & Nieto, J. J. (2006). Ectoines as compatible solutes and carbon and energy sources for the halophilic bacterium Chromohalobacter salexigens. *Journal of Applied Microbiology, 100*(1), 98–107. doi:10.1111/j.1365-2672.2005.02757.x PMID:16405689

Varma, V. S., & Kalamdhad, A. S. (2015). Evolution of chemical and biological characterization during thermophilic composting of vegetable waste using rotary drum composter. *International Journal of Environmental Science and Technology, 12*(6), 2015–2024. Advance online publication. doi:10.100713762-014-0582-3

Varshney, P., Mikulic, P., Vonshak, A., Beardall, J., & Wangikar, P. P. (2015). Extremophilic micro-algae and their potential contribution in biotechnology. *Bioresource Technology, 184*, 363–372. doi:10.1016/j.biortech.2014.11.040

Vashist, S., & Sharma, R. (2018). Why Settle for Mediocre, When Extremophiles Exist? In *Extremophiles in Eurasian Ecosystems: Ecology, Diversity, and Applications* (pp. 435–451). Springer. doi:10.1007/978-981-13-0329-6_16

Vasudevan, N., & Jayshree, A. (2020). Extremozymes and Extremoproteins in Biosensor Applications. Encyclopedia of Marine Biotechnology, 1711–1736. doi:10.1002/9781119143802.ch72

Vega, C. A., & Fernández, I. (1987). Mediating effect of ferric chelate compounds in microbial fuel cells with *Lactobacillus plantarum, Streptococcus lactis*, and *Erwinia dissolvens. Bioelectrochemistry and Bioenergetics, 17*(2), 217–222. doi:10.1016/0302-4598(87)80026-0

Veiga-Crespo, P., Blasco, L., Rosa-Dos-Santos, F., Poza, M., & Villa, T. G. (2005). Influence of culture conditions of Gordonia jacobaea MV-26 on canthaxanthin production. *International Microbiology : The Official Journal of the Spanish Society for Microbiology, 8*(1), 55–58. PMID:15906262

Venclovas, C. (2016). Structure of Csm2 elucidates the relationship between small subunits of CRISPR-Cas effector complexes. *FEBS Letters, 590*(10), 1521–1529. doi:10.1002/1873-3468.12179 PMID:27091242

Venkata Mohan, S., Lalit Babu, V., & Sarma, P. N. (2007a). Anaerobic biohydrogen production from dairy wastewater treatment in sequencing batch reactor (AnSBR): Effect of organic loading rate. *Enzyme and Microbial Technology, 41*(4), 506–515. doi:10.1016/j.enzmictec.2007.04.007

Venkata Mohan, S., Raghavulu, S., Srikanth, S., & Sarma, P. (2007b). Bioelectricity production by mediatorless microbial fuel cell under acidophilic condition using wastewater as substrate: Influence of substrate loading rate. *Current Science, 92*(12), 1720–1726.

Venugopal, V. (2016). Enzymes from seafood processing waste and their applications in seafood processing. *Advances in Food and Nutrition Research, 78*, 47–69. doi:10.1016/bs.afnr.2016.06.004 PMID:27452165

Vera, M., Schippers, A., & Sand, W. (2013). Progress in bioleaching: Fundamentals and mechanisms of bacterial metal sulfide oxidation—part A. *Applied Microbiology and Biotechnology, 97*(17), 7529–7541. doi:10.100700253-013-4954-2 PMID:23720034

Verhees, C., Kengen, S., Tuininga, J., Schut, G., Adams, M., de Vos, W., & Van der Oost, J. (2004). The unique features of glycolytic pathways in archaea. *Biochemistry, 377*(3), 819–822. doi:10.1042/bj3770819

Verma, P., Yadav, A. N., Khannam, K. S., Mishra, S., Kumar, S., Saxena, A. K., & Suman, A. (2019). Appraisal of diversity and functional attributes of thermotolerant wheat associated bacteria from the peninsular zone of India. *Saudi Journal of Biological Sciences, 26*(7), 1882–1895. doi:10.1016/j.sjbs.2016.01.042

Verma, S., Kumar, R., Kumar, P., Sharma, D., Gahlot, H., Sharma, P. K., & Meghwanshi, G. K. (2020). Cloning, Characterization, and Structural Modeling of an Extremophilic Bacterial Lipase Isolated from Saline Habitats of the Thar Desert. *Applied Biochemistry and Biotechnology, 192*(2), 557–572. doi:10.1007/s12010-020-03329-3

Verma, P., Yadav, A. N., Khannam, K. S., Kumar, S., Saxena, A. K., & Suman, A. (2016). Molecular diversity and multifarious plant growth promoting attributes of Bacilli associated with wheat (Triticum aestivum L.) rhizosphere from six diverse agro-ecological zones of India. *Journal of Basic Microbiology, 56*(1), 44–58. doi:10.1002/jobm.201500459 PMID:26567901

Vibrio vulnificus in aquariums is a novel threat to marine mammals and public health. (n.d.). Retrieved July 7, 2021, from https://www.ncbi.nlm.nih.gov/pmc/articles/PMC7169869/

Vibrio vulnificus: Review of Mild to Life-threatening Skin Infections. (n.d.). *MDedge Dermatology.* Retrieved July 7, 2021, from https://www.mdedge.com/dermatology/article/236136/infectious-diseases/*Vibrio-vulnificus*-review-mild-life-threatening-skin

Vieille, C., & Zeikus, G. J. (2001). Hyperthermophilic enzymes: Sources, uses, and molecular mechanisms for thermostability. *Microbiology and Molecular Biology Reviews, 65*(1), 1–43. doi:10.1128/MMBR.65.1.1-43.2001 PMID:11238984

Villarreal, P., Carrasco, M., Barahona, S., Alcaíno, J., Cifuentes, V., & Baeza, M. (2016). Tolerance to ultraviolet radiation of psychrotolerant yeasts and analysis of their carotenoid, mycosporine, and ergosterol content. *Current Microbiology, 72*(1), 94–101. doi:10.100700284-015-0928-1 PMID:26483082

Vincent, W. F., Hobbie, J. E., & Laybourn-Parry, J. (2008). Introduction to the limnology of high-latitude lake and river ecosystems. Polar lakes and rivers: limnology of Arctic and Antarctic aquatic ecosystems, 1-24. doi:10.1093/acprof:oso/9780199213887.003.0001

Vincent, S. J., Faber, E. J., Neeser, J.-R., Stingele, F., & Kamerling, J. P. (2001). Structure and properties of the exopolysaccharide produced by Streptococcus macedonicus Sc136. *Glycobiology, 11*(2), 131–139. doi:10.1093/glycob/11.2.131 PMID:11287400

Vincent, W. F. (2007). Cold tolerance in cyanobacteria and life in the cryosphere. In *Algae and cyanobacteria in extreme environments* (pp. 287–301). Springer. doi:10.1007/978-1-4020-6112-7_15

Vishniac, H. S. (1985). Cryptococcus socialis sp. nov. and Cryptococcus consortionis sp. nov., Antarctic basidioblastomycetes. *International Journal of Systematic and Evolutionary Microbiology, 35*(1), 119–122. PMID:11540820

Vishniac, H. S., & Kurtzman, C. P. (1992). Cryptococcus antarcticus sp. nov. and Cryptococcus albidosimilis sp. nov., basidioblastomycetes from Antarctic soils. *International Journal of Systematic and Evolutionary Microbiology, 42*(4), 547–553.

Voets, I. K. (2017). From ice-binding proteins to bio-inspired antifreeze materials. *Soft Matter, 13*(28), 4808–4823. doi:10.1039/C6SM02867E PMID:28657626

Vogler, M., Karan, R., Renn, D., Vancea, A., Vielberg, M.-T., Grötzinger, S. W., DasSarma, P., DasSarma, S., Eppinger, J., Groll, M., & Rueping, M. (2020). Crystal Structure and Active Site Engineering of a Halophilic γ-Carbonic Anhydrase. *Frontiers in Microbiology, 0*, 742. doi:10.3389/fmicb.2020.00742 PMID:32411108

Vögtlin, A., Fraefel, C., Albini, S., Leutenegger, C. M., Schraner, E., Spiess, B., Lutz, H., & Ackermann, M. (2002). Quantification of feline herpesvirus 1 DNA in ocular fluid samples of clinically diseased cats by real-time TaqMan PCR. *Journal of Clinical Microbiology, 40*(2), 519–523. doi:10.1128/JCM.40.2.519-523.2002 PMID:11825966

Voica, D. M., Bartha, L., Banciu, H. L., & Oren, A. (2016). Heavy metal resistance in halophilic Bacteria and Archaea. *FEMS Microbiology Letters, 363*(14), fnw146. Advance online publication. doi:10.1093/femsle/fnw146 PMID:27279625

Vyas, K. M., Jadeja, R. N., Patel, D., Devkar, R. V., & Gupta, V. K. (2014). Article. *Polyhedron, 80*, 20–33. doi:10.1016/j.poly.2013.12.037

Waditee-Sirisattha, R., Kageyama, H., Takabe, T., Waditee-Sirisattha, R., Kageyama, H., & Takabe, T. (2016). Halophilic microorganism resources and their applications in industrial and environmental biotechnology. *AIMS Microbiology, 2*(1), 42–54. doi:10.3934/microbiol.2016.1.42

Wadkar, D. V., Modak, P. R., & Chavan, V. S. (2013). Aerobic Thermophilic Composting of Municipal Solid Waste. *International Journal of Engineering Science and Technology, 5*(3), 716–718.

Wahlund, T. M., Woese, C. R., Castenholz, R. W., & Madigan, M. T. (1991). A thermophilic green sulfur bacterium from New Zealand hot springs, *Chlorobium tepidum* sp. nov. *Archives of Microbiology, 156*(2), 81–90. doi:10.1007/BF00290978

Wainwright, M., Wickramasinghe, N. C., Narlikar, J. V., & Rajaratnam, P. (2003). Microorganisms cultured from stratospheric air samples obtained at 41 km. *FEMS Microbiology Letters, 218*(1), 161–165. doi:10.1111/j.1574-6968.2003.tb11513.x PMID:12583913

Waksman, S. A., Schatz, A., & Reynolds, D. M. (2010). Production of antibiotic substances by actinomycetes. *Annals of the New York Academy of Sciences, 1213*(1), 112–124. doi:10.1111/j.1749-6632.2010.05861.x PMID:21175680

Waldhuber, S., Sattler, B., Semmler, J., Wille, A., & Psenner, R. (2002, November). Proceedings of the 2nd Eurpean Workshop on Exoas- trobiology. Comparative 16S rDNA analysis of extremophiles in the winter cover of a high mountain lake. In *Exo-Astrobiology* (Vol. 518, pp. 489-490). Academic Press.

Walker, J., Lanahan, A., Zheng, T., Toruno, C., Lynd, L., Cameron, J., Olson, D., & Eckert, C. (2020). Development of both type I–B and type II CRISPR/Cas genome editing systems in the cellulolytic bacterium *Clostridium thermocellum. Metabolic Engineering Communications, 10*, e00116. doi:10.1016/j.mec.2019.e00116 PMID:31890588

Walker, M., & Phillips, C. A. (2007). Alicyclobacillus acidoterrestris: An increasing threat to the fruit juice industry? *International Journal of Food Science & Technology, 43*(2), 250–260. doi:10.1111/j.1365-2621.2006.01427.x

Wallenstein, M. D., McMahon, S., & Schimel, J. (2007). Bacterial and fungal community structure in Arctic tundra tussock and shrub soils. *FEMS Microbiology Ecology, 59*(2), 428–435. doi:10.1111/j.1574-6941.2006.00260.x PMID:17313585

Wang, C. Y., Chang, C. C., Ng, C. C., Chen, T. W., & Shyu, Y. T. (2008). *Virgibacillus chiguensis* sp. nov., a novel halophilic bacterium isolated from Chigu, a previously commercial saltern located in southern Taiwan. *International Journal of Systematic and Evolutionary Microbiology*, *58*(2), 341–345. doi:10.1099/ijs.0.64996-0 PMID:18218928

Wang, J., Li, J., Zhao, H., Sheng, G., Wang, M., Yin, M., & Wang, Y. (2015). Structural and Mechanistic Basis of PAM-Dependent Spacer Acquisition in CRISPR-Cas Systems. *Cell*, *163*(4), 840–853. doi:10.1016/j.cell.2015.10.008 PMID:26478180

Wang, J., Salem, D. R., & Sani, R. K. (2021). Two new exopolysaccharides from a thermophilic bacterium *Geobacillus* sp. WSUCF1: Characterization and bioactivities. *New Biotechnology*, *61*, 29–39. doi:10.1016/j.nbt.2020.11.004 PMID:33188978

Wang, J., Zhao, C., Meng, B., Xie, J., Zhou, C., Chen, X., Zhao, K., Shao, J., Xue, Y., Xu, N., Ma, Y., & Liu, S. (2007). The proteomic alterations of *Thermoanaerobacter tengcongensis* cultured at different temperatures. *Proteomics*, *7*(9), 1409–1419. doi:10.1002/pmic.200500226 PMID:17469076

Wang, L., Wang, Y., Li, Q., Tian, K., Xu, L., Liu, G., & Guo, C. (2019). Exopolysaccharide, Isolated From a Novel Strain *Bifidobacterium breve* lw01 Possess an Anticancer Effect on Head and Neck Cancer–Genetic and Biochemical Evidences. *Frontiers in Microbiology*, *10*, 1044. doi:10.3389/fmicb.2019.01044 PMID:31143171

Wang, Q., Cen, Z., & Zhao, J. (2013). The survival mechanisms of thermophiles at high temperatures: An angle of omics. *Physiology (Bethesda, MD)*, *30*(2), 97–106. doi:10.1152/physiol.00066.2013 PMID:25729055

Wang, W. B., Chua, S.-L., Cao, B., Seviour, T., Nesatyy, V. J., Marsili, E., Kjelleberg, S., Givskov, M., Tolker-Nielsen, T., Song, H., Loo, J. S. C., & Yang, L. (2013). Engineering PQS Biosynthesis Pathway for Enhancement of Bioelectricity Production in *Pseudomonas aeruginosa* Microbial Fuel Cells. *PLoS One*, *8*(5), e63129. doi:10.1371/journal.pone.0063129 PMID:23700414

Wang, Z., Tong, W., Wang, Q., Bai, X., Chen, Z., Zhao, J., Xu, N., & Liu, S. (2012). The temperature dependent proteomic analysis of *Thermotoga maritima*. *PLoS One*, *7*(10), e46463. doi:10.1371/journal.pone.0046463 PMID:23071576

Wartiainen, I., Hestnes, A. G., & Svenning, M. M. (2003). Methanotrophic diversity in high arctic wetlands on the islands of Svalbard (Norway)-denaturing gradient gel electrophoresis analysis of soil DNA and enrichment cultures. *Canadian Journal of Microbiology*, *49*(10), 602–612. doi:10.1139/w03-080 PMID:14663494

Wei, N., Quarterman, J., & Jin, Y. S. (2013). Marine macro algae: An untapped resource for producing fuels and chemicals. *Trends in Biotechnology*, *31*(2), 70–77. doi:10.1016/j.tibtech.2012.10.009 PMID:23245657

Wei, W., Wang, K., Hu, X., Yang, S., Zeng, C., Hou, Z., Liu, S., Cui, H., & Zhu, L. (2021). *Phytoactinopolyspora limicola* sp. nov., an alkaliphilic actinomycete isolated from a soda alkali-saline soil. *Archives of Microbiology*, *203*(4), 1367–1374. doi:10.100700203-020-02134-0 PMID:33386868

Wei, X., Jiang, Y., Chen, X., Jiang, Y., & Lai, H. (2015). *Amycolatopsis flava* sp. nov., a halophilic actinomycete isolated from Dead Sea. *Antonie van Leeuwenhoek*, *108*(4), 879–885. doi:10.100710482-015-0542-z PMID:26233655

Welsh, D. T. (2000). Ecological significance of compatible solute accumulation by micro-organisms: From single cells to global climate. *FEMS Microbiology Reviews*, *24*(3), 263–290. doi:10.1111/j.1574-6976.2000.tb00542.x PMID:10841973

Werning, M. L., Notararigo, S., Nácher, M., Fernández de Palencia, P., Aznar, R., & López, P. (2012). Biosynthesis, purification and biotechnological use of exopolysaccharides produced by lactic acid bacteria. *Food Additives*, 83-114.

Whitaker, R. J., Grogan, D. W., & Taylor, J. W. (2003). Geographic barriers isolate endemic populations of hyperthermophilic archaea. *Science*, *301*(5635), 976–978. doi:10.1126cience.1086909 PMID:12881573

Whitman, W. B., Coleman, D. C., & Wiebe, W. J. (1998). Prokaryotes: The unseen majority. *Proceedings of the National Academy of Sciences of the United States of America, 95*(12), 6578–6583. doi:10.1073/pnas.95.12.6578 PMID:9618454

Wicklow, D. T., & Soderstrom, B. E. (1997). *Environmental and microbial relationships*. Springer.

Wiedenheft, B., Sternberg, S. H., & Doudna, J. A. (2012). RNA-guided genetic silencing systems in bacteria and archaea. *Nature, 482*(7385), 331–338. doi:10.1038/nature10886 PMID:22337052

Wilson, M., Tillery, L., Tabaie, E., Beery, G., Preusker, J., Bhola, I., & Frato, K. (2020). Alanine to serine substitutions drive thermal adaptation in a psychrophilic diatom cytochrome c6. *JBIC Journal of Biological Inorganic Chemistry, 25*(3), 489–500. doi:10.1007/s00775-020-01777-0

Windish, W. W., & Mhatre, N. S. (1965). Microbial amylases. *Advances in Applied Microbiology, 7*, 273–304. doi:10.1016/S0065-2164(08)70389-7 PMID:5321878

Woese, C. R., Kandler, O., & Wheelis, M. L. (1990). Towards a natural system of organisms: Proposal for the domains Archaea, Bacteria and Eucarya. *Proceedings of the National Academy of Sciences of the United States of America, 87*(12), 4576–4579. doi:10.1073/pnas.87.12.4576 PMID:2112744

Wong, J. X., & Rehm, B. H. (2018). Design of modular polyhydroxyalkanoate scaffolds for protein immobilization by directed ligation. *Biomacromolecules, 19*(10), 4098–4112. doi:10.1021/acs.biomac.8b01093 PMID:30199627

Wood, S. A., Rueckert, A., Cowan, D. A., & Cary, S. C. (2008). Sources of edaphic cyanobacterial diversity in the Dry Valleys of Eastern Antarctica. *The ISME Journal, 2*(3), 308–320. doi:10.1038/ismej.2007.104 PMID:18239611

Wu, S., Liu, B., & Zhang, X. (2006). Characterization of a recombinant thermostable xylanase from deep-sea thermophilic Geobacillus sp. MT-1 in East Pacific. *Applied Microbiology and Biotechnology, 72*(6), 1210–1216. doi:10.100700253-006-0416-4 PMID:16607523

Xavier, K. B., Martins, L. O., Peist, R., Kossmann, M., Boos, W., & Santos, H. (1996). High-affinity maltose/ trehalose transport system in the hyperthermophilic archaeon *Thermococcus litoralis*. *Journal of Bacteriology, 178*(16), 4773–4777. doi:10.1128/jb.178.16.4773-4777.1996 PMID:8759837

Xiao, Z., Zhang, Y., Xi, L., Huo, F., Zhao, J. Y., & Li, J. (2015). Thermophilic production of polyhydroxyalkanoates by a novel *Aneurinibacillus* strain isolated from Gudao oilfield, China. *Journal of Basic Microbiology, 55*(9), 1125–1133. doi:10.1002/jobm.201400843 PMID:25832555

Xia, Y., He, L., Mao, J., Fang, P., Ma, X., & Wang, Z. (2018). Purification, characterization, and gene cloning of a new cold-adapted β-galactosidase from Erwinia sp. E602 isolated in northeast China. *Journal of Dairy Science, 101*(8), 6946–6954. doi:10.3168/jds.2018-14605 PMID:29859695

Xue, T., Ni, J., Shang, F., Chen, X., & Zhang, M. (2015). Autoinducer-2 increases biofilm formation via an ica- and bhp-dependent manner in Staphylococcus epidermidis RP62A. *Microbes and Infection, 17*(5), 345–352. doi:10.1016/j.micinf.2015.01.003 PMID:25637952

Yadav & Patil. (2020). Microbial electroactive biofilms dominated by *Geoalkalibacter spp.* from a highly saline–alkaline environment. *NPJ Biofilms Microbiomes, 6*(1), 38.

Yadav, A. N. (2017). Beneficial role of extremophilic microbes for plant health and soil fertility. *Journal of Agricultural Science and Botany, 1*(1). doi:10.35841/2591-7897.1.1.30-33

Yadav, A. N. (2017). Extreme Cold Environments: A Suitable Niche for Selection of Novel Psychrotrophic Microbes for Biotechnological Applications. *Advances in Biotechnology & Microbiology, 2*(2). doi:10.19080/AIBM.2017.02.555584

Yadav, A. N., Kaur, T., Devi, R., Kour, D., & Yadav, N. (2021). Biodiversity and Biotechnological Applications of Extremophilic Microbiomes: Current Research and Future Challenges. Microbiomes of Extreme Environments, 278-290.

Yadav, A. N., Verma, P., Kumar, S., Kumar, V., Kumar, M., Sugitha, T. C. K., Singh, B. P., Saxena, A. K., & Dhaliwal, H. S. (2018). Actinobacteria from rhizosphere: molecular diversity, distributions, and potential biotechnological applications. In *New and future developments in microbial biotechnology and bioengineering* (pp. 13–41). Elsevier. doi:10.1016/B978-0-444-63994-3.00002-3

Yakovlev, A. G., Taisova, A. S., & Fetisova, Z. G. (2021). Utilization of blue-green light by chlorosomes from the photosynthetic bacterium *Chloroflexus aurantiacus*: Ultrafast excitation energy conversion and transfer. *Biochimica et Biophysica Acta (BBA) -. Bioenergetics, 1862*(6), 148396. doi:10.1016/j.bbabio.2021.148396 PMID:33581107

Yamamoto, Y., Takahashi, Y., Kawano, M., Iizuka, M., Matsumoto, T., Saeki, S., & Yamaguchi, H. (1999). *In vitro* digestibility and fermentability of levan and its hypocholesterolemic effects in rats. *The Journal of Nutritional Biochemistry, 10*(1), 13–18. doi:10.1016/S0955-2863(98)00077-1 PMID:15539245

Yamano, T., Zetsche, B., Ishitani, R., Zhang, F., Nishimasu, H., & Nureki, O. (2017). Structural Basis for the Canonical and Non-canonical PAM Recognition by CRISPR-Cpf1. *Molecular Cell, 67*(4), 633–645. doi:10.1016/j.molcel.2017.06.035 PMID:28781234

Yang, J. S., Yang, F. L., Yang, Y., Xing, G. L., Deng, C. P., Shen, Y. T., & Luo, L. Q. (2016). A proposal of core enzyme bioindicator in long-term Pb-Zn ore pollution areas based on topsoil property analysis. *Environmental Pollution, 213*, 760–769. doi:10.1016/j.envpol.2016.03.030

Yang, H., Ma, Y., Zhao, Y., Shen, W., & Chen, X. (2020). Systematic engineering of transport and transcription to boost alkaline α-amylase production in *Bacillus subtilis. Applied Microbiology and Biotechnology, 104*(7), 2973–2985. doi:10.100700253-020-10435-z PMID:32043188

Yang, S.-J., Ko, J.-A., Kim, H.-S., Jo, M.-H., Lee, H.-N., Park, B.-R., & Kim, Y. (2018). Biochemical Characterization of Alkaliphilic Cyclodextran Glucanotransferase from an Alkaliphilic Bacterium, *Paenibacillus daejeonensis. Journal of Microbiology and Biotechnology, 28*(12), 2029–2035. doi:10.4014/jmb.1810.10007 PMID:30394040

Ye, J., Huang, W., Wang, D., Chen, F., Yin, J., Li, T., Zhang, H., & Chen, G. Q. (2018). Pilot scale-up of poly (3-hydroxybutyrateco-4-hydroxybutyrate) production by *Halomonas bluephagenesis* via cell growth adapted optimization process. *Biotechnology Journal, 13*(5), e1800074. doi:10.1002/biot.201800074 PMID:29578651

Yi, H., & Chun, J. (2006). Flavobacterium weaverense sp. nov. and Flavobacterium segetis sp. nov., novel psychrophiles isolated from the Antarctic. *International Journal of Systematic and Evolutionary Microbiology, 56*(6), 1239–1244. doi:10.1099/ijs.0.64164-0 PMID:16738098

Yi, H., Oh, H. M., Lee, J. H., Kim, S. J., & Chun, J. (2005). Flavobacterium antarcticum sp. nov., a novel psychrotolerant bacterium isolated from the Antarctic. *International Journal of Systematic and Evolutionary Microbiology, 55*(2), 637–641. doi:10.1099/ijs.0.63423-0 PMID:15774636

Yi, H., Yoon, H. I., & Chun, J. (2005). Sejongia antarctica gen. nov., sp. nov. and Sejongia jeonii sp. nov., isolated from the Antarctic. *International Journal of Systematic and Evolutionary Microbiology, 55*(1), 409–416. doi:10.1099/ijs.0.63273-0 PMID:15653910

Yin, L., Sun, H., Zhang, H., He, L., Qiu, L., Lin, J., Xia, H., Zhang, Y., Ji, S., Shi, H., & Gao, M. (2019). Quantitatively Visualizing Tumor-Related Protease Activity in Vivo Using a Ratiometric Photoacoustic Probe. *Journal of the American Chemical Society, 141*(7), 3265–3273. doi:10.1021/jacs.8b13628 PMID:30689382

Yoneda, Y., Yoshida, T., Yasuda, H., Imada, C., & Sako, Y. (2013). A novel thermophilic, hydrogenogenic, and carboxydotrophic bacterium *Calderihabitans maritimus* gen. nov., sp. nov. from a marine sediment core of an undersea caldera. *International Journal of Systematic and Evolutionary Microbiology, 63*(Pt_10), 3602–3608. doi:10.1099/ijs.0.050468-0 PMID:23606483

Yosef, I., Goren, M. G., & Qimron, U. (2012). Proteins and DNA elements essential for the CRISPR adaptation process in *Escherichia coli. Nucleic Acids Research, 40*(12), 5569–5576. doi:10.1093/nar/gks216 PMID:22402487

Yoshida, T., Hatanaka, K., Uryu, T., Kaneko, Y., Suzuki, E., Miyano, H., Mimura, T., Yoshida, O., & Yamamoto, N. (1990). Synthesis and structural analysis of curdlan sulfate with a potent inhibitory effect in vitro of AIDS virus infection. *Macromolecules, 23*(16), 3717–3722. doi:10.1021/ma00218a001

You, H. J., Oh, D.-K., & Ji, G. E. (2004). Anticancerogenic effect of a novel chiroinositol-containing polysaccharide from *Bifidobacterium bifidum* BGN4. *FEMS Microbiology Letters, 240*(2), 131–136. doi:10.1016/j.femsle.2004.09.020 PMID:15522499

Youssef, N. H., Savage-Ashlock, K. N., McCully, A. L., Luedtke, B., Shaw, E. I., Hoff, W. D., & Elshahed, M. S. (2014). Trehalose / 2-sulfotrehalose biosynthesis and glycine-betaine uptake are widely spread mechanisms for osmoadaptation in the Halobacteriales. *The ISME Journal, 8*(3), 636–649. doi:10.1038/ismej.2013.165 PMID:24048226

Yuan, M., Chen, M., Zhang, W., Lu, W., Wang, J., Yang, M., Zhao, P., Tang, R., Li, X., Hao, Y., Zhou, Z., Zhan, Y., Yu, H., Teng, C., Yan, Y., Ping, S., Wang, Y., & Lin, M. (2012). Genome sequence and transcriptome analysis of the radioresistant bacterium Deinococcus gobiensis: Insights into the extreme environmental adaptations. *PLoS One, 7*(3), e34458. doi:10.1371/journal.pone.0034458 PMID:22470573

Yuan, M., Zhang, W., Dai, S., Wu, J., Wang, Y., Tao, T., Chen, M., & Lin, M. (2009). *Deinococcus gobiensis* sp. nov., an extremely radiation-resistant bacterium. *International Journal of Systematic and Evolutionary Microbiology, 59*(6), 1513–1517. doi:10.1099/ijs.0.004523-0 PMID:19502345

Yu, C., Liu, Y., Ma, T., Liu, K., Xu, S., Zhang, Y., Liu, H., La Russa, M., Xie, M., Ding, S., & Qi, L. S. (2015). Small molecules enhance CRISPR genome editing in pluripotent stem cells. *Cell Stem Cell, 16*(2), 142–147. doi:10.1016/j.stem.2015.01.003 PMID:25658371

Yue, H., Ling, C., Yang, T., Chen, X., Chen, Y., Deng, H., Wu, Q., Chen, J., & Chen, G. Q. (2014). A seawater-based open and continuous process for polyhydroxy alkanoates production by recombinant *Halomonas campaniensis* LS21 grown in mixed substrates. *Biotechnology for Biofuels, 7*(1), 108–119. doi:10.1186/1754-6834-7-108

Yu, F. B., Blainey, P. C., Schulz, F., Woyke, T., Horowitz, M. A., & Quake, S. R. (2017). Microfluidic-based mini-metagenomics enables discovery of novel microbial lineages from complex environmental samples. *eLife, 6*, e26580. doi:10.7554/eLife.26580 PMID:28678007

Yu, K. O., Jung, J., Ramzi, A. B., Choe, S. H., Kim, S. W., Park, C., & Han, S. O. (2013). Development of a *Saccharomyces cerevisiae* strain for increasing the accumulation of triacylglycerol as a microbial oil feedstock for biodiesel production using glycerol as a substrate. *Biotechnology and Bioengineering, 110*(1), 343–347. doi:10.1002/bit.24623 PMID:22886471

Yumoto, I., Hirota, K., Nodasaka, Y., Tokiwa, Y., & Nakajima, K. (2008). *Alkalibacterium indicireducens* sp. nov., an obligate alkaliphile that reduces indigo dye. *International Journal of Systematic and Evolutionary Microbiology, 58*(4), 901–905. doi:10.1099/ijs.0.64995-0 PMID:18398191

Yumoto, I., Hirota, K., Nodasaka, Y., Yokota, Y., Hoshino, T., & Nakajima, K. (2004a). *Alkalibacterium psychrotolerans* sp. nov., a psychrotolerant obligate alkaliphile that reduces an indigo dye. *International Journal of Systematic and Evolutionary Microbiology, 54*(6), 2379–2383. doi:10.1099/ijs.0.63130-0 PMID:15545487

Yumoto, I., Hishinuma-Narisawa, M., Hirota, K., Shingyo, T., Takebe, F., Nodasaka, Y., Matsuyama, H., & Hara, I. (2004b). *Exiguobacterium oxidotolerans* sp. nov., a novel alkaliphile exhibiting high catalase activity. *International Journal of Systematic and Evolutionary Microbiology, 54*(6), 2013–2017. doi:10.1099/ijs.0.63129-0 PMID:15545426

Yumoto, I., Yamazaki, K., Hishinuma, M., Nodasaka, Y., Suemori, A., Nakajima, K., Inoue, N., & Kawasaki, K. (2001). *Pseudomonas alcaliphila* sp. nov., a novel facultatively psychrophilic alkaliphile isolated from seawater. *International Journal of Systematic and Evolutionary Microbiology, 51*(2), 349–355. doi:10.1099/00207713-51-2-349 PMID:11321079

Yusof, N. A., Hashim, N. H. F., & Bharudin, I. (2021). Cold Adaptation Strategies and the Potential of Psychrophilic Enzymes from the Antarctic Yeast, Glaciozyma antarctica PI12. *Journal of Fungi, 7*(7), 528. doi:10.3390/jof7070528

Yu, Y., Xin, Y. H., Liu, H. C., Chen, B., Sheng, J., Chi, Z. M., Zhou, P.-J., & Zhang, D. C. (2008). Sporosarcina antarctica sp. nov., a psychrophilic bacterium isolated from the Antarctic. *International Journal of Systematic and Evolutionary Microbiology, 58*(9), 2114–2117. doi:10.1099/ijs.0.65838-0 PMID:18768614

Zaccai, G., Cendrin, F., Haik, Y., Borochov, N., & Eisenberg, H. (1989). Stabilization of halophilic malate dehydrogenase. *Journal of Molecular Biology, 208*(3), 491–500. doi:10.1016/0022-2836(89)90512-3 PMID:2795658

Zaghloul, A., Saber, M., Gadow, S., & Awad, F. (2020). Biological indicators for pollution detection in terrestrial and aquatic ecosystems. *Bulletin of the National Research Centre, 44*(1). doi:10.1186/s42269-020-00385-x

Zahid, N., Schweiger, P., Galinski, E., & Deppenmeier, U. (2015). Identification of mannitol as compatible solute in Gluconobacter oxydans. *Applied Microbiology and Biotechnology, 99*(13), 5511–5521. doi:10.1007/s00253-015-6626-x

Zainuddin, N., Abd Rahman, K. H., Razak Shaari, A., & Mohd Yusof, S. J. H. (2015). Isolation of Potential Alkaliphilic Cellulase-Producer Strains from Limestone Area of Perlis. *Applied Mechanics and Materials, 754–755*, 1054–1058. . doi:10.4028/www.scientific.net/AMM.754-755.1054

Zajc, J., Zalar, P., Plemenitaš, A., & Gunde-Cimerman, N. (2012). The mycobiota of the salterns. In *Biology of Marine Fungi* (pp. 133–158). Springer. doi:10.1007/978-3-642-23342-5_7

Zakhia, F., Jungblut, A. D., Taton, A., Vincent, W. F., & Wilmotte, A. (2008). Cyanobacteria in cold ecosystems. In R. Margesin, F. Schinner, J. C. Marx, & C. Gerday (Eds.), *Psychrophiles: from biodiversity to biotechnology* (pp. 121–135). Springer. doi:10.1007/978-3-540-74335-4_8

Zaldivar, J., Nielsen, J., & Olsson, L. (2001). Fuel ethanol production from lignocellulose: A challenge for metabolic engineering and process integration. *Applied Microbiology and Biotechnology, 56*(1–2), 17–34. doi:10.1007002530100624 PMID:11499926

Zannini, E., Waters, D. M., Coffey, A., & Arendt, E. K. (2016). Production, properties, and industrial food application of lactic acid bacteria-derived exopolysaccharides. *Applied Microbiology and Biotechnology, 100*(3), 1121–1135. doi:10.100700253-015-7172-2 PMID:26621802

Zarinviarsagh, M., Ebrahimipour, G., & Sadeghi, H. (2017). Lipase and biosurfactant from *Ochrobactrum intermedium* strain MZV101 isolated by washing powder for detergent application. *Lipids in Health and Disease, 16*(1), 1–13. doi:10.118612944-017-0565-8 PMID:28923075

Zeikus, J. G., & Wolfe, R. S. (1972). *Methanobacterium thermoautotrophicus* sp., an anaerobic, autotrophic, extreme thermophile. *Journal of Bacteriology, 109*(2), 707–715. doi:10.1128/jb.109.2.707-713.1972 PMID:4550816

Zeldes, B. M., Keller, M. W., Loder, A. J., Straub, C. T., Adams, M. W. W., & Kelly, R. M. (2015). Extremely thermophilic microorganisms as metabolic engineering platforms for production of fuels and industrial chemicals. *Frontiers in Microbiology*, *12*(6), 2015–2024. doi:10.3389/fmicb.2015.01209 PMID:26594201

Zelenkina, T., Eshinimayev, B. T., Dagurova, O., Suzina, N., Namsarayev, B. B., & Trotsenko, Y. (2009). Aerobic methanotrophs from the coastal thermal springs of Lake Baikal. *Microbiology*, *78*(4), 492–497. doi:10.1134/S0026261709040134 PMID:19827720

Zetsche, B., Gootenberg, J. S., Abudayyeh, O. O., Slaymaker, I. M., Makarova, K. S., Essletzbichler, P., Volz, S. E., Joung, J., van der Oost, J., Regev, A., Koonin, E. V., & Zhang, F. (2015). Cpf1 is a single RNA-guided endonuclease of a class 2 CRISPR-Cas system. *Cell*, *163*(3), 759–771. doi:10.1016/j.cell.2015.09.038 PMID:26422227

Zhai, L., Ma, Y., Xue, Y., Ma, Y. (2014). Bacillus alkalicola sp. nov., An Alkaliphilic, Gram-Positive Bacterium Isolated from Zhabuye Lake in Tibet, China. *Current Microbiology*, 311–316. Https://Doi.Org/10.1007/S00284-014-0576-X

Zhang, X., Li, Z., Che, X., Yu, L., Jia, W., Shen, R., Chen, J., Ma, Y., & Chen, G. Q. (2019). Synthesis and characterization of polyhydroxyalkanoate organo/hydrogels. *Biomacromolecules*, *20*(9), 3303-3312. .biomac.9b00479 doi:10.1021/acs

Zhang, X., Lin, Y., & Chen, G.-Q. (2018). Halophiles as Chassis for Bioproduction. *Advanced Biosystems*, *2*(11), 1800088. doi:10.1002/adbi.201800088

Zhang, D. C., Busse, H. J., Liu, H. C., Zhou, Y. G., Schinner, F., & Margesin, R. (2011). Sphingomonas glacialis sp. nov., a psychrophilic bacterium isolated from alpine glacier cryoconite. *International Journal of Systematic and Evolutionary Microbiology*, *61*(3), 587–591. doi:10.1099/ijs.0.023135-0 PMID:20382784

Zhang, D. C., Mörtelmaier, C., & Margesin, R. (2012). Characterization of the bacterial archaeal diversity in hydrocarbon-contaminated soil. *The Science of the Total Environment*, *421–422*, 184–196. doi:10.1016/j.scitotenv.2012.01.043 PMID:22386232

Zhang, G., huihua, G., & Yi, L. (2013). Stability of halophilic proteins: From dipeptide attributes to discrimination classifier. *International Journal of Biological Macromolecules*, *53*, 1–6. doi:10.1016/j.ijbiomac.2012.10.031 PMID:23142140

Zhang, J. W., & Zeng, R. Y. (2008). Purification and characterization of a cold-adapted α-amylase produced by *Nocardiopsis* sp. 7326 isolated from Prydz Bay, Antarctic. *Marine Biotechnology (New York, N.Y.)*, *10*(1), 75–82. doi:10.100710126-007-9035-z PMID:17934774

Zhang, X., Li, S. J., Li, J. J., Liang, Z. Z., & Zhao, C. Q. (2018). Novel natural products from extremophilic fungi. *Marine Drugs*, *16*(6), 194. doi:10.3390/md16060194 PMID:29867059

Zhang, Y. G., Lu, X. H., Ding, Y. B., Zhou, X. K., Li, L., Guo, J. W., Wang, H. F., Duan, Y. Q., & Li, W. J. (2016). *Phytoactinopolyspora alkaliphila* sp. Nov., an alkaliphilic actinomycete isolated from a saline-alkaline soil. *International Journal of Systematic and Evolutionary Microbiology*, *66*(5), 2058–2063. doi:10.1099/ijsem.0.000992 PMID:26920762

Zhang, Z., Cai, R., Zhang, W., Fu, Y., & Jiao, N. (2017). A novel exopolysaccharide with metal adsorption capacity produced by a marine bacterium *Alteromonas* sp. JL2810. *Marine Drugs*, *15*(6), 175. doi:10.3390/md15060175 PMID:28604644

Zhao, C.-C., & Eun, J.-B. (2020). Shotgun metagenomics approach reveals the bacterial community and metabolic pathways in commercial hongeo product, a traditional Korean fermented skate product. *Food Research International*, *131*. doi:10.1016/j.foodres.2020.109030

Zhao, L., Ma, Z., Luan, Y., Lu, A., Wang, J., & Pan, L. (2011). Molecular methods of studying microbial diversity in soil environments. In. *IFIP Advances in Information and Communication Technology International Conference on Computer and Computing Technologies in Agriculture*. Springer. 10.1007/978-3-642-18336-2_10

Zhao, B., Yan, Y., & Chen, S. (2014). How could haloalkaliphilic microorganisms contribute to biotechnology? *Canadian Journal of Microbiology, 60*(11), 717–727. doi:10.1139/cjm-2014-0233 PMID:25372346

Zheng, B., Jiang, X., Xu, Z., Fang, Y., & Li, L. (2016). Characterization of a novel metallo-β-lactamases fold hydrolase from *Pelagibacterium halotolerans*, a marine halotolerant bacterium isolated from East China Sea. *Extremophiles, 20*(1), 37–44. doi:10.100700792-015-0795-5 PMID:26560315

Zhou, Y., Sun, Z. Y., Li, H., Qian, C. J., Wu, X., Tang, H. Z., Ali, F., & di Liu, Y. (2017). Investigation of compatible solutes synthesis and transport of Virgibacillus halodenitrificans PDB-F2 with complete genome analysis. *International Biodeterioration & Biodegradation, 122*, 165–172. doi:10.1016/j.ibiod.2017.05.005

Zhu, D., Adebisi, W. A., Ahmad, F., Sethupathy, S., Danso, B., & Sun, J. (2020). Recent development of extremophilic bacteria and their application in biorefinery. *Frontiers in Bioengineering and Biotechnology, 8*, 483. doi:10.3389/fbioe.2020.00483 PMID:32596215

Zhu, J., Arsovska, B., & Kozovska, K. (2020). Ecology and diversity of Alkaliphilic Cyanobacteria from Soda Lake Lonar, India. *International Journal of Recent Scientific Research, 11*, 37471–37472. doi:10.24327/IJRSR

Zhu, S., Song, D., Gong, C., Tang, P., Li, X., Wang, J., & Zheng, G. (2013). Biosynthesis of nucleoside analogs via thermostable nucleoside phosphorylase. *Applied Microbiology and Biotechnology, 97*(15), 6769–6778. doi:10.100700253-012-4542-x PMID:23160980

Zinder, S. H. (1993). Physiological Ecology of Methanogens. *Methanogenesis*, 128-206. doi:10.1007/978-1-4615-2391-8_4

Zinger, L., Shahnavaz, B., Baptist, F., Geremia, R. A., & Choler, P. (2009). Microbial diversity in alpine tundra soils correlates with snow cover dynamics. *The ISME Journal, 3*(7), 850–859. doi:10.1038/ismej.2009.20 PMID:19322246

About the Contributors

Aparna Gunjal has completed her B.Sc. from Annasaheb Magar Mahavidyalaya, Hadapsar; M.Sc. from Modern College Arts, Commerce and Science College, Ganeshkhind and PhD in Environmental Sciences subject from Savitribai Phule Pune University, Pune, Maharashtra, India. She is working as an Assistant Professor in the Department of Microbiology at Dr. D.Y. Patil, Arts, Commerce and Science College, Pimpri, Pune, Maharashtra, India. Her research areas of expertise are solid waste management; plant growth-promoting rhizobacteria, e-waste management, bioremediation, etc. Aparna has 115 publications to her credit. She has received 10 Awards for the Best Paper presentations and also received travel grants. Aparna has also received Pune Municipal Corporation Award for excellent work in Environmental Sciences Research in 2015, The Elsevier Foundation - TWAS Sustainability Visiting Expert Programme" in 2018 and the Young Researcher award with Innovative Technology. She has worked on composting aspect as a Senior Researcher Assistant at Hongkong Baptist University, Hongkong. Aparna is an Editor Member of the International Journal of Microbial Science; Frontiers in Environmental Microbiology; Journal of Microbiology and Biotechnology Reports and Reviewer.

Rebecca Thombre is a London based Scientist with 19 years of experience in Microbiology and Molecular Biology. She has worked as a Scientist at Blue Marble Space Institute of Science, Seattle, Washington, USA (2020-2021), Visiting Scientist at University of Kent, UK (2019-2021), and as an Assistant Professor in Microbiology at Modern College, Pune, India (2005-2019). She has a rich teaching, research, and clinical laboratory experience in Microbiology, Bacteriology, Mycology, Antimicrobial Resistance (Testing and handling Drug-resistant strains-MDR, XDR, PDR, MRSA, VRSA, HG2,3 strains, Cancer cell lines), and Extremophilic bacteria. She was the recipient of Academic excellence award and DFF-Eureka Forbes fellowship from University of Mumbai for First rank in the University for M.Sc. Microbiology. She pursued her Ph.D. in Microbiology from Agharkar Research Institute, an institute under Department of Science and Technology, Government of India and her research focused on characterisation of the extremozyme cyclodextrin glycosyl transferase from alkaliphiles isolated from Lonar lake, a soda lake formed by a meteorite impact crater. She has received many postdoctoral grants/ awards/fellowships and has travelled to the USA, Europe and Taiwan for research. She has received funding as the Principal Scientific Investigator for projects on Microbiology and Astrobiology funded by Indian Space Research Organization (ISRO) - SPPU Space Technology cell, UGC Govt. of India and Europlanet (European Union). She was involved in Clinical research with hospitals to isolate and characterise oral microorganisms from patients with special reference to Streptococcus species. She has isolated and reported more than 200 extremophilic microorganisms and representative organisms have been deposited in the National Culture Collection, Centres. She has published more than 50 articles including popular science articles, peer-reviewed papers, book chapters and Edited Special issues and books published by Frontiers Publishing group and Elsevier, USA.

Javid Parray holds a Master's Degree in Environmental Science and has completed his Research Programme (MPhil and PhD) from the University of Kashmir after qualifying state level prestigious JKSLET examination. He has also done his Post Doctorate Research from the University of Kashmir. Dr Parray was also awarded a Fast Track Young Scientist Project by SERB – DST, GoI New Delhi. He currently teaches at the Department of Environmental Science, GDC Eidgah Srinagar affiliated to Cluster University Srinagar. He has attended many courses and conferences on environmental issues and other biotechnological aspects within and outside the country like Srilanka, Indonesia, Malaysia etc. His fundamental research interests include ecological and agricultural microbiology, climate change and microbial biotechnology, environmental microbiome etc. He has published more than 50 high-impact research papers and book chapters in reputed journals and publishing Hubs. Dr Parray has authored 08 books with international publishers like Elsevier, Springer, Callisto Reference USA and Wiley-Blackwell. Dr Parray is on the editorial board, and permanent reviewers of many journals and has been an invited speaker at various scientific meetings/conferences within India and abroad. He is also a Guest editor for a special issue on Environmental Biofilms with BioMed Research International, Hindwaii. He is member of many International and National scientific organizations and societies like Asian PGPR Society, IJMS Mumbai, Academy of Eco science, IAES Haridwar etc. Dr Javid was also awarded as "Emerging scientist year Gold Medal" for the year 2018 by Indian Academy of Environmental Science. Dr Parray is currently course coordinator for Two (02) CeC –Swayam Moocs programme on environmental Science.

***

Prashant S. Arya is currently a doctoral student at Department of Microbiology & Biotechnology Gujarat University Ahmedabad. He has been actively involved in enzyme research and studies on various aspects of enzyme production. He has published his work in referred journals and has made presentation at various national level scientific events. Recently has been awarded fellowship for pursuing his work under the "SHODH - ScHeme of Developing High-Quality Research" fellowship by Department of Higher Education, Government of Gujarat, Gujarat, India.

Nilaja Badodekar has worked on a project titled assessing the antimicrobial activity of silver and gold nanoparticles against antimicrobial resistant clinical isolates. She has designed protocols for testing the long-term efficiency of antimicrobial products.

Bidyut Bandyopadhyay is a professor in the Department of Biotechnology, Oriental Institute of Science and Technology, Burdwan.

Dhritiksha M. Baria is doctoral student at Department of Microbiology and Biotechnology, Gujarat University, Ahmedabad. Her area of interest includes extremophiles and their biopolymers. She has presented her work in national and international conferences. She co-authored few publications in referred journals and chapter in edited book. She is recipient of research fellowship under "SHODH - ScHeme of Developing High-Quality Research."

Aradhana Bavarva is pursuing M.Sc. in Microbiology. She got first prize in International Science Symposium 2021 organised by Christ College, Rajkot.

Shreays Bhatt is currently working as Head of the Department of Life Sciences, Hemchandracharya North Gujarat University. The author has 32 years of teaching and research experience. Dr. Bhatt has successfully supervised 15 Doctoral dissertations, more than 15 M. Phil. dissertations and 27 Master degree dissertations. He has more than 50 peer reviewed research publications in journals of national and international repute and has written four books as well as edited six books and four book chapters. He has completed five major and minor research and consultancy projects. He has served in all the authorities of the University viz., Executive Council, Academic Council, Board of University Teaching and Research. Dr. Bhatt is working as Chairman of Board of Studies in Microbiology and Environmental Science. Dr. Bhatt has also served in the Executive council of the Gujarat Science Academy. He has visited Germany, Canada, Mauritius and Maldives for presenting research papers and delivering invited lectures.

Pravin Dudhagara is a microbiologist, completed his post-graduation and doctorate in Microbiology from the Hemchandracharya North Gujarat University, Patan, India. He has started his research career as Research Fellow in 2007 at the Department of animal genetic and breeding at the Anand agricultural university, Anand, India. In 2009, he has shifted from research to academic field by joining as an assistant professor at the Department of Biotechnology, Veer Narmad South Gujarat University (VNSGU), Surat, India. Dr. dudhagara has awarded young scientist award in the FAOBMB conference Manila 2016, Philippines. His current research area includes (1) Metagenomics study of anthropogenic sites (2) Microbial enzymes (3) Extreme biosphere microbiome (4) Host-microbiome.

Kailas Fuldeore is working as Deputy Manager at Serum Institute of India.

Anjana Ghelani awarded Ph. D. in Microbiology from the HNGU, Patan, India. Presently, Dr. Anjana is working as a senior Assistant Professor in Microbiology at Shree Ramkrishna Institute of computer education and applied sciences, Surat, India. Currently, Dr. Anjana working in the area including (i) Actinomycetes bacterial diversity and its Plant growth promoting study (ii) Marine metagenomics (iii) Industrial enzymology. Dr. Anjana has contributed as an organizing member in more than ten seminar, conference and workshop.

Rakesh Goswami is pursuing PhD in Department of Botany, Raiganj University.

Sneha H. P. is a microbiologist with research experience in the field of coffee, cocoa fermentation and enzyme application with two research publications in esteemed international journals.

Chaitanya J. is an assistant professor in microbiology areas of interest are microbiology, agriculture microbiology, and PGPR having research and teaching experience of 13 years.

Kunal Jain has earned his post-graduate and doctorate degree from Sardar Patel University, Vallabh Vidyanagar, in Microbiology. He is currently working as an Assistant Professor (Adhoc), Department of Biosciences and Group Leader, RT-PCR (Covid-19) Testing Lab, Sardar Patel University, Vallabh Vidyanagar, Gujarat, India. He has been working in Environmental Microbiology for the last thirteen years, trying to understand microbial behaviour in an anthropogenic ecosystem. He was a Guest Co-Associate Editor for special issues in Frontiers in Bioengineering and Biotechnology and Frontiers in Microbiology. He has authored one book/compendium and currently has an h-index of 15.

Mukta Kothari currently working as an Assistant Professor at Department of Biotechnology, Modern College of Arts, Science & Commerce (Autonomous), Pune -5.

Amitsinh Mangrola awarded Ph. D. in Microbiology from the Hemchandracharya North Gujarat University, Patan, Gujarat, India. Presently, Dr. Amitsinh is working as a senior Assistant Professor in Biochemistry Department at Shri Alpesh N. Patel Post Graduate Institute of Science & Research, Anand, Gujarat, India. Dr. Amitsinh is currently working in the research area including (i) Extremophiles (ii) Cow urine and Panchagavya based research (iii) Enzyme Biotechnology (iv) Cancer and other therapeutic research using herbal formulations.

Pushpa S. Murthy holds PhD in Biotechnology from the University of Mysore and has over two decades of experience with combination of applied and basic science in the areas of Food and applied biotechnology. She has developed eight technology/processes, transferred to industry, and has three patents to her credit. She has received prestigious Har Gobind Khorana Best Scientist award for Food Biotechnology from Bose Science Society, Distinguished Biotechnologist Award, for research contributions in field of Biotechnology by the Board of Management of Venus International Foundation, Microbiologist award from Scire Science. She has fellow of UNU-Kirin, Japan, Society of Applied biotechnology, Fellow of Bose Science Society. She has guided more than 50 students for research investigation/ dissertation and PhD. She has over 50 publications in peer reviewed research journals, eight popular articles, seven Book chapters, 63 research posters/proceedings. She collaborated with other departments across CFTRI and outside the institutions for bringing out best research network. She has expertise is extraction bioactives from spice, plantation crops and microbial sources and application of Nutraceutical/functional molecules in food models and assessment of microbial Food safety. She is awarded more than 6 Grant in aid projects from Department of Science and engineering research Board, Ministry of Food processing, Govt of India. She is also involved and carrying out Industrially Sponsored (9) and consultancy projects(3) both nationally and internationally. She is life member of Indian science Congress, Association of food Scientist, The Biotech Research Society, Association of Microbiologists of India, Society for Educational & Scientific Research and Bose science society, India.

Amisha Panchal is an Assistant Professor in the Microbiology Department at Mehsana Urban Institute of Sciences, Ganpat University. She holds a Master of Science degree in Microbiology from the Kadi Sarva Vishwavidyalaya University of Gandhinagar. She is the committee member of various events organizes by Mehsana Urban Institute of Sciences, Ganpat University. At work, Amisha Panchal guides dissertation students and work pressure effectively. Her area of research interest is Fermentation technology, enzymology, and Microbial Diversity. she also remains engaged with co-curricular activities.

Khushbu Panchal is currently working as an assistant professor, Department of Microbiology, Mehsana Urban Institute of Sciences, Ganpat University, Gujarat. She has 04 years of teaching experience. Her area of interest is Environmental Microbiology, Agricultural Microbiology, Industrial Microbiology and Applied Microbiology.

Rakeshkumar R. Panchal presently is Associate Professor, Department of Microbiology and Biotechnology, Gujarat University, Ahmedabad. Dr. Panchal has completed his MPhil and PhD in microbiology from VNSGU. Surat, Gujarat. He has expertise in microbial fermentation both at lab scale and at industrial level. Dr. Panchal has vast experience in academics too and his research interests include microbial enzymes, fermentation and microbial secondary metabolites. He has been associated with industry and academics since past 25 years and has substantially contributed towards the development of skill based learning among the students.

Devendra Pandey is Principal Scientist & Head Division of Crop Improvement Central Institute for Subtropical Horticulture Lucknow, Uttar Pradesh, India.

Suneeta Panicker completed her Ph.D in Biotechnology through CSIR, JRF and SRF fellowship and then was awarded ICMR Post-doctoral fellowship at National Institute of Epidemiology, Chennai. Before her Ph.D she has worked as a Research assistant at National AIDS Research Institute (NARI), Pune and as a Senior research fellow at Haffkine research Institute, Mumbai. Currently she is working as an Assistant Professor in Dept of Microbiology at Dr. D. Y. Patil Arts, Commerce& Science College, Pimpri Pune. She has many peer reviewed research articles to her credit, few of which is published in Elsevier and Springer journals and has also authored two text books for the undergraduate Microbiology students. Last year Dr. Panicker was awarded ICMR international travel grant to present her work in a HBV meeting at Melbourne, Australia. She is a Member of Organization for Women in Science for the Developing World (OWSD) Strada Costiera, 11 34151 Trieste, Italy Dr. Suneeta is a mentor for the 1000 Girls, 1000 Futures and The Junior Academy program for New York Academy of Sciences She is on the editorial board of few prestigious journals and reviewed many manuscripts. She is actively involved in research related to virology and molecular biology.

Nafisa Patel has 18 years of teaching experience. Her core research areas are microbial technology and molecular biology.

Nidhi Yogeshbhai Patel is a doctoral student at Department of Microbiology & Biotechnology, University School of Sciences, Gujarat University, Navrangpura, Ahmedabad. She has completed here M.Phil. Microbiology and her major area of interest is Marine microbiology, Therapeutic enzymes from extremophiles. She is a young researcher and has presented her research work at various national and international conferences. She also has few good publications to her credit and is keen to contribute to research.

Inderjit Prakash, MSc in Biotechnology, has worked as Project Assistant for 1 year in CSIR-CFTRI. And recently worked on a Sponsored project on Coffee Fermentation and Pretreatments at the department of spices and flavour Sciences.

Kiransinh N. Rajput is currently associated with Department of Microbiology and Biotechnology, Gujarat University, Ahmedabad. Dr. Rajput is actively involved in microbial enzymes and his major focus is on cyclodextrin glucanotransferase. The other areas of research work includes microbial biosurfactant and bioremediation. Dr. Rajput has nearly two decades of experience in teaching various aspects of microbiology. He was been awarded the teachers fellowship (Faculty Improvement Program) from

Universities Grants Commission (UGC), New Delhi, India for pursuing doctoral research. Dr. Rajput has good number of publications in refereed journals and is been an active researcher in microbiology.

Vikram H. Raval is currently assistant professor at Department of Microbiology & Biotechnology, Gujarat University, Ahmedabad. Dr. Raval has been recipient of Junior and Senior Research Fellowships from Department of Biotechnology (DBT), New Delhi during his doctoral research work. His major works emphasis is on microbial enzymes from diverse ecological niches. Dr. Raval has also been selected for International Travel Fellowship and as Young Scientist by Scientific & Engineering Research Board (SERB), New Delhi, India. He has authored a good number of research papers, chapters in edited books and is actively involved in research with major focus on extremophiles, molecular biology, metagenomics and bioremediation. He has also been credited with two major research projects on Metagenomics and marine enzymes.

Sanjoy Sadhukhan is Assistant Professor in the Department of Botany, Raiganj University.

Hardik Shah is currently working as Associate Professor in Microbiology at Mehsana Urban Institute of Sciences, Faculty of Science, Ganpat University. He has completed bachelors, masters and doctorate in Microbiology, along with qualifying various eligibility tests such as CSIR-UGC NET (All India Rank 45), ICAR NET and GSET. With a rich academic experience of over 15 years in UG as well as PG courses, Dr. Shah has guided more than 30 post graduate dissertations and is also a recognized Ph.D. guide for Microbiology subject at Ganpat University. Dr. Shah has published one book and more than 8 publications in the journals of repute. His cardinal research interests include Microbial Fuel Cell, Extremophiles, Probiotics, Bacteriocins, Fermentative production of enzymes, and antimicrobial agents. Dr. Shah has a knack for technology and loves to do blogging. He's been operating a few blogs designed exclusively for students connected with allied subjects of life sciences.

Shivani M. Yagnik has been working as lecturer and is actively involved in teaching students at graduate and post graduate level at Christ College Rajkot. She has completed her MPhil in Microbiology and currently pursuing her doctoral research from Saurashtra University Rajkot. Her research area includes enzymology, bioremediaton, environmental microbiology and bioinformatics. She herself has excelled in academics and has publications in referred journals as well as presented her work at various national and international scientific platforms.

Index

F

G

H

I

M

N

O

P

R

S

T

www.ingramcontent.com/pod-product-compliance
Lightning Source LLC
LaVergne TN
LVHW060256011026
849161LV00011B/2725
* 9 7 8 1 7 9 9 8 9 1 4 4 4 *